Remote Sensing of the Environment

An Earth Resource Perspective

Prentice-Hall Series
in Geographic
Information Science

KEITH C. CLARKE,
Series Advisor

Arnold, *Interpretation of Airphotos and Remotely Sensed Imagery*

Avery/Berlin, *Fundamentals of Remote Sensing and Airphoto Interpretation*

Clarke, *Analytical and Computer Cartography*

Clarke, *Getting Started With Geographic Information Systems*, 2nd Edition

Foresman, *The History of Geographic Information Systems*

Jensen, *Introductory Digital Image Processing: A Remote Sensing Perspective*, 2nd Edition

Peterson, *Interactive and Animated Cartography*

Slocum, *Thematic Cartography and Visualization*

Vincent, *Fundamentals of Geological and Environmental Remote Sensing*

Remote Sensing of the Environment

An Earth Resource Perspective

John R. Jensen

University of South Carolina

Prentice Hall

Upper Saddle River, New Jersey 07458

Library of Congress Cataloging-in-Publication Data

Jensen, John R.
    Remote sensing of the environment: an earth resource perspective / John R. Jensen.
        p.   cm. — (Prentice Hall series in geographic information science)
    Includes bibliographical references and index.
    ISBN 0-13-489733-1
    1. Earth sciences — Remote sensing. I. Title. II. Series.
    QE33.2.R4 J46 2000
    550'.28 — dc21

                                         99-055656
                                       CIP

Executive Editor: Daniel Kaveney
Assistant Editor: Amanda Griffith
Editorial Assistant: Margaret Ziegler
Production Editors: Ed Thomas and Nicole Bush
Marketing Manager: Christine Henry
Art Director: Jayne Conte
Copy Editor: Barbara Booth
Production Assistant: Nancy Bauer
Manufacturing Manager: Trudy Pisciotti
Manufacturing Buyer: Michael Bell

Cover Credits: (top left) Solar and Heliospheric Observatory image of the Sun on September 14, 1999, courtesy of the SOHO EIT Consortium; SOHO is a joint ESA-NASA program; (top center) Dispersion of white light using a prism courtesy of David Parker, Photo Researchers, Inc.; (top right) Landsat 7 Enhanced Thematic Mapper Plus illustration courtesy of Lockheed Martin, Inc. and NASA Goddard Space Flight Center; (middle left and center) Landsat Thematic Mapper data of the Grand Canyon, AZ, and the Imperial Valley, CA, courtesy of Space Imaging, Inc.; (middle right) Calibrated Airborne Multispectral Scanner data of Pelican Key near Key West, FL, courtesy of NASA John C. Stennis Space Center; (bottom left) Airborne Visible/Infrared Imaging Spectrometer data of Sullivan's Island, SC, courtesy of NASA Jet Propulsion Laboratory; (bottom center) Color aerial photograph near Friar's Point, MS, courtesy of NASA John C. Stennis Space Center; (bottom right) Building infrastructure information and image of the capitol courtesy of LH Systems, LLC, and Marconi Integrated Systems, Inc., © SOCET SET.

Printed in the United States of America

10 9 8 7 6 5 4 3 2 1

ISBN 0-13-489733-1

Prentice-Hall International (UK) Limited, *London*
Prentice-Hall of Australia Pty. Limited, *Sydney*
Prentice-Hall Canada, Inc., *Toronto*
Prentice-Hall Hispanoamericana, S.A., *Mexico*
Prentice-Hall of India Private Limited, *New Delhi*
Prentice-Hall of Japan, Inc., *Tokyo*
Pearson Education Asia Pte. Ltd.
Editora Prentice-Hall do Brasil, Ltda., *Rio de Janeiro*

Brief Contents

1 *Remote Sensing of the Environment*   1

2 *Electromagnetic Radiation Principles*   29

3 *History of Aerial Photography and Aerial Platforms*   53

4 *Aerial Photography – Vantage Point, Cameras, Filters, and Film*   85

5 *Elements of Visual Image Interpretation*   119

6 *Photogrammetry*   137

7 *Multispectral Remote Sensing Systems*   181

8 *Thermal Infrared Remote Sensing*   243

9 *Active and Passive Microwave, and LIDAR Remote Sensing*   285

10 *Remote Sensing of Vegetation*   333

11 *Remote Sensing of Water*   379

12 *Remote Sensing the Urban Landscape*   407

13 *Remote Sensing of Soils, Minerals, and Geomorphology*   471

Index   531

Appendix A—Sources of Remote Sensing Information   541

Contents

Preface xiii
   Acknowledgments xvi

Chapter 1 — Remote Sensing of the Environment 1
   In Situ Data Collection 1
   Remote Sensing Data Collection 2
      Maximal/Minimal Definitions 3
      Remote Sensing Advantages and Limitations 7
   The Remote Sensing Process 8
      Statement of the Problem 9
      Identification of In Situ and Remote Sensing Data Requirements 10
      Remote Sensing Data Collection 12
      Remote Sensing Data Analysis 21
      Information Presentation 24
   Earth Resource Analysis Perspective 24
   Book Organization 24

Chapter 2 — Electromagnetic Radiation Principles 29
   Conduction, Convection, and Radiation 29
   Electromagnetic Radiation Models 30
      Wave Model of Electromagnetic Energy 30
      The Particle Model – Radiation from Atomic Structures 35
   Energy-Matter Interactions in the Atmosphere 39
      Refraction 39
      Scattering 41
      Absorption 42
      Reflectance 44
   Energy-Matter Interactions with the Terrain 44
      Hemispherical Reflectance, Absorptance, and Transmittance 44
      Radiant Flux Density 46
   Energy-Matter Interactions in the Atmosphere Once Again 48
   Energy-Matter Interactions at the Sensor System 48
      Target and Path Radiance 48
   Conclusion 51

Chapter 3 — History of Aerial Photography and Aerial Platforms 53
   History of Photography 53
      Light and Color 53
      The Camera Obscura 54
      Invention of Light-sensitive Emulsions and Methods of Permanently
         Fixing the Image 54
   Photography from Aerial Platforms 57
      Ornithopters 57

        Lighter-Than-Air Flight Using Balloons 58
        Lighter-Than-Air Flight Using Kites 61
        Heavier-Than-Air Flight Using Rockets 63
        Heavier-Than-Air Flight Using Pigeons, Gliders, and Aircraft 63
    Photo-reconnaissance in World War I and World War II .64
        Aerial Photography in World War I 67
        Aerial Photography in World War II 67
    Cold War Photo-Reconnaissance 69
        Genetrix Reconnaissance Balloons. 70
        U-2 Aircraft Reconnaissance Program 71
        The SR-71 72
        Stealth Technology 74
    Celestial Satellite Sentinels 74
        Corona 74
        Ongoing Satellite Sentinels 78
    Unmanned Aerial Vehicles 79
    Commercial, Nonmilitary Remote Sensing Platforms 79
    Use of National Technical Means Remotely Sensed Data by
        Nonmilitary Federal Agencies 80

Chapter 4 — Aerial Photography – Vantage Point, Cameras, Filters, and Film 85
    Vertical and Oblique Vantage Points 85
        Vertical Aerial Photography 85
        Oblique Aerial Photography 86
    Aerial Cameras 87
        Aerial Camera Components Compared with the Eye 87
        Types of Aerial Cameras. 91
    Aerial Photography Filtration. 97
        Types, Sizes, and Colors of Filters 99
    Aerial Photography Films. 102
        General Characteristics of Photographic Emulsions. 102
        Black-and-White Photographic Emulsions 103
        Color Photographic Emulsions 114
    Planning Aerial Photography Missions 116
        Time of Day— Sun Angle 116
        Weather 117
        Flightline Layout. 117

Chapter 5 — Elements of Visual Image Interpretation 119
    Introduction 119
        The Aerial/Regional Perspective. 120
        Three-Dimensional Depth Perception. 120
        Obtaining Knowledge Beyond Our Human Visual Perception. 121
        Historical Image Record and Change Detection Documentation 121
    Elements of Image Interpretation 121
    Methods of Search 133
        Using Collateral Information 133
        Convergence of Evidence 134
        The Multi-concept. 134
    Conclusion 135

Chapter 6 — Photogrammetry 137
    Flightlines of Vertical Aerial Photography 138

Fiducial Marks, Principal Points, and Conjugate Principal Points. 140
Geometry of Vertical Aerial Photography 140
Scale and Height Measurement on Single Vertical Aerial Photographs. 142
    Scale of a Vertical Aerial Photograph Over Level Terrain 144
    Scale of a Vertical Aerial Photograph Over Variable Terrain 147
    Height Measurement from Single Aerial Photographs. 148
Stereoscopic Measurement of Object Height or Terrain Elevation 151
    Fundamentals of Human Stereoscopy 152
    Stereoscopy Applied to Aerial Photography. 152
    Stereoscopic Aerial Photography – How Does It Work? .157
Orthophotos and Digital Elevation Models 164
    Advances in the Collection of Accurate Horizontal and Vertical Ground
        Control to Produce Orthophotography 166
    Advances in the Collection of Metric Aerial Photography and Other
        Remote Sensor Data Used in the Creation of Orthoimage Databases. 166
    Advances in Image Digitization Technology for the Creation of Orthoimage
        Databases 167
    Advances in Soft-Copy Photogrammetry for the Creation of Orthoimages. 167
    Advances in the Creation of the Digital Elevation Model 167
    Advances in the Creation of the Orthoimage 169
    Problems and Potential Solutions Associated with Digital Elevation Models
        Derived Using Soft-Copy Photogrammetry 169
    Problems and Potential Solutions Associated with Orthoimagery
        Derived Using Soft-Copy Photogrammetry 170
Area Measurement 174
    Area Measurement of Well-Known Geometric Shapes .175
    Area Measurement of Irregularly Shaped Polygons. 175

Chapter 7 — Multispectral Remote Sensing Systems 181
Multispectral Data Collection. 181
    Digital Image Terminology 182
Multispectral Imaging Using Discrete Detectors and Scanning Mirrors 184
    Earth Resource Technology Satellites (ERTS) and the Landsat Sensor Systems. . 184
    NOAA Multispectral Scanner Sensors 201
    ORBIMAGE and NASA Sea-viewing Wide Field of View Sensor (SeaWiFS) . . .208
    Aircraft Multispectral Scanners 209
Multispectral Imaging Using Linear Arrays. 212
    SPOT Sensor Systems. 212
    Indian Remote Sensing Systems. 220
    Advanced Spaceborne Thermal Emission and Reflection Radiometer (ASTER) .221
    Multi-angle Imaging Spectroradiometer (MISR). 222
    Very High-Resolution Linear Array Remote Sensing Systems 224
Imaging Spectrometry Using Linear and Area Arrays. 226
    Airborne Visible Infrared Imaging Spectrometer (AVIRIS). 228
    Compact Airborne Spectrographic Imager–2 (CASI-2) .230
    Moderate Resolution Imaging Spectrometer (MODIS) .231
Digital Frame Cameras. 231
    Digital Frame Camera Data-Collection 231
    Positive Systems, Inc. 233
    Litton Emerge Spatial, Inc. 235
Satellite Photographic Systems 235
    Russian SPIN-2 TK-350 and KVR-1000 Cameras. 235

U.S. Space Shuttle Photography 238
Digital Image Data Storage 238

Chapter 8 — Thermal Infrared Remote Sensing 243
    History of Thermal Infrared Remote Sensing 244
    Thermal Infrared Radiation Properties 246
        Kinetic Heat, Temperature, Radiant Energy and Radiant Flux 246
        Methods of Transferring Heat 246
    Thermal Infrared Atmospheric Windows 246
    Thermal Radiation Laws 247
        Stefan-Boltzmann Law 247
        Wien's Displacement Law 248
        Emissivity 248
        Kirchoff's Radiation Law 250
    Thermal Properties of Terrain 253
    Thermal Infrared Data Collection 254
        Thermal Infrared Multispectral Scanners 254
        Pushbroom Linear and Area-Array Charge-Coupled-Device (CCD) Detectors . . . 265
    Thermal Infrared Environmental Considerations 266
        Diurnal Temperature Cycle of Typical Materials 266
    Examples of Thermal Infrared Remote Sensing 271
        Nonpoint Source Pollution Monitoring 271
        Residential Thermal Imagery Energy Surveys 274
        Analysis of the Urban Heat Island Effect 278
        Use of Thermal Infrared Imagery for Forestry Applications 278

Chapter 9 — Active and Passive Microwave, and LIDAR Remote Sensing 285
    History of Active Microwave (RADAR) Remote Sensing 285
    Active Microwave System Components 288
        Sending and Receiving a Pulse of Microwave Energy – System Components . . . 288
        Slant-Range Versus Ground-Range RADAR Image Geometry 292
        Relief Displacement, Image Foreshortening, Layover, Shadows, and Speckle . . . 299
        Synthetic Aperture Radar Systems 302
    RADAR Environmental Considerations 308
        Surface Roughness Characteristics 308
        Electrical Characteristics (Complex Dielectric Constant)
            and the Relationship with Moisture Content 311
        Vegetation Response to Microwave Energy 312
        Water Response to Microwave Energy 317
        Urban Structure Response to Microwave Energy 317
    SAR Remote Sensing from Space 319
        Seasat 319
        Shuttle Imaging Radar SIR–A, SIR–B, SIR–C 319
        RADARSAT 319
        European Space Agency ERS–1 322
        JERS–1 322
        Almaz–1 323
    RADAR Interferometry 323
        Interferometric Topographic Mapping 323
        Interferometric Velocity Mapping 324
    Passive Microwave Remote Sensing 325
        Passive Microwave Radiometers 325
    Light Detection and Ranging (LIDAR) 326

LIDAR Sensor System 327
Accuracy of LIDAR Measurements 328
Canopy Penetration Capability 329

Chapter 10 — Remote Sensing of Vegetation 333
Photosynthesis Fundamentals. 333
Spectral Characteristics of Vegetation 334
Dominant Factors Controlling Leaf Reflectance 334
Temporal Characteristics of Vegetation 352
Natural Phenological Cycles. 353
Managed Phenological Cycles 354
Vegetation Indices. 361
Landscape Ecology Metrics 366
Landscape Indicators and Patch Metrics. 367
Biodiversity and GAP Analysis 368
Remote Sensing of Vegetation Change 370
Remote Sensing Inland Wetland Successional Changes. 370

Chapter 11 — Remote Sensing of Water 379
Remote Sensing Surface Water Biophysical Characteristics 380
Water Surface, Subsurface Volumetric, and Bottom Radiance. 380
Spectral Response of Water as a Function of Wavelength 381
Spectral Response of Water as a Function of Organic /Inorganic Constituents . . .385
Water Penetration (Bathymetry) 393
Water Surface Temperature. 393
Precipitation 395
Visible–Infrared Techniques 395
Active and Passive Microwave Techniques 395
Aerosols and Clouds. 396
Aerosols 396
Clouds. 397
Water Vapor 401
Snow. 401
Snow in the Visible Spectrum. 401
Snow in the Middle-Infrared and Microwave Regions. 401
Water Quality Modeling Using Remote Sensing and GIS 402
An Integrated Remote Sensing and GIS Water Quality Model 402

Chapter 12 — Remote Sensing the Urban Landscape. 407
Urban/Suburban Resolution Considerations 408
Urban/Suburban Temporal Resolution Considerations. 408
Urban/Suburban Spectral Resolution Considerations. 408
Urban/Suburban Spatial Resolution Considerations. .411
Remote Sensing Land Use and Land Cover. 413
Land Use/Land Cover Classification Schemes 413
Urban Land Use/Land Cover Classification (Levels I to IV) 414
Residential Land Use 418
Single-Family versus Multi-Family Residential 418
Building and Cadastral (Property Line) Infrastructure 422
Socioeconomic Characteristics Derived from Single- and Multi -Family
Residential-Housing Information 422
Energy Demand and Conservation 425
Commercial and Services Land Use. 426

The Central Business District 426
Commercial Land Use 427
Services (Public and Private) 439
Industrial Land Use 443
Industrial Land Use Classification Logic 443
Extraction Industries 443
Processing Industries 444
Fabrication Industries 453
Transportation Infrastructure 456
Communications and Utilities 460
Digital Elevation Model (DEM) Creation 463
Meteorological Data 464
Critical Environmental Area Assessment 465
Disaster Emergency Response 465
Observations 466

Chapter 13 — Remote Sensing of Soils, Minerals, and Geomorphology 471
Soil Characteristics and Taxonomy 472
Soil Horizons 472
Soil Grain Size and Texture 473
Soil Taxonomy 474
Remote Sensing of Soil Properties 474
Soil Texture and Moisture Content 476
Soil Organic Matter 478
Iron Oxide 478
Surface Roughness 479
Remote Sensing of Rocks and Minerals 479
Imaging Spectroscopy of Rocks and Minerals 480
Geology 483
Lithology 484
Structure 484
Drainage Density and Pattern 485
Geomorphology 490
Igneous Landforms 492
Landforms Developed on Horizontal Strata 494
Landforms Developed on Folded Strata 497
Fault-Controlled Landforms 500
Fluvial Landforms 503
Karst Landforms 509
Shoreline Landforms 511
Glacial Landforms 517
Eolian Landforms 522

Index 531

Appendix A—Sources of Remote Sensing Information 541

Preface

Remote sensing is defined as the art and science of obtaining information about an object without being in direct physical contact with the object. It is a scientific technology that can be used to measure and monitor important biophysical characteristics and human activities on Earth. *Remote Sensing of the Environment: An Earth Resource Perspective* is designed to introduce the reader to:

- the fundamental characteristics of electromagnetic radiation and how the energy interacts with Earth materials such as vegetation, soil, rock, water, and urban infrastructure,

- how the electromagnetic energy reflected or emitted from these materials is recorded using a variety of remote sensing instruments (e.g., cameras, multispectral scanners, hyperspectral instruments, RADAR, LIDAR), and

- how we can extract fundamental biophysical or land use/land cover information from the remote sensor data to solve important problems.

The book also introduces the principles of visual photo-interpretation and image analysis. It is a companion volume to the more advanced *Introductory Digital Image Processing: A Remote Sensing Perspective* (1996) published by Prentice-Hall, Inc., which introduces the fundamentals of digital image analysis.

This book was written for physical, natural, and social scientists interested in how remote sensing of the environment can be used to solve real-world problems. The reader should already have education or practical experience in some systematic body of knowledge (e.g., physical geography, soil taxonomy, biogeography, geology, hydrology, urban planning, agriculture, forestry, marine science) to which the remote sensing science will be applied. It is assumed that the reader has a background in college algebra. This remote sensing text can be used in undergraduate or graduate, one- or two-semester courses in remote sensing where the emphasis is on earth resource analysis.

The world has entered the electronic information age which, now more than ever before, includes *spatial* information. People responsible for managing the Earth's natural resources and planning future development recognize the importance of accurate, spatial information residing in a digital geographic information system (GIS). Many of the most important layers of biophysical, land use/land cover, and socioeconomic information in a GIS database are derived from an analysis of remotely sensed data. Thus, we now see a significant increase in the demand for remote sensing data. In fact, hundreds of public and private remote sensing systems sponsored by government agencies and others throughout the world now collect remote sensor data. Some remote sensor data are available at relatively reasonable cost per km^2 and may be obtained by the user over the Internet.

To utilize the remotely sensed data to its full potential, however, users must interpret it correctly. This requires that they understand how energy interacts with the atmosphere and terrain, how remote sensing systems function, and how to extract useful information from the remote sensor data. This book will assist the reader to understand remote sensing science.

The following features hopefully make the information in this book easy to comprehend and apply:

- Each chapter includes illustrations that were designed to make relatively complex principles easy to understand.

- Each chapter contains a substantive reference list.

- The larger 8.5 x 11 in. book format allows the remote sensing images and diagrams to be more readable and visually informative.

- Thirty-two pages of color are used to demonstrate various types of remote sensor data or biophysical information that may be extracted from remote sensor data.

- An appendix provides Internet addresses for some of the most important sources of remote sensing information and public and private sources of remotely sensed data.

Below is a brief summary of the topics discussed in each chapter.

Chapter 1. Remote Sensing of the Environment. This introductory chapter summarizes the remote sensing process. The various elements of the process are reviewed, including hypothesis testing procedures, data collection, data analysis (visual and digital), and information presentation alternatives. The chapter also introduces how remote sensing is used for a variety of national and international earth resource monitoring programs including the Global Change Research Program and NASA's Earth Science Enterprise. The chapter concludes with an overview of the contents of the book.

Chapter 2. Electromagnetic Radiation Principles. Using wave and particle models, this chapter introduces how electromagnetic energy is generated and transported through space. The reader is introduced to how electromagnetic energy interacts with the atmosphere, the terrain, and how the reflected or emitted energy is recorded by a remote sensing instrument. Emphasis is placed on minimizing the effects of atmospheric scattering to maximize the remote sensing signal from the terrain.

Chapter 3. History of Aerial Photography and Aerial Platforms. A substantive history of photography is provided, including the invention of specific types of emulsions and fixative agents. A history of the collection of aerial photography from various platforms is then reviewed, including: balloons, kites, rockets, pigeons, gliders, aircraft, and spacecraft. A history of photoreconnaissance in World War I, World War II, and the Cold War is presented. New developments in remote sensing platforms including unmanned aerial vehicles (UAV) are summarized.

Chapter 4. Aerial Photography – Vantage Point, Cameras, Filters, and Film. This chapter introduces the fundamental characteristics of vertical and oblique aerial photography. It then reviews the more important types of aerial cameras and band-pass filters that are used during data collection. The major black-and-white and color films (emulsions) used for aerial photography data collection are then identified, including color-infrared film. Methods of digitizing aerial photography are introduced. The chapter concludes with a brief discussion of flightline planning.

Chapter 5. Elements of Visual Image Interpretation. Unique characteristics of remote sensor data are reviewed, including: the power of an aerial regional perspective, the utility of three-dimensional depth perception, why it is important to obtain knowledge beyond our human visual perception, and the significance of being able to collect a historical spatial record of Earth resources to document change. The fundamental elements of image interpretation are introduced. The chapter concludes by summarizing various methods of search or inquiry involving collateral information, convergence-of-evidence, and use of the multi-concept in image analysis.

Chapter 6. Photogrammetry. The fundamental methods of extracting quantitative information from vertical aerial photography are presented. The chapter identifies the basic geometry of a single aerial photograph and how scale and height measurements are made. It then introduces stereoscopic principles and how they are used to extract three-dimensional information from overlapping, stereoscopic aerial photography. Much of the world today requires accurate orthophotos or orthophotomaps to monitor urban/suburban phenomena (e.g., cadastral property lines, tax mapping, utilities, floodplain management). A section summarizes how orthophotos and digital elevation models (DEM) are extracted from stereoscopic aerial photography. The final section describes how area measurements are made from vertical aerial photographs and other remote sensor data.

Chapter 7. Multispectral Remote Sensing Systems. The terminology of multispectral remote sensing is introduced. The major types of remote sensing instruments are then reviewed, including:

- multispectral imaging using discrete detectors and scanning mirrors (e.g., Landsat 1-7; AVHRR; SeaWiFS),

- multispectral imaging using linear arrays (e.g., Indian IRS-1C; SPOT, Inc.; Space Imaging, Inc., IKONOS),

- imaging spectrometry using linear and area arrays (e.g., AVIRIS, MODIS),

- digital frame cameras (e.g., Positive Systems, Inc., Litton Emerge Spatial, Inc.), and

- satellite photographic systems (e.g., Russian TK-350 and KVR-1000; U.S. Space Shuttle photography).

The major commercial high-resolution satellite systems are introduced (e.g., Space Imaging, Inc., EarthWatch, Inc., ORBIMAGE, Inc.) as well as the characteristics of the sensors onboard the NASA *Terra* satellite (e.g., ASTER, MODIS, MISR). The chapter concludes with a brief review of how remote sensor data are stored and transmitted.

Chapter 8. Thermal Infrared Remote Sensing. This chapter describes how surface temperature mapping is performed using thermal infrared remote sensing instruments. Thermal infrared radiation principles and atmospheric windows are summarized. Thermal radiation laws are reviewed. The diurnal temperature characteristics of vegetation, soil, rock, water, and urban phenomena are introduced. The characteristics of thermal infrared scanning instruments and forward looking infrared (FLIR) sensors are presented. Thermal infrared remote sensing case studies are provided.

Chapter 9. Active and Passive Microwave, and LIDAR Remote Sensing. Methods of sending and receiving a pulse of microwave energy are presented. Important system parameters such as frequency, polarization, pulse length, depression angle, look direction, and slant-range and ground-range image geometry are discussed. The major environmental factors influencing active microwave backscatter are presented, including: surface roughness, complex dielectric constant, and the cardinal effect. Active micro-

wave satellite remote sensing systems are reviewed. Principles of active microwave interferometry are explained. Passive microwave remote sensing is discussed. Finally, remote sensing using Light Detection and Ranging (LIDAR) technology is presented.

Chapter 10. Remote Sensing of Vegetation. This chapter reviews photosynthesis fundamentals. It then summarizes how vegetation interacts with incident electromagnetic energy in the visible, near-infrared, and middle-infrared portions of the spectrum. The bidirectional reflectance distribution function (BRDF) concept is introduced. The importance of understanding the temporal phenological cycle of the vegetation types under investigation is emphasized. Numerous vegetation indices derived from remote sensor data are identified. The use of remotely sensed data for computing landscape ecology metrics and for assisting in the assessment of biodiversity are introduced. A case study documents how vegetation characteristics may be inventoried through time to monitor vegetation regrowth.

Chapter 11. Remote Sensing of Water. This chapter introduces how incident electromagnetic energy interacts with the water surface, subsurface, and the bottom as a function of wavelength. The impact of organic (e.g., chlorophyll) and/or inorganic (e.g., suspended sediment) material in the water column on spectral reflectance is discussed. Remote sensing methods for monitoring precipitation, atmospheric aerosols and clouds, water vapor, and snow are presented.

Chapter 12. Remote Sensing the Urban Landscape. This chapter discusses the spatial, spectral, and temporal characteristics of urban and suburban phenomena. It introduces the major classification schemes used in urban research and how urban land use and/or land cover information are typically extracted from remote sensor data. Examples of residential, commercial, industrial, transportation, communications, and utilities land use are presented using examples from a variety of remote sensing systems. The increasing importance of obtaining a three-dimensional model of all the buildings and infrastructure in a city is discussed. The utility of high-resolution remote sensing data for monitoring sensitive environments and for urban disaster emergency response are presented.

Chapter 13. Remote Sensing of Soils, Minerals, and Geomorphology. The fundamental characteristics of soil taxonomy are introduced. The influence of soil texture, moisture content, organic matter, iron oxide, and surface roughness on the spectral reflectance of soils is presented. Rock and mineral discrimination using imaging spectroscopy techniques is discussed. The chapter identifies the geologic lithology, structure, drainage density and pattern of the Earth using remote sensor data. The chapter concludes with examples of how remote sensor data may be used to identify geomorphic features on the surface of the Earth.

Acknowledgments

The author wishes to thank the following people for their support and assistance in the preparation of this book. My wife, Marsha, provided patient encouragement during the years of manuscript preparation. Judith Berglund proof-read the manuscript and assisted in the preparation of the Appendix. Tim Ivey, a close friend, offered moral support. Dr. Susan Cutter provided resources and an environment within the Geography Department at the University of South Carolina that facilitated the creation of this book.

The author is indebted to the following scientists who reviewed selected chapters: Dr. John E. Estes, Dr. Glen Gustafson, Dr. Floyd Henderson, Dr. Chor Pang Lo, Dr. Jeff Luvall, Dr. Sunil Narumalani, Dr. Kevin Price, Dr. Dale Quattrochi, Dr. Don Rundquist, Judith Berglund, Jennifer Meisburger, and Steve Schill. Personnel at Prentice-Hall, Inc. were especially helpful including Dan Kaveney (Editor), Amanda Griffith (Assistant editor), and Ed Thomas (Production).

The American Society for Photogrammetry and Remote Sensing, the Association of American Geographers, Geocarto International Centre, Inc., American Elsevier Publishing Co., and Taylor & Francis, Inc. granted permission for the author to extract copyrighted material from his papers and other authors' works published in *Photogrammetric Engineering & Remote Sensing*, the *Manual of Remote Sensing, Manual of Color Aerial Photography, Annals* of the Association of American Geographers, *Geocarto International—A Multidisciplinary Journal of Remote Sensing & GIS, Professional Geographer, Remote Sensing of Environment, International Journal of Remote Sensing,* and *International Journal of Geographical Information Systems*.

John Pike of the Federation of American Scientists made available several useful remote sensing images. Large-scale kite aerial photography was provided by Dr. Cris Benton of the University of California Berkeley. Dr. Stefan Sandmeier of NASA Goddard provided much of the BRDF material.

Several historical photographs were made available by the Harry Ransom Humanities Research Center, University of Texas, Austin, TX, and the Deutsches Museum, Munich, Germany.

A number of federal agencies were helpful, including: personnel of the Smithsonian National Air and Space Museum, the Library of Congress, National Aeronautics and Space Administration (NASA) Observatorium, NASA Goddard Tropical Rainfall Measurement Mission Office (Alan Nelson), NASA Goddard Solar and Heliospheric Observatory (Joseph Gurman), NASA John C. Stennis Space Center Commercial Remote Sensing Program (Dr. Bruce A. Davis), NASA Johnson Space Center (Dr. Kamlesh Lulla), NASA Marshall Space Flight Center (Dr. Dale Quattrochi), NASA Jet Propulsion Laboratory (JPL), U.S. Geological Survey (Don Lauer), USGS Desert Processes Working Group, USGS National Aerial Photography Program (NAPP), USGS Imaging Spectroscopy Lab (Roger Clark), National Oceanic and Atmospheric Administration (NOAA) and the Geological Survey of Canada.

Commercial remote sensing data providers made available examples of their remote sensing instruments and image products. A special thanks to: Space Imaging, Inc. (Linda Lidov), SPOT Image, Inc. (Clark Nelson), Lockheed Martin, Inc. (Jeannie Duisenberg and Eric Schulzinger), Litton Emerge Spatial, Inc. (Don Light), Positive Systems, Inc. (Kim Hickman), ORBIMAGE, Inc. (Sue Hale), EarthWatch, Inc., Intermap, Inc. (Ron Birk), Sovinformsputnik and Aerial Images, Inc. (Michele Hane), Aeromap USA, Inc., Environmental Research Institute of Michigan (Eric Kasischke), RADARSAT, Inc., Eastman Kodak Company (Robert Lundquist and Mary Skerrett), and Hughes Santa Barbara Research Center, Inc. Photogrammetric engineering firms provided information and images of their instruments, including: L-H Systems, Inc. (Dr. A. Stewart Walker), Marconi Integrated Systems, Inc. (Bob Hayes), E. Coyote Enterprises and Z/I Imaging, Inc. (Marilyn O'Cuilinn), Vexcel, Inc., Carl Zeiss, Inc. (Anke B. Shimko), and Optem, Inc. Digital image processing was performed using ERDAS and ENVI software. GIS analysis was performed using ESRI Arc-Info software.

John R. Jensen
University of South Carolina

Remote Sensing of the Environment 1

Scientists are concerned with observing nature, making careful observations and measurements, and then attempting to accept or reject hypotheses concerning these phenomena. The data collection may take place directly in the field (referred to as *in situ* or *in-place* data collection), or at some remote distance from the subject matter.

 In Situ Data Collection

One form of *in situ* data collection involves the scientist going out in the field and questioning the phenomena of interest. For example, an enumerator for the decennial (10-year) census goes from door to door, asking people questions about their age, sex, education, income, etc. These data are recorded and used to document quantitatively the demographic characteristics of the population.

Conversely, a scientist may elect to use a *transducer* or other *in situ* measurement device at the study site to make measurements. Transducers are usually placed in direct physical contact with the object of interest. Many different types of transducers are available. For example, a scientist could use a thermometer to measure the temperature of the air, soil, or water; an anemometer to measure the speed of the wind; or a psychrometer to measure the humidity of the air. The data recorded by the transducers may be an analog electrical signal with voltage variations related to the intensity of the property being measured. Often these analog signals are transformed into digital values using analog-to-digital (A-to-D) conversion procedures. *In situ* data collection using transducers relieves the scientist of monotonous data collection often in inclement weather. Also, the scientist can distribute the transducers at important geographic locations throughout the study area, allowing the same type of measurement to be obtained at many locations at the same instant in time. Sometimes data from the transducers are telemetered electronically to a central collection point for rapid evaluation and archiving.

Two examples of *in situ* data collection are demonstrated in Figure 1-1. Leaf-area-index (LAI) measurements are being collected by a scientist at the study site using a handheld ceptometer in Figure 1-1a. Spectral reflectance measurements of the vegetation canopy are being obtained at the study site using a spectroradiometer in Figure 1-1b. LAI and spectral reflectance measurements obtained in the field may be used to calibrate LAI and spectral reflectance measurements collected by a remote sensing system located on an aircraft or satellite.

a.   b.

Figure 1-1   *In situ* (in-place) data are collected in the field at the study site. a) A scientist is collecting leaf-area-index (LAI) measurements of soybeans (*Glycine max L. Merrill*) using a ceptometer that measures the number of "sunflecks" that pass through the vegetation canopy. The measurements are typically made above the canopy and on the ground below the canopy. The *in situ* LAI measurements may be used to calibrate LAI estimates derived from remote sensor data. b) Spectral reflectance measurements from the vegetation canopy are being collected using a spectroradiometer located 1 m above the canopy. The *in situ* spectral reflectance measurements may be used to calibrate the spectral reflectance measurements obtained from a remote sensing system onboard an aircraft or satellite.

Data collection by scientists in the field or by instruments placed in the field provide much of the data for physical, biological, and social science research. However, it is important to remember that no matter how careful the scientist is, error may be introduced during the *in situ* data-collection process. First, the scientist in the field can be *intrusive*. This means that unless great care is exercised, the scientist can actually change the characteristics of the phenomenon being measured during the data-collection process. For example, a scientist could lean out of a boat to obtain a surface-water sample from a lake. Unfortunately, the movement of the boat into the area may have stirred up the water column in the vicinity of the water sample, resulting in an unrepresentative, or *biased,* sample. Similarly, a scientist collecting a ceptometer LAI reading could inadvertently step on the sample site, compacting the vegetation canopy prior to data collection. Scientists may also collect data in the field using biased procedures. This introduces *method-produced error.* It could involve the use of a biased sampling design or the systematic, improper use of a piece of equipment. Finally, the *in situ* data-collection measurement device may be calibrated incorrectly. This can result in serious measurement error.

Intrusive *in situ* data collection, coupled with human method-produced error and measurement-device miscalibration, all contribute to *in situ* data-collection error. Therefore, it is a misnomer to refer to *in situ* data as *ground truth data*. Instead, we should simply refer to it as *in situ ground reference data*, acknowledging that it contains error.

 Remote Sensing Data Collection

It is also possible to collect information about an object or geographic area from a distant vantage point (Figure 1-2) using specialized instruments (sensors). This remote data collection was originally performed using aerial cameras. *Photogrammetry* was defined in the early editions of the *Manual of Photogrammetry* as:

> "the art or science of obtaining reliable measurement by means of photography" (American Society of Photogrammetry, 1944; 1952; 1966).

Photographic interpretation is defined as:

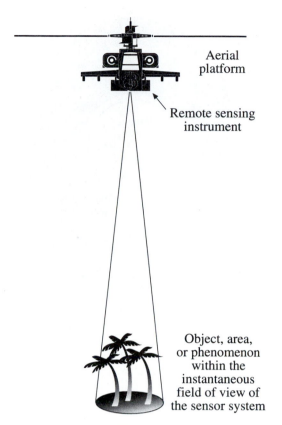

Aerial
platform

Remote sensing
instrument

Object, area,
or phenomenon
within the
instantaneous
field of view of
the sensor system

Figure 1-2   A remote sensing instrument collects information
about an object or phenomenon within the instantaneous field of view of the sensor system without being in direct physical contact.

"the act of examining photographic images for the
purpose of identifying objects and judging their significance" (Colwell, 1960).

Remote sensing was formally defined by the American Society for Photogrammetry and Remote Sensing (ASPRS) as:

ASPRS Definition: "the measurement or acquisition
of information of some property of an object or phenomenon, by a recording device that is not in physical or intimate contact with the object or
phenomenon under study" (Colwell, 1983).

In 1988, ASPRS adopted a combined definition of photogrammetry and remote sensing:

ASPRS Combined Definition: "photogrammetry and
remote sensing are the art, science, and technology
of obtaining reliable information about physical
objects and the environment, through the process of
recording, measuring and interpreting imagery and

digital representations of energy patterns derived
from noncontact sensor systems" (Colwell, 1997).

But where did the term *remote sensing* come from? The
actual coining of the term goes back to an unpublished paper
in the early 1960s by the staff of the Office of Naval
Research (ONR) Geography Branch (Pruitt, 1979; Fussell et
al., 1986). Evelyn L. Pruitt was the author of the paper. She
was assisted by staff member Walter H. Bailey. Aerial photo
interpretation had become very important in World War II.
The space age was just getting underway with the 1957
launch of *Sputnik* (U.S.S.R.), the 1958 launch of *Explorer 1*
(U.S.), and the collection of photography from the then
secret CORONA program initiated in 1960 (Table 1-1). In
addition, the Geography Branch of ONR was expanding its
research using instruments other than cameras (e.g., scanners, radiometers) and into regions of the electromagnetic
spectrum beyond the visible and near-infrared regions (e.g.,
thermal infrared, microwave). Thus, in the late 1950s it had
become apparent that the prefix "photo" was being stretched
too far in view of the fact that the root word, *photography*,
literally means "to write with [visible] light" (Colwell,
1997). Evelyn Pruitt (1979) wrote:

"The whole field was in flux and it was difficult for
the Geography Program to know which way to
move. It was finally decided in 1960 to take the
problem to the Advisory Committee. Walter H.
Bailey and I pondered a long time on how to present
the situation and on what to call the broader field
that we felt should be encompassed in a program to
replace the aerial photointerpretation project. The
term 'photograph' was too limited because it did not
cover the regions in the electromagnetic spectrum
beyond the 'visible' range, and it was in these nonvisible frequencies that the future of interpretation
seemed to lie. 'Aerial' was also too limited in view
of the potential for seeing the Earth from space."

The term *remote sensing* was promoted in a series of symposia sponsored by ONR at the Willow Run Laboratories of the
University of Michigan in conjunction with the National
Research Council throughout the 1960s and early 1970s and
has been in use ever since (Estes and Jensen, 1998).

Maximal/Minimal Definitions

Numerous other definitions of remote sensing have been
proposed. In fact, Colwell (1984) suggests that "one measure of the newness of a science, or of the rapidity with
which it is developing is to be found in the preoccupation of

its participating scientists with matters of terminology." Some have proposed an all-encompassing *maximal definition* where:

> *Maximal Definition:* "remote sensing is the acquiring of data about an object without touching it."

Such a definition is short, simple, general, and memorable. Unfortunately, it excludes little from the province of remote sensing (Fussell et al., 1986). It encompasses virtually all remote-sensing devices, including cameras, optical-mechanical scanners, linear and area arrays, lasers, radio-frequency receivers, radar systems, sonar, seismographs, gravimeters, magnetometers, and scintillation counters.

Others have suggested a more sharply focused, *minimalist definition* of remote sensing that adds qualifier after qualifier in an attempt to make certain that only legitimate functions are included in the term's definition. For example,

> *Minimal Definition:* "remote sensing is the noncontact recording of information from the ultraviolet, visible, infrared, and microwave regions of the electromagnetic spectrum by means of instruments such as cameras, scanners, lasers, linear arrays, and/or area arrays located on platforms such as aircraft or spacecraft, and the analysis of acquired information by means of visual and digital image processing."

Robert Green at NASA's Jet Propulsion Lab recently suggested we use the term *remote measurement* because data obtained using the new hyperspectral remote sensing systems are so accurate (Robbins, 1999). Each of the definitions are correct in an appropriate context. It is useful to briefly discuss components of these remote sensing definitions.

Remote Sensing: Art and/or Science?

Science: A *science* is defined as the broad field of human knowledge concerned with facts held together by *principles* (rules). Scientists discover and test these facts and principles by the scientific method, an orderly system of solving problems. Scientists generally feel that any subject that man can study by using the scientific method and other special rules of thinking may be called a science. The sciences include: 1) *mathematics* and *logic*, 2) the *physical sciences*, such as physics and chemistry, 3) the *biological sciences*, such as botany and zoology, and the 4) *social sciences*, such as geography, sociology, and anthropology (Figure 1-3). Interestingly, some persons do not consider mathematics and logic as sciences. But the fields of knowledge associated with

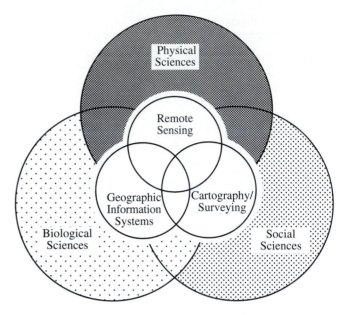

Figure 1-3   A three-way model of interaction between the mapping sciences of remote sensing, geographic information systems, and cartography/surveying as they are used in the physical, biological, and social sciences (after Dahlberg and Jensen, 1986; Fisher and Lindenberg, 1989).

mathematics and logic are such valuable *tools* for science that we cannot ignore them. Man's earliest questions were concerned with "how many" and "what belonged together." He struggled to count, to classify, to think systematically, and to describe exactly. In many respects, the state of development of a science is indicated by the use it makes of mathematics. A science seems to begin with simple mathematics to measure, then works toward more complex mathematics to explain.

Remote sensing is a tool or technique similar to mathematics. Using sophisticated sensors to measure the amount of electromagnetic energy exiting an object or geographic area from a distance and then extracting valuable information from the data using mathematically and statistically based algorithms is a *scientific* activity (Fussell et al., 1986). It functions in harmony with other *spatial* data-collection techniques or tools of the *mapping sciences*, including cartography and geographic information systems (GIS) (Fussell et al., 1986; Curran, 1987). In fact, Dahlberg and Jensen (1986) and Fisher and Lindenberg (1989) suggest a model where there is three-way interaction between remote sensing, cartography, and GIS, where no subdiscipline dominates and all are recognized as having unique yet overlapping areas of

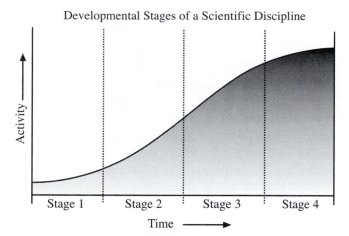

Developmental Stages of a Scientific Discipline

Stage 1 : Stage 2 : Stage 3 : Stage 4

Figure 1-4 The developmental stages of a scientific discipline (adapted from Wolter, 1975; Jensen and Dahlberg, 1983).

knowledge and intellectual activity as they are used in physical, biological, and social science research (Figure 1-3).

The theory of science suggests that scientific disciplines go through four classic developmental stages. Wolter (1975) suggested that the growth of a discrete scientific discipline, such as remote sensing, that has its own techniques, methodologies, and intellectual orientation seems to follow the sigmoid or logistic curve illustrated in Figure 1-4. The growth stages of a scientific field are: Stage 1 — a preliminary growth period with small absolute increments of literature; Stage 2 — a period of exponential growth when the number of publications doubles at regular intervals; Stage 3 — a period when the rate of growth begins to decline but annual increments remain constant; and Stage 4 — a final period when the rate of growth approaches zero. The characteristics of a scholarly field during each of the stages may be briefly described as follows (Wolter, 1975): Stage 1 — little or no social organization; Stage 2 — groups of collaborators and existence of invisible colleges, often in the form of ad hoc institutes, research units, etc.; Stage 3 — increasing specialization and increasing controversy; and Stage 4 — decline in membership in both collaborators and invisible colleges.

Using this logic, it may be suggested that remote sensing is in Stage 2 of a scientific field, experiencing exponential growth since the mid-1960s with the number of publications doubling at regular intervals (Colwell, 1983; Cracknell and Hayes, 1993). Empirical evidence is presented in Table 1-1, including: 1) the organization of many specialized institutes and centers of excellence associated with remote sensing, 2) the organization of numerous professional societies devoted to remote sensing research, 3) the publication of numerous

new scholarly remote sensing journals, 4) significant technological advancement such as improved sensor systems and methods of image analysis, and 5) intense self-examination. We are approaching Stage 3 with increasing specialization and theoretical controversy. However, the rate of growth of remote sensing has not begun to decline. In fact, there has recently been a tremendous surge in the numbers of persons specializing in remote sensing and commercial firms using remote sensing during the 1990s (Davis, 1999). Significant improvements in the spatial resolution of satellite remote sensing (e.g., more useful 1 x 1 m panchromatic data) is expected to bring even more social science GIS practictioners into the fold. Hundreds of new peer-reviewed remote sensing research articles are published every month.

Art: The process of visual photo or image interpretation brings to bear not only scientific knowledge but all of the background that a person has obtained through a lifetime. Such learning cannot be measured, programmed, or completely understood. The synergism of combining scientific knowledge with real-world analyst experience allows the interpreter to develop heuristic rules of thumb to extract valuable information from the imagery. It is a fact that some image analysts are much superior to other image analysts because they: 1) understand the scientific principles better, 2) are more widely traveled and have seen many landscape objects and geographic areas first-hand, and 3) they can synthesize scientific principles and real-world knowledge to reach logical and correct conclusions. Thus, remote sensing is both an art and a science.

Information About an Object or Area

Sensors can obtain very specific information about an object (e.g., the diameter of an oak tree crown) or the geographic extent of a phenomenon (e.g., the polygonal boundary of an entire oak forest). The electromagnetic energy emitted or reflected from an object or geographic area is used as a surrogate for the actual property under investigation. The electromagnetic energy measurements must be turned into information using visual and/or digital image processing techniques.

The Instrument (Sensor)

Remote sensing is performed using an instrument, often referred to as a *sensor*. The majority of remote sensing instruments described in this book record electromagnetic radiation (EMR) that travels at a velocity of 3×10^8 m s^{-1} from the source, directly through the vacuum of space or indirectly by reflection or reradiation to the sensor. The EMR represents an extremely efficient high-speed commu-

Table 1-1.   Major Milestones in Remote Sensing

1600 and 1700s
1687 - Sir Isaac Newton's *Principia* summarizes basic laws of mechanics

1800s
1826 - Joseph Nicephore Niepce takes first photographic image
1839 - Louis M. Daguerre invents positive print daguerrotype photography
1839 - William Henry Fox Talbot invents Calotype negative/positive process
1855 - James Clerk Maxwell postulates additive color theory
1858 - Gaspard Felix Tournachon takes first aerial photograph from a balloon
1860s- James Clerk Maxwell develops electromagnetic wave theory
1867 - The term *photogrammetry* is used in a published work
1873 - Herman Vogel extends sensitivity of emulsion dyes to longer wavelengths,
       paving the way for near-infrared photography

1900
1903 - Airplane invented by Wright Brothers (Dec 17)
1903 - Alfred Maul patents a camera used to obtain photographs from a rocket.

1910s
1910 - International Society for Photogrammetry (ISP) founded in Austria
1913 - First International Congress of ISP in Vienna
1914 - 1918 World War I photo-reconnaissance

1920s
1920 - 1930 Increase in civilian use of photointerpretation and photogrammetry
1926 - Robert Goddard launches liquid-powered rocket (Mar 16)

1930s
1934 - American Society for Photogrammetry (ASP) founded
1934 - *Photogrammetric Engineering* (ASP)
1938 - *Photogrammetria* (ISP)
1939 - 1945 World War II photo-reconnaissance advances

1940s
1940s - Radar invented
1940s - Jet aircraft invented by Germany
1942 - Kodak patents first false-color infrared film
1942 - Launch of German V-2 rocket by Wernher VonBraun (Oct 3)

1950s
1950s - Thermal infrared remote sensing invented by military
1950 - 1953 Korean War aerial reconnaissance
1953 - *Photogrammetric Record* (Photogrammetric Society, U.K.)
1954 - Westinghouse, Inc. develops side-looking airborne radar system
1955 - 1956 U.S. Genetrix balloon reconnaissance program
1956 - 1960 Central Intelligence Agency U-2 aerial reconnaissance program
1957 - Soviet Union launched *Sputnik* satellite (Oct 4)
1958 - United States launched *Explorer 1* satellite (Jan 31)

1960s
1960s - Emphasis primarily on visual image processing
1960s - Michigan Willow Run Laboratory active — evolved into ERIM
1960s- *Intl. Symp. on Remote Sensing of Environment* (ERIM)
1960s - Purdue Lab for Agricultural Remote Sensing (LARS) active
1960s - Forestry Remote Sensing Lab at U.C. Berkeley (Robert Colwell)
1960s - ITC– Delft, initiates photogrammetric education for students worldwide
1960s - Digital image processing initiated at LARS, Berkeley, Kansas, ERIM
1960s - Declassification of radar and thermal infrared sensor systems
1960 - 1972 United States Corona spy satellite program
1960 - *Manual of Photointerpretation* (ASP)
1960 - *Remote sensing* term introduced by Office of Naval Research personnel
1961 - Yuri Gagarin becomes first human to travel in space
1961 - 1963 Mercury space program
1962 - Cuban Missile Crisis — U-2 photo-reconnaissance shown to the public
1964 - SR-71 discussed in President Lyndon Johnson press briefing
1965 - 1966 Gemini space program
1965 - *ISPRS Journal of Photogrammetry & Remote Sensing*
1969 - *Remote Sensing of Environment*, Elsevier

1970s
1970s, 80s - Possible to specialize in remote sensing at universities
1970s -Digital image processing comes of age
1970s -Remote sensing integrated with digital geographic information systems
1972 - ERTS-1 launched (Earth Resource Technology Satellite)
1973 - 1979 Skylab program
1973 - *Canadian Journal of Remote Sensing* (Canadian RS Society)
1975 - ERTS-2 launched (renamed Landsat 2)
1975 - *Manual of Remote Sensing* (ASP)
1977 - European METEOSAT-1 launched
1978 - Landsat 3 launched
1978 - Nimbus 7 launched - Coastal Zone Color Scanner
1978 - TIROS-N launched with AVHRR sensor
1978 - SEASAT launched

1980s
1980s - AAG Remote Sensing Specialty Group > 500 members
1980s - Commercialization attempted – EOSAT, Inc.
1980 - ISP becomes Intl. Soc. for Photogrammetry & Remote Sensing
1980 - *Intl. Journal of Remote Sensing* (Remote Sensing Society)
1980 - European Space Agency (ESA) created (Oct 30)
1980 - *IEEE Trans. Geoscience and Remote Sensing* (GRSS Society)
1981 - First *Intl. Geoscience and Remote Sensing Symposium*
1981 - NASA Space Shuttle program initiated (STS-1)
1981 - NASA Space Shuttle Imaging Radar (SIR - A) launched
1982 - Landsat 4 - Thematic Mapper and MSS launched
1983 - *Manual of Remote Sensing*, 2nd Ed. (ASP)
1983 - *Remote Sensing Reviews*
1984 - Landsat 5 - Thematic Mapper launched
1984 - NASA Space Shuttle Imaging Radar (SIR-B) launched
1986 - SPOT Image, Inc., launched SPOT 1
1986 - *Geocarto International* (Geocarto International Center)
1989 - *The Earth Observer* (NASA Goddard Space Flight Center)

1990s
1990s - Digital soft-copy photogrammetry comes of age
1990s - University degree programs in remote sensing available
1990s - NASA assists commercial use of remote sensing (Stennis Space Center)
1990s - Increased use of hyperspectral and LIDAR sensors
1990 - *Backscatter* (Alliance for Marine Remote Sensing Association)
1990 - SPOT Image, Inc., launched SPOT 2
1991 - NASA initiates "Mission to Planet Earth" (Goddard Space Flight Center)
1991 - European ERS-1 launched
1992 - U.S. Land Remote Sensing Policy Act becomes law
1993 - EOSAT Inc., Landsat 6 did not achieve orbit
1993 - SPOT Image, Inc., launched SPOT 3
1993 - NASA Space Shuttle Imaging Radar (SIR-C) launched
1995 - Canadian RADARSAT-1 launched
1995 - ERS-2 launched
1995 - Indian IRS-1C launched (5 x 5 m)
1995 - Corona imagery declassified, transferred to National Archives
1995 - *The Earth Observer* (EOS-Goddard)
1996 - *Manual of Photographic Interpretation*, 2nd Ed. (ASPRS)
1997 - *Addendum to Manual of Photogrammetry* (ASPRS)
1997 - EarthWatch, Inc., lost contact with Earlybird satellite
1998 - NASA MTPE redefined as "Earth Science Enterprise"
1998 - *Manual of Remote Sensing - Radar* (ASPRS)
1998 - SPOT Image, Inc., launched SPOT 4
1999 - *Manual of Remote Sensing - Geosciences* (ASPRS)
1999 - NASA Landsat 7 Enhanced Thematic Mapper Plus launched (April 15)
1999 - Space Imaging, Inc., IKONOS did not achieve orbit (Apr 27)
1999 - Space Imaging, Inc., launched a second IKONOS (Sept 24)
1999 - NASA *Terra* Earth observing system launched

2000 – 2001
2000 - NASA to initiate New Millennium Program
2000 - OrbView 3,4 to be launched by ORBIMAGE, Inc.
2000 - Quickbird to be launched by EarthWatch, Inc.
2001 - European Space Agency to launch Envisat

nications link between the sensor and the remotely located phenomenon. In fact, we know of nothing that travels faster than the speed of light. Changes in the amount and properties of the EMR become, upon detection by the sensor, a valuable source of data for interpreting important properties of the phenomenon (e.g., temperature, color). Other types of force fields may be used in place of EMR, including sound waves (e.g., sonar). However, the majority of remotely sensed data collected for Earth resource applications are the result of sensors that record electromagnetic energy.

Distance: How Far Is Remote?

As the name implies, remote sensing occurs at a distance from the object or area of interest. Interestingly, there is no clear distinction about how great this distance should be. The distance could be 1 meter, 100 meters, or > 1 million meters from the object or area of interest. In fact, virtually all astronomy is based on remote sensing. Many of the most innovative remote sensing systems and visual and digital image processing methods were originally developed for remote sensing extraterrestrial landscapes such as the moon, Mars, Io, Saturn, Jupiter, etc. Remote sensing science conducted by the Jet Propulsion Laboratory at the California Institute of Technology is particularly noteworthy. This text, however, is concerned primarily with remote sensing of the terrestrial Earth, using sensors that are placed on suborbital air-breathing aircraft, or orbital satellite platforms placed in the vacuum of space.

Remote sensing techniques may also be used to analyze inner space. For example, an electron microscope and associated hardware may be used to obtain photographs of extremely small objects on the skin, in the eye, etc. Similarly, an X-ray device is a remote sensing instrument where the skin and muscle are equivalent to the atmosphere that must be penetrated, and the interior bone or other matter is often the object of interest.

Remote Sensing Advantages and Limitations

Remote sensing has several unique advantages as well as some limitations.

Advantages

Remote sensing is *unobtrusive* if the sensor is passively recording the electromagnetic energy reflected from or emitted by the phenomenon of interest. This is a very important consideration, as passive remote sensing does not disturb the object or area of interest.

Remote sensing devices are often programmed to collect data systematically, such as within a single 9 x 9 in. frame of vertical aerial photography or a matrix (raster) of Landsat image data. This systematic data collection can remove the sampling bias introduced in some *in situ* investigations.

Under carefully controlled conditions, remote sensing can provide fundamental biophysical data, including: x,y location, z elevation or depth, biomass, temperature, moisture content, etc. In this sense it is much like surveying, providing fundamental data that other sciences can use when conducting scientific investigations. However, unlike much of surveying, the remotely sensed data may be obtained systematically over very large geographic areas rather than just single point observations.

Remote sensing is also different from the other mapping sciences such as cartography or GIS because they rely on data produced elsewhere. Remote sensing science yields fundamental scientific information. For example, a properly calibrated thermal infrared remote sensing system can provide a geometrically correct map of land- or sea-surface temperature without any other intervening science. In fact, remote sensing-derived information is now critical to the successful modeling of numerous natural (e.g., water-supply estimation; eutrophication studies; nonpoint source pollution) and cultural processes (e.g., land-use conversion at the urban fringe; water-demand estimation; population estimation) (Walsh et al., 1999). A good example is the digital elevation model that is so important in many spatially distributed GIS models. Digital elevation models are now produced almost exclusively through the analysis of remotely sensed data.

Limitations

Remote sensing science has limitations. Perhaps the greatest limitation is that its utility is often oversold. *It is not a panacea* that will provide all the information needed for conducting physical, biological, or social science. It simply provides some spatial, spectral, and temporal information of value.

Human beings select the most appropriate sensor to collect the data, specify the resolution of the data, calibrate the sensor, select the platform that will carry the sensor, determine when the data will be collected, and specify how the data are processed. Thus, human method-produced error may be introduced as the various remote sensing instrument and mission parameters are specified.

Powerful *active* remote sensor systems, such as lasers or radars that emit their own electromagnetic radiation, can be intrusive and affect the phenomenon being investigated.

The Remote Sensing Process

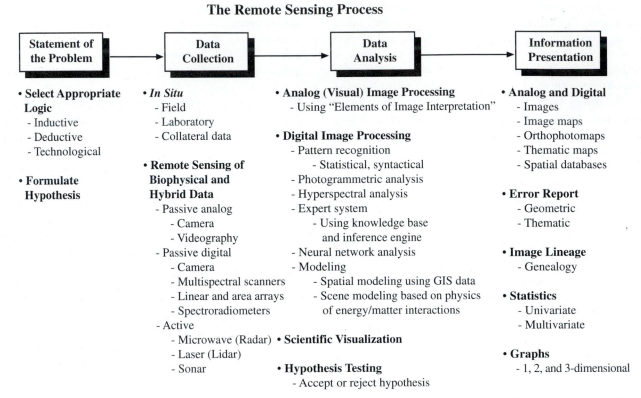

Statement of the Problem → **Data Collection** → **Data Analysis** → **Information Presentation**

- **Select Appropriate Logic**
  - Inductive
  - Deductive
  - Technological

- **Formulate Hypothesis**

- *In Situ*
  - Field
  - Laboratory
  - Collateral data

- **Remote Sensing of Biophysical and Hybrid Data**
  - Passive analog
    - Camera
    - Videography
  - Passive digital
    - Camera
    - Multispectral scanners
    - Linear and area arrays
    - Spectroradiometers
  - Active
    - Microwave (Radar)
    - Laser (Lidar)
    - Sonar

- **Analog (Visual) Image Processing**
  - Using "Elements of Image Interpretation"

- **Digital Image Processing**
  - Pattern recognition
    - Statistical, syntactical
  - Photogrammetric analysis
  - Hyperspectral analysis
  - Expert system
    - Using knowledge base and inference engine
  - Neural network analysis
  - Modeling
    - Spatial modeling using GIS data
    - Scene modeling based on physics of energy/matter interactions

- **Scientific Visualization**

- **Hypothesis Testing**
  - Accept or reject hypothesis

- **Analog and Digital**
  - Images
  - Image maps
  - Orthophotomaps
  - Thematic maps
  - Spatial databases

- **Error Report**
  - Geometric
  - Thematic

- **Image Lineage**
  - Genealogy

- **Statistics**
  - Univariate
  - Multivariate

- **Graphs**
  - 1, 2, and 3-dimensional

Figure 1-5   Scientists generally follow the remote sensing process when attempting to extract information from remotely sensed data.

Additional research is required to determine how intrusive these active sensors are.

Remote sensing instruments like *in situ* instruments often become uncalibrated, resulting in uncalibrated remote sensor data. Finally, remote sensor data may be expensive to collect and interpret or analyze. Hopefully, the information derived from the remote sensor data is of such value that the expense is warranted.

The Remote Sensing Process

Scientists have been developing procedures for collecting and analyzing remotely sensed data for more than 140 years. The first known photograph from an aerial platform (a tethered balloon) was obtained in 1858 by the Frenchman Gaspard Felix Tournachon (who called himself Nadar). Significant strides in aerial photography and other remote sensing data collection took place during World War I and II, the Korean conflict, the Cuban Missile Crisis, the Vietnam War, the Gulf War, and the war in Bosnia. Many of the

accomplishments are summarized in Table 1-1 and in Chapter 3 (History of Aerial Photography and Aerial Platforms). Basically, military contracts to commercial companies resulted in the development of sophisticated electro-optical multispectral remote sensing systems and thermal infrared and microwave (radar) sensor systems whose characteristics are summarized in Chapters 7, 8, and 9, respectively. While the majority of the remote sensing systems may have been initially developed for military reconnaissance applications, the systems are also heavily used for monitoring the Earth's natural resources.

The remote sensing data-collection and analysis procedures used for Earth resource applications are often implemented in a systematic fashion that can be termed the *remote sensing process*. The procedures in the remote sensing process are summarized in Figure 1-5:

- statement of the problem,

- data collection,

- data analysis, and

- presentation of the information so that informed decisions can be made.

It is useful to review the characteristics of these procedures.

Statement of the Problem

The average man or woman and some children can look at aerial photography or other remote sensor data and extract some useful information from it. Generally, they do not interpret the images with any particular plan or hypothesis to test. Unfortunately, it is likely that they may make serious interpretation errors because they do not understand the nature of the remote sensing system used to acquire the data or appreciate the vertical or oblique perspective of the terrain recorded in the imagery.

Scientists who use remote sensing, however, are usually trained in the *scientific method* — a way of thinking about problems and solving them. The formal plan has at least five elements, including: 1) stating the problem, 2) forming the hypothesis (i.e., a possible explanation), 3) observing and experimenting, 4) interpreting data, and 5) drawing conclusions. It is not necessary to follow this formal plan exactly.

Three methodologies or types of logic may be used to structure the problem, such as (Curran, 1987):

- inductive logic,

- deductive logic, and/or

- technological logic.

Inductive logic involves observation, classification, generalization, and theory formulation. Inductive logic is used to build an objective description of observed facts or phenomena, which are then shaped and ordered to derive theory and thereby knowledge. Using inductive logic, the procedure starts with a large number of observations (n) that are true and hopefully unbiased. For example, we may observe on thousands of aerial images that:

All healthy green mangrove forests appear red (magenta) on properly exposed color-infrared film.

Before such a theory can be considered legitimate, three conditions must be satisfied: 1) the number of observations must be large, 2) the observation must be repeated under a wide range of conditions, and 3) no accepted observation should *ever* conflict with the derived theory. The most seri-

ous limitation of inductive logic is that no number of apparently confirming observations can ever show that a theory is completely true.

The pursuit of knowledge using *deductive* logic has an emphasis not on observation but on the formulation of theory and the testing of hypotheses (Curran, 1987). Typically, the scientist states the problem and puts forth a speculative theory to solve that problem. He or she then puts forth a hypothesis (possible explanation). Observations are then made, often involving remote sensing imagery and *in situ* measurements made at unbiased locations. The null hypothesis is then tested at specific statistical confidence levels (e.g., 0.05 or 0.001). If the observations are such that the null hypothesis can be rejected, then the theory can be considered acceptable in the guarded sense that there is not an empirical basis for doubting its validity. If the observations do not support rejection (falsification) of the null hypothesis, then we must go back to our problem and evaluate other possible explanations that might lead to a falsifiable hypothesis.

For example, suppose we visited 100 mangrove forest sites in Florida and obtained color-infrared aerial photographs at each of the sites. We might develop the following null hypothesis:

There is no significant relationship between healthy green mangrove forest and a red (magenta) appearance on properly exposed color-infrared film.

If the healthy green mangrove forest at 99 of the 100 sites did indeed have a red (magenta) appearance in the corresponding color-infrared film, then it would be possible for us statistically to reject (falsify) the null hypothesis. We could then state that there appears to be a statistically significant relationship between healthy green vegetation and the red (magenta) appearance on properly exposed color-infrared film. Unfortunately, there is a tendency to suggest that whenever a null hypothesis is not rejected, that either the sensor was not working properly or the ground reference data were in error, i.e., the observations are in doubt, not the theory. Fortunately, scientists wait until a theory is tested numerous times before it is accepted or rejected (Curran, 1987).

Some scientists extract new thematic information directly from remotely sensed imagery without ever explicitly using inductive or deductive logic. They are just interested in extracting information from the imagery using appropriate methods and technology. This *technological* approach is not as rigorous, but it is common in what some call *applied remote sensing*. The approach can also generate new knowledge.

Remote sensing is used in both scientific (inductive and deductive) and technological approaches to gain knowledge. There is debate as to how the different types of logic used in the remote sensing process and in GIS yield new scientific knowledge (e.g., Fussell et al., 1986; Curran, 1987; Fisher and Lindenberg, 1989; Ryerson, 1989; Duggin and Robinove, 1990; Dobson, 1993; Wright et al., 1997).

Identification of In Situ and Remote Sensing Data Requirements

If a hypothesis is formulated using deductive logic, a list of variables or observations are identified that will be used to verify or falsify the hypothesis. *In situ* observation and/or remote sensing may be used to collect information on the most important variables.

In Situ Data Requirements

Scientists using remote sensing technology should be well trained in field and laboratory data-collection procedures. For example, if a scientist wants to measure the surface temperature of a lake, it is usually essential that some accurate *in situ* lake-temperature measurements be obtained at the same time the remote sensor data are collected. These *in situ* observations may be used to 1) calibrate the remote sensor data, and 2) perform an unbiased accuracy assessment of the final results (Congalton and Green, 1998). Remote sensing textbooks provide some information on field and laboratory sampling techniques. The *in situ* sampling procedures, however, are learned best through formal courses in the sciences (e.g., chemistry, biology, forestry, soils, hydrology, meteorology). In addition, methods of collecting socioeconomic and demographic information in urban environments is often essential (e.g., cultural geography, sociology).

Most *in situ* data are now collected in conjunction with accurate *x, y, z* global positioning system (GPS) data (Jensen and Cowen, 1999). Scientists should know how to collect the fundamental GPS data and then perform differential correction to obtain the most accurate *x, y, z* coordinate information. Sometimes *collateral* data (often called *ancillary* data), such as soil maps, political boundary files, and block population statistics, collected by other scientists are of value in the remote sensing process. Ideally, these spatial collateral data reside in a digital GIS.

Usually it is necessary to collect both *in situ* and remotely sensed data because each type of data may be used to calibrate the other. For example, some *in situ* studies sample the environment and obtain county or statewide data such as

population density. Remote sensing data can be used to disaggregate this sampled data. The spatial distribution of single- and multiple-family dwellings can be easily identified in high-resolution aerial photography. Such information may be used to disaggregate the population density within a county rather than assuming that the population is spread uniformly throughout the entire county. Similarly, a digital elevation model may be used to identify north-facing slopes and specific ranges of elevation (e.g., 1000 – 2000 m above sea level). Such information is very valuable when conducting a remote sensing vegetation study of mountainous terrain, where certain vegetation types only grow on north-facing slopes at an elevation of 1000 – 2000 m above sea level.

Remote Sensing Data Requirements

Once we have a list of variables, it is useful to determine those that can be remotely sensed. Remote sensing can provide information on two different classes of variables: *biophysical* and *hybrid*. Biophysical variables may be measured directly by the remote sensing system. This means that the remotely sensed data can provide fundamental biological and/or physical (*biophysical*) information directly, without having to use other surrogate or ancillary data. For example, a thermal infrared sensor can record the apparent temperature of a rock outcrop by measuring the radiant flux emitted from its surface. Similarly, it is possible to conduct remote sensing in a very specific region of the spectrum and identify the amount of water vapor in the atmosphere. It is also possible to measure soil moisture content directly using microwave remote sensing techniques (Engman and Chauhan, 1995). All three of these are true biophysical measurements. Such data are useful in physical science models.

Another example is the determination of the precise *x, y* location and height (*z*) of an object. Such information can be extracted directly from stereoscopic aerial photography, overlapping satellite imagery (e.g., SPOT) or interferometric radar imagery. A list of selected biophysical variables that can be remotely sensed and useful sensors to acquire such data are found in Table 1-2. The characteristics of most of these sensor systems are discussed in Chapters 7, 8, and 9. Great strides have been made in remotely sensing many of these biophysical variables (Eidenshink, 1992; ESA, 1992). They are important to the national and international effort under way to model the global environment (Lousma, 1993; Asrar and Dozier, 1994; Jones et al.,1997).

The second general group of variables that may be remotely sensed include *hybrid* variables, created by systematically analyzing more than one biophysical variable. For example,

Table 1-2. Biophysical and Hybrid Variables and Potentially Useful Remote Sensing Systems (proposed sensor systems are in italics)

Biophysical Variables	Potential Remote Sensing System
x, y Geographic location	Aerial photography, Landsat TM, SPOT HRV, Russian KVR-1000, IRS-1CD, ATLAS, Radarsat, ERS-1,2 microwave, Landsat 7 ETM$^+$, Space Imaging IKONOS, Terra *MODIS, ASTER, EarthWatch Quickbird, ORBIMAGE OrbView 3,4*
z Topographic/bathymetric	Aerial photography, TM, SPOT, IRS-1CD, Radarsat, LIDAR systems, ETM, IKONOS, *ASTER, Quickbird, OrbView 3,4*
Vegetation chlorophyll concentration biomass (green & dead) foliar water content Absorbed photosynthetically active radiation phytoplankton	 Air photos, TM, SPOT, IRS-1CD, ETM, IKONOS, *ASTER, MODIS, OrbView 3,4* Air photos, AVHRR, TM, SPOT, IRS-1CD, ETM, IKONOS, *MODIS, OrbView 3,4* Radarsat, ERS-1,2; TM Mid-IR, ETM, IKONOS, *MODIS, ASTER, OrbView 3,4* ETM, IKONOS, *MODIS, OrbView 3,4* SeaWiFS, TM, AVHRR, ETM, IKONOS, *MODIS, OrbView 3,4*
Surface temperature	GOES, SeaWiFS, AVHRR, TM, Daedalus, ATLAS, ETM, *ASTER, MODIS*
Soil moisture	ALMAZ, TM, ERS-1,2; Radarsat, Intermap Star 3i, IKONOS, *ASTER, OrbView 3,4*
Surface roughness	Air photos, ALMAZ, ERS-1,2; Radarsat, Star 3i, IKONOS, *ASTER, OrbView 3,4*
Evapotranspiration	AVHRR, TM, SPOT, CASI, ETM, *MODIS, ASTER*
Atmosphere tropospheric chemistry, temperature, water vapor, wind speed/direction, energy inputs, precipitation, cloud and aerosol properties	GOES, UARS, ATREM, *MODIS, MISR, CERES, MOPITT*
BRDF (bidirectional reflectance distribution function)	*MODIS, MISR, CERES*
Ocean color, phytoplankton, biochemistry, sea height	TOPEX/POSEIDON, SeaWiFS, ETM, IKONOS, *MODIS, MISR, ASTER, CERES, OrbView 3,4*
Snow and sea ice extent and characteristics	Aerial photography, AVHRR, TM, SPOT, Radarsat, SeaWiFS, IKONOS, ETM, *MODIS, ASTER, OrbView 3,4; Quickbird*
Volcanic effects temperature, gases	ATLAS, *MODIS, MISR, ASTER*
Selected Hybrid Variables	**Potential Remote Sensing System**
Land use urban infrastructure and land use	Aerial photography, AVHRR, TM, SPOT, Russian KVR-1000, IRS-1CD, Radarsat, Star 3i, ETM, IKONOS, *MODIS, ASTER, OrbView 3,4; Quickbird*
Vegetation stress	Aerial photography, Daedalus, ATLAS, AVHRR, TM, SPOT, IRS-1CD, IKONOS, SeaWiFS, ETM, *MODIS, ASTER, OrbView 3,4; Quickbird*

by remotely sensing a plant's chlorophyll absorption characteristics, temperature, and moisture content, it may be possible to model these data to detect vegetation stress, a hybrid variable. The variety of hybrid variables is large; consequently, no attempt is made to identify them. It is important to point out, however, that nominal-scale land-cover mapping is a hybrid variable. The land cover of a particular area on an image is usually derived by evaluating several of the fundamental biophysical variables at one time [e.g., object tone or color, location (x, y), height (z), and perhaps temperature]. So much attention has been placed on remotely sensing this hybrid *nominal*-scale variable that the *interval*- or *ratio*-scaled biophysical variables have been neglected until the last decade. Nominal-scale land-use mapping is an important capability of remote sensing technology and should not be minimized. In fact, many social and physical scientists routinely use such data in their research. However, we now see a dramatic increase in the extraction of interval- and ratio-scaled biophysical data that are so valuable when incorporated into quantitative models that can accept spatially distributed information (e.g., Asrar and Dozier, 1994; Moran et al., 1997; NASA, 1998).

Remote Sensing Data Collection

Remotely sensed data are collected using either passive or active remote sensing systems. *Passive* sensors record naturally occurring electromagnetic radiation that is reflected or emitted from the terrain. For example, cameras and video recorders may be used to record visible and near-infrared energy reflected from the terrain, and a multispectral scanner may be used to record the amount of thermal radiant flux emitted from the terrain. *Active* sensors such as microwave (radar) or sonar bathe the terrain in man-made electromagnetic energy and then record the amount of radiant flux scattered back toward the sensor system.

Remote sensing systems collect analog (e.g., hard-copy aerial photography or video data) and/or digital data [e.g., a matrix (raster) of brightness values obtained using a scanner, linear array, or area array]. A selected list of some of the most important current and proposed remote sensing systems is presented in Table 1-3.

Sensor Resolution

Each remote sensing system has four major resolutions associated with it. These resolutions should be understood by the scientist in order to extract meaningful biophysical or hybrid information from the remotely sensed imagery. *Resolution* (or resolving power) is defined as a measure of the ability of an optical system to distinguish between signals that are spatially near or spectrally similar.

Spectral Resolution: This refers to the number and dimension of specific wavelength intervals in the electromagnetic spectrum to which a remote sensing instrument is sensitive. The Landsat Multispectral Scanner (MSS) provided a tremendous amount of remotely sensed data of much of the Earth which is still of significant value for historical

studies. The bandwidths of the four MSS bands are displayed in Figure 1-6a (band 1 = $0.5 - 0.6\,\mu$m; band 2 = $0.6 - 0.7\,\mu$m; band 3 = $0.7 - 0.8\,\mu$m; and band 4 = $0.8 - 1.1\,\mu$m). The nominal size of a band may be large (i.e., coarse), as with the Landsat MSS near-infrared band 4 ($0.8 - 1.1\,\mu$m) or relatively smaller (i.e., finer), as with the Landsat MSS band 3 ($0.7 - 0.8\,\mu$m). Thus, Landsat MSS band 4 detectors record a relatively large range of reflected near-infrared radiant flux (300 nm) while the MSS band 3 detectors record a much reduced range of near-infrared radiant flux (100 nm).

The four spectral bandwidths associated with the Positive Systems, Inc. ADAR 5500 digital frame camera are shown for comparative purposes (Figure 1-6a). The frame camera's bands are refined to record information in more specific regions of the electromagnetic spectrum (band 1 = 450 – 515 nm; band 2 = 525 – 605 nm; band 3 = 640 – 690 nm; and band 4 = 750 – 900 nm). In fact, there are gaps in the spectral sensitivity of the detectors. Note that this digital camera system is also sensitive to reflected blue wavelength energy.

The aforementioned terminology is typically used to describe a sensor's *nominal spectral resolution*. Unfortunately, it is difficult to create a detector that has extremely sharp bandpass boundaries such as those shown in Figure 1-6a. Rather, the more precise method of stating bandwidth is to look at the typical Gaussian-shape of the detector sensitivity, such as the example shown in Figure 1-6b. The analyst then determines the Full Width at Half Maximum (FWHM). In this hypothetical example, the Landsat MSS near-infrared band 3 under investigation is sensitive to energy between 0.7 and 0.8 μm (700 – 800 nm).

Remote sensing systems may be configured to collect data in just a single band or region of the electromagnetic spectrum. For example, an ADAR 5500 band 4 near-infrared image is displayed in Figure 1-6c. *Multispectral* remote sensing takes place when radiant energy is recorded in multiple bands of the electromagnetic spectrum. The ADAR 5500 usually acquires four multispectral bands of imagery during a mission (Figure 1-6d). A *hyperspectral* remote sensing instrument acquires data in hundreds of spectral bands. For example, the Airborne Visible and Infrared Imaging Spectrometer (AVIRIS) has 224 bands in the region from $0.4 - 2.5\,\mu$m spaced just 10 nm apart based on the FWHM criteria (Clark, 1999). An AVIRIS hyperspectral datacube of Sullivan's Island, SC, is shown in Figure 1-7.

Certain regions or bands of the electromagnetic spectrum are optimum for obtaining information on biophysical parameters. The bands are normally selected to maximize the con-

Table 1-3. Selected Current and Proposed Remote Sensing Systems and their Major Characteristics

	Resolution								
	Spectral							Spatial (meters)	Temporal (days)
Remote Sensing Systems	Blue	Green	Red	Near-IR	Mid-IR	Thermal IR	Micro-wave		
Suborbital Sensors									
Panchromatic film (b&w)		0.5 ——— 0.7 µm						Variable	Variable
Color film	0.4 ————— 0.7 µm							Variable	Variable
Color-infrared film		0.5 ——— 0.9 µm						Variable	Variable
NASA Airborne Terrestrial Applications Sensor (ATLAS)	0.45 ——8 bands——2.35 µm					6	—	2.5 to 25	Variable
NASA Airborne Visible IR Imaging Spectrometer (AVIRIS)	0.41——224 bands ———2.5 µm							2.5 or 20	Variable
Intermap Star-3i X-band radar							1	Variable	Variable
Satellite Sensors									
NOAA-9 AVHRR LAC	—	—	1	1	—	3	—	1100	14.5/day
NOAA- K, L, M (proposed)	—	—	1	1	2	2	—	1100	14.5/day
Landsat Multispectral Scanner (MSS)	—	1	1	2	—	—	—	79	16–18
Landsat 4-5 Thematic Mapper (TM)	1	1	1	1	2	1	—	30 and 120	16
Landsat 7 Enhanced TM (ETM+) — Multispectral	1	1	1	1	2	1	—	30 and 60	16
— Panchromatic	—	0.52 ——— 0.9 µm			—	—	—	15	16
SPOT HRV — Multispectral	—	1	1	1	—	—	—	20	Pointable
— Panchromatic	0.51 ——— 0.73 µm			—	—	—		10	Pointable
GOES Series (East and West)	—	0.52 ——— 0.72 µm		—		4	—	700	0.5/hr
European Remote Sensing Satellite (ERS-1,2)	VV polarization C-band (5.3 GHz)						1	26 – 28	—
Canadian RADARSAT (several modes)	HH polarization C-band (5.3 GHz)						1	9 to 100	1–6 days
Shuttle Imaging Radar (SIR-C)	—	—	—	—	—	—	3	30	Variable
Sea-Viewing Wide Field-of-View Sensor (SeaWiFS)	3	2	1	2	—	—	—	1130	1
Terra Moderate Resolution Imaging Spectrometer (*MODIS*)	0.405 ——— 36 bands ——— 14.385 µm						—	250, 500, 1000	1-2
Terra Advanced Spaceborne Thermal Emission and Reflection Radiometer (*ASTER*)	0.52 — 3 bands — 0.86 µm							15	5
				1.6 – 6 bands – 2.43 µm				30	16
				8.12 – 5 bands – 11.6 µm				90	16
Terra Multiangle Imaging SpectroRadiometer (*MISR*)	Nine CCD cameras in four bands (440, 550, 670, 860 nm)							275 and 1100	
NASA Topex/Poseidon — TOPEX radar altimeter	(18, 21, 37 GHz)							315 km	10
— POSEIDON single-frequency radiometer	(13.65 GHz)								
NASA Upper Atmosphere Research Satellite (UARS): includes 9 sensors.									
Space Imaging IKONOS — Multispectral	1	1	1	1			—	4	Pointable
— Panchromatic	0.45 ——— 0.9 µm				—	—		1	
ORBIMAGE Orbview 3 — Multispectral	1	1	1	1			—	4	Pointable
— Panchromatic	0.45 ——— 0.9 µm				—	—	—	1	

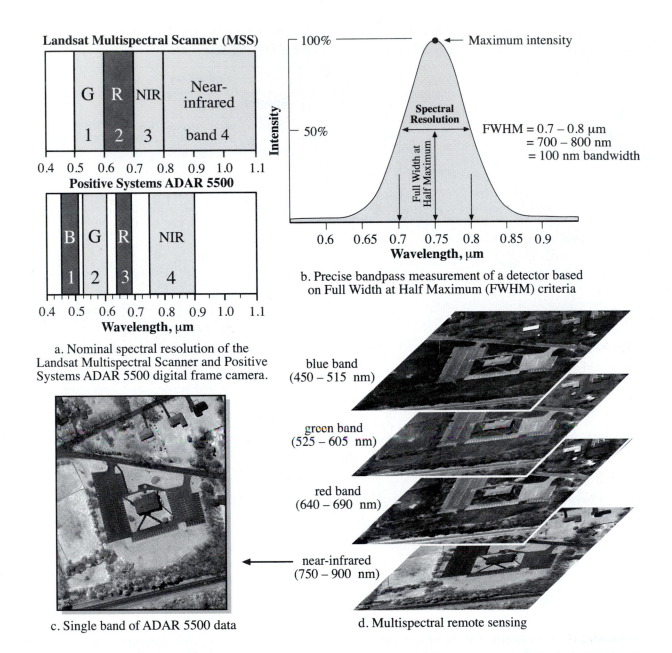

a. Nominal spectral resolution of the Landsat Multispectral Scanner and Positive Systems ADAR 5500 digital frame camera.

b. Precise bandpass measurement of a detector based on Full Width at Half Maximum (FWHM) criteria

c. Single band of ADAR 5500 data

d. Multispectral remote sensing

Figure 1-6   a) The spectral bandwidths of the four Landsat Multispectral Scanner (MSS) bands (green, red, and two near-infrared) compared with the bandwidths of the Positive Systems ADAR 5500 digital frame camera. b) The true spectral bandwidth is the width of the Gaussian-shaped spectral profile at Full Width at Half Maximum (FWHM) intensity (after Clark, 1999). This example has a spectral bandwidth of 0.1 μm (100 nm) between 700 and 800 nm. c) If desired, it is possible to collect reflected energy in a single band of the electromagnetic spectrum (e.g., 750 – 900 nm). d) Multispectral remote sensing instruments such as the ADAR 5500 collect data in multiple bands of the electromagnetic spectrum (images courtesy of Positive Systems, Inc.).

**Airborne Visible Infrared Imaging Spectrometer (AVIRIS) Datacube
of Sullivan's Island, SC, Obtained on October 26, 1998**

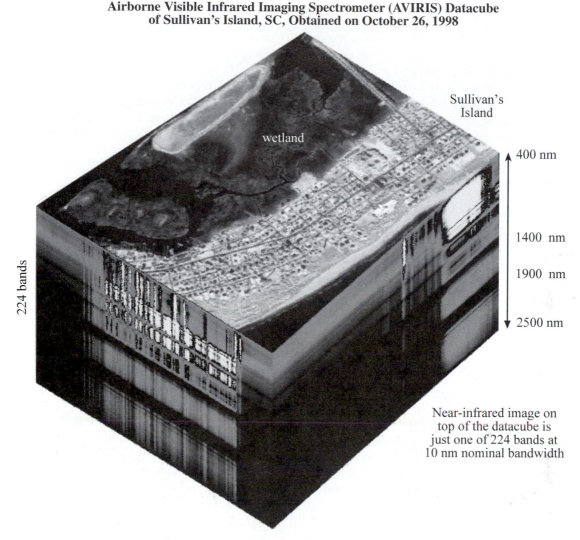

Figure 1-7   Hyperspectral remote sensing of Sullivan's Island, SC, on October 26, 1998, using NASA's Airborne Visible Infrared Imaging
Spectrometer (AVIRIS). The spatial resolution is 2.5 x 2.5 m. The atmosphere absorbs most of the electromagnetic energy
near 1.4 and 1.9 μm (Meisburger and Jensen, 1999; datacube courtesy of NASA Jet Propulsion Laboratory).

trast between the object of interest and its background (i.e., object-to-background contrast). Careful selection of the spectral bands may improve the probability that a feature will be detected and identified and biophysical information extracted.

Spatial Resolution: There is a relationship between the size of a feature to be identified and the spatial resolution of the remote sensing system. *Spatial resolution* is a measure of the smallest angular or linear separation between two objects that can be resolved by the sensor. The spatial resolution of aerial photography may be measured by 1) placing carefully calibrated, parallel black-and-white lines on tarps that are placed in the field, 2) obtaining aerial photography of the

study area, and 3) analyzing the photography and computing the number of resolvable *line pairs per millimeter* in the photography. It is also possible to determine the spatial resolution of imagery by computing its modulation transfer function, which is beyond the scope of this text.

Many satellite remote sensing systems operate in fixed orbits with fixed optical systems that have a constant instantaneous-field-of-view (IFOV). For practical purposes, therefore, we define a sensor system's nominal spatial resolution as simply the dimension in meters (or feet) of the ground-projected IFOV. For example, the SPOT panchromatic band has a nominal spatial resolution of 10 x 10 m, the Landsat Thematic Mapper has a nominal spatial resolution of 30 x 30 m for six of its bands, and the Landsat MSS has a nominal spatial resolution of 79 x 79 m. Generally, the smaller the spatial resolution, the greater the resolving power of the sensor system. Figure 1-8 depicts digital camera imagery of an area in Mechanicsville, N.Y. at resolutions ranging from 0.5 x 0.5 m to 80 x 80 m. Note that there is not a significant difference in the interpretability of 0.5 x 0.5 m data, 1 x 1 m data, and even 2 x 2 m data. However, the urban information content decreases rapidly when using 5 x 5 m imagery and is practically useless for urban analysis at spatial resolutions larger than 10 x 10 m. Landsat MSS data is particularly useless (79 x 79 m) for most urban applications.

Another useful rule is that in order to detect a feature, the spatial resolution of the sensor system should be less than one-half the size of the feature measured in its smallest dimension. For example, if we want to identify the location of all oak trees within a city park, the minimum acceptable spatial resolution would be approximately one-half the diameter of the smallest oak tree crown. Even this spatial resolution, however, will not guarantee success if there is no difference between the spectral response of the oak tree (the object) and the soil or grass surrounding it (i.e., its background).

Temporal Resolution: The *temporal resolution* of a remote sensing system refers to how often it records imagery of a particular area. For example, the temporal resolution of the sensor system shown in Figure 1-9a is every 16 days. Ideally, the sensor obtains data repetitively to capture unique discriminating characteristics of the object under investigation (Haack et al., 1997). For example, agricultural crops have unique crop calendars in each geographic region. To measure specific agricultural variables, it is necessary to acquire remotely sensed data at critical dates in the phenological cycle. Analysis of multiple-date imagery provides information on how the variables are changing through time. Change information provides insight into processes influencing the development of the crop (Steven, 1993). Fortunately, several satellite sensor systems such as SPOT are pointable, meaning that they can acquire imagery off-nadir (*nadir* is the point directly beneath the spacecraft) if necessary. This dramatically increases the probability that imagery might be obtained during a growing season or during an emergency. However, the off-nadir oblique viewing also introduces bidirectional reflectance distribution function (BRDF) issues that are addressed in Chapter 10 (Remote Sensing of Vegetation).

Radiometric Resolution: This is defined as the sensitivity of a remote sensing detector to differences in signal strength as it records the radiant flux reflected or emitted from the terrain. It defines the number of just discriminable signal levels; consequently, it can have a significant impact on our ability to measure the properties of scene objects. For example, the original multispectral scanner (MSS) onboard Landsat 1 recorded the reflected radiant energy with a precision of 6-bits (values ranging from 0 to 63). Landsat 4 and 5 Thematic Mapper recorded data in 8-bits (values from 0 to 255). Thus, the Landsat TM sensor had improved radiometric resolution when compared with the original MSS. Several new sensor systems have 12-bit radiometric resolution (values ranging from 0 to 4095) (Figure 1-9b).

Improvements in resolution generally increase the probability that phenomena may be remotely sensed more accurately. The trade-off is that any improvement in resolution will usually require additional data-processing capability for either human or computer-assisted analysis.

Suborbital (Airborne) Remote Sensing Systems

High-quality metric cameras mounted onboard aircraft continue to provide aerial photography for many Earth resource applications. For example, the U.S. Geological Survey's National Aerial Photography Program (NAPP) systematically collects 1:40,000-scale black-and-white or color-infrared aerial photography of much of the United States every 5 to 10 years. In addition, sophisticated remote sensing systems are routinely mounted on aircraft to provide high spatial and spectral resolution multispectral remotely sensed data. Examples include the Compact Airborne Spectrographic Imager (CASI), Daedalus multispectral scanners, and NASA's Airborne Terrestrial Applications Sensor (ATLAS) (Table 1-3). These sensors can collect data on demand when disaster strikes (e.g., oil spills or floods) if cloud-cover conditions permit. There are also numerous radars, such as Intermap's Star-3i radar, that can be flown on aircraft day and night and in inclement weather. Unfortunately, suborbital remote sensor data are usually expensive to acquire per km^2. Also, atmospheric turbulence can cause the data to have severe geometric distortions that can be quite difficult to correct.

Current and Proposed Satellite Remote Sensing Systems

Remote sensing systems onboard satellites provide high-quality, relatively inexpensive data per km^2. For example, the European Remote Sensing Satellite (ERS-1,2) collects 26 × 28 m spatial resolution C-band active microwave (radar) imagery of much of Earth, even through clouds. Similarly, the Canadian Space Agency RADARSAT obtains C-band active microwave imagery. The United States has pro-

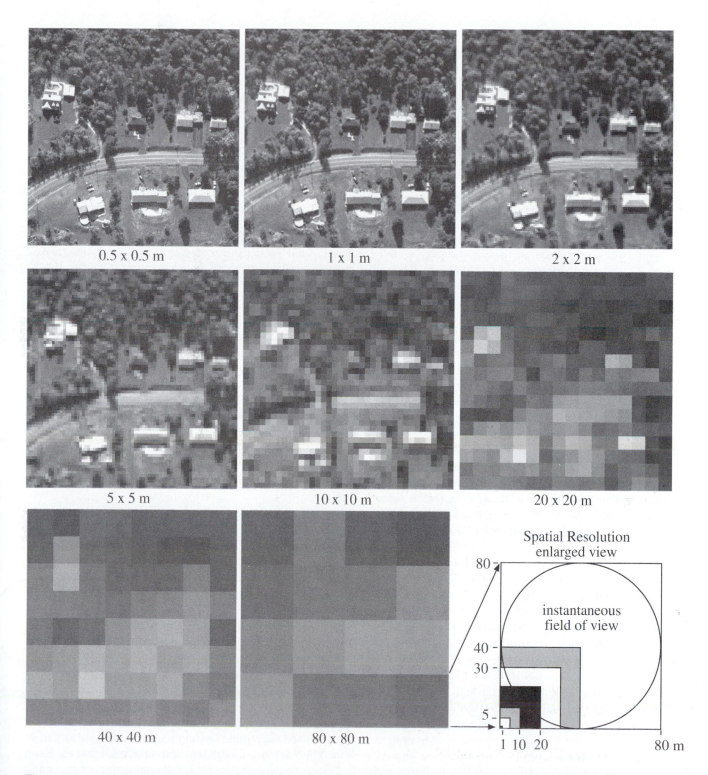

Figure 1-8 Imagery of residential housing near Mechanicsville, NY. The data were obtained on June 1, 1998, at a nominal spatial resolution of 0.3 x 0.3 m (approximately 1 x 1 ft) using a digital camera (courtesy of Litton Emerge, Inc.). The original data were resampled to derive the imagery with the simulated spatial resolutions shown.

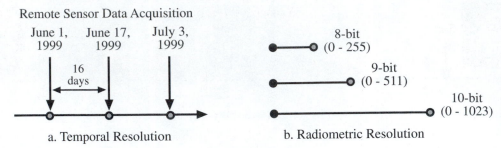

Figure 1-9   a) The temporal resolution of a remote sensing system refers to how often it records imagery of a particular area. This example depicts the systematic collection of data every 16 days, presumably at approximately the same time of day. Landsat Thematic Mapper 4 and 5 had 16-day revisit cycles. b) The radiometric resolution of a remote sensing system is defined as the sensitivity of remote sensing detectors to differences in signal strength as it records the radiant flux reflected or emitted from the terrain. The energy is normally quantized during the analog-to-digital (A-to-D) conversion process to 8, 9, 10, or 12-bits. Think of radiometric resolution as being like a ruler. If you had to measure something very precisely, would you rather have a ruler with just 256 subdivisions (8-bit) or one with 1024 subdivisions (10-bit)? Several new sensor systems record data in 12-bits (0-4095).

gressed from multispectral scanning systems (Landsat MSS, 1972 to present) to more advanced scanning systems (Landsat Thematic Mapper, 1982 to present). The Land Remote Sensing Policy Act of 1992 specified the future of satellite land remote sensing programs in the United States (Asker, 1992; Jensen, 1992). Unfortunately, Landsat 6, with its Enhanced Thematic Mapper (ETM), did not achieve orbit when launched on October 5, 1993. Landsat 7 was launched on April 15, 1999, to relieve the United States' land remote sensing data gap (Henderson, 1994). Meanwhile, the French have pioneered the development of linear array remote sensing technology with the launch of SPOT 1-4 High Resolution Visible (HRV) sensors in 1986, 1990, 1993, and 1998.

The International Geosphere–Biosphere Program (IGBP) and the U. S. Global Change Research Program (USGCRP) call for scientific research to:

> describe and understand the interactive physical, chemical, and biological processes that regulate the total Earth system (CEES, 1991).

Space-based remote sensing is an integral part of these research programs because it provides the only means of observing global ecosystems consistently and synoptically. NASA's Earth Science Enterprise (formerly Mission to Planet Earth) is the name given to the coordinated international plan to provide the necessary satellite platforms and instruments, an Earth Observing System Data and Information System (EOSDIS), and related scientific research for IGBP. In particular, new satellite remote sensing instruments will include 1) a series of near-term Earth probes to address discipline-specific measurement needs, 2) a series of multi-

purpose polar orbiting platforms, initiated in 1988, to acquire 15 years of continuous Earth observations, called the Earth Observing System (EOS), and 3) a series of geostationary platforms carrying advanced multidisciplinary instruments to fly sometime after the year 2000, called the Geostationary Earth Observing System (Price et al., 1994; NASA, 1998). Not everyone is convinced that NASA's Earth Science Enterprise is the most economic way of obtaining the required environmental information (e.g., Hudgins, 1997).

The first of the National Aeronautics and Space Administration Mission to Planet Earth sensors placed in orbit was the Upper Atmosphere Research Satellite (UARS) launched in 1991 (Luther, 1992). The UARS sensors collected information on upper atmospheric chemistry, temperature, wind speed, direction, and energy inputs. The TOPEX/POSEIDON satellite launched in 1992 uses radar altimetry to measure sea-surface height over 90 percent of the world's ice-free oceans. The system acquires global maps of ocean topography (barely perceptible hills and valleys of the sea surface), which scientists use to calculate the speed and direction of ocean currents (Jones, 1992).

The EOS Science Plan: Asrar and Dozier (1994) conceptualized the remote sensing science conducted as part of the Earth Science Enterprise. They suggested that the Earth consists of two subsystems, 1) the physical climate, and 2) biogeochemical cycles, linked by the global hydrologic cycle, as shown in Figure 1-10.

The *physical climate* subsystem is sensitive to fluctuations in the Earth's radiation balance. Human activities have caused

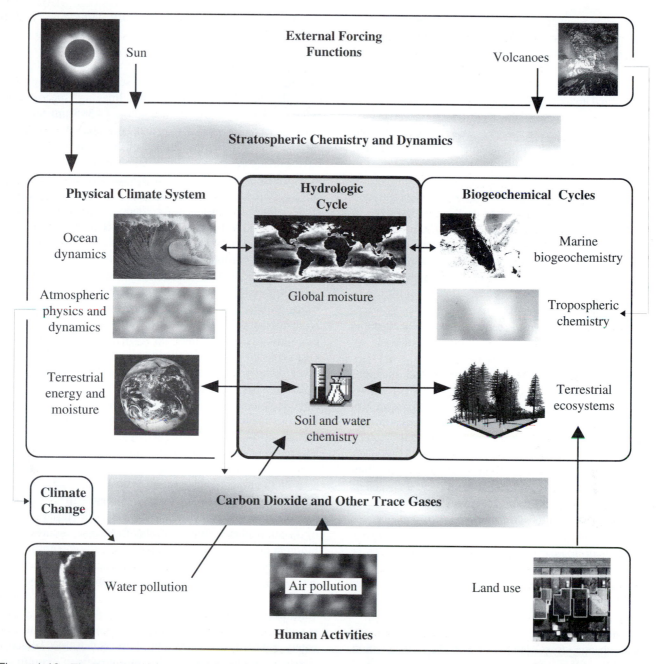

Figure 1-10   The Earth system may be subdivided into two subsystems — the physical climate system and biogeochemical cycles — that are linked by the global hydrologic cycle. Significant changes in the external forcing functions and human activities can have a dramatic impact on the physical climate system, biogeochemical cycles, and the global hydrologic cycle. Examination of these subsystems and their linkages defines the critical questions that the NASA Earth Observing System (EOS) is attempting to answer (after Asrar and Dozier, 1994).

changes to the planet's radiative heating mechanism that rival or exceed natural change. For example, increases in greenhouse gases between 1765 and 1990 have caused a radiative forcing of 2.5 W m^{-2}. If this rate is sustained, it could result in global mean temperatures increasing about 0.2 to 0.5°C per decade during the next century. Volcanic eruptions and the ocean's ability to absorb heat may impact the projections. Nevertheless, the following important questions are being addressed using remote sensing (Asrar and Dozier, 1994):

- How do clouds, water vapor, and aerosols in the Earth's radiation and heat budgets change with increased atmospheric greenhouse-gas concentrations?

- How do the oceans interact with the atmosphere in the transport and uptake of heat?

- How do land-surface properties such as snow and ice cover, evapotranspiration, urban/suburban land use, and vegetation influence circulation?

The Earth's *biogeochemical cycles* have also been changed by man. Atmospheric carbon dioxide has increased by 30 percent since 1859, methane by more than 100 percent, and ozone concentrations in the stratosphere have decreased, causing increased levels of ultraviolet radiation to reach the Earth's surface. Global change research is addressing the following questions:

- What role do the oceanic and terrestrial components of the biosphere play in the changing global carbon budget?

- What are the likely effects on natural and managed ecosystems of increased carbon dioxide, acid deposition, shifting patterns of precipitation, and changes in soil erosion, river chemistry, and atmospheric ozone concentrations?

The *hydrologic cycle* links the physical climate and bio-geochemical cycles. The phase change of water between its gaseous, liquid, and solid states involves storage and release of latent heat, so it influences atmospheric circulation and globally redistributes both water and heat (Asrar and Dozier, 1994). The hydrologic cycle is the integrating process for the fluxes of water, energy, and chemical elements among components of the Earth System. Important questions to be addressed include:

- How will atmospheric variability, human activities, and climate change affect patterns of humidity, precipitation, evapotranspiration, and soil moisture?

- How does soil moisture vary in time and space?

- Can we predict changes in the global hydrologic cycle using present and future observation systems and models?

EOS AM-1 (now referred to as the *Terra* satellite) houses five remote sensing instruments designed to address many of the previously mentioned research topics (NASA, 1998). The spatial, spectral, and temporal characteristics of three of the main *Terra* sensor systems (*MODIS*, *ASTER*, and *MISR*) are summarized briefly in Table 1-3, with additional detail provided in Chapter 7.

The *Terra* sensors use new remote sensing technology. For example, the Moderate Resolution Imaging Spectrometer (*MODIS*) has 36 bands from 0.405 – 14.385 μm that will collect data at 250- and 500-m and 1-km spatial resolutions. *MODIS* views the entire surface of the Earth every 1 to 2 days, making observations in 36 coregistered spectral bands, at moderate resolution (0.25 to 1 km), of land- and ocean-surface temperature, primary productivity, land-surface cover, clouds, aerosols, water vapor, temperature profiles, and fires (NASA, 1998). The Advanced Spaceborne Thermal Emission and Reflection Radiometer (*ASTER*) has five bands in the thermal infrared region between 8 and 12 μm with 90-m pixels. It also has three broad bands between 0.5 and 0.9 μm with 15-m pixels and stereo capability, and six bands in the shortwave infrared region (1.6 – 2.5 μm) with 30-m spatial resolution. *ASTER* is the highest spatial resolution sensor system on the EOS *Terra* platform and provides information on surface temperature that can be used to model evapotranspiration. The Multi-angle Imaging SpectroRadiometer (*MISR*) has nine separate CCD pushbroom cameras to observe Earth in four spectral bands and at nine separate view angles. It provides data on clouds, atmospheric aerosols, and multiple-angle views of the Earth's deserts, vegetation, and ice cover. The Clouds and the Earth's Radiant Energy System (*CERES*) consists of two scanning radiometers that measure the Earth's radiation balance and provide cloud property estimates to assess their role in radiative fluxes from the surface of the earth to the top of the atmosphere. Finally, *MOPITT* (Measurements of Pollution in the Troposphere) is a scanning radiometer that provides information on the distribution, transport, sources, and sinks of carbon monoxide and methane in the troposphere.

Commercial Vendors: EOSAT, Inc., launched Landsat 6 with its Enhanced Thematic Mapper in 1993. Unfortunately, it failed to achieve orbit. EarthWatch, Inc. launched Earlybird in December, 1997. Unfortunately, all communication with the satellite was lost. Space Imaging, Inc. launched IKONOS on April 27, 1999 and it failed to achieve orbit. A second IKONOS satellite was launched on September 24, 1999. The IKONOS sensor system has a 1 x 1 m panchromatic band as well as four 4 x 4 m multispectral bands (Table 1-3). Similar sensor systems are scheduled to be launched by EarthWatch, Inc. (*Quickbird*) and ORBIMAGE, Inc. (*Orbview 3*) in 2000. ORBIMAGE also plans to launch a hyperspectral satellite remote sensing system in 2000 (*Orbview 4*).

The analysis of remotely sensed data is performed using a variety of image processing techniques (Figure 1-11), including:

- analog (visual) image processing of image data and

- digital image processing of digital data.

Both analog and digital image processing should allow the analyst to perform *scientific visualization,* defined as "visually exploring data and information in such a way as to gain understanding and insight into the data" (Pickover, 1991). First, however, it is instructive to ask two questions: Why process the remotely sensed data digitally at all? Isn't visual image analysis sufficient?

Human beings are exceptionally adept at visually interpreting images produced by certain types of remote sensing devices, especially cameras. We could ask, Why try to mimic or improve on this capability? First, there are certain thresholds beyond which the human interpreter cannot detect "just noticeable differences" in the imagery. For example, it is commonly known that an analyst can discriminate only about nine shades of gray when interpreting continuous-tone black-and-white aerial photography. If the data were originally recorded with 256 shades of gray, there may be more subtle information present in the image than the interpreter can extract visually. Furthermore, the interpreter brings to the task all the pressures of the day, making the interpretation generally unrepeatable. Conversely, the results obtained by computer are repeatable (even when wrong!). Also, when it comes to keeping track of a great amount of detailed quantitative information, such as the spectral characteristics of a vegetated field throughout a growing season for crop identification purposes, the computer is very adept at storing and manipulating such tedious information and possibly making a more definitive conclusion as to what crop is being grown. This is not to say that digital image processing is superior to visual image analysis. This is certainly not the case. Rather, there may be times when a digital approach is better suited to the problem at hand.

But what about the actual processes of analog (visual) versus digital image processing? Are there similarities between the goals and methods of both procedures? Estes et al. (1983) suggest that there exist several image-analysis tasks and basic elements of image interpretation that the visual and digital image processing approaches share (Figure 1-11). First, both manual and digital analysis of remotely sensed data seek to detect and identify important phenomena in the scene. Once identified, the phenomena are usually measured, and the information is used in problem solving. Thus, both manual and digital analysis have the same general goals. However, the attainment of these goals may follow significantly different paths.

Analog (Visual) Image Processing

Most of the fundamental elements of image interpretation identified in Figure 1-11 are used in visual image analysis, including size, shape, shadow, color (tone), parallax, pattern, texture, site, and association. The human mind is amazingly adept at recognizing these complex elements in an image or photograph because we constantly process profile views of Earth features every day and continually process images in books and magazines and on television. Furthermore, we are adept at bringing to bear all the knowledge in our personal background and collateral information. We then converge all this evidence to identify phenomena in images and/or to judge their significance. Precise measurement of objects (location, height, width, etc.) may be performed using optical photogrammetric techniques applied to either monoscopic (single-photo) or stereoscopic (overlapping) images. Numerous books have been written on how to perform visual image interpretation and photogrammetric measurement. Chapter 5 summarizes the use of the fundamental elements of image interpretation. Chapter 6 introduces photogrammetry principles.

Interestingly, there is a resurgence in the art and science of visual photointerpretation as the digital remote sensor systems provide higher spatial resolution imagery. For example, Indian IRS-1C panchromatic data (5.8×5.8 m) is often photointerpreted and used as a base map in GIS projects. The new 1 x 1 m panchromatic data provided by commercial companies (e.g., Space Imaging IKONOS, ORBIMAGE Orbview 3) will cause even more visual image interpretation to take place.

Digital Image Processing

Scientists have made significant advances in digital image processing of remotely sensed data for scientific visualization and hypothesis testing. The methods are summarized in the companion book by Jensen (1996) and others (e.g., Wolff and Yaeger, 1993; Nadler and Smith, 1993; Schott, 1997). The major types of digital image processing include statistical and syntactical pattern recognition, photogrammetric image processing of stereoscopic imagery, hyperspectral data analysis, and expert system and neural network image analysis.

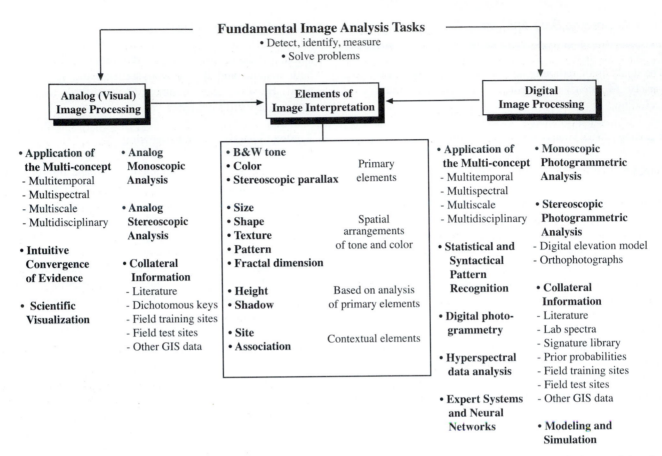

Figure 1-11   This conceptual diagram identifies analog (visual) and computer-assisted digital image processing of remotely sensed data that rely on the analysis of the fundamental elements of image interpretation. Visual analysis at the present time incorporates many more of the complex elements in the analysis of remote sensing images.

Pattern Recognition: Fundamental statistical methods of rectifying remotely sensed data to a map projection, enhancing the data, classifying the data into land use and land cover, and identifying change between dates of imagery are now performed routinely with reasonable precision. Interestingly, most of the computer-assisted image processing to date has involved the use of only a few of the basic elements of image interpretation. In fact, the overwhelming majority of all digital image analysis appears to be dependent primarily on just the *tone* or *color* of individual pixels in the scene using fundamental statistical pattern recognition techniques (Figure 1-11).

Various techniques have been used to incorporate additional elements of image interpretation into the image analysis process. For example, numerous studies have synthesized *texture* information from the spectral data in the imagery. Some have computed the fractal dimension of images and found it to be a valuable element of image interpretation (Emerson et al., 1999). *Contextual* classification has been performed, which makes use of neighboring pixel values, thus incorpo-

rating some level of *association* information (Gong and Howarth, 1992). Some image processing now takes into account *fuzzy set logic*, which attempts to model the imprecision in the real world (Ji and Jensen, 1996).

Photogrammetry: Significant advances have been made in the analysis of stereoscopic remote sensor data using computer workstations and digital image processing photogrammetric algorithms (Li, 1998). Soft-copy photogrammetric workstations can be used to extract accurate digital elevation models (DEMs) and differentially corrected orthophotography from the triangulated aerial photography or imagery (Ackerman, 1994; Jensen, 1995). The technology is revolutionizing the way DEMs are collected, especially for developing countries and how orthophotos are produced for rural and urban-suburban applications.

Hyperspectral: Analysts should be aware that special software is required to process the hyperspectral data from spectroradiometer remote sensor systems (e.g., AVIRIS, *MODIS*). Kruse et al. (1992), Landgrebe (1999) and ENVI

(1999) have pioneered the development of hyperspectral image analysis software. The software reduces the dimensionality of the data (number of bands) to a manageable degree, while still retaining the essence of the data. Under certain conditions the software can be used to compare the remotely sensed spectral reflectance curves with a library of spectral reflectance curves. Analysts are also able to identify the type and proportion of different materials within an individual picture element (referred to as end-member analysis). This is an exciting area of digital image processing.

Expert Systems and Neural Networks: Humans are very successful at visually interpreting aerial photographs because they focus their real-world knowledge about the study area and their years of visual processing experience on the task. It is difficult to make a computer understand and use the heuristic rules of thumb and knowledge that a human expert uses when interpreting an image (Moller-Jensen, 1990). Nevertheless, there has been considerable success in the use of artificial intelligence (AI) to try to make computers do things that, at the moment, people do better. One area of AI that has great potential in remote sensing image analysis is the use of expert systems. Expert systems can be used to 1) interpret an image, and/or 2) place all the information contained within an image in its proper context with other ancillary data and extract more valuable information (Bolstad and Lillesand, 1992). In the first case, collateral data and rules specified by an expert might be used by novices to more accurately interpret a remotely sensed image (Huang and Jensen, 1998). In the second case, an expert system might be used to produce geological engineering maps from input datasets (bedrock geology, agricultural soils, topography), some of which were derived using remote sensing (Usery et al., 1988). Scientists with good training in an Earth science discipline, an understanding of remote sensing, and expert system skills (e.g., how to create a knowledge base and query it with an inference engine) will make significant contributions in this area.

Neural networks have also been used to analyze remotely sensed data (Hepner et al., 1990; Jensen and Qiu, 1999). Neural networks do not require the input data to be normally distributed. Furthermore, they can be programmed to learn.

Modeling Remote Sensing Data Using A GIS Approach

Remotely sensed data should not be analyzed in a vacuum without the benefit of other collateral information, such as soils, hydrology, and topography (Price et al., 1994; Ramsey et al., 1995). Unfortunately, many scientists promoting the integration of remote sensing and GIS assume that the flow of data should be unidirectional — that is, from the remote sensing system to the GIS. Actually, the backward flow of ancillary data from the GIS to the remote sensing system is very valuable (Stow, 1993). For example, land-cover mapping using remotely sensed data has been significantly improved by incorporating topographic information from digital terrain models and other GIS data (Franklin and Wilson, 1992). Basically, the interface between GIS and remote sensing systems is functional but weak (Lunetta et al., 1991). Each technology suffers from a lack of critical support that could be provided by the other. GIS needs timely, accurate updating of the spatially distributed variables in the database that remote sensing can provide. Remote sensing can benefit from access to accurate ancillary information to improve classification accuracy and other types of modeling (Jensen et al., 1994). Such synergy is critical if successful expert system and neural network analyses are to be performed.

Scene Modeling

Strahler et al. (1986) describe a framework for modeling in remote sensing. Basically, a remote sensing model has three components: 1) a scene model, which specifies the form and nature of the energy and matter within the scene and their spatial and temporal order; 2) an atmospheric model, which describes the interaction between the atmosphere and the energy entering and being emitted from the scene; and 3) a sensor model, which describes the behavior of the sensor in responding to the energy fluxes incident on it and in producing the measurements that constitute the image. They suggest that "the problem of scene inference, then, becomes a problem of model inversion in which the order in the scene is reconstructed from the image and the remote sensing model." For example, Li and Strahler (1985) modeled the optical-geometric properties of a coniferous forest canopy that has been tested extensively (Franklin and Turner, 1992; Woodcock et al., 1997).

Basically, successful remote sensing modeling predicts how much radiant flux in certain wavelengths should exit a particular object (e.g., a conifer canopy) even without actually sensing the object. When the model's prediction is the same as the sensor's measurement, the relationship has been modeled correctly. The scientist then has a greater appreciation for energy-matter interactions in the scene and may be able to extend the logic to other regions or applications with confidence. The remote sensor data can then be used more effectively in physical deterministic models (e.g., watershed runoff, net primary productivity, and evapotranspiration models), which are so important for large ecosystem modeling. Recent work allows one to model the utility of sensors with different spatial resolutions for particular applications such as urban analysis (Collins and Woodcock, 1999).

Information Presentation

Information derived from remote sensor data are usually summarized as an enhanced image, image map, orthophotomap, thematic map, spatial database file, statistic, or graph (Figure 1-5). Thus, the final output products often require knowledge of remote sensing, cartography, GIS, and spatial statistics as well as the systematic science being investigated (e.g., soils, agriculture, forestry, wetland, urban studies). Scientists who understand the rules and synergistic relationship between the technologies can produce output products that communicate effectively. Conversely, those who violate fundamental rules (e.g., cartographic theory or database topology design) often produce poor output products that do not communicate effectively.

Image maps offer scientists an alternative to line maps for many cartographic applications. Thousands of satellite image maps have been produced from Landsat MSS (1:250,000 and 1:500,000 scale), TM (1:100,000 scale) and AVHRR data (Vickers, 1993). Image maps at scales of >1:24,000 are possible with the improved resolution of 1 x 1 m data (Li, 1998). Because image map products can be produced for a fraction of the cost of conventional line maps, they provide the basis for a national map series oriented toward the exploration and economic development of the less developed areas of the world, most of which have not been mapped at scales of 1:100,000 or larger.

Remote sensor data that has been geometrically rectified to a standard map projection is becoming indispensable in most sophisticated GIS databases. This is especially true of orthophotomaps that have the metric qualities of a line map and the information content of an aerial photograph or other type of image (Jensen, 1995).

Unfortunately, *error* is introduced at various stages in the remote sensing process and must be identified and reported. Innovations in error reduction include: 1) recording the genealogy or lineage of the various operations applied to the original remote sensor data (Lanter and Veregin, 1992), 2) documenting the geometric (spatial) error and thematic (attribute) error of the individual source materials, 3) improving legend design, especially for change detection map products derived from remote sensing, and 4) precise error evaluation statistic reporting (Khorram et al., 1999). Many of these concerns have not been adequately addressed. The remote sensing and GIS community should incorporate technologies that carefully track all types of error entering final map and image products (Goodchild and Gopal, 1992). This will result in more accurate information being used in the decision-making process.

 Earth Resource Analysis Perspective

Remote sensing may be used for numerous applications, including weapon guidance systems (e.g., the cruise missile), medical image analysis (e.g., X-raying a broken arm), nondestructive evaluation of machinery and products (e.g., on an assembly line), and analysis of Earth's resources. Earth resource information is defined as any information concerning terrestrial vegetation, soils, minerals, rocks, water, and urban infrastructure as well as certain atmospheric characteristics. *This book focuses on the art and science of applying remote sensing for the extraction of useful Earth resource information.* Such information may be useful for modeling the global carbon cycle, the biology and biochemistry of ecosystems, aspects of the global water and energy cycle, climate variability and prediction, atmospheric chemistry, characteristics of the solid Earth, and natural hazards (Paylor et al., 1999).

 Book Organization

This chapter defined important terms and provided a perspective on how remote sensing science can be useful for Earth resource investigations. Chapter 2 introduces the fundamental principles of electromagnetic radiation and how this radiation is used to perform remote sensing of the environment. Chapter 3 reviews the history of photography, and aerial and satellite platforms. Chapter 4 introduces the fundamental characteristics of aerial photography, filtration, and film. Chapter 5 presents the fundamental elements of visual image interpretation. Chapter 6 reviews principles of photogrammetry used to extract quantitative information from aerial photography. Chapter 7 presents the characteristics of optical-mechanical remote sensing systems. Chapter 8 introduces thermal infrared remote sensing. Chapter 9 presents active microwave remote sensing. Chapter 10 reviews how remote sensing may be used to extract fundamental biophysical characteristics of terrestrial and aquatic vegetation. Chapter 11 provides insight into remote sensing of terrestrial water, ice, and snow as well as atmospheric water vapor and temperature. Chapter 12 demonstrates how remote sensing can provide unique urban/suburban infrastructure information using a variety of remote sensing systems. Chapter 13 describes how selected soil and mineral characteristics may be remotely sensed and how major geomorphic features on the surface of the Earth may be identified.

 References

Ackerman, F., 1994, "Digital Elevation Models: Techniques and Application, Quality Standards, Development," *Proceedings, Symposium on Mapping and Geographic Information Systems,* Athens, GA: International Society for Photogrammetry & Remote Sensing, 30(4):421–432.

American Society of Photogrammetry, 1944, 1952, 1966, *Manual of Photogrammetry,* Falls Church: ASP, multiple editions.

Asker, J. R., 1992, "Congress Considers Landsat 'Decommercialization' Move," *Aviation Week and Space Technology,* May 11, 18–19.

Asrar, G. and J. Dozier, 1994, *EOS: Science Strategy for the Earth Observing System,* Woodbury: American Institute of Physics.

Bolstad, P. V. and T. M. Lillesand, 1992, "Rule-based Classification Models: Flexible Integration of Satellite Imagery and Thematic Spatial Data," *Photogrammetric Engineering & Remote Sensing,* 58(7):965–971.

Budge, A. and S. A. Morain, 1995, "Access Remote Sensing Data for GIS," *GIS World,* 8(2):45–49.

Canadian Space Agency, 1999, *RADARSAT,* Saint-Hubert: RADARSAT Program.

CEES, 1991, *Our Changing Planet: The FY 1992 U.S. Global Change Research Program,* Committee on Earth and Environmental Sciences, Office of Science & Technology, Washington, DC, 21 pp.

Clark, R. N., 1999, *Spectroscopy of Rocks and Minerals, and Principles of Spectroscopy,* Denver: U.S. Geological Survey, http://speclab.cr.usgs.gov, 58 pp.

Collins, J. B. and C. E. Woodcock, 1999, "Geostatistical Estimation of Resolution-Dependent Variance in Remotely Sensed Images," *Photogrammetric Engineering and Remote Sensing,* 65(1):41–50.

Colwell, R. N. (Ed.), 1960, *Manual of Photographic Interpretation,* Falls Church, VA: American Society for Photogrammetry & Remote Sensing.

Colwell, R. N. (Ed.), 1983, *Manual of Remote Sensing,* 2nd. Ed., Falls Church, VA: American Society of Photogrammetry.

Colwell, R. N., 1984, "From Photographic Interpretation to Remote Sensing," *Photogrammetric Engineering and Remote Sensing,* 50(9):1305.

Colwell, R. N., 1997, "History and Place of Photographic Interpretation," *Manual of Photographic Interpretation,* W. R. Philipson (Ed.), 2nd Ed., Bethesda: American Society for Photogrammetry & Remote Sensing, 33–48.

Congalton, R. G. and K. Green, 1998, *Assessing the Accuracy of Remotely Sensed Data,* Boca Raton: Lewis, 137 pp.

Cracknell, A. P. and L. W. B. Hayes, 1993, *Introduction to Remote Sensing,* London: Taylor & Francis, 293 pp.

Curran, P. J., 1987, "Remote Sensing Methodologies and Geography," *International Journal of Remote Sensing,* 8:1255–1275.

Dahlberg, R. W. and J. R. Jensen, 1986, "Education for Cartography and Remote Sensing in the Service of an Information Society: The United States Case," *The American Cartographer,* 13(1):51–71.

Davis, B. A., 1999, "An Overview of NASA's Commercial Remote Sensing Program," *Earth Observation Magazine,* 8(3):58–60.

Dobson, J. E., 1993, "Commentary: A Conceptual Framework for Integrating Remote Sensing, Geographic Information Systems, and Geography," *Photogrammetric Engineering & Remote Sensing,* 59(10):1491–1496.

Duggin, M. J. and C. J. Robinove, 1990, "Assumptions Implicit in Remote Sensing Data Acquisition and Analysis," *International Journal of Remote Sensing,* 11(10):1669–1694.

Eidenshink, J. C., 1992, "1990 Conterminous United States AVHRR Data Set," *Photogrammetric Engineering & Remote Sensing,* 58(6):809–813.

Emerson, C. W., N. Lam, and D. A. Quattrochi, 1999, "Multi-scale Fractal Analysis of Image Texture and Pattern," *Photogrammetric Engineering & Remote Sensing,* 65(1):51–61.

Engman, E. T. and N. Chauhan, 1995, "Status of Microwave Soil Moisture Measurements with Remote Sensing," *Remote Sensing of Environment,* 51:189–198.

ENVI, 1999, *ENVI User's Guide: Environment for Visualizing Images,* Boulder: Research Systems, Inc., 500 pp.

ESA, 1992, "The ERS-1 Spacecraft and Its Payload," *European Space Agency Bulletin,* 65:27–48.

Estes, J. R., 1966, *Geographic Applications of Multi-Image Correlation Remote Sensing Techniques,* Los Angeles, CA: University of California at Los Angeles, unpublished dissertation.

Estes, J. E., 1992, "Remote Sensing and Geographic Information System Integration: Research Needs, Status, and Trends," *ITC Journal,* 1992(1):2–10.

Estes, J. E., E. J. Hajic and L. Tinney, 1983, "Fundamentals of Image Analysis: Visible and Thermal Infrared Data," *Manual of Remote Sensing,* R. N. Colwell, (Ed.), Falls Church, VA: American Society for Photogrammetry & Remote Sensing, 987–1125.

Estes, J. E. and J. R. Jensen, 1998, "Development of Remote Sensing Digital Image Processing Systems and Raster GIS," *The History of Geographic Information Systems,* T. Foresman (Ed.), New York: Longman, Inc., 163–180.

Fisher, P. F. and R. E. Lindenberg, 1989, "On Distinctions among Cartography, Remote Sensing, and Geographic Information Systems," *Photogrammetric Engineering & Remote Sensing,* 55(10):1431–1434.

Franklin, J. and D. L. Turner, 1992, "Application of a Geometric Optical Canopy Reflectance Model to Semiarid Shrub Vegetation," *IEEE Transactions on Geoscience Remote Sensing,* 30:293–301.

Franklin, S. E. and B. A. Wilson, 1992, "A Three-stage Classifier for Remote Sensing of Mountain Environments," *Photogrammetric Engineering & Remote Sensing,* 58(4):449–454.

Fritz, L., 1996, "The Era of Commercial Earth Observation Satellites," *Photogrammetric Engineering & Remote Sensing,* 62(1):39–45.

Fussell, J., D. Rundquist and J. A. Harrington, 1986, "On Defining Remote Sensing," *Photogrammetric Engineering & Remote Sensing,* 52(9):1507–1511.

Gong, P. and P. Howarth, 1992, "Frequency Based Contextual Classification and Gray Level Vector Reduction for Land Use Identification," *Photogrammetric Engineering & Remote Sensing,* 58(4):423–437.

Goodchild, M. and S. Gopal, 1992, *Accuracy of Spatial Databases,* New York: Taylor & Francis, 290 pp.

Haack, B., S. C. Guptill, R. K. Holz, S. M. Jampoler, J. R. Jensen and R. A. Welch, 1997, "Urban Analysis and Planning," *The Manual of Photographic Interpretation,* Bethesda, MD: American Society for Photogrammetry & Remote Sensing, 517–553.

Henderson, F., 1994, "The Landsat Program—Life After Divorce?", *Earth Observation Magazine,* April, p. 8.

Hepner, G. F., T. Logan, N. Ritter and N. Bryant, 1990, "Artificial Neural Network Classification Using a Minimal Training Set: Comparison to Conventional Supervised Classification," *Photogrammetric Engineering & Remote Sensing,* 56(4):469–473.

Huang, X. and J. R. Jensen, 1998, "A Machine Learning Approach to Automated Construction of Knowledge Bases for Image Analysis Expert Systems that Incorporate Geographic Information System Data," *Photogrammetric Engineering & Remote Sensing,* 63(10):1185-1194.

Hudgins, E. L., 1997, "NASA and Mission to Planet Earth," Testimony to the U.S. House of Representatives Committee on Science Subcommittee on Space and Aeronautics (March 19), http://www.cato.org/testimony/ct-eh031997.html.

Jensen, J. R., 1992, "Testimony on S. 2297, The Land Remote Sensing Policy Act of 1992," Senate Committee on Commerce, Science, and Transportation, *Congressional Record,* (May 6):55–69.

Jensen, J. R., 1995, "Issues Involving the Creation of Digital Elevation Models and Terrain Corrected Orthoimagery Using Soft-copy Photogrammetry," *Geocarto International,* 10(1): 5–21.

Jensen, J. R., 1996, *Introductory Digital Image Processing: A Remote Sensing Perspective,* Upper Saddle River: Prentice-Hall, 318 pp.

Jensen, J. R. and D. C. Cowen, 1999, "Remote Sensing of Urban/Suburban Infrastructure and Socioeconomic Attributes," *Photogrammetric Engineering & Remote Sensing,* 65(5):611–622.

Jensen, J. R., F. Qiu and M. Ji, 1999, "Predictive Modeling of Coniferous Forest Age Using Statistical and Artificial Neural Network Approaches Applied to Remote Sensing Data," *International Journal of Remote Sensing,* 20(14):2805-2822.

Jensen, J. R. and R. E. Dahlberg, 1983, "Status and Content of Remote Sensing Education in the United States," *International Journal of Remote Sensing,* 4(2):235–245.

Jensen, J. R., E. W. Ramsey, B. Savitsky and B. Davis, 1990, "Environmental Sensitivity Index (ESI) Mapping for Oil Spills Using Remote Sensing and Geographic Information System Technology," *International Journal of Geographic Information Systems*, 4(2):181–201.

Jensen, J. R., D. J. Cowen, J. Halls, S. Narumalani, N. Schmidt, B. A. Davis and B. Burgess, 1994, "Improved Urban Infrastructure Mapping and Forecasting for Bell South Using Remote Sensing and Geographic Information System Technology," *Photogrammetric Engineering & Remote Sensing*, 60(3):339–346.

Ji, M. and J. R. Jensen, 1996, "Fuzzy Training in Supervised Image Classification," *Journal of Geographic Information Sciences*, 2(2):1–12.

Jones, L., 1992, *TOPEX/POSEIDON—Oceanography from Space: The Oceans and Climate*, Washington, DC: NASA, 22 pp.

Jones, K. B. et al., 1997, *An Ecological Assessment of the United States Mid-Atlantic Region*, Washington, DC: Environmental Protection Agency, 106 pp.

Khorram, S., G. Biging, N. Chrisman, D. Colby, R. Congalton, J. Dobson, R. Ferguson, M. Goodchild, J. Jensen and T. Mace, 1999, *Accuracy Assessment of Land Cover Change Detection*, Bethesda, MD: American Society for Photogrammetry & Remote Sensing, 64 pp.

Kruse, F. A., A. B. Lefkoff, J. W. Boardman, K. B. Heidebrecht, A. T. Shapiro, P. J. Barloon and A. F. H. Goetz, 1992, "The Spectral Image Processing System (SIPS)—Interactive Visualization and Analysis of Imaging Spectrometer Data," *Proceedings*, International Space Year Conference, Pasadena, CA, 10 pp.

Landgrebe, D., 1999, *An Introduction to MULTISPEC*, W. Lafayette, IN: Purdue University, 50 pp.

Lanter, D. P. and H. Veregin, 1992, "A Research Paradigm for Propagating Error in Layer-based GIS," *Photogrammetric Engineering & Remote Sensing*, 58(6):825–833.

Li, R., 1998, "Potential of High-Resolution Satellite Imagery for National Mapping Products," *Photogrammetric Engineering & Remote Sensing*, 64(12):1165–1169.

Li, X. and A. H. Strahler, 1985, "Geometric-optical Modeling of a Conifer Forest Canopy," *IEEE Transactions on Geoscience Remote Sensing*, 23:70-5-721.

Lillesand, R. M. and R. W. Kiefer, 1994, *Remote Sensing and Image Interpretation*, 3rd Ed. NY: John Wiley & Sons, 750 pp.

Lousma, J. R., 1993, "Rising to the Challenge: The Role of the Information Sciences," *Photogrammetric Engineering & Remote Sensing*, 59(6):957–959.

Lunetta, R. S., R. G. Congalton, L. K. Fenstermaker, J. R. Jensen, K. C. McGwire and L. R. Tinney, 1991, "Remote Sensing and GIS Data Integration: Error Sources and Research Issues," *Photogrammetric Engineering & Remote Sensing*, 57(6):677–687.

Luther, M. R., 1992, *UARS — Upper Atmosphere Research Satellite: A Program to Study Global Ozone Change.* Washington, DC: NASA, 29 pp.

Meisburger, J. and J. R. Jensen, 1999, *Quantification of Biomass and Leaf-Area-Index in a Charleston, SC, Estuary Using Low-Altitude AVIRIS Imagery*, Washington: NASA, 35 pp.

Moller-Jensen, L., 1990, "Knowledge-based Classification of an Urban Area Using Texture and Context Information in Landsat-TM Imagery," *Photogrammetric Engineering & Remote Sensing*, 56(6):899–904.

Moran, M. S., Y. Inoue and E. H. Barnes, 1997, "Opportunities and Limitations for Image-based Remote Sensing Precision Crop Management," *Remote Sensing of Environment*, 61:319–346.

Nadler, M. and E. Smith, 1993, *Pattern Recognition Engineering.* NY: John Wiley & Sons, 588 pp.

NASA, 1988, *Earth System Science: A Closer View.* Washington, DC: NASA, 36 pp.

NASA, 1998, *NASA's Earth Observing System —EOS AM-1*, Greenbelt: NASA Goddard Space Flight Center, 34 pp.

NASA Stennis, 1999, *ATLAS: Performance Verification Test Report*, SSC, MS: NASA Stennis Space Center, 43 pp.

Paylor, E. D., Kaye, J. A., Johnson, A. R. and N. G. Maynard, 1999, "Earth Science Enterprise Science and Technology for Society," *Earth Observation Magazine*, 8(3):8–12.

Pickover, C. A., 1991, *Computers and the Imagination: Visual Adventures Beyond the Edge*, New York: St. Martin's Press, 424 pp.

Price, R. D., et al., 1994, "Earth Science Data for All: EOS and the EOS Data and Information System," *Photogrammetric Engineering & Remote Sensing*, 60(3):469–473.

Pruitt, E. L., 1979, "The Office of Naval Research and Geography," *Annals,* Association of American Geographers, 69(1):106.

Ramsey, R. D., A. Falconer and J. R. Jensen, 1995, "The Relationship Between NOAA-AVHRR Normalized Difference Vegetation Index and Ecoregions in Utah," *Remote Sensing of Environment*, 53:188–198.

Rivard, B. and R. E. Arvidson, 1992, "Utility of Imaging Spectrometry for Lithologic Mapping in Greenland," *Photogrammetric Engineering & Remote Sensing*, 58(7):945–949.

Robbins, J., 1999, "High-Tech Camera Sees What Eye Cannot," *New York Times*, Science Section, September 14, D5.

Ryerson, R., 1989, "Image Interpretation Concerns for the 1990s and Lessons from the Past," *Photogrammetric Engineering & Remote Sensing*, 55(10):1427–1430.

Schott, J. R., 1997, *Remote Sensing —the Image Chain Approach*, New York: Oxford University Press, 394 pp.

Steven, M. D., 1993, "Satellite Remote Sensing for Agricultural Management: Opportunities and Logistic Constraints," *ISPRS Journal of Photogrammetry & Remote Sensing*, 48(4):29–34.

Stow, D. A., 1993, "The Role of GIS for Landscape Ecological Studies," *Landscape Ecology and GIS*, R. Haines-Young, D. Green, and S. Cousins, Eds., NY: Taylor & Francis, 11–21.

Strahler, A. H., C. E. Woodcock and J. A. Smith, 1986, "On the Nature of Models in Remote Sensing," *Remote Sensing of Environment*, 20:121–139.

Usery, E. L., P. Altheide, R. R. Deister and D. J. Barr, 1988, "Knowledge-based Geographic Information System Techniques Applied to Geological Engineering," *Photogrammetric Engineering & Remote Sensing*, 54(11):1623–1628.

Vickers, E. W., 1993, "Production Procedures for an Oversize Satellite Image Map," *Photogrammetric Engineering & Remote Sensing*, 59(2):247–254.

Walsh, S. J., T. P. Evans, W. F. Welsh, B. Entwisle and R. R. Rindfuss, 1999, "Scale-dependent Relationships Between Population and Environment in Northeastern Thailand," *Photogrammetric Engineering & Remote Sensing*, 65(1):97–105.

Wolff, R. S. and L. Yaeger, 1993, *Visualization of Natural Phenomena*, Santa Clara, CA: Telos Springer-Verlag, 374 pp.

Wolter, J. A., 1975, *The Emerging Discipline of Cartography*, Minneapolis: University of Minnesota, Department of Geography, unpublished dissertation.

Woodcock, C. E., J. B. Collins, V. Jakabhazy, X. Li, S. Macomber and Y. Wu, 1997, "Inversion of the Li-Strahler Canopy Reflectance Model for Mapping Forest Structure," *IEEE Transactions on Geoscience and Remote Sensing*, 35(2):405–414.

Wright, D. J., M. F. Goodchild and J. D. Procter, 1997, "GIS: Tool or Science; Demystifying the Persistent Ambiguity of GIS as Tool versus Science," *The Professional Geographer*, 87(2):346–362.

Electromagnetic Radiation Principles 2

Energy recorded by remote sensing systems undergoes fundamental interactions that should be understood to properly interpret the remotely sensed data. For example, if the energy being remotely sensed comes from the Sun, the energy

- is radiated by atomic particles at the source (the Sun),

- propagates through the vacuum of space at the speed of light,

- interacts with the Earth's atmosphere,

- interacts with the Earth's surface,

- interacts with the Earth's atmosphere once again, and

- finally reaches the remote sensor, where it interacts with various optical systems, filters, film emulsions, or detectors.

It is instructive to examine each of these fundamental interactions that electromagnetic energy undergoes as it progresses from its source to the remote sensing system detector.

 Conduction, Convection, and Radiation

Energy is the ability to do work. In the process of doing work, energy is often transferred from one body to another or from one place to another. The three basic ways in which energy can be transferred include conduction, convection, and radiation (Figure 2-1). Most people are familiar with *conduction* that occurs when one body (molecule or atom) transfers its kinetic energy to another by colliding with it. This is how a metal pan is heated by a hot burner on a stove. In *convection*, the kinetic energy of bodies is transferred from one place to another by physically moving the bodies. A good example is the heating of the air near the ground in the morning hours. The warmer air near the surface rises, setting up convectional currents in the atmosphere, which may produce cumulus clouds. The transfer of energy by electromagnetic *radiation* is of primary interest to remote sensing science because it is the only form of energy transfer that can take place in a vacuum such as the region between the Sun and the Earth.

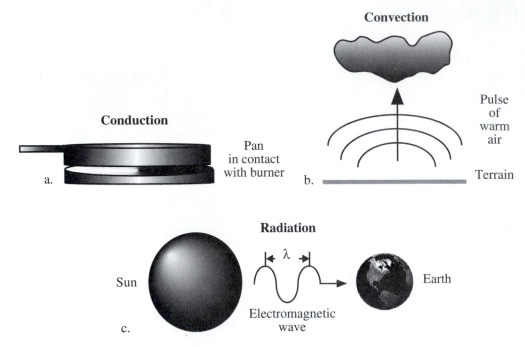

Figure 2-1   Heat may be transferred in three ways: conduction, convection, and radiation. a) Heat may be conducted directly from one object to another as when a pan is in direct physical contact with a hot burner. b) The Sun bathes the Earth's surface with radiant energy causing the air near the ground to increase in temperature. The less dense air rises creating convectional currents in the atmosphere. c) Electromagnetic energy in the form of electromagnetic waves may be transmitted through the vacuum of space from the Sun to the Earth.

 ### Electromagnetic Radiation Models

To understand how electromagnetic radiation is created, how it propagates through space, and how it interacts with other matter, it is useful to describe the processes using two different models: the *wave* model and the *particle* model (Englert et al., 1994).

Wave Model of Electromagnetic Energy

In the 1860s, James Clerk Maxwell (1831–1879) conceptualized electromagnetic radiation (EMR) as an electromagnetic wave that travels through space at the speed of light, c, which is 3 x 10^8 meters per second (hereafter referred to as m s^{-1}) or 186,282.03 miles s^{-1} (Trefil and Hazen, 1995). A useful relation for quick calculations is that light travels about 1 ft per nanosecond (10^{-9} s) (Rinker, 1999). The *electromagnetic wave* consists of two fluctuating fields — one electric and the other magnetic (Figure 2-2). The two vectors

are at right angles (orthogonal) to one another, and both are perpendicular to the direction of travel (Bolemon, 1985).

But how is an electromagnetic wave created? *Electromagnetic radiation* is generated whenever an electrical charge is accelerated. The wavelength (λ) of the electromagnetic radiation depends upon the length of time that the charged particle is accelerated. Its frequency (v) depends on the number of accelerations per second. *Wavelength* is formally defined as the mean distance between maximums (or minimums) of a roughly periodic pattern (Figure 2-2) and is normally measured in micrometers (μm) or nanometers (nm). *Frequency* is the number of wavelengths that pass a point per unit time. A wave that sends one crest by every second (completing one cycle) is said to have a frequency of one cycle per second, or one *hertz*, abbreviated 1 Hz. Frequently used measures of wavelength and frequency are found in Table 2-1.

The relationship between the wavelength (λ) and frequency (v) of electromagnetic radiation is based on the following formula, where c is the speed of light (Egan, 1985):

$$c = \lambda v \qquad\qquad (2\text{-}1)$$

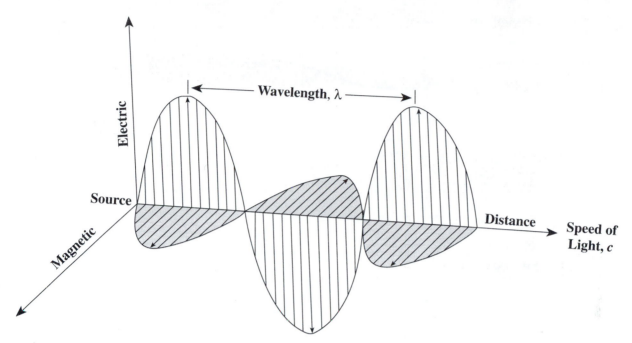

Figure 2-2   An electromagnetic wave is composed of both electric and magnetic vectors that are orthogonal (at 90° angles) to one another. The waves travel from the source at the speed of light (3 x 10^8 m s^{-1}).

$$v = \frac{c}{\lambda} \qquad (2\text{-}2)$$

and

$$\lambda = \frac{c}{v}. \qquad (2\text{-}3)$$

Note that frequency is *inversely* proportional to wavelength. This relationship is shown diagrammatically in Figure 2-3, where the longer the wavelength, the lower the frequency; the shorter the wavelength, the higher the frequency. When electromagnetic radiation passes from one substance to another, the speed of light and wavelength change while the frequency remains the same.

All objects above absolute zero (-273°C or 0 K) emit electromagnetic energy, including water, soil, rock, vegetation, and the surface of the Sun. The Sun represents the initial source of most of the electromagnetic energy recorded by remote sensing systems (except radar and sonar) (Figure 2-4; Color Plate 2-1). We may think of the Sun as a 6,000 K *blackbody* (a theoretical construct that absorbs and radiates energy at the maximum possible rate per unit area at each wavelength (λ) for a given temperature). The total emitted radiation from a blackbody (M_λ) measured in Watts per m^{-2} is proportional to the fourth power of its absolute temperature (T) measured

in degrees Kelvin. This is known as the *Stefan-Boltzmann law* and is expressed as:

$$M_\lambda = \sigma T^4 \qquad (2\text{-}4)$$

where σ is the Stefan-Boltzmann constant, 5.6697 x 10^{-8}W m^{-2}K^{-4}. The important thing to remember is that the amount of energy emitted by an object such as the Sun or the Earth is a function of its temperature. The greater the temperature, the greater the amount of radiant energy exiting the object. The actual amount of energy emitted by an object is computed by summing (integrating) the area under its curve (Figure 2-5). It is clear from this illustration that the total emitted radiation from the 6,000 K Sun is far greater than that emitted by the 300 K Earth.

In addition to computing the total amount of energy exiting a theoretical blackbody such as the Sun, we can determine its dominant wavelength (λ_{max}) based on *Wien's displacement law*:

$$\lambda_{max} = \frac{k}{T} \qquad (2\text{-}5)$$

where k is a constant equaling 2898 μm K, and T is the absolute temperature in degrees Kelvin. Therefore, as the Sun

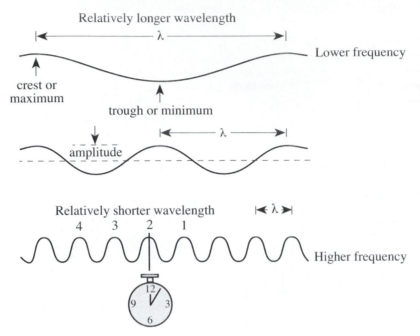

Figure 2-3   This cross-section of several electromagnetic waves illustrates the inverse relationship between wavelength (λ) and frequency (v). The longer the wavelength, the lower the frequency; the shorter the wavelength, the higher the frequency. The amplitude of an electromagnetic wave is the height of the wave crest above the undisturbed position. Successive wave crests are numbered 1, 2, 3, and 4. An observer at the position of the clock records the number of crests that pass by in a second. This frequency is measured in cycles per second, or *hertz*.

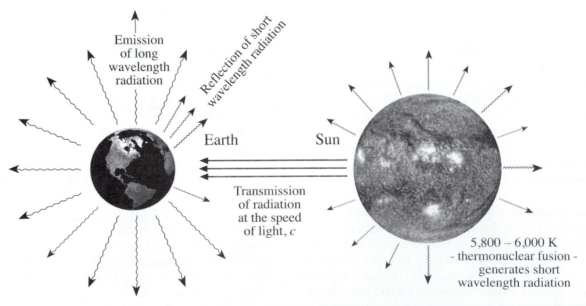

Figure 2-4   The thermonuclear fusion taking place on the surface of the Sun yields a continuous spectrum of electromagnetic energy. The 5,800 – 6,000 K temperature of this process produces a large amount of relatively short wavelength energy that travels through the vacuum of space at the speed of light. Some of this energy is intercepted by the Earth, where it interacts with the atmosphere and surface materials. The Earth reflects some of the energy directly back out to space or it may absorb the short wavelength energy and then reemit it at a longer wavelength (after Strahler and Strahler, 1989). This process takes place continually.

Table 2-1.   Wavelength and Frequency Standard Units of Measurement

Wavelength (λ)	
kilometer (km)	1,000 m
meter (m)	1.0 m
centimeter (cm)	$0.01 \text{ m} = 10^{-2} \text{ m}$
millimeter (mm)	$0.001 \text{ m} = 10^{-3} \text{ m}$
micrometer (µm)	$0.000001 = 10^{-6} \text{ m}$
nanometer (nm)	$0.000000001 = 10^{-9} \text{ m}$
Angstrom (A)	$0.0000000001 = 10^{-10} \text{ m}$
Frequency (cycles per second)	
hertz (Hz)	1
kilohertz (kHz)	$1,000 = 10^{3}$
megahertz (MHz)	$1,000,000 = 10^{6}$
gigahertz (GHz)	$1,000,000,000 = 10^{9}$

approximates a 6,000 K blackbody, its dominant wavelength (λ_{max}) is 0.48 µm:

$$0.483 \mu m = \frac{2898 \ \mu m \ K}{6000 \ K}.$$

Electromagnetic energy from the Sun travels in eight minutes across the intervening 93 million miles (150 million kilometers) of space to the Earth. As shown in Figure 2-5, the Earth approximates a 300 K (27°C) blackbody and has a dominant wavelength at approximately 9.66 µm:

$$9.66 \mu m = \frac{2898 \ \mu m \ K}{300 \ K}.$$

Although the Sun has a dominant wavelength at 0.48 µm, it produces a continuous spectrum with electromagnetic radiation ranging from very short, extremely high frequency gamma and cosmic waves to long, very low frequency radio waves (Figures 2-6 and 2-7). The Earth only intercepts a very small portion of the electromagnetic energy produced by the Sun.

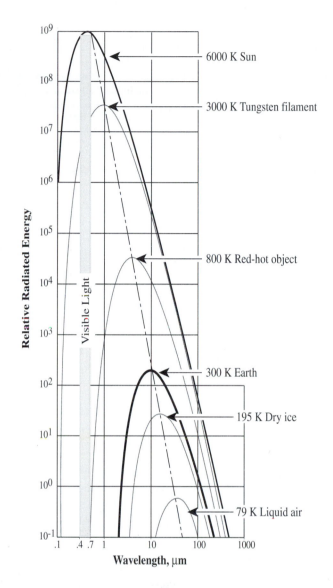

Figure 2-5   Blackbody radiation curves for several objects, including the Sun and the Earth, which approximate 6,000 K and 300 K blackbodies, respectively. The area under each curve may be summed to compute the total radiant energy (M_λ) exiting each object (Equation 2-4). Thus, the Sun produces more radiant exitance than the Earth because its temperature is greater. As the temperature of an object increases, its dominant wavelength (λ_{max}) shifts toward the shorter wavelengths of the electromagnetic spectrum.

As mentioned in Chapter 1, in remote sensing research we often specify a particular region of the electromagnetic spectrum (e.g., red light) by identifying a beginning and ending wavelength (or frequency) and then attaching a description.

Table 2-2.   Methods of Describing the Color Spectrum (after Nassau, 1983)

	Wavelength Descriptions				Energy Descriptions	
Color[a]	Angstrom (A)	Nanometer (nm)	Micrometer (μm)	Frequency Hz (x 1014)	Wave Number[c] (ψ cm^{-1})	Electron Volt (eV)
Ultraviolet, sw	2,537	254	0.254	11.82	39,400	4.89
Ultraviolet, lw	3,660	366	0.366	8.19	27,300	3.39
Violet (limit)[b]	4,000	400	0.40	7.50	25,000	3.10
Blue	4,500	450	0.45	6.66	22,200	2.75
Green	5,000	500	0.50	6.00	20,000	2.48
Green	5,500	550	0.55	5.45	18,200	2.25
Yellow	5,800	580	0.58	5.17	17,240	2.14
Orange	6,000	600	0.60	5.00	16,700	2.06
Red	6,500	650	0.65	4.62	15,400	1.91
Red (limit)[b]	7,000	700	0.70	4.29	14,300	1.77
Infrared, near	10,000	1,000	1.0	3.00	10,000	1.24
Infrared, far	300,000	30,000	30.00	0.10	333	0.041

[a]Typical values only; lw = long wavelength; sw = short wavelength. [b]Exact limit depends on the observer, light intensity, eye adaptation, and other factors. [c]The wave number (ψ) is the number of waves in a unit length (usually per cm). Therefore, $\psi = 1 / \lambda$ (cm) = 10,000 / λ (μm) =100,000,000 / λ (A) in cm^{-1}.

This wavelength (or frequency) interval in the electromagnetic spectrum is commonly referred to as a *band*, *channel*, or *region*. The major subdivisions of visible light are presented diagrammatically in Figure 2-7 and summarized in Table 2-2. For example, we generally think of visible light as being composed of energy in the blue (0.4 – 0.5 μm), green (0.5 – 0.6 μm), and red (0.6 – 0.7 μm) bands of the electromagnetic spectrum (Sagan, 1994). Similarly, reflected near-infrared energy in the region from 0.7 – 1.3 μm is commonly used to expose black-and-white and color-infrared sensitive film.

The middle-infrared region includes energy with a wavelength of 1.3 – 3 μm. The thermal infrared region has two very useful bands at 3 – 5 μm and 8 – 14 μm. The microwave portion of the spectrum consists of much longer wavelengths (1 mm – 1 m). The radio-wave portion of the spectrum may be subdivided into UHF, VHF, Radio (HF), LF, and ULF frequencies.

The spectral resolution of most remote sensing systems is described in terms of bands of the electromagnetic spectrum. For example, the spectral dimensions of the four bands of the Landsat Multispectral Scanner (MSS) and SPOT High Resolution Visible (HRV) sensors are shown in Figure 2-8, along with the spatial resolution of each band for comparison. The exact Landsat MSS and SPOT HRV band specifications are found in Chapter 7.

Electromagnetic energy may be described not only in terms of wavelength and frequency but also in photon energy units such as Joules (J) and electron volts (eV), as shown in Figure 2-7. Several of the more important mass, energy, and power conversions are summarized in Table 2-3.

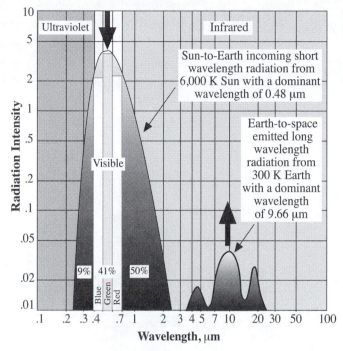

Figure 2-6   The Sun approximates a 6,000 K blackbody with a dominant wavelength of about 0.48 μm. The Earth approximates a 300 K blackbody with a dominant wavelength of about 9.66 μm. The 6,000 K Sun produces approximately 41 percent of its energy in the visible region from 0.4 – 0.7 μm (blue, green, and red light). The other 59 percent of the energy is in wavelengths shorter than blue light (<0.4 μm) and longer than red light (>0.7 μm). Our eyes are only sensitive to light from 0.4 to 0.7 μm (after Strahler and Strahler, 1989). Fortunately, it is possible to make remote sensor detectors that are sensitive to energy in these nonvisible regions of the spectrum.

Table 2-3.   Mass, Energy and Power Conversions

Conversion from English to SI Units

To get:	Multiply:	By:
newtons[a]	pounds	4.448
joules[b]	BTUs[c]	1055
joules	calories[d]	4.184
joules	kilowatt-hours[e]	3.6×10^6
joules	foot-pounds[f]	1.356
joules	horsepower[g]	745.7

Conversion from SI to English Units

To get:	Multiply:	By:
BTUs	joules	0.00095
calories	joules	0.2390
kilowatt-hours	joules	2.78×10^{-7}
foot-pounds	joules	0.7375
horsepower	watts	0.00134

[a]newton: force needed to accelerate a mass of 1 kg by 1 m s^{-2}
[b]joule: a force of 1 newton acting through 1 meter.
[c]British Thermal Unit, or BTU: energy required to raise the temperature of 1 pound of water by 1 degree Fahrenheit.
[d]calorie: energy required to raise the temperature of 1 kilogram of water by 1 degree Celsius.
[e]kilowatt-hour: 1000 joules per second for 1 hour.
[f]foot-pound: a force of 1 pound acting through 1 foot.
[g]horsepower: 550 foot-pounds per second.

The Particle Model - Radiation from Atomic Structures

In *Opticks* (1704), Sir Isaac Newton stated that light was a stream of particles, or corpuscles, traveling in straight lines. He also knew that light had wavelike characteristics based on his work with glass plates. Nevertheless, during the hundred years before 1905, light was thought of primarily as a smooth and continuous wave. Then, Albert Einstein (1879–1955) found that when light interacts with electrons, it has a different character. He concluded that when light interacts with matter, it behaves as though it is composed of many individual bodies called *photons*, which carry such particle-like properties as energy and momentum (Bolemon, 1985; Meadows, 1992). As a result, most physicists today would

answer the question, What is light? by saying that light is a *particular* kind of matter (Feinberg, 1985). Thus, we sometimes describe electromagnetic energy in terms of its wave-like properties. But when the energy interacts with matter, it is useful to describe it as discrete packets of energy, or *quanta*. It is practical to review how electromagnetic energy is generated at the atomic level, as this provides insight as to how light interacts with matter.

Electrons are the tiny negatively charged particles that move around the positively charged nucleus of an atom (Figure 2-9). Atoms of different substances are made up of varying numbers of electrons arranged in different ways. The inter-

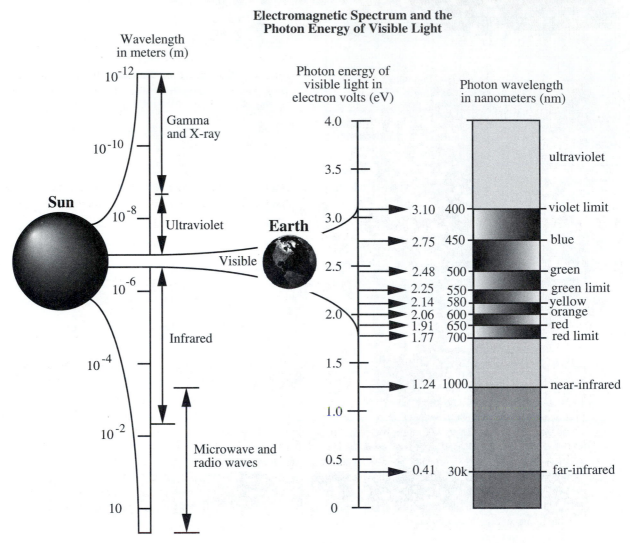

Figure 2-7   The electromagnetic spectrum and the photon energy of visible light. The Sun produces a continuous spectrum of energy from gamma rays to radio waves that continually bathe the Earth in energy. The visible portion of the spectrum may be measured using wavelength (measured in micrometers or nanometers, i.e., μm or nm) or electron volts (eV). All units are interchangeable.

action between the positively charged nucleus and the negatively charged electron keeps the electron in orbit. While its orbit is not explicitly fixed, each electron's motion is restricted to a definite range from the nucleus. The allowable orbital paths of electrons moving around an atom might be thought of as energy classes or levels (Figure 2-9a). In order for an electron to climb to a higher class, work must be performed. However, unless an amount of energy is available to move the electron up at least one energy level, it will accept no work. If a sufficient amount of energy is received, the electron will jump to a new level and the atom is said to be *excited* (Figure 2-9b). Once an electron is in a higher orbit, it possesses potential energy. After about 10^{-8} seconds, the

electron falls back to the atom's lowest empty energy level or orbit and gives off radiation (Figure 2-9c). The wavelength of radiation given off is a function of the amount of work done on the atom, i.e., the quantum of energy it absorbed to cause the electron to become excited and move to a higher orbit.

Electron orbits are like the rungs of a ladder. Adding energy moves the electron up the energy ladder; emitting energy moves it down. However, the energy ladder differs from an ordinary ladder in that its rungs are unevenly spaced. This means that the energy an electron needs to absorb, or to give up, in order to jump from one orbit to the next may not be the

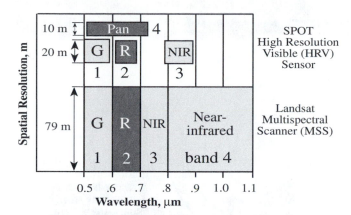

Figure 2-8 The nominal spectral bandwidths of the individual bands of the Landsat MSS and SPOT HRV sensor systems are summarized along the abscissa. The spatial resolution of each individual band is shown on the ordinate axis.

same as the energy change needed for some other step. Furthermore, an electron does not necessarily use consecutive rungs. Instead, it follows what physicists call *selection rules*. In many cases, an electron uses one sequence of rungs as it climbs the ladder and another sequence as it descends (Nassau, 1983). The energy that is left over when the electrically charged electron moves from an excited state (Figure 2-9b) to a de-excited state (Figure 2-9c) is emitted by the atom as a single packet of electromagnetic radiation; a particle-like unit of light called a *photon*. Every time an electron jumps from a higher to a lower energy level, a photon moves away at the speed of light.

Somehow an electron must disappear from its original orbit and reappear in its destination orbit without ever having to traverse any of the positions in between. This process is called a *quantum leap* or *quantum jump*. If the electron leaps from its highest excited state to the ground state in a single leap, it will emit a single photon of energy. It is also possible for the electron to leap from an excited orbit to the ground state in a series of jumps, e.g., from 4 to 2 to 1. If it takes two leaps to get to the ground state, then each of these jumps will emit photons of somewhat less energy. The energies emitted in the two different jumps must sum to the total of the single large jump (Trefil and Hazen, 1995).

Niels Bohr (1885–1962) and Max Planck recognized the discrete nature of exchanges of radiant energy and proposed the *quantum theory* of electromagnetic radiation. This theory states that energy is transferred in discrete packets called quanta or photons as discussed. The relationship between the frequency of radiation expressed by wave theory and the quantum is (Bolemon, 1985):

$$Q = hv \tag{2-6}$$

where Q is the energy of a quantum measured in Joules (J), h is the Planck constant (6.626×10^{-34} J s), and v is the frequency of the radiation. Referring to Equation 2-3, we can multiply the equation by h/h, or 1, without changing its value:

$$\lambda = \frac{hc}{hv}. \tag{2-7}$$

By substituting Q for hv (from Equation 2-6) we can express the wavelength associated with a quantum of energy as:

$$\lambda = \frac{hc}{Q} \tag{2-8}$$

or

$$Q = \frac{hc}{\lambda}. \tag{2-9}$$

Thus, we see that the energy of a quantum is inversely proportional to its wavelength, i.e., the longer the wavelength involved, the lower its energy content. This inverse relationship is important to remote sensing because it suggests that it is more difficult to detect longer wavelength energy being emitted at thermal infrared wavelengths than those at shorter visible wavelengths. In fact, it might be necessary to have the sensor look at or dwell longer on the parcel of ground if we are trying to measure the longer wavelength energy. The energy of quanta (photons) ranging from gamma rays to radio waves is summarized in Figure 2-10.

Substances have color because of differences in their energy levels and the selection rules. For example, consider energized sodium vapor that produces a bright yellow light that is used in some street lamps. When a sodium vapor lamp is turned on, several thousand volts of electricity energize the vapor. The outermost electron in each energized atom of sodium vapor climbs to a higher rung on the energy ladder and then returns down the ladder in a certain sequence of rungs, the last two of which are 2.1 eV apart (Figure 2-11). The energy released in this last leap appears as a photon of yellow light with a wavelength of 0.58 μm with 2.1 eV of energy (Nassau, 1983).

Matter can be heated to such high temperatures that electrons that normally move in captured, nonradiating orbits break free (Figure 2-9d). When this happens, the atom remains with a positive charge equal to the negatively charged electron that escaped. The electron becomes a free

Creation of Light from Atomic Particles and the Photoelectric Effect

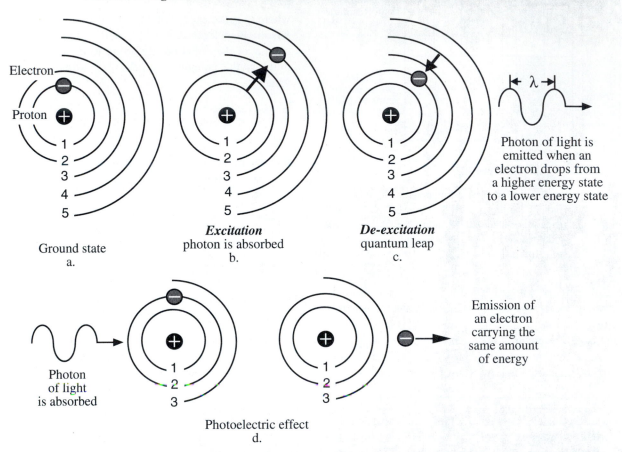

Figure 2-9   a–c) A photon of electromagnetic energy is emitted when an electron in an atom or molecule drops from a higher energy state to a lower energy state. The light emitted (i.e., its wavelength) is a function of the changes in the energy levels of the outer, valence electron. For example, yellow light is produced from a sodium vapor lamp in Figure 2-11. d) Matter can also be subjected to such high temperatures that electrons that normally move in captured, nonradiating orbits are broken free. When this happens, the atom remains with a positive charge equal to the negatively charged electron which escaped. The electron becomes a free electron, and the atom is called an ion. If another free electron fills the vacant energy level created by the free electron, then radiation from all wavelengths is produced, i.e., a continuous spectrum of energy. The intense heat at the surface of the Sun produces a continuous spectrum in this manner.

electron and the atom is called an *ion*. In the ultraviolet and visible (blue, green, and red) parts of the electromagnetic spectrum, radiation is produced by changes in the energy levels of the outer valence electrons. The wavelengths of energy produced are a function of the particular orbital levels of the electrons involved in the excitation process. If the atoms absorb enough energy to become ionized and if a free electron drops in to fill the vacant energy level, then the radiation given off is unquantized and a *continuous spectrum* is produced rather than a band or a series of bands. Every encounter of one of the free electrons with a positively charged nucleus causes rapidly changing electric and magnetic fields, so that radiation at all wavelengths is produced.

The hot surface of the Sun is largely a *plasma* in which radiation of all wavelengths is produced. As previously shown in Figure 2-7, the spectra of a plasma like the Sun is a continuous spectrum.

In atoms and molecules, electron orbital changes produce the shortest wavelength radiation, molecule vibrational motion changes produce near- and/or middle-infrared energy, and rotational motion changes produce long wavelength infrared or microwave radiation. More will be said about how thermal infrared radiation is produced and recorded by remote sensing systems in Chapter 8 (Thermal Infrared Remote Sensing).

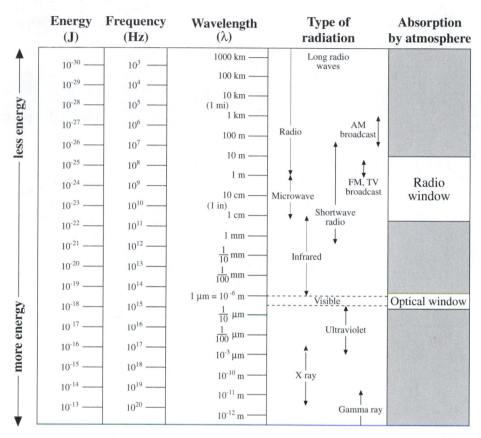

Figure 2-10 The energy of quanta (photons) ranging from gamma rays to radio waves in the electromagnetic spectrum.

Energy-Matter Interactions in the Atmosphere

Radiant energy is the capacity of radiation within a spectral band to do work (Colwell, 1983). Once electromagnetic radiation is generated, it is propagated through the Earth's atmosphere almost at the speed of light in a vacuum. Unlike a vacuum in which nothing happens, however, the atmosphere may affect not only the speed of radiation but also its wavelength, its intensity, and its spectral distribution. The electromagnetic radiation may also be diverted from its original direction due to refraction.

Refraction

The speed of light in a vacuum is 3×10^8 m s^{-1}. When electromagnetic radiation (EMR) encounters substances of different density, like air and water, refraction may take place. *Refraction* refers to the bending of light when it passes from one medium to another. Refraction occurs because the media are of differing densities and the speed of EMR is different in each. The *index of refraction (n)* is a measure of the optical density of a substance. This index is the ratio of the speed of light in a vacuum, c, to the speed of light in a substance such as the atmosphere or water, c_n (Mulligan, 1980):

$$n = \frac{c}{c_n}. \qquad (2\text{-}10)$$

The speed of light in a substance can never reach the speed of light in a vacuum. Therefore, its index of refraction must always be greater than 1. For example, the index of refraction for the atmosphere is 1.0002926 and 1.33 for water. Light travels more slowly through water.

Refraction can be described by Snell's law, which states that for a given frequency of light (we must use frequency since, unlike wavelength, it does not change when the speed of light changes), the product of the index of refraction and the sine of the angle between the ray and a line normal to the interface is constant:

**Creation of Light from Atomic Particles
The Sodium Vapor Lamp**

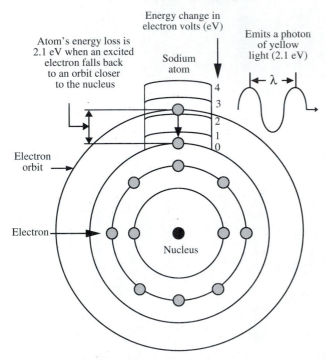

Figure 2-11   The creation of light from atomic particles in a sodium vapor lamp. After being energized by several thousand volts of electricity, the outermost electron in each energized atom of sodium vapor climbs to a high rung on the energy ladder and then returns down the ladder in a predictable fashion. The last two rungs in the descent are 2.1 eV apart. This produces a photon of yellow light which has 2.1 eV of energy (refer to Figure 2-7 and Table 2-2).

$$n_1 \sin\theta_1 = n_2 \sin\theta_2. \qquad (2\text{-}11)$$

From Figure 2-12 we can see that a nonturbulent atmosphere can be thought of as a series of layers of gases, each with a different density. Anytime energy is propagated through the atmosphere for any appreciable distance at any angle other than vertical, refraction occurs.

The amount of refraction is a function of the angle made with the vertical (θ), the distance involved (in the atmosphere the greater the distance, the more changes in density), and the density of the air involved (air is usually more dense near sea level). Serious errors in location due to refraction can occur in images formed from energy detected at high

Atmospheric Refraction

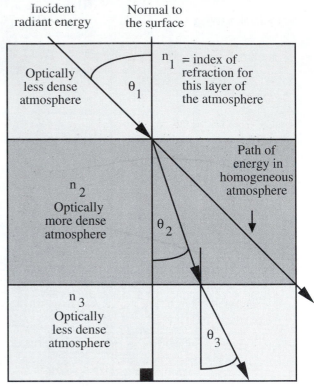

Path of energy affected
by atmospheric refraction

Figure 2-12   Refraction in three nonturbulent atmospheric layers. The incident energy is bent from its normal trajectory as it travels from one atmospheric layer to another. Snell's law can be used to predict how much bending will take place, based on a knowledge of the angle of incidence (θ) and the index of refraction of each atmospheric level, n_1, n_2, n_3.

altitudes or at acute angles. However, these location errors are predictable by Snell's law and thus can be removed. Notice that

$$\sin\theta_2 = \frac{n_1 \sin\theta_1}{n_2}. \qquad (2\text{-}12)$$

Therefore, if one knows the index of refraction of medium n_1 and n_2 and the angle of incidence of the energy to medium n_1, it is possible to predict the amount of refraction that will take place ($\sin\theta_2$) in medium n_2 using trigonometric relationships.

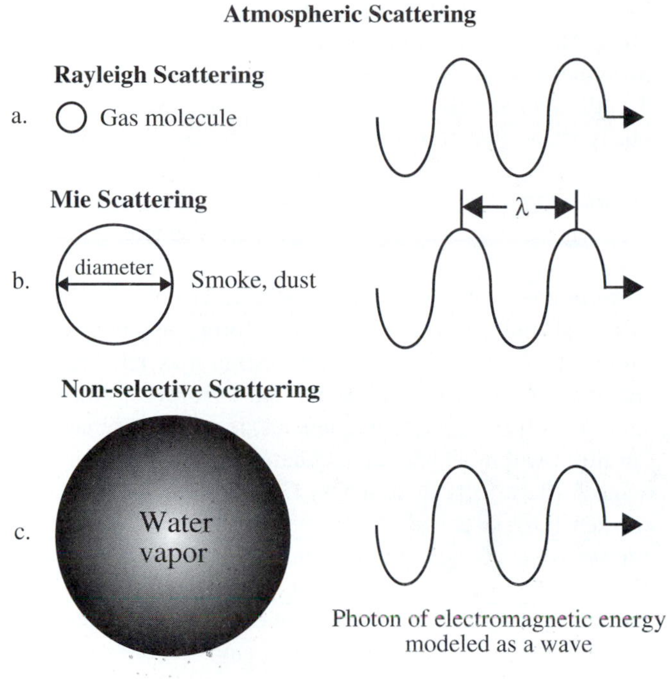

Atmospheric Scattering

Rayleigh Scattering

a. ○ Gas molecule

Mie Scattering

b. (diameter) Smoke, dust

Non-selective Scattering

c. Water vapor

λ

Photon of electromagnetic energy
modeled as a wave

Figure 2-13   Types of scattering encountered in the atmosphere. The type of scattering is a function of 1) the wavelength of the incident radiant energy, and 2) the size of the gas molecule, dust particle, and/or water vapor droplet encountered.

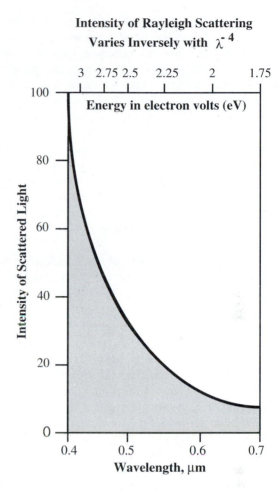

Intensity of Rayleigh Scattering Varies Inversely with λ^{-4}

Figure 2-14   The intensity of Rayleigh scattering varies inversely with the fourth power of the wavelength (λ^{-4}).

Scattering

One very serious effect of the atmosphere is the scattering of radiation by atmospheric particles. *Scattering* differs from reflection in that the direction associated with scattering is unpredictable, whereas the direction of reflection (to be defined shortly) is predictable. There are essentially three types of scattering: Rayleigh, Mie, and Non-selective scattering. The relative size of the wavelength of the incident electromagnetic radiation, the diameter of the gases, water vapor, and/or dust with which the energy interacts, and the type of scattering which should occur are summarized in Figure 2-13.

Rayleigh scattering (sometimes referred to as molecular scattering) occurs when the effective diameter of the matter (usually air molecules such as oxygen and nitrogen in the atmosphere) are many times smaller (usually < 0.1) than the wavelength of the incident electromagnetic radiation (Figure 2-13a). Rayleigh scattering is named after the English physicist Lord Rayleigh who offered the first coherent explana-

tion for it (Sagan, 1994). All scattering is accomplished through absorption and reemission of radiation by atoms or molecules in the manner described in the section on radiation from atomic structures. It is impossible to predict the direction in which a specific atom or molecule will emit a photon, hence scattering. The energy required to excite an atom is associated with powerful short-wavelength, high-frequency radiation. The amount of scattering is inversely related to the fourth power of the radiation's wavelength. For example, ultraviolet light at 0.3 μm is scattered approximately 16 times more than red light at 0.6 μm, i.e., [(0.6/0.3)^4 = 16]. Blue light at 0.4 μm is scattered about 5 times that of red light at 0.6 μm, i.e., [(0.6/0.4)^4 = 5.06]. The amount of Rayleigh scattering expected throughout the visible spectrum (0.4 – 0.7 μm) is shown in Figure 2-14.

Most Rayleigh scattering takes place in the upper 4.5 km of the atmosphere. It is responsible for the blue appearance of the sky. The shorter violet and blue wavelengths are more efficiently scattered than the longer orange and red wavelengths. When we look up on a cloudless day and admire the blue sky, we are witnessing the preferential scattering of the short-wavelength sunlight. Rayleigh scattering is also responsible for red sunsets. Since the atmosphere is a thin shell of gravitationally bound gas surrounding the solid Earth, sunlight must pass through a longer slant path of air at sunset (or sunrise) than at noon. Since the violet and blue wavelengths are scattered even more during their now longer path through the air than when the Sun is overhead, what we see when we look toward the sunset is the residue — the wavelengths of sunlight that are hardly scattered away at all, especially the oranges and reds (Sagan, 1994).

Mie scattering (sometimes referred to as nonmolecular scattering) takes place in the lower 4.5 km of the atmosphere, where there may be many essentially spherical particles present with diameters approximately equal to the size of the wavelength of the incident energy (Figure 2-13b). The actual size of the particles may range from 0.1 to 10 times the wavelength of the incident energy. For visible light, the main scattering agents are dust and other particles ranging from a few tenths of a micrometer to several micrometers in diameter. The amount of scatter is greater than Rayleigh scatter, and the wavelengths scattered are longer. Pollution also contributes to beautiful sunsets and sunrises. The greater the amount of smoke and dust particles in the atmospheric column, the more that violet and blue light will be scattered away and only the longer orange and red wavelength light will reach our eyes.

Non-selective scattering takes place in the lowest portions of the atmosphere where there are particles greater than 10 times the wavelength of the incident electromagnetic radiation (Figure 2-13c). This type of scattering is non-selective, i.e., all wavelengths of light are scattered, not just blue, green, or red. Thus, the water droplets and ice crystals that make up clouds and fog banks scatter all wavelengths of visible light equally well, causing the cloud to appear white. Non-selective scattering of approximately equal proportions of blue, green, and red light always appears as white light to the casual observer. This is the reason why putting our automobile high beams on in fog only makes the problem worse as we non-selectively scatter even more light into our visual field of view.

Scattering is a very important consideration in remote sensing investigations. It can severely reduce the information content of remotely sensed data to the point that the imagery loses contrast and it becomes difficult to differentiate one object from another. More will be said about how to minimize the effects of scattering in this chapter in the section on Terrain Interactions as well as in Chapter 4, where the various types of filters for aerial photography are described.

Absorption

Absorption is the process by which radiant energy is absorbed and converted into other forms of energy. The absorption of the incident radiant energy may take place in the atmosphere or on the terrain. An *absorption band* is a range of wavelengths (or frequencies) in the electromagnetic spectrum within which radiant energy is absorbed by a substance. The effects of water (H_2O), carbon dioxide (CO_2), oxygen (O_2), ozone (O_3), and nitrous oxide (N_2O) on the transmission of light through the atmosphere are summarized in Figure 2-15a. The cumulative effect of the absorption by the various constituents can cause the atmosphere to close down completely in certain regions of the spectrum. This is very bad for remote sensing because no energy is available to be sensed. Conversely, in the visible portion of the spectrum (0.4 – 0.7 μm), the atmosphere does not absorb all of the incident energy but transmits it rather effectively. Portions of the spectrum that transmit radiant energy effectively are called *atmospheric windows*.

Absorption occurs when incident energy of the same frequency as the resonant frequency of an atom or molecule is absorbed, producing an excited state. If instead of reradiating a photon of the same wavelength, the energy is transformed into heat motion and is subsequently reradiated at a longer wavelength, absorption occurs. When dealing with a medium like air, absorption and scattering are frequently combined into an *extinction coefficient*. Transmission is inversely related to the extinction coefficient times the thickness of the layer. Certain wavelengths of radiation are affected far more by absorption than by scattering. This is particularly true of infrared and wavelengths shorter than visible light. The combined effects of atmospheric absorption, scattering, and reflectance (from cloud tops) can dramatically reduce the amount of solar radiation reaching the Earth's surface at sea level as shown in Figure 2-15b.

Chlorophyll in vegetation absorbs much of the incident blue and red light for photosynthetic purposes. Chapter 10 describes the importance of these chlorophyll absorption bands and their role when remotely sensing vegetation. Similarly, water is an excellent absorber of energy (Chapter 11).

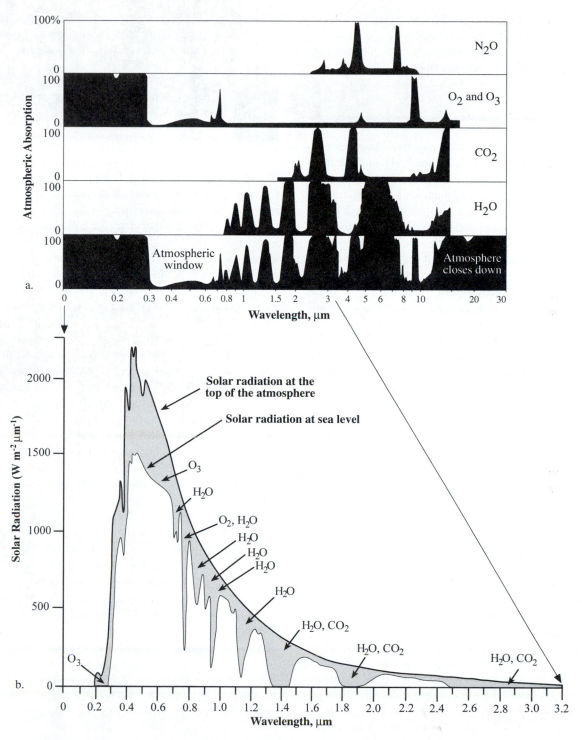

Figure 2-15   a) The absorption of the Sun's incident electromagnetic energy in the region from 0.1 – 30 μm by various atmospheric gases. The first four graphs depict the absorption characteristics of N_2O, O_2 and O_3, CO_2, and H_2O, while the final graphic depicts the cumulative result of having all these constituents in the atmosphere at one time. The atmosphere essentially "closes down" in certain portions of the spectrum while "atmospheric windows" exist in other regions that transmit incident energy effectively to the ground. It is within these windows that remote sensing systems must function. b) The combined effects of atmospheric absorption, scattering and reflectance reduce the amount of solar irradiance reaching the Earth's surface at sea level (after Slater, 1980).

Many minerals have unique absorption characteristics in very specific portions of the electromagnetic spectrum that allow us to use remote sensing to identify them (Chapter 13), assuming there is no overlying vegetation (Clark, 1999).

Reflectance

Reflectance is the process whereby radiation "bounces off" an object like the top of a cloud, a water body, or the terrestrial Earth. Actually, the process is more complicated, involving reradiation of photons in unison by atoms or molecules in a layer approximately one-half wavelength deep. Reflection exhibits fundamental characteristics that are important in remote sensing. First, the incident radiation, the reflected radiation, and a vertical to the surface from which the angles of incidence and reflection are measured all lie in the same plane. Second, the angle of incidence and the angle of reflection (exitance) are approximately equal as shown in Figure 2-16.

There are various types of reflecting surfaces. *Specular reflection* occurs when the surface from which the radiation is reflected is essentially smooth (i.e., the average surface-profile height is several times smaller than the wavelength of radiation striking the surface). Several features, such as calm water bodies, act like *near-perfect specular reflectors* (Figure 2-16a,b). If there are very few ripples on the surface, the incident energy will leave the water body at an angle equal and opposite to the incident energy. We know this occurs from our personal experience. If we shine a flashlight at night on a tranquil pool of water, the light will bounce off the surface and into the trees across the way at an angle equal to and opposite from the incident radiation.

If the surface has a large surface height relative to the size of the wavelength of the incident energy, the reflected rays go in many directions, depending on the orientation of the smaller reflecting surfaces. This *diffuse reflection* does not yield a mirror image, but instead produces diffused radiation (Figure 2-16c). White paper, white powders, and other materials reflect visible light in this diffuse manner. If the surface is so rough that there are no individual reflecting surfaces, then unpredictable scattering may occur. Lambert defined a perfectly diffuse surface; hence, the commonly designated *Lambertian surface* is one for which the radiant flux leaving the surface is constant for any angle of reflectance to the surface (Figure 2-16d).

A considerable amount of incident radiant flux from the Sun is reflected from the tops of clouds and other materials in the atmosphere. A substantial amount of this energy is reradiated back to space. As we shall see, the specular and diffuse reflection principles that apply to clouds also apply to the terrain.

 Energy-Matter Interactions with the Terrain

The amount of radiant energy onto, off of, or through a surface per unit time is called *radiant flux* (Φ) and is measured in Watts (W). The characteristics of the radiant flux and what happens to it as it interacts with the Earth's surface is of critical importance in remote sensing. In fact, this is the fundamental focus of much remote sensing research. By carefully monitoring the exact nature of the incoming (incident) radiant flux in selective wavelengths and how it interacts with the terrain, it is possible to learn important information about the terrain.

Various radiometric quantities have been identified that allow us to keep a careful record of the incident and exiting radiant flux (Table 2-4). We begin with the simple *radiation budget equation,* which states that the total amount of radiant flux in specific wavelengths (λ) incident to the terrain (Φ_{i_λ}) must be accounted for by evaluating the amount of energy reflected from the surface (r_λ), the amount of energy absorbed by the surface (α_λ), and the amount of radiant energy transmitted through the surface (τ_λ):

$$\Phi_{i_\lambda} = r_\lambda + \tau_\lambda + \alpha_\lambda. \qquad (2\text{-}13)$$

It is important to note that these radiometric quantities are based on the amount of radiant energy incident to a surface from any angle in a hemisphere (i.e., a half of a sphere).

Hemispherical Reflectance, Absorptance, and Transmittance

Hemispherical reflectance (r_λ) is defined as the dimensionless ratio of the radiant flux reflected from a surface to the radiant flux incident to it (Table 2-4):

$$r_\lambda = \frac{\Phi_{reflected}}{\Phi_{i_\lambda}}. \qquad (2\text{-}14)$$

Hemispherical transmittance (τ_λ) is defined as the dimensionless ratio of the radiant flux transmitted through a surface to the radiant flux incident to it:

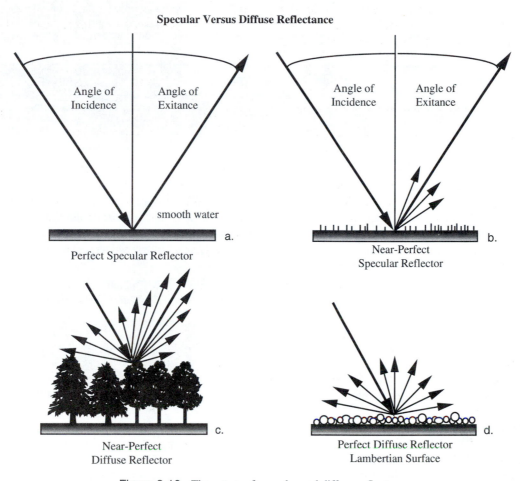

Specular Versus Diffuse Reflectance

Figure 2-16   The nature of specular and diffuse reflectance.

$$\tau_\lambda = \frac{\Phi_{transmitted}}{\Phi_{i_\lambda}}. \qquad (2\text{-}15)$$

Hemispherical absorptance (α_λ) is defined by the dimensionless relationship:

$$\alpha_\lambda = \frac{\Phi_{absorbed}}{\Phi_{i_\lambda}} \qquad (2\text{-}16)$$

or

$$\alpha_\lambda = 1 - (r_\lambda + \tau_\lambda). \qquad (2\text{-}17)$$

These definitions imply that radiant energy must be conserved, i.e., it is either returned back by reflection, transmitted through a material, or absorbed and transformed into some other form of energy inside the terrain. The net effect of absorption of radiation by most substances is that the

energy is converted into heat, causing a subsequent rise in the substance's temperature.

These radiometric quantities are useful for producing general statements about the spectral reflectance, absorptance, and transmittance characteristics of terrain features. In fact, if we take the simple hemispherical reflectance equation and multiply it by 100, we obtain an expression for percent reflectance (p_{r_λ}):

$$p_{r_\lambda} = \frac{\Phi_{reflected}}{\Phi_{i_\lambda}} \times 100 \qquad (2\text{-}18)$$

which is often used in remote sensing research to describe the spectral reflectance characteristics of various phenomena. Examples of spectral reflectance curves for selected urban-suburban phenomena are shown in Figure 2-17. Spectral reflectance curves typically provide no information

Table 2-4.   Summary of Radiometric Concepts (adapted from the American Society for Photogrammetry and Remote Sensing *Manual of Remote Sensing*; Colwell, 1983)

Name	Symbol	Units	Concept
Radiant energy	Q_λ	Joules, J	Capacity of radiation within a specified spectral band to do work
Radiant flux	Φ_λ	Watts, W	Time rate of flow of energy onto, off of, or through a surface
Radiant flux density at the surface			
Irradiance	E_λ	Watts per square meter $W\ m^{-2}$	Radiant flux incident upon a surface per unit area of that surface
Radiant exitance	M_λ	Watts per square meter $W\ m^{-2}$	Radiant flux leaving a surface per unit area of that surface
Radiance	L_λ	Watts per square meter per steradian $W\ m^{-2}\ sr^{-1}$	Radiant intensity per unit of projected source area in a specified direction
Hemispherical reflectance	r_λ	dimensionless	$\dfrac{\Phi_{reflected}}{\Phi_{i_\lambda}}$
Hemispherical transmittance	τ_λ	dimensionless	$\dfrac{\Phi_{transmitted}}{\Phi_{i_\lambda}}$
Hemispherical absorptance	α_λ	dimensionless	$\dfrac{\Phi_{absorbed}}{\Phi_{i_\lambda}}$

about the absorption and transmittance of the radiant energy. But because many of the sensor systems such as cameras and some multispectral scanners only record reflected energy, this information is still quite valuable and can form the basis for object identification and assessment. For example, it is clear from Figure 2-17 that grass reflects only approximately 15 percent of the incident red radiant energy (0.6 – 0.7 μm) while reflecting approximately 50 percent of the incident near-infrared radiant flux (0.7 – 0.9 μm). If we wanted to discriminate between grass and artificial turf, the ideal portion of the spectrum to remotely sense in would be the near-infrared region because artificial turf reflects only about 5 percent of the incident near-infrared energy. This would cause a black-and-white infrared image of the terrain to display grass in bright tones and the artificial turf in darker tones.

Hemispherical reflectance, transmittance, and absorptance radiometric quantities do not provide information about the exact amount of energy reaching a specific area on the ground from a specific direction or about the exact amount of radiant flux exiting the ground in a certain direction. Remote sensing systems can only be located in space at a single point in time, and they usually only look at a relatively small portion of the Earth at a single instant. Therefore, it is important to refine our radiometric measurement techniques so that more precise radiometric information can be extracted from the remotely sensed data. This requires the introduction of several radiometric quantities that provide progressively more precise radiometric information.

Radiant Flux Density

A flat area (e.g., 1 x 1 m in dimension) being bathed in radiant flux (Φ) in specific wavelengths from the Sun is shown in Figure 2-18. The amount of radiant flux intercepted divided by the area of the plane surface is the average *radiant flux density*.

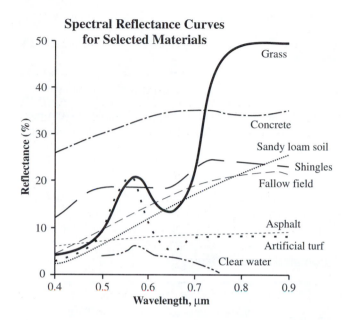

Spectral Reflectance Curves for Selected Materials

Figure 2-17 Typical spectral reflectance curves for urban-suburban phenomena in the region 0.4 – 0.9 μm (Jensen, 1989).

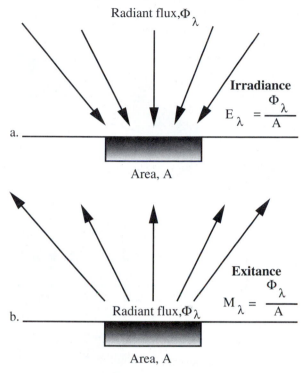

Concept of Radiant Flux Density

Figure 2-18 The concept of radiant flux density for an area on the surface of the Earth. a) *Irradiance* is a measure of the amount of incident energy in Watts m^{-2}. b) *Exitance* is a measure of the amount of energy leaving in Watts m^{-2}.

Irradiance and Exitance

The amount of radiant flux incident per unit area of a plane surface is called *irradiance* (E_λ):

$$E_\lambda = \frac{\Phi_\lambda}{A}. \tag{2-19}$$

The amount of radiant flux leaving per unit area of the plane surface is called *exitance* (M_λ):

$$M_\lambda = \frac{\Phi_\lambda}{A}. \tag{2-20}$$

Both quantities are usually measured in Watts per meter squared (W m^{-2}). Although we do not have information on the direction of either the incoming or outgoing radiant energy (i.e., the energy can come and go at any angle throughout the entire hemisphere), we have now refined the measurement to include information about the size of the study area of interest on the ground in m^2. Next we need to refine our radiometric measurement techniques to include information on what direction the radiant flux is leaving the study area.

Radiance

Radiance is the most precise remote sensing radiometric measurement. *Radiance* (L_λ) is the radiant flux per unit solid angle leaving an extended source in a given direction per unit of projected source area in that direction. It is measured in Watts per meter squared per steradian (W m^{-2} sr^{-1}). The concept of radiance is best understood by evaluating Figure 2-19. First, the radiant flux leaves the projected source area in a specific direction toward the remote sensor. We are not concerned with any other radiant flux that might be leaving the source area in any other direction. We are only interested in the radiant flux in certain wavelengths (L_λ) leaving the projected source area within a certain direction ($A \cos \theta$) and solid angle (Ω):

$$L_\lambda = \frac{\frac{\theta}{\Omega}}{A \cos \theta}. \tag{2-21}$$

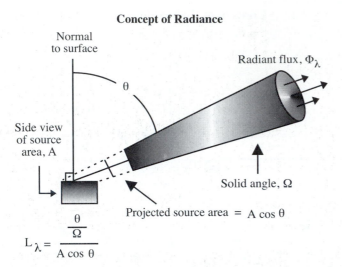

Concept of Radiance

$$L_\lambda = \frac{\dfrac{\theta}{\Omega}}{A \cos \theta}$$

Figure 2-19 The concept of radiance leaving a specific projected source area on the ground, in a specific direction, and within a specific solid angle. This is the most precise radiometric measurement used in remote sensing.

One way of visualizing the solid angle is to consider what you would see if you were in an airplane looking through a telescope at the ground. Only the energy that exited the terrain and came up to and through the telescope in a specific solid angle (measured in steradians) would be intercepted by the telescope and viewed by your eye. Therefore, the solid angle is like a three-dimensional cone (or tube) that funnels radiant flux from a specific point source on the terrain toward the sensor system. Hopefully, energy from the atmosphere or other terrain features does not become scattered into the solid angle field of view and contaminate the radiant flux from the area of interest on the ground. Unfortunately, this is not often the case as scattering in the atmosphere and from other nearby areas on the ground can contribute spurious spectral energy which enters into the solid angle field of view.

Energy-Matter Interactions in the Atmosphere Once Again

The radiant flux reflected or emitted from the Earth's surface once again enters the atmosphere, where it interacts with the various gases, water vapor, and particulates. Thus, atmospheric scattering, absorption, reflection, and refraction influence the radiant flux once again before the energy is recorded by the remote sensing system.

Energy-Matter Interactions at the Sensor System

Additional energy-matter interactions take place when the energy reaches the remote sensor. For example, if the remote sensing is being performed using a camera, then the radiance will interact with the camera filter, the optical glass lens, and finally the film emulsion with its light-sensitive silver-halide crystals. The emulsion must then be developed and printed before an analog copy is available for analysis. Similarly, an optical-mechanical detector will record the number of photons in very specific wavelength regions reaching the sensor.

Target and Path Radiance

Ideally, the radiant energy recorded by the camera or detector is a true function of the amount of radiance leaving the terrain within the instantaneous field of view (IFOV) at a specific solid angle. Unfortunately, other radiant energy may enter into the field of view from various other paths and introduce confounding noise into the remote sensing process. More refined radiometric variable definitions are needed to identify the major sources and paths of this energy (Table 2-5). The various paths and factors that determine the radiance reaching the satellite sensor are summarized in Figure 2-20:

- *Path 1* contains spectral solar irradiance (E_{o_λ}) that was attenuated very little before illuminating the terrain within the IFOV. Notice in this case that we are interested in the solar irradiance from a specific solar zenith angle (θ_o) and that the amount of irradiance reaching the terrain is a function of the atmospheric transmittance at this angle (T_{θ_o}). If all of the irradiance makes it to the ground, then the atmospheric transmittance (T_{θ_o}) equals one. If none of the irradiance makes it to the ground, then the atmospheric transmittance is zero.

- *Path 2* contains spectral diffuse sky irradiance (E_{d_λ}) that never even reaches the Earth's surface (the study area) because of scattering in the atmosphere. Unfortunately, such energy is often scattered directly into the IFOV of the sensor system. As previously discussed, Rayleigh scattering of blue light contributes much to this diffuse sky irradiance. That is why the blue band image produced by a remote sensor system is often much brighter than any of the other bands. It contains much unwanted diffuse sky irradiance that was inadvertently scattered into the IFOV of the

Table 2-5.   Fundamental Radiometric Variables

Radiometric Variables

E_o = solar irradiance at the top of the atmosphere (W m^{-2})

E_{o_λ} = spectral solar irradiance at the top of the atmosphere (W m^{-2} μm^{-1})

E_d = diffuse sky irradiance (W m^{-2})

E_{d_λ} = spectral diffuse sky irradiance (W m^{-2} μm^{-1})

E_g = global irradiance incident on the surface (W m^{-2})

E_{g_λ} = spectral global irradiance on the surface (W m^{-2} μm^{-1})

τ = normal atmospheric optical thickness

T_θ = atmospheric transmittance at an angle θ to the zenith

θ_o = solar zenith angle

θ_v = view angle of the satellite sensor (or scan angle)

μ = cos θ

r_λ = average target reflectance at a specific wavelength

r_{λ_n} = average reflectance from a neighboring area

L_s = total radiance at the sensor (W m^{-2} sr^{-1})

L_t = total radiance from the target of interest toward the sensor (W m^{-2} sr^{-1})

L_i = intrinsic radiance of the target (W m^{-2} sr^{-1}) (i.e., what a handheld radiometer would record on the ground without any intervening atmosphere)

L_p = path radiance from multiple scattering (W m^{-2} sr^{-1})

sensor system. Therefore, if possible, we want to minimize its effects.

- *Path 3* contains energy from the Sun that has undergone some Rayleigh, Mie, and/or Non-selective scattering and perhaps some absorption and reemission before illuminating the study area. Thus, its spectral composition and polarization may be somewhat different than the energy that reaches the ground from Path 1.

- *Path 4* contains radiation that was reflected or scattered by nearby terrain (r_{λ_n}) covered by snow, concrete, soil, water, and/or vegetation into the IFOV of the sensor system. The energy does not actually illuminate the study area of interest. Therefore, if possible, we would like to minimize its effects.

- *Path 5* is energy that was also reflected from nearby terrain into the atmosphere, but then scattered or reflected onto the study area.

Therefore, for a given spectral interval in the electromagnetic spectrum (e.g., λ_1 to λ_2 could be $0.5 - 0.6$ μm or green light), the total solar irradiance reaching the *Earth's surface*, E_{g_λ}, is an integration of several components:

$$E_{g_\lambda} = \int_{\lambda_1}^{\lambda_2} (E_{o_\lambda} T_{\theta_o} \cos \theta_o + E_{d_\lambda}) d\lambda \qquad (\text{W m}^{-2}\,\text{μm}^{-1}) \; (2\text{-}22)$$

It is a function of the spectral solar irradiance at the top of the atmosphere (E_{o_λ}) multiplied by the atmospheric transmittance (T_{θ_o}) at a certain solar zenith angle (θ_o) plus the contribution of spectral diffuse sky irradiance (E_{d_λ}).

Only a small amount of this irradiance is actually reflected by the terrain in the direction of the satellite sensor system. If we assume the surface of Earth is a diffuse reflector (a Lambertian surface), the total amount of radiance (L_T) exiting the study area toward the sensor is:

$$L_T = \frac{1}{\pi} \int_{\lambda_1}^{\lambda_2} r_\lambda T_{\theta_v} (E_{o_\lambda} T_{\theta_o} \cos \theta_o + E_{d_\lambda}) d\lambda \qquad (2\text{-}23)$$

The average target reflectance factor (r_λ) is included because the vegetation, soil, and water within the IFOV selectively absorb some of the incident energy. Therefore, not all of the energy incident to the IFOV (E_{g_λ}) leaves the IFOV. In effect, the terrain acts like a filter, selectively absorbing certain wavelengths of light while reflecting others. Note that the energy exiting the terrain is at an angle (θ_v), requiring the use of T_{θ_v} in the equation.

It would be wonderful if the total radiance recorded by the sensor, L_s, equaled the radiance returned from the study area of interest, L_t. Unfortunately, $L_s \neq L_t$ because there is some additional radiance from different *paths* that may fall within the IFOV of the sensor system detector (Figure 2-20). This is often called *path radiance*, L_p. Thus, the total radiance recorded by the sensor becomes:

$$L_s = L_t + L_p \quad (\text{W m}^{-2}\,\text{sr}^{-1}). \qquad (2\text{-}24)$$

We see from Equation 2-24 and Figure 2-20 that the path radiance (L_p) is an intrusive (bad) component of the total amount of radiance recorded by the sensor system (L_s). It is composed of radiant energy primarily from the diffuse sky irradiance (E_d) from path 2 as well as the reflectance from nearby

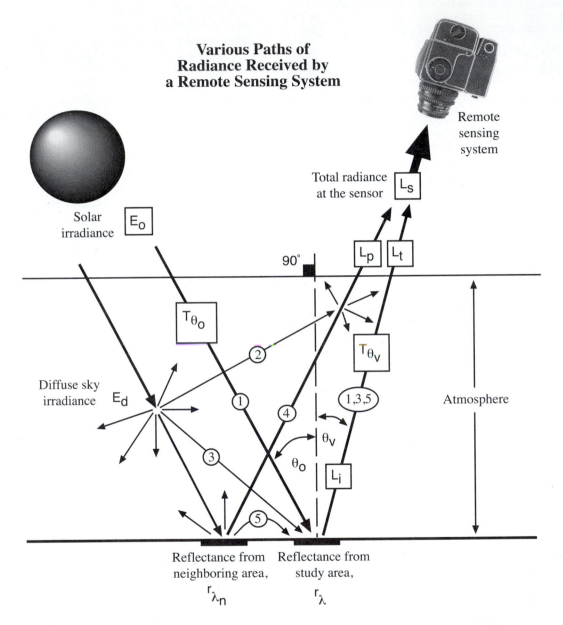

**Various Paths of
Radiance Received by
a Remote Sensing System**

Remote
sensing
system

Total radiance
at the sensor L_s

Solar
irradiance E_o

L_p L_t

90°

T_{θ_o}

Diffuse sky
irradiance E_d

T_{θ_v}

②

①   ④   ①,3,5   Atmosphere

③ θ_v

θ_o L_i

⑤

Reflectance from   Reflectance from
neighboring area,   study area,

r_{λ_n} r_λ

Figure 2-20   *Radiance* (L_t) *from paths 1, 3, and 5 contains intrinsic valuable spectral information about the target of interest. Conversely, the* path radiance (L_p) *from paths 2 and 4 includes diffuse sky irradiance or radiance from neighboring areas on the ground. This path radiance generally introduces unwanted radiometric noise in the remotely sensed data and complicates the image interpretation process.*

ground areas r_{λ_n} from path 4. Path radiance introduces error to the remote sensing data-collection process. It can impede our ability to obtain accurate spectral measurements.

A great deal of research has gone into developing methods to remove the contribution of path radiance (L_p). The methods usually require digital image processing and as such are beyond the scope of this text. However, several atmospheric correction methods are discussed in the chapter on Preprocessing in the companion book *Introductory Digital Image Processing* (Jensen, 1996). Also, commercial firms provide software that can be used to compute the atmospheric transmission and path radiance on a given day (Ontar, 1999). Such information can then be used to remove the path radiance (L_p) contribution to the remote sensing signal (L_S). The models require some atmospheric data that are hopefully available from nearby weather stations.

Conclusion

A scientist or lay person using remote sensing technology should understand the nature of the interactions taking place from the time that electromagnetic energy is created until it is recorded in analog or digital form by the remote sensing system (Duggin and Robinove, 1990). This knowledge increases the probability of accurate image interpretation and the extraction of valuable biophysical information from the imagery.

References

Bolemon, J., 1985, *Physics: An Introduction,* Englewood Cliffs, NJ: Prentice-Hall, Inc., 628 pp.

Clark, R. N., 1999, "Spectroscopy of Rocks and Minerals, and Principles of Spectroscopy," *Manual of Remote Sensing - Remote Sensing for the Earth Sciences,* A. N. Rencz (Ed.), NY: John Wiley & Sons, 672 pp.

Colwell, R. N., (Ed.), 1983, *Manual of Remote Sensing,* 2nd Ed., Falls Church: American Society of Photogrammetry, 2440 pp.

Duggin, M. J. and C. J. Robinove, 1990, "Assumptions Implicit in Remote Sensing Data Acquisition and Analysis," *International Journal of Remote Sensing,* 11(10):1669–1694.

Egan, W. G., 1985, *Photometry and Polarization in Remote Sensing,* NY: Elsevier, 503 pp.

Englert, B., M. O. Scully and H. Walther, 1994, "The Duality in Matter and Light," *Scientific American,* 271(6):86–92.

Feinberg, G., 1985, "Light," *The Surveillant Science: Remote Sensing of the Environment,* 2nd Ed., R. K. Holz (Ed.), NY: John Wiley & Sons, 2–11.

Jensen, J. R., (Ed.), 1989, "Remote Sensing in America," *Geography in America,* NY: Bobs Merrill, Inc., 746–775.

Jensen, J. R., 1996, *Introductory Digital Image Processing: A Remote Sensing Perspective,* Saddle River, NJ: Prentice-Hall, 318 pp.

Lousma, J. R., 1993, "Rising to the Challenge: The Role of the Information Sciences," *Photogrammetric Engineering & Remote Sensing,* 59(6):957–959.

Meadows, J., 1992, *The Great Scientists,* NY: Oxford University Press, 248 pp.

Mulligan, J. R., 1980, *Practical Physics: The Production and Conservation of Energy,* NY: McGraw-Hill, Inc., 526 pp.

Nassau, K., 1983, *The Physics and Chemistry of Color: The Fifteen Causes of Color,* NY: John Wiley & Sons.

Nassau, K., 1984, "The Physics of Color," in *Science Year 1984,* Chicago: World Book, 126–139.

Ontar, *1999 Product Catalog: PcModWin,* (commercial version of the U.S. Air Force Research Lab's MODTRAN), Andover: Ontar Inc., 23 pp.

Rinker, J. N., 1999, *Introduction to Spectral Remote Sensing,* Alexandria, VA: U.S. Army Topographic Engineering Center, http://www.tec.army.mil/terrain/desert/tutorial.

Sagan, C., 1994, *Pale Blue Dot,* NY: Random House, 429 pp.

Slater, P. N., 1980, *Remote Sensing: Optics and Optical Systems,* Reading, Pennsylvania: Addison-Wesley Publishing Company, 575 pp.

Strahler, A. N. and A. H. Strahler, 1989, *Elements of Physical Geography,* 4th Ed., NY: John Wiley & Sons, 562 pp.

Trefil, J. and R. M. Hazen, 1995, *The Sciences: An Integrated Approach,* NY: John Wiley & Sons, 634 pp.

Wolff, R. S. and L. Yaeger, 1993, *Visualization of Natural Phenomena,* Santa Clara, CA: Springer-Verlag, 374 pp.

History of Aerial Photography and Aerial Platforms

3

E lectromagnetic energy reflected from the Earth's surface may be recorded by a variety of remote sensing systems. The *camera* is still one of the most reliable and useful remote sensing instruments. This chapter first reviews the history of photography. It then documents the development of suborbital platforms used to obtain aerial photography. Finally, a brief history of orbital aerial photography data collection is presented.

History of Photography

Photography was an invention waiting to happen from 1833 to 1839. Two of the technological components had been around for centuries, waiting for the right inventor to put everything together, including: 1) a correct theory of light and color, and 2) a recording instrument (the camera obscura). All that was lacking was the invention of a light-sensitive emulsion that could be made permanent (fixed).

Light and Color

For more than a thousand years India's astrologers have taught that the Sun's white light is composed of all colors. Unfortunately, Aristotle's belief that all colors are created by mixing black and white prevailed well into the seventeenth century (Wolinsky, 1999). Even Leonardo da Vinci could not make up his mind, declaring on different occasions that there were six primary colors — or eight. It took the genius of Sir Isaac Newton to put forth the correct concept of light and color. In 1672, when making public his *New Theory about Light and Colours*, Newton wrote,

> In the begining of the yeare 1666... I procured me a Triangular glasse Prisme, to try therewith the Celebrated Phaenemena of Colours.

Newton found that in using the prism, he could disperse white light into a spectrum of colors — red, orange, yellow, green, blue, indigo, and violet. Utilizing a second prism, he found that he could recombine the colors into white light. He was the first to scientifically document the spectral nature of light. His ideas helped launch the era of modern optics.

The Camera Obscura

Fundamental camera principles were known as early as the 4th century BC in China and in Greece. A device known as the *camera obscura* (a dark chamber) had long been utilized by artists as a drawing aid (Figure 3-1). The camera obscura uses a simple lens inserted into a box to focus images of the outside world onto a mirror and subsequently onto a plate of glass. The person then sketches the desired information onto relatively translucent paper (e.g., thin onionskin paper). In this way, the relative proportions and shape of objects can be recorded. The problem facing those who would invent photography was how to capture and preserve the image on the glass (Fanton, 1989).

Invention of Light-sensitive Emulsions and Methods of Permanently Fixing the Image

M. Joseph Nicephore Niepce (1763 – 1833) lived in France. He produced the world's first photographic image of nature, circa 1826 (Gernsheim and Gernsheim, 1952). It was obtained from an upper window at his estate "Le Gras" in the village of Saint Loup de Varenne, near Charlon-sur-Saone, France, and records the buildings and vegetation in the courtyard (Figure 3-2). On the left of the photograph is the upper loft of the Niepce family house; to the right is a pear tree with a patch of sky showing through an opening in the branches; in the middle is the slanting roof of the barn; to the far right is another wing of the house.

This is believed to be the first successful *permanent* photograph (remotely sensed image). But how was the first photograph produced? Niepce dissolved bitumen of Judea (a kind of asphalt) in lavender oil (a solvent used in varnishes), then coated a sheet of pewter (metal) with the mixture. He placed the sheet in a camera obscura and aimed the camera through an open window at his courtyard. He exposed the sheet for eight hours. The shading of the buildings is odd because of the diverse lighting caused by the Sun traversing across the sky from east to west during the eight-hour exposure. Incident light from the Sun hardened the bitumen of Judea in bright areas and left it soft in dark areas. Niepce then washed the plate with a mixture of lavender oil and white petroleum. This time it removed the soft bitumen that had not been struck by light. "The result was a permanent direct positive picture, in which the lights were represented by bitumen and the shades by bare metal" (Gernsheim and Gernsheim, 1952). Niepce had found a way to remove the unexposed and still light-sensitive material so that the image stayed visible. Niepce called it a *heliograph,* from the Greek *helios* for

Figure 3-1   A portable camera obscura focused reflected light from people or landscapes through a lens (A), onto a mirror (B) and subsequently onto a clear plate of glass (C). The person then sketched the relative proportions and shape of objects onto transparent paper. The camera obscura was an important technological stepping stone in the development of cameras and photography (after Ganot, 1855).

"Sun" and *graphos* for "drawing" (London and Upton, 1994). Niepce died in 1833 and was never able to appreciate the impact that photography would have on society.

Louis Jacques Mande Daguerre (1787 – 1851) was a scenic artist who created stage sets for operas and theater. He and Niepce corresponded often and became business partners in 1829. After Niepce's death in 1833, Daguerre continued experimenting with various emulsions and image "fixers." On January 7, 1839, he presented the *daguerreotype* process to the French Academy of Sciences. The daguerreotype process involved the following steps (Quackenbush, 1960; Newhall, 1993; London and Upton, 1994):

• A polished surface of silver was plated onto a sheet of copper.

• The plate was made light-sensitive by exposing it to the vapors from iodine crystals in a box. The vapor reacted with the silver, producing the light-sensitive compound silver iodide.

• The photo-sensitive plate was placed in a camera obscura, and an exposure was made. The exposure of the plate often lasted several minutes, during which time the participants

Figure 3-2   Joseph Nicephore Niepce took the world's first photographic image in 1826 of his estate courtyard. He used a camera obscura and a metal plate of pewter coated with an emulsion of bitumen of Judea (a kind of asphalt). The exposure lasted eight hours. The parts of the plate exposed to light hardened. Areas that did not receive light remained soft. He removed the unexposed soft bitumen by washing the plate in lavender oil and white petroleum. This resulted in the first permanent photographic image (Gernsheim and Gernsheim, 1952; © Gernsheim Collection, Harry Ransom Humanities Research Center, University of Texas, Austin; used with permission).

(or landscape) had to remain very still. During exposure in the camera, the plate recorded a *latent image* of the scene, which was chemically present but undetectable to the human eye.

• The latent image was developed by putting the exposed plate in a box that had a dish of heated mercury at the bottom. Where many photons of light had impacted the plate, the mercury vapor formed a frosty amalgam, or alloy, with the silver. This amalgam made up the bright areas of the image. Where few photons of light impacted the plate, no amalgam was formed. The plate was then placed in a bath of common salt (sodium chloride), which caused any unexposed silver iodide to become insensitive to further light action. The plate was then washed in water and dried. The result was a one-of-a-kind positive image. A few years later sodium thiosulphate (hypo), which Herschel had discovered in 1819, was used instead of sodium chloride as the fixative agent.

This direct positive process yielded a single positive print called a "Sun-drawn miniature," as the word "photograph" had not been coined. Daguerre made it clear that he considered the invention to be his own, but agreed to transfer it to the partnership with Isidore Niepce (Joseph's son) "on condition that this new process shall bear the name of Daguerre alone; it may, however, only be published simultaneously with the first process, in order that the name of M. Joseph Nicephore Niepce may always figure, as it should in this invention" (Fouque, 1867).

The secretary of the French Academy of Sciences marveled at Daguerre's invention and stated that, "It upsets all scientific theories of light and optics, and will revolutionize the art of drawing." Other imminent members of the French Academy of Sciences, such as Alexander Von Humboldt (the geographer who conceived of the "isoline"), verified the importance of the invention. The daguerreotype process was purchased by the French government and both Daguerre and

Isidore Niepce were granted lifetime pensions. Daguerre wrote a 79 page booklet describing the daguerreotype process, which was published in more than 30 editions and distributed throughout the world. A serious problem for early photographers making daguerreotypes, however, was that some of them went insane; a photographer exposed to the mercury fumes over a prolonged period of time could obtain mercury poisoning.

Daguerreotypes recorded very fine detail, and the public loved them. Unfortunately, there was no way of producing multiple copies except by rephotographing the original. What was needed was a process where any number of positive images could be made from a single exposure. In 1839, an Englishman named William Henry Fox Talbot (1800 – 1877), who was a member of the Royal Society (equivalent to the French Academy of Sciences), discovered that an image formed on paper coated with silver iodide, though barely visible, could be developed or strengthened with gallic acid and silver nitrate. He used waxed, transparent paper negatives, fixed in hypo, to make positive copies on silver chloride paper. Photographs made in this way did not equal daguerreotypes in brilliancy or sharpness, but the *calotype* (Greek *kalos* for "beautiful" and *typos* for "impression") process made it possible to produce a number of positive copies from the negative. Talbot's *Pencil of Nature*, published in 1844, was the first book to be illustrated with photographs.

Talbot's negative-positive process is basically the same one we use today in terrestrial and aerial photography. Talbot contended that his process predated Daguerre's process and presented his case to the Royal Society. Interestingly, when the famed astronomer Sir John F. W. Herschel of England was asked to come to France and compare Daguerre's process with that of Talbot's, he replied, "I must tell you that compared to these masterpieces of Daguerre, Monsieur Talbot produces nothing but vague, foggy things. There is as much difference between these two products as there is between the moon and the sun" (Newhall, 1993). Herschel also introduced the word *photography* instead of Talbot's term "photogenic drawing" and coined the terms *positive* and *negative*.

Photography was considered miraculous at the time because prior to 1839 it was impossible to know exactly what any person or place looked like without actually knowing the person or visiting the location. Portraiture was available only to the wealthy and was subject to embellishment by the artist. Even the best descriptions of faraway wonders of the world were still only words. Photography allowed the common man and woman to reach into the flow of time, stop it for an instant, and then preserve that moment for posterity.

In 1851, the Englishman Frederick Scott Archer found that a glass plate coated with *collodion* (nitrocellulose dissolved in ether and alcohol) could be used as an emulsion. He developed a way to coat a glass plate with the substance (Newhall, 1969). The plate was made sensitive by dipping it in silver nitrate. The silver ions combined with the iodine ions to form light-sensitive iodide within the collodion. The plate was exposed while still moist and then developed in pyrogallic acid, fixed in hypo, and washed. The *collodion wet-plate* process had to be performed on location. This required a complete mobile darkroom if photography was acquired in the field. From 1851 until 1888 this was the most popular form of photography, completely replacing the daguerreotype and calotype processes.

In the *Proceedings* of the Royal Society in 1855, the noted Scottish physicist James Clerk Maxwell made what is believed to be the earliest suggestion that objects could be reproduced in color by photography (Estes, 1966). His paper consists of a basic discussion on the theory of color vision. Referring to early works by Thomas Young, whom he says seems to have been the first to understand Newton's suggestion on the mixing of colors, Maxwell makes the following statement:

> This theory of colour may be illustrated by a supposed case taken from the art of photography. Let it be required to ascertain the colours of a landscape, by means of Impressions taken on a preparation equally sensitive to rays of every colour.

> Let a plate of red be placed before the camera and an impression taken. The positive of this will be transparent wherever the red light has been abundant in the landscape, and opaque where it has been wanting. Let it now be put in a magic lantern [i.e., a projector] along with the red glass, and a red picture will be thrown on the screen.

> Let this operation be repeated with a green and violet glass and by means of three magic lanterns let the three images be superimposed on the screen. The colour of any point on the screen will then depend on that of the corresponding point of the landscape; and, by properly adjusting the intensities of the light, etc., a complete copy of the landscape as far as visible colour is concerned, will be thrown on the screen (Niven, 1890).

On May 17, 1861, with the help of photographer Thomas Sutton, Maxwell demonstrated this *additive color combining* technique. He photographed a bow of multicolored ribbon four times using black-and-white film. Each exposure was made through a different filter: red (sulfo-cyanide of iron), green (copper chloride), blue (ammoniacal sulfate of copper), and yellow (lemon-colored glass). He then projected light through the red, green, and blue filtered black-and-white images and was able to recreate an image of the bow of multicolored ribbon. He found that the yellow filter was unnecessary (Watt, 1925). Chapter 4 provides additional information on additive color combining principles.

Great improvements in photography were realized after Richard L. Maddox, a London physician and photo-micrographer, invented the dry-plate process in 1871, which used gelatin as the medium for suspending light-sensitive silver salts (Maddox, 1871). This was greatly superior to the collodion wet-plate process in that the emulsion was 1) much more sensitive (approximately 60 times faster than collodion), and 2) could be developed when the emulsion was dry. It was no longer necessary to have a wet laboratory nearby. The more sensitive emulsion was also important because it stopped action more rapidly, greatly improving the quality of detail in terrestrial and aerial photographs. Building on this logic, roll film was invented by Leon Warnerke in 1875.

In 1873, Herman Vogel discovered that by soaking silver halide emulsions in various dyes, he could extend their sensitivity to longer wavelengths. Subsequent investigations revealed that dyes could be rendered that extend emulsion sensitivity into the infrared portion of the spectrum (Thompson, 1966).

Prior to 1888, the general public could buy a camera and film, but they had to know how to develop the film and print the photographs. In 1888, George Eastman revolutionized photography. Building upon the work by Warnerke, he used a gelatin emulsion in which light-sensitive silver salts were suspended. This material was then coated on paper and produced in a roll format. This clear, flexible film support did much to advance and increase the usefulness of photography (Thompson, 1966). George Eastman founded the Eastman Kodak Company and developed a photofinishing system that lived up to the slogan "You press the button, we do the rest." The general public bought the inexpensive box cameras, exposed 100 negatives, and sent the exposed film still inside the camera to Eastman's company, where it was processed. The company then sent the 100 prints back to the customer, along with the camera and a new roll of film inside. This sounds very much like the recyclable cardboard or plastic cameras sold today!

Figure 3-3　A man-powered ornithopter. Flapping the arms was supposed to generate sufficient aerodynamic lift to sustain flight. Experimentation with ornithopters often resulted in serious accidents.

Finally, in 1924, Mannes and Godousky patented the first part of their work on multilayer film, which led to the marketing of the three-layer color film Kodachrome in 1935 (McCamy, 1960; Estes, 1966).

 ## Photography from Aerial Platforms

Mankind has had the desire to fly like a bird and have an aerial perspective or bird's-eye view of the Earth for a long time. Even our earliest mythology is consumed with flight. For example, in Greek mythology Daedalus built wings of feathers and wax for himself and his son, Icarus. They escaped from prison using the wings. Unfortunately, Icarus became caught up in the joy of flying and flew too close to the Sun. The wings melted, and he fell to Earth. The Greek hero Perseus used winged sandals when he flew to rescue Andromeda (Lopez, 1995). After dreaming about flying for centuries, mankind's creativity eventually resulted in a series of inventions that allowed us to escape gravity and obtain an aerial perspective. Some of the aerial platforms invented were more useful and safe than others.

Ornithopters

Not surprisingly, the first real attempts at flight involved people trying to imitate birds. They built flapping devices called *ornithopters* (Figure 3-3). They jumped from high places strapped to these devices and often died. For example, in 1010 a monk named Eilmer, equipped with an ornithopter, took off from an abbey in England and broke his legs. Denis Bolori of France in 1536 tried to fly using wings flapped by a spring. He died when the spring broke. Leonardo da Vinci's notebooks written between 1488 and 1514 contain designs of ornithopters and simple helicopters. We do not know if da Vinci actually built the designs. He did suggest that the devices should be tested over water for safety.

Lighter-Than-Air Flight Using Balloons

The hot-air balloon was invented by Joseph and Etienne Montgolfier in 1783 in France (Lopez, 1995). They burned straw and wool to produce the less dense air that lifted the balloon and basket off the ground. The first balloon passengers were animals. The first humans to ride in a Montgolfier balloon were J. F. Pilatre de Rozier and the Marquis d'Arlandes, who flew over Paris. People who ventured into the sky in balloons often referred to themselves as *aeronauts*.

The first-known aerial photograph was obtained in 1858 by the Parisian portrait photographer and passionate aeronaut Gaspard Felix Tournachon (he called himself *Nadar*) (Figure 3-4). Tournachon had a vision of what aerial photography might contribute in the future, and on October 23, 1858, he applied for a patent for what we now call an aerial survey — the mapping of the land from a series of overlapping aerial photographs (Newhall, 1969). Later in the month he ascended in a captive balloon over Paris. He suffered many failures attempting to obtain an aerial photograph, because the gas escaping from the mouth of the balloon desensitized the collodion-coated glass plates he was using. He finally achieved success over the Val de Bievre on the outskirts of Paris when he went aloft in a tethered balloon only 80 m (264 ft) above the ground, exposed a photographic plate, and then was rapidly hauled back down to Earth, where he dashed into an inn and developed the picture. The world's first aerial photograph no longer exists but Nadar described its contents:

> I develop my picture... Good luck! There is something!... It cannot be denied: here right under me are all of the three houses in the little village: the farm, the inn and the police station... You can distinguish perfectly a delivery van on the road whose driver has stopped short before the balloon, and on the roof-tiles two white pigeons who have just landed there. Thus I was right! (*Le Monde Illustre*, October 30, 1858).

Nadar himself referred to the photograph as "a simple positive upon glass," made with "detestable materials." Nevertheless, aerial photography was born.

The first successful aerial photographs that we have a record of were recorded two years later on October 13, 1860 from the tethered balloon *Queen of the Air* in the United States by James W. Black and Samuel A. King from a height of 1,200 ft over Boston, Massachusetts (Figure 3-5). The photographs were obtained using wet collodion plates (Tennant, 1903). Black was a professional photographer from the firm of Black & Bathelder. King had his own photography business — King & Allen. At the conclusion of the days' aerial photography, King (1860) commented,

> This is only the precursor, no doubt, of numerous other experiments; for no one can look upon these pictures, obtained by aid of the balloon, without being convinced that the time has come when what has been used only for public amusement can be made to serve some practical end.

In July, 1863, Sir Oliver Wendell Holmes (a photographic student of Samuel King) photointerpreted the contents of this photograph for *The Atlantic Monthly* (Newhall, 1969):

> Boston, as the eagle and the wild goose see it, is a very different object from the same place as the solid citizen looks up at its eaves and chimneys. The Old South and Trinity Church are two landmarks not to be mistaken. Washington Street slants across the picture as a narrow cleft. Milk Street winds as if the cowpath which gave it a name had been followed by the builders of its commercial palaces. Windows, chimneys, and skylights attract the eye in the central parts of the view, exquisitely defined, bewildering in numbers.... As a first attempt it is on the whole a remarkable success; but its greatest interest is in showing what we may hope to see accomplished in the same direction.

During the Civil War in America, Union General McClellan used balloons to observe Confederate Army positions and movement. For example, in June, 1862, the Union Army used balloons tethered at an altitude of 1,400 ft to gather intelligence and draw maps of the defenses of Richmond, Virginia. The balloon *Intrepid* is shown tethered by Union troops on the ground at the battle of Fair Oaks on June 1, 1862, in Figure 3-6 (Quackenbush, 1960; Fischer, 1975). The *Intrepid* was inflated using a special hydrogen-producing system (Figure 3-7). It is believed that some aerial photographs were obtained by observers using tethered balloons. However, not a single Civil War (1861 – 1865) aerial photograph has survived (Colwell, 1997).

Meanwhile, in Europe, Gaspard Felix Tournachon (Nadar) was still actively involved in balloon aerial photography. In

Figure 3-4   Photograph of Gaspard Felix Tournachon (1820-1910), the famous Parisian photographer. He called himself Nadar. Here he is seen kneeling in a fragile balloon gondola. He obtained the first aerial photograph from a balloon in 1858 near Paris, France, and patented the aerial survey as we know it today. Unfortunately, the first aerial photograph did not survive (© Roger-Viollet, Paris, France; used with permission).

Figure 3-5   A portion of an aerial photograph of downtown Boston, MA, obtained by aeronauts James W. Black and Samuel A. King from a tethered balloon at an altitude of 1,200 ft on October 13, 1860. It is believed to be the first aerial photograph taken from a captive balloon in the United States and the earliest aerial photograph still in existence. It was obtained using a wet collodion plate (used with permission of the Smithsonian Institution, Washington, DC; #3B-15472).

Figure 3-6   The *Intrepid* balloon being tethered by Union troops at the battle of Fair Oaks on June 1, 1862 (used with permission of the Smithsonian Institution, Washington, DC; #2A-03710).

The balloon *Le Geant* will be employed in various aerostatic photography... the results of which will be so valuable for all planispheric, cadastral, strategical, and other surveys.

Unfortunately, in October 1863, the gigantic balloon descended too rapidly. Nine passengers aboard, including Ms. Nadar, were almost killed when the balloon was dragged 25 miles across the French countryside. No aerial photographs obtained from *Le Geant* have survived. However, Nadar did successfully obtain oblique aerial photographs using perhaps the first aerial multiple-lens camera from the *Hippodrome Balloon* tethered 1,700 ft above Paris in 1868 (Figure 3-8).

The dry-plate process pioneered by Richard Maddox in 1871 was a real boon to obtaining quality aerial photographs from balloons because no wet laboratory was required in the balloon gondola or nearby on the ground, and the emulsion was much faster, resulting in less image blur. This resulted in a great deal of balloon aerial photography taking place during the last few decades of the nineteenth century. Gaston Tissandier published the first manual on air photography from balloon platforms, *La Photographie en Balloon* (Newhall, 1969).

Aerial photography from tethered and free-flying balloons continues today. Stabilizing the balloon during the instant of exposure continues to be a major problem. Also, untethered balloons are at the mercy of the prevailing wind, making it difficult to navigate the balloon over the intended landscape. Nevertheless, many scientists and groups continue to obtain aerial photography using balloons. A few of the most active organizations are identified in Appendix A.

Lighter-Than-Air Flight Using Kites

Cameras carried aloft by large kites have also been used to obtain aerial photography. Frenchman Arthur Batut pioneered the development of kite aerial photography. In 1890, he published *La Photographie Aerienne par Cerf-volant,* in which he outlined the uses of kite aerial photography for the explorer (to view the land beyond), the archaeologist (to locate ruins), the military (to reconnoiter), and the agronomist (to locate vine diseases). Batut felt that the kite would bring aerial photography within the grasp of the common man (Batut, 1890; Tennant, 1903).

George R. Lawrence of Chicago was almost killed twice trying to obtain aerial photographs from captive balloons. On one occasion he fell 228 feet in a gondola from a balloon

fact, he designed and built a tremendous balloon called *Le Geant* (The Giant) in 1863. The balloon held 210,000 ft³ of gas, and its gondola consisted literally of a two-story house fitted with three-decker beds, a bathroom, and even a printing press. The gondola could comfortably house 12 passengers (Newhall, 1969). Nadar wrote:

Figure 3-7   The balloon *Intrepid* being inflated by using Thaddeus Lowe's portable hydrogen generating system during the Civil War battle of Fair Oaks on June 1, 1862 (used with permission of the Smithsonian Institution, Washington, DC; #2A-03675).

suspended over the Chicago stockyards and was only saved
because the free-falling gondola landed on telegraph wires.
Therefore, it is not surprising that he switched to kite aerial
photography. He often used as many as 17 kites in a connected "train" (although five to 10 usually sufficed) to carry
aloft cameras weighing as much as 2,000 lbs. He called this
collection of tethered kites a *Captive Airship* (Baker, 1994).
A mount hanging below the lowest kite in the series allowed
him to fix the camera in any direction before sending it up. A
system of booms, lines, and lead weights prevented the camera from turning horizontally while at the same time
decreased the camera's tendency to swing. Lawrence tripped
the shutter by incorporating an insulated wire into the steel
kite line, which was used to carry an electric current to the
camera (Newhall, 1969; Baker, 1994). Some of Lawrence's
most celebrated aerial photographs were acquired six weeks
after the April 18, 1906, San Francisco earthquake using 17
of his kites at one time (Figure 3-9). At first, rain soaked the
line and shorted-out the command to trip the shutter. Then,
the Sun came out and dried the line, allowing the shutter to
function. Panoramic negatives were acquired using a 49-lb
panoramic camera that held a celluloid-film plate 18.75 by
48 inches. Contact prints created from the negatives at the
time were the largest ever produced from an airborne platform. The camera covered 160° in a single sweep of the lens,

Figure 3-8   Oblique photograph of Paris obtained by Gaspard
Felix Tournachon (Nadar) from the *Hippodrome
Balloon* tethered 1,700 ft above the ground in 1868
(after Newhall, 1969).

Figure 3-9   Aerial photograph of "San Francisco in Ruins" obtained by George R. Lawrence after the April 18, 1906 earthquake using a 49-lb camera attached to a *captive airship* consisting of 17 kites tethered from a ship in San Francisco Bay. The kites achieved an altitude of 2,000 ft above sea level (courtesy Library of Congress; #LC-USZ 62-16440).

producing extremely detailed wide-angle photographs of the ruined city (Baker, 1994).

The major problem with kites and balloons is that they are not navigable in the strict sense of the word. Still, many excellent aerial photographs have been obtained using them. There is an Internet user group devoted to obtaining aerial photography using kites (Appendix A).

Heavier-Than-Air Flight Using Rockets

Even before airplanes were invented, aerial photographs were obtained by rocketing a camera into the sky and then retrieving the camera and film. In 1888, *La Nature* described a "photo rocket" invented by Amadee Denisse. In 1891, Ludwig Rahrmann received a German patent for a photographic system that was shot into the air using a rocket or large-caliber gun. The camera returned to the ground via parachute (Newhall, 1969). In 1903, Alfred Maul patented a rocket camera. Aerial photographs from a rocket that achieved an altitude of 2,600 ft were published in the *Illustrated London News* on December 7, 1912.

Heavier-Than-Air Flight Using Pigeons, Gliders, and Aircraft

All heavier-than-air birds, gliders, and airplanes achieve sustained flight based on the principle of *aerodynamic lift*. Wind blowing above and below a wing will cause the wing to achieve aerodynamic lift if the wing has the proper shape. A flat wing shape fights airflow, causing drag (resistance), while a curved wing shape (Figure 3-10) allows air to flow smoothly around it. A wing that is curved on the top and almost flat on the bottom creates aerodynamic lift. The molecules of air passing over the top of the wing surface have a longer distance to travel and therefore must move more rapidly, creating less pressure than the slower air flowing below the wing. The higher pressure air below the wing exerts pressure upward, causing the wing to lift. Tilting the wing upward will increase the aerodynamic lift even more. However, if it is tilted too much, the aerodynamic lift will be lost and the wing (and aircraft!) will stall and fall (Lopez, 1995).

Pigeons

In 1903, Julius Neubronner patented a breast-mounted aerial camera for carrier pigeons that weighed only 2.5 oz (Figure 3-11a,b). An article of the day described the pigeon data collection process:

> As a carrier pigeon, after starting, at first describes a spiral line, it is quite easy to take a number of views of a given portion of the ground from different points of view. After once determining the position of its cote (which it recognizes from a distance upward of 20 miles) the pigeon flies towards its goal in a straight line and at the uniform speed of an express train, so that

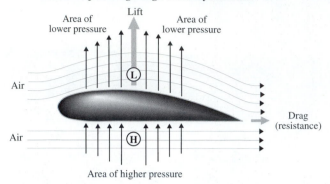

Relationship of Wing Design to Aerodynamic Lift

Figure 3-10   A wing that is curved on the top and relatively flat on the bottom creates *aerodynamic lift* (after Lopez, 1995).

the route to be recorded photographically can be readily determined in advance (*Scientific American*, 1909).

Exposures were made automatically at 30-second intervals (Quackenbush, 1960; Fischer, 1975). For obvious reasons, pigeons are not ideal remote sensing platform.

Gliders

George Cayley and John Stringfellow dramatically improved the aerodynamics and control mechanisms of heavier-than-air gliders. Cayley's coachman was the first person to successfully fly in a glider in 1853. From 1891 to 1896, a German civil engineer named Otto Lilienthal completed nearly 2,000 glider flights. Sometimes he would glide for more than 1,000 ft. He died in a glider flight in 1896 (Lopez, 1995). An American engineer name Octave Chanute built on Lilienthal's designs and dramatically improved glider control. He was also a great supporter of the aviation work of Orville and Wilbur Wright.

The Wright Brothers were absolutely fascinated with the concept of flight. Initially they experimented with gliders at Kitty Hawk, N.C., where a consistent wind could be found and the only obstructions were sand dunes and the surf. They built on the glider designs of Lilienthal and Chanute and eventually designed their own gliders in 1900. A photograph of one of the Wright Brothers lying prone in a glider in 1902 is shown in Figure 3-12a.

Aircraft

In 1903, the Wright Brothers built a 12-horsepower engine and the first operational aircraft propeller. By the flip of a coin, Wilbur Wright was given the chance to fly the first heavier-than-air motor-driven machine on December 14, 1903. He stalled and crashed without injury. On December 17, 1903, at Kill Devil Hills at Kitty Hawk, N.C. Orville Wright flew for 12 seconds and 120 ft (Figure 3-12b). Both brothers flew several times that day, with Wilbur flying the longest distance of 852 ft. Manned heavier-than-air powered flight had begun.

The original Wright Flyer was controlled by the pilot lying prone on the lower wing (Figure 3-12b). Turns were made by swinging the body from one side to another. This caused the rudder to move and the wings to become warped. The pilot's left hand controlled the up and down motion (ascent and descent) of the aircraft using an "elevator." The pilot held on firmly to the aircraft with the right hand. Not surprisingly, no aerial photographs were obtained using the original Wright Flyer because all hands were required just to maintain control and keep from crashing. The Wright Brothers built a much improved airplane in 1907 that allowed the pilot and a passenger to sit upright on the lower wing.

The first time that an airplane was used as a platform to obtain aerial photography was in 1908, when Pathe motion-picture photographer L. P. Bonvillain accompanied Wilbur Wright. He took motion pictures over Camp d'Auvours, near Le Mans, France. The original motion-picture film has not been found (Newhall, 1969). But, an enlargement of one of the frames of the motion picture was published in a French magazine in 1908.

 Photo-reconnaissance in World War I and World War II

Numerous new aircraft companies came into existence from 1907 to 1930. Many of the founder's names are still associated with United States' aircraft companies today (Lopez, 1995):

- Glenn Curtiss in 1907 — Curtiss

- Wright Brothers in 1909 — eventually Curtiss-Wright

- Glenn Martin in 1912 — now Martin Marietta

- William Boeing in 1916 — Boeing

- Donald Douglas in 1920 — now McDonnell Douglas

- Alan Lockheed in 1926 — now Lockheed-Martin

a.

b.

Figure 3-11 a) A squadron of pigeons equipped with lightweight (approximately 2.5 oz) 70-mm aerial cameras. b) Portion of an oblique aerial photograph obtained from a camera carried by a pigeon. The pigeon's wings are visible (© Deutsches Museum, Munich, Germany; used with permission).

a.

Figure 3-12   a) One of the Wright Brothers in an un-powered heavier-than-air glider at Kitty Hawk, N.C., in 1902. b) The Wright Flyer built by Orville and Wilbur Wright had a 12-horsepower engine and was equipped with specially designed propellers. Orville took the first successful heavier-than-air engine-powered flight on December 17, 1903. It lasted 12 seconds and covered 120 ft (used with permission of the Smithsonian Institution, Washington, DC; #3B-03541 and 3B-03300, respectively).

b.

• John Northrop in 1929 — Northrop

• LeRoy Grumman in 1929 — now Northrop/Grumman

Aircraft produced by these companies and others in Germany, France, and Britain were used for military photo-reconnaissance and other purposes in World War I and World War II.

Aerial Photography in World War I

In order to take an aerial photograph in the early years of World War I, it was necessary for the airman to point the handheld camera out the side of the plane, make an exposure, and then change the plate before taking another photograph. Cameras were initially strapped to the photographer's chest or attached to the side of the plane (Figure 3-13ab.) Aerial photography was very dangerous duty, especially in the early formative years.

The wartime use of aerial photography was not appreciated in the beginning of the war. In fact, World War I trench maps derived from photographs taken by the pilots on their own initiative were at first dismissed as being "a most disgraceful thing to have attempted" (Newhall, 1969). Eventually, however, commanders learned the value of aerial reconnaissance and it completely changed the military tactics used in World War I. First, aerial photographs were used to make relatively accurate maps for planning military strategy over poorly mapped terrain. A knowledge of existing and destroyed roads, barriers, and constriction points was very valuable for planning the movement of troops and materials. Second, a vast amount of military material was simply impossible to conceal from the aerial camera lens (Quackenbush, 1960). Troop movements and the stockpiling of arms and supplies could be documented. For example, on August 22, 1914, World War I British reconnaissance aerial photography revealed a major change in direction of the German forces advancing on Paris. This timely information allowed the Allied army to fortify its position on the Marne River and hold off the German advance to Paris (Lopez, 1995). World War I often consisted of trench warfare. Battlefield trenches dug in World War I for an unknown area are displayed in Figure 3-14. Trained photointerpreters routinely analyzed stereoscopic photography of trenches to locate gun emplacements and ammunition dumps. By 1918, French aerial units were developing and printing as many as 10,000 photographs each night during periods of intense activity. During the Meuse-Argonne offensive, 56,000 aerial photographs were made and delivered to American Expeditionary Forces in four days.

At the end of World War I, Lieutenant Edward Steichen of the Photographic Section of the American Expeditionary Forces stated,

> The consensus of expert opinion, as expressed at the various inter-Allied conferences on aerial photography, is that at least two-thirds of all military information is either obtained or verified by aerial photography.

Aerial Photography in World War II

Aerial photo-reconnaissance played a significant role in World War II. In 1938, General Werner von Fritsch, Chief of the German General Staff, made a prophetic statement: "The nation with the best photo-reconnaissance will win the next war." By 1940, Germany led the world in military photo-reconnaissance. Fortunately, German photointerpretation stagnated as the war progressed. The United States had almost no capability in military photointerpretation when it entered World War II (Fischer, 1975). However, British and American aerial photography data-collection and photo-interpretation capabilities improved quickly. After the retreat from the mainland at Dunkirk in 1940, the British were almost completely cut off from their normal sources of military intelligence and had to rely on aerial photography as their chief source of information. The photo-identification of German invasion barges in canals near the coast of France in the summer of 1940 constituted the major evidence that an invasion of England would take place in less than 48 hours. The British launched such an effective air attack on the invasion forces that Germany was forced to postpone the invasion and finally to abandon it (Quackenbush, 1960). Aerial photography also assisted in the destruction of German V-2 rocket facilities late in the war (Figure 3-15).

Most reconnaissance World War II aircraft had little defense capability. Therefore, they were often escorted by fighter aircraft. Millions of reconnaissance aerial photographs were obtained during World War II by modified aircraft such as the P-38. In addition, bombers often obtained aerial photographs during bombing raids that were used in subsequent bomb-damage assessment. For example, consider the rare accident documented during a bombing and aerial reconnaissance mission over Berlin, Germany (Figure 3-16ab). A B-17 Flying Fortress of the U.S. 8th Air Force in a lower group flew under an upper group just as bombs were released. The first aerial photograph reveals details about the B-17 Flying Fortress, such as several bombs being released, clouds, and the Berlin countryside below. The next exposure documents that a bomb crashed through the port horizontal

a.

b.

Figure 3-13   a) An aerial photographer and pilot in a Curtiss AH-13 airplane with Graflex camera in 1915. b) Close-up view of a World War I Curtiss JN-4 Jenny with the camera mounted on the right side (used with permission of the Smithsonian Institution, Washington, DC; #2B-41742 and 2B-41668, respectively).

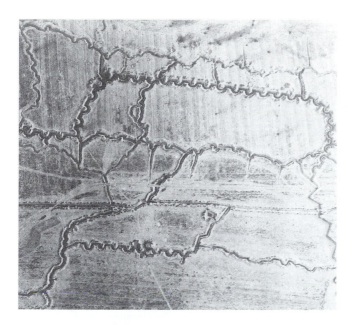

Figure 3-14   Vertical photography of World War I trenches in Europe. Examination of stereoscopic photography revealed the location of men, gun emplacements, and ammunition bunkers. Millions of men died in trench warfare during World War I (used with permission of the Smithsonian Institution, Washington, DC; #2B-41711).

Figure 3-15   Vertical aerial photograph of a V-2 rocket launching facility at Peenemunde in World War II. Note the large circular earth revetment around the launch facilities designed to deflect the blast during liftoff or to minimize the destruction during an accident. These facilities were eventually destroyed (courtesy of Federation of American Scientists, Intelligence Research Program).

stabilizer of the lower aircraft, clearing the way for subsequent bombs. It is unlikely that the damaged B-17 could have returned safely. Germany introduced jet aircraft platforms near the end of World War II (Boyne, 1993).

 ## Cold War Photo-Reconnaissance

After World War II, the arms race between the United States and the Soviet Union caused the U.S. government to think that there was a tremendous bomber and/or missile gap. Great emphasis was placed at the highest levels on developing technology to collect information. Similarly, the Soviet Union desired information about U.S. activities. Any aerial reconnaissance overflight of another state without authorization was considered to be an illegal and hostile act unless national leaders agreed to it beforehand. But, such overflights were absolutely essential if the United States was to know what was taking place in the Soviet Union and other countries. United States President Dwight Eisenhower approved the U-2 aerial reconnaissance program in December 1954 (Brugioni and Doyle, 1997). Subsequently, President Eisenhower attended the Four-Power Summit

Conference in Geneva, Switzerland, on July 21, 1955, and presented what the press eventually called the historic Open Skies Doctrine. He suggested that the absence of trust and the presence of "terrible weapons" among states provoked in the world "fears and dangers of surprise attack." To eliminate those fears, he urged that the Soviet Union and the United States provide "facilities for aerial photography to the other country" and conduct mutually supervised reconnaissance overflights. First Secretary of the Communist Party Khruschev privately rejected the Open Skies Doctrine the same day (Hall, 1996). Nevertheless, the concept had been presented.

Just a few days later on July 29, 1955, President Eisenhower publicly announced plans for launching "small unmanned, Earth-circling satellites as part of the U.S. participation in the 1957–1958 International Geophysical Year." The real purpose, however, was to establish the principle in international law of "freedom of space" with all that it implied for strategic reconnaissance conducted at altitudes above the "airspace" to which the states beneath claimed exclusive sovereignty (Hall, 1996). In effect, the President was laying the groundwork for the already conceived Genetrix Reconnaissance Balloon Project, a U-2 suborbital aircraft project, and subsequent satellite orbital reconnaissance.

a.   b.

Figure 3-16   a) A B-17 Flying Fortress of the U.S. 8th Air Force in a lower group flew under an upper group just as bombs were released. b) The next exposure documents that the bomb crashed through the port horizontal stabilizer of the lower aircraft (used with permission of the Smithsonian Institution, Washington, DC; #3A-20850 and 20851, respectively).

Genetrix Reconnaissance Balloons

Considerable research was conducted by the U.S. Air Force and the Central Intelligence Agency (CIA) during 1954 and 1955 on the utility of using unmanned balloons operating at high altitudes with cameras in their gondolas to obtain reconnaissance information over vast geographic areas. This was called the Grayback Program. A trial balloon was launched by the CIA from Scotland which drifted across the Soviet Union and was retrieved near South Korea. Similar balloons were also launched and recovered in the United States (Peebles, 1997).

Based on these successes, the Genetrix Reconnaissance Balloon Project was initiated in October 1955. The goal of the program was to obtain complete photographic coverage of the Soviet landmass based on the proposed launch and recovery of 2,500 unmanned high-altitude balloons. President Eisenhower gave approval for the launches to begin on December 27, 1955. From January 10, 1956, through February 6, 1956, approximately 448 balloons were launched from Scotland, Norway, West Germany, and Turkey (Peebles, 1997). The gondolas were to be retrieved by aircraft stationed in a large arc from Okinawa through Japan to Alaska. Many of the early balloons were recovered. However, Eastern European, Soviet, and Communist Chinese air defenses responded quickly and were able to stop many of the balloons.

On February 4, 1956, Soviet Deputy Foreign Minister Andrei A. Gromyko protested to the U. S. Ambassador. On

February 7, 1956, anticipating the Soviet response, President Eisenhower suggested to Secretary of State John Foster Dulles that the operation be suspended and "we should handle it so it would not look as though we had been caught with jam on our fingers" (Day et al., 1998). On February 9, 1956, the Soviets held a press conference and displayed about 50 balloons and instrument containers. The United States said they were weather balloons. It was a major embarrassment to the United States (Day et al., 1998). Only 44 of the 448 balloons were recovered. However, the cameras on these balloons obtained 13,813 aerial photographs covering 1,116,449 mi^2 of Soviet and Chinese terrain, including the very significant nuclear refining facility at Dononovo in Siberia (Peebles, 1997). This marked the end of this early U.S. balloon reconnaissance program. The U-2 aircraft reconnaissance program was initiated just a few months later, in July 1956.

U-2 Aircraft Reconnaissance Program

In December 1954, President Eisenhower instructed the CIA to contract with Lockheed (in cooperation with the U.S. Air Force) to develop a photo-reconnaissance jet aircraft that could fly above the Soviet Union at will to document their military capability. This led to the creation of the secret Lockheed Skunk Works (named after a smelly location in the Lil' Abner cartoon strip) in Burbank, and later Palmdale, CA, where Kelly Johnson and his engineers developed the aluminum U-2 aircraft (Rich and Janos, 1994).

The original U-2s were painted dull black to prevent them from glinting in the Sun (Figure 3-17a). It could fly at 70,000 ft above sea level and expose 3,600 ft of high-resolution film on a single mission. The camera was developed by Dr. Edwin Land, inventor of the Polaroid camera. The original U-2 payload consisted of two high-resolution cameras — one a special long-focal-length spotting camera able to resolve objects two to three feet across from a height of 70,000 ft, and the other a tracking camera that could produce a continuous strip of film of the whole flight path (Rich and Janos, 1994). The first Russian overflight occurred on July 4, 1956. It took off from Weisbaden, West Germany to survey Soviet naval shipyards and submarine construction. It overflew Poland, Belorussia, Moscow, Leningrad, and the Soviet Baltic states. Much to the consternation of the Americans, the Soviet radar detected and tracked the first U-2 at its design altitude of 70,000 ft (Hall, 1996).

Soviet surface-to-air missiles and aircraft could not touch the U-2 for four years. Information provided from the U-2 overflights had a tremendous impact on our knowledge about the Soviet Union during the 1950s and early 1960s. However, missile technology continued to improve, and on May 1, 1960, a cluster of 14 SA-2 air-to-air missiles shot down U-2 pilot Francis Gary Powers (Figure 3-17b). He was interrogated for 10 to 16 hours a day for 61 days. In August 1960, Soviet authorities staged a widely publicized, open trial that was designed to embarrass the United States. They sentenced Francis Gary Powers to 10 years in a Soviet prison; however, he was exchanged after 21 months for Soviet spy Rudolph Abel, who was being held in the United States (Powers, 1997).

After the Francis Gary Powers incident, the U-2 still provided important strategic reconnaissance information throughout the world. For example, in July 1962, the director of the CIA sent a memo to President John F. Kennedy saying that he believed the Soviets would deploy medium-range ballistic missiles (MRBM) just 90 miles from the U.S. mainland on Cuba; exactly when was unclear (Walter, 1992; Goldberg, 1993). On October 9, 1962, President Kennedy approved reconnaissance flights over western Cuba using high-altitude U-2s. Hurricane Ella kept the planes on the ground until October 14, 1962. The photographs were interpreted on October 15 and presented to President Kennedy on October 16, 1962. A portion of one of the U-2 photographs shown to President Kennedy is found in Figure 3-18a. An oblique aerial photograph obtained by an RF-101 aircraft recorded MRBM facilities near San Cristobal, Cuba, on October 25, 1962 (Figure 3-18b).

On October 25, 1962, United Nations U.S. Ambassador Adlai Stevenson challenged the Soviet Ambassador, Valerian A. Zorin, to deny the U.S. charge that the Russians had installed offensive missile bases on Cuba. "In due course, sir, you will have your reply. Do not worry," replied Zorin. Stevenson responded, "I am prepared to wait until hell freezes over, if that's your decision. I am also prepared to present the evidence in this room." Aerial photographs on easels were then brought in, and the peaceful Cuban countryside in August was shown to contain missile facilities by mid-October. Twenty-four hours later there were unmistakable signs of missile installations. President Kennedy then initiated a naval blockade of Cuba. After much political brinkmanship, Soviet Premier Khruschev relented on October 28, 1962, and the world returned from the brink of nuclear war.

Interestingly, the detail in the photographs was largely lost on the untrained presidential staff. Attorney General Robert F. Kennedy wrote about the CIA's explanation of the photographs at the first emergency meeting: "I, for one, had to take their word for it. I examined the pictures carefully, and what

a.

b.

c.

Figure 3-17   a) The Lockheed U-2 high altitude reconnaissance aircraft was developed at the Skunk Works in Burbank and Palmdale, CA, for President Eisenhower. The program was supervised by the Central Intelligence Agency. b) Francis Gary Powers in front of a U-2 aircraft (courtesy Lockheed Martin, Inc.). c) Many U-2s are still in service as Earth resource observation aircraft. In this example there is a U-2 in the foreground and an ER-2 in the background in flight near San Francisco, CA (courtesy NASA and Lockheed Martin, Inc.).

I saw appeared to be no more than the clearing of a field for a farm or the basement of a house. I was relieved to hear later that this was the same reaction of virtually everyone at the meeting, including President Kennedy. Even a few days later, when more work had taken place on the site, he remarked that it looked like a football field" (Kennedy, 1969).

Other Presidents have used the suborbital U-2 platform as an intelligence asset. For example, President Reagan ordered U-2 overflights of Nicaragua to identify Soviet arms buildup in support of rebel forces. President Bush used U-2 assets in the Gulf War in 1991. President Clinton used them in the war in Bosnia in 1998–1999. Specially modified U-2 aircraft are also used to support various NASA Earth resource (designated ER) remote sensing projects (Brugioni, 1985).

The SR-71

"We knew in 1958, two years before it happened, that the Russians were going to shoot down a U-2. The U-2 could fly higher than any other plane, but it was slow. We needed a replacement that could fly higher, farther, and faster," said Kelly Johnson (Hartz, 1981). The Skunk Works subsequently developed the SR-71 (Figure 3-19). It was delivered to the U.S. Air Force in 1965. In 1976, it set the world speed and altitude records of 2,193 m.p.h. in level flight at 85,126 ft, which exceeds the velocity of a 30.06 rifle bullet. It is still the fastest, highest-flying aircraft in the world as far as we know. At Mach-3 the glass of the cockpit is blistering hot, over 6,000°F. The SR-71's first mission was in 1965 over Hanoi, the capital of North Vietnam. It flew daily reconnais-

a.

b.

Figure 3-18   a) A portion of the U-2 aerial photograph of San Cristobal, Cuba, shown to President John F. Kennedy by the U.S.'s top photointerpreter Arthur C. Lundahl on October 16, 1962. The photograph was obtained on October 14, 1962, at an altitude of 21,300 m. It depicts missile trailers, a few launchers, and tents that were used to prepare missiles and warheads for launch. President Kennedy asked, "Are you sure?" "Mr. President," Lundahl replied, "I am as sure of this as a photointerpreter can be sure of anything. And I think, sir, you might agree that we have not misled you on anything we have reported to you. Yes, I am convinced they are missiles" (Brugioni, 1996) (U.S. Air Force photograph). b) Low-oblique aerial photograph of Medium Range Ballistic Missile (MRBM) Launch Site 1 near San Cristobal, Cuba, obtained on October 25, 1962, by low-flying RF-101 aircraft. Long missile shelter tents and missile transporters are visible (U.S. Air Force photograph; courtesy Federation of American Scientists).

a.   b.

Figure 3-19   Two views of the Lockheed SR-71 reconnaissance aircraft. It can fly at greater than 70,000 ft above sea level and achieve air-
speeds greater than 2,000 m.p.h. (courtesy Lockheed Martin, Inc.).

sance missions over North Vietnam and was never touched. The SR-71 detected the Soviet combat brigade in Cuba in 1979 (Hartz, 1981). The SR-71 carries only remote sensing and signal intelligence equipment. Several are still in use although they have been officially retired numerous times.

Stealth Technology

From 1975 to the present, the Lockheed Skunk Works and other aerospace companies throughout the world have developed *stealth* technology. A stealth aircraft is constructed of computer-designed facets of metal that have special shapes and orientations designed to deflect incoming radar waves away from the source of transmission on the ground or from another plane in the air. In addition, special composite material is usually applied to the aircraft, which absorbs incident radar energy (Rich and Janos, 1994). These conditions make the aircraft relatively invisible to radar detection. The radar cross-section (signature) of a non-stealth airplane looks as big as an 18-wheel tractor trailer. Conversely, the radar cross-section of an F-117 is equivalent to that of a small steel marble. This stealth capability was dramatically unveiled to the world in the 1990 Gulf War when previously secret F-117 aircraft (Figure 3-20) were used to fly night reconnaissance and tactical missions over Iraq and Kuwait. Stealth technology was also used extensively in the conflict in Bosnia in 1998–1999.

Celestial Satellite Sentinels

The first man-made satellite to orbit the Earth was *Sputnik I* launched by the Soviet Union on October 24, 1957. In a single moment it effectively confirmed a worldwide "open skies" policy for objects launched into orbit. The United States initiated its *Corona* orbital satellite reconnaissance program in the late 1950s and began launch operations on February 28, 1959. It was managed jointly by the CIA and the U.S. Air Force. The first eight reconnaissance missions did not produce any imagery (McDonald, 1997b). However, on August 18, 1960, Mission 9009 was successful. This ninth mission was recovered in midair and became not only the first Corona to return from space with reconnaissance film but also the first object to return from space and be recovered in midair (McDonald, 1997a). The age of space reconnaissance had begun. In just one mission it provided more photographic coverage of the Soviet Union than all previous U-2 missions.

Corona

Keyhole (KH) was the codeword assigned to the United States' space reconnaissance activities for the Corona, Argon, and Lanyard programs. The KH abbreviation was

Figure 3-20 The Lockheed F-117 stealth aircraft. The unique fuselage geometry and energy-absorbent materials on the plane make it difficult to detect on conventional radars (courtesy Lockheed Martin, Inc.).

used as a descriptor to refer to the camera systems associated with these reconnaissance programs, e.g., KH-1, KH-2, KH-3, and KH-4 were Corona sensors; Argon's camera was KH-5, and Lanyard's camera was KH-6. The initial Corona spatial resolution was 40 ft for Corona KH-1 in 1960 and eventually 4.5 to 6 ft for KH-4B in 1972 (Ruffner, 1995; McDonald, 1997b).

The first Corona photograph was of the Mys Shmidta Air Field (Figure 3-21a). All of the KH cameras had a focal length of 24 inches. All KH-4 satellites (1962–1972) contained twin panoramic cameras that could obtain stereoscopic photographs. An artist's rendition of the KH-4B platform and internal components is shown in Figure 3-21b. The details of each Corona launch from June 25, 1959, to the final mission on May 25, 1972, are found in McDonald (1997b). A Corona KH-4 panchromatic photograph obtained on October 20, 1964, four days after China's first aboveground nuclear test at Lop Nor is shown in Figure 3-22a. Corona played a major role monitoring nuclear proliferation.

Despite their great value, the early reconnaissance satellites were far from perfect. They returned their images by parachuting the film back to Earth in a capsule, sometimes days or even weeks after they were taken (Figure 3-22b). That delay could be crippling. Both the 1967 Six-Day War in Israel and the Soviet invasion of Czechoslovakia in 1968 ended before the United States could obtain satellite imagery of the trouble spots (Richelson, 1992).

It is important to point out that both the Soviet Union and the United States initiated their orbital space reconnaissance systems in 1956. However, due to its technological superiority, the United States was able to attempt launching Corona beginning in 1959 while the Soviet's Zenit-2 program was first launched in 1961. Both the Zenit and Corona platforms met their design objectives as reliable reconnaissance systems operating in an entirely new orbital dimension (Gorgin, 1997). The last Corona image was obtained on May 31, 1972. Over 800,000 images were acquired consisting of approximately 2.1 million feet of film in 39,000 cans (Clarke, 1999).

In 1967, President Lyndon Johnson addressed a group of educators,

> I wouldn't want to be quoted on this, but we've spent thirty-five or forty billion dollars on the space program. And if nothing else had come out of it except the knowledge we've gained from space photography, it would be worth ten times what the whole program has cost. Without satellites, I'd be operating by guess. But tonight we know how many missiles the enemy has, and it turned out our guesses were way off. We were doing things we didn't need to do. We were building things we didn't need to build. We were harboring fears we didn't need to harbor (Richelson, 1992; Walter, 1992).

President Johnson was responding to critics that said he had spent too much on the space program and not enough on poverty.

Only recently has it been made known that imagery from the Corona program were also used for domestic map compilation and updating. For example, the U.S. Geological Survey has used the data to update the 1:250,000 and 1:24,000-scale map series (McDonald, 1997a; Clarke, 1999). A domestic Corona image of the Pentagon in Washington, DC is shown in Figure 3-22c.

Executive Order Number 12951, issued by President William Clinton on February 22, 1995, changed the world of photo-satellite reconnaissance. The order directed:

> Imagery acquired by the space-based national intelligence reconnaissance systems known as the Corona, Argon, and Lanyard Missions shall, within 18 months of the date of this order be declassified.

The Web site for browsing and obtaining duplicates of Corona film is found in Appendix A. An excellent overview

First Corona Satellite Reconnaissance Photograph
Mys Shmidta Air Field, U.S.S.R. on August 18, 1960

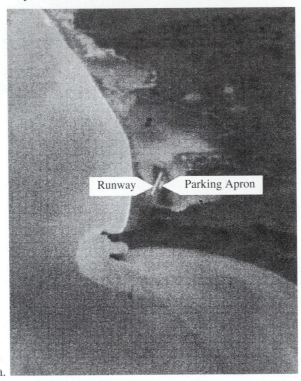

Runway   Parking Apron

a.

Artist's Rendition of the Corona KH-4B Camera in Flight

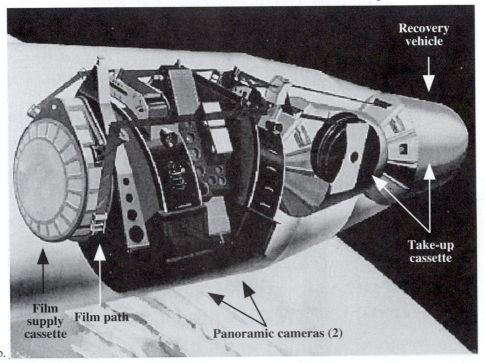

Recovery
vehicle

Take-up
cassette

Film
supply
cassette

Film path

Panoramic cameras (2)

b.

Figure 3-21   a) The first photograph by Corona KH-1. The spatial resolution was approximately 40 ft. b) Artist's rendition of the internal components of the Corona KH-4B orbital platform (courtesy National Reconnaissance Office).

Corona Photography of Lop Nor, China on October 20, 1964

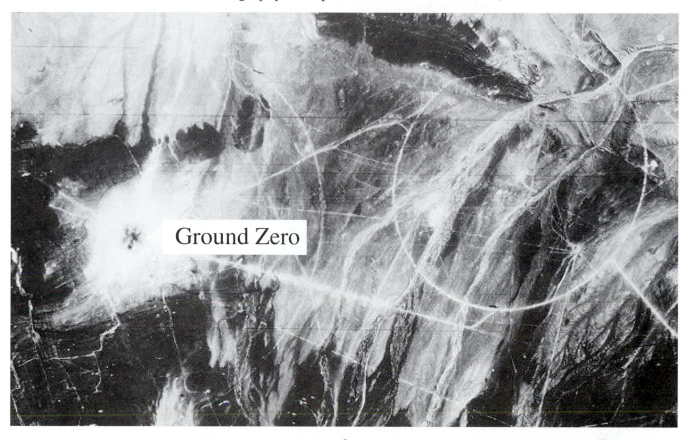

Ground Zero

a.

b.

c.

Figure 3-22 a) Corona KH-4 vertical panchromatic photograph of Lop Nor, China, aboveground nuclear test site on October 20, 1964. b) Photograph of the parachute of a Corona reconnaissance satellite capsule being captured. The film in the capsule was then developed and photointerpreted (courtesy Federation of American Scientists, Intelligence Research Program). c) Domestic image of the Pentagon in Washington, DC, recorded by Corona Mission 1101 on September 25, 1967 (courtesy USGS).

of the Corona program and the potential utility of the data for Earth resource analysis is found at Clarke (1999).

Ongoing Satellite Sentinels

Bailey and Kearney wrote in *Defense News* (1991):

> Satellite data and airborne radars have replaced the cavalry scout and the foot patrol as the commander's eyes.... Although the fog of war was not eliminated, General Schwarzkopf's view of the battlefield exceeded anything before possible. There was far less uncertainty regarding the enemy's vulnerabilities.

They called the Gulf War a *hyperwar* because so many decisions were based on satellite and aircraft remote sensing and signal intelligence (Walter, 1992). An example of Gulf War image intelligence is found in Figure 3-23.

In the United States, at least four types of satellites now gather intelligence information. LaCrosse bathes the Earth in microwaves (radar) and can resolve through cloud-cover objects less than three feet across, augmenting the work of the optical remote sensing satellites when bad weather blocks their view. Optical satellite remote sensing systems are far more powerful than the earlier Corona systems. Optical sensors resolve objects as small as 5 in. across using area array charge-coupled-device (CCD) technology discussed in Chapter 7, and the sensors telemeter the data directly to the ground (Richelson, 1992).

Ferrets (signal intelligence satellites — SIGINT) listen with very sensitive receivers to radio and microwave transmissions. Even in the early 1970s, a ferret named Rhyolite could record signals that aimed Soviet and Chinese missiles and at the same time monitor as many as 11,000 conversations going on between telephones and walkie-talkies. Tracking and Data Relay Satellite Systems (TDRS) orbit the globe at 22,300 miles above the equator in geostationary orbit just like Arthur C. Clarke, the noted author, wrote about in *Wireless World* in 1945 (Walter, 1992). They may be used to transmit remote sensor data from satellites to the ground.

Treaty compliance, arms-control agreements, and border disputes continue to be monitored by nations using high-resolution imagery obtained from satellite platforms (Richelson, 1990). For example, consider the KH-12 panchromatic images of Zhawar Kili, a suspected terrorist training camp

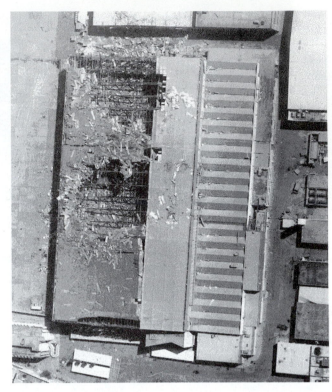

Figure 3-23　A warehouse south of Kuwait City suspected of housing Iraqi aircraft (U. S. Navy TARPS image released under the Freedom of Information Act to William M. Arkin).

support complex in Afghanistan (Figure 3-24a), and the Shifa pharmaceutical plant in the Sudan, suspected of producing chemical weapons (Figure 3-24b). These satellite images were used by Secretary of State William S. Cohen in his debriefing after cruise missile attacks by the United States on August 20, 1998.

Fortunately, humankind has also devoted some of its resources to placing remote sensing platforms in space for peaceful, Earth-resource applications. The most noteworthy include *Skylab*, the Landsat Multispectral Scanner (MSS) from 1972 to present, the Landsat Thematic Mapper series of satellites (1984 to present), the Landsat Enhanced Thematic Mapper+ (1999), the Space Shuttle Photography Program (ongoing), and numerous meteorological satellites (e.g., U.S. GOES and AVHRR, European METEOSAT). The Earth Observing System is scheduled to launch its first Earth-resource oriented EOS-AM (*Terra*) satellite in 1999 or early 2000. Chapter 7 provides details about many of these remote sensing platforms and sensor systems.

a.

b.

Figure 3-24 a) Satellite platform Keyhole 12 (KH-12) imagery of the Zhawar Kili suspected terrorist training camp in Afghanistan (spatial resolution was approximately 10 x 10 cm). b) Shifa pharmaceutical plant in the Sudan suspected of producing chemical weapons. These satellite images were used by Secretary of State William S. Cohen in his debriefing after cruise missile attacks on August 20, 1998 (courtesy of Federation of American Scientists, Intelligence Research Program).

Unmanned Aerial Vehicles

It is sad that many of the greatest improvements in aerial photography data collection and in the art and science of photointerpretation have taken place during World War I, World War II, the Korean War, Vietnam War, Gulf War, and Bosnia. One of the most important recent advancements are unmanned aerial vehicles (UAV) that have been under development since the Vietnam War (McDaid and Oliver, 1997). Scientists have equipped model airplanes and helicopters with light-weight cameras to obtain quality, high spatial resolution aerial photography (Walker, 1993). Similarly, the United States' Predator UAV was used in surveillance missions over sensitive areas of Bosnia in 1995 (Figure 3-25a,b,c). Lockheed Martin Inc. has been developing the stealthy Darkstar UAV (Figure 3-25d), a hi-tech surveillance UAV designed for the twenty-first century. Micro UAVs as small as 12 to 18 in. are now available, with hummingbird-size UAVs on the horizon (McDaid and Oliver, 1997). An artist's rendition of a potential micro UAV proposed by the Naval Research Lab is shown in Figure 3-26. UAVs may use optical, thermal infrared, as well as synthetic aperture radar sensor systems for all-weather capability. They can stay aloft virtually undetectable due to their size for hours on end, constantly monitoring the same geographic area. Some have SAR moving-target indicator capability that automatically detects whether anything is moving in the environment.

UAV technology represents a double-edged sword. On the one hand, UAV platforms and sensors could be especially useful for low-cost remote sensing data collection for law enforcement, Earth-resource analysis, urban planning, etc. UAVs are also extremely useful for military reconnaissance. "The soldier of the future will have his own mechanical bird which will provide him with pictures, sound and even smells of what is inside a building or bunker" (McDaid and Oliver, 1997). However, the same technology could be used to spy on the general public. The technology could also be used to conduct illegal activities such as monitoring the conditions around a bank preparatory to a robbery or for industrial espionage, where one company spies on another. Our generation of remote sensing scientists must be careful how it utilizes this new remote sensing technology (Slonecker et al., 1998).

Commercial, Nonmilitary Remote Sensing Platforms

The majority of nonmilitary commercial aerial photography is collected using single-wing aircraft (Figure 3-27). More affluent photogrammetric engineering and remote sensing

Predator Unmanned Aerial Vehicle (UAV) Imagery of Vogosca Ammunition Plant

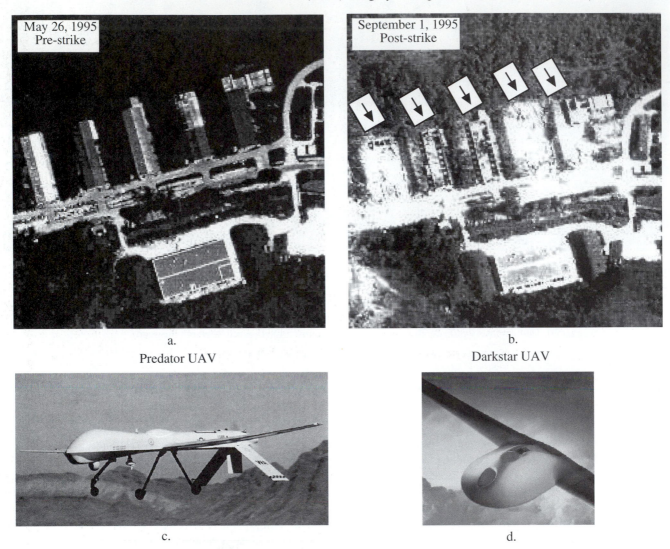

Figure 3-25   Pre-strike (a) and post-strike (b) Predator unmanned aerial vehicle (UAV) imagery of the Vogosca Ammunition Plant in Bosnia (courtesy John Pike, Federation of American Scientists, Intelligence Research Program). c) A Predator unmanned aerial vehicle in flight. d) The Darkstar unmanned aerial vehicle (courtesy Lockheed Martin, Inc.).

companies and nonmilitary government agencies have access to jet platforms. These types of platforms normally cannot fly as high or as fast as military platforms, but they are sufficient for acquiring quality metric aerial photography and other types of remote sensor data for managing our cities and natural resources. Most of the aircraft now have onboard GPS that keep track of the aircraft's exact location during remote sensing data collection.

 Use of National Technical Means Remotely Sensed Data by Nonmilitary Federal Agencies

As previously discussed, certain "classified" or "National Technical Means" U.S. government imagery is used for nonmilitary intelligence "civil applications." The major initiatives include (Hodgson, 1999):

Prototype Micro Unmanned Aerial Vehicle (UAV)

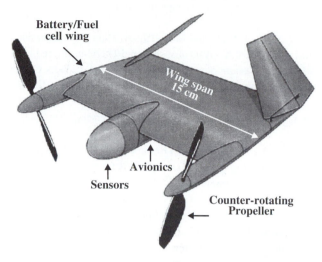

Battery/Fuel cell wing

Wing span 15 cm

Avionics

Sensors

Counter-rotating Propeller

Figure 3-26   A prototype unmanned aerial vehicle (UAV) developed by the Naval Research Lab. It is supposed to weigh only 40 g and have an airspeed of 10 m s^{-1} (adapted from McDaid and Oliver, 1997).

- supervision by the Advanced Systems Center within the National Mapping Program at the U.S. Geological Survey,

- the Government Applications Task Force (GATF), and

- the Federal Emergency Management Agency's (FEMA) use of remotely sensed imagery for emergency response applications.

Advanced Systems Center

Most government agencies using classified remotely sensed data are required to work through the USGS Advanced Systems Center and the Civilian Applications Committee (CAC). Requests for access and use of the imagery must be stated in a request to the CAC that includes the *Proper Use Statement,* which states the type of data to be used and for what purpose the data will be used. The Proper Use Statement must explicitly define the agencies and units within each agency that will have access to the data and the time period for which the data are used. The CAC is composed of representatives from each federal agency. The request is reviewed and approved or declined by the CAC. If approved, the requesting agency works with the CAC in 1) tasking the reconnaissance satellites for image collection, and 2) analyzing the data at the Advanced Systems Center or processing the data in a classified facility.

Figure 3-27   A single-wing, dual propeller aircraft. Such aircraft are ideal for obtaining sub-orbital aerial photography and other types of remote sensor data.

Government Applications Task Force (GATF) Projects

Several projects have been funded by intelligence agencies and the government agency with a need (e.g., NOAA, DOE, etc.) for the use of classified assets for managing federal resources. These projects were referred to as GATF projects and were made possible through the down-classification of imagery. Previously, access to any of the classified remotely sensed imagery required either a "top secret" or "SCI" clearance. Certain data forms have now been classified at only the "secret" level, thereby allowing individuals with more easily obtainable clearances to work with the data. This down-classification of imagery allowed for the GATF projects.

FEMA's Use of National Technical Means Imagery

Although several agencies use classified remotely sensed imagery such as the USGS to update maps, FEMA's use is the most publicly known. Currently, the National Imagery and Mapping Agency (NIMA) acquires and processes the imagery and provides FEMA with what is known as a "derived product." The derived product may be tabular or cartographic (digital or analog). The cartographic products provided to FEMA illustrate the spatial distribution of damage from a natural disaster (e.g., hurricane, tornado). Damage polygons are categorized in levels of severity (Hodgson, 1999).

References

Adams, A. 1985, *The Camera*, Boston: Little, Brown & Co., 1–19.

Anderson, J. R., E. Hardy, J. Roach and R. Witmer, 1976, *A Land Use and Land Cover Classification System for Use with Remote*

Sensor Data, Washington: U.S. Geological Survey Professional Paper #964.

Avery, T. E. and G. L. Berlin, 1985, *Interpretation of Aerial Photographs*, Minneapolis: Macmillan., 470 pp.

Bailey, R. and T. Kearney, 1991, "Combat Enters Hyperwar Era," *Defense News*, July 22, 1991.

Baker, S., 1994, "San Francisco in Ruins: The 1906 Aerial Photographs of George R. Lawrence," *Landscape*, 10(10):9–14.

Batut, A., 1890, *La Photographie Aerienne par Cerf-volant*, Paris: Gauthier-Villars.

Boyne, W. J., 1993, *Silver Wings: A History of the United States Air Force*, New York: Simon & Schuster, 366 pp.

Brugioni, D., 1985, "New Roles for Recce," *Air Force Magazine*, October, 94–101.

Brugioni, D. A., 1996, "The Art and Science of Photoreconnaissance," *Scientific American*, 274(3):78–85.

Brugioni, D. A. and F. J. Doyle, 1997, "Arthur C. Lundahl: Founder of the Image Exploitation Discipline," *Corona: Between the Sun & the Earth*, R. A. McDonald (Ed.), Bethesda: American Society for Photogrammetry & Remote Sensing, 159–166.

Clarke, A. C., 1945, "Extra-terrestrial Relays: Can Rocket Stations Give World-wide Radio Coverage?" *Wireless World*, October, 305–309.

Clarke, K., 1999, *Project Corona*, sponsored by the National Science Foundation, Santa Barbara: U.C, Santa Barbara, http://www.geog.ucsb.edu/~kclarke/Corona/Corona.html.

Colwell, R. N., 1997, "History and Place of Photographic Interpretation," *Manual of Photographic Interpretation*, W. Philipson (Ed.), 2nd Ed., Bethesda: American Society for Photogrammetry & Remote Sensing, 3–48.

Day, D. A., J. M. Logsdon and B. Latell, Eds., 1998, *Eye in the Sky: The Story of the Corona Spy Satellites*, History of Aviation Series, Washington: Smithsonian Institution, p. 128.

de Vries, G., 1999, "Jules Verne Collection: Striking Similarities," http://www.phys.uu.nl/~gdevries/sim/sim.html.

Eder, J. M., 1945, *History of Photography*, translated by Edward Epstean, New York: Columbia University Press, 318–324.

Estes, J. E., 1966, *Geographic Applications of Multi-Image Correlation Remote Sensing Techniques*, Los Angeles: University of California at Los Angeles, unpublished dissertation.

Estes, J. E., E. J. Hajic and L. R. Tinney, 1983,"Fundamentals of Image Analysis: Analysis of Visible and Thermal Infrared Data," *Manual of Remote Sensing*, R. N. Colwell, Ed., Falls Church, VA: American Society of Photogrammetry, 1: 1039–1040.

Fanton, B., 1989, "Photography: 150th Anniversary," *Sky*, 84–91.

Fischer, W. A., 1975, "History of Remote Sensing," in R. G. Reeves, (Ed.), *Manual of Remote Sensing*, Falls Church, VA: American Society of Photogrammetry, 27-50.

Fouque, V., 1867, *La Verite sur l'invention de la photographie: Nicephore Niepce, sa vie, ses essais, ses travaux*, Paris: Libraire des Auteurs et de l'Academie des Bibliophiles, p. 61.

Ganot, A., 1855, *Traite elementaire de physique*, Paris.

Gernsheim, H. and A. Gernsheim, 1952, "Re-discovery of the World's First Photograph," *The Photographic Journal*, Section A (May): 118–121.

Goldberg, V., 1993, *The Power of Photography*, NY: Abbeville, 287.

Gorgin, P. A., 1997, "Zenit: Corona's Soviet Counterpart," *Corona: Between the Sun & the Earth*, R. A. McDonald (Ed.), Bethesda: American Society for Photogrammetry & Remote Sensing, 85–107.

Hall, R. C, 1996, "Post War Strategic Reconnaissance and the Genesis of Project Corona," *Corona: Between the Sun & the Earth*, R. A. McDonald (Ed.), Bethesda: American Society for Photogrammetry & Remote Sensing, 25–58.

Hartz, 1981, "Alone, Unarmed, Untouchable: The Amazing SR-71," *Readers Digest* (January), 133–136.

Hodgson, M. E., 1999, personal communication, Dept. of Geography, University of South Carolina.

Jensen, J. R., 1983, "Biophysical Remote Sensing," *Annals of the Association of American Geographers*, 73(1):111–132.

Kennedy, R. F., 1969, *Thirteen Days: A Memoir of the Cuban Missile Crisis*, New York: Mentor Co., 192 pp.

King, S. A., 1860, "The Late Balloon Photographic Experiment," *Boston Herald*, October 16, 1860.

Kodak, 1981, *Applied Infrared Photography*, Publication M-28, Rochester: Eastman Kodak Co., 86 pp.

Lillesand, T. M. and R. W. Kiefer, 1994, *Remote Sensing and Image Interpretation*, 3rd. Ed., New York: John Wiley & Sons, 750 pp.

London, B. and J. Upton, 1994, *Photography*, 5th Ed. New York: Harper Collins, 422 pp.

Lopez, D. S., 1995, *Aviation*, New York: Macmillan, 256 pp.

Lyon, J. G. and E. Falkner, 1995, "Estimating Cost for Photogrammetric Mapping and Aerial Photography," *Journal of Surveying Engineering*, 121(2):63–86.

Maddox, R. L., 1871, *British Journal of Photography*, 18:422–423.

McCamy, C. S., 1960, "A Demonstration of Color Perception with Abridged Color Projections Systems," *Photographic Science and Engineering*, 4(3):156.

McDaid, H. and D. Oliver, 1997, *Smart Weapons: Top Secret History of Remote Controlled Airborne Weapons*, New York: Barnes & Noble, 208 pp.

McDonald, R. A. (Ed.), 1997a, *CORONA: Between the Sun and the Earth: The First NRO Reconnaissance Eye in Space*, Bethesda: American Society for Photogrammetry & Remote Sensing, 400.

McDonald, R. A., 1997b, "Corona, Argon, and Lanyard: A Revolution for U.S. Overhead Reconnaissance," *Corona: Between the Sun & the Earth*, Bethesda: American Society for Photogrammetry & Remote Sensing, 61–74.

Newhall, B., 1969, *Airborne Camera: The World from the Air and Outer Space*, New York: Hastings House, 144 pp.

Newhall, B., 1993, *The History of Photography*, New York: Museum of Modern Art, 319 pp.

Niven, W. D., Ed., 1890, *The Scientific Papers of James Clerk Maxwell*, Cambridge: Cambridge University Press, 1:136–137.

Peebles, C., 1997, *The Corona Project: America's First Spy Satellites*, ISBN 1-55750-688-4, 31–33.

Pike, J., 1998-1999, *Intelligence Research Program*, Washington: Federation of American Scientists, www.FAS.org.

Powers, F. G., Jr., 1997, "Foreward: From the U-2 to Corona," *CORONA: Between the Sun and the Earth: The First NRO Reconnaissance Eye in Space*, R. A. McDonald, Ed., Bethesda: American Society for Photogrammetry & Remote Sensing, vii–ix.

Quackenbush, R. S., 1960, "Development of Photo Interpretation," *Manual of Photographic Interpretation*, Falls Church, VA: American Society of Photogrammetry, 1–18.

Rich, B. R. and L. Janos, 1994, *Skunk Works — A Personal Memoir of My Years at Lockheed*, New York: Little Brown, 370 pp.

Richelson, J. T., 1990, "From CORONA to LACROSSE: A Short History of Satellites," *The Washington Post*, February 25, 1990, B1–B4.

Richelson, J. T., 1992, "Spies in Space," *Air & Space*, 6(5):75–80.

Ruffner, K. C., 1995, *CORONA: America's First Satellite Program*, Washington: Central Intelligence Agency, 360 pp.

Sabins, F. F., Jr., 1997, *Remote Sensing — Principles and Interpretation*, New York: W. H. Freeman, 494 pp.

Scientific American, 1909, "Pigeon Cameras," *Scientific American*, 100(4):27.

Sewall, E. D., 1957, "Fifty Years of Aerial Photography – Beginning in 1880," *Photogrammetric Engineering*, 23:835–850.

Slonecker, E. T., D. M. Shaw and T. M. Lillesand, 1998, "Emerging Legal and Ethical Issues in Advanced Remote Sensing Technology," *Photogrammetric Engineering & Remote Sensing*, 64(6):589–595.

Taylor, M. J. H. (Ed.), 1995, *Jane's Encyclopedia of Aviation*, New York: Crescent Books, 964 pp.

Tennant, J. A., 1903, "Aerial Photography," *The Photo-Miniature*, 5(52):144–173.

Thompson, M. M., 1966, *Manual of Photogrammetry*, Menasha, Wisconsin: George Banta Co. and the American Society of Photogrammetry, 3rd Ed., 1:244–245.

Walker, J. W., 1993, *Low Altitude Large Scale Reconnaissance: A Method of Obtaining High Resolution Vertical Photographs for Small Areas*, Denver: National Park Service, 127 pp.

Walter, W. J., 1992, *Space Age*, New York: Random House, 335 pp.

Watt, E. J., 1925, *History of Three-Color Photography*, Boston: American Photographic Publishing Co., 3–4.

Wolff, R. S. and L. Yaeger, 1993, *Visualization of Natural Phenomena*, Santa Clara: Telos Springer-Verlag, 374 pp.

Wolinsky, C., 1999, "The Quest for Color," *National Geographic*, 196(1):72–93.

Aerial Photography — Vantage Point, Cameras, Filters, and Film

4

At first glance it might appear that obtaining and interpreting aerial photography is a routine task because the technology has been available for more than a century. This is not the case. Important decisions must be made to ensure that quality aerial photography is collected and interpreted properly. Many of these decisions are based on principles of optics and how energy interacts with light-sensitive materials. Some of the most important issues to be addressed include:

- vantage point (e.g., vertical, low-oblique, high-oblique),

- camera (e.g., metric, panoramic),

- filtration (e.g., haze, minus-blue, band-pass filters), and

- film emulsion (black-and-white, black-and-white infrared, color, and color-infrared).

The more that an image analyst knows about how an aerial photograph was collected, the better image interpreter he or she will be. This chapter also describes how imagery is digitized.

 Vertical and Oblique Vantage Points

Aerial photography may be obtained from vertical or oblique vantage points depending upon project requirements.

Vertical Aerial Photography

A *vertical photograph* is obtained when the camera's optical axis is within $\pm$ 3° of being vertical (perpendicular) to the Earth's level surface (Figure 4-1a). A portion of a vertical black-and-white photograph of the Goosenecks of the San Juan River in Utah is shown in Figure 4-1b. This aerial photograph can be visually interpreted to extract detailed qualitative information about the surface geology, geomorphology, and hydrology within the vertical field of view. This aerial photograph in conjunction with another overlapping vertical aerial photograph obtained from a slightly different viewing position may be analyzed using quantitative photogrammetric principles (discussed in Chapter 6) to derive the following types of information:

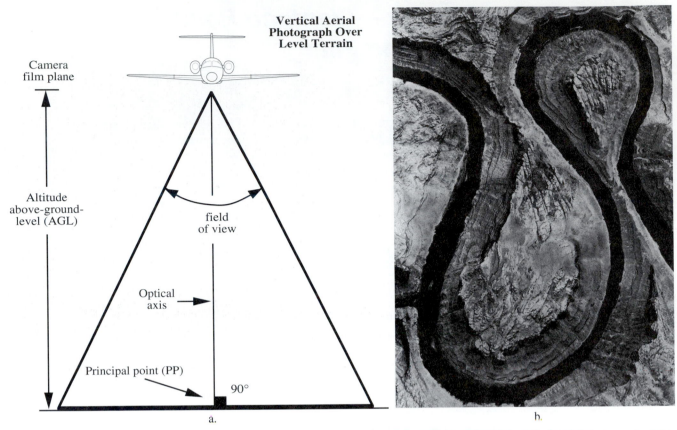

Figure 4-1   a) A vertical aerial photograph has $\leq 3°$ of tilt from a ray perpendicular to the terrain. b) A vertical aerial photograph of the Goosenecks of the San Juan River in Utah.

- accurate *planimetric* (x,y location) base maps of natural (e.g., stream network, rock outcrops) and man-made features (e.g., trails, roads, buildings),

- *topographic* (z-elevation above sea level) base maps,

- raster *digital elevation models* (DEMs), and

- accurate *orthophotographs* (aerial photographs that are geometrically accurate in x,y).

Resource managers often forget that the planimetric, topographic, and orthophotomaps used in almost all geographic information systems (GIS) are created using fundamental photogrammetric principles applied to near-vertical aerial photography. For example, the U.S. Geological Survey's 7.5-minute 1:24,000 map series and digital elevation models are derived photogrammetrically. Unfortunately, the general public is not used to viewing the tops of objects such as buildings, trees, roads, etc. It takes considerable practice and

experience to efficiently and accurately interpret a vertical aerial photograph of the terrain (Cavanaugh, 1990).

Oblique Aerial Photography

An *oblique aerial photograph* is obtained if the camera's optical axis deviates more than 3° from vertical. It is called a *low-oblique aerial photograph* if the horizon is not visible (Figure 4-2a). A low-oblique photograph of a bridge over the Congaree River in South Carolina is shown in Figure 4-2b. It was acquired at an altitude of 600 ft above-ground-level. Visual interpretation of the photograph yields detailed information about the bridge superstructure.

A *high-oblique aerial photograph* is obtained if the horizon is visible (Figure 4-3a). A high-oblique photograph of the Grand Coolee Dam in Washington is shown in Figure 4-3b. This impressive 1940 oblique photograph represents a valuable historical record of the dam and the cultural landscape

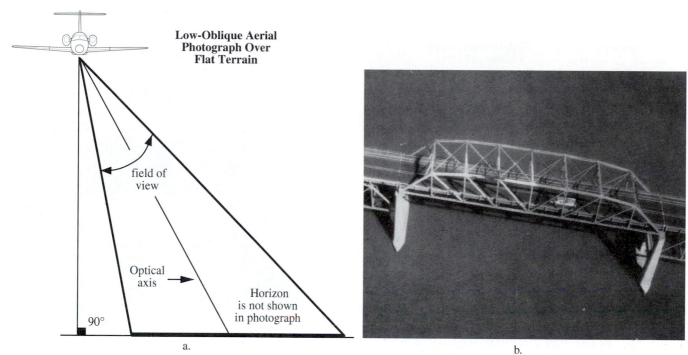

Low-Oblique Aerial Photograph Over Flat Terrain

field of view

Optical axis →

Horizon is not shown in photograph

90°

a.

b.

Figure 4-2   a) A low-oblique aerial photograph is tilted > 3° from vertical, and the horizon is not visible. b) A low-oblique aerial photograph of a bridge on the Congaree River near Columbia, SC. The photograph was obtained at an elevation of approximately 600 ft above-ground-level. One should always view an oblique aerial photograph so that the important features such as the bridge fall away from the viewer.

that was created adjacent to the dam during its construction to house the workers.

People are generally better able to photointerpret oblique aerial photography than vertical aerial photography because they have a lifetime of experience looking at the façade (side) of objects as they navigate in their daily environment. Oblique aerial photographs record a familiar façade or side view of the terrain. During interpretation, the analyst should always orient the oblique photograph so that the features of interest appear in natural perspective. Usually this will mean that objects with height appear to lean away from the person viewing the image.

Scientists and/or resource managers may live their entire lives and never extract quantitative information from oblique aerial photography. Therefore, oblique aerial photography photogrammetric principles will not be presented here. If it is necessary to extract quantitative information from oblique photography, consult the photogrammetric algorithms in the *Manual of Photogrammetry* and the *Addendum to the Manual of Photogrammetry* (Slama, 1980; Greve, 1996). This discussion will focus on obtaining near-vertical aerial photography and extracting both qualitative and quantitative information.

Aerial Cameras

One of the very first box cameras made for commercial purchase was developed for Louis Daguerre in France by Samuel F. B. Morse, who invented the Morse code (Figure 4-4). While modern cameras are much more sophisticated than this simple box camera, they nevertheless share certain fundamental characteristics. A good way to understand how a modern camera functions is to compare its components with those of the human eye (Figure 4-5).

Aerial Camera Components Compared with the Eye

The light-sensitive *retina* in the human eye is analogous to the light-sensitive *film* located at the *film plane* at the back of the camera. Both the eye and the camera use a *lens* to focus reflected light from the real world onto the retina or film. In the eye, the amount of light allowed to illuminate the retina is controlled by the iris, which can expand or contract in dark or light conditions and by the eyelid, which acts as a shutter. In the camera, the amount of light reaching the film plane is controlled by 1) the *size of the lens aperture opening*, and 2)

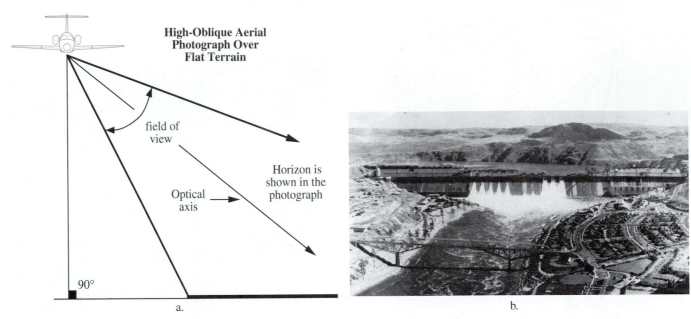

Figure 4-3   a) A high-oblique aerial photograph is tilted > 3° from vertical, and the horizon is visible. b) A high-oblique aerial photograph of the Grand Coulee Dam in Washington taken in 1940. The homes were built to house the workers during construction of the dam.

Figure 4-4   One of the first commercially available box cameras was created for Louis Daguerre by Samuel F. B. Morse, inventor of the Morse code.

a *shutter*, which controls the length of time – *exposure* – that the lens aperture remains open and allows light to pass through. The shutter can be placed in the camera lens assembly or just in front of the film at the back of the camera, in which case it is referred to as a *focal plane shutter*.

Focal Plane and Focal Length

The area in which the film is held flat during an exposure is called the *focal plane* or *film plane*. When a camera is focused at infinity, the distance from the rear nodal point of the lens to the film plane is known as the *focal length* (Adams, 1985). Some focal-length lenses used for aerial mapping include 88 mm (3.5 in.), 152 mm (6 in.), 210 mm (8.25 in.), and 305 mm (12 in.). Zoom lenses have a constantly changing focal length. Conversely, most metric aerial cameras have a fixed focal length, such as 152 mm (6 in.). Longer focal-length lenses, such as the 305 mm (12 in.) lens, are especially useful for high-altitude aerial photography data collection. Military photo-reconnaissance operations commonly employ lenses of 3 to 6 ft to obtain detailed photographs from extremely high altitudes or to acquire oblique photographs across borders.

f/**stop**

The ratio of the camera lens focal length (*f*) to the diameter of the lens opening (*d*) is known as the *f*/stop:

$$f/\text{stop} = \frac{\text{lens focal length}}{\text{diameter of lens opening}} = \frac{f}{d} \qquad (4\text{-}1)$$

The *f*/stop ratio is often used to designate the relative aperture setting or *speed* of the camera lens system. For example, a camera with a focal length of 100 mm and a maximum lens diameter opening of 10 mm (called *full aperture*) would have an *f*/10 speed lens:

Eye

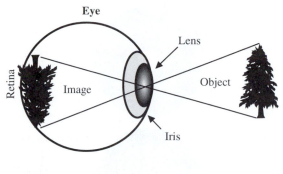

Camera

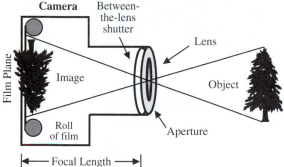

Figure 4-5   A comparison of the optical components of the simple camera with those of the human eye.

$$f/10 \ = \ \frac{100 \ \text{mm}}{10 \ \text{mm}}.$$

If the lens diameter at full aperture were twice as large, say 20 mm instead of 10 mm, the lens rating would be *f*/5. Basically, the *smaller* the *f*-rating, the faster the lens, i.e., the more light admitted through the lens opening per unit of time. In the previous example, the *f*/5 lens is said to be faster than the *f*/10 lens. The aircraft is always moving during aerial photography data collection (except when using a helicopter or tethered balloon or kite). Therefore, it is important to be able to acquire the photograph very rapidly. We do not want to have the image of a building smeared across the camera film plane as the aircraft moves forward. Therefore, aerial photography cameras generally have very fast lenses that can in effect "stop action." The fast lenses are often coupled with very sensitive film (to be discussed).

A complete system of apertures beginning at *f*/1 can be produced by multiplying any aperture by 1.4142136 (the square root of 2), which yields the succeeding smaller aperture. For example, the sequence of full-stop increments is *f*/1, *f*/1.4142, *f*/2, *f*/2.828, *f*/4, *f*/5.656, *f*/8, *f*/11, *f*/16, *f*/22, *f*/32, etc., as shown in Table 4-1. Each lens opening in the series transmits one-half as much light as the preceding lens opening (i.e., *f*/8 transmits one-half as much light as *f*/5.6). For exam-

Table 4-1.   International Series of *f*/stops and Shutter Speeds

Relative Aperture of f/stop		
Larger lens openings --------------- Smaller lens openings		
f/1---1.4---2---2.8---4---5.6---8---11---16---22---32---etc.		
Shutter Speeds		
Slower --- Faster		
1-1/2-1/4-1/8-1/16-1/30-1/60-1/125-1/250-1/500-1/1000-1/2000		

ple, if we "stop down" from *f*/11 to *f*/16, we decrease both the aperture of the lens opening and the amount of light reaching the film plane by two. Conversely, if we "stop up" from *f*/16 to *f*/11, we would increase the diameter of the lens diaphragm and the amount of light illuminating the film plane by a factor of two. This is shown diagrammatically in Figure 4-6. The *lens speed* is the *f*-number of the maximum effective diameter of the lens when the diaphragm is wide open (full aperture). Thus, we have an *f*/2.8 lens in Figure 4-6 (London and Upton, 1994).

Shutter Speed

To ensure that the film emulsion receives the correct amount of light during an exposure, the aerial photographer must select the correct relationship between the size of the lens aperture opening (i.e., the *f*/stop) and how long the light is allowed to illuminate the film plane. The length of time the shutter is open is called *exposure time*. It is controlled by the shutter mechanism. Shutter speeds on a camera usually range from "bulb," which lasts as long as the photographer pushes the exposure button, to 1 second, 0.75, 0.5, 0.25, 1/100, 1/200, 1/400, 1/500, 1/1000, and 1/2,000 seconds. For example, perhaps an ideal aerial photography exposure at 1:00 p.m. in the afternoon in clear Sunlight is *f*/16 at 1/200 second. If the sky above the aircraft suddenly becomes completely overcast and allows much less light to illuminate the terrain below, the aerial photographer might have to increase the amount of light illuminating the film plane. He could do this in one of two ways: 1) by increasing the size of the *f*/stop, e.g., from *f*/16 to *f*/11, which would double the area of the aperture opening, or 2) by allowing a longer exposure, e.g., a 1/100-sec exposure would double the length of time the aperture was open. Hopefully, either of these adjustments would result in an acceptable aerial photograph. But what if the photographer already had a good exposure at *f*/16 at 1/200-sec, but for some reason the pilot said that the plane's speed would need to be increased? The photographer would then want to increase the shutter speed to perhaps 1/400-sec so that blurring at the film plane would not take place. This would effectively cut the exposure time in half.

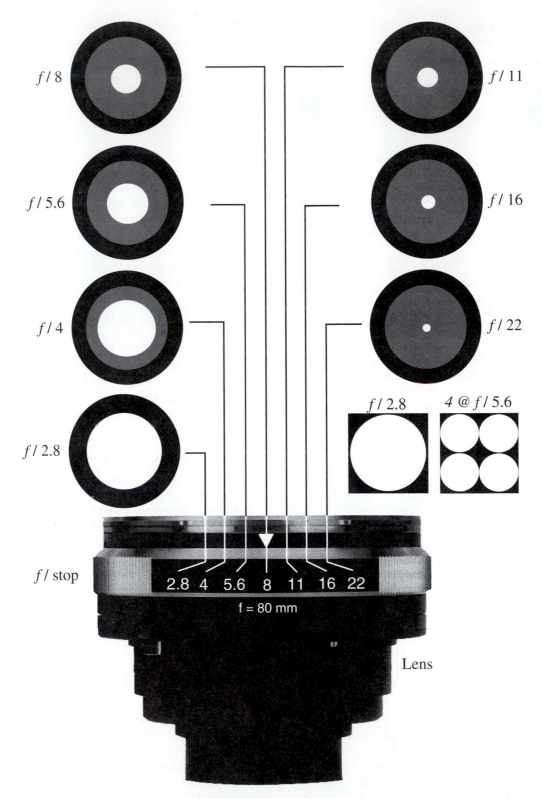

Figure 4-6   The *f*/stops for a camera lens and the size of their aperture openings. In this example, the *f*/stops range from *f*/2.8 to *f*/22, with the size of the apertures shown by the white circles. The lowest *f*/stop has the largest opening and lets in the greatest amount of light. The greater the *f*/stop, the less light admitted. Each *f*/stop lets in half (or double) the light of the next setting. For example, it takes four circles the size of an *f*/5.6 aperture to equal the size of an *f*/2.8 aperture (after London and Upton, 1994).

The photographer would then have to compensate for this by doubling the aperture of the lens by going from $f/16$ to $f/11$. Hopefully, the new 1/400-sec exposure at $f/11$ would produce the desired photographic results.

Why is it important to know about $f/$stops and shutter speeds when most of the aerial photography you will ever analyze will be acquired by engineers working for photogrammetric engineering firms who pay meticulous attention to these parameters? First, high-stratus clouds, cloud shadows, and Sun elevation differences throughout the day can cause light to fall off or increase, resulting in underexposed or overexposed aerial photography, respectively. Proper adjustment of the $f/$stop and shutter speed can maintain proper exposure throughout the day. Sometimes it is necessary to reject a batch of aerial photography if it has been improperly exposed. Second, most persons who use aerial photography will at some time desire to acquire their own photography even if it is obtained by pointing a 35-mm camera out the window of a light plane. At this time you will need to select an optimum combination of $f/$stop and shutter speed to obtain a good exposure of the terrain on film. It is also likely that relatively inexpensive unmanned aerial vehicles (UAVs) specifically designed for obtaining aerial photography soon will be available. For example, Hinckley and Walker (1993) provide information on how to acquire aerial photography using low-altitude remotely controlled aircraft.

Types of Aerial Cameras

There are several types of cameras used to obtain aerial photography, including:

- single-lens mapping (metric) cameras,

- multiple-lens (multiple-band) cameras,

- panoramic cameras,

- digital cameras, and

- miscellaneous cameras.

Single-Lens Mapping (Metric) Cameras

Single-lens metric cameras obtain most of the aerial photography used to map the planimetric (x,y) location of features and to derive topographic (contour) maps. These *cartographic* cameras are calibrated to provide the highest geometric and radiometric quality aerial photography. They usually consist of a camera body, lens cone assembly, shutter, film feed and uptake motorized transport assembly at the film plane, and an aircraft mounting platform. Filter(s) placed in front of the lens determine the wavelengths of light that are allowed to illuminate the film plane. An artist's rendition of the internal components of a metric camera is shown in Figure 4-7a. A profile view of a typical metric camera is shown in Figure 4-7b.

In the United States, Federal Aviation Administration (FAA) approval is required to cut a hole in an airplane's fuselage to accommodate an aerial camera. An example of a single camera mounted in the floor of an aircraft fuselage is shown in Figure 4-7c. If two aerial cameras are mounted in the aircraft (Figure 4-7d), it is possible to expose two types of emulsions at the same time (e.g., color and color-infrared) by synchronizing the camera shutter release mechanisms.

The lens cone assembly is the most important part of the camera. It usually consists of a single, expensive *multiple-element lens* that projects undistorted images of the real world onto the film plane. The multiple-element lens is focused at infinity because the aircraft typically flies at thousands of meters above-ground-level during data collection. Metric mapping cameras use various lenses with different angular fields of view, depending on the mission requirements. *Narrow* camera lenses have an angular field of view of $< 60°$, *normal* $60° - 75°$, *wide-angle* $75° - 100°$, and *super-wide-angle* $> 100°$. The wider the angular field of view, the greater the amount of Earth recorded on the film at a given altitude above-ground-level. The higher the altitude, the greater the amount of Earth recorded on the film by each lens. These relationships are summarized in Figure 4-8.

An *intervalometer* is used to expose the photographic film at specific intervals of time (dependent upon the aircraft altitude above-ground-level and speed) that will result in the proper amount of endlap to be obtained for overlapping (stereoscopic) coverage.

Aerial cameras usually expose film that is 24 cm (9.5 in.) wide in rolls ≥ 100 to 500 ft in length, depending upon the thickness of the film. Individual exposures are typically 9 x 9 in. (23 x 23 cm). At the instant of exposure, the film is held in place against a flat platen surface located at the focal plane. Vacuum pressure is applied to the film via the platen just prior to the instant of exposure to remove any bubbles, bumps, or irregularities in the unexposed film. After exposure, the vacuum is released and the drive mechanism moves the exposed film onto the take-up reel in preparation for the next exposure.

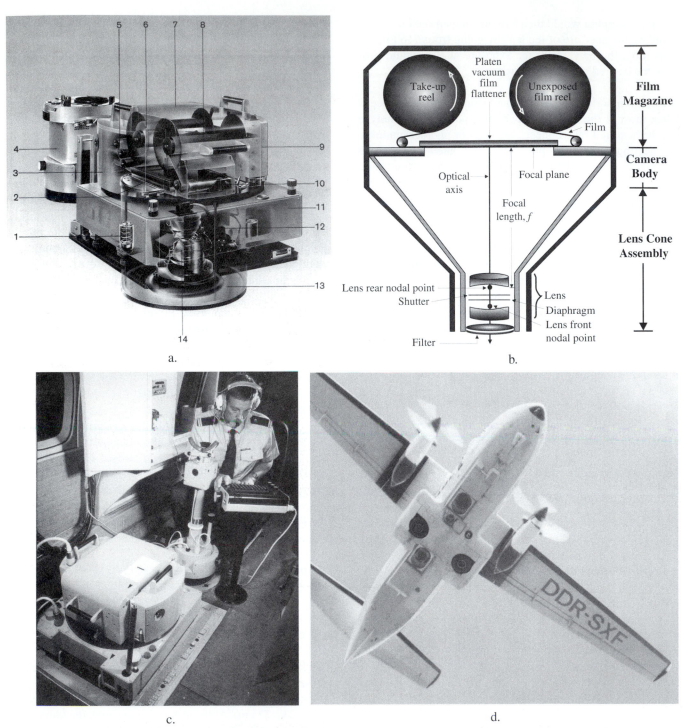

Figure 4-7   a) System components of a metric camera consisting of 1) vibration isolation elements, 2) film platen pressure plate, 3) drive unit, 4) control unit, 5) film feed rollers, 6) film take-up reel, 7) film magazine, 8) unexposed film supply reel, 9) drive unit, 10) forward-motion compensation device, 11) mount, 12) exchangeable universal shutter, 13) lens cone assembly, and 14) lens. b) Profile view of a metric camera and system components. c) Example of a metric camera installed in the floor of an aircraft complete with operator console and a terrain bore-sighting instrument. d) A plane with two camera ports can expose two types of emulsions at exactly the same time, e.g., color and color-infrared film. Note that in this photograph two camera ports and two retracted wheels are present (courtesy E. Coyote Enterprises, Inc., Marilyn O'Cuilinn; Z/I Imaging).

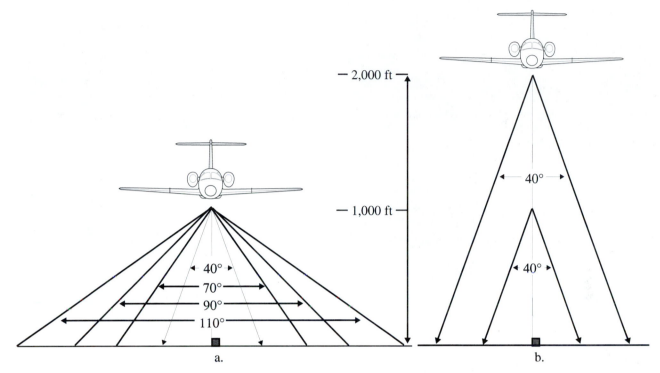

Figure 4-8   a) The greater the camera lens angle of view, the greater the amount of terrain photographed at a constant altitude above-ground-level. Narrow (40°), normal (70°), wide-angle (90°), and super-wide-angle (110°) lenses are portrayed. b) Using the same narrow (40°) angle-of-view camera lens at two different altitudes results in dramatically different amounts of terrain being photographed.

Depending on the velocity of the aircraft (v) and the aircraft altitude above-ground-level (h), the film might be advanced slightly during exposure to compensate for image motion. Special-purpose *image motion compensation* (IMC) magazines move the film across the focal plane anywhere from 0 to 64 mm per second. Correction is achieved by shifting the platen pressure plate with the film attached via vacuum in the flight direction in accordance with a velocity-to-height ratio (v/h) and the focal length of the lens. This greatly increases the quality of the aerial photography. An example of the effectiveness of utilizing image motion compensation and stabilization is shown in Figure 4-9. Notice the improved detail in the aerial photograph that was obtained using image motion compensation.

Most modern metric cameras provide detailed *image annotation* around the 9 x 9 in. image area of the film. For example, numerous types of ancillary information are displayed around the perimeter of the vertical aerial photograph of Mineral Wells, TX, shown in Figure 4-10. A programmable light-emitting diode inside the camera exposed text information onto the film. Important information present includes: 1) a grayscale step wedge used to determine if a proper

exposure has been obtained, 2) a notepad where the aerial photographer can enter mission critical notes in pencil if necessary, 3) altimeter, 4) white cross-hair fiducial marks, 5) clock, 6) lens cone serial number, 7) focal length in mm, 8) project frame number, 9) mission name and date, and 10) navigation data (not visible). Fiducial marks are discussed in Chapter 6.

Sometimes we analyze aerial photographs that are many years old. Having detailed image annotation information is critical to successful information extraction, especially if sophisticated photogrammetric instruments have to be calibrated or if the photography will be used in computer soft-copy photogrammetry applications discussed in Chapter 6.

Multiple-Lens (Multiple-band) Cameras

More information can usually be obtained about the environment from a study of photographs taken simultaneously in several regions (bands) of the electromagnetic spectrum than from photographs taken in any single band. When conducting multiband spectral reconnaissance (Colwell, 1997), each of the cameras simultaneously records photographs of the

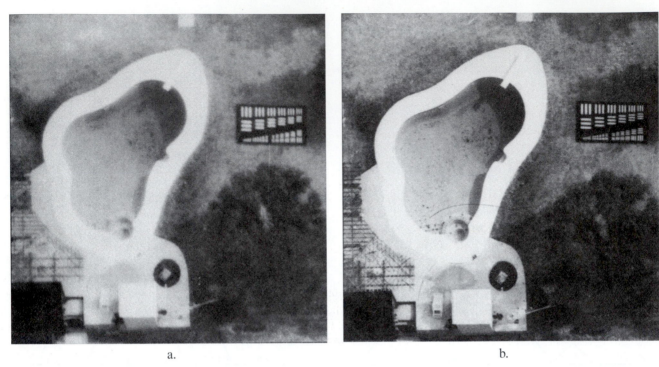

a.   b.

Figure 4-9   a) Vertical aerial photograph of a backyard swimming pool obtained without forward image motion compensation. Note the resolution target lying on the ground. b) Photograph obtained with forward image motion compensation. More elements of the resolution target can be discerned (courtesy E. Coyote Enterprises, Inc., Marilyn O'Cuilinn; Z/I Imaging).

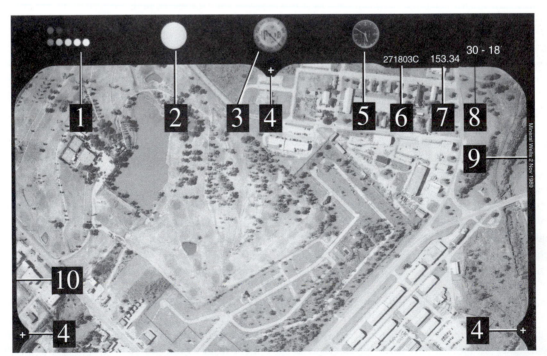

Figure 4-10   Annotation around the perimeter of an aerial photograph is very important. This particular metric camera provides the following information: 1) a grayscale step wedge, 2) notepad, 3) altimeter, 4) fiducial marks, 5) clock, 6) lens cone serial number, 7) focal length in mm, 8) frame number, 9) mission name and date, and 10) navigation data (not visible). Only one-half of this aerial photograph of Mineral Wells, TX is visible (courtesy E. Coyote Enterprises, Inc., Marilyn O'Cuilinn; Z/I Imaging).

same geographic area, but using different film and/or filter combinations. For example, a four-camera Hasselblad 70-mm configuration is shown in Figure 4-11a. By carefully selecting the film and filter combinations, specific wavelengths of light reflected from the scene can be recorded by each of the cameras.

Multiple-band aerial photography of Century City, Los Angeles, CA, is shown in Figure 4-11b. Notice the different information recorded in the individual blue, green, red, and near-infrared photographs. A natural-looking color composite of these photographs can be produced by simultaneously projecting blue light through the blue photograph onto a screen, green light through the green photograph, and red light through the red photograph. A color-infrared color composite could be created by simultaneously projecting blue light through the green photograph, green light through the red photograph, and red light through the near-infrared photograph. More will be said about color composites later in this chapter.

Panoramic Cameras

A panoramic camera uses a rotating lens (or prism) to produce a narrow strip of imagery *perpendicular* to the flight-line. Each of these long, narrow exposures will typically be vertical in the center and more oblique toward the ends. A panoramic camera intended for low-altitude use will typically pan across the flightline from one horizon to the other, giving rise to a 180° field of view in the resulting airphoto. Panoramic cameras are very common in military photo-reconnaissance, but much less so in the civilian world. Because the panoramic images are produced by dynamic motion during the exposure, the resulting airphoto does not have a rigid geometry like that resulting from a standard frame camera. Nevertheless, all types of measuring and mapping are done from panoramic imagery by organizations with the required systems and capabilities (Hooper and Gustafson, 1983).

Digital Cameras

A high-quality digital camera is shown in Figure 4-12a. It uses an area array of solid-state charge-coupled-device (CCD) detectors. The detectors are arranged in a matrix format with 1524 columns and 1024 rows.

Digital cameras also utilize a lens with its associated diaphragm to control the *f*/stop, the shutter to control the length of exposure, and a focusing device. However, the major difference is that instead of using film, a CCD area array is located at the film plane. The lens focuses the light from the outside world onto the bank of detectors. The photons of light illuminating each of the detectors cause an electrical charge to be produced that is directly related to the amount of incident radiant energy. This analog signal is then typically sampled electronically and converted into a digital brightness value ranging from 8-bit (values from 0 – 255) to 10-bit (values from 0 to 1023). The brightness values obtained from the analog-to-digital (A-to-D) conversion may be stored within the camera in computer memory or on small flash cards that can be read by standard computer systems. The charge-coupled-devices are actually more sensitive to spectral reflectance changes in the scene than the silver halide crystals used in conventional photography.

Digital cameras such as the one shown in Figure 4-12a can also produce color images. Inside the camera there is a small blue, green, and red filter wheel (or a dichroic grate). At the instant of exposure, the camera rapidly records three versions of the scene using the three filters. The result is one image based solely on blue light reflected from the terrain, another based on only green light reflected from the terrain, and a final image produced only from reflected red light. The three individual black-and-white images can be color-composited together using additive color theory to produce a natural-looking color photograph. It is also possible to make the detectors sensitive to near-infrared light.

One of the drawbacks of using certain types of digital cameras for aerial remote sensing is that it is necessary to register the three individual images using digital image processing techniques. Unlike terrestrial photography, where the photographer is normally holding the camera very still during the exposure, the aircraft is moving during the acquisition of digital aerial photography. This causes each successive exposure (e.g., blue, green, and red) to be acquired from a slightly different vantage point, which may amount to hundreds of feet in a fast-moving aircraft. Thus, the geographic area recorded on each of the individual images is different and even the common geographic area among the three images is not automatically registered.

In addition to the registration problem, there is also the issue of spatial resolution. Light (1996) found that to replicate the spatial resolution of standard 9 x 9 in. metric aerial cameras and high-quality aerial film, a digital camera would require approximately 20,000 rows by 20,000 columns of detectors. Color photography in certain digital cameras would require three banks of 20,000 by 20,000 detectors. Currently, the highest quality digital cameras available to the general public have about 2,000 by 3,000 picture elements (pixels), and these digital cameras are very expensive. Fortunately, the cost of digital cameras continues to decline. It is likely that

a.

Near-infrared (0.7 – 1.0 μm)

Red (0.6 – 0.7 μm)

Green (0.5 – 0.6 μm)

Blue (0.4 – 0.5 μm)

b.

Figure 4-11   a) Four 70-mm Hasselblad cameras arranged in a mount that may be installed in a specially designed hole in the bottom of an aircraft. All four cameras are exposed electronically at exactly the same instant. Different film and filter combinations may be used in each camera. b) Multiple-band aerial photography of Century City, Los Angeles, CA. The large panels in the parking lot were used to calibrate the photography. Notice how bright the vegetation is around the curvilinear hotel in the center of the black-and-white near-infrared photograph and how dark it appears in the blue, green, and red photographs.

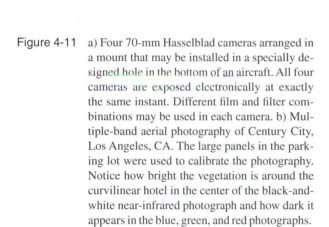

a.   b.

Figure 4-12   a) Kodak Professional DCS 420 digital camera with a Nikon camera lens and body (courtesy Eastman Kodak Company). b) A high-quality 70-mm camera is ideal for obtaining oblique aerial photography. Cameras such as this were used by the Apollo astronauts to obtain photography through portholes (note the moon insignia on the camera). Newer versions continue to be used to obtain handheld photography in space.

digital cameras will be heavily used in remote sensing of the environment in the future once the image registration issue is resolved and there are 20,000 by 20,000 pixel CCD camera systems. Examples of imagery from digital cameras are provided in Chapter 7 (Multispectral Remote Sensing Systems).

Miscellaneous Cameras

There are a variety of relatively simple, inexpensive cameras used both commercially and in a research mode that provide high-quality aerial photography. For example, extensive use is made of 35-mm cameras mounted inside a plane or hand-held by the scientist to obtain aerial photography of small research areas (Warner et al., 1996). Such systems can provide excellent, inexpensive aerial photography if properly mounted, exposed, and processed. A more expensive 70-mm camera that is ideal for handheld operation is shown in Figure 4-12b.

 Aerial Photography Filtration

In 1666, Sir Isaac Newton, while experimenting with a prism, found that he could disperse white light into a *spectrum* of colors (red, orange, yellow, green, blue, indigo, and violet — ROYGBIV) (Figure 4-13a). A color example of what takes place when white light is directed through a prism is provided in Color Plate 4-1a. Utilizing a second prism, Newton found that he could recombine the colors into white light (Figure 4-13b). He published his research on the properties of light in *Opticks* in 1704 (Figure 4-14).

Additive color theory is based on what happens when light is mixed. Color Plate 4-1b reveals that white light consists of all the colors of the visible spectrum. Black is the absence of all these colors. Blue, green, and red are the *primary colors*. Additively combining green and red light creates yellow light. Additively combining blue and green light creates cyan. Additively combining blue and red light creates magenta. Yellow, magenta, and cyan are referred to as *complementary colors* because, when paired, they produce white light, e.g., yellow plus blue, magenta plus green, and cyan plus red. Color relationships are summarized in Table 4-2.

Additive color theory is used to display images on television screens and on computer monitors. Each picture element (pixel) on a monitor screen (there are usually 480 rows by 525 lines of pixels) consists of three color guns — blue, green, and red. Each color gun's intensity in each pixel is modulated according to the amount of primary color present in the scene being transmitted from the TV station or from

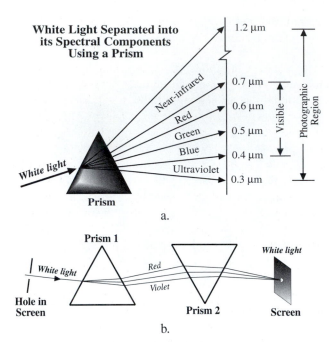

Figure 4-13   a) White light can be separated into its spectral components using a prism, as originally discovered by Sir Isaac Newton in 1666 and published in *Opticks* in 1704. Only blue, green, and red light are visible to the human eye. b) A diagrammatic representation of Newton's classic experiment in which white light is decomposed into its spectral components using one prism and then recombined back into white light using a second prism, before being projected onto a screen.

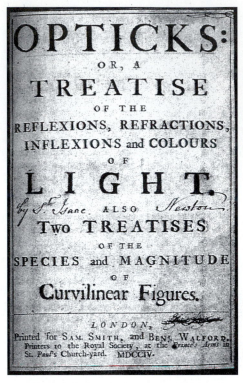

Figure 4-14   Sir Isaac Newton published *Opticks* in 1704. This work eventually led to the invention of the reflecting telescope.

Table 4-2.   The Color of an Object as Seen in White Light

Color as Seen in White Light	Colors of Light Absorbed
Blue	Green and red
Green	Blue and red
Red	Blue and green
Yellow (green and red)	Blue
Magenta (blue and red)	Green
Cyan (blue and green)	Red
Black	Blue, green, and red
White	None
Gray	Equal proportions of blue, green, and red

the computer central processing unit (Jensen, 1996). The result is a color visual display of the phenomena based on additive color theory.

Subtractive color theory is based on the use of pigments or dyes and not light. We use subtractive color theory when we paint or work with filters. For example, we know that if we mix equal proportions of blue, green, and red paint, we do not get white but obtain some dark gray color. Subtractive color theory is based on the use of the complementary color dyes — yellow, magenta, and cyan. An example of subtractive color theory is found in Color Plate 4-1c. If we projected white light onto a translucent filter made of yellow dye, the filter would subtract the blue light and allow the green and red light to pass through. Similarly, a magenta dye filter subtracts the green light and allows the blue and red to be transmitted. A cyan dye filter subtracts the red light and allows the blue and green light to be transmitted. But what if we superimposed two of these dye filters? If we superimposed the magenta dye filter and the cyan dye filter, everything but blue light would be subtracted. This is because the magenta filter subtracts the green light while the cyan filter subtracts the red light. If we superimposed the yellow and cyan filters only green light would be perceived. Finally, if we superimposed the yellow

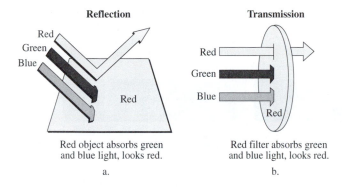

| Reflection | Transmission |

Red object absorbs green and blue light, looks red.

a.

Red filter absorbs green and blue light, looks red.

b.

Figure 4-15 a) A red object absorbs most of the blue and green incident light and reflects most of the red light toward our eyes or a remote sensing instrument. b) A red filter looks red because it selectively absorbs blue and green light while allowing red light to be transmitted through it.

and magenta dye filters, only red light would be perceived. If all three of the subtractive dyes (yellow, magenta, and cyan) were used, they would subtract (filter out) all of the white light and we would not perceive any color; it would appear black. It will be demonstrated how subtractive color dyes are used to create the colors we see in 1) developed color negatives, and 2) color positive prints.

A scientist can selectively record very specific wavelengths of light reflected from the scene onto a photographic emulsion. This is done using specific filter and film combinations. We will first investigate various filter options.

Types, Sizes, and Colors of Filters

A filter placed in front of the camera lens does exactly what the name implies, i.e., it filters out certain wavelengths of light before they can reach the film plane and expose the film. To understand how filters work, consider why a red piece of paper looks red (Figure 4-15). White light, consisting of equal proportions of blue, green, and red wavelength energy, illuminates a piece of red paper. The paper absorbs the blue and green light but reflects the red light. Therefore, anything that absorbs (subtracts) both blue and green light and reflects red light will look red. Refer to Table 4-2 for a summary of these relationships. Conversely, if we hold a red filter up to a white light, it will appear red to us because it lets through (i.e., transmits) only the red light while it absorbs both the blue and green light. This is the key to an understanding of photographic filtration. *Filters* subtract some of the light reflected from a scene before the light reaches the film in the camera. Get into the habit of thinking

of a red filter not so much as one that looks red, but as one that absorbs (subtracts) blue and green light.

Using this logic, a yellow filter absorbs blue light and allows green and red light to be transmitted (Table 4-2). Our eyes see a mixture of red and green light as yellow (i.e., the lack of blue). Due to Rayleigh scattering (Chapter 2), blue light is scattered in the atmosphere to a much greater degree than either green or red light and can therefore cause aerial photography to record considerable unwanted, scattered blue light. Therefore, it is common to use a yellow filter to selectively remove some of the scattered path radiance (especially ultraviolet and some blue light) before it ever reaches the emulsion. This *minus-blue filter* will be shown to be particularly important when collecting near-infrared aerial photography.

Most aerial photography is acquired using at least one standard filter. The spectral-transmittance characteristics of selected Kodak Wratten filters over the wavelength interval 200 to 1,100 nm (0.2 – 1.1 μm) are shown in Figure 4-16. In addition, a transmittance curve for Kodak filters HF3 and Wratten 12 are shown in Figures 4-17a and 4-17b, respectively (Kodak, 1999). These filters are important to aerial photography. When collecting natural color aerial photography, it is desirable to eliminate much of the scattering of ultraviolet radiation caused by atmospheric haze. For this purpose and to obtain a more satisfactory color balance, *haze filters* (HF) were developed that absorb light shorter than 400 nm. Similarly, when collecting color-infrared aerial photography, a yellow filter is used, which subtracts almost all of the blue light (wavelengths shorter than 500 nm). This minus-blue filter reduces the effects of atmospheric scattering and allows the proper energy to interact with each of the film's layers, to be discussed shortly. If desired, it is possible to configure a camera film/filter combination so that it selectively records a very specific band of reflected electromagnetic energy on the film. This is called spectral *band-pass filtering*. For example, if one wanted to photograph only reflected green light for a specific aerial photography project, a Kodak Wratten 58 filter could be used (refer to Figure 4-16).

In addition to the normal filters, *polarizing filters* may also be used with aerial photography. To understand how polarizing filters work, it is necessary to briefly review the nature of light. Remember that an electromagnetic quanta of light discussed in Chapter 2 vibrates in all directions perpendicular to its direction of travel as shown in Figure 4-18. When a quanta of light hits a nonmetallic surface, the vibration in only one direction, or plane, is reflected completely. Conversely, all vibrations are reflected by a bare metallic sur-

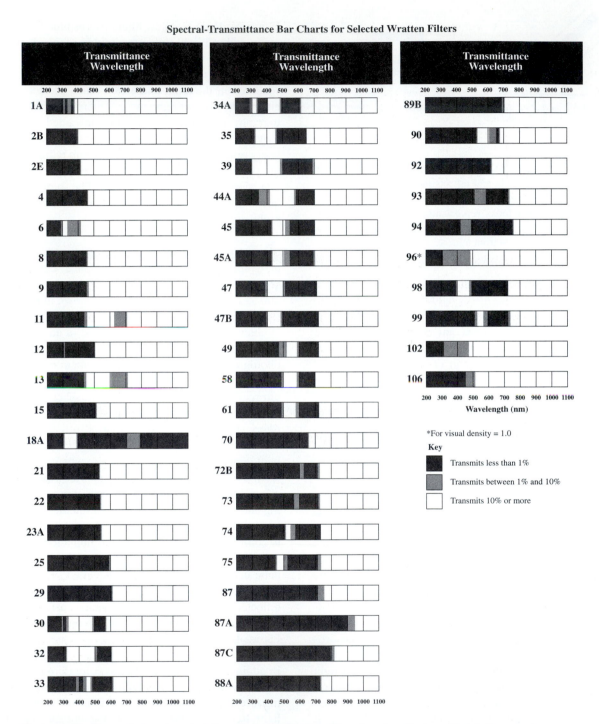

Figure 4-16   Transmission characteristics of selected Wratten filters (courtesy Eastman Kodak Company).

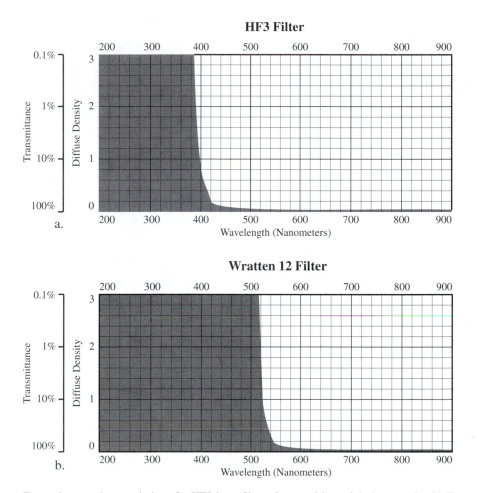

Figure 4-17 a) Transmittance characteristics of a HF3 haze filter often used in aerial photography. b) Transmittance characteristics of a Wratten 12 (minus-blue) filter used when acquiring color-infrared photography (courtesy Eastman Kodak Company).

face. Now, depending upon the angle at which the camera or our eyes are viewing the object, vibrations in other planes are reduced or eliminated completely. This reflected light — vibrating in only one plane — is called *polarized light*. The light from a blue sky is polarized because it is reflected from nonmetallic particles in the atmosphere. Likewise, light reflected from a wheat field or a body of water into the field of view of a camera is polarized since wheat and water are nonmetallic.

Sometimes it is useful to only view a certain "angle" of polarized light leaving the scene. For example, Figure 4-19 demonstrates how a polarizing filter placed in front of a camera lens will pass the vibration of a light ray in just one plane at a time. Some filters even have handles with which to adjust the filters so that they pass the light vibration in a plane parallel to the handle.

One of the most widely used applications of polarizing filters is when photographing waterbodies. By manipulating the polarizing filter, it is possible to filter out the unwanted reflections of some types of polarized light reflected from the water surface. This can improve our ability to see further into the water column. Polaroid glasses worn by people are simply polarized filters that perform much the same function. Interestingly, no one ever asks us whether we want vertically or horizontally polarized glasses, which provide different views of the world to our eyes. Polarized energy will also be important in active microwave (radar) remote sensing discussed in Chapter 9.

Thus, filters allow us to selectively filter out certain types of unwanted light while allowing very specific wavelengths of light to pass through the filter. The light that the filter transmits through it is then allowed to pass through the lens sys-

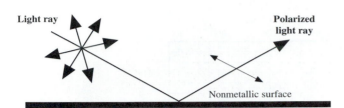

Figure 4-18   Light entering the upper portion of the atmosphere is unpolarized. It is polarized when it is reflected from nonmetallic surfaces such as atmospheric particles, water, trees, concrete, etc.

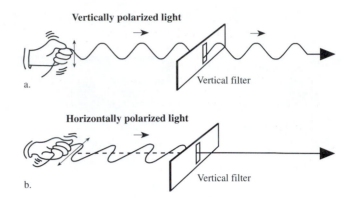

Figure 4-19   a) Vertically polarized light is allowed to pass through a vertically polarized filter. b) Horizontally polarized light is not allowed to pass through the vertically polarized filter.

tem and illuminate the light-sensitive silver halide crystals in the film emulsion.

Aerial Photography Films

Film is usually exposed in a camera mounted in an aircraft that is moving relatively fast (e.g., > 150 m.p.h.). Fortunately, films made especially for aerial photography can stop action and yield high-quality aerial photography if 1) the proper film/filter combination is used, 2) the aperture opening (*f*/stop) is appropriate, and 3) the length of the exposure is correct. Unfortunately, analysts sometimes are forced to interpret underexposed (too dark; not enough light was allowed to create an exposure) or overexposed (too bright; an excessive amount of light was allowed to create an exposure) aerial photography. Therefore, it is important for photointerpreters to understand the fundamental nature of film emulsions and their speed in order to properly interpret aerial photography or to understand why problems occur in some aerial photography.

General Characteristics of Photographic Emulsions

Cross-sections through four typical films (Figure 4-20) reveal that they consist of:

- an emulsion layer(s) containing light-sensitive silver halide crystals in a gelatin suspension,

- a base or support material composed of cellulose acetate or polyester, and

- an anti-halation layer that absorbs light that passes through the emulsion and the base to prevent reflection back to the emulsion.

The sensitivity of a photographic emulsion is a function of the size, shape, and number of *silver halide crystals* in the emulsion per unit area and the wavelengths of light to which the grains are sensitive. This is real silver, and although it is present in relatively small quantities in a roll of film, it does account for the relatively high cost of film.

Silver Halide Crystal Grain Size, Density, and Shape

To understand the relationship between grain size and density, we will depict a piece of photographic negative film as being composed of simply silver halide crystals and support backing (Figure 4-21a). Film A has 10 silver halide crystals per unit distance while Film B has only 7 crystals. As grain size increases, the total number of grains in the emulsion per unit distance decreases. If an emulsion is composed of grains one-half as large as those in another emulsion, it will require approximately twice as much light to expose it. Therefore, Film B is said to be faster than Film A because it requires less light for proper exposure. Under the scanning electron microscope, conventional silver halide crystals appear as cubes, octahedra (eight-sided solids), or irregularly shaped pebbles (Figure 4-21b). Also, note that the crystals are not all exactly the same size.

Faster films can be used advantageously when photographing objects that are moving rapidly across the film plane, as in aerial photography. Unfortunately, as sensitivity and grain size increase, the resulting image becomes more coarse, and resolution (sharpness or crispness of the image) may be reduced. Film resolution can be tested by photographing a standard test pattern that consists of groups of line pairs (parallel lines of varying thickness separated by spaces equal to

Generalized Cross-Sections of Black-and-White Panchromatic, Black-and-White Infrared, Color, and Color-Infrared Film

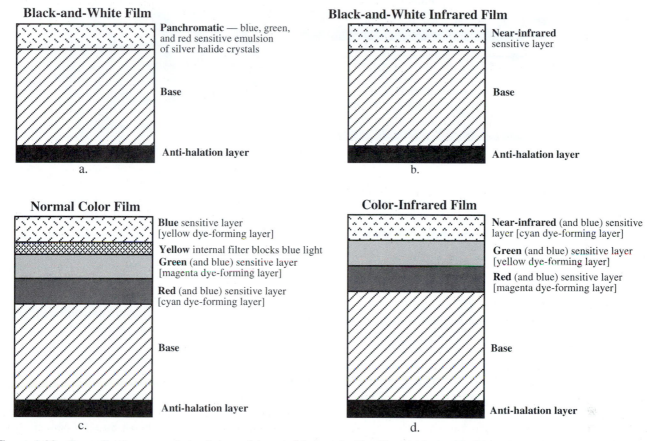

Black-and-White Film

Panchromatic — blue, green, and red sensitive emulsion of silver halide crystals

Base

Anti-halation layer

a.

Black-and-White Infrared Film

Near-infrared sensitive layer

Base

Anti-halation layer

b.

Normal Color Film

Blue sensitive layer [yellow dye-forming layer]

Yellow internal filter blocks blue light
Green (and blue) sensitive layer [magenta dye-forming layer]

Red (and blue) sensitive layer [cyan dye-forming layer]

Base

Anti-halation layer

c.

Color-Infrared Film

Near-infrared (and blue) sensitive layer [cyan dye-forming layer]

Green (and blue) sensitive layer [yellow dye-forming layer]

Red (and blue) sensitive layer [magenta dye-forming layer]

Base

Anti-halation layer

d.

Figure 4-20   Generalized cross-sections of a) panchromatic black-and-white film, b) black-and-white infrared film, c) normal or natural color film, and d) color-infrared film.

the line thickness). An example of a resolution test pattern target was shown previously in Figure 4-9. The number of lines per millimeter in the smallest line pattern that can clearly be discerned on the developed film is the resolution of the film. It is common to be able to resolve 25 to 100 line pairs per millimeter on 1:10,000 scale aerial photography positive prints. This type of spatial resolution results in aerial photography with very high spatial detail.

Black-and-White Photographic Emulsions

Just as the retina of the human eye is sensitive to different wavelengths of light ranging from blue through red (0.4 – 0.7 μm), it is possible for film manufacturers to create black-and-white photographic emulsions that have the following sensitivities:

- *orthochromatic* emulsions are sensitive to blue and green light;

- *panchromatic* emulsions are sensitive to ultraviolet, blue, green, and red light; and

- *near-infrared* emulsions are sensitive to blue, green, red, and near-infrared light.

The spectral sensitivity of standard black-and-white printing paper, panchromatic film, and near-infrared film are summarized in Figure 4-22. The diagram reveals why it is possible to use a red "safe light" in a photographic darkroom. The printing paper is simply not sensitive to red light. Conversely, black-and-white panchromatic film records ultraviolet, blue, green, and red reflected light. Therefore, a haze filter (such as the Kodak HF3 previously discussed) is often

**Two Films with Different Sizes and Densities
of Silver Halide Crystals**

**Electron Microscope Photograph
of Silver Halide Crystals**

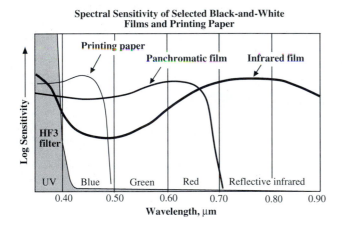

a.

b.

Figure 4-21   a) Films A and B represent hypothetical profiles of two films with different sizes and densities of silver halide crystals. b) An electron microscope photograph of actual silver halide crystals. Note that the crystals are not perfectly uniform in size, shape, or spacing (courtesy Eastman Kodak Company).

**Spectral Sensitivity of Selected Black-and-White
Films and Printing Paper**

Figure 4-22   Spectral sensitivity of black-and-white film and paper emulsions over the wavelength interval 0.35 – 0.9 µm (courtesy Eastman Kodak Company).

used to keep ultraviolet and blue light from exposing the film. Panchromatic film is the most widely used black-and-white aerial film, as it produces gray tones that are expected and recognized by human beings, i.e., water is dark, sand is white, concrete is gray, clouds are white. Much of the aerial photography acquired for photogrammetric purposes to make planimetric and topographic maps is panchromatic aerial photography. Typical panchromatic aerial photography films marketed by Eastman Kodak Company are summarized in Table 4-3.

A black-and-white panchromatic aerial photograph is shown in Figure 4-23a. A black-and-white near-infrared photograph of the same area is shown in Figure 4-23b. Care must be exercised when interpreting near-infrared photography because the camera has recorded energy that is beyond our human perception. Consequently, certain tones might not represent what we would anticipate in the image. For example, healthy green vegetation is dark on panchromatic photography because it absorbs almost 80 – 90 percent of the incident blue, green, and red radiant energy for photosynthetic purposes. Conversely, healthy vegetation is bright on the black-and-white near-infrared photography because it reflects 40 – 70 percent of the incident near-infrared energy. Human beings have very little idea how much near-infrared radiant energy is reflected from commonplace materials such as vegetation, water, and concrete. Therefore, we should constantly refer to spectral reflectance curves of these phenomena that summarize how much green, red, and near-infrared energy these objects typically reflect (refer to Chapter 2). In this manner we can understand and even predict how certain objects in the real world will look on infrared photography. It is instructive to review the process whereby simple black-and-white aerial photography negatives and positive prints are produced.

**Creating a Black-and-White Aerial
Photography Negative**

Consider the simple binary terrain depicted in Figure 4-24 that consists of calm, non-turbid ocean water and a sandy beach with no vegetation on it. Assume that the silver halide

Table 4-3.   Eastman Kodak Aerial Photography Films (courtesy Eastman Kodak Company; all film names are trademarked)

Black-and-White Aerial Films	Characteristics
Plus-X AEROGRAPHIC 2402	Medium speed, high dimensional stability for aerial mapping and reconnaissance.
Plus-X AERECON II 3404	Medium speed, fine-grain, medium- to high-altitude reconnaissance film.
Tri-X AEROGRAPHIC 2403	High speed, high dimensional stability for mapping under low illumination.
Tri-X AERECON SO-50	Similar to 2403; thin base for increased spool capacity.
Double-X AEROGRAPHIC 2405	Medium- to high-speed film for mapping and charting, dimensional stability.
Aero LX 2408	Intermediate speed, very fine-grain, medium- to high-altitude mapping and reconnaissance.
Panatomic-X AEROGRAPHIC II 2412	Intermediate speed, very fine-grain, medium- to high-altitude mapping and reconnaissance.
Panatomic-X AEROGRAPHIC II 3412	Similar to 2412; thin base; greater spool capacity for medium- to high-altitude mapping.
AERECON High Altitude 3409	Thin base, extremely fine-grain, high definition film for high-altitude, stabilized platform aerial cameras with high quality optical systems.
Infrared AEROGRAPHIC 2424	Reduces haze effects in vegetation surveys, highlights water, multispectral photography.
Color Aerial Films	
AEROCOLOR II Negative 2445	Medium speed for mapping and reconnaissance.
AEROCHROME II MS 2448	Color reversal film for low- to medium-altitude aerial mapping and reconnaissance.
AEROCHROME HS SO-358	High-speed color negative film for low-altitude aerial photography.
AEROCHROME HS SO-359	High-speed color reversal film for low- to medium-altitude mapping and reconnaissance.
AEROCHROME II Infrared 2443	False-color reversal film, high dimensional stability for vegetation surveys, camouflage detection, and Earth resource investigations.
AEROCHROME II Infrared NP SO-134	Similar to 2443; greater infrared response, suitable for altitudes above 15,000 ft.

crystals in the film in the aerial camera are sensitive to blue, green, and red light from 0.4 – 0.7 μm. The non-turbid ocean water would absorb much of the incident blue, green, and red radiant flux from the Sun. Conversely, the sandy beach would reflect much of the incident radiant flux. Some portion of this energy would be collected by the optics of the camera lens and focused onto the silver halide crystals at the film plane. When the required amount of light exposes a silver halide crystal suspended in the emulsion, the entire crystal becomes exposed, regardless of its size. Basically, the bond between the silver and the halide is weakened when silver halide crystals are exposed to light. Notice in our example that there was sufficient energy to expose the silver halide crystals on the left side of the film but insufficient energy to expose the crystals on the right side of the film. An emulsion that has been exposed to light contains an invisible image of the object called the *latent image*. To turn the latent image on the emulsion into a negative, it must be *developed*.

When the latent image is developed with the proper chemicals, areas of the emulsion that were exposed to intense light turn to free silver and become black (dense or opaque), as shown. Areas that received no light become clear if the support is the typical transparent plastic film. The degree of darkness of the developed negative is a function of the total exposure (product of illuminance and exposure time), which caused the emulsion to form the latent image.

Creating a Positive Aerial Photographic Print from a Black-and-White Negative

What good is a negative? Most people do not photointerpret *negatives*, because they are a reversal of both the tone and geometry of the real world, as shown in Figure 4-24. Therefore, it is customary to produce a positive print of the scene from the negative (Figure 4-25). A *positive print* is produced by placing the developed negative in an enlarger with the emulsion side of the negative facing the light source. White light is transmitted through the negative, passed through the enlarger lens, and onto photographic paper that has its own silver halide crystal sensitivity and film speed. In this case, the dense (dark) beach area on the negative allows very little radiant flux to pass through the negative while the clear (ocean) area on the negative allows a lot of radiant flux to

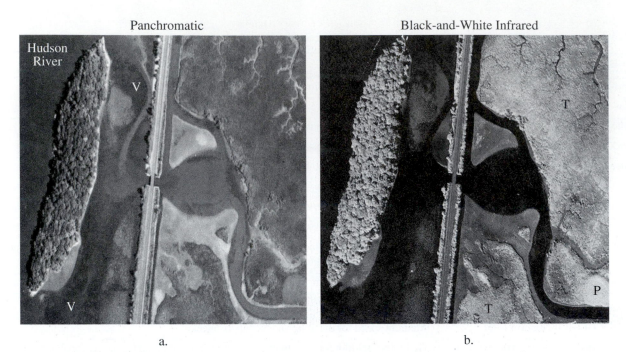

Panchromatic

Black-and-White Infrared

Hudson River

a.

b.

Figure 4-23 a) Black-and-white panchromatic photograph of Tivoli North Bay on the Hudson River, NY. Vegetation is recorded in relatively dark tones because it absorbs much of the incident green and red energy. Some green and red energy passes through the water column and provides information on suspended sediment patterns and submerged aquatic vegetation (V = wild celery, *Vallisneria americana*). b) A black-and-white near-infrared photograph depicts the land-water interface well, because water absorbs most of the incident near-infrared energy causing it to appear dark. Conversely, deciduous trees on the island and tidal wetland vegetation (T = cattail, *Typha angustifolia*; P = common reed, *Phragmites aus.*) are bright because healthy vegetation reflects much of the incident near-infrared energy (Berglund, 1999) (please refer to Color Plate 4-3ab).

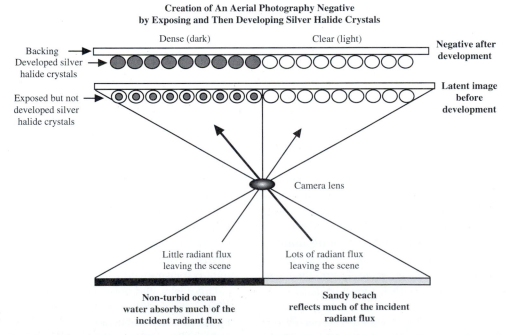

Figure 4-24 A diagrammatic representation of how a black-and-white negative of an ocean-beach scene is exposed and developed.

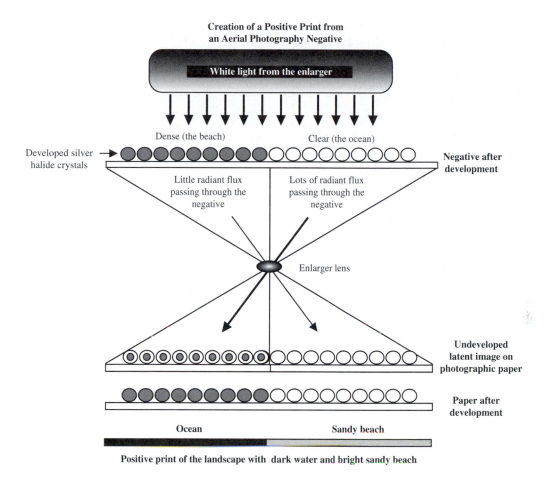

Creation of a Positive Print from
an Aerial Photography Negative

White light from the enlarger

Dense (the beach)   Clear (the ocean)

Developed silver halide crystals →   **Negative after development**

Little radiant flux passing through the negative   Lots of radiant flux passing through the negative

Enlarger lens

Undeveloped latent image on photographic paper

Paper after development

Ocean   Sandy beach

Positive print of the landscape with dark water and bright sandy beach

Figure 4-25   Diagrammatic representation of how a positive black-and-white print of the ocean-beach scene is produced from a developed negative.

pass through it. The energy is focused through the lens onto the undeveloped photographic paper, and once again the individual silver halide crystals are either completely exposed or unexposed, forming a latent image on the photographic paper. After development, the exposed ocean area becomes dark on the paper print and the sand beach area becomes light. Thus, we now have the correct tone and geometry of the terrain in the form of a positive paper print that can be visually interpreted. If we want a positive film transparency to view, then positive transparency material is substituted for the photographic paper. This typically creates an airphoto with improved fine detail and more gray tones.

Advanced Radiometric Characteristics of Black-and-White Films

Transmittance: The ability of a portion of a developed film to pass light is called its *transmittance* ($T_{i,j}$). A black portion of the film may transmit no light, while a clear por-

tion of the film may transmit almost 100 percent of the incident light. Therefore, the transmittance at location i,j in the photograph is:

$$T_{i,j} = \frac{\text{light passing through the film}}{\text{total incident light}} \tag{4-2}$$

Opacity: There is an inverse relationship between transmittance and how opaque an area on the film is. Something that is very opaque does not transmit light well. *Opacity* ($O_{i,j}$) is the reciprocal of transmittance:

$$O_{i,j} = \frac{1}{T_{i,j}} \tag{4-3}$$

Density: Transmittance and opacity are two good measures of the darkness of any portion of a developed negative. However, psychologists have found that the human visual system does not respond linearly to light stimulation, but rather we respond logarithmically. Therefore, it is common to use *density* $(D_{i,j})$ which is the common logarithm of opacity, as our measure of choice.

$$D_{i,j} = \log_{10} O_{i,j} = \log\left(\frac{1}{T_{i,j}}\right) \qquad (4\text{-}4)$$

If 10 percent of the light can be transmitted through a film at a certain *i,j* location, transmittance is 1/10, opacity is 1/0.10 or 10, and density is the common logarithm of 10 or 1.0. Table 4-4 summarizes the general relationship between transmittance, opacity, and density values.

Characteristic Curve: Numerous variables, such as the length of the exposure, length of film development, type of film emulsion, speed of the film, etc., can cause differences in how a given object is recorded on the film and its density. Therefore, it is important to understand the nature of the characteristic curve of films and how this impacts the creation of properly exposed areas on the photographic negative.

The *characteristic curves* of two hypothetical black-and-white negative films, X and Y, are shown in Figure 4-26. These are called *D log E* curves where the density of the portion of the film under investigation is plotted on the y-axis (already in logarithmic form, as previously discussed) and the total exposure of the film is plotted in logarithmic units on the x-axis. We will use relative log Exposure units instead of ergs/cm^2 or meter-candle-second used in photometric science. Different films have different D log E curves, but the shapes are generally similar, consisting of three parts — the *toe*, a *straight-line* or *linear section*, and the *shoulder*. The curves provide important information about the film emulsion.

Unfortunately, even if no exposure has been made (e.g., an aerial photograph has not even been taken yet), there is some density already present in the film. The density of the unexposed emulsion is called *fog*. There is also some density from the film base material. *Gross fog* is the sum of the unexposed density from the film base and the emulsion *fog*. The *gross fog* level is located in the lower position of the toe and labeled D_{min} in Figure 4-26. In order to begin to create an image of value in the photograph, we must have an exposure greater than D_{min}.

At this point it is useful to give a formal definition of *exposure* (E), which is a function of several factors previously discussed, including *f*/stop (i.e., *f/d*), the amount of radiant energy coming from the scene (*s*) measured in Joules mm^{-2} sec^{-1}, and the exposure time (*t*) in seconds:

Table 4-4.   Relationship Between Transmittance, Opacity, and Density

Percent Transmittance	Transmittance $T_{i,j}$	Opacity $O_{i,j}$	Density $D_{i,j}$
100	1	1	0.00
50	0.50	2	0.30
25	0.25	4	0.60
10	0.10	10	1.00
1.0	0.01	100	2.00
0.1	0.001	1000	3.00

$$E = \frac{st}{4 \cdot \left(\frac{f}{d}\right)^2}. \qquad (4\text{-}5)$$

As exposure begins, the density increases curvilinearly from point *a* to point *b*. This is called the *toe* of the curve. If objects are recorded with just this length of exposure, they may be underexposed. As the length of the exposure increases, there is a portion of the curve from *b* to *c* where the increase in density is nearly linear with changes in the log of exposure. This is called the *linear* or straight line part of the characteristic curve. High-quality aerial photography may be obtained if the exposure is made in the linear portion of the curve and perhaps a small part of the toe. As the length of the exposure time increases from *c* to *d*, density increases at a decreasing curvilinear rate. This is the *shoulder* of the curve. The maximum density of the film is D_{max}. Objects recorded in the shoulder of the curve are usually overexposed.

The slope of the linear section of the D log E curve is called *gamma* (γ) and provides valuable information about the *contrast* of the film. The greater the slope of the line, the higher the gamma and the higher the contrast of the film. Notice that films X and Y in Figure 4-26 have significantly different gammas. If the same length of exposure (e.g., 1.6 relative log Exposure units) were applied to both films, Film X would have a greater range of density (approximately 2.2) while Film Y would only yield a density of approximately 0.5 at this same exposure. Hopefully, Film X would provide a full continuum of grays from white to black, with plenty of gray tones in between at this exposure. Conversely, Film Y might not have such a full range of gray tones at this exposure. We often want to retain the subtle gray tones in aerial photography because most natural landscapes (soils, rocks, vegetation, water) are relatively low in contrast.

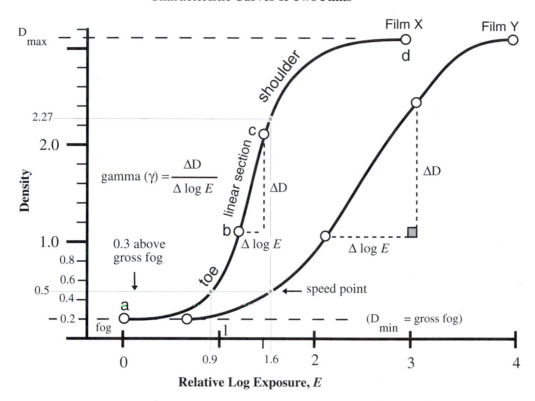

Figure 4-26 Characteristic curves of two hypothetical black-and-white films. To obtain densities in the developed negative of 0.5, Film X requires a relative log Exposure of 0.9 units while Film Y requires 1.6 units. Thus, Film X is *faster* than Film Y. This means that we could obtain the same densities in Film X using a shorter exposure time. This could be very important if we want to stop action when collecting aerial photography to obtain quality aerial photographs. Film X also has a much steeper gamma (γ), meaning that it produces a much greater range of density (i.e., contrast) than Film Y in the linear (straight-line) section of the D log E curve.

This brings us to *film speed* once again. Some films are more sensitive to light than others. In our example, it appears that Film X is the faster film. It requires a shorter exposure time to produce a wider range of density (and therefore contrast). This can be of value for aerial photography because we may want to "stop" the action as the aircraft flies above the terrain at hundreds of miles per hour. Unfortunately, fast films require larger silver halide crystals that can also reduce the spatial resolution of the film and impede our ability to resolve objects that are spectrally different yet spatially near to one another. Conversely, Film Y would require a longer exposure to produce the same range of density. It would probably be composed of smaller silver halide crystals. If Film Y were properly exposed, it might provide high spatial resolution and subtle grayscale information. But if Film Y was not fast enough to stop action, the result might be blurred aerial photography. Obviously there must be some quantitative way of comparing one film speed with another.

Every film speed system requires the selection of a *speed point* to compare the speed of two films. For aerial photography films, the speed point is "density = 0.3 + density above gross fog" (Kodak, 1988). Therefore, since gross fog is 0.2 for both hypothetical aerial films, the speed point for both films is 0.3 + 0.2 = 0.5, as shown in Figure 4-26. The 0.5 density line intersecting the two films can be used to determine how much faster one film is than another. For example, to produce a density of 0.5 on Film X, approximately 0.9 units of relative log Exposure are required. To produce the same density (0.5) on Film Y, approximately 1.6 units of relative log Exposure are required. Thus, Film X is faster than Film Y by approximately 0.7 relative log units of Exposure. For every 0.3 log E, there is one *f*/stop difference in exposure (Kodak, 1988). Thus, these two films differ in speed by greater than two *f*/stops (i.e., 0.7/0.3 = 2.33). If we required a great range of image density in our photographs such as a value of 2.0 (i.e., well-exposed blacks, whites, and intermediate gray tones) but we had to have short exposure times to stop action, as is common in aerial photography, then we might want to select Film X because it would provide the density range we desire with a much shorter relative log Exposure than Film Y. It is possible to compute the exact film speed (*FS*) of an aerial film using the formula (Kodak, 1988):

$$FS = \frac{3}{2E} \qquad (4\text{-}6)$$

where E is the exposure measured in meter-candle-seconds (MCS) at the point on the characteristic curve where the density is 0.3 units above D_{min} (i.e., gross fog), as discussed.

Sometimes we acquire handheld aerial photography using non-aerial photography films. The speed of these films is determined by the American Standards Association (ASA) film speed system that assigns a number to a film that is approximately equal to the inverse of the shutter speed (in seconds) required for proper exposure in good Sunlight for a lens opening of *f*/16. Therefore, if a film is properly exposed in good Sunlight at *f*/16 and 1/200 second, it is classified as an ASA 200 film (Wolf, 1983). The original Kodacolor Film introduced in 1942 had an ASA speed of just 25. That speed jumped to ASA 32 in 1955 and to ASA 64 in 1963 with the introduction of Kodacolor-X Film. In 1977, Kodak improved Kodacolor II Film to an ASA of 100 and introduced Kodacolor 400 Film. In 1983, the Kodacolor VR 1000 film was introduced with an ASA of 1000, the most significant single advance in silver halide technology in more than 50 years.

If one must plan an aerial photography mission, the *Kodak Aerial Exposure Computer* can be used to determine the most appropriate lens aperture (*f*/stop) and shutter speed once information is provided on the following variables: effective aerial film speed, haze factor, speed of aircraft, altitude above-ground-level, latitude, and time of day.

Digitizing Black-and-White (and Color) Film

The density (*D*) characteristics of a negative or positive transparency film can be measured using a *densitometer*. There are several types of densitometers, including flatbed and drum microdensitometers, video densitometers, and linear or area array charge-coupled-device densitometers.

Microdensitometer Digitization: The characteristics of a typical *flatbed microdensitometer* are shown in Figure 4-27. This instrument can measure the density characteristics of very small portions of a negative or positive transparency, down to just a few micrometers in size, hence the term

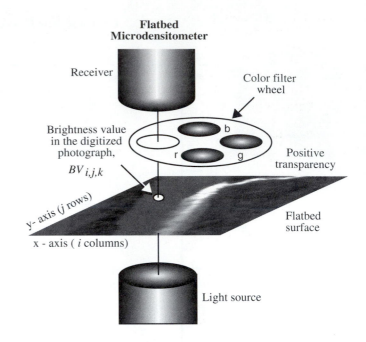

Figure 4-27   Schematic of a flatbed microdensitometer. A black-and-white negative or positive transparency is converted from an analog photographic image into a single matrix of digital brightness values, $BV_{i,j,k}$. A color negative or positive transparency is separated into three registered matrices based on the density of the three dyes (yellow, magenta, and cyan) found at each location in the photography. The spot size that is analyzed during the digitization process may be as small as a few micrometers; hence the term *microdensitometer*.

microdensitometer. Basically, a known quantity of light is sent from the light source toward the receiver. If the light encounters a very dense portion of the film, very little light will be transmitted to the receiver. If the light encounters a very clear portion of the film, then much of the light will be transmitted to the receiver. The densitometer can output the characteristics at each *i,j* location in the photograph in terms of transmittance, opacity, or density, as previously discussed. The amount of light recorded by the receiver is more commonly converted into a digital brightness value designated, $BV_{i,j,k}$ which refers to the location in the photograph at row *i* and column *j* and band *k*. At the end of each scan line, the light source steps in the *y* direction some Δy to scan along a line contiguous and parallel to the previous one. As the light source is scanned across the image, the continuous output from the receiver is converted to a series of discrete numerical values on a pixel-by-pixel basis. This analog-to-digital (A-to-D) conversion process results in a matrix of values that is usually recorded in 8-bit bytes (values ranging

Table 4-5. Relationship Between Digitizer Instantaneous Field of View (IFOV) Measured in Dots per Inch or Micrometers, and the Pixel Ground Resolution at Various Scales of Photography

Digitizer Detector IFOV		Pixel Ground Resolution at Various Scales of Photography (meters)					
Dots per inch	Micrometers	1:40,000	1:20,000	1:9,600	1:4,800	1:2,400	1:1,200
100	254.00	10.16	5.08	2.44	1.22	0.61	0.30
200	127.00	5.08	2.54	1.22	0.61	0.30	0.15
300	84.67	3.39	1.69	0.81	0.41	0.20	0.10
400	63.50	2.54	1.27	0.61	0.30	0.15	0.08
500	50.80	2.03	1.02	0.49	0.24	0.12	0.06
600	42.34	1.69	0.85	0.41	0.20	0.10	0.05
700	36.29	1.45	0.73	0.35	0.17	0.09	0.04
800	31.75	1.27	0.64	0.30	0.15	0.08	0.04
900	28.23	1.13	0.56	0.27	0.14	0.07	0.03
1000	25.40	1.02	0.51	0.24	0.12	0.06	0.03
1200	21.17	0.85	0.42	0.20	0.10	0.05	0.03
1500	16.94	0.67	0.34	0.16	0.08	0.04	0.02
2000	12.70	0.51	0.25	0.12	0.06	0.03	0.02
3000	8.47	0.33	0.17	0.08	0.04	0.02	0.01
4000	6.35	0.25	0.13	0.06	0.03	0.02	0.008

Useful Scanning Conversions:

DPI = dots per inch;   µm = micrometers;   I = inches;   M = meters
From DPI to micrometers: µm = (2.54 / DPI)10,000
From micrometers to DPI: DPI = (2.54 / µm)10,000
From inches to meters: M = I × 0.0254
From meters to inches: I = M × 39.37

Computation of Pixel Ground Resolution:

PM = pixel size in meters;   PF = pixel size in feet;   S = photo scale
Using DPI: PM = (S/DPI)/39.37   PF = (S/DPI)/12
Using micrometers: PM = (S × µm) 0.000001   PF = (S × µm) 0.00000328
For example, if a 1:6,000 scale aerial photograph is scanned at 500 DPI, the pixel size will be (6000/500)/39.37 = 0.3048 meters per pixel or (6000/500)/12 = 1.00 foot per pixel. If a 1:9,600 scale aerial photograph is scanned at 50.8 µm, the pixel size will be (9,600 × 50.8)(0.000001) = 0.49 meters or (9,600 × 50.8)(0.00000328) = 1.6 feet per pixel.

from 0 to 255). These data are then stored on disk or tape for future analysis.

Scanning imagery at spot sizes <12 µm may result in noisy digitized data, because the spot size approaches the dimen-sion of the film's silver halide crystals. Table 4-5 summarizes the relationship between digitizer scanning spot size (IFOV) measured in dots per inch or micrometers and the pixel ground resolution at various scales of aerial photography or imagery.

A simple black-and-white photograph has only a single band, $k=1$. However, we may need to digitize color photography. In such circumstances, we use three specially designed filters that determine the amount of light transmitted by each of the dye layers in the film (Figure 4-20). The negative or positive transparency is scanned three times ($k = 1, 2$ and 3), each time with a different filter. This extracts spectral information from the respective dye layers found in color and color-infrared aerial photography and results in a registered three-band digital data set for subsequent image processing.

Video Digitization: It is possible to digitize hard-copy imagery by sensing it through a video camera and then performing an analog-to-digital conversion on the 525 lines by 512 rows of data that are within the standard field of view (as established by the National Television System Committee). Video digitizing involves freezing and then digitizing a frame of analog video camera input. A full frame of video input can be read as rapidly as 1/30-sec. A high-speed analog-to-digital converter, known as a *frame grabber*, digitizes the data and stores them in a buffer memory. The memory is then read by the host computer and the digital information stored on disk or tape.

Video digitization of hard-copy imagery is performed very rapidly, but the results are not always useful for digital image processing purposes. For example, there are dramatic differences in the radiometric sensitivity and repeatability of various video cameras. A serious problem is vignetting (light fall-off) away from the center of the image being digitized. This can affect the spectral signatures extracted from the scene. Also, any distortion in the vidicon optical system will be transferred to the digital remote sensor data, making it difficult to edge-match between adjacent images that have been digitized in this manner.

Linear and Area Array Charge-Coupled-Device (CCD) Digitization: Advances in the personal computer industry have spurred the development of flatbed, desktop linear array digitizers based on linear array charge-coupled-devices (CCDs) that can be used to digitize hard-copy negatives, paper prints, or transparencies at 300 to 3,000 pixels per inch (Figure 4-28ab). The hard-copy photograph is placed on the glass. The digitizer optical system illuminates an entire line of the hard-copy photograph at one time with a known amount of light. A linear array of detectors records the amount of light reflected from or transmitted through the photograph along the array and performs an A-to-D conversion. The linear array is stepped in the y direction, and another line of data is digitized. It is possible to purchase desktop color scanners for under $500. Many digital image processing laboratories use these inexpensive desktop digitizers to convert hard-copy remotely sensed data into a digital format. Desktop scanners provide surprising spatial precision and a reasonable characteristic curve when scanning black-and-white images. An optional "transilluminater" can be purchased for the backlighting of film (or 35-mm slide) to be scanned. Unfortunately, most desktop scanners are designed for 8.5×14 in. originals, and most aerial photographs are 9×9 in. Under such conditions, the analyst must digitize the 9×9 in. photograph in two sections (e.g., 8.5×9 in. and 0.5×9 in.) and then digitally *mosaic* the two pieces together. The mosaicking process can introduce both geometric and radiometric error.

An area array consisting of 2,048 by 2,048 charge-coupled-devices (CCDs) is shown in Figure 4-28c. CCD digital camera technology has been adapted specifically for remote sensing image digitization. For example, the scanner shown in Figure 4-28d digitizes from 160 dpi to 3,000 dpi (approximately 160 µm to 8.5 µm) over a 10×20 in. image area (254 mm $\times 508$ mm). The system scans the film (ideally the original negative) as a series of rectangular image segments, or tiles. It then illuminates and scans a *reseau grid*, which is an array of precisely located crosshatches etched into the glass of the film carrier. The reseau grid coordinate data are used to locate the exact orientation of the CCD camera during scanning and to geometrically correct each digitized tile of the image relative to all others. Radiometric calibration algorithms are then used to compensate for uneven illumination encountered in any of the tile regions. When scanning a color image, the scanner stops on a rectangular image section and captures that information sequentially with each of four color filters (blue, green, red, and neutral) before it moves to another section. Most other scanners digitize an entire image with one color filter and then repeat the process with the other color filters. This can result in color misregistration and loss of image quality. Area array digitizing technology has obtained geometric accuracy of ≤ 5 µm over 23×23 cm images when scanned at 25 µm per pixel, and repeatability of ≤ 3 µm.

Hopefully there is a relationship between the brightness value ($BV_{i,j,k}$) or density ($D_{i,j}$) at any particular location in the film and the energy reflected from the real-world object space ($O_{x,y}$) at the exact location. Scientists take advantage of this relationship by 1) making careful *in situ* observations in the field, such as the amount of biomass for a 1 x 1 m spot on the Earth located at $O_{x,y}$ and then 2) measuring the brightness value ($BV_{i,j,k}$) or density of the object at that exact location in the photograph using a *densitometer*. If enough samples are located in the field and in the photography, it may be possible to develop a correlation between the real-

Linear Array CCD

a.

Linear Array CCD Flatbed Digitizer

b.

Area Array CCD

c.

Area Array CCD Image Digitizer

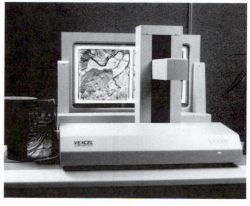

d.

Figure 4-28   a) Enlarged view of a 2,048 element charge-coupled-device (CCD) linear array. b) Inexpensive desktop linear array CCD digitizer. c) Enlarged view of a 2,048 x 2,048 area array CCD. d) An image digitizer based on area array CCD technology (courtesy Vexcel, Inc.).

world object space and the image space. This is an important use of digitized aerial photography.

Digitized National Aerial Photography Program (NAPP) Data: The NAPP was initiated in 1987 as a replacement for the National High Altitude Aerial Photography (NHAP) Program. The objective of the NAPP is to acquire and archive photographic coverage of the conterminous United States at 1:40,000 scale using either color-infrared or black-and-white film. The photography is acquired at an altitude of 20,000 ft above-ground-level (AGL) with a 6-in. focal-length metric camera. The photography is acquired ideally on a five-year cycle, resulting in a nationwide photographic database that is readily available through the EROS Data Center in Sioux Falls, SD, or the Aerial Photography Field Office in Salt Lake City, UT (USGS, 1994).

This high spatial resolution NAPP photography represents a wealth of information for on-screen photointerpretation and

can become a high-resolution base map upon which other GIS information (e.g., parcel boundaries, utility lines, and tax data) may be overlaid after it is digitized and rectified to a standard map projection. Light (1993) summarized the optimum methods for converting the NAPP data into a national database of digitized photography that meets National Map Accuracy Standards. Microdensitometer scanning of the photography, using a spot size of 15 µm, preserves the 27 resolvable line pair per millimeter (lp/mm) spatial resolution in the original NAPP photography. This process generally yields a digital dataset that has a ground spatial resolution of 1 x 1 m, depending on original scene contrast. This meets most land-cover and land-use mapping user requirements.

The digitized information can be color separated into separate bands of information if desired. The 15 µm scanning spot size will support most digital soft-copy photogrammetry for which coordinate measurements are made using a computer and the monitor screen (Light, 1993). Because the digitized NAPP data are so useful as a high spatial resolution GIS base map, many states are entering into cost-sharing relationships with the U.S. Geological Survey and having their NAPP coverage digitized and output as digital ortho-photomaps. A large amount of NAPP data have been digitized and converted into digital orthophotoquads. Much of the digital data is available and can be browsed and ordered at the USGS orthophotoquad Website in Appendix A.

Color Photographic Emulsions

Normal color and color-infrared photographic emulsions are heavily used in photogrammetry and photointerpretation.

Normal Color Aerial Photography

Normal color photography records energy in the region from 0.4 – 0.7 μm (blue, green, and red light) and depicts the terrain in the same hues (colors) as our eyes perceive the landscape. A haze filter is normally used to prevent ultraviolet light from exposing the film (e.g., Kodak HF3). Color film emulsions generally consist of three layers of silver halide crystals arranged as shown in Figure 4-20c. The top layer is sensitive to blue light, the second layer is sensitive to blue and green light, and the bottom layer is sensitive to red and blue light. To prevent blue light from exposing the bottom two layers, a yellow blue-blocking filter is placed in the emulsion between the top two layers. The spectral sensitivity of the blue, green, and red emulsion layers is shown in Figure 4-29a. Color aerial photography films marketed by Eastman Kodak Company are summarized in Table 4-3.

Blue, green, and red light reflected from the various objects in the scene interact with and activate the blue, green, and red-sensitive silver halide crystals in the film, forming a latent image. A color rendition of the process is shown in Color Plate 4-2a.

Color film development is exactly like the first step of black-and-white film development, i.e., the exposed silver halide crystals in each layer are turned into black crystals of silver. The remainder of the process, however, depends on whether the film is color negative or color reversal film. With *color negative film*, a negative is produced and color prints are made from the negative. *Color reversal film* produces a true color transparency directly on the film. This process is used for making slides or positive transparencies.

The second step in the creation of a positive normal color photograph from a color negative involves the use of special

Table 4-6.   Primary and Complementary Colors

Primary Colors	Complementary Colors
blue	yellow
green	magenta
red	cyan

dyes. Silver halides in each layer that turned black in the first step are replaced with *dyes* of the *complementary colors* of the layer (Color Plate 4-2a; Table 4-6). Dense black silver grains in the blue-sensitive layer are replaced with yellow dye (yellow is the complementary color of blue and is composed of green and red light). *However, the amount of dye placed at each location in the film layer is inversely proportional to the intensity of light that illuminated that location.* Therefore, the more blue light that illuminated the area, the less yellow dye that will be present. Black silver crystals in the green-sensitive layer are replaced with magenta dye. Black silver crystals in the red-sensitive layer are replaced with cyan dye. This process results in a negative that is composed of yellow, magenta, and cyan dyes. This is the reason that when we look at a color negative we see strange colors (yellow, magenta, and cyan) instead of the traditional blue, green, and red found in the landscape. We are not used to seeing nature in hues of yellow, magenta, and cyan; therefore, we rarely photointerpret color negatives.

To produce a color positive print from the negative that we can photointerpret, white light is projected through the color negative to expose a three-layered color printing paper emulsion. The color negative acts as a filter, exposing the three layers on the photographic paper to the three colors yellow, magenta, and cyan. When developing the positive print paper, however, the complementary colors of yellow, magenta, and cyan (i.e., blue, green, and red) are produced. This second color reversal yields the original colors of the scene on the positive print that are then suitable for photointerpretation (Color Plate 4-2a). A normal color aerial photograph of Tivoli North Bay, NY is shown in Color Plate 4-3a.

Color-Infrared Aerial Photography

Color-infrared film was originally developed during World War II, when there was great interest in increasing the sensitivity of films in the infrared region of the spectrum to detect camouflage. In 1941, S. Q. Duntley of the Massachusetts Institute of Technology headed up a study on optical and visual camouflage detection. The research concluded that the vegetation chlorophyll absorption band located in the red spectral region (0.60 – 0.68 μm) and the high amount of

Spectral Sensitivity of the Three Layers of Normal Color Film

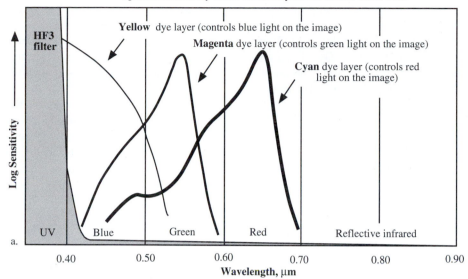

a.

Spectral Sensitivity of the Three Layers of Color-Infrared Film

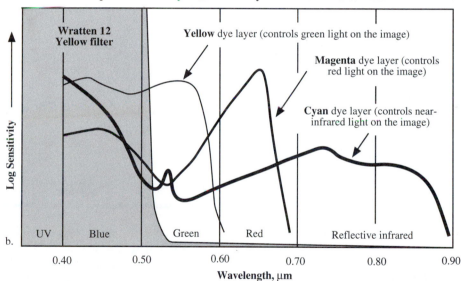

b.

Figure 4-29 a) The spectral sensitivity of the three dye layers of normal color film. A haze filter (e.g., HF3) may be used that filters out much of the light less than 0.4 μm. b) The spectral sensitivity of the three dye layers of color-infrared film. A Wratten 12 (yellow) filter is often used, which filters out much of the light shorter than 0.5 μm (courtesy Eastman Kodak Company).

near-infrared reflectance from vegetation in the region > 0.7 μm would be difficult to be replicated in a camouflage paint. Therefore, in 1942, Duntley requested that Eastman Kodak Company, produce a false-color camouflage-detection film. Dr. Mannes of the Eastman Kodak Company demonstrated the test results on October 19, 1942. It was during this time that the decision was made for foliage to appear red in the finished color-infrared aerial photography. It was made available to the military under the name "Aero Kodacolor Reversal Film, Camouflage Detection" (Colwell, 1997). The

film was successful in the war for locating targets covered with camouflage netting or paint. It has also proven to be of significant value for monitoring the condition and location of vegetation in Earth resource surveys. Color-infrared aerial photography films marketed by Eastman Kodak Company, are summarized in Table 4-3. The most popular is Kodak AEROCHROME II Infrared Film 2443.

Color-infrared film records reflected energy in the region from just below 0.3 μm to just above the 0.9 μm (ultraviolet, blue, green, red, and near-infrared) and portrays the energy on the film in false colors (Rundquist and Sampson, 1988; Kodak, 1999). Like normal color film, color-infrared film also has three emulsion layers, each sensitive to a different part of the spectrum (Figure 4-20d). The top layer has a sensitivity from ultraviolet through near-infrared. The middle layer is most sensitive to green and blue light. The bottom layer is most sensitive to blue and red light.

The color-infrared film is exposed through a dark yellow filter (usually a Kodak Wratten 12 minus-blue filter) that allows the green, red, and near-infrared light from the terrain to expose the film while preventing light shorter than 0.5 μm from reaching the film. Color Plate 4-2b depicts how the dyes introduced into each of the film layers are different when compared to the development of a standard color photo. Basically, they are "offset by one," meaning that the green-sensitive layer receives yellow dye, the red-sensitive layer receives magenta dye, and the infrared-sensitive layer receives cyan dye. Thus, when white light is eventually projected through this negative and a positive print is produced, anything that reflected *only* green light in the scene shows up as bright blue, anything that reflected *only* red light in the scene will appear in bright shades of green, and anything in the scene that reflected *only* near-infrared energy will be recorded in bright shades of red. If something reflected equal proportions of red and near-infrared light but no green light, then it would show up in shades of yellow on the color-infrared film. Similarly, equal amounts of green, red, and near-infrared reflected from the scene, such as from a concrete surface, will yield a medium gray tone.

Thus, color-infrared film exhibits a *color balance shift* when compared to traditional color aerial photography. What can be confusing to interpreters is the fact that as human beings we have no idea how much near-infrared energy is reflected by certain phenomena since we cannot see or appreciate energy in this region. For example, a color-infrared aerial photograph is shown in Color Plate 4-3b. This photograph was obtained using Kodak AEROCHROME II Infrared 2443 film with a 6 in. focal-length lens. Deep, clear water absorbs almost all of the near-infrared radiant flux incident to it while reflecting somewhat more green and red light. Therefore, if the water is deep and free of suspended sediment or any organic matter, it will appear dark or almost black on color-infrared film. However, if there is substantial suspended sediment in the water, it may appear in relatively dark shades of blue and green on the near-infrared photography. Conversely, vegetation absorbs approximately 80 percent of the green and red radiant flux incident to it while it reflects approximately 40 – 70 percent of the incident near-

infrared radiant flux. This condition causes healthy green vegetation to be recorded in shades of red because the near-infrared energy is dominant. Concrete streets or gravel railroad beds reflect approximately equal proportions of green, red, and near-infrared energy and are recorded as a "steel gray" signature on the color-infrared photography.

Similar color reversals are evident in the example of a building at the University of Nebraska photographed using color and color-infrared film (Color Plate 4-4ab). Objects that reflect blue light exclusively are recorded in black on the color-infrared photograph (e.g., the car on the left). The red carpet on the steps of the building is recorded in shades of green. The tile roof is recorded in shades of yellow, meaning that the tile absorbs much of the green light while reflecting heavily in the red and near-infrared portions of the spectrum.

Color-infrared aerial photography will continue to be one of the most important remote sensing tools. In fact, it is the emulsion of choice for the USGS National Aerial Photography Program. It will be interesting to see if the new high spatial resolution ($\leq$ 1 x 1 m panchromatic and 4 x 4 m multispectral) satellite remote sensor data from commercial vendors reduces our dependency on large-scale color-infrared photography (Fritz, 1996).

Planning Aerial Photography Missions

Time of Day — Sun Angle

The ideal time of day to obtain aerial photography is when the Sun is between 30° and 52° above the horizon, i.e., within two hours of solar noon. A Sun angle less than 30° may not allow proper illumination or adequate reflectance from the scene. This can result in underexposed aerial photography that is difficult to interpret. Also, photography taken at low Sun angles is usually unacceptable because of the extremely long shadows cast by trees and buildings that can obscure valuable information. However, some scientists (especially geologists) prefer low Sun angle photography to enhance terrain representation (Lyon and Falkner, 1995).

A Sun angle greater than 52° may yield *hotspots* on the photography. These are unusually bright areas on the photograph, which are a function of Sun and camera orientation. They become especially bothersome if there is a lot of water in the study area.

Photographic flightlines can be oriented in any direction as long as terrain and safety factors are considered. The angle of the Sun at a particular latitude and specific time of day may be obtained from solar ephemeris tables.

Weather

Aerial photography is ideally collected a few days after the passage of a frontal system, which removes water vapor and particulates (especially smog) from the air and allows the wind associated with the frontal system to decrease. Aerial photography obtained in very humid conditions is degraded because of atmospheric scattering of unwanted light into the field of view of the camera. Also, the water vapor may absorb some of the signal we are trying to record on film.

Strong winds can cause the plane to drift off course, resulting in poor flightline coverage, loss of stereoscopic overlap, and sometimes oblique rather than vertical photography. Clouds in aerial photography, of course, are unacceptable. Not only is a portion of the terrain obscured by a cloud, but the cloud shadow causes tonal variations that might be interpreted incorrectly. Aerial photography obtained prior to the passage of a frontal system may be ideal if the relative humidity is low and the winds are not strong.

Flightline Layout

To obtain the desired aerial photography, it is necessary to know the desired photo scale (e.g., 1:10,000), the scale of the base map on which all information is to be plotted (e.g., 1:12,000), the x,y coordinates of the four corner points of the study area (latitude, longitude or UTM), the size of the geographic area to be photographed (e.g., 100 km^2), the average forward overlap of each frame of photography (e.g., 60 percent), the average sidelap of each frame (e.g., 20 percent), the film format (e.g., 9 x 9 in.), and camera focal-length. Once this information is known, it is possible to compute 1) the necessary flight altitude above-ground-level (AGL), 2) the number of flightlines required, 3) map distance between flightlines, 4) ground distance between exposures, 5) map distance between exposures, and 6) total number of exposures required. Many of these variables are described in Chapter 6 (Photogrammetry).

References

Adams, A. 1985, *The Camera*, Boston: Little, Brown & Co., 1–19.

Berglund, J. A., 1999, "Evaluating the Use of RADARSAT SAR and JERS-1 SAR for Estimating LAI and Delineating Upland Land Cover and Tidal Freshwater Wetland Vegetation Species in the Tivoli North Bay, Area, New York," Columbia, SC: Department of Geography, unpublished masters thesis, 310 pp.

Cavanaugh, J., 1990, "My Point-of-View," *Kodak Technical Bits*, 3:22.

Caylor, J. A. and H. M. Lachowski, 1988, *How to Use Aerial Photographs for Natural Resource Applications*, Washington: USDA Forest Service, 350 pp.

Colwell, R. N., 1997, "History and Place of Photographic Interpretation," Philipson, W., (Ed.), *Manual of Photographic Interpretation*, Bethesda: American Society for Photogrammetry & Remote Sensing, 2nd Ed., 3–47.

Estes, J. E., Hajic, E. J. and L. R. Tinney, 1983. "Fundamentals of Image Analysis: Analysis of Visible and Thermal Infrared Data," *Manual of Remote Sensing*, R. N. Colwell, (Ed.), Falls Church, VA: American Society for Photogrammetry & Remote Sensing, 1:1039–1040.

USGS, 1994, *Digital Orthophotos Fact Sheet*, Washington, DC: U.S. Department of the Interior, U.S. Geological Survey, 2 pp.

Fanton, B., 1989, "Photography: The 150th Anniversary," *Sky*, 84–91.

Fritz, L. W., 1996, "The Era of Commercial Earth Observation Satellites," *Photogrammetric Engineering & Remote Sensing*, 62(1):39–45.

Greve, C., 1996, *Digital Photogrammetry: An Addendum to the Manual of Photogrammetry*, Bethesda: American Society for Photogrammetry & Remote Sensing, 247 pp.

Hinckley, T. K. and J. W. Walker, 1993, "Obtaining and Using Low Altitude, Large-Scale Imagery," *Photogrammetric Engineering & Remote Sensing*, 59(3):310–318.

Hooper, N. J. and G. C. Gustafson, 1983, "Automation and Recording of the Image Interpreter's Mensuration Tasks for Man-made Objects," *SPIE Proceedings*, 424:56–67.

Jensen, J. R., 1996, *Introductory Digital Image Processing: A Remote Sensing Perspective*, Upper Saddle River: Prentice-Hall, Inc., 318 pp.

Kodak, 1981, *Applied Infrared Photography*, Publication M-28, Rochester, NY: Eastman Kodak Co., 86 pp.

Kodak, 1988, "Kodak Technical Information (Part III): The Characteristic Curve," *Kodak Technical Bits*, Summer:12–16.

Kodak, 1999, Kodak Home Page, Rochester, NY: Eastman Kodak, Co., http://www.kodak.com.

Light, D. L., 1990, "Characteristics of Remote Sensors for Mapping and Earth Science Applications," *Photogrammetric Engineering & Remote Sensing*, 56(12):1613–1623.

Light, D. L., 1993, "The National Aerial Photography Program as a Geographic Information System Resource," *Photogrammetric Engineering & Remote Sensing*, 48(1):61–65.

Light, D. L., 1996, "Film Cameras or Digital Sensors? The Challenge Ahead for Aerial Imaging," *Photogrammetric Engineering & Remote Sensing*, 62(3):285–291.

London, B. and J. Upton, 1994, *Photography*, 5th Ed., NY: Harper Colling, 422 pp.

Lunetta, R. S. and C. D. Elvidge, 1998, *Remote Sensing Change Detection*, Ann Arbor, MI: Ann Arbor Press, 318 pp.

Lyon, J. G. and E. Falkner, 1995, "Estimating Cost for Photogrammetric Mapping and Aerial Photography," *Journal of Surveying Engineering*, 121(2):63–86.

Meadows, J., 1992, *The Great Scientists*, NY: Oxford University Press, 250 pp.

Philipson, W., 1997, *Manual of Photographic Interpretation*, Bethesda, MD: American Society for Photogrammetry & Remote Sensing, 2nd Ed., 689 pp.

Rasher, M. E. and W. Weaver, 1990, *Basic Photo Interpretation*, Washington, DC: U.S. Department of Agriculture Soil Conservation Service, 320 pp.

Rundquist, D. C. and S. A. Sampson, 1988, *A guide to the Practical Use of Aerial Color-Infrared Photography in Agriculture*, Lincoln, NB: Conservation and Survey Division, 27 pp.

Slama, C. C., 1980, *Manual of Photogrammetry*, 4th Ed., Falls Church, VA: American Society for Photogrammetry & Remote Sensing, 1056.

Warner, W. S., R. W. Graham and R. E. Read, 1996, *Small Format Aerial Photography*, Scotland: Whittles Publishing Co., 348 pp.

White, L., 1995, *Infrared Photography Handbook*, Amherst, NY: Amherst Media, 108 pp.

Wolf, P. R., 1983, *Elements of Photogrammetry*, 2nd Ed., NY: John Wiley & Sons.

Elements of Visual Image Interpretation 5

Human beings are adept at interpreting images of objects. After all, they have been doing this all their lives. With some instruction they can become excellent image analysts. *Photo* or *image interpretation* is defined as

> the examination of images for the purpose of identifying objects and judging their significance (Philipson, 1997).

This chapter introduces the fundamental concepts associated with the visual interpretation of images of objects recorded primarily by remote sensing systems operating in the optical blue, green, red, and reflective near-infrared portions of the electromagnetic spectrum. The imagery that is interpreted may be acquired using a variety of sensors, including traditional analog cameras (e.g., Leica RC 30), digital cameras (e.g., Kodak DCS), multispectral scanners (e.g., Landsat Thematic Mapper), and current and proposed linear or area-array sensor systems (e.g., SPOT, IRS-1C, IKONOS, *MODIS*, *OrbView 3, Quickbird*). Methods of extracting quantitative information from remotely sensed images obtained in the optical portion of the spectrum are presented in Chapter 6 (Photogrammetry). Distinctive elements of image interpretation associated with thermal infrared (temperature) and active microwave (RADAR) imagery are presented in Chapters 8 and 9, respectively.

 Introduction

There are a number of important reasons why photo or image interpretation is such a powerful scientific tool, including:

- the aerial/regional perspective,

- three-dimensional depth perception,

- knowledge beyond our human visual perception, and

- the ability to obtain a historical image record to document change.

This chapter discusses these considerations and then introduces the fundamental elements of image interpretation used by image analysts to implement them. Various methods of search are also presented, including the use of collateral (ancillary) information, convergence of evidence, and application of the multi-concept in image analysis.

The Aerial/Regional Perspective

A vertical or oblique aerial photograph or other type of visible/near-infrared image records a detailed but much reduced version of reality. A single image usually encompasses much more geographic area than we as human beings could possibly traverse or really appreciate in a given day. For example, consider the photograph obtained by the astronauts through a porthole in *Apollo 17* that captures one-half of the entire Earth (a hemisphere) at one time (Figure 5-1). Much of Africa is visible from the arid Sahara to the dark vegetation of the Congo to the Cape of Good Hope shrouded in clouds. A single 9 x 9 in. 1:63,360-scale vertical aerial photograph records approximately 81 mi^2 of geography at one time.

Examination of the Earth from an aerial perspective allows scientists and the general public to identify objects, patterns, and man-land interrelationships that may never be completely understood if we were constrained to a terrestrial, Earth-bound vantage point. It does not matter whether the aerial perspective is from the top of a tall building, an elevated hillside, a light plane, a high-altitude jet, or a satellite platform (Warner et al., 1996). The resultant remotely sensed image provides spatial terrain information that we would not be able to acquire and appreciate in any other manner. This is why remote sensing image interpretation is so important for military reconnaissance and peaceful Earth resource scientific investigation.

Care must be exercised, however, when interpreting vertical and oblique imagery. Human beings are accustomed to looking at the façade (side) of objects from a terrestrial vantage point and do not normally have an appreciation for what objects look like when they are recorded from a vertical or oblique perspective (Haack et al., 1996). In addition, we definitely are not used to looking at and interpreting the significance of many square kilometers of terrain at one time. Our line of sight on the ground is usually less than a kilometer. Therefore, the regional analysis of vertical and oblique remote sensor data requires training and practice.

Three-Dimensional Depth Perception

We can view a single aerial photograph or image with our eyes and obtain an appreciation for the geographic distribution of features in the landscape. However, it is also possible to obtain a *three-dimensional view* of the terrain as if we were actually in an airborne balloon or aircraft looking out the window. The three-dimensional effect is accomplished

Figure 5-1   A photograph of the Earth obtained by the astronauts onboard *Apollo 17*, shooting through a porthole of the spacecraft. Almost the entire continent of Africa is visible as well as Saudi Arabia and part of Iraq and India. Note the arid Sahara and the dark, vegetated terrain of the rain forest along the equator in central Africa. Antarctica is especially apparent at the South Pole. Photographs like this helped mankind to realize how vulnerable and precious the Earth is as it rests like a multicolored jewel in the blackness of space (courtesy NASA).

by obtaining two photographs or images of the terrain from two slightly different vantage points. We can train our eyes to view the two images of the terrain at the same time. Our mind fuses this stereoscopic information into a three-dimensional model of the landscape that we perceive in our minds as being real (Caylor and Lachowski, 1988).

We know from life that it is important to not only know the size and shape of an object but that its height, depth, and volume are also very diagnostic. Analysis of stereoscopic imagery in three dimensions allows us to appreciate the three-dimensional nature of the undulating terrain and the slope and aspect of the land. In addition, the stereoscopic analysis process usually exaggerates the height or depth of the terrain, allowing us to appreciate very subtle differences in object height and terrain slope and aspect that we might never appreciate from a terrestrial vantage point. Chapter 6 (Photogrammetry) introduces the principles of stereoscopy and how they are used to extract three-dimensional information from remote sensor data.

Obtaining Knowledge Beyond our Human Visual Perception

Our eyes are sensitive primarily to blue, green, and red light. Therefore, we sample a very limited portion of the electromagnetic energy that is actually moving about in the environment and interacting with soil, rock, water, vegetation, the atmosphere, and urban structure. Fortunately, ingenious sensors have been invented that can measure the activity of X-rays, ultraviolet, near-infrared, middle-infrared, thermal infrared, microwave, and radiowave energy. Carefully calibrated remote sensor data provides entirely new information about an object that we as humans might never be able to appreciate in any other manner.

Historical Image Record and Change Detection Documentation

A single aerial photograph or image captures the Earth's surface and atmosphere at a unique moment in space and time, not to be repeated again. These photographs or images are valuable historical records of the spatial distribution of natural and man-made phenomena. When we acquire multiple images of the Earth, we can compare the historic imagery with the new imagery to determine if there are any subtle, dramatic, or particularly significant changes (Cowen and Jensen, 1998). The study of change usually increases our understanding about the natural and human-induced *processes* at work in the landscape. Knowledge about the spatial and temporal dynamics of phenomena allows us to develop predictive models about what has happened in the past and what may happen in the future (Lunetta and Elvidge, 1998). Predictive modeling is one of the major goals of science. Remote sensing image interpretation is playing an increasingly important role in predictive modeling and simulation (Vane and Goetz, 1993; Moran et al., 1997).

 Elements of Image Interpretation

To perform regional analysis, view the terrain in three dimensions, interpret images obtained from multiple regions of the electromagnetic spectrum, and perform change detection, it is customary to use principles of image interpretation that have been developed through empirical experience for more than 150 years (Rabben, 1960; Estes et al., 1983; Schott, 1997). The most fundamental of these principles are the *elements of image interpretation* that are routinely used when photointerpreting an aerial photograph or analyzing a

Table 5-1.   Elements of Image Interpretation

Element	Common Adjectives (quantitative and qualitative)
***x,y* location**	• x,y coordinates: longitude and latitude or meters easting and northing in a UTM map grid
Size	• length, width, perimeter, area (m²) • small, medium (intermediate), large
Shape	• an object's geometric characteristics: linear, curvilinear, circular, elliptical, radial, square, rectangular, triangular, hexagonal, pentagonal, star, amorphous, etc.
Shadow	• a silhouette caused by solar illumination from the side
Tone/color	• gray tone: light (bright), intermediate (gray), dark (black) • color: IHS = intensity, hue (color), saturation; RGB = red, green, and blue; Munsell
Texture	• characteristic placement and arrangement of repetitions of tone or color: smooth, intermediate (medium), rough (coarse), mottled, stippled
Pattern	• the spatial arrangement of objects on the ground: systematic, unsystematic or random, linear, curvilinear, rectangular, circular, elliptical, parallel, centripetal, serrated, striated, braided
Height/depth volume/ slope/aspect	• z-elevation (height), depth (bathymetry), volume (m³), slope °, aspect °
Site/ situation/ association	• Site: elevation, slope, aspect, exposure, adjacency to water, transportation, utilities • Situation: objects are placed in a particular order or orientation relative to one another • Association: related phenomena are usually present

photolike image. The elements include location, size, shape, shadow, tone and color, texture, pattern, height and depth, and site, situation, and association. Some of the adjectives associated with each of these elements of image interpretation are summarized in Table 5-1. A well-trained image interpreter uses many of the elements of image interpretation during analysis without really thinking about them. However, a novice interpreter may have to systematically force himself or herself to consciously evaluate an unknown object with respect to these elements to finally identify it and judge its significance in relationship to all the other phenomena in the scene.

Figure 5-2   A global positioning system (GPS) unit located on the grounds of the State Capitol in Columbia, SC. This survey grade instrument can be used to obtain x,y,z coordinate information to within ± 3 cm, with differential correction and phase processing.

x,y Location

There are two primary methods of obtaining precise x,y coordinate information about an object: 1) survey it in the field using traditional surveying techniques or global positioning system (GPS) instruments, or 2) collect remote sensor data of the object, register (rectify) the image to a basemap, and then extract the x,y coordinate information directly from the rectified image.

If option one is selected, most scientists now use relatively inexpensive GPS instruments in the field (Figure 5-2) to obtain a precise measurement of an object's location in degrees of longitude and latitude on the Earth's graticule or in meters easting and northing in a map grid such as the Universal Transverse Mercator (UTM). Scientists must then transfer the coordinates of the point (e.g., a specific tree location) or polygon (e.g., the perimeter of a small lake) onto

accurate planimetric maps. In the United States we normally use the U.S. Geological Survey's 7.5-minute quadrangle map. In Britain they use the Ordinance Survey map.

Most modern aircraft or spacecraft used to collect remote sensor data now have a GPS receiver. This allows the remote sensing instrument onboard to obtain exact x,y,z GPS coordinates at each photograph exposure station or the center of each line scan. In the case of aerial photography, this means that we can obtain information about the exact location of the center of each aerial photograph (i.e., the *principal point*) at the instant of exposure. We can use the GPS information collected by the sensor (and perhaps some collected on the ground) to register (rectify) the uncontrolled photo or image to a UTM or other map projection. If we also correct for the relief displacement of the topography, then the photo or image becomes an *orthophoto* or *orthoimage* with all the metric qualities of a line map. Geographic coordinates (x,y) of points and polygons can then be extracted directly from the rectified image. Chapter 6 (Photogrammetry) discusses how orthophotos are created. Jensen (1996) describes methods used to digitally rectify remote sensor data to a standard map projection.

Size — Length, Width, Perimeter, and Area

The size of an object is one of its most distinguishing characteristics and one of the most important elements of image interpretation. The most commonly measured parameters are length, width, perimeter, area (m^2) and occasionally volume (m^3). The analyst should routinely measure the size of unknown objects. To do this it is necessary to know the scale of the photography (e.g., 1:24,000) and its general unit equivalent or verbal scale (i.e., 1" = 2,000'). In the case of digital imagery it is necessary to know the ground spatial resolution of the sensor system (e.g., 1 x 1 m).

Measuring the size of an unknown object allows the interpreter to rule out many possible alternatives. One must be careful, however, because all of the objects in remote sensor data are at a scale less than 1:1, and we are not used to looking at a miniature version of an object that may measure only a few centimeters in length and width on the image. Measuring the size of a few well-known objects in an image such as car length, road and railroad width, size of the typical single-family house in the area, etc., allows us to understand the size of unknown features in the image and eventually to identify them. There are also several subjective, relative size adjectives, including small, medium, and large. These adjectives should be used sparingly.

Size

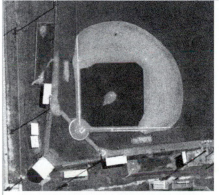

a. Automobiles: ≈ 15 ft in length, 6 ft wide.

b. Railroad: ≈ 4.71 ft between the rails and ≈ 8-in. between the railroad ties.

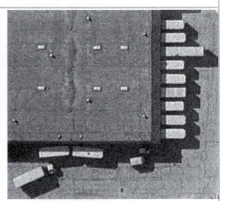

c. Single trailer on a tractor-trailer rig ≈ 45 to 50 ft in length.

d. Baseball: 90 ft between bases, 60 ft from home plate to the pitcher's mound.

e. Diving board: ≈ 12 ft in length.

f. Cars and trucks can be used to scale the size of the air conditioning units.

Figure 5-3   Elements of Image Interpretation — Size

Objects that have relatively unique sizes can be used to judge the size of other objects in the scene. For example, midsize cars are approximately 15 ft long and 6 ft wide in the United States (Figure 5-3a). They may be two-thirds that size in Europe, Asia, etc. Notice that it is possible to differentiate between automobiles and pickup trucks. Also note that the 6-in. white line separating parking spaces is quite visible, giving some indication as to the high spatial resolution of this aerial photography. The distance between regular-gauge railroad tracks is 4.71 ft in the United States (Figure 5-3b). This provides diagnostic information about the length of the individual railroad cars. The average length of a trailer on a tractor-trailer rig (Figure 5-3c) is 45 to 50 ft allowing us to appreciate the size of the adjacent warehouse.

Field dimensions of major sports such as soccer, baseball (Figure 5-3d), football, and tennis are standardized worldwide. The distance between the bases on a baseball diamond is 90 ft while the distance from the pitcher's mound to home plate is 60 ft. Most swimming pool diving boards (Figure 5-3e) are 12 ft long. Additional examples are found in Chapter 12 (Remote Sensing the Urban Landscape).

If these objects are visible within an image, it is possible to determine the size of other objects in the scene by comparing their dimensions with those of the known object dimension. For example, the diameter of the two rooftop air-conditioning units shown in Figure 5-3f are at least the length of the car and truck also visible in the image.

It is risky to measure the precise length, perimeter, and area of objects on unrectified aerial photography or other types of unrectified remote sensor data. The terrain is rarely completely flat within the instantaneous-field-of-view of an aerial photograph or other type of image. This causes points that are higher than the average elevation to be closer to the sensor and points that are lower than the average elevation to be farther away from the sensor system. Thus, different parts

Shape

a. Triangular (delta) shape of
a typical passenger jet airliner.

b. Rectangular single- and
double-wide mobile homes for sale.

c. Black-and-white infrared image of the
Pentagon (courtesy Positive Systems, Inc.).

d. A curvilinear cloverleaf highway
intersection in the United States.

e. The curvilinear shape of carefully engineered rice field levees in Louisiana.

f. Radial palm tree fronds
in San Diego, CA.

Figure 5-4   Elements of Image Interpretation — Shape

of the image have different scales. Tall buildings, hills, and depressions may have significantly different scales than those at the average elevation within the photograph.

Therefore, the optimum situation is where the aerial photography or other image data have been geometrically rectified and terrain-corrected to become, in effect, an orthophotograph or orthoimage where all objects are in their proper planimetric x,y location. It is then possible to measure the length, perimeter, and area of features using several methods, including: polar planimeter, tablet digitization, dot-grid analysis, or digital image analysis. These size measurement methods are discussed in Chapter 6 (Photogrammetry).

Shape

It would be wonderful if everything had a unique shape that could be easily discerned from a vertical or oblique perspec-

tive. Unfortunately, novice interpreters sometimes have difficulty even identifying the shape of the building they are in, much less appreciating the planimetric (x,y) shape of natural and man-made objects recorded in aerial photography or other imagery. Nevertheless, many features do have very unique shapes. There are numerous shape adjectives such as (Table 5-1) linear, curvilinear, circular, elliptical, radial, square, rectangular, triangular, hexagonal, star, elongated, and amorphous (no unique shape).

There are an infinite variety of uniquely shaped natural and man-made objects in the real world. Unfortunately, we can only provide a few examples (Figure 5-4). Modern jet aircraft (Figure 5-4a) typically have a triangular (delta) shaped outline and distinctively shaped shadows. Mankind's residential housing and public/commercial buildings may range from very simple rectangular mobile homes for sale (Figure 5-4b) to complex geometric patterns such as the Pentagon in Washington, DC (Figure 5-4c). The 0.5 x 0.5 m black-and-

Shadow

a. People and benches recorded in kite photography (courtesy Cris Benton).

b. Shadows cast from La Gloriette – Arch of Glory – in Vienna, Austria.

c. Bridge and sign shadows provide valuable information.

d. Pyramids of Giza (courtesy of Sovin-formsputnik and Aerial Images, Inc.).

e. Shadows provide information about object height (Litton Emerge, Inc.).

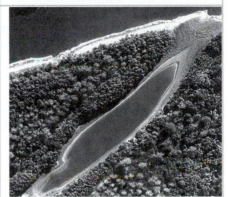

f. Orient images so that shadows fall toward the viewer during image analysis.

Figure 5-5   Elements of Image Interpretation — Shadow

white infrared image of the Pentagon was obtained using a digital camera. Man's transportation systems (Figure 5-4d) in developed countries usually have a curvilinear shape and exhibit extensive engineering.

Man modifies nature in a tremendous variety of ways, some of them very interesting. For example, Figure 5-4e depicts the curvilinear shape of carefully engineered levees (rising just 2 ft above the ground) that direct water continuously through a rice field in Louisiana. An adjacent field has been systematically plowed. But nature designs the most beautiful shapes, patterns and textures, including the radial frond pattern of palm trees shown in Figure 5-4f.

The best image interpreters spend a great amount of time in the field viewing and appreciating natural and man-made objects and their shapes. He or she is then in a good position to understand how these shapes appear when recorded on vertical or oblique imagery.

Shadow

Most remote sensor data is collected within ± 2 hours of solar noon to avoid extensive shadows in the imagery. This is because shadows from objects can obscure other objects that might otherwise be detected and identified. On the other hand, the shadow or *silhouette* cast by an object may be the only real clue to an object's identity. For example, consider the shadows cast by two people standing on a pier and the shadows cast by benches in Figure 5-5a. The shadows in the image actually provide more information than the objects themselves. La Gloriette — Arch of Glory — in Vienna, Austria, has unique statues on top of it (Figure 5-5b). Through careful evaluation of the shadows in the vertical photograph, it is possible to determine the location of the statues on top of the building. Similarly, shadows cast by signs or bridges (Figure 5-5c) are often more informative than the objects themselves in vertical aerial photography.

Very small scale photography or imagery usually does not contain shadows of objects unless they protrude a great distance above surrounding terrain such as mountains, extremely tall buildings, etc. For example, consider the shadows cast by the great pyramids of Giza in Egypt (Figure 5-5d). The distinctive shadows are very diagnostic during image interpretation.

In certain instances, shadows can provide clues about the height of an object when the image interpreter does not have access to stereoscopic imagery. For example, the building shadows in Figure 5-5e provide valuable information about the relative height of the building above the ground, i.e., that it is a one-story single-family residence. Chapter 6 (Photogrammetry) describes how shadow information can be used to measure the height of objects.

When interpreting imagery with substantial shadows, it is a good practice to orient the imagery so that the shadows fall toward the image interpreter such as those shown in Figure 5-5f. This keeps the analyst from experiencing *pseudoscopic illusion* where low points appear high and vice versa. For example, it is difficult to interpret the photograph of the forest and wetland shown in Figure 5-5f when it is viewed with the shadows falling away from the viewer. Please turn the page around 180° and see how difficult it is to interpret correctly. Unfortunately, most aerial photography of the northern hemisphere is obtained during the leaf-off spring months when the Sun casts shadows northward. This can be quite disconcerting. The solution is to reorient the photographs so that south is at the top. Unfortunately, if we have to make a photomap or orthophotomap of the study area, it is cartographic convention to orient the map with north at the top. This can then cause some problems when lay persons interpret the photomap because they do not know about pseudoscopic illusion.

Shadows on radar imagery are completely black and contain no information. Fortunately, this is not the case with shadows on aerial photography. While it may be relatively dark in the shadow area, there may still be sufficient light scattered into the area by surrounding objects to illuminate the terrain to some degree and enable careful image interpretation to take place.

Tone and Color

Real-world surface materials such as vegetation, water, and bare soil often reflect different proportions of energy in the blue, green, red, and near-infrared portions of the electromagnetic spectrum. We can plot the amount of energy reflected from each of these materials at specific wavelengths and create a spectral reflectance curve, sometimes called a *spectral signature*. For example, generalized spectral reflectance curves for objects found in a south Florida mangrove ecosystem are shown in Figure 5-6a (Davis and Jensen, 1998). Spectral reflectance curves of selected materials provide insight as to why they appear as they do on black-and-white or color imagery. We will first consider why objects appear in certain grayscale tones on black-and-white images.

Tone

A band of electromagnetic energy (e.g., green light from 0.5 – 0.6 μm) recorded by a remote sensing system may be displayed in shades of gray ranging from black to white. These shades of gray are usually referred to as *tone*. We often say, "This part of an image has a 'bright' tone, this area has a 'dark' tone, and this feature has an intermediate 'gray' tone." Of course, the degree of darkness or brightness is a function of the amount of light reflected from the scene within the specific wavelength interval (band). For example, consider three black-and-white images of a south Florida mangrove ecosystem (Figure 5-6b–d). The three images record the amount of green, red, and near-infrared energy reflected from the scene, respectively.

Incident green light (0.5 – 0.6 μm) penetrates the water column farther than red and near-infrared energy and is reflected off the sandy bottom or the coral reef (Figure 5-6b). Therefore, the green band provides subsurface detail about the reef structure surrounding the mangrove islands. Mangrove vegetation absorbs approximately 86 percent of the incident green light for photosynthetic purposes and reflects approximately 14 percent. This causes mangroves to appear relatively dark in a single-band green image. Sand reflects high, equal proportions of blue, green, red, and near-infrared incident energy, so it appears bright in all images.

Mangroves reflect approximately 9 percent of the incident red energy (0.6 – 0.7 μm) while absorbing approximately 91 percent of the incident energy for photosynthetic purposes. This causes the mangroves to appear very dark in the red photograph (Figure 5-6c). Red light does not penetrate as well into the water column so the water has a slightly darker tone, especially in the deeper channels. As expected, sandy areas have bright tones.

In the black-and-white image recording only near-infrared energy (0.7 – 0.92 μm), vegetation is displayed in bright tones (Figure 5-6d). Healthy vegetation reflects much of the incident near-infrared energy (approximately 28 percent).

Tone and Color

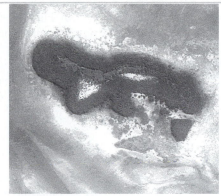

a. Spectral reflectance curves for sand, mangrove, and water in Florida.

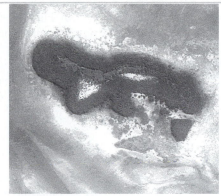

b. Black-and-white photograph of green reflected energy from Florida mangroves.

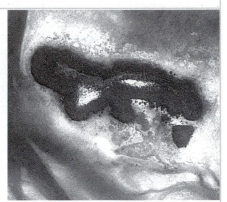

c. Black-and-white photograph of red reflected energy.

d. Black-and-white photograph of near-infrared reflected energy.

e. Stand of pine (evergreen) surrounded by hardwoods (courtesy Litton Emerge, Inc.).

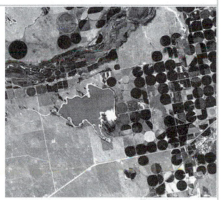

f. Vegetation is dark, fallow is bright, and turbid water is gray (Space Imaging, Inc.).

g. U-2 photograph of a Russian *Sputnik* launch site (courtesy John Pike, FAS).

h. High-contrast terrestrial photograph of a Dalmatian.

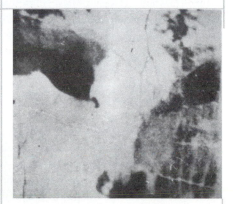

i. High-contrast terrestrial photograph of a cow.

Figure 5-6   Elements of Image Interpretation — Tone and Color

Generally, the brighter the tone from a vegetated surface, the greater the amount of biological matter (biomass) present (Jensen et al., 1999). Conversely, water absorbs most of the incident near-infrared energy, causing the water to appear dark. There is a great contrast between the bright upland consisting of vegetation and sand, and the dark water. Therefore, it is not surprising that the near-infrared region is considered to be the best for discriminating between the upland and water interface.

A black-and-white infrared image of an evergreen pine stand surrounded by deciduous hardwood forest is shown in Figure 5-6e. The tonal contrast makes it easy to discriminate between the two major species.

One must be careful, however, when interpreting individual band black-and-white images. For example, consider the Landsat Thematic Mapper band 3 (red) image of a Colorado agricultural area (Figure 5-6f). As expected, the greater the amount of vegetation, the greater the absorption of the incident red light by the vegetation, and the darker the vegetated area within the center-pivot irrigation system. Conversely, fallow fields and areas not in agricultural production show up in much brighter tones. Unfortunately, the lake also shows up as an intermediate shade of gray. This is because the lake has received a substantial amount of suspended sediment, causing it to reflect more red radiant energy than it normally would if it were deep, nonturbid water. If we had additional blue, green, and perhaps near-infrared bands to analyze, it would probably be clear that this is a water body. However, when viewing only a single band of imagery displayed in black-and-white tones, we could come to the conclusion that the lake is simply a large vegetated field perhaps in the early stages of development when some of the sandy bare soil is still visible through the canopy. In fact, it has approximately the same gray tone as several of the adjacent fields within the center-pivot irrigation systems.

Human beings can differentiate between approximately 40-50 individual shades of gray in a black-and-white photograph or remote sensor image. However, it takes practice and skill to extract useful information from broad-band panchromatic black-and-white images or black-and-white images of individual bands. For example, consider the U-2 photograph of a Russian *Sputnik* launching site shown in Figure 5-6g. Careful examination of the gray tones and the shadows by a trained analyst reveals that the excavated earth from the blast area depression was deposited in a large mound nearby. Human beings simply are not used to viewing the tops of objects in shades of gray. They must be trained. Furthermore, humans often have a very difficult time identifying features if the scene is composed of very high contrast infor-

mation. This is exemplified by viewing terrestrial photographs of two very well known objects: a Dalmatian and a cow in Figure 5-6h,i, respectively. Many novice analysts simply cannot find the Dalmatian or the cow in the photographs. This suggests that extremely high contrast aerial photographs or images are difficult to interpret and that it is best to acquire and interpret remotely sensed imagery that has a continuum of grayscale tones from black to gray to white, if possible.

Color

We may use additive color-combining techniques to create color composite images from the individual bands of remote sensor data as previously discussed in Chapter 4. This introduces *hue* (color) and *saturation* in addition to grayscale tone (*intensity*). A color composite of the green, red, and near-infrared bands of the mangrove study area is found in Color Plate 5-1. Notice how much more visual information is present in the color composite. Humans can discriminate among thousands of subtle colors. In this false-color image, all vegetation is depicted in shades of red (magenta), sand is bright white, and the water is in various shades of blue. Most scientists prefer to acquire some form of multispectral data so that color composites can be made. This may be the collection of natural color aerial photography, color-infrared aerial photography, or multispectral data, where perhaps many individual bands are collected and a select few are additively color-combined to produce color images.

Unfortunately, some peoples' color perception is impaired. This means that they do not experience the same mental impression of a color (e.g., green) as does the vast majority of the population. While this may be somewhat of a disadvantage when selecting a shirt or tie to wear, many excellent image analysts have some color perception disorder. There are special tests such as those shown in Color Plate 5-2 that can be viewed to determine if color blindness is present.

Texture

Texture is the characteristic placement and arrangement of repetitions of tone or color in an image. In an aerial photograph, it is created by tonal repetitions of groups of objects that may be too small to be discerned individually. Sometimes two features that have very similar spectral characteristics (e.g., similar black-and-white tones or colors) exhibit different texture characteristics that allow a trained interpreter to distinguish between them. We often use the textural adjectives smooth (uniform, homogeneous), intermediate, and rough (coarse, heterogeneous).

Texture

a. Relatively coarse-texture avocado field. The grass and road have a smooth texture.

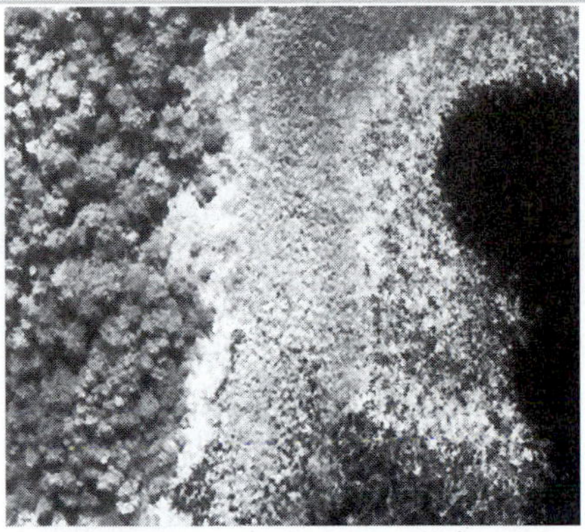

b. Pine has coarse texture, cattails and waterlilies have intermediate texture.

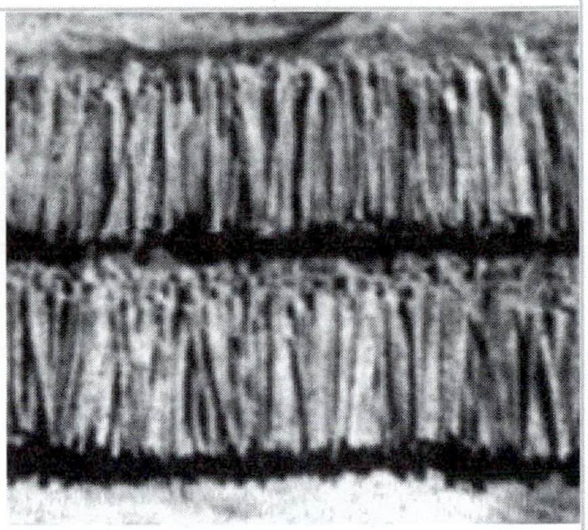

c. Coarse texture of freshly cut pine logs at a sawmill in Georgia.

d. Coarse-texture South Carolina corn field interspersed with circular marijuana plants.

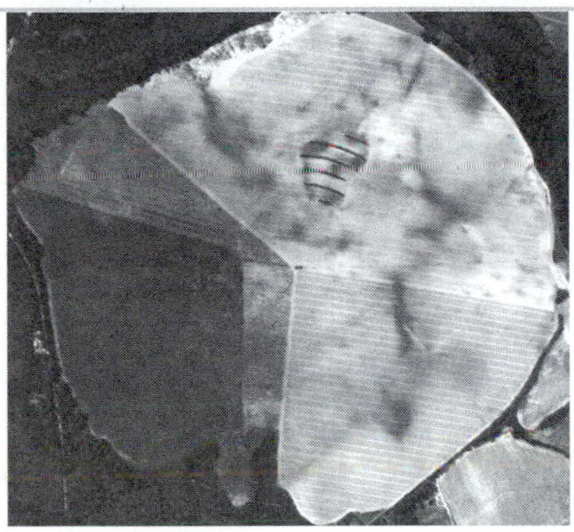

e. Mottled texture on fallow soil in a Georgia center-pivot irrigation system.

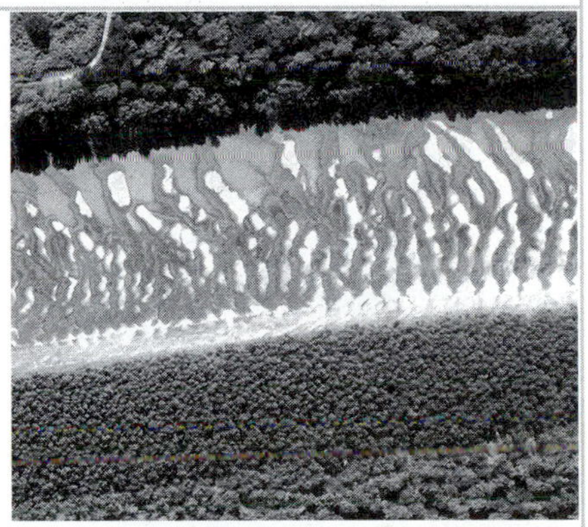

f. A variety of textures along a tributary of the Mississippi.

Figure 5-7   Elements of Image Interpretation — Texture

It is important to understand that the texture in a certain portion of a photograph is strictly a function of scale. For example, in a very large scale aerial photograph (e.g., 1:500) we might be able to actually see the leaves and branches in the canopy of a stand of trees and describe the area as having a coarse texture. However, as the scale of the imagery becomes smaller (e.g., 1:5,000), the individual leaves and branches and even the tree crowns might coalesce, giving us the impression that the stand now has an intermediate texture, i.e., it is not smooth but definitely not rough. When the same stand of trees is viewed at a very small scale (e.g., 1:50,000), it might appear to be a uniform forest stand with smooth texture. Thus, texture is a function of the scale of the imagery and the ability of the interpreter to perceive and describe it.

Several other texture adjectives are often used, including mottled, stippled, etc. It is difficult to define exactly what is meant by each of these textures. It is simply better to present a few examples, as shown in Figure 5-7. Both the avocado orchard and the trees in the courtyard have a coarse texture on this large-scale photograph (Figure 5-7a). Conversely, the concrete road and much of the grass yard have a smooth texture. Just behind the pool, the soil exhibits varying degrees of moisture content, causing a mottled texture.

In Figure 5-7b, the pine forest on the left has a relatively coarse texture as the individual tree crowns are visible. The bright sandy beach has a smooth texture. Both cattails near the shore and waterlilies farther out into L-Lake on the Savannah River Site exhibit intermediate to rough textures. Finally, the waterlilies give way to dark, smooth textured water.

Two piles of 50 ft pine logs at a sawmill in Georgia are shown in Figure 5-7c. The logs exhibit a coarse, heterogeneous texture with a linear pattern. The shadow between the stacks has a smooth texture.

Pattern

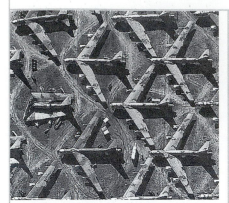

a. Systematic, triangular pattern of B-52s being dismantled (courtesy USGS).

b. Seven circular grain silos exhibit a curvilinear pattern on this southeastern farm.

c. Random, sinuous braided stream pattern on sandy soil at Pen Branch, SC.

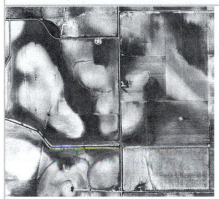

d. Township & Range survey pattern on mottled soil in Texas.

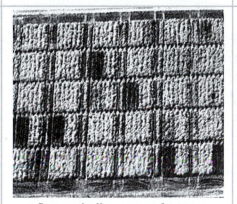

e. Systematic, linear rows of potatoes with some damaged (dark) by late blight.

f. Radiating road pattern in Paris (Sovinformsputnik and Aerial Images, Inc.).

Figure 5-8   Elements of Image Interpretation — Pattern

Figure 5-7d is an interesting photograph of systematically placed circular marijuana plants interspersed in a field of corn. The physiology (structure) of the two types of plants, their spacing, and orientation combine to produce a coarse-textured agricultural field. The shadows produced by the marijuana plants contribute substantially to the texture of the area. Interestingly, the goal of the farmers appears to be working. Few novice interpreters appreciate the subtle differences in texture visible in the field.

Part of the agricultural field in Figure 5-7e is being cultivated while the remainder is in fallow. The vegetated southwest portion of the center-pivot irrigation system has a relatively smooth texture. However, the remaining fallow portion of the field appears to have areas with varying amounts of soil moisture or different soil types. This causes this area to have a mottled texture. One part of the mottled texture region still bears the circular scars of six wheels of the center-pivot irrigation system.

Various vegetation and sand bar textures are present in the large scale photograph of a tributary to the Mississippi River in Figure 5-7f. A dense stand of willow parallels the lower shoreline, creating a relatively fine texture when compared with the hardwood behind it with its more coarse texture. The sand bars interspersed with water create a unique, sinuous texture as well as a serrated pattern. Some of the individual tree crowns in the upper portion of the image are spaced well apart, creating a more coarse texture.

Pattern

Pattern is the spatial arrangement of objects in the landscape (Figure 5-8). The objects may be arranged randomly or systematically. They may be natural, as with a drainage pattern, or man-made, as with the Township and Range land tenure system present in the western United States. Pattern is a very diagnostic characteristic of many features. Typical pattern

Height and Depth

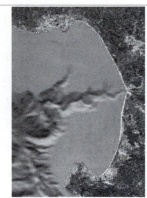

a. Relief displacement is an important monoscopic cue about object height.

b. Shadows and masking in downtown San Francisco (courtesy Space Imaging, Inc.)

c. Bathymetry of Monterey Bay, CA (courtesy HJW, Inc.; © SPOT Image, Inc.).

Figure 5-9   Elements of Image Interpretation — Height and Depth

adjectives include: random, systematic, circular, centripetal, oval, curvilinear, linear, radiating, rectangular, hexagonal, pentagonal, octagonal, etc.

Examples of typical patterns captured on remote sensor data are shown in Figure 5-8. The first example depicts the systematic, triangular pattern of B-52s being dismantled at Montham Air Force Base (Figure 5-8a). A large metal blade cuts the fuselage into a specific number of parts. The parts must remain visible for a certain number of days so that foreign countries can use their own aerial reconnaissance technology to verify that the specified number of B-52s have been removed from service as part of the strategic arms limitation process. Heavy equipment moving between the aircraft creates a unique curvilinear transportation pattern.

Seven large silos used to store agricultural grain are seen in Figure 5-8b. The individual silos are circular but they are situated in a curvilinear pattern on the landscape. Numerous rectangular farm buildings oriented north-south are arranged in a random fashion.

A random, sinuous braided stream pattern is present at the mouth of Pen Branch, SC, in Figure 5-8c. This particular pattern resembles braided hair, hence the terminology. The major drainage patterns visible in remotely sensed data are discussed in Chapter 13.

Figure 5-8d depicts the systematically surveyed Township & Range cadastral system superimposed on an agricultural region in Texas. The NAPP photograph reveals small farmsteads separated by large tracts of agricultural land. The soil

moisture and soil type differences in the fields combine to create an unsystematic, mottled soil texture.

Potatoes arranged in systematically spaced linear rows are shown in Figure 5-8e. Various rows are arranged in a rectangular pattern. This near-infrared photograph reveals that some of the fields that appear dark have experienced late blight damage.

A KVR-1000 Russian satellite photograph reveals the systematic, radiating road pattern centered on the Arch de Triumph in Paris (Figure 5-8f).

Height and Depth

The ability to visually appreciate and measure the height (elevation) or depth (bathymetry) of an object or landform is one of the most diagnostic elements of image interpretation (Figure 5-9). Stereoscopic parallax is introduced to remotely sensed data when the same object is viewed from two different vantage points along a flightline. Viewing these overlapping photographs or images using *stereoscopic* instruments is the optimum method for visually appreciating the three-dimensionality of the terrain and for extracting accurate x,y, and z topographic and/or bathymetric information. Techniques of stereoscopic analysis are presented in Chapter 6 (Photogrammetry).

However, there are also *monoscopic* cues that we can use to appreciate the height or depth of an object. For example, any object such as a building or utility pole that protrudes above the local datum will exhibit radial relief displacement out-

Site, Situation, and Association

a. Thermal electric Haynes Steam Plant in Long Beach, CA.

b. A sawmill with its associated piles of raw and finished lumber.

c. Vogtle Nuclear Power Plant near Augusta, GA.

Figure 5-10   Elements of Image Interpretation — Site, Situation, and Association

ward from the principal point of a typical aerial photograph. In effect, we are able to see the side of the feature, as demonstrated in Figure 5-9a. Also, all objects protruding above the local datum cast a shadow that provides diagnostic height or elevation information such as the various buildings in San Francisco shown in Figure 5-9b. Also, masking takes place in some images where tall objects obscure objects behind them, making it clear that one object has more elevation than another. For example, the building at the top of Figure 5-9b is masking the buildings behind it, suggesting that it has greater height.

The optimum method of obtaining bathymetric measurements is to use a sonar remote sensing device which sends out a pulse of sound and measures how long it takes for the sound to pass through the water column, bounce off the bottom, and be recorded by the sensor. The image of Monterey Bay, CA, in Figure 5-9c was obtained using sonar and merged with a SPOT image of the terrestrial landscape.

Site, Situation, and Association

Site, situation, and *association* characteristics are very important when trying to identify an object or activity. A *site* has unique physical and/or socioeconomic characteristics. The physical characteristics might include elevation, slope, aspect, and type of surface cover (e.g., bare soil, grass, shrub/scrub, rangeland, forest, water, asphalt, concrete, housing, etc.). Socioeconomic site characteristics might include the value of the land, the land-tenure system at the site (metes and bounds versus Township and Range), adja-

cency to water, and/or adjacency to a certain type of population (professional, blue-collar, retired, etc.).

Situation refers to how certain objects in the scene are organized and oriented relative to one another. Often, certain raw materials, buildings, pipelines, and finished products are situated in a logical, predictable manner.

Association refers to the fact that when you find a certain phenomena or activity, you almost invariably encounter related or associated features or activities. Site, situation, and association elements of image interpretation are rarely used independently when analyzing an image. Rather, they are used synergistically to arrive at a logical conclusion. For example, sewage disposal plants are almost always located on flat sites situated near a water source so they can dispose of the treated water, and they exist relatively close to the producing community. Large commercial shopping malls typically have multiple large buildings on level sites, massive parking lots, and are ideally situated near major transportation arteries and population centers.

Thermal electric power plants such as the Haynes Steam Plant in Long Beach, CA, shown in Figure 5-10a are usually located on flat, well-engineered sites with large tanks of petroleum (or other type of natural resource) nearby that is burned to create steam to propel the electric generators. Man-made levies, called "revetments," often encompass the tank farm to contain the petroleum in the event of an accident. Thermal electric power plants are often associated with some type of recirculating cooling ponds. The water is used to cool critical steam-generating components.

Sawmills such as the one shown in Figure 5-10b are usually sited on flat terrain within 20 km of many stands of trees and associated with large piles of raw timber, well-organized piles of finished lumber, a furnace to dispose of wood waste products, and an extensive processing facility. Railroad spurs are often used to transport the finished lumber or wood chip products to market.

Nuclear power plants exist on extremely well-engineered level sites. They have large concrete reactor containment building(s). The site may contain large recirculating cooling water ponds or enormous cooling towers such as those under construction at the Vogtle Nuclear Power Plant near Augusta, GA, in Figure 5-10c. Power-generating plants do not need to be adjacent to the consuming population as electricity can be transported economically great distances.

Expert image analysts bring to bear site, situation, and association knowledge to an image interpretation problem. Such knowledge is gained best by observing phenomena in the field. The best image analysts have "been around" and have seen and appreciate a diverse array of natural and man-made environments. It is difficult to identify an object in an image if one has never seen the object in the real world and does not appreciate its site, situation, and association characteristics.

 Methods of Search

Photointerpretation has been taking place since Gaspard Felix Tournachon (Nadar) took the first successful aerial photograph in France in 1858. Over the years, scientists have developed some valuable approaches to interpreting remotely sensed data, including: 1) utilizing collateral (ancillary) information, 2) converging the evidence, and 3) applying the multi-concept in image analysis.

Using Collateral Information

Trained image interpreters rarely interpret aerial photography or other remote sensor data in a vacuum. Instead, they collect as much *collateral* (often called *ancillary*) information about the subject and the study area as possible. Some of the major types of collateral information are summarized in Table 5-2, including: the use of a variety of maps for orientation, political boundary information, property line cadastral data, geodetic control (x,y,z), forest stand data, geologic data, hazard information, surface and subsurface hydrologic data, socioeconomic data, soil taxonomy, topographic and bathymetric data, transportation features, and wetland infor-

Table 5-2. Collateral Information Often Used in the Interpretation of Aerial Photography and Other Remotely Sensed Data in the United States

Topic	Collateral Information
General orientation	International Map of the World 1:1,000,000 National Imagery and Mapping Agency (NIMA) 1:100,000; 1:250,000 USGS 7.5-min 1:24,000 USGS 15-min 1:63,360 Image browsing systems: Space Imaging, EarthWatch, SPOT, CORE
Boundaries or districts	USGS 7.5-min 1:24,000 USGS 15-min 1:63,360 Boards – state, county, city, school, fire, voting, water/sewer
Cadastral	City and county tax maps
Geodetic control	USGS digital line graph – elevation NGS – nautical and bathymetric charts
Forestry	USFS – forest stand information
Geology	USGS – surface and subsurface
Hazards	FEMA – flood insurance maps USCG – environmental sensitivity index
Hydrology	USGS digital line graph – surface hydrology NGS – nautical and bathymetric charts USGS – water supply reports USGS – stream gauge reports
Socio-economic	Bureau of the Census – demographic data – TIGER block data – census tracts
Soils	SCS, NRCS – soil taxonomy maps
Topography/ bathymetry	USGS digital line graph – elevation USGS – digital elevation model NIMA - digital terrain elevation data (DTED) USCG – nautical and bathymetric charts
Trans-portation	USGS digital line graph – transportation County and state transportation maps
Weather/ atmosphere	National Weather Service – NEXRAD
Wetland	USGS – National Wetland Inventory maps NOAA – Coastal change analysis program

mation. Ideally, these data are stored in a geographic information system (GIS) for easy retrieval and overlay with the remote sensor data.

It is useful to contact the local National Weather Service to obtain quantitative information on the meteorological conditions that occurred on the day the remote sensor data were collected (cloud cover, visibility, and precipitation). USGS water-supply and stream-gauge reports are also useful.

Scientists also obtain local street maps, terrestrial photographs, local and regional geography books, and journal and popular magazine articles about the locale or subject matter. They talk with local experts. Well-trained image analysts get out into the field to appreciate first-hand the lay of the land, its subtle soil and vegetation characteristics, the drainage and geomorphic conditions, and man's cultural impact.

Often much of this collateral spatial information is stored in a GIS. This is particularly useful since the remote sensor data can be geometrically registered to the spatial information in the GIS database and important interrelationships evaluated.

Convergence of Evidence

It is generally a good idea to work from the known to the unknown. For example, perhaps we are having difficulty identifying a particular type of industry in an aerial photograph. Careful examination of what we do know about things surrounding and influencing the object of interest can provide valuable clues that could allow us to make the identification. This might include a careful interpretation of the building characteristics (length, width, height, number of stories, type of construction), the surrounding transportation pattern (e.g., parking facilities, railroad spur to the building, adjacent to an interstate), site slope and aspect, site drainage characteristics, unique utilities coming into or out of the facility (pipelines, water intake or output), unusual raw materials or finished products in view outside the building, and methods of transporting the raw and finished goods (tractor trailers, loading docks, ramps, etc.). We bring all the knowledge we can to the interpretation problem and *converge our evidence* to identify the object or process at work.

The Multi-concept

Robert Colwell of the Forestry Department at the University of California at Berkeley put forth the *multiconcept* in image interpretation in the 1960s (Philipson, 1997; Colwell, 1997). He suggested that the most useful and accurate method of scientific image interpretation consisted of performing the following types of analysis: *multispectral, multidisciplinary, multiscale,* and *multitemporal.* The multiconcept has been further elaborated upon by Estes et al. (1983) and Teng (1997).

Colwell pioneered the use of *multiband* aerial photography and *multispectral* remote sensor data. He documented that in agriculture and forest environments, measurements made in multiple discrete wavelength regions (bands) of the electromagnetic spectrum were usually more valuable than acquiring single broad-band panchromatic-type imagery. For example, Figure 5-6 documented the significant difference in information content found in green, red, and near-infrared multispectral images of mangrove.

Colwell also suggested that *multiscale* (often called multistage) photography or imagery of an area was very important. Smaller scale imagery (e.g.,1:80,000) was useful for placing intermediate scale imagery (e.g., 1:40,000) in its proper regional context. Then, very large-scale imagery (e.g., > 1:10,000) could be used to provide detailed information about local phenomena. *In situ* field investigation is the largest scale utilized and is very important. Each scale of imagery provides unique information that can be used to calibrate the others.

Professor Colwell was a great believer in bringing many *multidisciplinary* experts together to focus on a remote sensing image analysis or information extraction problem. The real world consists of soils, surface and subsurface geology, vegetation, water, atmosphere, and man-made urban structure. In this age of increasing scientific specialization, it is difficult for any one person to be able to understand and extract all the pertinent and valuable information present within a remote sensor image. Therefore, Colwell strongly suggested that image analysts embrace the input of other multidisciplinary scientists in the image interpretation process. This philosophy and process often yields synergistic, novel, and unexpected results as multidisciplinary scientists each bring their expertise to bear on the landscape appreciation problem. Table 5-3 lists the disciplines of colleagues that often collaborate when systematically studying a certain topic.

While single-date remote sensing investigations can yield important information, they do not always provide information about the processes at work. Conversely, a *multitemporal* remote sensing investigation obtains more than one image of an object. Monitoring the phenomena through time allows us to understand the processes at work and to develop predictive models (Lunetta and Elvidge, 1998; Schill et al., 1999). Colwell pioneered the concept of developing crop phenological calendars in order to monitor the spectral changes that take place as plants progress through the growing season. Once crop calendars are available, they may be used to select the optimum dates during the growing season to acquire remote sensor data. Many other phenomena, such

Table 5-3. Multidisciplinary Scientists Bring their Unique Training to the Image Interpretation Process

Topic	Disciplines
Agriculture	Agronomy, agricultural engineering, biology, biogeography, geology, landscape ecology
Biodiversity, habitat	Biology, zoology, biogeography, landscape ecology, marine science
Database and algorithm preparation	Cartography, GIS, computer science, photogrammetry, programming, analytical modeling
Forestry, Rangeland	Forestry, agronomy, rangeland ecology, landscape ecology, biogeography
Geodetic control	Geodesy, surveying, photogrammetry
Geology Soils	Geology, geomorphology, agronomy, geography
Hazards	Geology, hydrology, urban and physical geography
Hydrology	Hydrology, chemistry, geology, geography
Topography/ bathymetry	Geodesy, surveying, photogrammetry
Transportation	Transportation engineering, city planning, urban geography
Urban studies	Urban, economic, and political geography, city planning, transportation engineering, civil engineering, landscape ecology
Weather/ atmosphere	Meteorology, climatology, physics, chemistry
Wetland	Biology, landscape ecology, biogeography

as residential urban development, have been found to undergo predictable cycles that can be monitored using remote sensor data. A trained image analyst understands the phenological cycle of the phenomena he or she is interpreting and uses this information to acquire the optimum type of remote sensor data on the optimum days of the year.

Conclusion

We now have an understanding of the fundamental elements of image interpretation. We can utilize the elements of image interpretation to carefully analyze aerial photography or other types of optical (blue, green, red, and near-infrared wavelength) remote sensor data. Based on this foundation, we are prepared to progress to more sophisticated image analysis techniques, including the extraction of quantitative information from remote sensor data using principles of photogrammetry.

References

Caylor, J. A. and H. M. Lachowski, 1988, *How to Use Aerial Photographs for Natural Resource Applications*, Washington: USDA Forest Service, 350 pp.

Colwell, R. N., 1997, "History and Place of Photographic Interpretation," *Manual of Photographic Interpretation*, W. Philipson (Ed.), 2nd Ed., Bethesda, MD: American Society for Photogrammetry & Remote Sensing, 3–48.

Cowen, D. J. and J. R. Jensen, 1998, "Extraction and Modeling of Urban Attributes Using Remote Sensing Technology," *People and Pixels*, D. Liverman, Ed., Washington, DC: National Academy Press, 164–188.

Davis, B. A. and J. R. Jensen, 1998, "Remote Sensing of Mangrove Biophysical Characteristics," *Geocarto International: A Multidisciplinary Journal of Remote Sensing and GIS*, 13(4):55–64.

Estes, J. E., E. J. Hajic and L. R. Tinney, 1983, "Fundamentals of Image Analysis: Analysis of Visible and Thermal Infrared Data," *Manual of Remote Sensing*, R. N. Colwell, Ed., Falls Church, VA: American Society of Photogrammetry, 1:1039–1040.

Haack, B., S. Guptill, R. Holz, S. Jampoler, J. R. Jensen and R. A. Welch, 1996, "Urban Analysis and Planning," *Manual of Photographic Interpretation*, W. Philipson (Ed.), 2nd Ed., Bethesda, MD: American Society for Photogrammetry & Remote Sensing, 517–547.

Jensen, J. R., 1996, *Introductory Digital Image Processing: A Remote Sensing Perspective*, Upper Saddle River, NJ: Prentice-Hall, Inc., 319 pp.

Jensen, J. R., C. Coombs, D. Porter, B. Jones, S. Schill and D. White, 1999, "Extraction of Smooth Cordgrass (*Spartina alterniflora*) Biomass and Leaf Area Index Parameters from High Resolution Imagery," *Geocarto International: A Multidisciplinary Journal of Remote Sensing and GIS*, 13(4):25–34.

Lunetta, R. S. and C. D. Elvidge, 1998, *Remote Sensing Change Detection: Environmental Monitoring Methods and Applications*, Ann Arbor, MI: Ann Arbor Press, 318 pp.

McDonnell Douglas, 1982, *Reconnaissance Handy Book for the Tactical Reconnaissance Specialist*, St. Louis: McDonnell Douglas, Inc., 183 pp.

Moran, M. S., Y. Inoue and E. H. Barnes, 1997, "Opportunities and Limitations for Image-based Remote Sensing Precision Crop Management," *Remote Sensing of Environment*, 61:319–346.

Philipson, W., 1997, *Manual of Photographic Interpretation*, 2nd Ed., Bethesda, MD: American Society for Photogrammetry & Remote Sensing, 555 pp.

Rabben, E. L., 1960, "Fundamentals of Photo Interpretation," in *Manual of Photographic Interpretation*, Falls Church: American Society for Photogrammetry, 99–168.

Rasher, M. E. and W. Weaver, 1990, *Basic Photo Interpretation*, Washington, DC: USDA Soil Conservation Service, 320 pp.

Schill, S., J. R. Jensen and D. C. Cowen, 1999, "Bridging the Gap Between Government and Industry: the NASA Affiliate Research Center Program," *Geo Info Systems*, 9(9):26–33.

Schott, J. R., 1997, *Remote Sensing: The Image Chain Approach,* Oxford: Oxford University Press, 394 pp.

Teng, W. L., 1997, "Fundamentals of Photographic Interpretation," *Manual of Photographic Interpretation*, W. Philipson (Ed.), 2nd Ed., Bethesda, MD: American Society for Photogrammetry & Remote Sensing, 49–113.

Vane, G. and A. F. H. Goetz, 1993, "Terrestrial Imaging Spectrometry: Current Status, Future Trends," *Remote Sensing of Environment*, 44:117–126.

Warner, W. S., R. W. Graham and R. E. Read, 1996, *Small Format Aerial Photography*, Scotland: Whittles Publishing, 348 pp.

Photogrammetry 6

Photogrammetry is the art and science of making accurate measurements by means of aerial photography. *Analog photogrammetry* is performed visually by humans and is used when the data are in a hard-copy format such as a 9 x 9-in. aerial photographic print or positive transparency. *Digital* or *analytical photogrammetry* is performed with the aid of a computer using digitized aerial photography. Neither method is superior to the other. Each approach has an appropriate role depending upon the nature of the remote sensor data and the image analysis goals. This chapter focuses on methods to extract *quantitative* information from aerial photography using both analog and digital photogrammetric techniques. Sometimes scientists use the term *photo mensuration* to describe the variety of methods used by photointerpreters to collect numerical information from aerial images. The methods are generally based on simplified photogrammetric principles (Gustafson, 1999).

Below are some of the more important measurements that can be obtained from a *single vertical aerial photograph* using either analog or digital photogrammetric techniques:

- scale of the photography,

- object height,

- object length,

- area of an object or polygon,

- perimeter of an object or polygon, and

- the grayscale tone or color of an object.

The following quantitative measurements may be made using *multiple (overlapping) stereoscopic aerial photographs* and analog or digital measurement of stereoscopic parallax (to be defined):

- precise planimetric (x,y) object location of buildings, streets, hydrology, and shorelines in a standard map projection, and

- precise object height (z).

In addition, digital photogrammetric techniques applied to stereoscopic aerial photography can yield:

Flightline of Aerial Photography

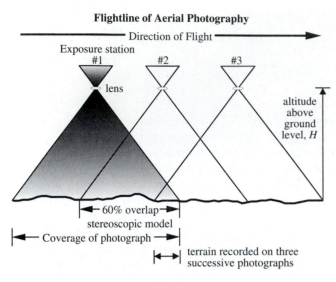

Figure 6-1   A single flightline of vertical aerial photography with 60 percent overlap obtained at three exposure stations *H* meters above-ground-level.

Block of Aerial Photography

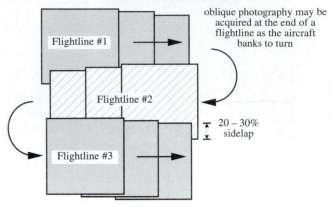

Figure 6-2   A block of aerial photography is produced by photographing multiple flightlines that have 20 – 30 percent sidelap between them. If the aerial photographer does not turn off the camera during a turn, some oblique aerial photography may be acquired at the end of a flightline.

- planimetrically accurate (x,y) orthophotographs,

- digital elevation models (DEM),

- bathymetric models, and

- slope and aspect information derived from the digital elevation or bathymetric models.

 ## Flightlines of Vertical Aerial Photography

It is necessary to understand how individual aerial photographs are acquired at various *exposure stations* along a flightline before we can correctly apply photogrammetric techniques to extract quantitative measurements.

A *flightline* of vertical aerial photography is obtained by mounting a camera in an aircraft, helicopter, or spacecraft and obtaining photographs of the terrain directly beneath the platform at specific exposure stations. For example, the geometry of three hypothetical vertical aerial photographs taken in succession over relatively level terrain are depicted in Figure 6-1. The time between individual exposures along a flightline is determined by setting the camera *intervalometer*. The aerial photographer takes into account the speed of the aircraft and the scale of the desired photography and sets the intervalometer so that each vertical aerial photograph

overlaps the next photograph in the flightline by approximately 60 percent (referred to as *stereoscopic overlap*). This *overlap* is very important because it provides at least two and sometimes three photographic views of each object in the real world along a flightline (Figure 6-1). Sometimes aerial photographs are acquired with >80 percent overlap in very mountainous terrain.

Most aerial photography projects, such as mapping a watershed, county, or entire city, require multiple flightlines of photography to cover the geographic area of interest. When this occurs, the flightlines are normally overlapped by 20 – 30 percent, commonly referred to as *sidelap* (Figures 6-2 and 6-3). To acquire multiple flightlines, a pilot must make a 180° turn at the end of a flightline and then fly in the opposite direction. Care should be exercised when analyzing vertical aerial photographs acquired at the end of a flightline because some photography obtained while the plane was banking during the turn may be oblique. The algorithms presented in this chapter work best with near-vertical aerial photographs (≤ 3° from nadir). Also, sometimes wind or pilot error will cause the plane to drift to the left or right of the desired line of flight. When this occurs, the footprint of each photograph or the entire flightline may be offset, as shown in Figure 6-3b. This is not a serious condition as long as it does not become too severe and the 60 percent overlap and 20 – 30 percent sidelap are maintained.

Multiple flightlines with 20 – 30 percent sidelap are commonly referred to as a *block* of aerial photography (Figure 6-

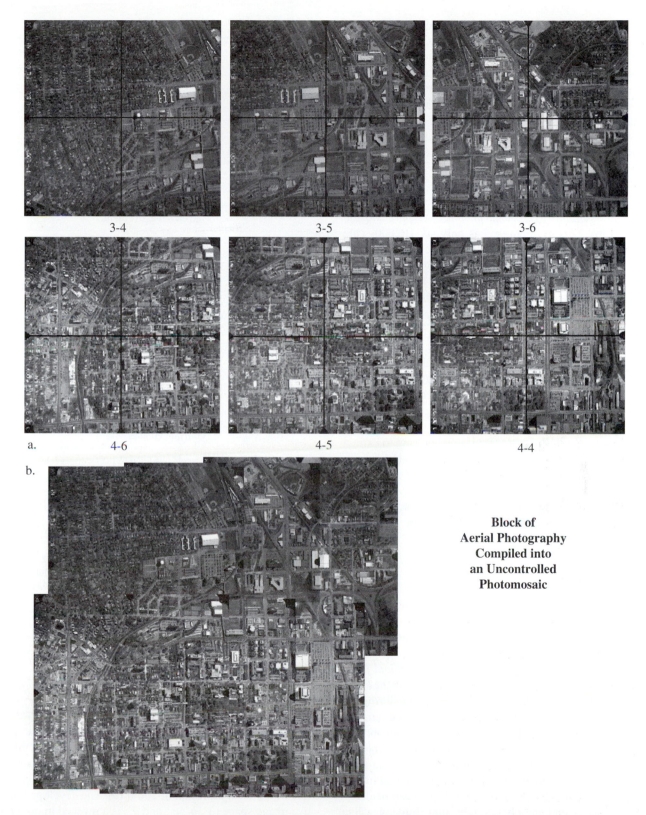

3-4　　　　　　　　　3-5　　　　　　　　　3-6

a.　　　　4-6　　　　　　　　　4-5　　　　　　　　　4-4

b.

**Block of
Aerial Photography
Compiled into
an Uncontrolled
Photomosaic**

Figure 6-3　　a) Two flightlines (#3 and #4) of 1:6,000-scale vertical aerial photography of Columbia, SC, obtained at 3,000 ft above local ground level using a 6-in. (152.82 mm) focal-length lens. b) An uncontrolled photomosaic of the block of aerial photography.

2). For example, consider a portion of a block of vertical aerial photography of downtown Columbia, SC, shown in Figure 6-3a. The block consists of three photographs in each of two flightlines (#3 and #4). The photography was acquired on March 30, 1993, at approximately 3,000 ft above-ground-level with a 6-in. (152.82 mm) focal-length lens and a Wild RC 10 camera yielding approximately 1:6,000-scale photography. There is approximately 60 percent overlap along each flightline and approximately 20 percent sidelap between the two flightlines.

It is possible to combine several vertical photographs in the block of photography to create an *uncontrolled photomosaic* (Figure 6-3b). This example depicts only six 1:6,000-scale photographs. Acquiring photography of a county, state or country sometimes requires thousands of photographs, depending upon the size of the country and the scale of the photography. Table 6-1 provides several metric and English unit equivalents that can be of significant value when measuring the length and width of flightlines.

Much of the aerial photography acquired each year throughout the world is obtained in the early spring leaf-off period when snow is not present and deciduous trees are still dormant. This makes it much easier for the image analyst to see through the branches of the canopy to the ground surface below. This is especially important if one is trying to extract topographic information from stereoscopic aerial photography (to be discussed).

Fiducial Marks, Principal Points, and Conjugate Principal Points

A typical 9 x 9 in. panchromatic vertical aerial photograph is shown in Figure 6-4 (scaled to be 7 x 7-in. for page-printing purposes). Note the eight *fiducial marks* located in the four corners and in the centers of the four sides. Some metric cameras have only four fiducial marks. Drawing a line between opposite fiducial marks locates the *principal point (PP)* of the photograph, which is the exact point on the Earth where the optical axis of the camera was pointing during the instant of exposure (Figure 6-5). By carefully examining the vicinity of the principal point on an airphoto, its location can be visually transferred to each of the two adjacent photos in the flightline. The transferred principal point is commonly called a *conjugate principal point (CPP)*. The actual line of flight (as opposed to the desired line of flight) can be determined by laying out the photography and drawing a line through the principal points and conjugate principal points, as shown in Figure 6-5.

Table 6-1.   Selected Units and Their Equivalents

Metric	English	Conversion
Linear Measurement		
centimeter (cm)	inch (in.)	1 cm = 0.3937 in. 1 in. = 2.54 cm
meter (m)	foot (ft)	1 m = 3.28 ft 1 ft = 0.305 m
kilometer (km)	mile (mi)	1 km = 0.621 mi 1 mi = 1.61 km
Area Measurement		
square centimeter (cm^2)	square inch (sq in.)	1 cm^2 = 0.155 sq in. 1 sq in. = 6.4516 cm^2
square meter (m^2)	square foot (sq ft)	1 m^2 = 10.764 sq ft 1 sq ft = 0.0929 m^2
square kilometer (km^2)	square mile (sq mi)	1 km^2 = 0.3861 sq mi 1 sq mi = 2.59 km^2
hectare (ha)	acre	1 ha = 10,000 m^2 1 ha = 2.471 acres 1 acre = 43,560 sq ft 1 acre = 0.4047 ha
Volume Measurement		
cubic centimeter (cm^3)	cubic inch (cu in.)	1 cm^3 = 0.061 cu in. 1 cu in. = 16.387 cm^3
cubic meter (m^3)	cubic foot (cu ft)	1m^3 = 35.315 cu ft 1 cu ft = 0.02832 m^3
liter	gallon	1 liter = 1,000 cm^3 1 liter = 0.264 gallons 1 gallon = 231 cu in. 1 gallon = 3.7853 liter

Geometry of Vertical Aerial Photography

A diagrammatic representation of the geometry of a single vertical aerial photograph of downtown Columbia, SC, is shown in Figure 6-6. In this example, a 9 x 9-in. negative was obtained at exposure station *L* at 3,200 ft above sea level (*H*) with a local elevation of 200 ft (*h*). The developed negative image space (*a'*, *b'*, *c'*, and *d'*) is a reversal in both *tone* and *geometry* of the Earth object space (*A*, *B*, *C*, and *D*) and is situated a distance equal to the focal length *f* (distance *o'L*)

Figure 6-4   Flightline #4, Photo #5 of vertical panchromatic aerial photography of Columbia, SC, obtained on March 30, 1993, at approximately 3,000 ft above-ground-level using a Wild RC 10 camera with a 6-in. focal-length lens. The photography depicts the state capitol in the lower right corner and the University of South Carolina campus in the center of the image. Eight fiducial marks are present. The "titling" ancillary data contains very valuable information, including a verbal scale (1" = 500') but unfortunately does not show the exact altitude, time of day, amount of tilt, etc., that many metric cameras now routinely provide. This is a 7 x 7-in. reduction of the original 9 x 9-in. 1:6,000-scale photograph. Several photogrammetric measurements will be obtained within the area bounded by the white dashed box.

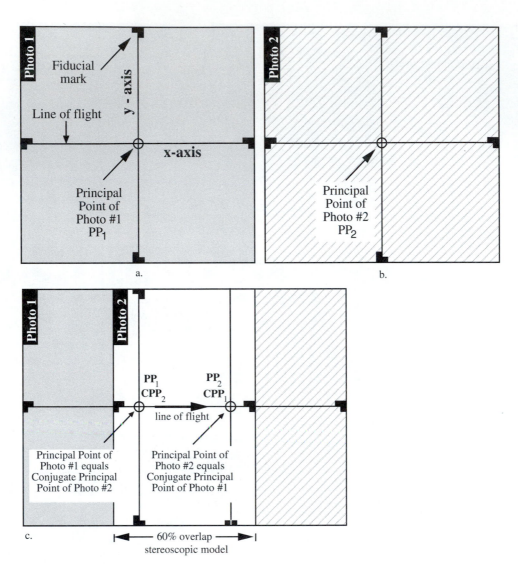

Figure 6-5   a) Geometry of a single aerial photograph (Photo #1). The principal point (PP) is located at the intersection of lines drawn through opposing fiducial marks. b) Geometry of Photo #2. c) Superposition of Photo #1 and Photo #2. The 60 percent overlap area can be viewed stereoscopically. Note the location of the principal point (PP) and conjugate principal points (CPP) on each photograph. A line drawn between the principal points identifies the line of flight.

behind the rear nodal point of the camera lens. A positive version of the scene can be produced by contact-printing the negative with either photographic paper or positive transparency material. The result is a positive print or positive transparency with tone and geometry the same as that of the original object space. The reversal in geometry from negative to positive space is seen by comparing the locations of negative image points *a', b', c',* and *d'* with their corresponding positive image points *a, b, c,* and *d* in Figure 6-6.

The photographic coordinate axes *x* and *y* radiate from the principal point in the positive contact print. These axes are important for photogrammetric measurements, which will

be discussed shortly. Because photointerpreters normally work with positive prints or positive transparencies, they are primarily interested in the positions of images in the positive plane. Consequently, most of the line drawings in this chapter depict the positive rather than the negative plane.

Scale and Height Measurement on Single Vertical Aerial Photographs

The scale of an aerial photograph may be expressed as a *verbal scale* or as a *representative fraction* (dimensionless). For

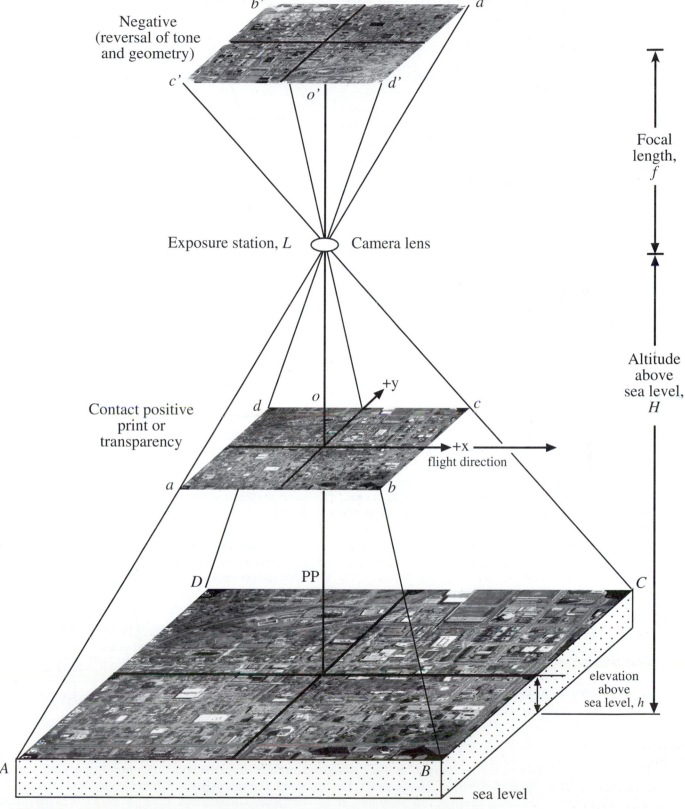

Negative
(reversal of tone
and geometry)

Exposure station, L Camera lens

Contact positive
print or
transparency

+y

+x
flight direction

D PP C

A B

sea level

Focal
length,
f

Altitude
above
sea level,
H

elevation
above
sea level, h

Figure 6-6   The geometry of a vertical aerial photograph obtained over flat terrain (Columbia, SC).

example, if 1 in. on an aerial photograph represents 2,000 ft (24,000 in.) on the ground, the scale may be expressed as a:

- *Verbal Scale:* 1 in. = 2,000 ft

- *Representative Fraction:* $\frac{1}{24,000}$ or 1:24,000.

When comparing two scales, it is important to remember that the larger the number in the scale expression, the *smaller* the scale. For example, compare the scale expressions 1/11,000 and 1/12,000. The image of a given object on the aerial photograph will actually be larger on the larger scale 1/11,000-scale photograph, i.e., a 100-yard football field will be larger on the 1/11,000-scale photo than on the 1/12,000-scale photograph. Table 6-2 is a useful scale conversion chart.

Quite often a verbal scale is found in the titling ancillary information on the border of an aerial photograph. While this scale may be correct, it is much better to calculate the exact scale of every photograph that will be used to extract photogrammetric measurements. Several methods are available for computing the scale of aerial photographs obtained over level or variable relief terrain.

Scale of a Vertical Aerial Photograph Over Level Terrain

There are two main methods of determining the scale of single aerial photographs obtained over level terrain. One involves comparing the size of objects measured in the real world or from a map (e.g., the length of a section of road) with the same object measured on the aerial photograph. The second method involves computing the relationship between camera focal length and altitude of the aircraft above-ground-level.

Computing Scale by Comparing Real World Object Size versus Photographic Image Size

The scale, s, of a vertical aerial photograph obtained over nearly level terrain is the ratio of the size of the object as measured on the aerial photograph, ab, compared to the actual measured length of the object in the real world, AB:

$$s = \frac{ab}{AB}. \tag{6-1}$$

This relationship is based on the geometry of similar triangles *Lab* and *LAB* in Figures 6-6 and 6-7. To compute scale using Equation 6-1, the image analyst first identifies an object in the vertical aerial photograph whose length (*AB*)

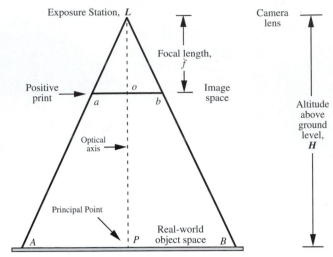

Figure 6-7 The geometry of a vertical aerial photograph collected over relatively flat terrain.

was measured on the ground or, more commonly, was measured from a reference map. The analyst then measures the corresponding distance on the photograph (*ab*) and solves for *s*.

For example, consider the vertical aerial photograph of downtown Columbia, SC, previously shown in Figure 6-4. The titling information says the original photograph was obtained at a nominal scale of 1 in. = 500 ft (1:6,000). But is this correct? Many subsequent computations will make use of the scale parameter, so it is important to make sure that we know the actual scale of the photograph and not just a nominal scale.

In this example, we will first determine the scale of the photograph using the road shown in Figure 6-8. The road width from curb to curb was 56.1 ft (*AB*) as measured in the field using a surveyor's tape. The width of the same road was 0.113 in. (*ab*) on the vertical aerial photograph (Figure 6-8). Using Equation 6-1, the scale of the photograph is:

$$s = \frac{ab}{AB} = \frac{0.113''}{56.1'} = \frac{0.113''}{673.2''} = \frac{1''}{5957.52''}$$

Representative fraction: $\frac{1}{5957}$

Verbal Scale: 1 in. = 496.46 ft.

Just to be sure, the width of a nearby sidewalk was measured. The sidewalk was 6 ft in the real world and 0.012 in. on the aerial photograph. If we use the width of the sidewalk to compute the scale of the photography, we get 1:6,000:

Table 6-2.   Scale Conversion Chart (after Rasher and Weaver, 1990)

Scale	feet per in. (a)	in. per mile	miles per in.	acres per sq. in. (b)	hectares per sq. in.	meters per cm	cm per km	km per cm
1:500	41.67	126.72	0.008	0.03986	0.0161	5.00	200	0.005
1:600	50.00	105.60	0.009	0.05739	0.0232	6.00	166.66	0.006
1:1,000	83.33	63.36	0.016	0.15940	0.0645	10.00	100	0.010
1:2,000	166.67	31.68	0.032	0.63771	0.2581	20.00	50.00	0.020
1:3,000	250.00	21.12	0.047	1.4348	0.5807	30.00	33.33	0.030
1:4,000	333.33	15.84	0.063	2.5507	1.0323	40.00	25.00	0.040
1:5,000	416.67	12.67	0.079	3.9856	1.6129	50.00	20.00	0.050
1:6,000	500.00	10.56	0.095	5.7392	2.3226	60.00	16.66	0.060
1:10,000	833.33	6.336	0.158	15.9421	6.4517	100	10.00	0.100
1:12,000	1000.00	5.280	0.189	22.9568	9.29058	120.00	8.33	0.120
1:15,000	1250.00	4.224	0.237	35.8700	14.5164	150.00	6.66	0.150
1:20,000	1666.67	3.168	0.316	63.7692	25.8070	200.00	5.0	0.200
1:24,000	2000.00	2.640	0.379	91.8273	37.1620	240.00	4.166	0.240
1:25,000	2083.33	2.534	0.395	99.6387	40.3234	250.00	4.00	0.250
1:48,000	4000.00	1.320	0.758	367.309	148.6481	480.00	2.08	0.480
1:50,000	4166.67	1.267	0.789	398.556	161.2935	500.00	2.00	0.500
1:63,360	5280.00	1	1	640.000	259.0056	633.60	1.58	0.634
1:100,000	8333.33	0.634	1.578	1594.00	645.174	1,000.00	1	1
1:250,000	20833.33	0.2534	3.946	9963.90	4032.338	2,500.00	0.40	2.50
1:500,000	41666.67	0.1267	7.891	39855.63	16129.35	5,000.00	0.2	5.00
1:1,000,000	83333.33	0.0634	15.783	159422.49	64517.41	10,000.00	0.100	10.00
	$\dfrac{scale}{12}$	$\dfrac{63360}{scale}$	$\dfrac{scale}{63360}$	$\dfrac{(a)^2}{43560}$	$\dfrac{(b)}{2.471}$	$\dfrac{scale}{100}$	$\dfrac{100000}{scale}$	$\dfrac{scale}{100000}$

$$s = \frac{ab}{AB} = \frac{0.012''}{6'} = \frac{0.012''}{72''} = \frac{1''}{6000''}$$

Representative fraction: $\dfrac{1}{6000}$

Verbal scale: 1 in. = 500 ft.

When computing scale, the goal is to modify the input values so that the numerator has a value of one and both the numerator and denominator are in the same units, e.g., in inches, feet, or meters. Sometimes this requires multiplying both the numerator and denominator by the reciprocal of the numerator. The best way to conceptualize scale is as a representative fraction. Therefore, in this last example 1 in. on the photograph equals 6,000 in. in the real world; 1 ft on the photograph equals 6,000 ft; and 1 m on the photograph equals 6,000 m. With a representative fraction, we can work in any units we desire.

Figure 6-8 An extreme enlargement of a portion of the aerial photograph shown in Figure 6-4. The graphic shows the width of the road from curb to curb (56.1') and the sidewalk (6') as measured using a surveyor's tape and measured on the aerial photograph (0.113" and 0.012", respectively). These measurements may be used to compute the scale of the photograph in this portion of the photograph.

To offset the effects of aircraft tilt at the instant of exposure, the measurement of scale using this method should be based on the average of several scale checkpoints in the photograph. When possible, some analysts like to compute the scale by selecting objects along lines such as roads that intersect approximately at right angles and are centrally located on the photograph. They then compute the mean of the multiple-scale measurements.

Vertical aerial photography scale can also be determined if objects whose lengths are already known appear on the photograph. A baseball diamond, football field, or soccer field may be measured on the photograph and the photo scale determined as the ratio of the photo distance to the known ground distance. For example, what is the scale of a vertical aerial photograph on which the distance between home plate and first base equals 0.5"? The distance from home plate to first base is 90 ft. Therefore,

$$s = \frac{ab}{AB} = \frac{0.5''}{90'}$$

$$s = \frac{0.5''}{90' \times 12''}$$

$$s = \frac{0.5''}{1080''}$$

$$s = \frac{1''}{2160''}$$

Representative Fraction: $\frac{1}{2160}$

Examples of sports field dimensions found throughout the world are summarized in Chapter 12 (Remote Sensing the Urban Landscape).

Computing Scale by Relating Focal Length to Altitude Above-Ground-Level (AGL)

Scale may also be expressed in terms of camera focal length, f, and flying height above the ground, H, by equating the geometrically similar triangles Loa and LPA in Figure 6-7:

$$s = \frac{f}{H}. \tag{6-2}$$

From Equation 6-2 it is evident that the scale of a vertical aerial photograph is directly proportional to camera focal length (image distance) and inversely proportional to flying height above-ground-level (object distance). This means that if the altitude above-ground-level is held constant, increasing the size of the focal length will result in larger images of objects at the film plane. Conversely, if focal length is held constant, the images of objects will be smaller as we gain additional altitude above-ground-level. If the focal length of the camera and the altitude of the aircraft above-ground-level are unknown, the image analyst will have to consult the ancillary information appearing on the edges of the photograph, the flight roll, or the aerial photographer's mission summary.

For example, a vertical aerial photograph is obtained over flat terrain with a 12-in. focal-length camera lens from an altitude of 60,000 ft above-ground-level. Using Equation 6-2, the scale is:

$$s = \frac{f}{H} = \frac{12''}{60000'}$$

$$s = \frac{1'}{60000'}$$

Representative fraction: $= \dfrac{1}{60000}$

Verbal scale: 1 in. = 5000 ft .

By transposing Equation 6-2 to read,

$$H = \frac{f}{s} \qquad (6\text{-}3)$$

or

$$f = H \times s \qquad (6\text{-}4)$$

it is possible to determine the altitude above-ground-level, H, of the photograph at the instant of exposure, or the focal length, f, of the camera if the scale of the photograph, s, is known along with one of the other variables. For example, in the case of the Columbia, SC, photograph (Figure 6-4) we know that the scale of the photograph based on the measurement of road width is 1:5,957 or 1" = 5,957", and the focal length is 6 in. Using Equation 6-3, we can determine the altitude of the aircraft above local datum at the instant of exposure:

$$H = \frac{6''}{\left(\dfrac{1''}{5957''}\right)} = 35742'' = 2978.5' .$$

Similarly, if we only knew the altitude above datum (2,978.5') and the scale (1:5,957) we could compute the focal length of the camera system using Equation 6-4:

$$f = 2978.5' \times \frac{1'}{5957'} = 0.5' = 6'' .$$

Scale of a Vertical Aerial Photograph Over Variable Terrain

One of the principal differences between a near-vertical aerial photograph and a planimetric map is that for photographs taken over variable terrain, there are an infinite number of different scales present in the photograph. If topographic elevation decreases within a certain portion of the aerial photograph relative to other areas, then that portion of the photograph will have a smaller scale than the rest of the photograph because the land will have "moved away" from the aerial camera that is flown at a constant altitude. Conversely, if a topographic feature such as a mountain or a building protrude above the average elevation of the local terrain, then the scale in that area of the photograph will be

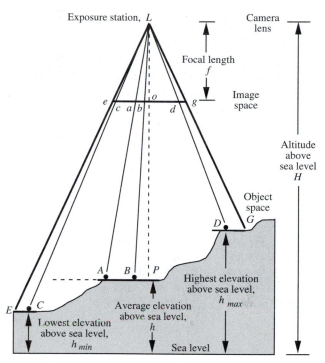

Figure 6-9 Geometry of a vertical aerial photograph obtained over terrain with variable relief.

larger because the land will have "moved closer" to the aerial camera.

The geometry of a single vertical aerial photograph taken over terrain with variable local relief from exposure station L is shown in Figure 6-9. Points A and B in the real world object space are located on level terrain and recorded on the positive print image space at a and b, respectively. Points A and B are located at an elevation of h units above sea level. As previously discussed, the scale of the aerial photograph at location ab in the positive print is equal to the ratio of the photo distance ab to ground distance AB. By similar triangles Lab and LAB, the photo scale at location ab in the vertical aerial photograph may be computed using several relationships, including:

$$s_{ab} = \frac{ab}{AB} = \frac{Lb}{LB} . \qquad (6\text{-}5)$$

Also, by similar triangles Lob and LPB,

$$\frac{Lb}{LB} = \frac{Lo}{LP} = \frac{f}{H-h} . \qquad (6\text{-}6)$$

Substituting Equation 6-6 into Equation 6-5,

$$s_{ab} = \frac{ab}{AB} = \frac{f}{H-h}. \qquad (6\text{-}7)$$

If the line between points a and b in the vertical aerial photograph is considered to be infinitely small, Equation 6-7 reduces to an expression for photo scale at a specific point. Dropping subscripts, the scale at any point whose elevation above sea level is h and whose altitude above sea level is H, may be expressed as (Estes et al., 1983):

$$s = \frac{f}{H-h}. \qquad (6\text{-}8)$$

In our example in Figure 6-9, different scale values would be computed at locations c and d in the aerial photograph. One scale value would be a function of the minimum elevation above sea level within the photograph (s_{min}) while the other would be a function of the maximum elevation above sea level within the photograph (s_{max}):

$$s_{min} = \frac{f}{H-h_{min}} \qquad (6\text{-}9)$$

$$s_{max} = \frac{f}{H-h_{max}}. \qquad (6\text{-}10)$$

Usually an average or *nominal* scale is computed to define the overall scale of a vertical aerial photograph taken over variable terrain:

$$s_{avg} = \frac{f}{H-h_{avg}}. \qquad (6\text{-}11)$$

It should be remembered that the average scale is only at those points which lie at average elevation, and it is only an approximate scale for all other locations on the photograph.

To demonstrate these relationships, let us consider the maximum elevation, h_{max}, average elevation, h_{avg}, and minimum elevation, h_{min}, of the terrain in Figure 6-9 to be 10,000, 8,000, and 6,000 ft above sea level, respectively. The flying height of the aircraft above sea level is 20,000 ft, and the camera focal length is 6 in. (152.82 mm). The maximum, minimum, and average scale of the vertical aerial photograph would be:

$$s_{max} = \frac{f}{H-h_{max}}$$

$$s_{max} = \frac{6''}{20000' - 10000'}$$

$$s_{max} = \frac{6''}{10000'}$$

$$s_{max} = 1:20000$$

$$s_{min} = \frac{f}{H-h_{min}}$$

$$s_{min} = \frac{6''}{20000' - 6000'}$$

$$s_{min} = \frac{6''}{14000'}$$

$$s_{min} = 1:28000$$

$$s_{avg} = \frac{f}{H-h_{avg}}$$

$$s_{avg} = \frac{6''}{20000' - 8000'}$$

$$s_{avg} = \frac{6''}{12000'}$$

$$s_{avg} = 1:24000$$

Height Measurement from Single Aerial Photographs

There are two primary methods of computing the heights of objects on single vertical aerial photographs. The first involves the measurement of image relief displacement, and the second is based on the measurement of shadow length.

Height Measurement Based on Relief Displacement

The image of any object lying above or below the horizontal plane passing through the elevation of the principal point is displaced on a truly vertical aerial photograph from its true planimetric (x,y) location. The *relief displacement* is outward from the principal point for objects whose elevations are above the local datum, and inward or toward the principal point for objects whose elevations are below the local datum. The direction of relief displacement is radial from the principal point of the photograph.

The amount of the relief displacement, d, is:

- *directly proportional* to the difference in elevation, h, between the top of the object whose image is displaced and the local datum, i.e., the greater the height of the object above the local datum, the greater its displacement.

- *directly proportional* to the radial distance, r, between the top of the displaced image and the principal point, i.e., the farther the object is from the principal point, the greater the displacement.

- *inversely proportional* to the altitude, H, of the camera above the local datum. Therefore, a reduction in relief displacement of an object can be achieved by increasing the flying height.

These relationships are depicted in Figure 6-10. Notice from similar triangles in the diagram that:

$$\frac{h}{H} = \frac{d}{r}. \qquad (6\text{-}12)$$

Rearranging the relationship, we can see that the amount of displacement, d, is directly proportional to the height of the object, h, and its distance from principal point, r, and inversely proportional to the altitude above local datum, H:

$$d = \frac{h \times r}{H}. \qquad (6\text{-}13)$$

If we solve for the height of the object, h, the equation becomes:

$$h = \frac{d \times H}{r}. \qquad (6\text{-}14)$$

Therefore, we can compute the height of an object from its relief displacement characteristics on a single vertical aerial photograph. It is important that both the top and the bottom of the object being measured are clearly visible and that the base is on level terrain.

For example, let us compute the height of the Senate Condominium shown in Figure 6-10a. The photograph has been rotated 90° counterclockwise to facilitate viewing (shadows now fall toward the viewer, and the displaced façade of the building leans comfortably away from the viewer). Based on previous measurements, the altitude of the camera above local datum, H, is known to be 2,978.5 ft, the radial distance from the principal point to the top of the building, r, is measured as 2.23 in., and building relief displacement, d, is

0.129 in. The photogrammetrically computed height, h, of the condominium is:

$$h = \frac{0.129'' \times 2978.5'}{2.23''} = 172.3'.$$

The actual height of the building measured with a surveyor's tape is 172.75 ft. To obtain accurate object height measurements using this technique, it is imperative that the altitude of the aircraft above the local datum be as precise as possible. Also, great care should be exercised when measuring r and d on the photograph. Keep in mind that r is measured from the principal point to the top of the object.

Height Measurement Based on Shadow Length

The height of an object, h, may be computed by measuring the length of the shadow cast, L, on vertical aerial photography. Because the rays of the Sun are essentially parallel throughout the area shown on vertical aerial photographs, the length of an object's shadow on a horizontal surface is proportional to its height. Figure 6-11 illustrates the trigonometric relationship involved in determining object heights from shadow measurements. Notice that the tangent of angle a would be equal to the opposite side, h, over the adjacent side, which is the shadow length, L, i.e.,

$$\tan a = \frac{h}{L}. \qquad (6\text{-}15)$$

Solving for height yields:

$$h = L \times \tan a. \qquad (6\text{-}16)$$

The Sun's elevation angle, a, above the local horizon can be predicted using a solar ephemeris table. This requires a knowledge of the approximate geographic coordinates of the site (longitude and latitude), and the acquisition date and time of day. Alternatively, the solar altitude may be empirically computed if sharply defined shadows of known height are formed on the photograph. For example, we know from previous discussion that the height of the Senate Condominium in Figure 6-12 is 172.75 ft. It casts a shadow onto level ground that is 0.241" in length on the photograph. The scale of the photography is 1:5,957 or 1" = 496.46'. Therefore, the shadow length on the photograph is 119.65 ft. The tangent of angle a can be found using Equation 6-15:

$$\tan a = \frac{h}{L} = \frac{172.75'}{119.65'} = 1.44$$

Other shadow lengths on the same aerial photograph can be measured and their lengths multiplied by 1.44 to determine

a.

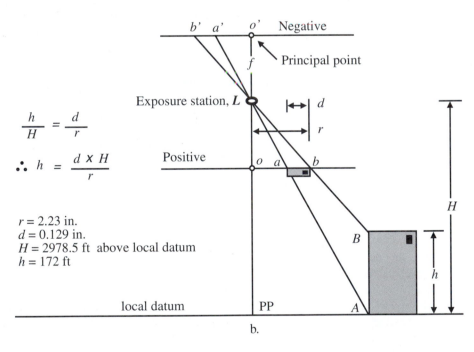

$$\frac{h}{H} = \frac{d}{r}$$

$$\therefore \quad h = \frac{d \times H}{r}$$

$r = 2.23$ in.
$d = 0.129$ in.
$H = 2978.5$ ft above local datum
$h = 172$ ft

b.

Figure 6-10   Measurement of object height (in this case, the Senate Condominium in Columbia, SC) from a single vertical aerial photograph based on relief displacement principles. a) The radial distance from the principal point (PP) and the top of the building is r. The distance from the base of the building to the top of the building is d. b) The height of the building (h) above local datum is computed based on the relationship between similar triangles.

Measurement of the Height of Objects Based on Shadow Length

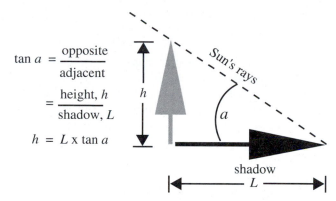

$$\tan a = \frac{\text{opposite}}{\text{adjacent}}$$

$$= \frac{\text{height, } h}{\text{shadow, } L}$$

$$h = L \times \tan a$$

Figure 6-11   The height of objects can be measured from vertical aerial photography based on the length of shadows cast on level terrain.

their heights. For example, the tower on the nearby parking structure in Figure 6-12 casts a shadow that is 0.119", or 59.1' long onto level ground in the photograph. Therefore, the height of the tower is:

$$h = L \times \tan a = 59.1' \times 1.44 = 85.10'$$

The actual height of the tower is 86 ft.

Care must be exercised when computing the height of objects based on shadow length in aerial photography. Important factors to be considered include shadows falling on nonlevel terrain, shadows produced from leaning objects, shadows not cast from the true top of the object, and snow or other types of groundcover obscuring the true ground level.

 ### Stereoscopic Measurement of Object Height or Terrain Elevation

A single aerial photograph captures a precise record of the positions of objects in the scene at the instant of exposure. If we acquire multiple photographs along a flightline, we record images of the landscape from different vantage points. For example, the top of a tall building might be on the left side of photo #1 and in the middle of overlapping photo #2 because the aircraft has moved hundreds of meters between exposures. If we opened up the back of the aerial camera, held the shutter wide open, and looked at the groundglass at the focal plane while the aircraft was flying along a flightline, we would literally see the tall building first

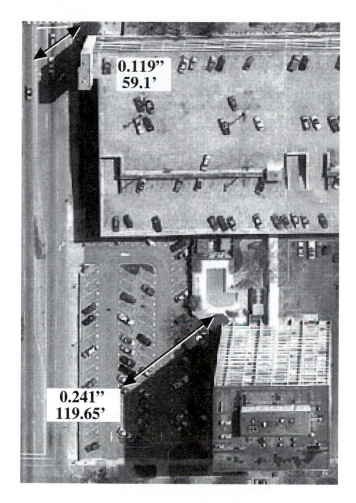

Figure 6-12   The height of the parking garage tower may be computed by determining the relationship between its shadow length and the length of the shadow cast by a building of known height such as the nearby Senate Condominium. It is important that the shadow falls on level ground and that both the beginning and end of the shadow are visible.

enter the groundglass field of view and then traverse across the groundglass until it eventually leaves the camera's field of view.

The change in position of an object with height, from one photograph to the next relative to its background, caused by the aircraft's motion, is called *stereoscopic parallax*. *Parallax* is the apparent displacement in the position of an object, with respect to a frame of reference, caused by a shift in the position of observation. Parallax is a normal characteristic of aerial photography and is the basis for three-dimensional stereoscopic viewing. Differences in the parallax of various objects of interest (called *differential parallax*) can be used

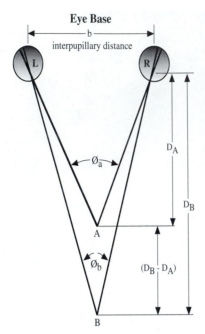

Figure 6-13 Parallactic angles are formed when our eyes focus on objects in the real world. The mind associates differences in parallactic angles with differences in distance to the various objects. This allows us to have very sensitive depth perception.

to measure the heights of objects and to extract topographic information such as contour lines from aerial photographs by means of stereoscopic instruments. The measurement of stereoscopic parallax is the basis for the creation of almost all topographic maps (x, y, and z) and planimetric maps (x,y) and is a very important element of photogrammetry and remote sensing.

Fundamentals of Human Stereoscopy

Stereoscopy is the science of perceiving depth using two eyes. When a human being's two eyes (*binocular vision*) are focused on a certain point, the optical axes of the eyes converge on that point, forming a *parallactic angle (ϕ)*. The nearer the object, the greater the parallactic angle. For example, in Figure 6-13 the optical axes of the left and right eyes, L and R, are separated by the *eye base* or *interpupillary distance*. The eye base of the average adult is between 63 and 69 mm (approximately 2.5 to 2.76 in.). When the eyes are focused on point A, the optical axes converge, forming parallactic angle ϕ_a. Similarly, when looking at point B, the optical axes converge, forming parallactic angle ϕ_b. The brain has learned to associate distances D_A and D_B with corresponding parallactic angles ϕ_a and ϕ_b and gives the viewer

the visual and mental impression that object A is closer than object B. This is the basis of *depth perception*. If both objects were exactly the same distance from the viewer, then $\phi_a = \phi_b$ and the viewer would perceive them as being the same distance away.

When we walk outside, the maximum distance at which distinct stereoscopic depth perception is possible is approximately 1000 m for the average adult. Beyond that distance, parallactic angles are extremely small, and changes in parallactic angle necessary for depth perception may not be discerned. This is why humans have trouble determining whether one house is behind another house, or one car is behind another car when these objects are thousands of meters away from us. Conversely, if we could somehow stretch our eyes to be a meter or even hundreds of meters apart, then we would be able to resolve much more subtle differences in parallactic angles and determine which objects are closer to us over much greater distances. Such *hyperstereoscopy* depth perception would be ideal for hunting and sports activities, but it would require a substantial modification of the human head. Fortunately, there is a simple method that we can use to obtain a hyperstereoscopy condition when collecting and interpreting stereoscopic aerial photography.

Stereoscopy Applied to Aerial Photography

Overlapping aerial photography (usually 60 percent endlap) obtained at exposure stations along a flightline contain stereoscopic parallax. The exposure stations are separated by large distances. Nevertheless, it is possible to let our eyes view the photographs as if our eyes were present at the two exposure stations at the instant of exposure (Figure 6-14a). This results in a *hyperstereoscopy* condition that allows us to view the terrain in three dimensions. We normally view the stereoscopic aerial photography using a lens or mirror stereoscope with magnifying lenses. These instruments enhance the three-dimensional nature of the stereoscopic model. There are other stereoscopic viewing alternatives.

Methods of Stereoscopic Viewing

The photointerpreter can view the vertically exaggerated stereoscopic model of the 60 percent endlap area of two successive aerial photographs using one of four methods: 1) keeping the lines of sight parallel with the aid of a stereoscope, 2) keeping the lines of sight parallel without the aid of a stereoscope, 3) crossing the eyes and reversing the order of the stereoscopic images, or 4) using anaglyphic or polarizing glasses.

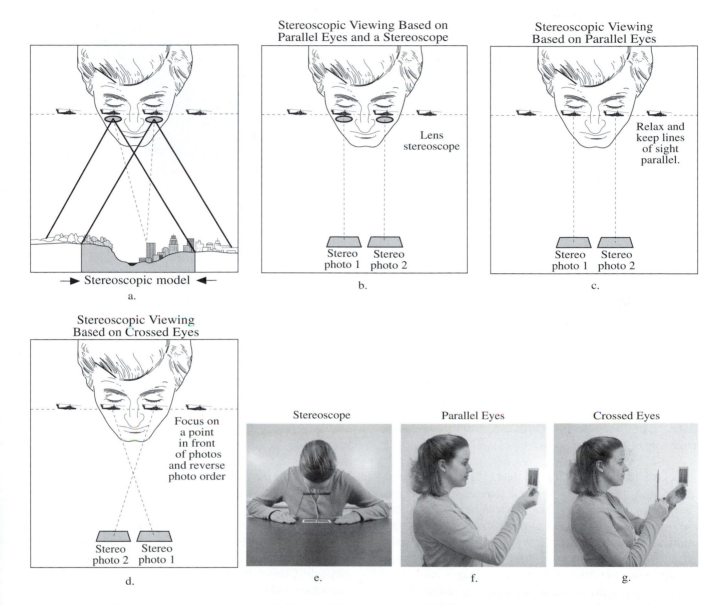

a. Stereoscopic model

b. Stereoscopic Viewing Based on Parallel Eyes and a Stereoscope — Lens stereoscope — Stereo photo 1, Stereo photo 2

c. Stereoscopic Viewing Based on Parallel Eyes — Relax and keep lines of sight parallel. — Stereo photo 1, Stereo photo 2

d. Stereoscopic Viewing Based on Crossed Eyes — Focus on a point in front of photos and reverse photo order — Stereo photo 2, Stereo photo 1

e. Stereoscope

f. Parallel Eyes

g. Crossed Eyes

Figure 6-14 a) A hyperstereoscopy condition can be achieved by taking overlapping vertical aerial photographs along a flightline. The analyst then views one photo of the stereopair with the left eye and the adjacent photo with the right eye. When the viewer focuses both eyes on a single feature within the stereoscopic model, such as the top or base of a building, the mind subconsciously associates differences in the parallactic angle between the base and the top of the building as depth-perception information. We perceive a realistic three-dimensional model of the terrain in our minds. This means that we can literally view any landscape in the world from the comfort of our office in very realistic three dimensions if we have stereoscopic aerial photography and the proper viewing equipment. b,e) A lens or mirror stereoscope assists our eyes' lines of sight to remain parallel rather than converging, which is our natural tendency. This analyst is looking at a specially prepared stereopair of photographs. c,f) It is possible to obtain a stereoscopic view by training the eyes' lines of sight to remain parallel without using a stereoscope. This is very useful in the field. d,g) Some people can see stereo by focusing on a point in front of the stereoscopic photographs and letting their lines of sight cross. This is not a natural thing to do and can cause eye strain.

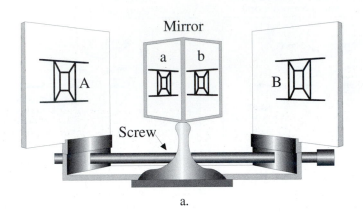

Wheatstone's Mirror Stereoscope

Figure 6-15 a) Wheatstone's mirror stereoscope used two 45° angled mirrors (a and b) to reflect the left and right images of photographs (A and B) toward the viewer. A screw mechanism adjusted the views to accommodate various interpupillary distances. b) An example of Brewster's lens stereoscope.

The vast majority of image analysts prefer to use a simple lens pocket stereoscope or mirror stereoscope that assists the eyes in keeping parallel lines of sight and in addition, usually magnifies the photographs (Figure 6-14b,e). This produces some eye strain. It is suggested that novice interpreters only view photographs in stereo for 10 to 15 minutes at a time in the beginning while the eyes become accustomed to using a stereoscope.

Some people are adept at forcing their eyes to remain parallel and thus do not need to use a stereoscope (Figure 6-14c,f). They simply situate the overlapping portion of two stereo photographs adjacent to one another, position their head approximately 8 in. from the photographs, and then let their eyes relax as if they were looking at infinity. Gradually, the mind will fuse the two stereoscopic images into a third image directly in the middle of the two stereo photos. This is a good skill to acquire since one can then easily view stereoscopic photographs without a stereoscope whenever the need arises. It is particularly useful when conducting field work. However, this is unnatural for the eyes and may be uncomfortable and cause eye strain.

Some image analysts are able to reverse the order of the stereo photographs (Figure 6-14d,g) and then fixate on a point directly in front of the photos. This causes the eyes to cross. This produces a true stereoscopic impression, but it is very strenuous on the eyes and is not recommended.

Another way of making sure that the left and right eyes view distinct images is to use *anaglyphic* or *polarizing* glasses in conjunction with specially prepared image materials. It is possible to produce aerial photography where the left image is depicted in shades of blue and the right image is projected in shades of red. The analyst then wears anaglyphic glasses with blue (left) and red (right) lenses. The mind fuses the two views and creates a three-dimensional scene. Similarly, it is possible to view the left and right photographs through specially prepared polarizing glasses that accomplish the same goal.

Lens and Mirror Stereoscopes and Stereo Cameras

The *stereoscope* is a binocular viewing system specially developed to analyze terrestrial stereoscopic photographs (*not* aerial photographs). It was invented by the English physicist Charles Wheatstone in 1833, although he did not describe it publicly until 1838. Wheatstone used a pair of mirrors, one before each eye, oriented at 45° to allow pictures placed at either side to be fused by the eyes, as shown in Figure 6-15a. David Brewster invented an alternative stereoscopic system using lenses in 1849 (Figure 6-15b). The pictures were smaller, but they were enlarged by the lenses. This became the parlor stereoscope so popular during the 1800s and early 1900s for viewing specially prepared stereoscopic photographs.

Throughout most of the 1800s, stereo photographs were obtained by taking one exposure and then picking up the camera on a tripod and moving it a certain distance to the left or right and taking another picture. This introduced stereoscopic parallax between the two photographs. For example, consider the 1899 stereogram of the Salt Lake Temple, in

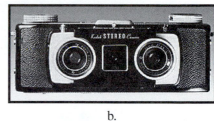

a.

b.

Figure 6-16   a) A terrestrial stereogram of the temple in Salt Lake City, UT, obtained in 1899. b) A vintage stereo camera.

Salt Lake City, Utah, in Figure 6-16a. A wealth of historical information about the temple is available when the stereopair is viewed using a stereoscope.

Stereoscopic photographs may also be acquired using special stereoscopic cameras that contain two identical lenses that are situated 60 to 70 mm apart (2.36 – 2.75") that take two photographs of the scene at exactly the same time. The cameras obtain photographs of objects from slightly different vantage points, introducing stereoscopic parallax. Note the two separate lenses on the vintage stereoscopic camera in Figure 6-16b. Companies continue to produce stereoscopic cameras.

The development of the concept of stereoscopic photographs and the invention of stereoscopic cameras allowed people to view marvelous stereoscopic views of local and foreign landscapes in three dimensions as if they were really at the location. There continues to be significant interest in the collection of stereoscopic photography by the public. The National Stereoscopic Association promotes the study and collection of stereographs, stereo cameras, and related materials and publishes *Stereo World*.

Viewing Stereoscopic Aerial Photographs

The same stereoscopic principles used in the original stereoscopes are used in our current photogrammetric stereoscopes. The simple *pocket lens stereoscope* consists of two convex lenses mounted on a rigid metal or plastic frame (Figure 6-17a). The distance between the lenses can be varied to accommodate various eye bases (interpupillary distances). The special lenses help keep the viewer's lines of sight parallel and also magnify the photography. The proper method of arranging stereoscopic photographs for analysis using a pocket lens stereoscope is demonstrated in Figure 6-18ab. First, the principal point and conjugate principal points are located (PP and CPP, respectively) on each photograph. A line is then drawn through them on each photograph. This is the line of flight previously discussed. The flightlines on each of the photographs are oriented so that they form a continuous line (i.e., they become *colinear*). The analyst then slides one of the photographs left or right so that a portion of the stereoscopic overlap area is visible. Then, the stereoscope is placed above the overlap area and stereoscopic viewing takes place. The common overlap area of a pair of 9 x 9-in. aerial photographs taken with 60 percent overlap is about 5.4 in. that can be viewed in stereo. Unfortunately, when the photographs are aligned for stereovision using the pocket stereoscope, not all of the 5.4 in. of the stereo model can be seen at one time. When this occurs, the interpreter can gently lift up the edge of the top photograph to see what is underneath.

A *mirror stereoscope* (Figure 6-17b) permits the entire stereoscopic model of the two overlapping aerial photographs to be viewed. Mirror stereoscopes often have magnification options (e.g., 2x, 3x, 6x) available that greatly increase the interpreter's ability to magnify and interpret fine detail in the stereo model. For example, a more sophisticated and expensive mirror stereoscope with zoom magnification is shown attached to a light table in Figure 6-17c. This configuration allows the image analyst to view stereoscopically successive overlapping aerial photographs on a roll of film without having to cut the roll of film. Still more sophisticated instruments, based on the camera-lucida principle, allow the

a. Lens stereoscope

b. Mirror stereoscope

c. Zoom stereoscope

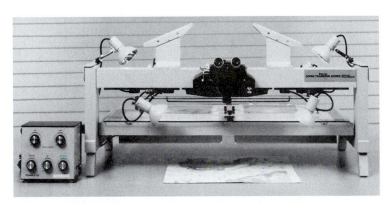

d. Analog stereoscopic zoom-transfer-scope

Figure 6-17   a) Lens stereoscope with attached parallax bar, situated over stereoscopic aerial
photographs having 60 percent overlap.
Only a portion of the entire stereoscopic
model can be viewed at one time using a
lens stereoscope. b) An analyst using a
mirror stereoscope can view the entire
stereoscopic model at one time. c) A
zoom stereoscope functions like a mirror
stereoscope. In this example, positive
transparency photographs are being
viewed on a light table. d) An analog stereoscopic zoom-transfer-scope can be
used to view stereoscopic aerial photographs and a map at the same time (courtesy Image Interpretation Systems, Inc.).
e) A digital stereoscopic zoom-transfer-
scope (courtesy Optem, Inc.).

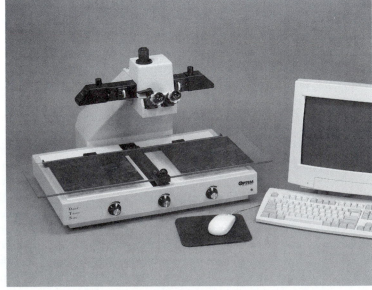

e. Digital stereoscopic zoom-transfer-scope

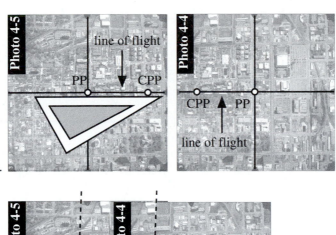

Initial stereoscopic
airphoto alignment

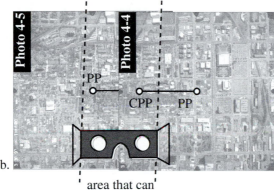

Final alignment

area that can
be viewed
using a lens
stereoscope

Figure 6-18   Preparation of a stereopair for viewing using a lens stereoscope. a) The principal point and conjugate principal point are identified on each photograph, and a line is drawn through them with a straight edge. This locates the line of flight. b) A portion
of the common overlap area that is of interest to the interpreter is moved parallel with the line of flight. The interpreter then
views the overlap area using the lens stereoscope.

analyst to view stereoscopic imagery while simultaneously viewing a map in superposition (Figure 6-17d). This is commonly called a *zoom-transfer-scope* and is often used to transfer information interpreted from aerial photography onto a map base. Digital stereoscopic zoom-transfer-scopes are now available (Figure 6-17e).

Stereoscopic Aerial Photography – How Does It Work?

A hypothetical example will demonstrate how the image analyst perceives the third dimension in stereoscopic aerial photography using a stereoscope. First, consider the profile view of two stereoscopic aerial photographs taken from two exposure stations, L_1 and L_2 in Figure 6-19. Let us evaluate the characteristics of the top of a very tall building (which we will designate as object A in the real-world object space) and the top of a smaller building (object B in the real-world object space) lying on extremely flat ground just above sea level. The distance between the two exposure stations, L_1

and L_2, is called the *air base (A-base)*. The air base in effect becomes a stretched interpupillary distance. This condition is responsible for the exaggerated third dimension when analyzing the photographs stereoscopically.

In our example, objects A and B are recorded on the right photograph (L_1) at locations a and b, respectively (Figure 6-19b). The aircraft then continues along its designated line of flight. The forward progress of the aircraft causes the *image* of objects a and b to move across the camera's focal plane parallel to the line of flight. When exposure L_2 is made, it captures the image of objects A and B at locations a' and b', on the left photograph (Figure 6-19a). This change in position of an image of an object from one photograph to the next caused by the aircraft's motion is referred to as stereoscopic *x-parallax* (Wolf, 1983).

There are some very interesting properties associated with *x-parallax* that allow us to obtain precise photogrammetric measurements from vertical aerial photography. To under-

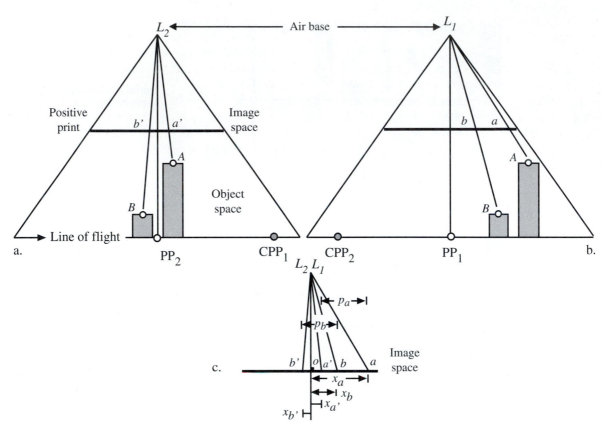

Figure 6-19   Stereoscopic parallax is introduced when an object is viewed from two different vantage points. a,b) Objects *A* and *B* in the real world are recorded as points *a* and *b* in exposure L_1 and at *a'* and *b'* in exposure L_2. c) When the locations of *a* and *b* and *a'* and *b'* are placed in superposition, the image of object *A* has greater x-parallax (p_a) than the image of object *B* (p_b). This difference in parallax (called *differential parallax, dp*) can be related to the actual height of the objects.

stand these relationships, consider Figure 6-19c, which depicts the two photographs taken at exposure stations L_1 and L_2 in *superposition*. Superposition means that we adjust the profile views of photos L_1 and L_2 so that the vertical lines running through each of the photos' principal points (PP_1 and PP_2) are superimposed on top of one another. This allows us to determine how much the objects have moved across the film plane from one exposure to the next.

Notice that the image of object *A* moved from *a* to *a'* on the two successive aerial photographs. The parallax of point *a* is $p_a = x_a - x_{a'}$. Similarly, the image of object *B* moved from *b* to *b'*. The parallax of point *b* is $p_b = x_b - x_{b'}$. The important thing to recognize is that *a* (the taller building) moved a greater distance across the film plane than *b* (the shorter building) because *a* had greater stereoscopic parallax than *b*.

It is also very important to understand that *all objects in the scene at exactly the same elevation will have an identical*

amount of x-parallax. For example, any point located on the top of building *A* will have exactly p_a x-parallax when the stereoscopic photographs are analyzed. Similarly, any other building in the entire stereoscopic model that had the same height as building *A* would have the same x-parallax.

Another way of thinking about x-parallax is to place a beach ball (which we will call object *A)* anywhere on a very flat plane, such as the Bonneville Salt Flats in Utah or on a lake, and then acquire stereoscopic photographs. No matter where the beach ball was in the stereoscopic model, its image, *a*, on the first photograph would travel exactly the same distance along the image coordinate system to *a'* on the adjacent overlapping photograph because the ball would have exactly the same amount of x-parallax everywhere in the stereo model! Therefore, the x-parallax of any point is:

• directly related to the elevation of the point above mean terrain, and

- greater for high points (e.g., those closer to the aerial camera) than for low points.

The ability to measure small differences in the parallax (i.e., the *differential parallax*) between any two points using two overlapping photographs allows us to determine elevation differences using stereoscopic parallax equations. This is how almost all topographic maps are made and how very precise tree and building height measurements can be obtained using stereo photographs. In order to do this, we use the fundamental stereoscopic parallax equation.

Measurement of Absolute and Differential Parallax

To compute the height of an object, h_o, using stereoscopic photogrammetric techniques, we use the *parallax equation*:

$$h_o = (H - h) \times \frac{dp}{(P + dp)} \qquad (6\text{-}17)$$

where $H - h$ is the altitude of the aircraft above-ground-level (AGL); P is the *absolute stereoscopic parallax* at the base of the object being measured (we usually use the air base for this measurement), and dp is the differential parallax.

This equation may yield incorrect results unless the following conditions are met:

- the vertical aerial photographs have $\leq 3°$ tilt,

- the adjacent photographs are exposed from almost exactly the same altitude above-ground-level,

- the principal points (PPs) of both photographs lie at approximately the same elevation above-ground-level, and

- the base of the objects of interest are at approximately the same elevation as that of the principal points.

If these conditions are met, then we only need to obtain three measurements on the stereoscopic photographs to compute the absolute height of an object found within the stereoscopic overlap portion of a stereopair. For example, consider the stereoscopic characteristics of Photo$_{4\text{-}4}$ and Photo$_{4\text{-}5}$ of Columbia, SC, shown previously in Figure 6-3 and shown diagrammatically in Figure 6-20. Remember that the aerial photography was acquired at an altitude of 2,978.5 ft above-ground-level with a 6-in. focal-length lens yielding a nominal scale of approximately 1:5,957. We will compute the height of the Senate Condominium. The computation of the height of the condominium is accomplished by:

1) determining the altitude of the aircraft above-ground-level $(H - h)$ which we know to be 2,978.5 ft. When using the parallax equation, it is necessary to express the aircraft height above-ground-level $(H - h)$ in the units desired. In this case we will use feet.

2) locating the *principal point* (PP) on each of the photographs by drawing lines through the opposing four fiducial marks on each photograph.

3) finding the *conjugate principal point* (CPP) on both photos by locating the position of each photo's principal point on the other photograph.

4) positioning the photographs along the flightline by aligning the PP and CPP of each photograph so that they are in a straight line. This represents the *line of flight*. Note that the line of flight is from right to left in Figure 6-20. Novice interpreters often think that flightlines must progress from left to right. They do not.

5) determining the average photo air base (absolute stereoscopic parallax, P). First, measure the distance between the principal point (PP$_{4\text{-}4}$) and the conjugate principal point (CPP$_{4\text{-}5}$) on Photo$_{4\text{-}4}$, which we will call A-base$_{4\text{-}4}$ (Figure 6-20b,d). This was 3.41 in. Do the same thing for Photo$_{4\text{-}5}$. In this case A-base$_{4\text{-}5}$ was 3.39 in. (Figure 6-20a,c). The mean of these two measurements is the average photo air base (P) between the two exposure stations (i.e., absolute stereoscopic parallax, $P = 3.4$ in.). Absolute stereoscopic parallax (P) and differential parallax (dp) must be measured in the same units, e.g., either millimeters or thousands of an inch.

6) measuring the differential parallax (dp) between the base of the building and the top of the building. This measurement is best obtained using a *parallax bar* (often called a *stereometer*), which will be discussed shortly. However, it may also be possible to compute the difference in parallax between the objects using a quality ruler. In this example, the top corner of the building is designated as being *Object A* and the base of the building is *Object B*. Notice in Figure 6-20ab that the building is just visible in the lower left-hand portion of Photo$_{4\text{-}4}$ but that it has moved dramatically across the film plane and is located near the vertical line through the principal point in Photo$_{4\text{-}5}$. The location of the top of the building is labeled a in Photo$_{4\text{-}4}$ and a' in Photo$_{4\text{-}5}$. The location of the base of the building is labeled b in Photo$_{4\text{-}4}$ and b' in Photo$_{4\text{-}5}$. The x-parallax of the top (from a to a') and the x-parallax of the base of the building (from b to b') can be measured using a ruler in one of three ways: 1) measurement using fiducial lines, 2) measurement based on superposition, or 3) measurement using a parallax bar (stereometer).

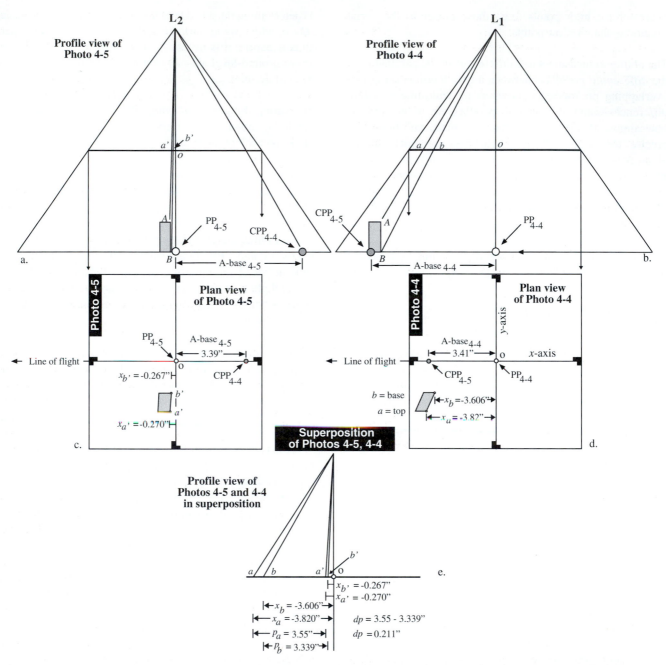

Figure 6-20 Computing the height of the Senate Condominium in Columbia, SC, by measuring the stereoscopic x-parallax of the top and base of the building from Photo $_{4\text{-}4}$ and Photo$_{4\text{-}5}$. When placed in superposition (e), the differential parallax (*dp*) of the building is determined and can be used to compute the height of the building. Refer to the next illustration to view the actual aerial photographs used in the computation.

X-Parallax Measurement Using Fiducial Lines

Figure 6-20c,d depict a diagrammatic plan view of the information content of Photo$_{4\text{-}4}$ and Photo$_{4\text{-}5}$. Note that the location of the principal points (PP), conjugate principal points (CPP), base of the building, and top of the building are located on each photograph. Also note that the diagram depicts the building leaning away from the principal point due to relief displacement, as previously discussed.

In conjunction with the planimetric map, it is useful to actually view the image of the Senate Condominium as recorded on Photo$_{4-4}$ and Photo$_{4-5}$. There is a much enlarged view in Figure 6-21. Note that the principal point line (fiducial line) is visible near the building in Photo$_{4-5}$ (Figure 6-21a), but that the principal point line is far away from the building in Photo$_{4-4}$, necessitating a break in the line (Figure 6-21b). Using the real aerial photography, it is possible to measure the x-parallax of the top of the building on Photo$_{4-4}$ (designated x_a = -3.82 in.) and x-parallax of the base of the building (designated x_b = -3.606 in.) from the fiducial line. Similarly, it is possible to measure the x-parallax of the top of the building on Photo$_{4-5}$ (designated $x_{a'}$ = -0.270 in.) and the base of the building (designated $x_{b'}$ = -0.267 in.) from the fiducial line. The absolute value of the x-parallax of the top of the building is p_a = 3.55 in. and the x-parallax of the base of the building is p_b = 3.339 in. The differential parallax (dp) is the difference between the two values, dp = 0.211 in.

Substituting the measured values into Equation 6-17

$$h_o = (H - h) \times \frac{dp}{(P + dp)}$$

yields

$$h_o = 2978.5' \times \frac{0.211''}{(3.4'' + 0.211'')}$$

$$h_o = 174' \quad .$$

The actual height of the building is 172 ft.

X-Parallax Measurement Based on Superposition

It is possible to orient the two overlapping photographs as shown in Figures 6-20e and 6-21c where the base and top of the building are aligned parallel with the line of flight. It does not matter how much distance separates the two photographs. The x-parallax of the top of the building is measured as the distance from the top of the building on Photo$_{4-5}$ to the top of the same corner of the building on Photo$_{4-4}$ (p_a = 0.30 in.). The x-parallax of the base of the building is the distance from the base of the building on Photo$_{4-5}$ to the base on Photo$_{4-4}$ (p_b = 0.511 in.). The differential parallax between the top and base of the building is identical to what was computed using the fiducial line method, dp = 0.211 in., and would yield the same building height estimate. Many image analysts prefer this simple parallax measurement technique.

Now that we have the altitude of the aircraft above-ground-level (H – h) and the absolute stereoscopic parallax (P) computed for these two photographs, it is a simple task to mea-

sure the bottom and top of other features within the stereoscopic model and compute their differential parallax measurements (dp). While it is simple and useful to be able to measure the amount of differential parallax using a ruler as just described, it is also possible to use the parallax bar or stereometer that was designed specifically for measuring stereoscopic parallax.

X-Parallax Measurement Using the Parallax Bar (Stereometer)

The parallax of specific objects in the stereopair can be computed accurately and quickly using the *parallax bar* (*stereometer*). The stereometer consists of a bar with two clear plastic or glass plates attached to it (Figure 6-22). One of the plates is usually fixed, while the other plate can be moved back and forth along the length of the bar by adjusting a vernier mechanism. There is usually a bright red dot etched into each of the plastic (glass) plates. These are called the *measuring marks*.

When using a simple lens stereoscope, the image analyst arranges the photographs so that the object of interest (such as the corner of the top of a building) in the left photograph is viewed by the left eye and the same object in the right-hand photograph is viewed by the right eye. Next, the analyst adjusts the stereometer so that 1) the left measuring mark is placed exactly on the corner of the top of the building in the left-hand photograph, and 2) the right measuring mark is adjusted so that it is placed on the same corner of the top of the building in the right-hand photograph. The measuring marks are shifted in position on the stereometer using the vernier screw device until they visually fuse together into a single mark, which appears to exist in the stereo model as a three-dimensional red ball that lies at the particular elevation of the corner of the top of the building. This produces the same effect as if a red ball had actually existed at the corner of the top of the building at the time the two photographs were exposed. If the measuring marks are successfully fused and hover exactly at the corner of the top of the building, then the stereoscopic x-parallax of this point may be easily read from the vernier scale on the parallax bar. We have just used the principle of the *floating mark*.

If we wish to measure the parallax of the base of the corner of the same building, we simply focus our stereoscopic vision on the base and adjust the two (left and right) measuring marks until they once again fuse as a three-dimensional ball that just touches the ground at the base of the building. We can then read the vernier scale to obtain the x-parallax measurement of the base of the building.

**Methods of Measuring
Stereoscopic *x*-parallax from
Overlapping Aerial Photographs:**
- Measurement Using Fiducial Lines (a,b)
- Measurement Based on Superposition (c)

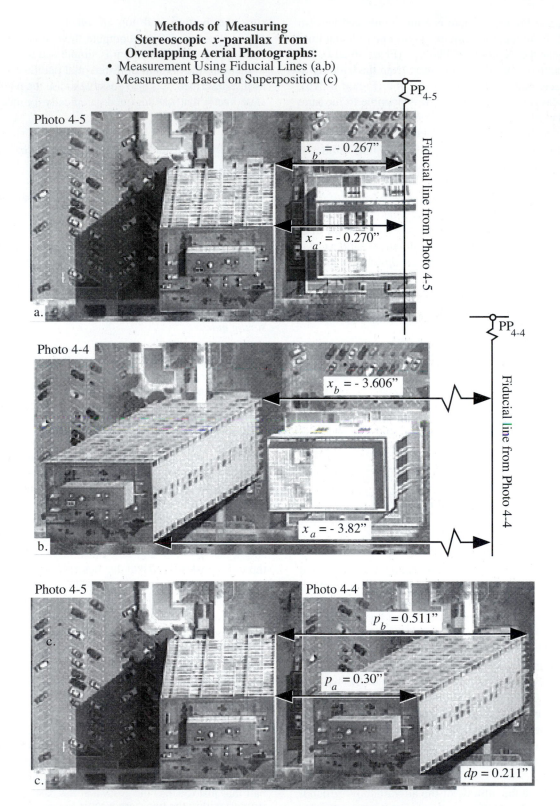

Figure 6-21   a,b) The measurement of stereoscopic x-parallax using the fiducial lines for $Photo_{4-5}$ and $Photo_{4-4}$, respectively. c) Measurement of stereoscopic x-parallax using superposition.

Figure 6-22 Close-up view of a parallax bar with the floating marks and vernier measuring mechanism. It is attached to the lens stereoscope and placed over the stereopair to make parallax measurements.

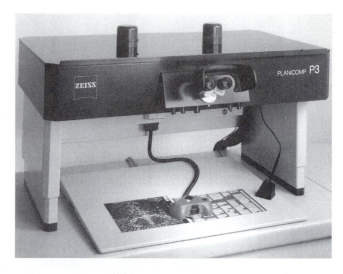

Figure 6-23 A Zeiss P3 Planicomp analytical stereoplotter. The analyst views the stereo model through the binocular lens system and can adjust the floating mark using the cursor lying on the table (courtesy Carl Zeiss).

The significance of the floating mark for photogrammetric measurement is that it can be used to very accurately measure the parallax at any two points over the stereo model. It can be placed at the top of an object such as a building and the parallax of the point read, e.g., p_a. Then, the floating mark can be placed at the bottom of the object and the parallax read, e.g., p_b. The difference $(p_a - p_b)$ results in the differential parallax (dp) measurement required as input to Equation 6-17. It makes no difference if the object is man-made with vertical sides or if it is a terrain feature, i.e., it can be the top and base of a building or tower, the peak of a mountain or a river floodplain. From the *differential parallax* between the two measurement points, the elevation difference may be derived.

The principle of the floating mark can be used to map lines of constant elevation in the terrain. For example, if the floating mark (the fused, red three-dimensional ball) were moved around the stereo model so that it maintained contact with the terrain, i.e., it was not allowed to float above the terrain or go down below the terrain (which is possible when viewing in stereo!), then a line of constant stereoscopic *x*-parallax would be identified. In such instances, the floating-mark mechanism is usually physically connected to a *pantograph* which marks in pencil on a manuscript map the exact location of the floating mark at all times. If this were done along the side of a mountain, the line of constant x-parallax could be determined, which could then be converted directly into elevation above sea level. This line of constant parallax is related to actual elevation through the use of horizontal (x,y) and vertical (z) ground-control markers that have been sur-

veyed in the field. This is exactly how USGS topographic maps with contour lines are produced. The stereoplotter operator places the floating mark on the terrain (representing a specific elevation) and then moves the floating mark about the terrain in the stereo model while keeping the mark firmly on the ground. One can convert the x-parallax measurement to actual elevation by locating horizontal-vertical ground-control points (e.g., 100 ft above sea level at location x,y) in the stereoscopic model and determining its x-parallax. Any other point within the stereoscopic model with the same x-parallax must then lie at 100 ft above sea level. It is also possible to map building perimeters by placing the three-dimensional floating mark so that it just rests on the rooftop, and then tracing the mark around the edge of the rooftop. This results in a rudimentary map of the planimetric (x,y) location of the building.

The simple parallax bar (stereometer) is the least expensive and one of the least accurate of all instruments that are based on the concept of the floating mark. Very expensive analog and digital stereoscopic plotting instruments are available that allow precise parallax measurements to be made. A Zeiss analytical stereoplotter is shown in Figure 6-23. When used in conjunction with x,y,z ground-control information collected in the field, the system analytically correlates and computes the amount of x-parallax for each picture element (pixel) within the two stereoscopic aerial photographs inside the system. A stereoplotter operator interacts with the analytical instrument to produce planimetric and topographic

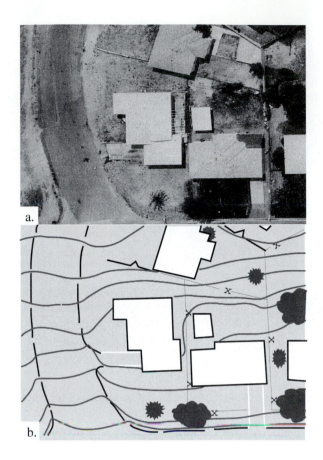

Figure 6-24 a) One aerial photograph of a stereopair of Covina, CA. b) A line drawing of the planimetric (x,y) location of the roads, buildings, fences, and landscape elements (trees, shrubs), as well as 1-ft contours derived from the stereo model.

maps. In addition, they can be used to produce raster digital elevation models. An example of typical x,y planimetry and contours (z) derived from a stereoplotter is shown in Figure 6-24. For a detailed discussion of how analytical stereoplotters function, refer to the *Manual of Photogrammetry* published by the American Society for Photogrammetry and Remote Sensing (1983) and the *Addendum to the Manual of Photogrammetry* (Greve, 1996).

Stereoscopic photography may also be acquired from a terrestrial vantage point rather than from an aerial vantage point. The stereoscopic photographs can be analyzed using the principle of the floating mark to produce a detailed three-dimensional representation of the façade of buildings if desired (of course, field x,y,z ground-control measurements are also required to scale the stereo model). This is called *close-range photogrammetry* (Warner et al., 1996). Stereo-

scopic photogrammetric techniques can be applied to terrestrial photography to restore historical buildings, restore certain terrain characteristics, or even reconstruct a human's arm, leg, or face.

Orthophotos and Digital Elevation Models

More than 24 years ago, Thrower and Jensen (1976) reviewed how analog (hard-copy) orthophotography was created, and identified numerous cartographic applications. They stated:

> Ortho-photomapping represents a technique by which spatially arrayed data might be both more accurately measured and communicated because of the special attributes of the orthophoto map, namely, the image of an aerial photograph and the metric qualities of a controlled line map.

As predicted, the last two decades have seen a steady increase in the creation and utilization of orthoimage products. But in the last decade, there has been a tremendous increase in the use of digital orthoimagery and digital elevation models in remote sensing and GIS applications (Morain and Baros, 1996). The general public, GIS practitioners, cartographers, and the media (e.g., newspapers, magazines, television) are using orthoimages as cartographic backdrops upon which thematic information is overlaid (e.g., property lines, utility lines, drainage networks, contours, troop deployment). *More importantly, many image analysts are now using desktop soft-copy photogrammetric systems to generate their own personal DEMs and orthophotographs for input to GIS.* This section describes how DEMs and orthophotographs are created and important issues that should be considered when creating and using them.

Orthoimages are created from remotely sensed images. The geometry of an unrectified digital photograph is changed from that of a conical bundle of rays to parallel rays that are orthogonal to the ground and to the image plane (Keating, 1993). Therefore, instead of having a perspective center, the viewing perspective is modeled as being an infinite distance from the ground. Figure 6-25 demonstrates the change in image geometry and the removal of terrain-induced displacement (Δs). The point P that lies at a specific elevation above mean sea level on the ground is found at p in the image plane coordinate system when it should be located at c. The correction of the shift from p to c (i.e., Δs) is the goal

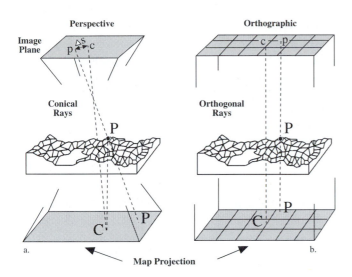

a. Uncorrected vertical aerial photograph

b. Orthophotograph

Figure 6-25 a) A perspective projection aerial photograph displaces the image of objects away from their true planimetric position. For example, P is located at p in the photograph when it should be located at *c*. b) An orthographic projection places all images of objects in the correct planimetric position. The goal when creating an orthophotograph is to move objects recorded on an aerial photograph specific distances and directions, Δs, to their proper planimetric position.

Figure 6-26 a) An uncorrected aerial photograph of a power transmission line in rugged terrain. b) Planimetrically accurate orthophotograph after correction to remove roll, pitch, and yaw errors introduced by the aircraft at the instant of exposure and the effects of relief displacement.

of orthorectification (Michael, 1992). Each part of the terrain is independently corrected during the rectification process.

The effects of topographic relief displacement and camera altitude variations are removed in orthoimagery. The result is a planimetrically correct orthoimage. For example, Figure 6-26 depicts a power transmission line in rugged terrain. The powerline was surveyed to be very straight in the real world, but the substantial local relief causes relief displacement to take place in the photograph. This causes the transmission line to not be in its proper planimetric position. The process of differential bit-by-bit orthographic rectification removes distortions caused by tilt of the camera at the instant of exposure and the effects of relief displacement. The powerline is now in its proper planimetric position, and its shape and geometry are correct. The photograph is now an orthophotograph with the accuracy of a traditional line map.

This *planimetric* accuracy allows scientists to use orthoimages like maps for making direct measurements of geographic location, distances, angles, and area. On unrectified imagery such measurements can only be approximated because of image displacement and scale change caused by variations in local relief (i.e., relief displacement) and small amounts of tilt in the airphoto.

The increased use of terrain-corrected orthoimagery and derivative digital elevation models can be explained by comparing and contrasting how the orthoimagery and DEMs were produced traditionally versus how they are created today using digital photogrammetric techniques. This involves a discussion of advances in the following technologies:

• ground-control point collection,

• collection of aerial photography and/or digital remote sensor data,

• image digitization technology,

• hardware and software associated with the photogrammetric derivation of orthoimages and DEMs,

• the generation of the DEM, and

• the creation of the digital orthoimage.

A concluding section identifies problems and possible solutions associated with the creation of DEMs and derivative orthoimages.

Advances in the Collection of Accurate Horizontal and Vertical Ground Control to Produce Orthophotography

The creation of digital elevation models and orthoimagery has always required the collection of horizontal (x,y) and vertical (z) ground-control points (GCPs). Historically, only registered surveyors were capable of obtaining the required x,y,z GCP data. Only government laboratories and photogrammetric engineering firms could afford this type of GCP data collection. Fortunately, the situation has changed dramatically. The availability of survey-grade GPS makes it possible to collect accurate GCP information with x,y,z root-mean-square-errors (RMSE) of <3 cm when the data are *differentially corrected* (Keating, 1993).

The use of GPS for ground-control point collection "... has become everyday practice, with considerable gain in time and economy, especially as intervisibility between points is not any more required" (Ackermann, 1994b). Some members of the surveying community are alarmed that nonsurveyors, i.e., geographers, foresters, geologists, and agronomists are collecting such information (Ayers, 1994; Perkins, 1994). Nevertheless, the x,y,z ground-control data necessary to orthographically rectify a block of aerial photography to a certain level of precision using soft-copy photogrammetry instruments can now be collected economically by nonsurveyors in a day or two for relatively large regions. Thus, a major stumbling block to the creation of *personal* orthoimagery and DEMs has been removed. This of course assumes that the analyst uses the GPS and photogrammetric equipment correctly. The nonsurveyor analyst should record the *metadata* about the GCP points, including methods of data capture and accuracy, for future reference using federal metadata standards (Ayers, 1994).

Advances in the Collection of Metric Aerial Photography and Other Remote Sensor Data Used in the Creation of Orthoimage Databases

In previous decades, medium-scale (1:20,000 to 1:40,000) metric aerial photography was acquired primarily by government agencies (e.g., U.S. Geological Survey NAPP data,

U.S. Coast and Geodetic Survey, Bureau of Land Management). Private photogrammetric engineering firms collected most of the large-scale (>1:20,000) photography. The data were expensive to acquire, and only a fraction of such data were ever converted into orthophoto products or DEMs.

Scientists still have difficulty obtaining inexpensive large-scale data and it is difficult to obtain current NAPP data of a region. Therefore, there is a significant demand for some type of remotely sensed data that can meet many of the orthoimage mapping demands.

Some satellite remotely sensed data suitable for general photogrammetric applications are available. For example, it is possible to obtain digital stereo panchromatic remote sensor data from space with a nominal ground spatial resolution of 10 x 10 m from the SPOT satellites. When very precise ephemeris information is available concerning the sensor and its position at the exact instant of data collection, SPOT panchromatic data may be used to derive good quality digital elevation models and orthoimages (Dowman and Neto, 1994; Trinder et al., 1994).

Orthoimagery is often used as a background image upon which thematic information is overlaid in many urban applications. This is the reason so much 10 x 10 m SPOT and 5 x 5 m Indian IRS-1C panchromatic data are used as local and regional backdrop images, despite the fact that individual houses and small buildings cannot be resolved. Imagery with a spatial resolution of 1 to 4 m is required to resolve trailers, houses, small buildings, narrow roads, and drainage networks, so important in many urban applications (Jensen et al., 1994; Haack et al., 1996; Jensen and Cowen, 1999). Fortunately, several companies received permission to use previously classified reconnaissance technology to build high spatial resolution satellite sensor systems to fill this important niche. Simulations of the information content of 1, 2, 3, and 5 m spatial resolution satellite imagery of San Jose, CA, are shown in Figure 6-27. Space Imaging, Inc., launched IKONOS into orbit on September 24, 1999, containing 1 x 1 m panchromatic and 4 x 4 m multispectral sensors.

While such digital satellite remote sensor data may never replace the demand for quality large-scale metric aerial photography (Jensen and Cowen, 1999), there will be many applications where the orthographically rectified satellite data are sufficient. Thus, another major stumbling block is being overcome as relatively high spatial resolution remote sensor data becomes available for use in the creation of personal orthoimages and DEMs.

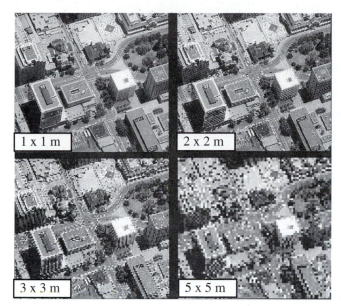

1 x 1 m   2 x 2 m

3 x 3 m   5 x 5 m

Figure 6-27 Simulation of 1, 2, 3, and 5 m satellite data for a portion of San Jose, CA.

Advances in Image Digitization Technology for the Creation of Orthoimage Databases

Only large government and university research laboratories or private photogrammetric firms traditionally had access to high-resolution optical-mechanical image-scanning systems to digitize aerial photography or other remote sensor data. Today, linear and area array digitization technology based on charge-coupled-devices (CCDs) have revolutionized image digitization. Scientists now have access to desktop systems that will digitize black-and-white imagery to 12 bits (values from 0 to 4095) and color imagery to 30 bits (>1 billion colors) at repeatable spatial resolutions approaching <10 μm. Scientists can now inexpensively scan images of their choosing (e.g., perhaps just a single airphoto of a flightline) at repeatable high spatial resolutions for photogrammetric projects.

Advances in Soft-Copy Photogrammetry for the Creation of Orthoimage Databases

All orthoimagery traditionally were derived using specially modified analog stereoplotting instruments called *orthophotoscopes*. They were used to create hard-copy orthophotographs based on *differential (bit-by-bit) correction* of the stereo model in a line-by-line manner (Thrower and Jensen, 1976). In 1958, the analytical stereoplotter was invented by Helava. By 1994, there were approximately 1,500 analytical

stereoplotters in existence compared to approximately 5,000 analog instruments. Analytical stereoplotters can be used to extract digital elevation models and orthophotographs from the stereoscopic model. Considerable human interaction is required, and the systems are still expensive (usually >$50,000), basically out of the reach of many users of remote sensing data. However, in the military photointelligence field, and in some remote sensing laboratories, analytical plotters are commonly used by photointerpreters (Gustafson, 1989).

James Case (1982) developed the first photogrammetric soft-copy system (Leberl, 1994). The term *soft-copy* means that a digital version of the imagery is analyzed rather than a hard-copy of it. Current photogrammetric soft-copy systems allow the analyst to view the stereo model in three dimensions on a CRT screen, extract contours (lines of equal elevation), vectors, and planimetric information (e.g., elevation, building outlines, road centerlines, drainage network), extract a digital elevation model, and produce terrain-corrected orthophotographs.

The typical procedures used to extract DEMs and orthoimagery using a soft-copy photogrammetric system are summarized in Figure 6-28. The soft-copy photogrammetry systems usually cost <$50,000 and some function with limited capability for <$5,000. Using proper ground control and orientation (interior and exterior), a scientist can perform *aero-triangulation* of the stereo model. Once this is accomplished, contours may be extracted, a digital elevation model created, and orthoimagery produced. Many systems allow color imagery to be viewed (Helpke, 1995). Thus, the ability to perform true photogrammetric operations can now be performed on the scientist's desktop using soft-copy photogrammetric systems. The scientist can produce *on-demand* planimetric maps, DEMs, and orthoimages for specific projects rather than being at the mercy of slow, tedious government agencies.

Advances in the Creation of the Digital Elevation Model

A *digital elevation model* is a regular array of terrain elevations (x,y,z) normally obtained in a grid or hexagonal pattern (Petrie and Kennie, 1990). DEMs may be created using 1) ground survey data, 2) cartographic digitization of contour data, and/or 3) photogrammetric measurements.

Ground surveys may obtain accurate x,y,z information. These data may then be interpolated to a grid or triangular irregular network (TIN) model. Unfortunately, each point is usually expensive to acquire.

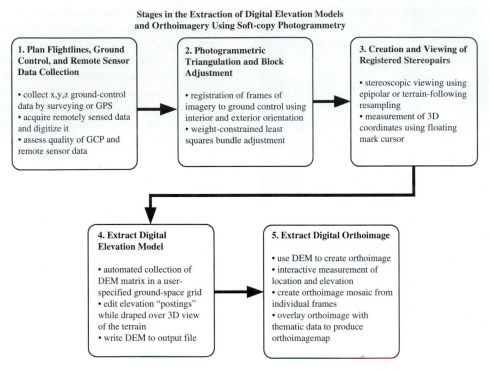

**Stages in the Extraction of Digital Elevation Models
and Orthoimagery Using Soft-copy Photogrammetry**

**1. Plan Flightlines, Ground
Control, and Remote Sensor
Data Collection**

• collect x,y,z ground-control
data by surveying or GPS
• acquire remotely sensed data
and digitize it
• assess quality of GCP and
remote sensor data

**2. Photogrammetric
Triangulation and Block
Adjustment**

• registration of frames of
imagery to ground control using
interior and exterior orientation
• weight-constrained least
squares bundle adjustment

**3. Creation and Viewing of
Registered Stereopairs**

• stereoscopic viewing using
epipolar or terrain-following
resampling
• measurement of 3D
coordinates using floating
mark cursor

**4. Extract Digital
Elevation Model**

• automated collection of
DEM matrix in a user-
specified ground-space grid
• edit elevation "postings"
while draped over 3D view
of the terrain
• write DEM to output file

5. Extract Digital Orthoimage

• use DEM to create orthoimage
• interactive measurement of
location and elevation
• create orthoimage mosaic from
individual frames
• overlay orthoimage with
thematic data to produce
orthoimagemap

Figure 6-28   The stages in the creation of a digital elevation model and orthoimagery using soft-copy photogrammetric techniques.

When using the cartographic contour digitization approach, "...the quality of the derived DEM is generally poor, unless great efforts are made to also extract characteristic features and break lines. In that case efforts and costs may be as high as for a new photogrammetric DEM generation" (Ackermann, 1994a). The accuracy of the resulting DEM will be much lower than that achievable with field survey or photogrammetric instrumentation, but it is acceptable to a considerable body of users concerned with the modeling of large areas of terrain (Petrie and Kennie, 1990).

**Photogrammetric Methods of Terrain Data
Collection**

The x,y,z accuracy of the digital elevation model that can be extracted using photogrammetric methods is a function of the scale and resolution of the remote sensor data, the flying height at which the imagery were acquired, the base/height ratio (i.e., geometry) of the stereoscopic imagery, and the accuracy of the stereoplotting equipment used for the measurements (Petrie and Kennie, 1990).

If quality aerial photography is available, stereoplotting instruments may be used to create an exact three-dimensional stereo model of the terrain. The analyst measures the x,y,z coordinates of individual points in the stereo model very accurately using the "principle of the floating mark"

previously discussed instead of going out to the terrain to obtain the elevation measurement. Traditionally, the terrain models measured in the stereoplotting machine were either optical or mechanical. Analytical stereoplotting machines replace the optical or mechanical models by equivalent purely mathematical or numerical solutions that are executed by a computer. In such systems, the x,y,z coordinates of a point measured by the operator are passed to the computer, which computes the corresponding position of the same point on each photograph of the stereopair in real time. The differential parallax at the point is determined and the elevation of the point is computed and often stored off-line as a digital elevation model. Digital elevation data may be collected randomly for very specific points of interest, systematically (e.g., grid-based sampling), or a combination of the two. These analytical stereoplotters produce very accurate DEMs but at significant expense (primarily due to the cost of the instrument).

Using soft-copy photogrammetric systems and suitable ground-control and camera-calibration information, it is now possible to perform aero-triangulation of the photography (or imagery) and utilize information obtained during the process to generate a lattice (grid) of elevation values within each stereo model of a flightline. The software automatically performs stereo correlation on orthorectified patches of imagery and computes the parallax (and related elevation) associated with each new point in the stereo model. The stereo correlation is performed for all points in a user-specified grid. Typical systems can process thousands of points in the stereo model per second.

Scientists can resample and mosaic the final DEM to whatever spatial resolution is considered appropriate (e.g., 5 x 5 m, 20 x 20 m). In this manner the DEM can be brought into geometric congruence with data in a GIS. *Thus, Earth scientists now have the capability of creating very accurate DEMs on demand for site-specific projects using desktop soft-copy photogrammetric workstations.* However, there are problems associated with the DEMs, to be discussed.

Advances in the Creation of the Orthoimages

Soft-copy photogrammetric instruments can produce terrain-corrected orthoimages on demand at minimal expense. The general procedure is summarized in Figure 6-28. The user selects any triangulated image in the block of aerial photography (or SPOT data) and its associated DEM to generate a digital orthoimage. During the orthorectification process, the effects of elevation upon the image perspective are removed to produce a geocoded data set with an even pixel spacing in map space. For each orthoimage pixel of known latitude and longitude, the algorithm first uses the DEM to determine the height of the point. The rational functions for the image are then used to determine the pixel in the triangulated image that corresponds to the point in ground space. The intensity of this point (resampled based on its neighbors using a bilinear interpolation scheme) is then assigned to the output orthoimage pixel. This process typically runs at a rate >100,000 pixels/sec until the rectified output orthoimage is completely filled. The result is a terrain-corrected orthoimage that can be used as a cartographic map.

Problems and Potential Solutions Associated with Digital Elevation Models Derived Using Soft-Copy Photogrammetry

Users sometimes feel that a DEM derived using soft-copy photogrammetry will be error free. This is not the case.

Tall Structures and Trees Impact the Creation of a Photogrammetrically Derived Digital Elevation Model

The automatic stereo correlation used to create the DEMs works well when the terrain is devoid of trees, buildings, overpasses, bridges, etc., which extend above the nominal terrain. When such objects are present, however, the algorithms assume these objects are terrain and compute the differential parallax and resultant height of such surfaces. The heights are then placed in the DEM. For example, Figure 6-29a depicts a DEM derived for a four-block region of the University of South Carolina campus derived from 1:6,000-scale photography. The blocky appearance of the DEM is due to detailed height information for each building and tree in the study area.

Methods Used to Edit a DEM Impact its Accuracy

Soft-copy photogrammetric systems allow users to view the grid of DEM elevation "postings" superimposed on a stereopair in three-dimensions. The analyst can edit individual elevation "postings" by moving them so that they come in contact with the ground using the "principle of the floating mark," whereby each posting becomes a floating mark. The analyst may 1) correct individual postings, 2) select a polygon of postings and change all of them to the same elevation, or 3) select a polygon of postings along a slope and have them scaled to lie between the highest and lowest points encountered within the polygon. When carefully used, the analyst can correct most problems encountered in the DEM. For example, the DEM in Figure 6-29b was edited so that the top of each building was at the correct elevation. DEMs that include elevation information about buildings and trees may be of use if the analyst desires to drape an orthophoto on top of the DEM and perhaps eventually do a "fly-by" through the city. However, if the analyst wanted a DEM of just the nominal ground terrain in the four-block region, this is certainly not present in Figures 6-29ab.

To create a DEM of the region that does not have building and tree information in it, the analyst must usually manually edit the elevation "postings" in the DEM that correspond with the buildings and trees and effectively drive or push them to the nominal terrain height in the area. This can be difficult if a building or stand of trees is extremely large. However, if the buildings and trees are not too large it is possible to identify the general trend of the terrain between buildings and large trees such that the "postings" of buildings and trees can be moved to the nominal terrain elevation. Careful editing of the original DEM in this manner can produce a revised DEM that depicts just the local relief of the area, without buildings and trees, as shown in Figure 6-29c. A percent slope database of the region (important in many environmental studies) cannot accurately be computed from the DEM with buildings and trees in it. It can be produced from the DEM with buildings and trees removed, as shown in Figure 6-29d.

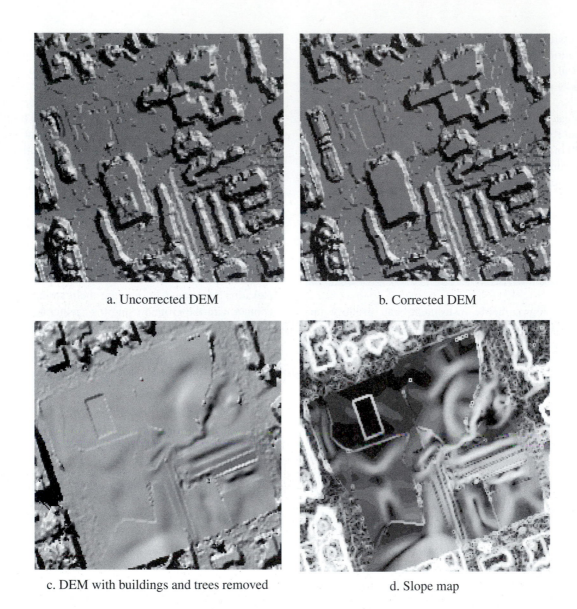

a. Uncorrected DEM   b. Corrected DEM

c. DEM with buildings and trees removed   d. Slope map

Figure 6-29   a) Uncorrected digital elevation model derived using soft-copy photogrammetric techniques. b) Corrected DEM. c) DEM with buildings and trees removed. d) Slope map produced from DEM with buildings and trees removed.

DEMs are most accurate and require the least editing when produced for rural areas that do not have significant "leaf-on" tree cover or man-made structures (buildings, bridges, etc.). Urbanized areas with buildings and trees may have to be manually edited to obtain a DEM of just the nominal terrain. This can be a laborious process and is subject to error being introduced by the analyst. Research is under way by many scientists to automate the DEM creation process. The larger the scale of aerial photography and the greater the height of the buildings and structures in the study area, the greater the difficulty of obtaining an accurate DEM of the terrain.

Problems and Potential Solutions Associated With Orthoimagery Derived Using Soft-Copy Photogrammetry

The accuracy of a digital orthoimage is a function of the quality of the imagery, the ground control, the photogrammetric triangulation, and the DEM used to create it (Figure 6-30). An orthoimage may be produced from the original DEM, a DEM with building rooftops cleaned up, or even a DEM with buildings and trees removed. A DEM produced from the collection of field surveying or even digitized con-

a. Orthophoto derived from uncorrected DEM

b. Orthophoto derived from corrected DEM

Figure 6-30   The quality of the DEM influences the quality of the orthophoto. a) Orthophoto generated from uncorrected DEM. b) Orthophoto generated from corrected DEM.

a. Orthophoto draped over an uncorrected DEM

b. Orthophoto draped over a corrected DEM

Figure 6-31   The quality of the DEM influences the quality of the orthophoto. a) Orthophoto draped over an uncorrected DEM. b) Orthophoto draped over a corrected DEM. The impact of trees is minimized, and the buildings are shaped correctly.

Figure 6-32   A quality orthophotograph of Rosslyn, VA draped over a carefully edited digital elevation model (courtesy L-H Systems, Inc., and Marconi, Inc.; © SOCET set).

tours may also be used to create the orthoimage. Therefore, the analyst should always have access to the metadata (history) of how the DEM was created. In this way only the most appropriate DEM data will be used in the creation of the orthoimagery.

Large-scale (e.g., 1:6,000) urban orthoimages derived from uncorrected DEMs often exhibit severe distortion of building edges (Joffe, 1994; Nale, 1994). For example, the orthoimage in Figure 6-30a was derived using an uncorrected DEM while Figure 6-30b was produced using a DEM with buildings and tree elevations corrected. The roof edges at "a" and "b" are correct in Figure 6-30b.

The importance of the quality of the DEM can be appreciated even more when the orthoimage is draped back onto the DEM used in its creation. For example, consider the two orthophotos in Figure 6-31ab created and draped over uncorrected and corrected DEMs, respectively. The edges of buildings are "smeared" due to the effects of adjacent trees in Figure 6-31a. Conversely, when the rooftop and tree elevations are corrected in the DEM, the resultant orthoimage depicts the same buildings with sharp, distinct edges (Figure

6-31b). This is the type of orthoimagery that should be used in simulated fly-bys through urban environments. More sophisticated soft-copy photogrammetric systems can produce extremely accurate digital elevation models and very clean orthophotography, such as the three-dimensional view of Rosslyn, VA, in Figure 6-32.

Traditional Ortho-rectification Does Not Eliminate Radial and Relief Displacement of Tall Structures

Most users have the mistaken impression that the tops of tall buildings or extremely tall trees in an orthoimage are in their proper planimetric location, i.e., that the roofs of tall buildings are over their foundations. This is only the case when the sensor acquires the data from a great altitude and the relief displacement of the buildings or other structures caused by radial distortion is minimal. Unfortunately, when using large-scale aerial photography with significant building and tree relief displacement, only the base of the buildings or trees are in their proper planimetric location in an orthoimage. A traditional digital orthophoto does not correct for "building lean" caused by radial distortion. Indeed, the

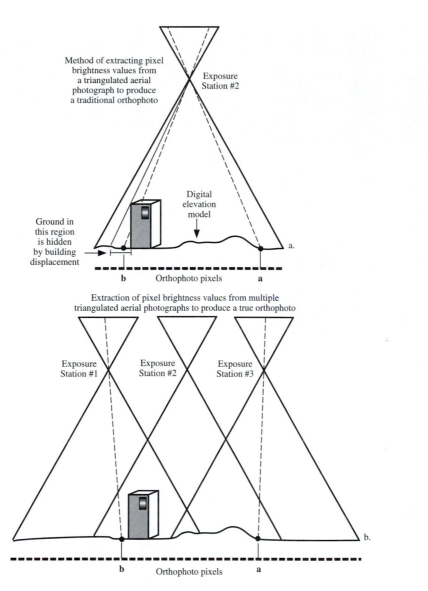

Figure 6-33 a) The brightness value of pixel *b* in a traditional orthophoto would be extracted from the top of the building due to relief displacement. Pixel *a* would have the correct gray shade because it is not obscured from view. Note that both pixels are extracted from a single photo obtained at a single exposure station. b) Pixels *a* and *b* in a true orthophoto are derived from the triangulated photo in the flightline that is most appropriate. For example, pixel *a* is derived from the photo obtained at Exposure Station #3. The tone of *b* is extracted from Exposure Station #1, where it is not obscured by building displacement (Southard, 1994; courtesy of L-H Systems, Inc., and Marconi, Inc.; © SOCET set).

"building lean" of tall buildings or other structures (bridges, overpasses, etc.) may completely obscure the image of the Earth's surface for several hundred feet. If we attempted to digitize buildings from a traditional digital orthoimage, we would discover that the building locations are misplaced on the Earth's surface relative to the building heights (Nale, 1994).

Recently, an accurate method of creating *true orthophotos* has been developed. In order to understand how it is derived,

it is first useful to review how a traditional orthophoto is produced. The brightness value of pixel *a* is obtained by starting at *a*'s ground x,y position, interpolating the elevation from the DEM, tracing up through the math model to the image, interpolating the proper shade of gray from the image, and assigning the resulting gray shade to pixel *a* in the new orthophoto (Figure 6-33a). The process is acceptable for pixel *a* because there is no obstruction (e.g., building) between the ground x,y,z location at *a* that obscures the view of the ground from the exposure station. The problem that

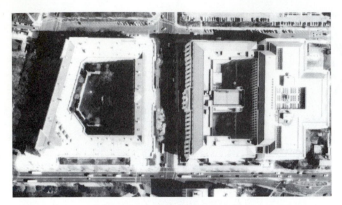

a. Conventional orthophoto

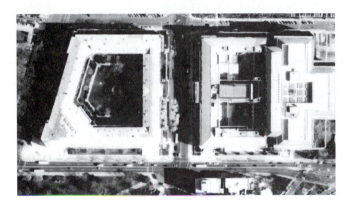

b. True orthophoto

Figure 6-34   The difference between a conventional orthophotograph (a) and a true orthophoto (b) of the U.S. Capitol. Building lean has been removed and all building rooftops are in their proper planimetric position (Southard, 1994; courtesy of L-H Systems, Inc., and Marconi, Inc.; © SOCET set).

arises is illustrated by pixel *b*. In the original image, the ground "behind" the building at *b* is not depicted because of the height of the building. Pixel *b* will have the tone or color of the roof of the building. The final traditional orthophoto will show the building leaning away from the center of the photograph as in Figure 6-34a. The roof will be shown where there should be ground at the back of the building, and the side of the building will be shown where the roof should be. The ground behind the building will not be shown at all. These displacements are related to the height of the building and the position of the building in the original photo. The greater the height of the building and the closer it is to the side of the original photograph, the worse the displacement will be.

An elegant solution to the above orthophoto problems was developed (Southard, 1994; Walker, 1994). In Figure 6-33b

we see three triangulated aerial photographs and a DEM covering the entire footprint of the project area. Using traditional three-dimensional stereoscopic feature extraction tools, the outlines of buildings, bridges, and other obstructions are identified. However, the brightness value or gray tone for pixel *a* is interpolated from the most nadir (directly overhead) Exposure Station (#3 in Figure 6-33b) that has the best view of the ground at location *a*. The algorithm then examines the DEM and feature data and determines that the view of the ground for pixel *b* is obscured by the building at Exposure Station #2 and automatically selects imagery from Exposure Station #1 to obtain the proper pixel color for pixel *b*. The application of these algorithms results in a *true orthophoto* where:

- building rooftops are shown in their correct planimetric x,y location,

- sides of buildings are not shown,

- the ground on all sides of all buildings is shown in its proper location,

- tops and bottoms of overpasses are shown in their proper locations, and

- orthophotos and map sheets can be made that are larger than any of the input images.

A comparison between a traditional orthophotograph and a true orthophoto of the U.S. Capitol is shown in Figure 6-34ab. Building lean (displacement) has been removed, and all building rooftops are in their proper planimetric location. Creation of true orthoimages should accelerate the application of orthoimagery in GIS.

 Area Measurement

It is possible to obtain area measurements directly from unrectified vertical aerial photography if the terrain is very level. However, if the local topographic elevation within the field of view of the photograph varies by more than about 5 percent of the flying height above-ground-level (e.g., 200 m if $H = 4,000$ m), then serious area measurement error could occur (Avery and Berlin, 1992). If the local relief varies by more than 5 percent, the image interpreter could carefully stratify the photograph into geographic areas that have approximately the same scale. For example, if a single aerial photograph consisted primarily of the image of a flat plateau at one elevation (e.g., 400 m above sea level) and a gently sloping river valley at another elevation (e.g., 150 m above sea level), the interpreter could stratify the photograph into two parts. He would then compute the exact scale of the plateau portion of the photograph and the exact scale of the river valley portion of the photograph. These two different scales would be used when measuring the area of selected features found in these two different landscapes. The optimum solution is to convert the aerial photography into orthoimagery as previously discussed. The area measurements will then be accurate because the orthophotograph has a uniform scale.

Area Measurement of Well-Known Geometric Shapes

It is a straightforward task to determine the area of well-known geometric shapes such as rectangles, squares, and circles in aerial photography. The analyst first measures the length, width, side, or diameter of the feature on the image. He or she then converts this measurement to a ground distance based on the scale of the photography. The ground distance is then used with the appropriate mathematical relationship to compute the area:

$$\text{Area of a circle} = \pi r^2$$

$$\text{Area of a square} = s^2$$

$$\text{Area of a rectangle} = l \times w.$$

For example, if the diameter of an agricultural center-pivot irrigation system is 0.5 in. when measured on 1:24,000-scale (1 in. = 2,000 ft) aerial photography, then the ground distance diameter (d) is 1000 ft and the radius (r) is 500 ft. The center-pivot irrigation system would contain 3.1416 x 500^2 or 785,400 ft^2. An acre contains 43,560 ft^2. A hectare (ha) contains 10,000 m^2 (Table 6-1). One acre = 0.4047 ha; one hectare = 2.471 acres. Therefore, the center-pivot irrigation system contains 18.03 acres (7.3 ha).

Area Measurement of Irregularly Shaped Polygons

There are several methods for obtaining accurate area measurements from irregularly shaped polygonal features such as lakes, forest stands, urbanized areas, etc. found in vertical aerial photography. The most popular methods include the use of compensating polar planimeters, dot grids, and on-screen digitization using digital image processing techniques.

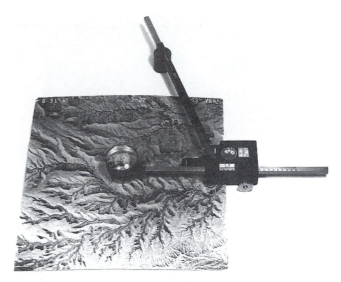

Figure 6-35   A compensating polar planimeter may be used to measure the area of annotated polygons in aerial photography or other remotely sensed data. Measurements from unrectified imagery can yield inaccurate results. It is better if the photography or other imagery have been orthorectified.

Compensating Polar Planimeter

A *compensating polar planimeter* is shown in Figure 6-35. The weighted polar arm is usually placed outside the single aerial photograph as shown. A tracing arm containing a magnifying tracer lens with a cross-hair is moved by the analyst around the polygon of interest in a clockwise direction. Both the weighted polar arm and the tracing arm are connected to the vernier mechanism that sits on top of a wheel. The vernier dial records the area of the polygon in square inches or square centimeters as the interpreter circumscribes the polygon. The interpreter then consults a table that summarizes the number of hectares or acres per square inch or square centimeter found at the scale of the photography or map (Table 6-2). It is wise to measure each polygon several times and take the mean polar planimeter reading.

Whenever possible, it is best *not* to measure directly from the aerial photography but rather to transfer the polygon of interest from the photograph to a base map (e.g., USGS 7.5-minute 1:24,000 scale) using a camera-lucida device such as a zoom-transfer-scope previously discussed, and then use the planimeter on the base map. Modern digital compensating polar planimeters may have built-in microprocessors and digital displays.

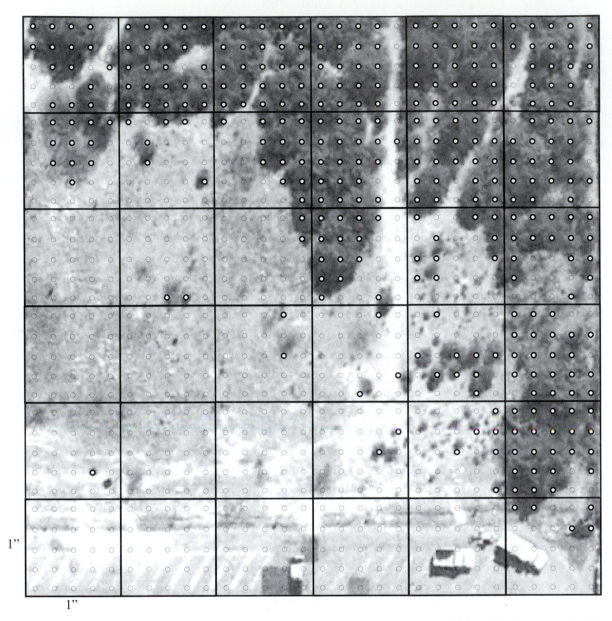

1"

1"

Figure 6-36   A dot grid consisting of 25 dots per square inch overlaid on 1:600-scale (1" = 50 ft) aerial photography. In this example, the goal is to determine the acres (hectares) of terrain covered by forest.

Dot Grids

A dot grid is simply a piece of clear acetate that has a user-specified number of dots scribed onto it per sq. in. (or cm^2), e.g., 25 dots per sq. in. (Figure 6-36). The goal is to use a dot grid that has a dot density sufficient to capture the essence of the subject matter you want to measure (e.g., forest cover acreage) yet not so dense that a tremendous number of dots must be counted. It is useful to create a grid that is most suitable for the scale of the photography or map under investiga-

tion. For example, if an analyst was going to use a dot grid on 1:24,000-scale aerial photography, it would be wise to make a dot grid with nine dots per in^2 because at 1:24,000 scale, 1 in^2 equals approximately 9 acres. If more precision were desired, 18 dots could be systematically placed in each square inch of the grid. The analyst then randomly drops the grid over a polygon of interest and counts the number of dots falling within it. An alternative method of increasing the accuracy of the measurement is, after the first count, to lift up the clear plastic dot grid and randomly drop it again and

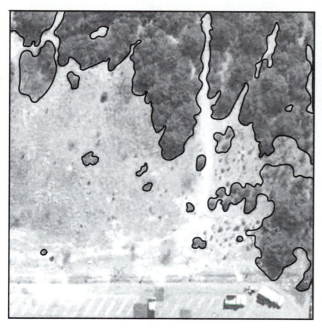

a. On-screen digitization

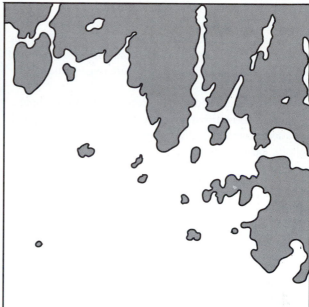

b. Individual polygons

Figure 6-37   a) Example of on-screen digitization using a digital image processing system. b) Polygon attribute tables for each of the individual polygons contain information about the perimeter and area.

recount. When the results from the two countings are averaged, the answer will generally be closer to the true value. This can be repeated any number of times. Again, the best results are obtained when the photointerpreted polygons have been transferred onto a base map and the dot grid is overlaid on the map.

The 6 x 6-in. orthophoto in Figure 6-36 is 1:600 scale (1" = 50 ft). Therefore, it contains 90,000 ft^2 of terrain (2.066 acres or 0.836 ha). It is overlaid with a dot grid consisting of 25 dots per in^2. At this scale, each square inch contains 2500 ft^2. Each dot therefore represents 100 ft^2. The interpreter analyzes the image and highlights those dots that appear to coincide with the location of forest (they are exaggerated in this example to improve visibility). The analyst simply counts the number of dots representing forest, 336, which equates to 33,600 ft^2 or 0.771 acre (0.316 ha). Forest occupies 37.3 percent of the study area (0.771 / 2.066 acres).

It is obvious from the example that there are some small, dark vegetated forest features that did not fall neatly below a dot and were not counted. Another interpreter might have included the dots closest to these features. Thus, the use of a dot grid involves "method-induced error" that is primarily a function of the methodology (logic) used by the image analyst as he or she allocates the dots. The important thing to

remember when using a dot grid is to be consistent. A variety of dot grids are available from forestry-supply companies.

On-screen Digitization

Perhaps the most widely used method of obtaining polygon area estimates from aerial photography or other remote sensor data is to analyze the image using a digital image processing system or GIS (Figure 6-37). While it is easy to scan an unrectified 9 x 9-in. aerial photograph and perform area calculations using the system, it is better if the aerial photography is scanned and then geometrically rectified to a map base, in effect turning it into an orthophotograph. All major digital image processing systems (ERDAS, PCI, ER Mapper, ENVI, etc.) and GIS (ESRI, Autodesk, etc.) have this fundamental capability. Details on how the image rectification is performed are found in Jensen (1996).

With the aerial photograph rectified to a map base, it is a straightforward task to utilize the system software and the cursor to place a "rubberband" polygon around the area of interest. For example, Figure 6-37a depicts hand-drawn polygons around the forest land cover. The vertices of the polygon can be easily edited. The beauty of this on-screen measurement approach is that each of the individual polygons has a record in a "polygon attribute table" associated with it that summarizes the attribute type (e.g., forest or non-forest), perimeter, and area (in hectares or acres). For example, there are only 14 forest polygons in Figure 6-37b. The attributes for all 14 polygons (e.g., hectares) can be summed by the image processing or GIS software to generate totals.

Again, it is not wise to digitize directly from aerial photography unless it is rectified orthophotography. For example, it is not good practice to digitize polygons from unrectified soil photomaps.

 References

Ackermann, F., 1994a, "Digital Elevation Models — Techniques and Application, Quality Standards, Development," *Proceedings, Symposium on Mapping and Geographic Information Systems*, Athens, GA: International Society for Photogrammetry & Remote Sensing, 30(4): 421–432.

Ackermann, F., 1994b, "On the Status and Accuracy Performance of GPS Photogrammetry," *Proceedings, Mapping and Remote Sensing Tools for the 21st Century*, Washington, DC: American Society for Photogrammetry & Remote Sensing, 80–90.

Avery, T. E. and G. L. Berlin, 1992, *Fundamentals of Remote Sensing and Airphoto Interpretation*, Upper Saddle River: Prentice-Hall, Inc., 472 pp.

Ayers, L., 1994, "Redefining the Role of Surveyors and GIS," *The Professional Surveyor*, 15(5):10–12.

Case, J., 1982, "The Digital Stereo Comparator/Compiler DSCC," *Proceedings*, International Society for Photogrammetry & Remote Sensing, Commission II, Canadian Institute of Surveying, Box 5378, Station F., Ottawa, 23–29.

Cowen, D. J. and J. R. Jensen, 1998, "Extraction and Modeling of Urban Attributes Using Remote Sensing Technology," in *People and Pixels*, Washington, DC: National Academy Press,164–188.

Dowman, I and F. Neto, 1994, "The Accuracy of Along Track Stereoscopic Data for Mapping: Results from Simulations and JERS OPS," *Proceedings, Symposium on Mapping and Geographic Information Systems*, Athens, GA: International Society for Photogrammetry & Remote Sensing, 30(4): 216–201.

Estes, J. E., E. J. Hajic and L. R. Tinney, 1983, "Fundamentals of Image Analysis: Analysis of Visible and Thermal Infrared Data," *Manual of Remote Sensing*, R. N. Colwell, Ed., Falls Church, VA: American Society of Photogrammetry, 1:1039–1040.

Greve, C., 1996, *Addendum to the Manual of Photogrammetry*, Bethesda: American Society for Photogrammetry & Remote Sensing, 318 pp.

Gustafson, G. C., 1989, "The Computer Brings Photogrammetry to the Photo Interpreter," University of Munich, *Muenchener Geographische Abhandlungen*, Vol. A-41, 61–74.

Gustafson, G. C., 1999, George Mason University, personnel correspondence.

Haack, B. et al., 1996, "Urban Analysis," *Manual of Photographic Interpretation*, 2nd Ed., Bethesda: American Society for Photogrammetry & Remote Sensing, 517–554.

Helava, U., 1958, "New Principle for Photogrammetric Plotters," *Photogrammetria*, 14(2):89–96.

Helpke, C., 1995, "State-of-the-Art of Digital Photogrammetric Workstations for Topographic Applications," *Photogrammetric Engineering & Remote Sensing*, 61(1):49–56.

Jensen, J. R., 1995, "Issues Involving the Creation of Digital Elevation Models and Terrain Corrected Orthoimagery Using Soft-Copy Photogrammetry," *Geocarto International*, 10(1):5–21.

Jensen, J. R., 1996a, *Introductory Digital Image Processing: A Remote Sensing Perspective*, Upper Saddle River, NJ: Prentice-Hall, 318 pp.

Jensen, J. R., 1996b, "Confluence of Mapping and Resource Management," *Addendum to the Manual of Photogrammetry*, C. Greve (Ed.), Bethesda: American Society for Photogrammetry & Remote Sensing, 167-179.

Jensen, J. R. and D. J. Cowen, 1999, "Remote Sensing of Urban/Suburban Infrastructure and Socio-Economic Attributes," *Photogrammetric Engineering & Remote Sensing*, 65(5):611–622.

Jensen, J. R., D. J. Cowen, J. Halls, S. Narumalani, N. J. Schmidt, B. A. Davis and B. Burgess, 1994, "Improved Urban Infrastructure Mapping and Forecasting for BellSouth Using Remote Sensing and GIS Technology," *Photogrammetric Engineering & Remote Sensing*, 60(3):339–346.

Joffe, B., 1994, "Better and Cheaper: Technical Innovation Builds Palo Alto's High Quality Map Base," *Geo Info Systems*, 4(6):47–51.

Keating, J. B., 1993, *The Geo-Positioning Selection Guide for Resource Management:* Technical Note 389, Cheyenne, WY: Bureau of Land Management, 72.

Leberl, F. W., 1994, "Practical Issues in Softcopy Photogrammetric Systems," *Proceedings*, Mapping and Remote Sensing Tools for the 21st Century, Washington, DC: American Society for Photogrammetry & Remote Sensing, 223–230.

Michael, J. H., 1992, "Digital Orthoimagery: the Map of the Future," *Northpoint*, 29(1): Survey Discipline of the Ontario Association of Certified Engineering Technicians and Technologists, 10 pp.

Morain, S. and S. L. Baros, 1996, *Raster Imagery in Geographic Information Systems*, Santa Fe: Onword Press, 495 pp.

Nale, D. K., 1994, "Digital Orthophotography: "What It Is and Isn't," *GIS World*, 7(6):22.

Perkins, S., 1994, "Photogrammetry and Surveying: Working Together as Science and Business," *The Professional Surveyor*, 15(5):20–29.

Petrie, G. and T. J. M. Kennie, 1990, *Terrain Modeling in Surveying and Civil Engineering*, London,: Whittles Publishing Company, 351 pp.

Rasher, M. E. and W. Weaver, 1990, *Basic Photo Interpretation*, Washington, DC: U. S. Department of Agriculture, Soil Conservation Service, 320 pp.

Southard, G. W., 1994, *True Orthophoto Capabilities of SOCET SET*, Englewood, CO: Leica Helava, Inc., 8 pp.

TeSalle, G., J. Plasker, A. Mikuni and K. Wortman, 1994, "A National Digital Orthophoto Program," *Proceedings*, GIS/LIS '94, Bethesda, MD: American Society for Photogrammetry & Remote Sensing, 741–757.

Thrower, N. J. W. and J. R. Jensen, 1976, "The Orthophoto and Orthophotomap: Characteristics, Development and Application," *The American Cartographer*, 3(1):39–56.

Trinder, J. C., A. Vuillemin, B. E. Donnelly and V. K. Shettigara, 1994, "A Study of Procedures and Tests on DEM Software for SPOT Images," *Proceedings*, Symposium on Mapping and Geographic Information Systems, Athens, GA: International Society for Photogrammetry & Remote Sensing, 30(4): 449–456.

Walker, A. S., 1994, Correspondence concerning the SOCET SET (TM), San Diego, CA: Leica AG Photogrammetry and Metrology, Inc., December 20.

Warner, W. S., R. W. Graham and R. E. Read, 1996, *Small Format Aerial Photography*, Scotland: Whittles Publishing, 348 pp.

Wolf, P. R., 1983, *Elements of Photogrammetry*, New York: John Wiley & Sons.

Multispectral Remote Sensing Systems 7

Multispectral remote sensing is defined as the collection of reflected, emitted, or backscattered energy from an object or area of interest in multiple bands (regions) of the electromagnetic spectrum. *Hyperspectral remote sensing* involves data collection in hundreds of bands (Logicon, 1997). Most multispectral and hyperspectral remote sensing systems collect data in a digital format. This chapter first introduces the fundamental structure of a digital image. The generic types of multispectral and hyperspectral remote sensing systems are then introduced. Finally, the characteristics of the current and proposed multispectral and hyperspectral remote sensing systems are presented.

 Multispectral Data Collection

Chapters 3 – 6 devoted attention to the analysis of aerial photography data using both visual (analog) and digital soft-copy photogrammetric techniques. Most of the data sets consisted of natural color or color-infrared aerial photography. Both color and color-infrared aerial photography may be considered to be three-band multispectral data sets based on the sensitivity of their emulsions as previously discussed in Chapter 4. While these three-band multispectral data are sufficient for many applications, there are times when more spectral bands located at optimum locations throughout the electromagnetic spectrum might be much better for a specific application. Fortunately, engineers have developed detectors that are sensitive to hundreds of bands in the electromagnetic spectrum. The measurements made by the detectors are almost always stored in a digital format.

An overview of how these digital remote sensing data are turned into useful information is shown in Figure 7-1. The remote sensor system first detects electromagnetic energy that exits from the phenomena of interest and passes through the atmosphere. The energy detected is recorded as an analog electrical signal, which is usually converted into a digital value through an analog-to-digital (A-to-D) conversion. If an aircraft platform is used, the digital data are simply returned to Earth. However, if a spacecraft platform is used, the digital data are telemetered to Earth receiving stations directly or indirectly via tracking and data relay satellites (TDRS). In either case, it may be necessary to perform some radiometric and/or geometric preprocessing of the digital remotely sensed data to improve its interpretability. The data may then be enhanced for subsequent human visual analysis or processed further using digital image processing algorithms. Biophysical and/or land-cover information extracted using either a visual or digital image processing approach is distributed and used to make decisions.

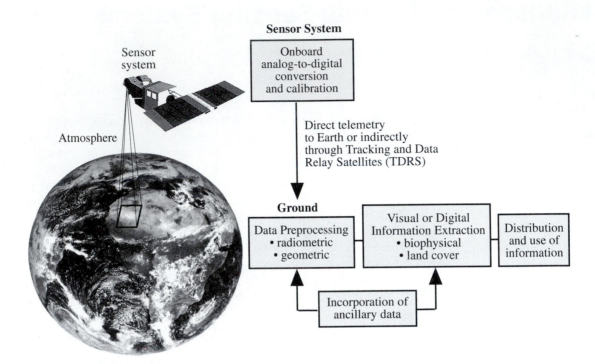

Figure 7-1   Overview of how digital remotely sensed data are turned into useful information. The data recorded by the detectors are often converted from an analog electrical signal to a digital value and calibrated. Ground preprocessing removes geometric and radiometric distortions. This may involve the use of ephemeris and/or ancillary data such as map x,y coordinates, a digital elevation model, etc. The data are then ready for visual or digital analysis to extract biophysical or land-use/land-cover information.

Digital Image Terminology

The digital remote sensor data are usually stored as a matrix (array) of numbers. Each digital value is located at a specific row (i) and column (j) in a matrix, as shown in Figure 7-2. A *pixel* is defined as "a two-dimensional picture element that is the smallest non-divisible element of a digital image." The pixel has a brightness value (BV) associated with each row (i) and column (j) in the image. The dataset may consist of n individual multispectral bands (k). Thus, it is possible to identify the brightness value (BV) of a particular picture element (pixel) in the multispectral dataset by specifying its row (i), column (j), and band (k) coordinate, i.e., $BV_{i,j,k}$. It is important to understand that the n multispectral bands are all geometrically registered to one another. Therefore, a road intersection in band 1 at row 4, column 4 (i.e., $BV_{4,4,1}$) should be located at the same row and column coordinate in the tenth band (i.e., $BV_{4,4,10}$). Hopefully, the brightness values at the two locations are different; otherwise, the information content of the two images is redundant.

The analog-to-digital conversion that takes place onboard the sensor system usually creates pixels with a brightness value range of 8 to 12 bits. This is generally called the *quantization* level of the sensor system. Remote sensor data quantized to 8 bits have brightness values that range from 0 to 255. Data quantized to 12 bits range from 0 to 1023, etc. The greater the range of possible brightness values, the more precise we may be able to measure the amount of radiance recorded by the detector. One can think of quantization as if it were a ruler. We can obtain much more accurate measurements of an object with a ruler that has 1024 subdivisions than with a ruler that only has 256 subdivisions.

There are a tremendous variety of digital multispectral and hyperspectral remote sensing systems. It is beyond the scope of this book to provide detailed information on each of them. However, it is possible to review selected remote sensing systems that are or will be of significant value for Earth resource investigations. They are organized according to the type of remote sensing technology used, as summarized in Figure 7-3, including:

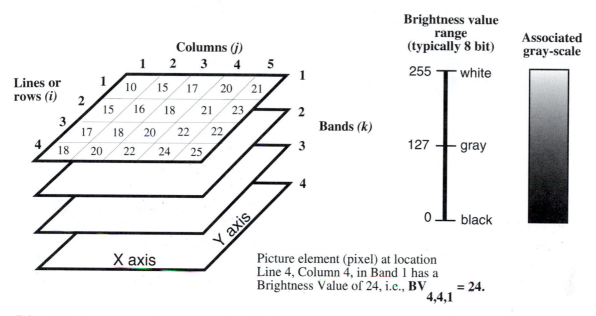

Figure 7-2 Digital remote sensor data are stored in a matrix (array or raster) format. Picture element (pixel) brightness values *(BV)* are located at row *i*, column *j*, and band *k* in the multispectral dataset. The digital remote sensor data are normally stored as 8-bit bytes, with values ranging from 0 to 255. However, several sensor systems now routinely collect 10- and 12-bit data.

Multispectral Imaging Using Discrete Detectors and Scanning Mirrors

- Landsat Multispectral Scanner (MSS)

- Landsat Thematic Mapper (TM)

- Landsat 7 Enhanced Thematic Mapper Plus (ETM[+])

- NOAA Geostationary Operational Environmental Satellite (GOES)

- NOAA Advanced Very High Resolution Radiometer (AVHRR)

- NASA and ORBIMAGE, Inc., Sea-viewing Wide field of view Sensor (SeaWiFS)

- Daedalus, Inc., Aircraft Multispectral Scanner (AMS)

- NASA Airborne Terrestrial Applications Sensor (ATLAS)

Multispectral Imaging Using Linear Arrays

- SPOT 1, 2, and 3 High Resolution Visible (HRV) sensors and SPOT 4 High Resolution Visible Infrared (HRVIR) and Vegetation sensor

- Indian Remote Sensing System (IRS) Linear Imaging Self-scanning Sensor (LISS)

- Space Imaging, Inc., (IKONOS)

- ORBIMAGE, Inc., OrbView-3 and OrbView-4 (Warfighter)

- EarthWatch, Inc., (Quickbird)

- NASA *Terra* Advanced Spaceborne Thermal Emission and Reflection Radiometer (ASTER)

- NASA *Terra* Multi-angle Imaging Spectroradiometer (MISR)

Imaging Spectrometry Using Linear and Area Arrays

- NASA Jet Propulsion Lab (JPL/CALTECH) Airborne Visible-Infrared Imaging Spectrometer (AVIRIS)

- Canadian Compact Airborne Spectrographic Imager-2 (CASI-2)

- NASA *Terra* Moderate Resolution Imaging Spectrometer (MODIS)

Types of Detector Configurations Used for Multispectral and Hyperspectral Remote Sensing

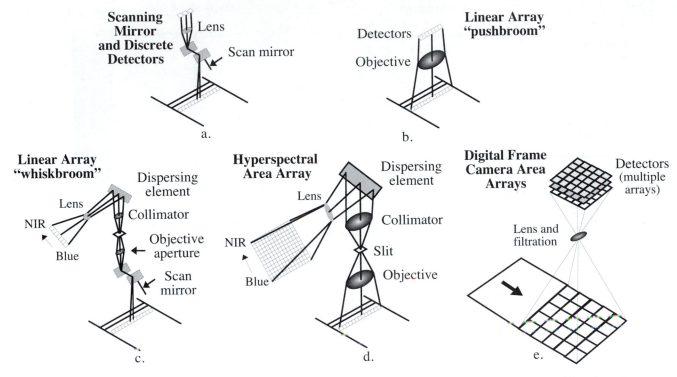

Figure 7-3   Five types of remote sensing systems used for multispectral and hyperspectral imaging: a) multispectral imaging using a scanning mirror and discrete detectors, b) multispectral imaging with linear arrays (often referred to as "pushbroom" technology), c) imaging with a scanning mirror and linear arrays (often referred to as "whiskbroom" technology), d) imaging spectrometry with area arrays, and e) data collection using a digital frame camera equipped with multiple area arrays.

- NASA Earth Observer (EO-1) Advanced Land Imager (ALI), Hyperion, and LEISA Atmospheric Corrector (LAC)

Digital Frame Cameras

- Positive Systems, Inc.

- Litton Emerge Spatial, Inc.

Satellite Photographic Systems

- Russian SPIN-2 TK-350 and KVR-1000

- NASA Space Shuttle Photography

This discussion identifies the spatial, spectral, temporal, and radiometric characteristics of the remote sensing systems. It will be clear that some sensors were designed to collect very specific types of biophysical information.

 Multispectral Imaging Using Discrete Detectors and Scanning Mirrors

The collection of multispectral remote sensor data using discrete detectors and scanning mirrors has been with us since the mid-1960s. Despite the technology's age, several new remote sensing systems still utilize it.

Earth Resource Technology Satellites (ERTS) and the Landsat Sensor Systems

In 1967, the National Aeronautics and Space Administration (NASA), encouraged by the U.S. Department of the Interior, initiated the Earth Resource Technology Satellite (ERTS) program. This program resulted in the deployment of five satellites carrying a variety of remote sensing systems designed primarily to acquire Earth resource information. The most noteworthy sensors were the Landsat Multispec-

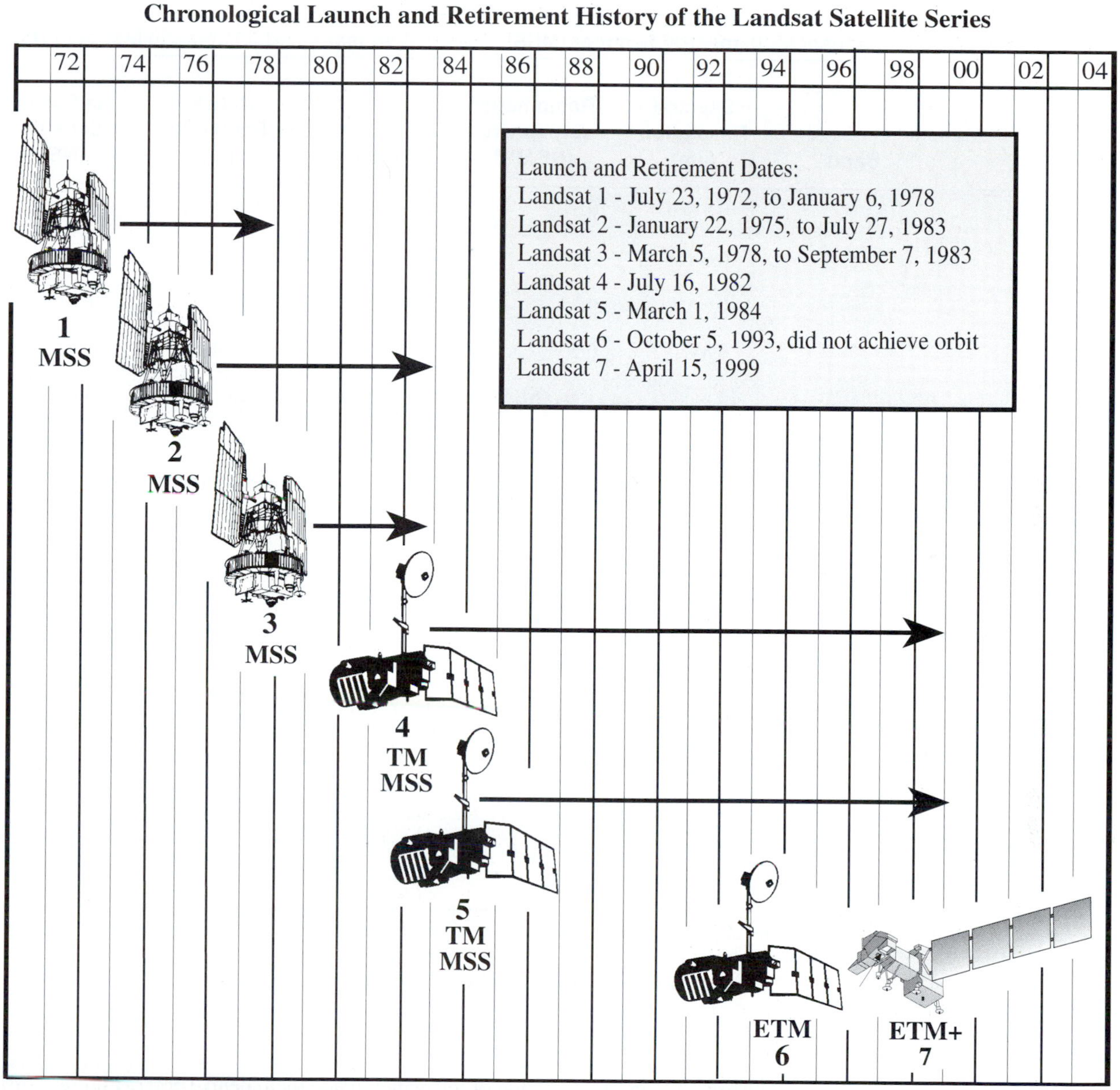

Chronological Launch and Retirement History of the Landsat Satellite Series

Launch and Retirement Dates:
Landsat 1 - July 23, 1972, to January 6, 1978
Landsat 2 - January 22, 1975, to July 27, 1983
Landsat 3 - March 5, 1978, to September 7, 1983
Landsat 4 - July 16, 1982
Landsat 5 - March 1, 1984
Landsat 6 - October 5, 1993, did not achieve orbit
Landsat 7 - April 15, 1999

Figure 7-4   Chronological launch and retirement history of the Landsat series of satellites (1 through 7) from 1972 to 1999.

tral Scanner (MSS) and the Landsat Thematic Mapper (Table 7-1). The Landsat Program is the United States' oldest land-surface observation satellite system, having obtained data since 1972. It has had a tumultuous history of management and funding sources.

The chronological launch and retirement history of the satellites is shown in Figure 7-4 (NASA, 1999). The ERTS-1 satellite launched on July 23, 1972, was the first experimental system designed to test the feasibility of collecting Earth resource data by unmanned satellites. Prior to the launch of ERTS-B on January 22, 1975, NASA renamed the ERTS program *Landsat*, distinguishing it from the *Seasat* oceanographic satellite launched on June 26, 1978. At this time, ERTS-1 was retroactively named Landsat-1 and ERTS-B became Landsat-2 at launch. Landsat-3 was launched March 5, 1978; Landsat-4 on July 16, 1982; and Landsat-5 on March 1, 1984. A variety of mechanical failures prompted the retirement of some of the Landsat satellites (EOSAT, 1992a).

Table 7-1.   Landsat Multispectral Scanner (MSS) and Thematic Mapper (TM) Sensor System Characteristics

	Landsat Multispectral Scanner (MSS)			Landsat 4 and 5 Thematic Mapper (TM)		
Band	Spectral Resolution (μm)	Radiometric Sensitivity (NEΔP)[a]		Band	Spectral Resolution (μm)	Radiometric Sensitivity (NEΔP)
4[b]	0.5–0.6	0.57		1	0.45–0.52	0.8
5	0.6–0.7	0.57		2	0.52–0.60	0.5
6	0.7–0.8	0.65		3	0.63–0.69	0.5
7	0.8–1.1	0.70		4	0.76–0.90	0.5
8[c]	10.4–12.6	1.4K (NEΔT)		5	1.55–1.75	1.0
				6	10.40–12.5	0.5 (NEΔT)
				7	2.08–2.35	2.4
IFOV at nadir	79 × 79 m for bands 4 to 7 240 × 240 m for band 8			30 × 30 m for bands 1 to 5, 7 120 × 120 m for band 6		
Data rate	15 Mb/s			85 Mb/s		
Quantization levels	6 bit (values from 0 to 63)			8 bit (values from 0 to 255)		
Earth coverage	18 days Landsat 1, 2, 3 16 days Landsat 4, 5			16 days Landsat 4, 5		
Altitude	919 km			705 km		
Swath width	185 km			185 km		
Inclination	99°			98.2°		

[a] The radiometric sensitivities are the noise-equivalent reflectance differences for the reflective channels expressed as percentages (NEΔP) and temperature differences for the thermal infrared bands (NEΔT).
[b] MSS bands 4, 5, 6, and 7 were renumbered bands 1, 2, 3, and 4 on Landsats 4 and 5.
[c] MSS band 8 was present only on Landsat 3.

The Earth Observation Satellite Company (EOSAT) obtained control of the Landsat satellites in September, 1985. Unfortunately, Landsat 6 with its Enhanced Thematic Mapper (ETM) (a 15 × 15 m panchromatic band was added) failed to achieve orbit on October 5, 1993 (Silvestrini, 1993). For a detailed history of the Landsat program, refer to the *Landsat Data User Notes* published by the EROS Data Center (NOAA, 1975–1984), Space Imaging, Inc., *Imaging Notes*, and the NASA Landsat 7 home page on the Internet (Appendix A).

Landsats 1 to 3 were launched into circular orbits around Earth at a nominal altitude of 919 km (570 mi). The platform is shown in Figure 7-5a. The satellites had an orbital inclina-

tion of 99° which made them nearly polar (Figure 7-5b) and caused them to cross the equator at an angle of approximately 9° from normal. The satellites orbited Earth once every 103 min, resulting in 14 orbits per day (Figure 7-5c). This Sun-synchronous orbit meant that the orbital plane precessed around Earth at the same angular rate at which Earth moved around the Sun. This characteristic caused the satellites to cross the equator at approximately the same local time (9:30 to 10:00 a.m.) on the illuminated side of Earth.

Figures 7-5c and 7-6 illustrate how repeat coverage of a geographic area was acquired. From one orbit to the next, a position directly below the spacecraft moved 2875 km (1785 mi) at the equator as Earth rotated beneath it. The next day, 14

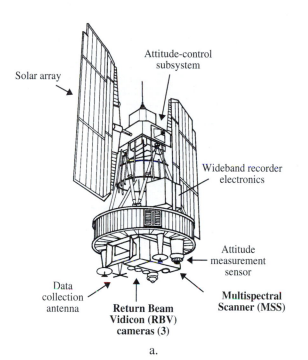

Solar array

Attitude-control
subsystem

Wideband recorder
electronics

Attitude
measurement
sensor

**Multispectral
Scanner (MSS)**

**Return Beam
Vidicon (RBV)
cameras (3)**

Data
collection
antenna

a.

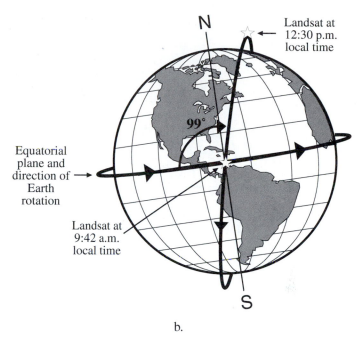

N

Landsat at
12:30 p.m.
local time

99°

Equatorial
plane and
direction of
Earth
rotation

Landsat at
9:42 a.m.
local time

S

b.

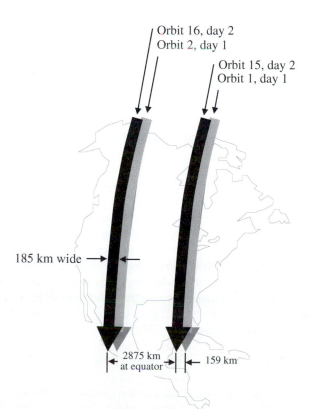

Orbit 16, day 2
Orbit 2, day 1

Orbit 15, day 2
Orbit 1, day 1

185 km wide →

2875 km
at equator

159 km

c.

Figure 7-5   a) Nimbus-style platform used for Landsats 1, 2, and 3 and associated sensor and telecommunication systems. b) Inclination of the Landsat orbit to maintain a Sun-synchronous orbit. c) From one orbit to the next, the position directly below the satellite moved 2875 km (1785 mi) at the equator as Earth rotated beneath it. The next day, 14 orbits later, it was approximately back to its original location, with orbit 15 displaced westward from orbit 1 by 159 km (99 mi). This is how repeat coverage of the same geographic area was obtained.

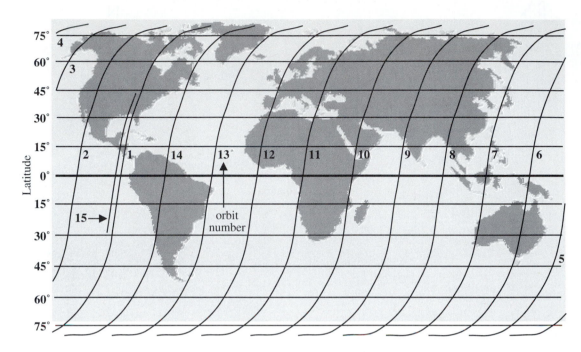

Figure 7-6   Orbital tracks of Landsat 1, 2, or 3 during a single day of coverage. The satellite crossed the equator every 103 minutes, during which time the Earth rotated a distance of 2,875 km under the satellite at the equator. Every 14 orbits, 24 hours elapsed.

orbits later, it was back to its original location, with orbit 15 displaced westward from orbit 1 by 159 km (99 mi) at the equator. This continued for 18 days, after which orbit 252 fell directly over orbit 1 once again. Thus, the Landsat sensor systems had the capability of observing the entire globe (except poleward of 81°) once every 18 days, or about 20 times a year. There were approximately 26 km (16 mi) of overlap between successive orbits. This overlap was a maximum at 81° north and south latitudes (about 85 percent) and a minimum at the equator (about 14 percent). This has proven useful for some stereoscopic analysis applications.

The nature of the orbiting Landsat system has given rise to a Path and Row *Worldwide Reference System* (WRS) for locating and obtaining Landsat imagery for any area on Earth. The WRS has catalogued the world's landmass into 57,784 scenes. Each scene is approximately 185 km wide by 170 km long and consists of approximately 3.8 gigabits of data (NASA, 1999). Figure 7-7 depicts a small portion of the 1993 WRS map for the southeastern United States with one scene highlighted. The user locates the area of interest on the path and row map (e.g., Path 16, Row 37 is the nominal Charleston, SC, scene) and then requests information from Space Imaging, Inc., or the EROS Data Center in Sioux Falls, SD, about Landsat imagery available for this path and

row. If no path and row map is available, a geographic search can be performed by specifying the longitude and latitude at the center of the area of interest or by defining an area of interest with the longitude and latitude coordinates of each corner.

In the context of this section on data acquisition, we are interested in the type of sensors carried aloft by the Landsat satellites and the nature and quality of remote sensor data provided for Earth resource investigations. The sensors included the Return-Beam-Vidicon camera (RBV), Multispectral Scanner (MSS), and Thematic Mapper (TM). The focus will be only on the Multispectral Scanner and Thematic Mapper systems because they continue to provide a significant part of all remotely sensed imagery used for Earth resource studies.

Landsat Multispectral Scanner (MSS)

This sensor was placed on Landsat satellites 1 through 5. The MSS multi-detector array and the scanning system are shown diagrammatically in Figure 7-8a. Sensors such as the Landsat MSS (and Thematic Mapper to be discussed) are optical-mechanical systems in which a mirror scans the terrain perpendicular to the flight direction. While it scans, it

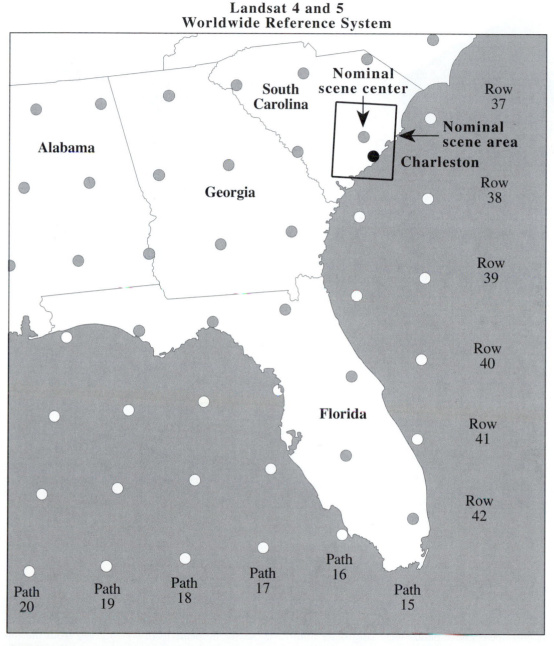

Figure 7-7 Portion of the 1993 Landsat Worldwide Reference System. Path 16, Row 37 is near Charleston, SC.

focuses radiant flux from the terrain onto discrete detector elements. The detectors convert the reflected solar radiant flux measured within each instantaneous field of view (IFOV) in the scene into an electronic signal (refer to Figure 7-3a). The detector elements are placed behind filters that pass broad portions of the spectrum. The MSS has four sets of filters and detectors, whereas the TM has seven. The primary limitation of this approach is the short residence time of the detector in each IFOV. To achieve adequate signal-to-

noise ratio without sacrificing spatial resolution, such a sensor must operate in broad spectral bands of ≥100 nm or must use optics with unrealistically small ratios of focal-length to aperture (*f*/stop).

The MSS scanning mirror oscillates through an angular displacement of ±5.78° off-nadir. This 11.56° field of view resulted in a swath width of approximately 185 km (115 mi) for each orbit. Six parallel detectors sensitive to four spectral

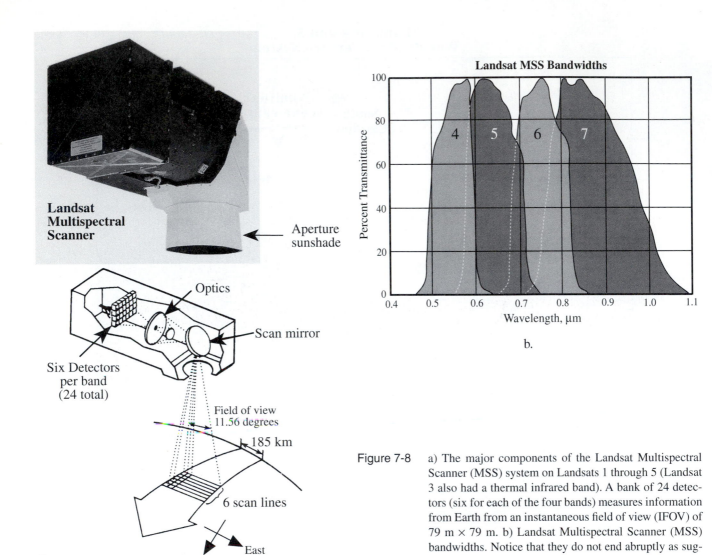

Figure 7-8   a) The major components of the Landsat Multispectral Scanner (MSS) system on Landsats 1 through 5 (Landsat 3 also had a thermal infrared band). A bank of 24 detectors (six for each of the four bands) measures information from Earth from an instantaneous field of view (IFOV) of 79 m × 79 m. b) Landsat Multispectral Scanner (MSS) bandwidths. Notice that they do not end abruptly as suggested by the usual nomenclature.

bands (channels) in the electromagnetic spectrum viewed the ground simultaneously: 0.5 to 0.6 μm (green), 0.6 to 0.7 μm (red), 0.7 to 0.8 μm (reflective infrared), and 0.8 to 1.1 μm (reflective infrared). These bands were originally numbered 4, 5, 6, and 7, respectively, because a Return-Beam-Vidicon (RBV) sensor system also onboard the satellite recorded energy in three bands labeled 1, 2, and 3.

When not viewing the Earth, the MSS detectors were exposed to internal light and Sun calibration sources. The spectral sensitivity of the bands is summarized in Table 7-1 and shown diagrammatically in Figure 7-8b. Note that there is some spectral overlap between the bands.

Prior to the launch of the satellite, the engineering model of the ERTS MSS was tested by viewing the scene behind the

Santa Barbara Research Center in Goleta, CA. Bands 4 and 6 (green and near-infrared, respectively) of the area are shown in Figure 7-9. Note the spatial detail present when the sensor is located only 1 to 2 km from the mountains. The spatial resolution is much lower when the sensor is placed 919 km above Earth in orbit.

The IFOV of each detector was square and resulted in a ground resolution element of approximately 79 × 79 m (67,143 ft²). The voltage analog signal from each detector was converted to a digital value using an onboard A-to-D converter. The data were quantized to 6 bits with a range of values from 0 to 63. These data were then rescaled to 7 bits (0 to 127) for three of the four bands in subsequent ground processing (i.e., bands 4, 5, and 6 were decompressed to a range of 0 to 127). It is important to remember that the early

a.

b.

Figure 7-9   Two terrestrial images acquired by the engineering model of the Landsat MSS on March 4, 1972, at the Santa Barbara Research Center of Hughes Aircraft, Inc. The top image (a) was acquired using the MSS band 4 detectors (0.5 – 0.6 μm), and the bottom image (b) was acquired using band 6 detectors (0.7 – 0.8 μm). Note the high spatial fidelity of the images, which is possible when the terrain is close and not 919 km away.

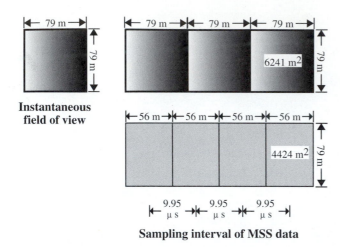

Instantaneous field of view

Sampling interval of MSS data

Figure 7-10   Relationship between the original 79 × 79 m IFOV of the Landsat MSS and the rate at which it was re-sampled (i.e., every 9.95 µs). This resulted in picture elements (pixels) that were 56 × 79 m in dimension on tapes acquired from the EROS Data Center at Sioux Falls, SD.

1970s Landsat MSS data were quantized to 6 bits when comparing MSS data collected in the late 1970s and 1980s which were collected at 8 bits.

During each scan, the voltage produced by each detector was sampled every 9.95 µs. For one detector, approximately 3,300 samples were taken along a 185-km line. Thus, the IFOV of 79 m × 79 m became about 56 m on the ground between each sample (Figure 7-10). The 56 × 79 m area is called a Landsat MSS picture element. Thus, although the measurement of landscape brightness was made from a 6241-m² area, each pixel was reformatted as if the measurement were made from a 4424-m² area (Figure 7-10). Note the overlap of the areas from which brightness measurements were made for adjacent pixels.

The MSS scanned each line across-track from west to east as the southward orbit of the spacecraft provided the along-track progression. Each MSS scene represents a 185 × 170 km parallelogram extracted from the continuous swath of an orbit and contains approximately 10 percent overlap. A typical scene contains approximately 2,340 scan lines with about 3,240 pixels per line or about 7,581,600 pixels per channel. All four bands represent a data set of more than 30 million brightness values. Landsat MSS images provide an unprecedented ability to observe large geographic areas while viewing a single image. For example, approximately 5,000 conventional vertical aerial photographs obtained at a

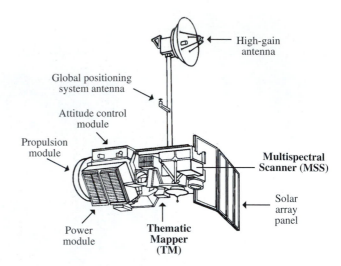

Figure 7-11   Landsat 4 and 5 platform with associated sensor and telecommunication systems.

scale of 1:15,000 are required to equal the geographic coverage of a single Landsat MSS image. This allows regional terrain analysis to be performed using one data source rather than a multitude of aerial photographs.

Landsat Thematic Mapper (TM)

Landsat Thematic Mapper sensor systems were launched on July 16, 1982 (Landsat 4), and on March 1, 1984 (Landsat 5). The TM is a scanning optical-mechanical sensor (Figure 7-3a) that records energy in the visible, reflective-infrared, middle-infrared, and thermal infrared regions of the electromagnetic spectrum. It collects multispectral imagery that has higher spatial, spectral, temporal, and radiometric resolution than the Landsat MSS (SBRC, 1994).

Detailed descriptions of the design and performance characteristics of the TM can be found in EOSAT (1992a). The Landsat 4 and 5 platform is shown in Figure 7-11. The Thematic Mapper sensor system configuration is shown in Figure 7-12. A telescope directs the incoming radiant flux obtained along a scan line through a scan line corrector to 1) the visible and near-infrared primary focal plane, or 2) the middle-infrared and thermal infrared cooled focal plane. The detectors for the visible and near-infrared bands (1 to 4) are four staggered linear arrays, each containing 16 silicon detectors. The two middle-infrared detectors are 16 indium-antimonide cells in a staggered linear array, and the thermal infrared detector is a four-element array of mercury-cadmium-telluride cells.

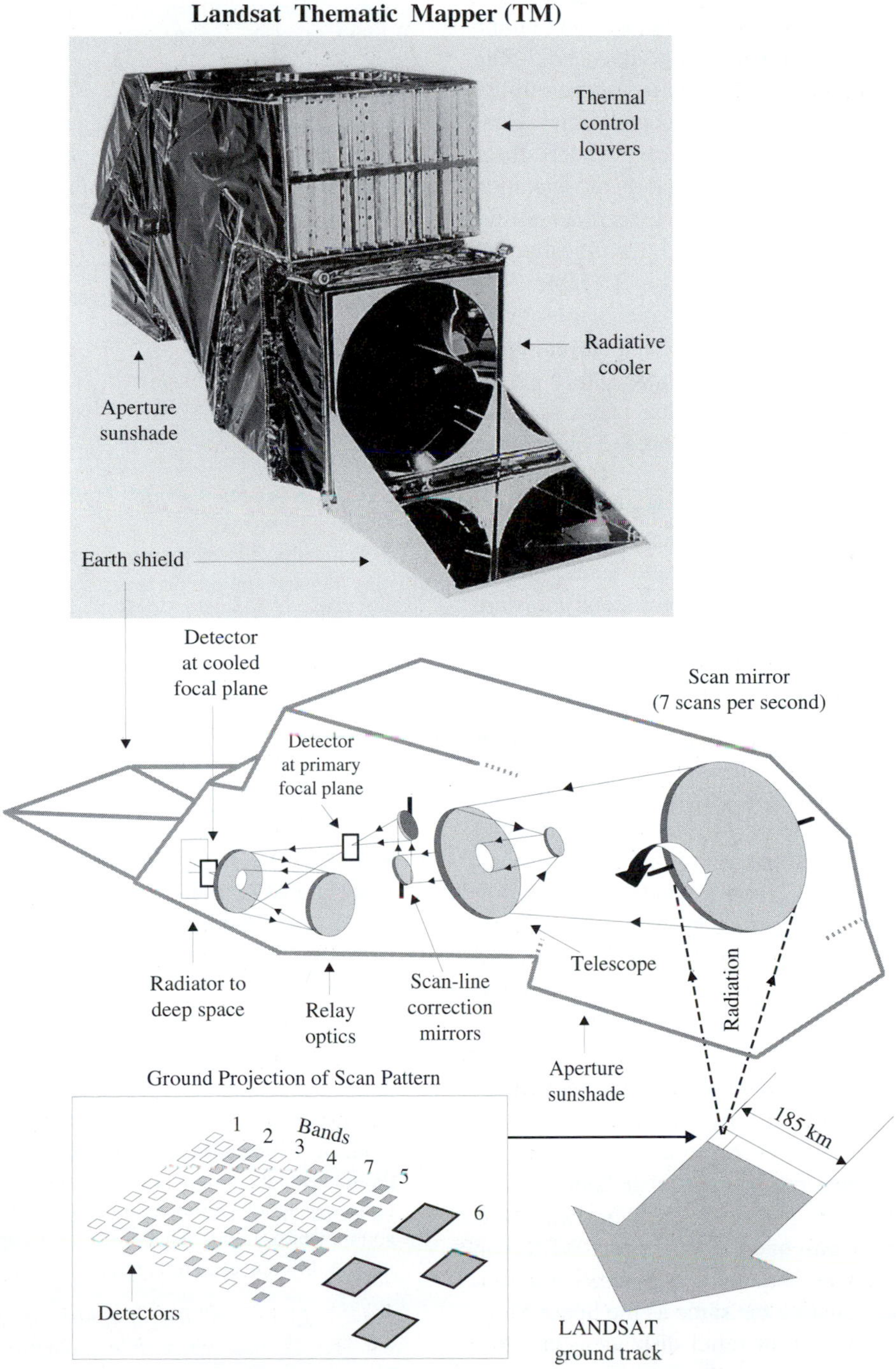

Landsat Thematic Mapper (TM)

Figure 7-12 Major components of the Landsat 4 and 5 Thematic Mapper sensor system. The sensor is sensitive to the seven bands of the electromagnetic spectrum summarized in Table 7-1. Six of the seven bands have a spatial resolution of 30×30 m, while the thermal infrared band has a spatial resolution of 120×120 m.

Landsat TM data have a ground-projected IFOV of 30×30 m for bands 1 – 5 and 7. The thermal infrared band 6 has a spatial resolution of 120×120 m. The TM spectral bands represent important departures from the bands found on the traditional MSS also carried onboard Landsats 4 and 5. The original MSS bandwidths were selected based on their utility for general vegetation inventories and geologic studies. Conversely, the TM bands were chosen after years of analysis for their value in water penetration, discrimination of vegetation type and vigor, plant and soil moisture measurement, differentiation of clouds, snow, and ice, and identification of hydrothermal alteration in certain rock types (Table 7-2). The refined bandwidths and improved spatial resolution of the Landsat TM versus the Landsat MSS and several other sensor systems (Landsat 7 and SPOT 1-4) are shown graphically in Figure 7-13. Examples of individual bands of Landsat Thematic Mapper imagery of Charleston, SC, obtained in 1994 are provided in Figure 7-14.

The Landsat TM bands were selected to make maximum use of the dominant factors controlling leaf reflectance, such as leaf pigmentation, leaf and canopy structure, and moisture content, as demonstrated in Figure 7-15. Band 1 (blue) provides some water-penetration capability. Vegetation absorbs much of the incident blue, green, and red radiant flux for photosynthetic purposes; therefore, vegetated areas appear dark in TM band 1 (blue), 2 (green), and 3 (red) images, as seen in the Charleston, SC, TM data (Figure 7-14). Vegetation reflects approximately half of the incident near-infrared radiant flux, causing it to appear bright in the band 4 (near-infrared) image. Bands 5 and 7 both provide more detail in the wetland because they are sensitive to soil and plant moisture conditions. The band 6 (thermal) image provides limited information of value. Landsat Thematic Mapper color composites of Charleston, SC, are shown in Color Plate 7-1a,b.

The equatorial crossing time was 9:45 a.m. for Landsats 4 and 5 with an orbital inclination of 98.2°. The transition from an approximately 919-km orbit to a 705-km orbit for Landsats 4 and 5 disrupted the continuity of Landsat 1, 2, and 3 MSS path and row designations in the Worldwide Reference System (WRS). Consequently, a separate WRS map is required to select images obtained by Landsats 4 and 5. The lower orbit (approximately the same as the Space Shuttle) also increased the amount of relief displacement introduced into the imagery obtained over mountainous terrain. The new orbit also caused the period between repetitive coverage to change from 18 to 16 days for both the MSS and TM data collected by Landsats 4 and 5.

There was a substantial improvement in the level of quantization from 6 to 8 bits per band (Table 7-1). This, in addition

Table 7-2.   Characteristics of the Landsat Thematic Mapper (TM) Spectral Bands

Band 1: 0.45 to 0.52 μm (blue). Provides increased penetration of water bodies, as well as supporting analyses of land use, soil, and vegetation characteristics. The shorter-wavelength cutoff is just below the peak transmittance of clear water, while the upper-wavelength cutoff is the limit of blue chlorophyll absorption for healthy green vegetation. Wavelengths below 0.45 μm are substantially influenced by atmospheric scattering and absorption.

Band 2: 0.52 to 0.60 μm (green). This band spans the region between the blue and red chlorophyll absorption bands and therefore corresponds to the green reflectance of healthy vegetation.

Band 3: 0.63 to 0.69 μm (red). This is the red chlorophyll absorption band of healthy green vegetation and represents one of the most important bands for vegetation discrimination. It is also useful for soil-boundary and geological-boundary delineations. This band may exhibit more contrast than bands 1 and 2 because of the reduced effect of atmospheric attenuation. The 0.69-μm cutoff is significant because it represents the beginning of a spectral region from 0.68 to 0.75 μm, where vegetation reflectance crossovers take place that can reduce the accuracy of vegetation investigations.

Band 4: 0.76 to 0.90 μm (reflective infrared). For reasons discussed, the lower cutoff for this band was placed above 0.75 μm. This band is especially responsive to the amount of vegetation biomass present in a scene. It is useful for crop identification and emphasizes soil/crop and land/water contrasts.

Band 5: 1.55 to 1.75 μm (mid-infrared). This band is sensitive to the turgidity or amount of water in plants. Such information is useful in crop drought studies and in plant vigor investigations. In addition, this is one of the few bands that can be used to discriminate between clouds, snow, and ice, which are so important in hydrologic research.

Band 6: 10.4 to 12.5 μm (thermal infrared). This band measures the amount of infrared radiant flux emitted from surfaces. The apparent temperature is a function of the emissivities and the true or kinetic temperature of the surface. It is useful for locating geothermal activity, thermal inertia mapping for geologic investigations, vegetation classification, vegetation stress analysis, and soil moisture studies. The sensor often captures unique information on differences in topographic aspect in mountainous areas.

Band 7: 2.08 to 2.35 μm (mid-infrared). This is an important band for the discrimination of geologic rock formations. It has been shown to be particularly effective in identifying zones of hydrothermal alteration in rocks.

to a greater number of bands and a higher spatial resolution, increased the data rate from 15 to 85 Mb/s. Ground receiving stations were modified to process the increased data flow.

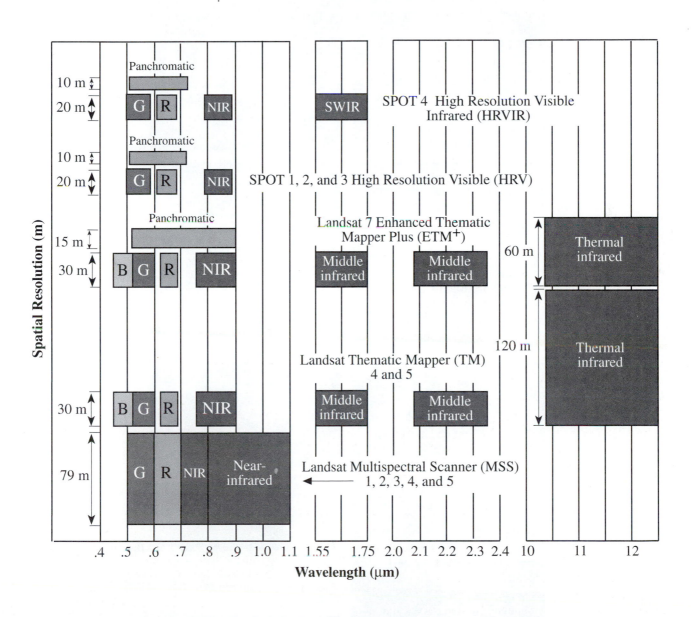

Figure 7-13 Spectral and spatial resolution of the Landsat Multispectral Scanner (MSS), Landsat 4 and 5 Thematic Mapper (TM), Landsat 7 Enhanced Thematic Mapper Plus (ETM+), SPOT 1, 2, and 3 High Resolution Visible (HRV), and SPOT 4 High Resolution Visible Infrared (HRVIR) sensor systems. The SPOT 4 Vegetation sensor characteristics are not shown (it consists of four 1.15 x 1.15 km bands).

Landsat Thematic Mapper Data of Charleston, SC

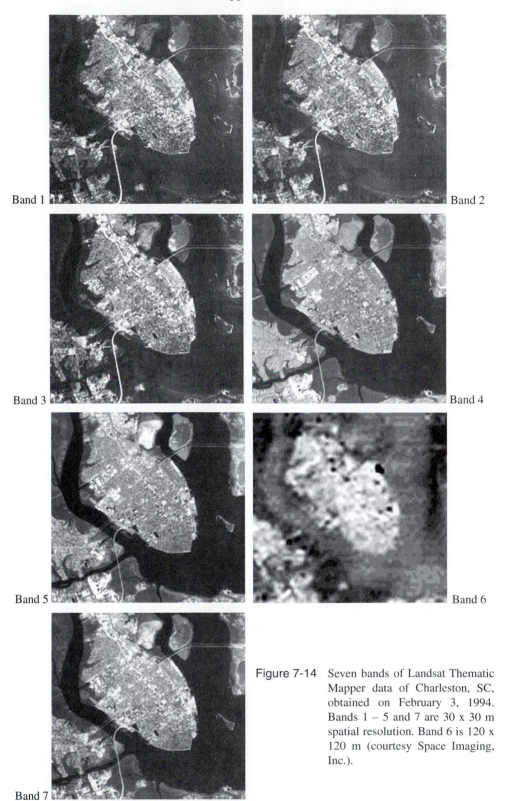

Band 1

Band 2

Band 3

Band 4

Band 5

Band 6

Band 7

Figure 7-14   Seven bands of Landsat Thematic Mapper data of Charleston, SC, obtained on February 3, 1994. Bands 1 – 5 and 7 are 30 x 30 m spatial resolution. Band 6 is 120 x 120 m (courtesy Space Imaging, Inc.).

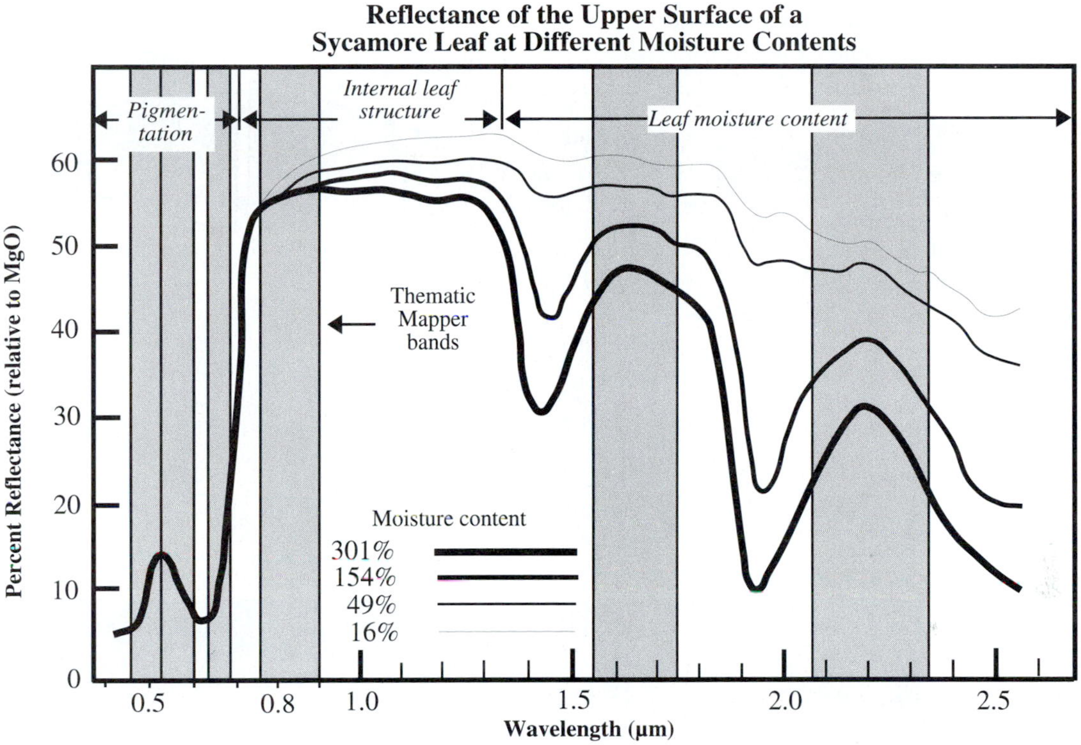

Reflectance of the Upper Surface of a
Sycamore Leaf at Different Moisture Contents

Figure 7-15 Progressive changes in percent reflectance for a Sycamore leaf at varying oven dry weight moisture contents. The dominant factors controlling leaf reflectance and the location of six of the Landsat Thematic Mapper bands are superimposed.

Based on the improvements in spectral, spatial, and radiometric resolution, Solomonson (1984) suggested that "it appears that the TM can be described as being twice as effective in providing information as the Landsat MSS. This is based on its ability to provide twice as many separable classes over a given area as the MSS, numerically provide two more independent vectors in the data or demonstrate through classical information theory that twice as much information exists in the TM data." Remarkably, Landsat 5 is still acquiring images. Efforts to move the Landsat Program into the commercial sector began under the Carter Administration in 1979 and resulted in legislation passed in 1984 that charged the National Oceanic and Atmospheric Administration to transfer the program to the private sector. The Earth Observing Satellite Company (EOSAT) took over operation in 1985 and was given the rights to market Landsat data.

Landsat 7 Enhanced Thematic Mapper Plus (ETM+)

On October 28, 1992, President Clinton signed the Land Remote Sensing Policy Act of 1992 (Public Law 102-555). The law authorized the procurement of Landsat 7 and called for its launch within five years of the launch of Landsat 6. In parallel actions, Congress funded Landsat 7 procurement and stipulated that data from publicly funded remote sensing satellite systems like Landsat must be sold to United States government agencies and their affiliated users at the cost of fulfilling user requests (Asker, 1992; EOSAT, 1992b). Unfortunately, Landsat 6 did not achieve orbit on October 5, 1993, when the rocket's upper stage failed to fire.

With the passage of the Land Remote Sensing Policy Act of 1992, oversight of the Landsat program began to be shifted from the commercial sector back to the federal government. NASA is responsible for the design, development, launch and on-orbit checkout of Landsat 7, installation of the ground system, and operation of it until October 2000. The USGS is responsible for capture, processing, and distribution of the Landsat 7 data, mission management, and maintaining an archive of Landsat data. In October 2000, USGS will take over Landsat 7 satellite operations.

Landsat 7 was officially integrated into NASA's Earth Observing System (EOS) in 1994. It was launched on April 15, 1999 from Vandenburg Air Force Base, CA, using a

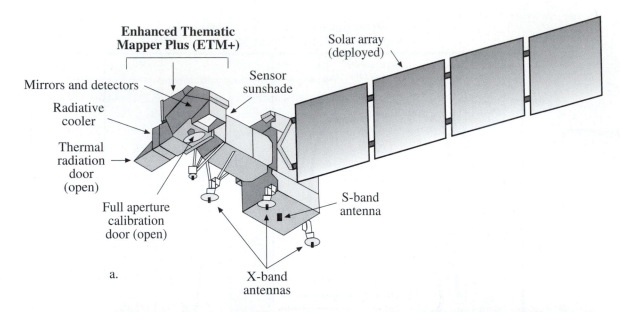

Figure 7-16   a) Schematic diagram, and b) artist's rendition of the Landsat 7 satellite with its Enhanced Thematic Mapper Plus (ETM+) sensor system (courtesy of Lockheed Martin, Inc.).

Delta-II Expendable Launch Vehicle into a Sun-synchronous orbit (Figure 7-16). Landsat 7 was designed to work in harmony with NASA's EOS *Terra* satellite which uses a suite of relatively coarse spatial resolution sensors (except for ASTER). Landsat 7 provides a unique suite of high-resolution observations of the terrestrial environment. It was designed to achieve three main objectives (NASA, 1999):

• maintain data continuity by providing data that are consistent in terms of geometry, spatial resolution,

calibration, coverage characteristics, and spectral characteristics with previous Landsat data;

• generate and periodically refresh a global archive of substantially cloud-free, sunlit landmass imagery; and

• continue to make Landsat-type data available to U.S. and international users at the cost of fulfilling user requests (COFUR) and to expand the use of such data for global-change research and commercial purposes.

Figure 7-17 The first Landsat 7 ETM⁺ image was obtained over Sioux Falls, SD. This is a small subscene of the 15 x 15 m panchromatic image obtained on April 18, 1999 (courtesy U.S. Geological Survey National Mapping Division, Sioux Falls, SD).

Landsat 7 is a three-axis stabilized platform carrying a single nadir-pointing instrument, the ETM⁺ (Figure 7-16). The ETM⁺ instrument is a derivative of the Landsat 4 and 5 Thematic Mapper sensors. Therefore, it is possible to refer to Figure 7-12 for a review of its mirror and detector design. Interestingly, it is still based on scanning technology despite the fact that linear array "pushbroom" technology has been commercially available since the launch of the French SPOT 1 satellite in 1986. Nevertheless, the ETM⁺ instrument is an exceptional sensor with several notable improvements over its Landsat 4 and 5 predecessors.

The characteristics of the Landsat 7 ETM⁺ are summarized in Table 7-3. The ETM⁺ bands 1 – 5 and 7 are identical to those found on Landsat 4 and 5 and have the same 30 x 30 m spatial resolution. The thermal infrared band 6 now has 60 x 60 m spatial resolution (instead of 120 x 120 m). Perhaps most notable is the new 15 x 15 m panchromatic band (0.52 – 0.90 µm). The first Landsat 7 ETM⁺ panchromatic image was obtained over Sioux Falls, SD. A portion of the image is shown in Figure 7-17. Landsat 7 color composites of Palm Springs, CA, are displayed in Color Plate 7-1c,d.

Like its predecessors, Landsat 7 is in orbit at 705 km above the Earth, collects data in a swath 185 km wide, and cannot view off-nadir. Its revisit interval is 16 days. It has a 378 gigabit solid-state recorder that can hold 42 minutes of sensor data and 29 hours of housekeeping telemetry data. This is necessary because the ETM⁺ obtains 150 megabits of data each second. Landsat 7 can transmit data directly to ground receiving stations in the United States at the EROS Data Center in Sioux Falls, SD, or to Fairbanks, AK. Landsat 7 international data may by acquired by retransmission using TDRS satellites or by international receiving stations.

The Landsat 7 ETM⁺ has significantly better radiometric calibration, which is accomplished using several schemes. The ETM⁺ is equipped with an internal calibration paddle that moves into the path of the incoming radiation once each scan line. This paddle has calibration lamps with known energy signatures that when turned on are sensed by the detectors and supply calibration data at the end of every scan line. In addition, the ETM⁺ was built to perform two types of calibration that use the Sun as a radiation source. Partial Aperture Solar Calibration (PASC) is performed once every day during normal operations and is used to calibrate bands 1 – 5, 7, and 8. Approximately one minute after the spacecraft passes into sunlight and while its *subsatellite point* is still in darkness (i.e., the point that it would be viewing on the ground), the ETM⁺ is commanded to acquire an image. However, instead of viewing the terrain, a small device housed inside the ETM⁺ sunshade is moved into position and reflects sunlight toward the detectors. In effect, it creates an image of the Sun. The operation lasts just over two minutes and provides valuable detector calibration information.

A Full Aperture Solar Calibration (FASC) is performed once every four to six weeks. It involves positioning the FASC paddle in front of the ETM⁺ aperture to once again reflect sunlight into the field of view. The ETM⁺ then images 10 scans in low gain, 10 scans in high gain, and then 10 scans in low gain. This operation takes place over the North Pole terminator. Spectral information from these procedures is then used to radiometrically adjust the sensor system. Also, Ground Look Calibration (GLC) is performed by acquiring images of certain Earth landmass calibration targets. Biophysical and atmospheric characteristics of these targets are especially well-instrumented on the ground.

Data acquisition by the ETM⁺ is directed by the mission goal of acquiring and updating periodically a global archive of daytime, substantially cloud-free images of land areas. In addition to the periodic global archive acquisitions, Landsat 7 acquires every daytime scene on every pass over the

Table 7-3.   Landsat Enhanced Thematic Mapper Plus (ETM[+]) Compared with the Proposed Earth Observer (EO-1) Advanced Land Imager (ALI), Hyperion, and LEISA Atmospheric Corrector (LAC) Sensors (NASA, 1999)

Landsat 7 Enhanced Thematic Mapper Plus (ETM[+])			EO-1 Advanced Land Imager (ALI)		
Band	Spectral Resolution (μm)	Spatial Resolution (m) at Nadir	Band	Spectral Resolution (μm)	Spatial Resolution (m) at Nadir
1	0.450–0.515	30 x 30	MS-1'	0.433–0.453	30 x 30
2	0.525–0.605	30 x 30	MS-1	0.450–0.510	30 x 30
3	0.630–0.690	30 x 30	MS-2	0.525–0.605	30 x 30
4	0.750–0.900	30 x 30	MS-3	0.630–0.690	30 x 30
5	1.55–1.75	30 x 30	MS-4	0.775–0.805	30 x 30
6	10.40–12.50	60 x 60	MS-4'	0.845–0.890	30 x 30
7	2.08–2.35	30 x 30	MS-5'	1.20–1.30	30 x 30
8 (pan)	0.52–0.90	15 x 15	MS-5	1.55–1.75	30 x 30
			MS-7	2.08–2.35	30 x 30
			Pan	0.480–0.690	10 x 10
			EO-1 Hyperion Hyperspectral Sensor 220 bands from 0.4 – 2.4 μm at 30 x 30 m		
			EO-1 LEISA Atmospheric Corrector (LAC) 256 bands from 0.9 – 1.6 μm at 250 x 250 m		
Sensor Technology	Scanning mirror spectrometer		Advanced Land Imager is a pushbroom radiometer Hyperion is a pushbroom spectroradiometer LAC uses area arrays		
Swath Width	185 km		ALI = 37 km Hyperion = 7.5 km LAC = 185 km		
Data Rate	250 images per day @ 31,450 km^2		—		
Revisit	16 days		16 days		
Orbit and Inclination	705 km, sun-synchronous Inclination = 98.2° Equatorial crossing 10:00 a.m. ±15 min.		705 km, sun-synchronous Inclination = 98.2° Equatorial crossing = Landsat 7 + 1 min		
Launch	April 15, 1999; 6 year duration		1999/2000; 1-year experimental		

United States. Approximately 250 images are processed by the EROS Data Center each day.

Landsat 7 data are processed to several product levels, including (NASA, 1999):

- Level 0R Data Product: reformatted, raw data including reversing the order of the reverse scan data, aligning the staggered detectors, nominal alignment of the forward and reverse scans, and metadata. Data are not radiometrically corrected.

- Level 1R Data Product: Radiometrically corrected data and metadata describing calibration parameters, payload correction data, mirror scan correction data, a geolocation table, and internal calibration lamp data.

- Level 1G Data Product: The data are radiometrically and geometrically corrected. Metadata describes calibration parameters and a geolocation table. Data are resampled to a user-specified map projection.

The Level 0R data are distributed for no more than $500 per scene. The Level 1R and 1G products are more expensive. All users are charged standard prices.

Follow-on Landsat satellites may be designed based on the analysis of the next-generation Earth Observer (EO-1) mission scheduled for launch in 2000. Preliminary specifications are summarized in Table 7-3. It contains a linear array Advanced Land Imager (ALI) with 10 bands from 0.4 – 2.35 μm at 30 x 30 m spatial resolution. It will also have onboard the Hyperion hyperspectral sensor that can record data in 220 bands from 0.4 – 2.4 μm at 30 x 30 m spatial resolution. The Linear Etalon Imaging Spectrometer Array (LEISA) Atmospheric Corrector (LAC) is a 256 band hyperspectral instrument sensitive to the region from 0.9 – 1.6 μm at 250 x 250 m spatial resolution. It is designed to correct for water vapor variations in the atmosphere. The EO-1 satellite is to be placed in orbit so that it covers the same ground track as Landsat 7, approximately one minute later.

National Atmospheric and Oceanic Administration (NOAA) Multispectral Scanner Sensors

NOAA operates two series of remote sensing satellites: the Geostationary Operational Environmental Satellites (GOES) and the Polar-Orbiting Operational Environmental Satellites (POES). Both are currently based on multispectral scanner technology. The U.S. National Weather Service uses data from these sensors to forecast the weather. We often see GOES images of North and South America weather patterns on the daily news and on the Weather Channel. While AVHRR data were developed for meteorological purposes, research on global climate change has also focused attention on the use of AVHRR data to map vegetation and sea-surface characteristics (Eidenshink and Faundeen, 1999).

Geostationary Operational Environmental Satellites (GOES)

The GOES system is operated by the National Environmental Satellite Data and Information Service (NESDIS) of NOAA. The system was developed by NESDIS in conjunction with NASA. In April, 1994, the first of NOAA's new generation of geostationary satellites, GOES-8 East, was launched. It is still operational. GOES-9 was launched on May 23, 1995, but is in standby mode due to mechanical difficulties. GOES-10 West was launched April 25, 1997 and is operational (Figure 7-18a).

The GOES system consists of several observing subsystems:

- GOES Imager (provides multispectral image data),

- GOES Sounder (provides hourly 19-channel soundings), and

- a data-collection system (DCS) that relays data from *in situ* sites at or near the Earth's surface to other locations.

The GOES spacecraft is a three-axis (x,y,z) stabilized design capable of continuously pointing the optical line of sight of the imaging and sounding radiometers toward the Earth. GOES are placed in geostationary orbits 35,790 km (22,240 statute miles) above the equator. The satellites remain at a stationary point above the equator and rotate at the same speed and direction as Earth. This enables the sensors to stare at the Earth from the geosynchronous orbit and thus more frequently obtain images of clouds, monitor the Earth's surface temperature and water vapor characteristics, and sound the Earth's atmosphere for its vertical thermal and water vapor structures.

GOES-8 East (Figure 7-18a) is normally situated at 75° W longitude and GOES-10 West is at 135° W longitude. The geographic coverage of GOES East and GOES West is summarized in Figure 7-18a. These sensors view most of the Earth from approximately 20° W to 165° E longitude. Poleward coverage is between approximately 77° N and S latitude. GOES East and West view the contiguous 48 states, South America, and major portions of the central and eastern Pacific Ocean and the central and western Atlantic Ocean areas. Pacific coverage includes the Hawaiian Islands and Gulf of Alaska, the latter known to weather forecasters as "the birthplace of North American weather systems" (Loral Space Systems, 1996).

GOES Imager: The Imager is a five-channel multispectral scanner. The bandwidths and spatial resolution are summarized in Table 7-4. By means of a two-axis gimballed mirror in conjunction with a 31.1-cm (12.2-in.) diameter Cassegrain telescope, the Imager's multispectral channels can simultaneously sweep an 8-km (5-statue mile) north-to-south swath along an east-to-west/west-to-east path, at a rate

GOES East and West Coverage

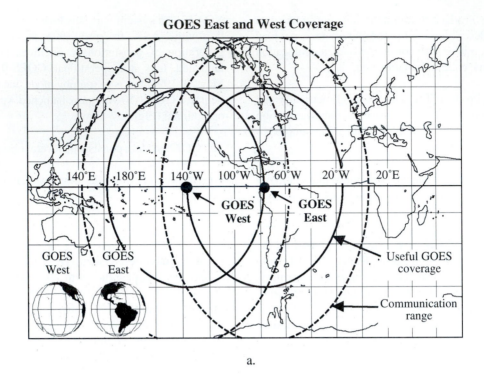

a.

GOES Imager Optical Elements

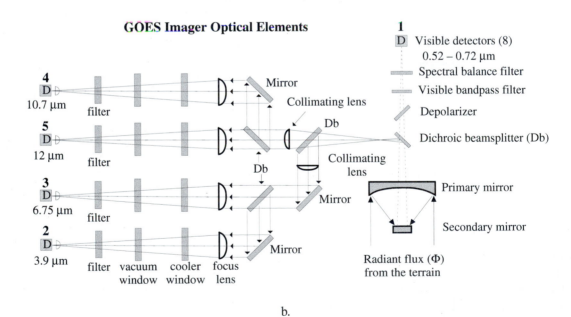

b.

Figure 7-18   a) Geographic coverage of GOES East (75° W) and GOES West (135° W). b) Radiant flux from the terrain is reflected off a scanning mirror (not shown) onto the primary and secondary mirrors. A dichroic beamsplitter separates the visible light from the thermal infrared energy. Subsequent beamsplitters separate the thermal energy into specific bands (after Loral Space Systems, 1996).

Table 7-4.   NOAA Geostationary Operational Environmental Satellite (GOES) Imager Sensor System Characteristics

GOES-8,10 Band	Spectral Resolution (µm)	Spatial Resolution (km)	Band Characteristics
1	0.52 – 0.72	1 x 1	cloud, pollution, haze detection and severe storm identification
2	3.78 – 4.03	4 x 4	fog detection, discriminates between water clouds and snow or ice clouds during daytime, detects fires and volcanoes, and nighttime mapping of sea surface temperature
3	6.47 – 7.02	8 x 8	estimates mid- and upper-level water vapor, detects advection, and tracks mid-level atmospheric motion
4	10.2 – 11.2	4 x 4	cloud-drift winds, severe storm identification, cloud-top heights, and location of heavy rainfall
5[a]	11.5 – 12.5	4 x 4	identification of low-level water vapor, sea surface temperature, and airborne dust and volcanic ash

[a] This band may be used to remove the path radiance contributed by atmospheric water vapor when temperature mapping.

of 20° (optical) per second (Loral Space Systems, 1996). The telescope concentrates both the visible and thermal radiant flux from the terrain onto a 5.3-cm secondary mirror (Figure 7-18b). Dichroic beamsplitters separate the incoming scene radiance and focus it onto 22 detectors (8 visible and 14 thermal). The visible energy passes through the initial beamsplitter and is focused onto eight silicon visible detector elements. Each of the eight visible detectors has an IFOV of approximately 1 x 1 km at the satellite's suborbital point on the Earth.

All thermal infrared energy is deflected to the specialized detectors within the radiative cooler. The thermal infrared energy is further separated into the 3.9, 6.75, 10.7, and 12 µm channels. Each of the four infrared channels has a separate set of detectors: four-element indium-antimonide (InSb) detectors for band 2; two-element mercury-cadmium-telluride (Hg:Cd:Te) detectors for band 3; and four-element mercury-cadmium-telluride (Hg:Cd:Te) detectors for both bands 4 and 5.

The GOES-8 channels have 10 bit radiometric precision. The primary utility of the visible band 1 (1 x 1 km) is in the daytime monitoring of thunderstorms, frontal systems, and tropical cyclones. Band 2 (4 x 4 km) responds to both emitted terrestrial radiation and reflected solar radiation. It is useful for identifying fog and discriminating between water and ice clouds, between snow and clouds, and for identifying large or very intense fires. It can be used at night to track low-level clouds and infer near-surface wind circulation. Band 3 (8 x 8 km) responds to mid- and upper-level water vapor and clouds. It is especially useful for identifying the jet stream,

upper-level wind fields, and thunderstorms. Energy recorded by band 4 (4 x 4 km) is not absorbed to any significant degree by atmospheric gases. It is ideal for measuring cloud top heights, identifying cloud-top features, assessing the severity of some thunderstorms, and tracking clouds and frontal systems at night. Thermal band 5 (4 x 4 km) is similar to band 4 except that this wavelength region has a unique sensitivity to low-level water vapor. GOES-8 East visible, thermal infrared, and water vapor images of Hurricane Bonnie on August 25, 1998, are shown in Figure 7-19c-e.

The Imager scans the continental United States every 15 min; scans most of the hemisphere from near the North Pole to approximately 20° S latitude every 30 min; and scans the entire hemisphere once every three hours in "routine" scheduling mode. Optionally, special imaging schedules are available, which allow data collection at more rapid time intervals over reduced geographic areas. During Rapid Scan Operations (RSO) and Super Rapid Scan Operations (SRSO), images are collected over increasingly reduced-area sectors at 7.5-min intervals (RSO) and at either 1-min or at 30-sec intervals (SRSO). A typical image acquired at 30-second intervals covers a rectangle of about 10° of latitude and 15° of longitude. The 1-min SRSO collects 22 images per hour with 2 segments of 1-min interval images, allowing for regularly scheduled 15-min interval operational scans.

GOES Sounder: The GOES Sounder utilizes 1 visible and 18 infrared sounding channels to record data in a north-to-south swath across an east-to-west path. The Sounder and Imager both provide full Earth imagery, sector imagery, and area scans of local regions. The 19 bands yield the prime

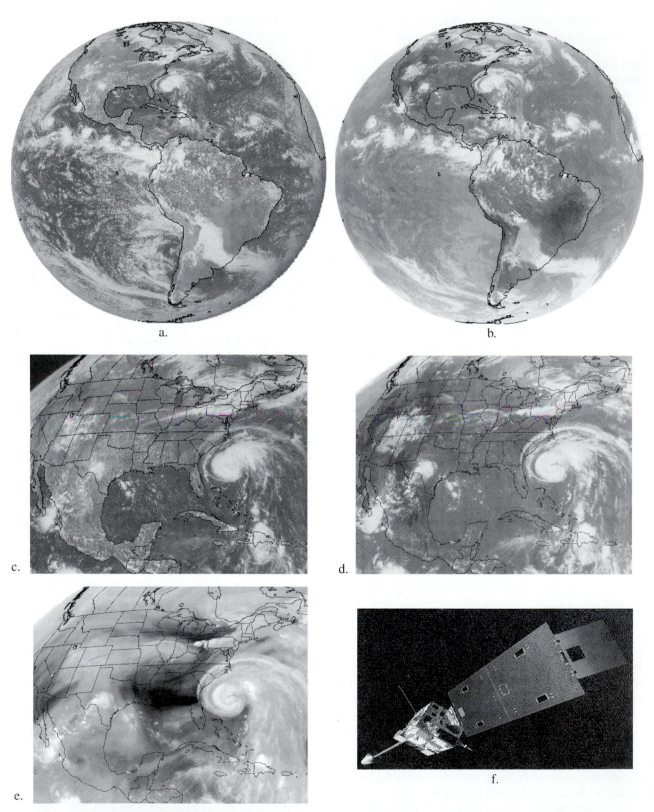

Figure 7-19   GOES-8 East global visible (a) and thermal infrared (b) images obtained on August 25, 1998, at 1745 GMT. GOES-8 East visible (c), thermal infrared (d), and water vapor (e) images acquired on August 25, 1998, at 1845. f) The GOES-8 satellite (images courtesy of NOAA).

sounding products of vertical atmospheric temperature profiles, vertical moisture profiles, atmospheric layer mean temperature, layer mean moisture, total precipitable water, and the lifted index (a measure of stability). These products are used to augment data from the Imager to provide information on atmospheric temperature and moisture profiles, surface and cloud-top temperatures, and the distribution of atmospheric ozone (Loral Space Systems, 1996).

Advanced Very High Resolution Radiometer (AVHRR)

The Satellite Services Branch of the National Climatic Data Center, under the auspices of NESDIS, has established a digital archive of data collected from the NOAA Polar-Orbiting Operational Environmental Satellites (POES) (Kidwell, 1998). This series of satellites commenced with TIROS-N (launched in October 1978) and continued with NOAA-A (launched in March, 1983 and renamed NOAA-8) to the current NOAA-15. These Sun-synchronous polar-orbiting satellites carry the Advanced Very High Resolution Radiometer (AVHRR). Substantial progress has been made in using AVHRR data for land-cover characterization and the mapping of daytime and nighttime clouds, snow, ice, and surface temperature (Loveland et al., 1991; Kidwell, 1998; Eidenshink and Faundeen, 1999). Unlike the Landsat TM and Landsat 7 ETM$^+$ sensor systems with nadir revisit cycles of 16 days, the AVHRR sensors acquire images of the entire Earth two times each day. This high frequency of coverage enhances the likelihood that cloud-free observations can be obtained for specific temporal windows and makes it possible to monitor change in land-cover conditions over short periods, such as a growing season. Moreover, the moderate resolution (1.1 x 1.1 km) of the AVHRR data makes it feasible to collect, store, and process continental or global data sets. For these reasons, NASA and NOAA initiated the AVHRR Pathfinder Program to create universally available global long-term remotely sensed data sets that can be used to study global climate change.

The AVHRR satellites orbit at approximately 833 km above Earth at an inclination of 98.9° and continuously record data in a swath 2700 km wide at 1.1 × 1.1 km spatial resolution at nadir. Normally, two NOAA-series satellites are operational at one time (one odd, one even). The odd-numbered satellite typically crosses the equator at approximately 2:30 p.m. and 2:30 a.m., while an even-numbered satellite crosses the equator at 7:30 p.m. and 7:30 a.m. local time. Each satellite orbits Earth 14.1 times daily (every 102 min) and acquires complete global coverage every 24 hours. Because the number of orbits per day is not an integer, the suborbital tracks do not repeat on a daily basis, although the local solar time of

the satellite's passage is essentially unchanged for any latitude. However, the satellite's orbital drift over time causes a systematic change of illumination conditions and local time of observation, which is a major source of nonuniformity when analyzing multidate AVHRR data.

The AVHRR is a cross-track scanning system. The scanning rate of the AVHRR is 360 scans per minute. A total of 2,048 samples (pixels) are obtained per channel per Earth scan, which spans an angle of $\pm$ 55.4° off-nadir. The IFOV of each band is approximately 1.4 milliradians leading to a resolution at the satellite subpoint of 1.1 x 1.1 km (Figure 7-20a). The more recent AVHRR systems have five channels (Table 7-5; Figure 7-20b).

The infrared channels are calibrated in-flight using an internal blackbody and outer space as a reference. Radiometric calibration of the AVHRR visible and near-infrared bands (1 and 2) is difficult because there is not always reliable preflight calibration, no onboard calibration, and difficulty with in-flight calibration. Radiometric calibration to radiance and reflectance is typically performed using techniques summarized in Teillet and Holben (1993).

The full resolution AVHRR data obtained at 1.1 × 1.1 km is referred to as *local area coverage* (*LAC*) data. It may be resampled to 4 × 4 km *global area coverage* (*GAC*) data. The GAC data contains only one out of three original AVHRR lines and the data volume and resolution are further reduced by starting with the third sample along the scan line, averaging the next four samples, and skipping the next sample. The sequence of average four, skip one is continued to the end of the scan line. Some studies use GAC data while others use the full-resolution LAC data.

The AVHRR provides regional information on vegetation condition and sea-surface temperature. For example, a portion of an AVHRR image of the South Carolina coast obtained on May 13, 1993, at 3:00 p.m. is shown in Figure 7-21. Band 1 is approximately equivalent to Landsat TM band 3. Vegetated land appears in dark tones due to chlorophyll absorption of red light. Band 2 is approximately equivalent to TM band 4. Vegetation reflects much of the near-infrared radiant flux, yielding bright tones, while water absorbs much of the incident energy. The land–water interface is usually quite distinct. The three thermal bands provide information about Earth's surface and water temperature. The gray scale is inverted for the thermal infrared data with cold, high clouds in black and warm land and water in lighter tones. This particular image captured a large lobe of warm Gulf Stream water. A color-infrared composite AVHRR image of the conterminous United States is shown in Color Plate 7-2a.

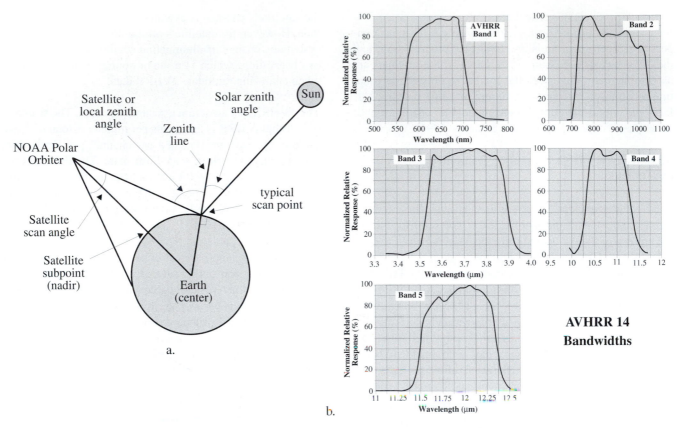

Figure 7-20   a) Relationship between the Earth, Sun, and the NOAA Polar Orbiter. The satellite subpoint lies at nadir. b) The NOAA-14 AVHRR bandwidths for bands 1 – 5.

Scientists often compute a normalized difference vegetation index (NDVI) from the AVHRR data using the visible ($AVHRR_1$) and near-infrared ($AVHRR_2$) bands to map the condition of vegetation on a regional and national level. It is a simple transformation based on the following ratio:

$$NDVI = \frac{AVHRR_2 - AVHRR_1}{AVHRR_2 + AVHRR_1} \qquad (7\text{-}1)$$

The NDVI equation produces values in the range of -1.0 to 1.0, where increasing positive values indicate increasing green vegetation, and negative values indicate nonvegetated surfaces such as water, barren land, ice, and snow or clouds. To obtain the most precision, the NDVI is derived from calibrated, atmospherically corrected AVHRR channel 1 and 2 data in 16-bit precision, prior to geometric registration and sampling. The final NDVI results from -1 to 1 are normally scaled from 0 to 200 (Eidenshink and Faundeen, 1999).

NDVI data obtained from multiple dates of AVHRR data may be composited to provide summary seasonal information. The *n*-day NDVI composite is produced by examining each NDVI value pixel by pixel for each observation during the compositing period to determine the maximum value. The retention of the highest NDVI value reduces the number of cloud-contaminated pixels. A 10-day composite image depicting the maximum NDVI of the northern hemisphere centered on Canada and the U.S. is shown in Color Plate 7-2b. Such images allow scientists to watch the green wave move northward during spring and vegetation senescence in the fall anywhere in the world (Loveland et al., 1991).

There is an ongoing effort by six major participants (NOAA, ESA, USGS, NASA, CSIRO, and SMC) to compile a 1 x 1 km AVHRR Global Land Data Set for the period from April 1, 1992, through September 30, 1996. A data set of more than 40,000 AVHRR images has been archived and made available for distribution by the U.S. Geological Survey, EROS Data Center and the European Space Agency (Eidenshink and Faundeen, 1999). In addition, NOAA routinely uses AVHRR data to produce Global Vegetation Index (GVI) products, including the (NOAA, 1999):

• GVI Normalized Density Vegetation Index,

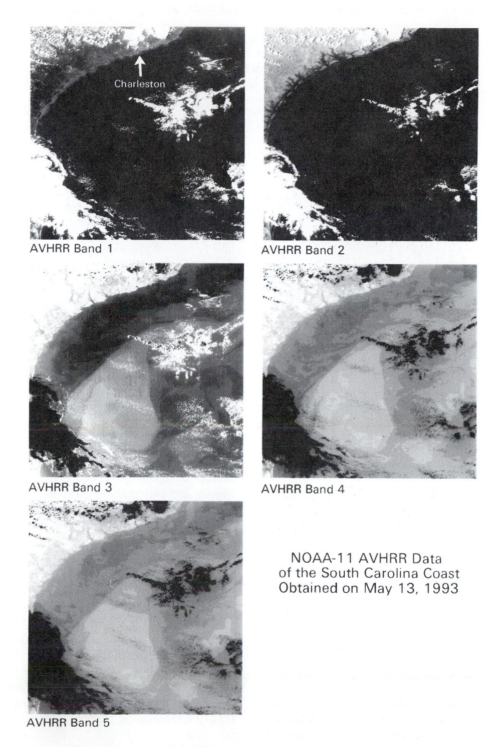

AVHRR Band 1

AVHRR Band 2

AVHRR Band 3

AVHRR Band 4

NOAA-11 AVHRR Data
of the South Carolina Coast
Obtained on May 13, 1993

AVHRR Band 5

Figure 7-21 Portion of an AVHRR image of the South Carolina coast obtained on May 13, 1993, at 3:00 p.m. (refer to Table 7-5 for band specifications). Vegetated land appears dark in band 1 due to chlorophyll absorption of red light. Vegetation appears bright in band 2 because it reflects much of the near-infrared radiant flux. Water absorbs much of the incident energy; therefore, the land–water interface is usually distinct. The three thermal bands (3, 4, and 5) provide surface-temperature information. The gray scale is inverted with cold, high clouds in black and warm land and water in lighter tones. A large lobe of warm gulf stream water is easily identified (images courtesy of NOAA).

Table 7-5.   NOAA Advanced Very High Resolution Radiometer (AVHRR) Sensor System Characteristics

Band	NOAA 6, 8, 10 Spectral Resolution (μm)[a]	NOAA 7, 9, 11, 12, 13, 14, 15 Spectral Resolution (μm)[a]	Band Characteristics
1	0.580 – 0.68	0.580 – 0.68	Daytime cloud, snow, ice, and vegetation mapping; used to compute NDVI
2	0.725 – 1.10	0.725 – 1.10	Land–water interface delineation; snow, ice, and vegetation mapping; used to compute NDVI
3	3.55 – 3.93	3.55 – 3.93	Monitoring hot targets (volcanoes, forest fires), nighttime cloud mapping
4	10.50 – 11.50	10.30 – 11.30	Day-and-night cloud and surface-temperature mapping
5	None	11.50 – 12.50	Cloud and surface temperature, day and night cloud mapping; removal of atmospheric water vapor path radiance
IFOV at nadir	1.1 × 1.1 km		
Swath width	2700 km at nadir		

[a] TIROS-N was launched on October 13, 1978; NOAA-6 on June 27, 1979; NOAA-7 on June 23, 1981; NOAA-8 on March 28, 1983; NOAA-9 on December 12, 1984; NOAA-10 on September 17, 1986; NOAA-11 on September 24, 1988; NOAA-12 on May 14, 1991; NOAA-13 on August 9, 1993 (to August 21, 1993); NOAA-14 on December 30, 1994.

- GVI Fractional Vegetation, and

- GVI Precipitable Water Index.

The NDVI and other vegetation indexes (refer to Chapter 10) have been used extensively with AVHRR data to monitor natural vegetation and crop condition, identify deforestation in the tropics, and monitor areas undergoing desertification and drought (Eidenshink, 1992; Sampson, 1993; Eidenshink and Faundeen, 1999).

ORBIMAGE, Inc., and NASA Sea-viewing Wide Field of view Sensor (SeaWiFS)

The oceans cover more than two-thirds of the Earth's surface and play an important role in the global climate system. The *SeaWiFS* (Sea-viewing Wide Field of view Sensor) onboard the SeaStar spacecraft is an advanced scanning system designed specifically for ocean monitoring. The SeaStar spacecraft (also referred to as *OrbView-2*), developed by ORBIMAGE, Inc., in conjunction with NASA, carried the SeaWiFS into orbit using a Pegasus rocket on August 1, 1997. Interestingly, the Pegasus rocket was flown aloft under the body of a modified Lockheed L-1011 and released at an altitude of about 39,000 ft, whereupon the rocket was ignited, and the spacecraft was lifted into orbit. The final orbit was 705 km above the Earth. The equatorial crossing time is 12 p.m.

SeaWiFS builds on all that was learned about ocean remote sensing using the Nimbus-7 satellite Coastal Zone Color Scanner (CZCS) launched in 1978. CZCS ceased operation in 1986. The SeaWiFS instrument consists of an optical scanner with a 58.3° total field of view. Incoming scene radiation is collected by a telescope and reflected onto the rotating half-angle mirror. The radiation is then relayed to dichroic beamsplitters that separate the radiation into eight wavelength intervals (Table 7-6). SeaWiFS has a spatial resolution of 1.13 x 1.13 km (at nadir) over a swath of 2800 km. It has a revisit time of one day.

SeaWiFS records energy in eight spectral bands with very narrow wavelength ranges (Table 7-6) tailored for the detection and monitoring of very specific ocean phenomena, including: ocean primary production and phytoplankton processes, ocean influences on climate processes (heat storage and aerosol formation), and the cycles of carbon, sulfur, and

Table 7-6.   Characteristics of the Sea-viewing Wide Field of view Sensor (SeaWiFS)

SeaWiFS Band	Bandcenter (nm)	Bandwidth (nm)	Utility
1	412	402 – 422	identify yellow substances
2	443	433 – 453	chlorophyll concentration
3	490	480 – 500	increased sensitivity to chlorophyll concentration
4	510	500 – 520	chlorophyll concentration
5	555	545 – 565	Gelbstoffe (yellow substance)
6	670	660 – 680	chlorophyll concentration
7	765	745 – 785	surface vegetation, land-water interface, atmospheric correction
8	865	845 – 885	surface vegetation, land-water interface, atmospheric correction

nitrogen. In particular, SeaWiFS has specially designed bands centered at 412 nm (to identify yellow substances through their blue wavelength absorption), at 490 nm (to increase sensitivity to chlorophyll concentration), and in the 765 and 865 nm near-infrared (to assist in the removal of atmospheric attenuation).

SeaWiFS observations help scientists understand the dynamics of ocean and coastal currents, the physics of mixing, and the relationships between ocean physics and large-scale patterns of productivity. The data fill the gaps in ocean biological observations between those of the test-bed CZCS and the Moderate Resolution Imaging Spectrometer (MODIS). OrbView-2 also provides laser water penetration depth imagery for naval operations. Examples of SeaWiFS imagery are provided in Chapter 11.

Other satellites of particular value for marine remote sensing include the Japanese Marine Observation Satellites (MOS-1, 1b) launched in 1987 and 1990, respectively, and the Japanese Ocean Colour and Temperature Scanner (OCTS), launched on August 17, 1996.

Aircraft Multispectral Scanners

Orbital sensors such as the Landsat MSS, TM, and ETM⁺ collect data on a repetitive cycle and at set spatial and spectral resolutions. Often it is necessary to acquire remotely sensed data at specific times that do not coincide with the scheduled satellite overpasses and at perhaps different spatial and spectral resolutions. Rapid collection and analysis of high-resolution remotely sensed data may be required for specific studies and locations. When such conditions occur

or when a sensor configuration different from the Landsat or SPOT sensors is needed, agencies and companies often use a multispectral scanner (MSS) placed onboard an aircraft to acquire remotely sensed data. There are several commercial and publicly available MSS that can be flown onboard aircraft, including the Daedalus and the NASA Airborne Terrestrial Applications Sensor (ATLAS).

Daedalus DS-1260, DS-1268, and Airborne Multispectral Scanner (AMS)

Numerous remote sensing laboratories and/or government agencies in 25 countries have purchased Daedalus DS-1260, DS-1268, or the Airborne Multispectral Scanner (AMS) over the last 30 years. These relatively expensive sensor systems have provided much of the useful high spatial and spectral resolution multispectral scanner data (including thermal infrared) for monitoring the environment. The DS-1260 records data in 10 bands spanning the region from the ultraviolet through near-infrared (0.38 – 1.10 μm), plus a thermal infrared channel (8.5 – 13.5 μm). The DS-1268 incorporates the Thematic Mapper middle-infrared bands. The AMS contains a hot-target, thermal infrared detector (3.0 – 5.5 μm) in addition to the standard thermal infrared detector (8.5 – 12.5 μm) (England, 1994). Table 7-7 summarizes the characteristics of the AMS sensor system.

The basic principles of operation and components of the airborne multispectral scanner (AMS) are shown in Figure 7-22. Radiant flux reflected or emitted from the terrain is collected by the scanning optical system and projected onto a dichroic grating. The grating separates the reflected radiant flux from the emitted radiant flux. Energy in the reflective part of the spectrum, including ultraviolet (optional), blue,

Table 7-7.   Daedalus Airborne Multispectral Scanner (AMS) and NASA Airborne Terrestrial Applications Sensor (ATLAS) Characteristics

	Daedalus Airborne Multispectral Scanner (AMS)			**NASA Airborne Terrestrial Applications Sensor (ATLAS)**		
	Band	**Spectral Resolution (μm)**	**Spatial Resolution (m)**	**Band**	**Spectral Resolution (μm)**	**Spatial Resolution (m)**
	1	0.42 – 0.45	variable,	1	0.45 – 0.52	2.5 to 25 m
	2	0.45 – 0.52	depending	2	0.52 – 0.60	depending
	3	0.52 – 0.60	upon altitude-	3	0.60 – 0.63	upon altitude-
	4	0.60 – 0.63	above-ground-	4	0.63 – 0.69	above-ground-
	5	0.63 – 0.69	level (AGL)	5	0.69 – 0.76	level (AGL)
	6	0.69 – 0.75		6	0.76 – 0.90	
	7	0.76 – 0.90		7	1.55 – 1.75	
	8	0.91 – 1.05		8	2.08 – 2.35	
	9	3.00 – 5.50		9	removed	
	10	8.50 – 12.5		10	8.20 – 8.60	
				11	8.60 – 9.00	
				12	9.00 – 9.40	
				13	9.60 – 10.20	
				14	10.20 – 11.20	
				15	11.20 – 12.20	
IFOV	2.5 mrad			2.0 mrad		
Quantization	8 – 12 bits			8 bits		
Altitude	variable			variable		
Swath width	714 pixels			800 pixels		

green, red, and reflective-infrared, is directed from the grating to a prism (or refraction grating) that further separates the energy into specific bands. At the same time, the emitted thermal incident energy is separated from the reflective incident energy. The independent bands of energy are focused onto a bank of discrete detectors situated behind the grating and the prism. The detectors that record the emitted energy are usually cooled by a dewar of liquid nitrogen or some other substance. The signals recorded by the detectors are amplified by the system electronics and recorded on a multi-channel tape recorder.

If the emphasis is on visual analysis of the MSS data, they may be recorded on an analog recorder and converted directly into hard-copy imagery during or after the flight. If the data are to be digitally processed, it is necessary to convert the analog electrical scanner output into a numerical (digital) format. This A-to-D conversion ensures that the data collected in the several bands are precisely synchronized. Such data are recorded on high-density digital tape (HDDT) onboard the aircraft. Later, the HDDT data are converted into a computer-compatible tape (CCT) format suit-

able for digital image processing. The flight altitudes for aircraft MSS surveys are determined by evaluating the size of the desired ground-resolution element (or pixel) and the size of the study area. Basically, the diameter of the circular ground area viewed by the sensor, D, is a function of the instantaneous field of view, β, of the scanner and the altitude-above-ground-level, H, where

$$D = H \times \beta. \tag{7-2}$$

For example, if the IFOV of the scanner is 2.5 mrad, the ground size of the pixel in meters is a product of the IFOV (0.0025) and the altitude AGL in meters. Table 7-8 presents flight altitudes and corresponding pixel sizes at nadir for an IFOV of 2.5 mrad.

The following factors should be considered when collecting aircraft MSS data:

• The field of view of the MSS optical system and the altitude AGL dictate the width of a single flightline of coverage. At lower altitudes, the high-spatial resolution

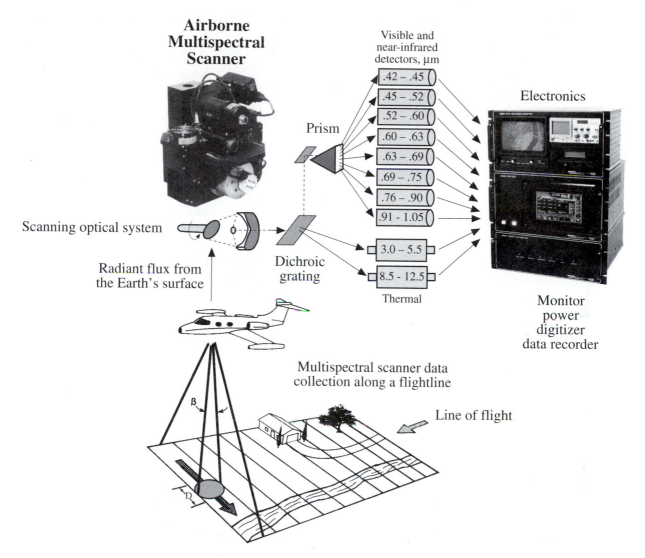

Figure 7-22 Characteristics of the Daedalus airborne multispectral scanner (AMS) and associated electronics that are carried onboard the aircraft during data collection. The diameter of the circular ground area viewed by the sensor, D, is a function of the instantaneous field of view, β, of the scanner and the altitude above-ground-level of the aircraft, H, at the time of data collection. Radiant flux from Earth's surface is passed from the optical system onto a dichroic grate, which sends the various wavelengths of light to detectors that are continuously monitored by the sensor system electronics (after Daedalus Enterprises, Inc.).

may be outweighed by the fact that more flightlines are required to cover the area compared to more efficient coverage at higher altitudes with larger pixels. The pixel size and the geographic extent of the area to be surveyed are considered, objectives are weighed, and a compromise is reached. Multiple flightlines of aircraft MSS data are difficult to mosaic.

- Even single flightlines of aircraft MSS data are difficult to rectify to a standard map series because of aircraft roll, pitch, and/or yaw during data collection (Jensen et al.,

1987). Notches in the edge of a flightline of data are indicative of aircraft roll. Such data require significant human and machine resources to make the data planimetrically accurate (Ramsey et al., 1992). Several agencies have placed GPS units on the aircraft to obtain precise flightline coordinates, which are useful when rectifying the aircraft MSS data (Fisher, 1991).

A Daedalus DS-1260 near-infrared band 10 image of the Four Mile Creek delta on the Savannah River Site in South Carolina is shown in Color Plate 7-3a. A color composite of

Table 7-8.   Aircraft Multispectral Scanner Flight Altitude AGL and Pixel Size Assuming an Instantaneous field of view of 2.5 milliradians (mrad)

Flight Altitude AGL (m)	Pixel Size (m)
1,000	2.5
2,000	5.0
4,000	10.0
16,000	40.0
50,000	125.0

Daedalus DS-1260 bands 10, 6, and 4 (near-infrared, red, and green) is shown in Color Plate 7-3b. Near-infrared band 10 imagery of the same region collected on April 23, 1992, is shown in Color Plate 7-3c. Color Plate 7-3d is a color composite of bands 10, 6, and 4. Thermal effluent was not allowed to enter Four Mile Creek after 1985. Examination of the imagery reveals that revegetation has taken place in many of the wetland sloughs. These two datasets were registered together and are the basis of a wetland change detection study documenting revegetation in the swamp (Jensen et al., 1994a).

Airborne multispectral scanning systems operating at relatively low altitudes are one of the few sensors that can acquire high spatial resolution multispectral and temperature information for a variety of environmental monitoring purposes on demand, weather permitting.

NASA Airborne Terrestrial Applications Sensor (ATLAS)

NASA's ATLAS is a multispectral scanner. It is operated by the Commercial Remote Sensing Program at NASA Stennis Space Center, MS (Spiering, 1998). ATLAS has 14 channels with a spectral range from 0.45 – 12.2 μm. There are six visible and near-infrared bands, two short-wavelength infrared bands (identical to Thematic Mapper bands 5 and 7) and six thermal infrared bands. The bandwidths are summarized in Table 7-7. The sensor has a total field of view of 72° and an IFOV of 2.0 milliradians. ATLAS is flown on a Learjet 23 aircraft from 6,000 ft to 41,000 ft above-ground-level yielding pixels with a ground resolution of approximately 2.5 x 2.5 m to 25 x 25 m, depending upon user specifications. There are normally 800 pixels per line plus three calibration source pixels. The data are quantized to 8 bits.

Calibration of the thermal data is performed using two internal blackbodies. Visible and near-infrared calibration is accomplished on the ground between missions using an integrating sphere. Onboard GPS documents the location in space of each line of data collected.

The ATLAS sensor is ideal for collecting high-spatial resolution data in the visible, near-infrared, middle-infrared, and thermal infrared regions so important for many commercial remote sensing applications. The Learjet is an ideal suborbital platform because of its stability in flight and its ability to travel to the study area quickly. It is particularly useful for obtaining data immediately after a disaster such as Hurricane Hugo or Andrew. An example of 2.5 x 2.5 m ATLAS band 6 (near-infrared) data obtained in 1998 for an area adjacent to Sullivans Island, SC, is shown in Figure 7-23.

 Multispectral Imaging Using Linear Arrays

Linear array sensor systems use very sensitive diodes or charge-coupled-devices (CCDs) to record the reflected or emitted radiance from the terrain. Linear array sensors are often referred to as *pushbroom* sensors because, like a single line of bristles in a broom, the linear array stares constantly at the ground while the spacecraft moves forward (Figure 7-3b). The result is usually a more accurate measurement of the reflected radiant flux because 1) there is no moving mirror, and 2) the linear array detectors are able to dwell longer on a specific portion of the terrain, resulting in a hopefully more accurate representation.

SPOT Sensor Systems

The sensors onboard the TIROS and NIMBUS satellites in the 1960s provided remotely sensed imagery with ground spatial resolutions of about 1000 × 1000 m and were the first to reveal the potential of space as a vantage point for Earth resource observation. The multispectral Landsat MSS and TM sensor systems developed in the 1970s and 1980s provided imagery with spatial resolutions from 79 × 79 m to 30 × 30 m. The first SPOT satellite was launched on February 21, 1986. It was developed by the French Centre National d'Etudes Spatiales (CNES) in cooperation with Belgium and Sweden, has a spatial resolution of 10 × 10 m (panchromatic mode) and 20 × 20 m (multispectral mode), and provides several other innovations in remote sensor system design. SPOT satellites 2 and 3 with identical payloads were launched on January 22, 1990, and September 25, 1993, respectively (Figure 7-24). SPOT 1 and 2 are still in service. SPOT 3 continued in service until November 14, 1996

Figure 7-23   Near-infrared band 6 (0.76 – 0.90 μm) Airborne Terrestrial Applications Sensor (ATLAS) image of a portion of the smooth cordgrass (*Spartina alterniflora*) marsh behind Sullivans Island, SC. The 2.5 x 2.5 m data were obtained on October 15, 1998.

(SPOT, 1999b). SPOT 4 with a modified payload was launched on March 24, 1998.

The SPOT Earth observation satellites are an unqualified success story. Since 1986 they have been the one consistent, dependable source of high-resolution Earth resource information. While many countries have seen their primary Earth resource monitoring sensor systems come and go depending upon the vagaries of politics, one could always count on SPOT Image, Inc., to provide quality imagery. In fact, as of January 1, 1998, SPOT 1 had collected 1,973,461 scenes, SPOT 2 collected 2,437,716 scenes, and SPOT 3 obtained 1,026,716 scenes (SPOT, 1998a). Unfortunately, the imagery has always been relatively expensive, usually >$2000 per panchromatic *or* multispectral scene, although it has been reduced in recent years. If you wanted both panchromatic and multispectral imagery of a study area, the cost was usually >$4000.

SPOT 1, 2, and 3

These satellites are all identical and consist of two parts: 1) the SPOT bus, which is a standard multipurpose platform, and 2) the sensor system instruments (Figure 7-25) consisting of two identical high-resolution visible (HRV) sensor systems and a package comprising two tape recorders and a telemetry transmitter. The satellite operates in a Sun-syn-

chronous, near-polar orbit (inclination of 98.2°) at an altitude of 832 km. The satellite passes overhead at the same solar time; the local clock time varies with latitude.

The HRV sensors may operate in two modes in the visible and reflective-infrared portions of the spectrum, a *panchromatic* mode corresponding to observation over a broad spectral band (similar to a typical black-and-white photograph) and a *multispectral* (color) mode corresponding to observation in three relatively narrower spectral bands (Table 7-9; Figure 7-13). Thus, the spectral resolution of SPOT 1 – 3 is not as good as the Thematic Mapper. The ground spatial resolution, however, is 10×10 m for the panchromatic band and 20×20 m for the three multispectral bands when the instruments are viewing at nadir, directly below the aircraft.

Radiant energy reflected from the terrain enters the HRV via a plane mirror and is then projected onto two CCD arrays. Each CCD array consists of 6,000 detectors arranged linearly. An electron microscope view of some of the individual detectors in the linear array is shown in Figure 7-26ab (SPOT, 1988). This linear array *pushbroom* sensor images a complete line of the ground scene in the cross-track direction in one look as the sensor system progresses downtrack (refer to Figure 7-25b). This capability breaks tradition with the Landsat MSS, Landsat TM, and Landsat 7 ETM⁺ sensors in that no mechanical scanning takes place. Engineers will

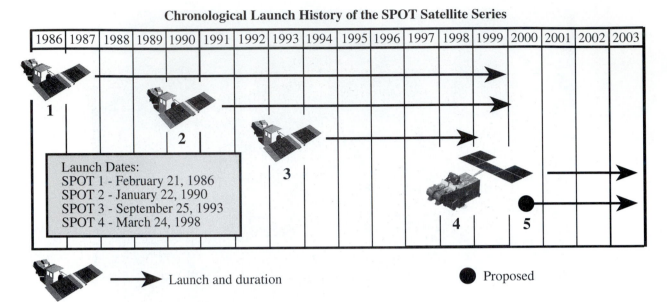

Chronological Launch History of the SPOT Satellite Series

| 1986 | 1987 | 1988 | 1989 | 1990 | 1991 | 1992 | 1993 | 1994 | 1995 | 1996 | 1997 | 1998 | 1999 | 2000 | 2001 | 2002 | 2003 |

Launch Dates:
SPOT 1 - February 21, 1986
SPOT 2 - January 22, 1990
SPOT 3 - September 25, 1993
SPOT 4 - March 24, 1998

Launch and duration   ● Proposed

Figure 7-24   Chronological launch history of the SPOT satellites.

tell you that a linear array sensor is superior because there is no mirror that must scan back-and-forth to collect data (mirror scan velocity is a serious issue) and this allows the detector to literally 'stare' at the ground for a longer time obtaining a more accurate record of the spectral radiant flux exiting the terrain. The SPOT satellites pioneered this linear array pushbroom technology in commercial Earth resource remote sensing as early as 1986.

When looking directly at the terrain beneath the sensor system, the two HRV instruments can be pointed to cover adjacent fields, each with a 60 km swath width (Figure 7-25c). In this configuration the total swath width is 117 km and the two fields overlap by 3 km. However, it is also possible to selectively point the mirror to off-nadir viewing angles through commands from the ground station. In this configuration it is possible to observe any region of interest within a 950-km-wide strip centered on the satellite ground track (i.e., the observed region may not be centered on the ground track) (Figure 7-27). The width of the swath actually observed varies between 60 km for nadir viewing and 80 km for extreme off-nadir viewing.

If the HRV instruments were only capable of nadir viewing, the revisit frequency for any given region of the world would be 26 days. This interval is often unacceptable for the observation of phenomena evolving on time scales ranging from several days to a few weeks, especially where the cloud cover hinders the acquisition of usable data. During the 26-day period separating two successive SPOT satellite passes over a given point on Earth and taking into account the steering capability of the instruments, the point in question could be observed on seven different passes if it were on the equator and on 11 occasions if at a latitude of 45° (Figure 7-28a). A given region can be revisited on dates separated alternatively by one to four (or occasionally five) days.

The SPOT sensors can also acquire cross-track stereoscopic pairs of images for a given geographic area (Figure 7-28b). Two observations can be made on successive days such that the two images are acquired at angles on either side of the vertical. In such cases, the ratio between the observation base (distance between the two satellite positions) and the height (satellite altitude) is approximately 0.75 at the equator and 0.50 at a latitude of 45°. Tests have shown that SPOT data with these base-to-height ratios may be used for topographic mapping. Toutin and Beaudoin (1995) applied photogrammetric techniques to SPOT data and produced maps with a planimetric accuracy of 12 m with 90 percent confidence for well-identifiable features and an elevation accuracy for a digital elevation model of 30 m with 90 percent confidence.

SPOT HRV multispectral and panchromatic data of Charleston, SC, are shown in Color Plate 7-4. Merging the panchromatic 10×10 m data with the multispectral 20×20 m data dramatically improves the visual interpretability of the region (Color Plate 7-4f).

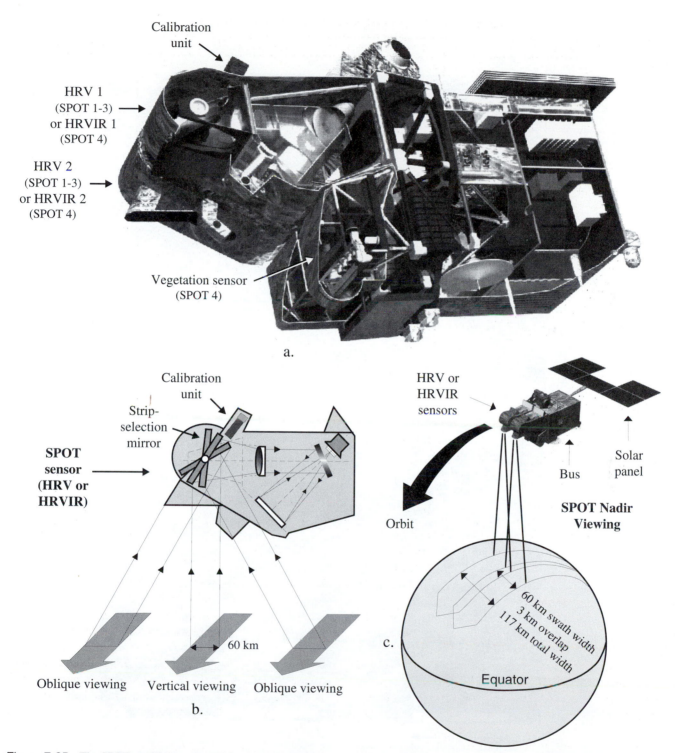

Calibration
unit

HRV 1
(SPOT 1-3)
or HRVIR 1
(SPOT 4)

HRV 2
(SPOT 1-3)
or HRVIR 2
(SPOT 4)

Vegetation sensor
(SPOT 4)

a.

Calibration
unit

Strip-
selection
mirror

**SPOT
sensor
(HRV or
HRVIR)**

HRV or
HRVIR
sensors

Solar
panel

Bus

**SPOT Nadir
Viewing**

Orbit

60 km

Oblique viewing   Vertical viewing   Oblique viewing

b.

60 km swath width
3 km overlap
117 km total width

c.

Equator

Figure 7-25 The SPOT satellite consists of the SPOT bus which is a multipurpose platform and the sensor system payload. Two identical high-resolution visible (HRV) sensors are found on SPOT 1, 2, and 3 and two identical high-resolution visible infrared (HRVIR) sensors on SPOT 4. Radiant energy from the terrain enters the HRV or HRVIR via a plane mirror and is then projected onto two CCD arrays. Each CCD array consists of 6,000 detectors arranged linearly. This results in a spatial resolution of 10×10 or 20×20 m, depending on the mode in which the sensor is being used (Table 7-9). The swath width at nadir is 60 km. The SPOT HRV and HRVIR sensors may also be pointed off-nadir to collect data. SPOT 4 carries a Vegetation sensor with 1.15 x 1.15 km spatial resolution and 2,250-km swath width (courtesy SPOT Image, Inc.).

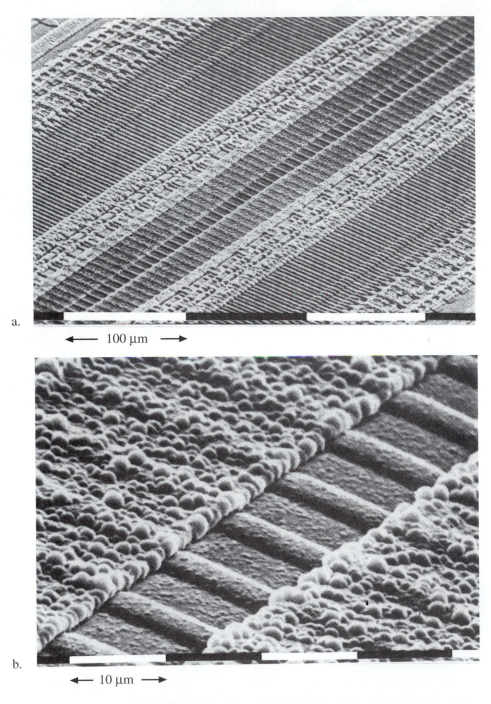

a.

←— 100 μm —→

b.

←— 10 μm —→

Figure 7-26   a) Scanning electron microscope images of the front surface of a CCD linear array like that used in the SPOT HRV sensor systems. Approximately 58 CCD detectors are visible, with rows of readout registers on either side. b) Seven detectors of a CCD linear array are shown at higher magnification (© Spot Image Corporation, Inc.).

Table 7-9.   SPOT 1, 2, and 3 High Resolution Visible (HRV), SPOT 4 High Resolution Visible and Infrared (HRVIR), and SPOT 4 Vegetation Sensor System Characteristics

SPOT 1, 2, and 3 HRV			**SPOT 4 HRVIR**			**SPOT 4 Vegetation**		
Band	Spectral Resolution (μm)	Spatial Resolution (m) at Nadir	Band	Spectral Resolution (μm)	Spatial Resolution (m) at Nadir	Band	Spectral Resolution (μm)	Spatial Resolution (km) at Nadir
1	0.50 – 0.59	20 x 20	1	0.50 – 0.59	20 x 20	0	0.43 – 0.47	1.15 x 1.15
2	0.61 – 0.68	20 x 20	2	0.61 – 0.68	20 x 20	2	0.61 – 0.68	1.15 x 1.15
			Pan		10 x 10			
3	0.79 – 0.89	20 x 20	3	0.79 – 0.89	20 x 20	3	0.78 – 0.89	1.15 x 1.15
Pan	0.51 – 0.73	10 x 10	SWIR	1.58 – 1.75	20 x 20	SWIR	1.58 – 1.75	1.15 x 1.15
Sensor	Linear array pushbroom		Linear array pushbroom			Linear array pushbroom		
Swath	60 km ±50.5°		60 km ±27°			2250 km ±50.5°		
Rate	25 Mb/s		50 Mb/s			50 Mb/s		
Revisit	26 days		26 days			1 day		
Orbit	832 km, Sun-synchronous Inclination = 98.2° Equatorial crossing 10:30 a.m.		832 km, Sun-synchronous Inclination = 98.2° Equatorial crossing 10:30 a.m.			832 km, Sun-synchronous Inclination = 98.2° Equatorial crossing 10:30 a.m.		

SPOT 10 x 10 m panchromatic data are of such high geometric fidelity that they can be photointerpreted like a typical aerial photograph in many instances. For this reason, SPOT panchromatic data are often commonly registered to topographic base maps and used as orthophotomaps. Such image maps are useful in GIS databases because they contain more accurate planimetric information (e.g., new roads, subdivisions) than out-of-date 7.5-min topographic maps (Jensen et al., 1994b). The improved spatial resolution available is demonstrated in Figure 7-29, which presents a TM band 3 image and a SPOT panchromatic image of Charleston, SC.

SPOT sensors collect data over a relatively small 60 × 60 km (3600 km²) area compared with Landsat MSS and TM image areas of 170 × 185 km (31,450 km²) (Figure 7-30). It takes 8.74 SPOT images to cover the same area as a single Landsat TM or MSS scene. This may be a limiting factor for extensive regional studies. However, SPOT does allow imagery to be purchased by the km² (e.g., for a watershed or school district) or by the linear km (e.g., along a highway).

SPOT 4 and 5

SPOT Image, Inc. launched SPOT 4 on March 24, 1998. Its characteristics are summarized in Table 7-9. The viewing

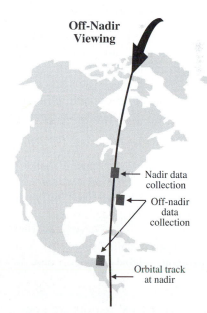

Off-Nadir Viewing

Nadir data collection

Off-nadir data collection

Orbital track at nadir

Figure 7-27   The SPOT HRV instruments are pointable and can be used to view areas that are not directly below the aircraft (i.e., off-nadir). This is very useful for collecting information in the event of a natural or man-made disaster, when the satellite track is not optimum or for collecting stereoscopic imagery (courtesy Spot Image, Inc.).

SPOT Off-Nadir Revisit Capabilities

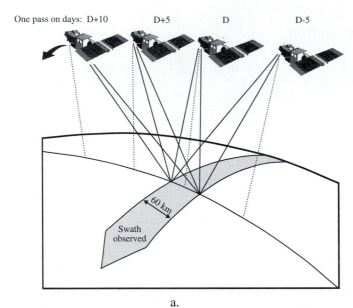

Stereoscopic Viewing Capabilities

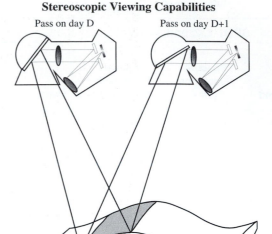

a.   b.

Figure 7-28   a) During the 26-day period separating two successive SPOT satellite overpasses, a point on Earth could be observed on seven different passes if it is at the equator and on 11 occasions if at a latitude of 45°. A given region can be revisited on dates separated alternatively by one, four, and occasionally five days. b) Two observations can be made on successive days such that the two images are acquired at angles on either side of the vertical, resulting in stereoscopic imagery. Such imagery can be used to produce topographic and planimetric maps (© Spot Image Corporation, Inc.).

angle may be adjusted to ±27° off-nadir. SPOT 4 has several notable new features of significant value for Earth resource remote sensing, including:

- the addition of a short-wavelength infrared (SWIR) band (1.58 – 1.75 μm) for vegetation and soil moisture applications;

- onboard registration of the spectral bands, achieved by replacing the original HRV panchromatic sensor (0.51 – 0.73 μm) with band 2 (0.61 – 0.68 μm) operating in both 10- *and* 20-m resolution mode; and

- an independent sensor called simply *Vegetation* for small-scale vegetation, global change, and oceanographic studies.

Because the SPOT 4 HRV sensors are now sensitive to a SWIR band, they are referred to as HRVIR 1 and HRVIR 2.

The SPOT Vegetation sensor is completely independent of the HRVIR sensors. It is a multispectral electronic scanning radiometer operating at optical wavelengths with a separate objective lens and sensor for each of the four spectral bands; (blue = 0.43 – 0.47 μm used primarily for atmospheric cor-

rection; red = 0.61 – 0.68 μm; near-infrared = 0.78 – 0.89 μm; and SWIR = 1.58 – 1.75 μm). Each sensor takes the form of a 1728 CCD linear array located in the focal plane of the corresponding objective lens. The spectral resolution of the individual bands is summarized in Table 7-9. The Vegetation sensor has a spatial resolution of 1.15 × 1.15 km. The objective lenses offer a field of view of ±50.5° which translates into a 2250-km swath width. The Vegetation sensor has several important characteristics:

- Multidate radiometric calibration accuracy better than 3 percent and absolute calibration accuracy better than 5 percent is superior to the AVHRR, making it more useful for repeatable global and regional vegetation surveys;

- Because it is based on pushbroom technology, pixel size is uniform across the entire swath width, with geometric precision better than 0.3 pixels and interband multidate registration better than 0.3 km;

- It has a 10:30 a.m. equatorial crossing time versus AVHRR's 2:30 p.m. crossing time;

- It has a short-wavelength infrared (SWIR) band for improved vegetation mapping;

Charleston, SC

a. Landsat Thematic Mapper Band 3 (30 x 30 m)　　　　February 3, 1994

b. SPOT HRV Panchromatic Band (10 x 10 m)　　　　January 10, 1996

Figure 7-29　Comparison of the detail in the 30 x 30 m Landsat TM band 3 data and SPOT 10 × 10 m panchromatic data of Charleston, SC (© Spot Image Corporation, Inc.).

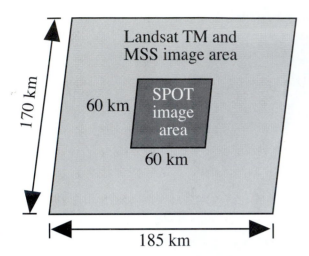

Figure 7-30 Geographic coverage of the SPOT HRV and Landsat Thematic Mapper remote sensing systems.

- It is straightforward to relate the HRVIR 20 x 20 m data nested within the Vegetation 2250 x 2250 km swath width data; and

- Individual images can be obtained, or data can be summarized over a 24-hr period (called a Daily Synthesis), or daily synthesis data can be compiled into *n*-day syntheses.

A Daily Synthesis vegetation index map of the whole Earth can be created using the SPOT 4 Vegetation sensor. These data are archived and used to generate Global Ten-Day Synthesis vegetation index maps. A portion of a Global Ten-Day Synthesis map centered on Europe, the Middle East, and Africa is shown in Color Plate 7-5.

With SPOT 1, 2, and 4 all in orbit and functioning, the constellation of satellites offers unrivaled ability to acquire imagery of almost any point on the globe daily. Stereopair acquisition capacity is also greatly enhanced. SPOT 5, to be launched in 2000, may include three spectral bands at 10 x 10 m, the SWIR 20 x 20 m band, a panchromatic band at 2.5 x 2.5 m, and perhaps the Vegetation sensor.

Indian Remote Sensing Systems

The Indian Space Research Organization (ISRO) has launched several Indian Remote Sensing (IRS) satellites: IRS-1A on March 17, 1988, IRS-1B on August 29, 1991, IRS-1C in 1995 and IRS-1D in September 1997 (Table 7-10). The sensors onboard the satellites utilize linear array

Figure 7-31 Indian Remote Sensing Satellite (IRS-1D) panchromatic image of downtown San Diego, CA. The 5.8 x 5.8 m image was resampled to 5 x 5 m (courtesy Indian Space Research Organization and Space Imaging, Inc.).

sensor technology. The IRS data (Figure 7-31) are marketed through Space Imaging Inc., in the United States.

IRS-1A, 1B, 1C, and 1D

The IRS 1A and 1B satellites acquire data with Linear Imaging Self-scanning Sensors (LISS-I and LISS-II) at spatial resolutions of 72.5 × 72.5 m and 36.25 × 36.25 m, respectively (Table 7-10). The data are collected in four spectral bands that are nearly identical to the TM visible and near-infrared bands. The satellites' altitude is 904 km, the orbit is Sun-synchronous, repeat coverage is every 22 days at the equator (11-day repeat coverage with two satellites), and orbital inclination is 99.5°. The swath width is 146 – 148 km.

The IRS-1C and 1D satellites carry three sensors (Table 7-10): the LISS-III multispectral sensor, a panchromatic sensor, and a Wide Field Sensor (WiFS). The LISS-III has four bands with the green, red, and near-infrared bands at 23 x 23 m spatial resolution and the short-wavelength infrared

Table 7-10. Indian Space Research Organization (ISRO) Indian Remote Sensing (IRS) Satellite Characteristics

IRS-1A and 1B			IRS-1C and 1D		
LISS-I and LISS-II Bands	Spectral Resolution (μm)	Spatial Resolution (m) at Nadir	LISS-III, Pan, and WiFS Bands	Spectral Resolution (μm)	Spatial Resolution (m) at Nadir
1	0.45–0.52	LISS-I @72.5 m LISS-II @36.25 m	1	-	-
2	0.52–0.59	LISS-I @72.5 m LISS-II @36.25 m	2	0.52 – 0.59	23 x 23
3	0.62–0.68	LISS-I @72.5 m LISS-II @36.25 m	3	0.62 – 0.68	23 x 23
4	0.77–0.86	LISS-I @72.5 m LISS-II @36.25 m	4	0.77 – 0.86	23 x 23
			5	1.55 – 1.70	70 x 70
			Pan	0.50 – 0.75	5.8 x 5.8
			WiFS 1	0.62 – 0.68	188 x 188
			WiFS 2	0.77 – 0.86	188 x 188
Sensor	Linear array pushbroom		Linear array pushbroom		
Swath width	LISS-I = 148 km; LISS-II = 146 km		LISS-III = 142 km for bands 2, 3, and 4; band 5 = 148 km Pan = 70 km; WiFS = 774 km		
Revisit	22 days at Equator		LISS-III is 24 days at Equator; Pan is 5 days with ±26° off-nadir viewing; WiFS is 5 days at Equator		
Orbit	904 km, Sun-synchronous Inclination = 99.5° Equatorial crossing 10:26 a.m.		817 km, Sun-synchronous Inclination = 98.69° Equatorial crossing 10:30 a.m. ± 5 min		
Launch	IRS-1A on March 17, 1988 IRS-1B on August 29, 1991		IRS-1C in 1995 IRS-1D in September 1997		

(SWIR) band at 70 x 70 m spatial resolution. The swath width is 142 km for bands 2, 3, and 4 and 148 km for band 5. Repeat coverage is every 24 days at the equator.

The panchromatic sensor has a spatial resolution of 5.8 x 5.8 m and stereoscopic imaging capability. The panchromatic band has a 70 km swath width with repeat coverage every 24 days at the equator and a revisit time of five days with ±26° off-nadir viewing. An example of the 5.8 x 5.8 m panchromatic data of downtown San Diego, CA, (resampled to 5 x 5 m) is shown in Figure 7-31.

The Wide Field Sensor (WiFS) has 188 x 188 m spatial resolution. The WiFS has two bands comparable to NOAA's AVHRR satellite (0.62 – 0.68 μm and 0.77 – 0.86 μm) with

a swath width of 774 km. Repeat coverage is 5 days at the equator.

Advanced Spaceborne Thermal Emission and Reflection Radiometer (ASTER)

The ASTER is an imaging instrument on the EOS *Terra* satellite. The satellite is a cooperative effort between NASA and Japan's Ministry of International Trade and Industry. ASTER obtains detailed information on surface temperature, emissivity, reflectance, and elevation. It is the only high-spatial resolution instrument on the *Terra* satellite. As such, it is used in conjunction with MODIS, MISR, and CERES sensors that monitor the Earth at moderate to coarse

Table 7-11.   Advanced Spaceborne Thermal Emission and Reflection Radiometer (ASTER) Characteristics (NASA, 1999)

Advanced Spaceborne Thermal Emission and Reflection Radiometer (ASTER)					
Band	VNIR Spectral Resolution (μm)	Band	SWIR Spectral Resolution (μm)	Band	TIR Spectral Resolution (μm)
1 (nadir)	0.52 – 0.60	4	1.600 – 1.700	10	8.125 – 8.475
2 (nadir)	0.63 – 0.69	5	2.145 – 2.185	11	8.475 – 8.825
3 (nadir)	0.76 – 0.86	6	2.185 – 2.225	12	8.925 – 9.275
3 (backward)	0.76 – 0.86	7	2.235 – 2.285	13	10.25 – 10.95
		8	2.295 – 2.365	14	10.95 – 11.65
		9	2.360 – 2.430		
Technology (detector)	pushbroom Si		pushbroom PtSi:Si		whiskbroom Hg:Cd:Te
Spatial Resolution (m)	15 x 15		30 x 30		90 x 90
Swath Width	60 km		60 km		60 km
Quantization	8 bits		8 bits		8 bits

spatial resolutions. In effect, ASTER serves as a zoom lens for the other *Terra* instruments and is particularly important for change detection and calibration/validation studies (Tan and Hook, 1999).

ASTER obtains data in 14 channels from the visible through thermal infrared regions of the electromagnetic spectrum. It consists of three separate instrument subsystems. Individual bandwidths and subsystem characteristics are summarized in Table 7-11.

The VNIR detector subsystem operates in three spectral bands at visible and near-infrared wavelengths with a spatial resolution of 15 x 15 m. It consists of two telescopes — one nadir-looking with a three-spectral-band CCD detector, and another backward-looking with a single-band CCD detector. The backward-looking telescope provides a second view of the study area in band 3 for stereoscopic observations. Across-track pointing to 24° is accomplished by rotating the entire telescope assembly.

The SWIR subsystem operates six spectral bands in the near-infrared region through a single nadir-pointing telescope that provides 30 x 30 m spatial resolution. Cross-track pointing (± 8.55°) is accomplished by a pointing mirror.

The TIR subsystem operates in five bands in the thermal infrared region using a single, fixed-position nadir-looking telescope with a spatial resolution of 90 x 90 m. Unlike the other subsystems, it has a whiskbroom scanning system instead of a pushbroom system. Each band uses 10 detectors in a staggered array with optical bandpass filters over each detector element. The scanning mirror functions both for scanning and cross-track pointing (±8.55°). During scanning, the mirror rotates 90° from nadir to view an internal blackbody.

Multi-angle Imaging Spectroradiometer (MISR)

MISR was built by NASA's Jet Propulsion Laboratory. It is one of the five *Terra* satellite instruments. The MISR instrument measures the Earth's brightness in four spectral bands, at each of nine look angles spread out in the forward and aft directions along the flightline. Spatial samples are acquired every 275 m. Over a period of 7 minutes, a 360 km wide swath of Earth comes into view at all nine angles.

An illustration of the nine look angles is shown in Figure 7-32. The digital pushbroom sensors image the Earth at 26.1°, 45.6°, 60°, and 70.5° forward and aft of the local vertical

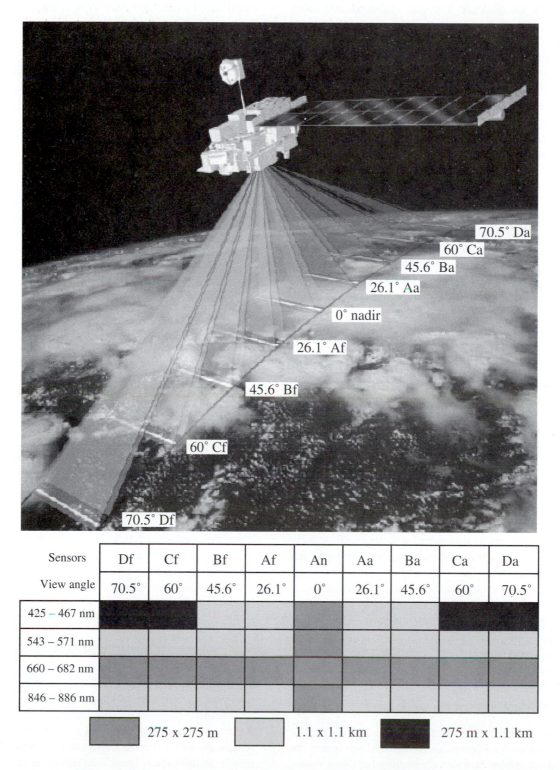

Sensors	Df	Cf	Bf	Af	An	Aa	Ba	Ca	Da
View angle	70.5°	60°	45.6°	26.1°	0°	26.1°	45.6°	60°	70.5°
425 – 467 nm									
543 – 571 nm									
660 – 682 nm									
846 – 886 nm									

275 x 275 m   1.1 x 1.1 km   275 m x 1.1 km

Figure 7-32   Artist's rendition of the Multi-angle Imaging Spectroradiometer (MISR) on EOS *Terra*. It uses linear array technology to acquire imagery of the terrain in four bands at nine different angles: at nadir (0°) and at 26.1°, 45.6°, 60°, and 70.5° forward and aft of nadir (courtesy NASA Jet Propulsion Laboratory).

(nadir 0°). Note that the fore and aft camera angles are the same, i.e., the cameras are arranged symmetrically about nadir. In general, large viewing angles provide enhanced sensitivity to atmospheric aerosol effects and to cloud reflectance effects, whereas more modest angles are required for land-surface viewing.

Each MISR camera sees instantaneously a single row of pixels at right angles to the ground track in a pushbroom format. It records data in four bands: blue, green, red, and near-infrared. The band center wavelengths are identified in Figure 7-32. Each camera has four independent linear CCD arrays (one per filter), with 1,504 active pixels per linear array.

The nadir-viewing camera (labeled An in Figure 7-32) provides imagery that is less distorted by surface topographic effects than any of the other cameras. It also is the least affected by atmospheric scattering. It provides 1) useful reference for navigating within all the MISR imagery, and 2) a base image to compare with images acquired at different angles of view. Such comparisons will provide important "bidirectional reflectance distribution function" information explained in Chapter 10. The nadir-viewing camera also offers an opportunity to compare observations with other nadir-viewing sensors such as Landsat TM or ETM$^+$. The nadir-viewing camera also facilitates calibration.

The fore and aft 26.1° view angle cameras (Af and Aa) provide useful stereoscopic information that can be of benefit for measuring topographic elevation and cloud heights. The fore and aft 45.6° view angle cameras (Bf and Ba) are positioned to be especially sensitive to atmospheric aerosol properties. The fore and aft 60° view angle cameras (Cf and Ca) provide observations looking through the atmosphere with twice the amount of air compared with the vertical view. This provides unique information about the hemispherical albedo of land surfaces. The fore and aft 70.5° view angle cameras (Df and Da) provide the maximum sensitivity to off-nadir effects. The scientific community is interested in obtaining quantitative information about clouds and the Earth surface from as many angles as possible.

Very High-Resolution Linear Array Remote Sensing Systems

In 1994, the United States government made a decision to allow civil commercial companies to market high spatial resolution remote sensor data (approximately 1 x 1 to 4 x 4 m). This resulted in the creation of a number of commercial consortiums that have the capital necessary to create, launch, and market high spatial resolution digital remote sensor data.

The most notable companies include, Space Imaging, Inc., ORBIMAGE, Inc., and EarthWatch, Inc. These companies are targeting the geographic information system (GIS) and cartographic mapping markets traditionally serviced by the aerial photogrammetric industries. Some estimate the growing Earth observation industry to be $2 billion to $10 billion dollars a year. The new commercial remote sensing firms hope to have a potentially crucial impact in markets as diverse as agriculture, natural resource management, local and regional government, transportation, emergency response, mapping, and eventually an array of average consumer applications as well (Space Imaging, 1999a).

All commercial vendors offer an Internet on-line ordering service, heretofore unknown in the remote sensing industry. All vendors offer a suite of standard and nonstandard products that can be tailored to user requirements, including the creation of digital elevation models from the remote sensor data. The commercial remote sensing companies typically price the imagery according to the type of product ordered and the amount of geographic coverage desired (km^2). The sensors used by these companies are based primarily on linear array CCD technology. The sensor system characteristics are summarized in Table 7-12.

EarthWatch, Inc., Earlybird and Quickbird

EarthWatch, Inc., launched *EarlyBird* in 1996 with a 3 x 3 m panchromatic band and three visible to near-infrared (VNIR) bands at 15 x 15 m spatial resolution. Unfortunately, EarthWatch subsequently lost contact with the satellite. EarthWatch, Inc. intends to launch *QuickBird* in 2000 into a 600 km orbit. Interestingly, it will be in a 66° non Sun-synchronous orbit. Revisit times range from one to five days, depending on latitude. It has a swath width of 20 to 40 km. Quickbird has a 1 x 1 m panchromatic band and four visible/near-infrared bands at 4 x 4 m spatial resolution (Table 7-12). The data are quantized to 11 bits (brightness values from 0 – 2047). The sensor may be pointed fore and aft and across-track to obtain stereoscopic data.

Space Imaging, Inc., IKONOS

Space Imaging, Inc., launched IKONOS on April 27, 1999. Unfortunately, contact was lost with the satellite 8 minutes after launch. The satellite never achieved orbit. Space Imaging, Inc., successfully launched a second IKONOS on September 24, 1999. The IKONOS satellite sensor has a 1 x 1 m panchromatic band (Figure 7-33) and four multispectral visible and near-infrared bands at 4 x 4 m spatial resolution (Space Imaging, 1999b). The sensor characteristics are summarized in Table 7-12. IKONOS is in a Sun-synchronous

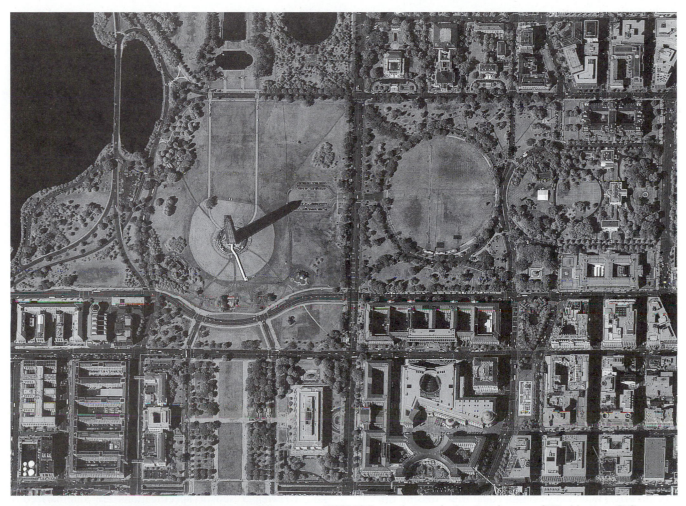

a. IKONOS panchromatic 1 x 1 m image of Washington, DC

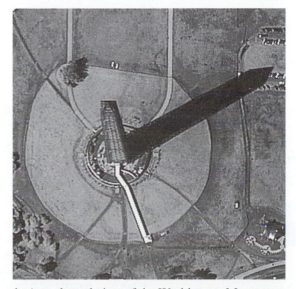

b. An enlarged view of the Washington Monument

Figure 7-33   a) Space Imaging, Inc., IKONOS 1 x 1 m panchromatic image of downtown Washington, DC, obtained on September 30, 1999. The Washington Monument and White House are clearly visible. b) An enlargement of the Washington Monument. Individual cars are discernible. The successful launch of the IKONOS satellite and the proper functioning of the remote sensing system onboard represent a remarkable achievement for mankind. City planners, Earth resource managers, utility companies, recreational users, and others can now obtain very high spatial resolution imagery of almost any cloud-free environment in the world (courtesy Space Imaging, Inc.).

Table 7-12.   Sensor Characteristics of Space Imaging, Inc., IKONOS Satellite; ORBIMAGE, Inc., *OrbView-3* Satellite; and EarthWatch, Inc., *Quickbird* Satellite

| Space Imaging, Inc. | | | ORBIMAGE, Inc. | | | EarthWatch, Inc. | | |
| IKONOS | | | OrbView-3 | | | Quickbird | | |
Band	Spectral Resolution (µm)	Spatial Resolution (m) at Nadir	Band	Spectral Resolution (µm)	Spatial Resolution (m) at Nadir	Band	Spectral Resolution (µm)	Spatial Resolution (m) at Nadir
1	0.45 – 0.52	4 x 4	1	0.45 – 0.52	4 x 4	1	0.45 – 0.52	4 x 4
2	0.52 – 0.60	4 x 4	2	0.52 – 0.60	4 x 4	2	0.52 – 0.60	4 x 4
3	0.63 – 0.69	4 x 4	3	0.625 – 0.695	4 x 4	3	0.63 – 0.69	4 x 4
4	0.76 – 0.90	4 x 4	4	0.76 – 0.90	4 x 4	4	0.76 – 0.890	4 x 4
Pan	0.45 – 0.90	1 x 1	Pan	0.45 – 0.90	1 x 1	Pan	0.45 – 0.90	1 x 1
Sensor	Linear array pushbroom			Linear array pushbroom			Linear array pushbroom	
Swath	11 km			8 km			20 to 40 km	
Rate	25 Mb/s			50 Mb/s			50 Mb/s	
Revisit	< 3 days			< 3 days			1 to 5 days depending on latitude	
Orbit	681 km, Sun-synchronous Equatorial crossing 10 – 11 am			470 km, Sun-synchronous Equatorial crossing 10:30 a.m.			600 km, not Sun-synchronous Equatorial crossing variable	
Launch	April 27, 1999 (failed) September 24, 1999			2000			2000	

681 km orbit, with a descending equatorial crossing time of between 10 and 11 a.m. It has both cross-track and along-track viewing instruments, which enables flexible data acquisition and frequent revisit capability: < 3 days at 1 x 1 m spatial resolution (for look angles < 26°) and 1.5 days at 4 x 4 m spatial resolution. The nominal swath width is 11 km. Data are quantized to 11 bits. The first IKONOS 1 x 1 m panchromatic image was obtained on September 30, 1999 of downtown Washington, DC (Figure 7-33). The image contains a wealth of high spatial resolution detail sufficient for many city planning and earth resource investigations.

ORBIMAGE, Inc., Orbview-3,4

ORBIMAGE, Inc., plans to launch *OrbView-3* in 2000 with 1 x 1 m panchromatic data and four visible and near-infrared multispectral bands at 4 x 4 m spatial resolution. It will have a Sun-synchronous orbit at 470 km above the Earth with a 10:30 a.m. equatorial crossing time. It will have an 8 km swath width. The sensor will revisit each location on Earth in less than three days, with an ability to turn from side to side

45°. *Orbview-3* sensor specifications are summarized in Table 7-12.

ORBIMAGE intends to launch *OrbView-4 (Warfighter)* in the future. It will acquire 1 x 1 m panchromatic and 4 x 4 m multispectral data similar to *OrbView-3*. In addition, *Orb-View-4* will be the first satellite to acquire hyperspectral imagery. It will obtain 200 channels (0.45 – 2.5 µm) of 8 x 8 m hyperspectral imagery with a swath width of 5 km. It will also have an off-nadir viewing capability and a revisit cycle of < 3 days. There is some debate at the present time as to whether ORBIMAGE, Inc., will be allowed to provide 8 x 8 m hyperspectral data to the public.

Imaging Spectrometry Using Linear and Area Arrays

This section describes a major advance in remote sensing, *imaging spectrometry*, defined as:

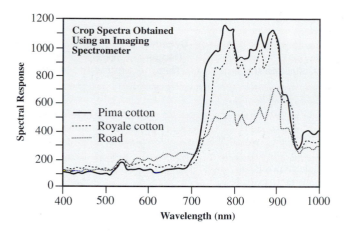

Figure 7-34   Imaging spectrometer crop spectra for Pima cotton, Royale cotton, and road surfaces extracted from 2 × 2 m data obtained near Bakersfield, CA (after SBRC, 1994).

The simultaneous acquisition of images in many relatively narrow, contiguous and/or non-contiguous spectral bands throughout the ultraviolet, visible and infrared portions of the spectrum.

In the past, most remotely sensed data were acquired in 4 to 12 spectral bands. Imaging spectrometry makes possible the acquisition of data in hundreds of spectral bands simultaneously (Goetz et al., 1985; Rubin, 1993; Kruse, 1994). Because of the very precise nature of the data acquired by imaging spectrometry, more Earth resource problems may be addressed in greater detail (Vane and Goetz, 1993).

The value of an imaging spectrometer lies in its ability to provide a high-resolution reflectance spectrum for each picture element in the image. The reflectance spectrum in the region from 0.4 − 2.5 μm may be used to identify a large range of surface cover materials that cannot be identified with broadband, low-spectral-resolution imaging systems such as the Landsat MSS, TM, or SPOT (Goetz et al., 1985). Many surface materials, although not all, have diagnostic absorption features that are only 20 − 40 nm wide. Therefore, spectral imaging systems that acquire data in contiguous 10 nm bands may produce data with sufficient resolution for the direct identification of those materials with diagnostic spectral features. For example, Figure 7-34 depicts high spectral resolution crop spectra over the interval from 400 − 1000 nm obtained using a Hughes Wedge Imaging Spectrometer for an agricultural area near Bakersfield, CA. The spectra for the Pima and Royale cotton differ from one another from about 725 nm, where the "red edge" is located to about 900 nm, leading to the possibility that species

within the same crop type may be distinguishable (SBRC, 1994). The Landsat scanners and SPOT HRV sensors, which have relatively large bandwidths, may not be able to resolve these spectral differences.

Simultaneous imaging in many contiguous spectral bands requires a new approach to remote sensor system design. One approach is to increase the residence time of a detector in each IFOV using a linear array of detector elements (Figure 7-3b). In this configuration, there is a dedicated detector element for each cross-track pixel, which increases the residence time to the interval required to move one IFOV along the flight direction. The French SPOT HRV sensor uses a linear array of detector elements. Despite the improved sensitivity of the SPOT detectors in the cross-track direction, they only record energy in three very broad green, red, and near-infrared bands. Thus, its major improvement is in the spatial domain and not in the spectral domain.

Two more practical approaches to imaging spectrometry are shown in Figures 7-3c and d. The whiskbroom scanner linear array approach (Figure 7-3c) is analogous to the scanner approach used for Landsat MSS or ETM+, except that radiant flux from within the IFOV is passed onto a spectrometer, where it is dispersed and focused onto a linear array of detectors. Thus, each pixel is simultaneously sensed in as many spectral bands as there are detector elements in the linear array. For high spatial resolution imaging (ground IFOVs of 10 to 30 m), this approach is suited only to an airborne sensor that flies slowly and when the readout time of the detector array is a small fraction of the integration time. Because of high spacecraft velocities, orbital imaging spectrometry may require the use of two-dimensional area arrays of detectors at the focal plane of the spectrometer. This eliminates the need for the optical scanning mechanism. In this situation, there is a dedicated column of spectral detector elements for each cross-track pixel in the scene.

Thus, traditional broadband remote sensing systems such as Landsat MSS or SPOT HRV *under-sample* the information content available from a reflectance spectrum by making only a few measurements in spectral bands up to several hundred nanometers wide. Conversely, imaging spectrometers sample at close intervals (bands on the order of tens of nanometers wide) and have a sufficient number of spectral bands to allow construction of spectra that closely resemble those measured by laboratory instruments. Analysis of imaging spectrometer data allows extraction of a detailed spectrum for each picture element in the image (Figure 7-35). Such spectra often allow direct identification of specific materials within the IFOV of the sensor based upon their reflectance characteristics, including minerals, atmospheric

Table 7-13.   Characteristics of the Airborne AVIRIS and CASI-2 Hyperspectral Remote Sensing Systems

Sensor	Technology	Spectral Resolution (nm)	Spectral Interval (nm)	Data Collection Mode	Dynamic Range (bits)	IFOV (mrad)	Total field of view (°)
AVIRIS	Whiskbroom linear array	400 - 2500	10	224 bands	12	1.0	30°
CASI-2	Area array CCD (512 x 288)	400 - 1000	1.9	Spatial: 19 bands @ 512 pixels Spectral: 288 bands @ 39 non-adjacent pixels, or 288 bands @ 101 pixels, or 48 bands @ 511 pixels	12	1.0	37.8°

gases, vegetation, snow and ice, and dissolved matter in water bodies (Kruse, 1994).

Analysis of hyperspectral data often requires the use of sophisticated digital image processing software (e.g., ENVI). This is because it is usually necessary to calibrate (convert) the raw hyperspectral radiance data to "percent reflectance" before it can be properly interpreted. This means removing the effects of atmospheric attenuation, topographic effects (slope, aspect), and any sensor anomalies. The Atmospheric Removal Program (ATREM) available from the University of Colorado performs many of these functions well. Similarly, to get the most out of the hyperspectral data it is usually necessary to use algorithms that 1) allow one to analyze a typical spectra to determine its constituent materials, and/or 2) compare the spectra with a library of spectra obtained using handheld spectroradiometers such as that provided by the U.S. Geological Survey.

Government agencies and commercial firms have designed hundreds of imaging spectrometers capable of acquiring hyperspectral data. It is beyond the scope of this book to list them all. Only three systems are summarized: NASA JPL's Airborne Visible Infrared Imaging Spectrometer (AVIRIS), the commercially available Compact Airborne Spectrographic Imager-2 (CASI), and NASA's Moderate Resolution Imaging Spectrometer (MODIS) onboard the *Terra* satellite.

Airborne Visible Infrared Imaging Spectrometer (AVIRIS)

The first airborne imaging spectrometer (AIS) was built to test the imaging spectrometer concept with infrared area arrays (Goetz et al., 1985; Vane and Goetz, 1993). It was operated in the mode shown in Figure 7-3d. The spectral coverage of the instrument was 1.9 – 2.1 μm in the *tree mode* and 1.2 – 2.4 μm in *rock mode* in contiguous bands that were

9.3 nm wide. This sampling interval was sufficient to describe absorption features for solids in this wavelength region. Continuous strip images, 32 pixels wide in 128 spectral bands, were acquired. The 128 spectral bands were acquired by stepping the spectrometer grating through four positions ($4 \times 32 = 128$) during the time it took to fly forward one pixel width on the ground. The area array was read out between each grating position, and the data were recorded on the aircraft with a high-density analog tape recorder. The IFOV of the AIS was 1.9 mrad, which produced a ground pixel size of approximately 8×8 m from an operating altitude of 4200 m.

To acquire data with greater spectral and spatial coverage, the *Airborne Visible Infrared Imaging Spectrometer* (AVIRIS) was developed at NASA's Jet Propulsion Laboratory in Pasadena, CA (Table 7-13). Using a whiskbroom scanning mirror and linear arrays of silicon (Si) and indium-antimonide (InSb) configured as in Figure 7-3c, AVIRIS acquires images in 224 bands, each 10 nm wide in the 400 to 2500 nm region (Green, 1994; Analytical Imaging and Geophysics, 1999). The sensor is typically flown onboard the NASA/ARC ER-2 aircraft at 20 km above-ground-level and has a 30° total field of view and an instantaneous field of view of 1.0 mrad, which yields 20×20 m pixels. The data are recorded in 12 bits (values from $0 - 4095$).

Many AVIRIS characteristics are summarized in Figure 7-35. It depicts a single band of AVIRIS imagery (band 30; 655.56 nm) obtained over the Kennedy Space Center, FL, and radiance data extracted for a single pixel of saw palmetto vegetation. Color Plate 7-6 depicts a portion of an AVIRIS data set acquired over Moffett Field, CA, the southern end of San Francisco Bay, and large evaporation ponds. Three of the 224 spectral bands of data were used to produce the color composite on top of the *hyperspectral datacube*. The black areas in the data cube represent atmospheric absorption bands at 1.4 and 1.9 μm.

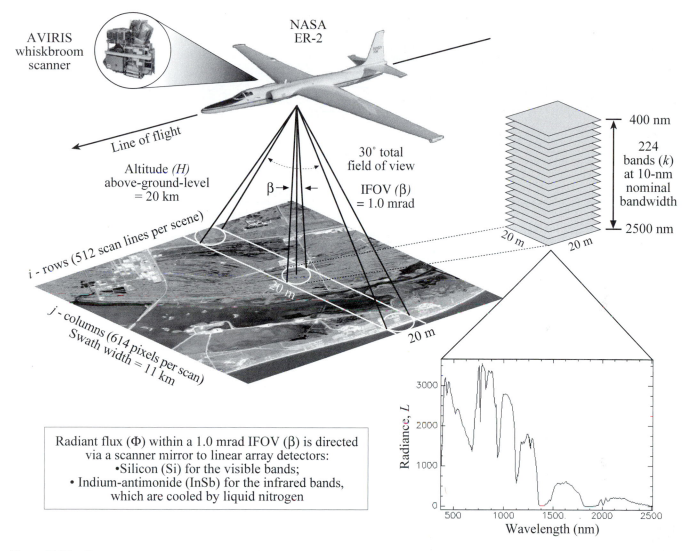

Figure 7-35 Conceptual representation of imaging spectroscopy as implemented by the NASA Jet Propulsion Lab Airborne Visible-Infrared Imaging Spectrometer (AVIRIS). The scanner mirror focuses radiant flux onto linear arrays that contain 224 detector elements with a spectral sensitivity ranging from 400 – 2500 nm. A spectra of radiance (L) or percent reflectance may be obtained for each picture element. The AVIRIS scene was acquired over the Kennedy Space Center, FL. Only band 30 (655.56 nm) is displayed (adapted from Filippi, 1999).

Numerous AVIRIS flights are made each year to support scientific experiments in the following areas (Green, 1994):

• Ecology: chlorophyll, leaf water, cellulose, lignin, nitrogen compounds;

• Oceanography and limnology: phytoplankton chlorophyll, dissolved organic compounds, suspended sediments, pigments of other planktonic organisms, marine plants and corals;

• Soils and geology: clay minerals, iron minerals, carbonates, sulfates;

• Snow and ice hydrology: ice absorption, water absorption, ice particle scattering;

• Atmosphere: water-vapor, aerosols, water clouds, ice clouds, smoke, oxygen, carbon dioxide, ozone, methane;

• Calibration of other satellite and aircraft sensor systems.

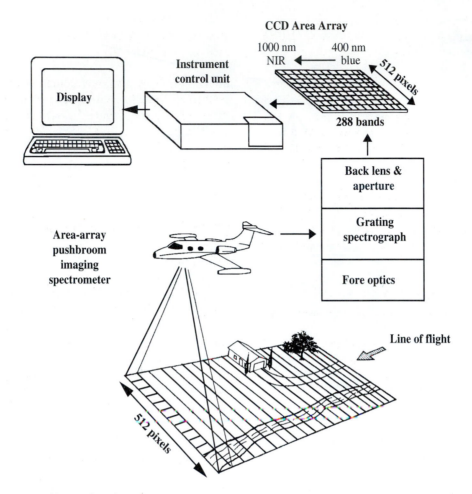

Figure 7-36   Area array pushbroom imaging spectrometer concept.

In 1998, AVIRIS was flown on a NOAA Twin Otter aircraft acquiring data at an altitude of 12,500 ft above sea level. This resulted in data with a spatial resolution of 3.5 x 3.5 m (JPL, 1999).

Compact Airborne Spectrographic Imager-2 (CASI-2)

ITRES Research Ltd. of Canada markets the CASI remote sensing system first introduced in 1989. It is a pushbroom imaging spectrometer based on the use of a 512 x 288 area array CCD. The instrument operates over a 545 nm spectral range from 400 nm – 1000 nm and has a 37.8° total field of view across-track (Figure 7-36).

The CASI is one of the first programmable imaging spectrometers, meaning that the user can program the sensor to collect specific types of hyperspectral data (ITRES, 1999). A single line of terrain 512 pixels wide perpendicular to the flight path is sensed by the spectrometer optics (Figures 7-3d and 7-36). The incoming radiant flux from each pixel is then spectrally dispersed along the other axis of the CCD so that the spectrum of energy (from blue through near-infrared) is obtained for each pixel across the swath. By repetitively reading the contents of the area array CCD as the aircraft moves along the flight path, a two-dimensional image at high spectral resolution is acquired. Since the radiant flux for all pixels in a particular swath are recorded simultaneously, spatial and spectral co-registration is assured. The across-track spatial resolution is determined by the altitude of the CASI above-ground-level and the IFOV, while the along-track resolution depends upon the velocity of the aircraft and the rate at which the CCD is read.

CASI may be programmed to collect 12 bit data in several modes (Table 7-13):

- Spatial Mode — where the full across-track resolution of 512 pixels are obtained for up to 19 non-overlapping spectral bands with programmable center wavelengths and bandwidths.

- Spectral Mode — where 288 bands of data at 1.9 nm intervals are obtained for a limited number of pixels across the swath (maximum is 39); or up to 288 spectral bands of data for 101 pixels; or up to 48 spectral bands for 511 across-track pixels.

The specific bandwidths are selected according to the application (e.g., bathymetric mapping, inventorying chlorophyll *a* concentration). The result is a powerful, programmable area array remote sensing system that may be the precursor of future satellite hyperspectral sensor systems.

Moderate Resolution Imaging Spectrometer (MODIS)

The Moderate Resolution Imaging Spectrometer is to be flown on NASA's EOS *Terra* satellite (equatorial crossing time of 10:30 a.m.) in late 1999 or early 2000 (Table 7-14). It is also scheduled for launch on the *EOS-PM* (afternoon crossing) satellite (Asrar and Dokken, 1993; NASA, 1999). MODIS provides long-term observations to derive an enhanced knowledge of global dynamics and processes occurring on the surface of Earth and in the lower atmosphere (SBRC, 1994). It yields simultaneous observations of high-atmospheric (cloud cover and associated properties), oceanic (sea-surface temperature and chlorophyll), and land-surface features (land-cover changes, land-surface temperature, and vegetation properties).

MODIS is in a 705 km Sun-synchronous orbit. It views the entire surface of the Earth every one to two days. It has a field of view of ±55° off-nadir, which yields a swath width of 2,330 km. MODIS obtains high radiometric resolution images (12 bit) of daylight-reflected solar radiation and day/night thermal emission over all regions of the globe. MODIS is a whiskbroom scanning imaging radiometer consisting of a cross-track scan mirror, collecting optics, and a set of linear detector arrays with spectral interference filters located in four focal planes. It collects data in 36 co-registered spectral bands: 20 bands from 0.4 – 3 μm and 16 bands from 3 – 15 μm. The bandwidths and their primary uses are summarized in Table 7-14.

MODIS's coarse spatial resolution ranges from 250 x 250 m (bands 1 – 2), to 500 x 500 m (bands 3 – 7) and 1 x 1 km (bands 8 – 36). MODIS provides daylight reflection and day/night emission spectral imaging of any point on Earth at

least every two days, with a continuous duty cycle. Daedalus, Inc., developed a 50-band MODIS Airborne Simulator used to simulate MODIS data.

MODIS has one of the most comprehensive calibration subsystems ever flown on a remote sensing instrument. The calibration hardware includes a solar diffuser, a solar diffuser stability monitor, a spectroradiometric calibration instrument, a blackbody for thermal calibration, and a space viewport. The calibration allows the raw brightness values to be converted into true percent reflectance or radiance measurements.

 Digital Frame Cameras

The CCD was invented in the late 1960s by scientists at Bell Labs. It was originally conceived as a new type of computer memory circuit, but it soon became apparent that it had many other applications, including image data collection, because of the sensitivity of silicon to light. A matrix of CCDs (often referred to by Kodak, Inc. as *photosites*) represents the heart and soul of digital frame cameras that are now being used to collect remote sensor data (McGarigle, 1997).

Digital Frame Camera Data-Collection

A digital frame camera draws many similarities to a regular camera. It consists of a camera body, a shutter mechanism to control the length of the exposure, and a lens to focus the incident energy onto a film or data-collection plane. However, this is where the similarity ends. Digital frame cameras use electronics rather than chemistry to capture, record, and process images. In a digital frame camera, the standard photographic silver-halide crystal emulsion at the film plane is replaced with a matrix (area array) of CCDs. There are usually thousands of light-sensitive photosites (pixels) that convert varying incident wavelengths of light into electrical signals (Kodak, 1999). An example of an area array CCD and digital frame camera is shown in Figure 7-37.

The CCD — *charged coupled devices*

Below is a step-by-step explanation of the CCD sensor's role in the digital image capture process (Kodak, 1999):

- Mechanical shutter opens, exposing the CCD sensor to light.

- Light is converted to a charge in the CCD.

Table 7-14.   Characteristics of the Moderate Resolution Imaging Spectrometer – MODIS (NASA, 1999)

Band	Spectral Resolution (μm)	Spatial Resolution	Primary Use
1	0.620–0.670	250 x 250 m	Land-cover classification and
2	0.841–0.876	250 x 250 m	chlorophyll absorption
3	0.459–0.479	500 x 500 m	Land, cloud, and aerosol properties
4	0.545–0.565	500 x 500 m	
5	1.230–1.250	500 x 500 m	
6	1.628–1.652	500 x 500 m	
7	2.105–2.155	500 x 500 m	
8	0.405–0.420	1 x 1 km	Ocean color, phytoplankton, biogeochemistry
9	0.438–0.448	1 x 1 km	
10	0.483–0.493	1 x 1 km	
11	0.526–0.536	1 x 1 km	
12	0.546–0.556	1 x 1 km	
13	0.662–0.672	1 x 1 km	
14	0.673–0.683	1 x 1 km	
15	0.743–0.753	1 x 1 km	
16	0.862–0.877	1 x 1 km	
17	0.890–0.920	1 x 1 km	Atmospheric water vapor
18	0.931–0.941	1 x 1 km	
19	0.915–0.965	1 x 1 km	
20	3.600–3.840	1 x 1 km	Surface–cloud temperature
21	3.929–3.989	1 x 1 km	
22	3.929–3.989	1 x 1 km	
23	4.020–4.080	1 x 1 km	
24	4.433–4.498	1 x 1 km	Atmospheric temperature
25	4.482–4.549	1 x 1 km	
26	1.360–1.390	1 x 1 km	Cirrus clouds
27	6.535–6.895	1 x 1 km	Water vapor
28	7.175–7.475	1 x 1 km	
29	8.400–8.700	1 x 1 km	
30	9.580–9.880	1 x 1 km	Ozone
31	10.780–11.280	1 x 1 km	Surface–cloud temperature
32	11.770–12.270	1 x 1 km	
33	13.185–13.485	1 x 1 km	Cloud top altitude
34	13.485–13.785	1 x 1 km	
35	13.785–14.085	1 x 1 km	
36	14.085–14.385	1 x 1 km	

- The shutter closes, blocking the light.

- The charge is transferred to the CCD output register and converted to a signal.

- The signal is digitized and stored in computer memory.

- The stored image is processed and displayed on the camera's liquid crystal display (LCD), on a computer screen, or used to make hard-copy prints.

An image is acquired when incident light in the form of photons falls on the array of pixels. The energy associated with

a.   b.

Figure 7-37   a) An area array CCD. b) The area array CCD is located at the film plane in the camera and used instead of film to record reflected light (courtesy Eastman Kodak Co.).

each photon is absorbed by the silicon, and a reaction takes place that creates an electron-hole charge pair (e.g., an electron). The number of electrons collected at each pixel is linearly dependent on the amount of light (photons) received per unit time and nonlinearly dependent on wavelength.

Like a traditional photographic system, the digital CCD area array captures a whole "frame" of terrain during a single exposure. The geographic area of the terrain recorded by the CCD area array is a function of 1) the dimension of the CCD array in rows and columns, 2) the focal length of the camera lens (the distance from the rear nodal point of the lens to the CCD array), and 3) the altitude of the aircraft above-ground-level. Moving the aircraft with the frame camera closer to the ground results in higher spatial resolution but less geographic area coverage. Increasing the actual size (dimension) of the CCD array will record more geographic area if all other variables are held constant. For example, doubling the size of the CCD array from 1000 x 1000 to 2000 x 2000 will effectively record four times as much geographic area during the same exposure. Area arrays used for remote sensing applications typically have 1500 x 1500 to 2000 x 3000 CCDs. Digital cameras based on CCD area array technology will not obtain the same resolution as traditional cameras until they contain approximately 20,000 x 20,000 detectors per band (Light, 1996). The general public does not have access to such technology at the present time. Fortunately, the cost of digital cameras continues to decrease.

Filtration

Because CCDs are inherently monochromatic, special filters called *color filter array* (*CFA*) patterns are used to capture the incident blue, green, and red photons of light. A number of CFAs have been invented, but only three are described.

The *RGB Filter Wheel CFA* is one of the simplest methods. A filter wheel is mounted in front of a monochromatic CCD sensor. The CCD makes three sequential exposures — one for each color. In this case, all photosites on the sensor capture red, blue, or green during the appropriate exposure. This method produces true colors, but it requires three exposures. Therefore, it is ideal for still photography, but not for collecting digital photography from a rapidly moving aircraft.

Three-chip Cameras use three separate full-frame CCDs, each one coated with a filter to make it red, green, or blue-sensitive. A beamsplitter inside the camera divides incoming energy into three distinct bands and sends the energy to the appropriate CCD. This design delivers high resolution and good color rendition of rapidly moving objects. It is the preferred method for remote sensing data collection. However, the cameras tend to be costly and bulky. Using this technology, it is also possible to send incident near-infrared light to an additional near-infrared sensitive CCD.

A *Single-chip Technology* filter captures all three colors with a single full-frame CCD. This is performed by placing a specially designed filter over each pixel, giving it the ability to capture red, green, and blue information. Obviously this reduces the cost and bulk of the camera since only one CCD is required. It acquires all the information instantaneously.

Timeliness

One of the most important characteristics of digital camera remote sensing is that the data are available as soon as they are collected. There is no need to send the imagery off for chemical processing and then wait for its return. If desired, the digital imagery can be downlinked electronically to the ground while the aircraft is still in the air. Furthermore, the cost of photographic processing is removed unless one wants to make hard copies.

Digital frame camera technology is used by companies such as Positive Systems, Inc., and Litton Emerge Spatial, Inc. to collect digital remote sensor data. It is instructive to review the characteristics of these two sensor systems.

Positive Systems, Inc.

Positive Systems, Inc., designs and markets the Airborne Data Acquisition and Registration (ADAR) system (Figure 7-38d). They configure the following components into an

a. Green band

b. Red band

c. Near-infrared band

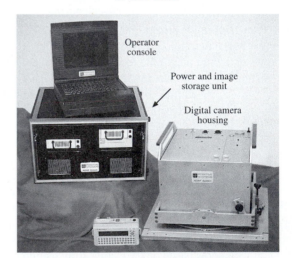

d. ADAR 5500 System Components

Figure 7-38   (a-c) Positive Systems, Inc., imagery of an area in Hartford, CT. The data were collected at a spatial resolution of 0.5 x 0.5 m. North is at the top to facilitate image interpretation. A color-infrared composite of the three bands is found in Color Plate 7-7b. d) ADAR 5500 System Components (courtesy of Positive Systems, Inc., ADAR System Imagery).

integrated digital frame camera remote sensing system: a camera housing that mounts over the aircraft camera port, an operator console that controls image and GPS information data collection, and a storage and power distribution unit. ADAR systems make extensive use of Eastman Kodak, Inc. color and color-infrared digital frame camera technology, including the Kodak Professional DCS 420 (1520 x 1012 pixel area array; 39° across-track field of view) and DCS 460 (3060 x 2036; 69° across-track field of view). The instantaneous field of view of each pixel in the CCD array is 0.44 mrad. The ADAR 5500 can be configured to acquire four bands of 8 bit visible and near-infrared digital frame photography in the spectral region from 400 – 900 nm (blue 450 - 515 nm; green 525 – 605 nm; red 640 – 690 nm; near-infrared 750 – 900 nm) at spatial resolutions ranging from 50 cm to 3 meters (Figure 7-38). A full frame is 1500 x 1000 pixels. Higher spatial resolutions are possible if a helicopter is used (Positive Systems, Inc., 1999).

The individual frames can be mosaicked together accurately if appropriate methods are employed. Positive Systems, Inc., was selected to provide 1 – 3 m digital multispectral remote sensing data to NASA scientists as part of the 1999 NASA Data Buy. Several examples of Positive Systems ADAR imagery are found throughout this book.

Litton Emerge Spatial, Inc.

Emerge Spatial, Inc., is a subsidiary of Litton TASC, Inc. The Emerge sensor system was designed initially by the U. S. Forest Service for forest/parks survey where high-resolution color-infrared imagery was required. It uses a proprietary configuration of Kodak DCS 460 and/or 560 digital cameras with 3072 x 2048 pixels. Each pixel in the array is 9 x 9 microns. Users can specify color (blue, green, and red bands) or color-infrared (green, red, and near-infrared) multiband imagery over the spectral region from 0.4 – 0.86 µm. In infrared mode, the Emerge sensor system has a spectral response similar to color-infrared film Kodak SO-134 or 2443, with a significantly higher dynamic range. The data are originally recorded at 12 bits per pixel (values from 0 – 4,095). Emerge collects real-time differentially corrected GPS data about each digital frame of imagery. These data are used to mosaic and ortho-rectify the imagery to the NAD 83 datum in the United States using photogrammetric techniques. The pixel placement accuracy meets national map accuracy standards (Emerge Spatial, 1999).

A variety of flying heights and different focal-length Nikon lenses are typically used to obtain imagery ranging from 0.3 m (1 ft) to 3 m. For example, using a 20 mm focal-length lens and flying at 2,200 ft above-ground-level results in imagery with a 1 x 1 ft ground resolved distance. Increasing the altitude to 6,660 ft results in 3 x 3 ft pixels.

Several examples of Litton Emerge Spatial, Inc., imagery are found throughout this book. Figure 7-39 depicts multispectral bands of 1 x 1 m imagery obtained over Dunkirk, NY. A color-infrared composite of the same area is found in Color Plate 7-8. Such high-resolution data can be collected on-demand in good weather. In emergencies, data can be collected below 4,000 ft above-ground-level.

 ## Satellite Photographic Systems

Despite the ongoing development of electro-optical remote sensing instruments, traditional optical camera systems continue to be used for space-survey purposes (Lavrov, 1997). For example, the Russian SOVINFORMSPUTNIK SPIN-2 TK-350 and KVR-1000 cameras provide cartographic quality photography suitable for making topographic and planimetric map products at 1:50,000 scale. Also, the U. S. Space Shuttle astronauts routinely obtain photography using Hasselblad and Linhof cameras.

Russian SPIN-2 TK-350 and KVR-1000 Cameras

The Russian Space Agency granted the Interbranch Association SOVINFORMSPUTNIK, Moscow, Russia, an exclusive right to use remote sensor data acquired by Russian defense satellites, to distribute these data commercially, and to produce and market value-added products. Most of the data marketed by SOVINFORMSPUTNIK are acquired by the KOMETA Space Mapping System onboard the KOSMOS series of satellites, which was designed to obtain high spatial resolution stereoscopic analog photography from space to produce 1:50,000-scale topographic maps. The data are also used to produce digital elevation models and 2 x 2 m orthorectified images. The high-resolution image archive contains global coverage acquired since 1981. The KOMETA system includes the TK-350 Camera and the KVR-1000 Panoramic Camera (Figure 7-40).

Basically, a rocket carries the KOMETA satellite into a 220 km near-circular orbit with an orbital duration of approximately 45 days. For example, one KOMETA mission lasted from February 17, 1998, through April 3, 1998. The total film capacity of the two cameras covers approximately 10.5 million km^2 of land area. The entire KOMETA system is retrieved from orbit at a predefined location. The film is then

a. Green band

b. Red band

c. Near-infrared band

Figure 7-39   Litton Emerge Spatial, Inc., multiband digital imagery of Dunkirk, NY. The data were collected on December 12, 1998, at a spatial resolution of 1 x 1 m. North is at the top to facilitate image interpretation. A color-infrared composite of the three bands is found in Color Plate 7-8a (courtesy of Litton Emerge Spatial, Inc.).

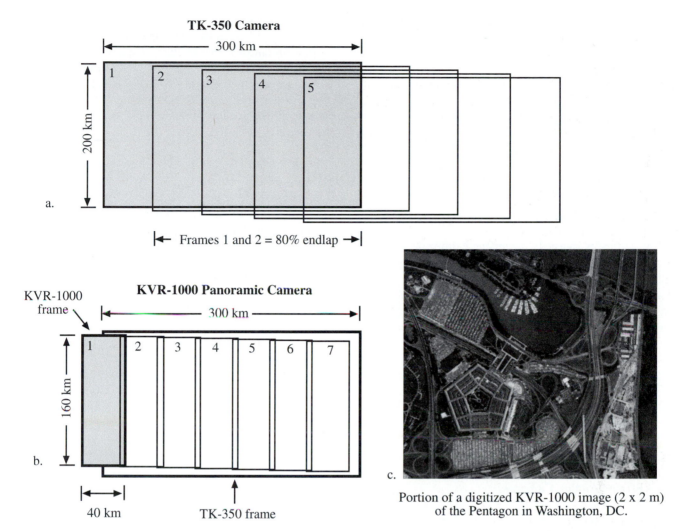

TK-350 Camera

300 km

200 km

1 2 3 4 5

a.

|← Frames 1 and 2 = 80% endlap →|

KVR-1000 Panoramic Camera

KVR-1000 frame

300 km

160 km

1 2 3 4 5 6 7

b.

40 km

TK-350 frame

c.

Portion of a digitized KVR-1000 image (2 x 2 m) of the Pentagon in Washington, DC.

Figure 7-40   a) Geographic coverage of the SPIN-2 SOVINFORMSPUTNIK TK-350 Camera system. b) Geographic coverage of the KVR-1000 Panoramic Camera. c) A small portion of a KVR-1000 frame of imagery depicting the Pentagon (courtesy of SOVINFORMSPUTNIK and Aerial Images, Inc.).

processed and digitized. SOVINFORMSPUTNIK has an agreement with Aerial Images, Inc., Raleigh, NC, and Central Trading Systems, Inc., Huntington Bay, NY, to market the data. Much of the U.S. coverage is served interactively on the Internet using Microsoft's Terra Server, a revolutionary system capable of retrieving quicklooks of much of the SOVINFORMSPUTNIK KVR-1000 archive for selected locations. The imagery can be ordered directly through the Terra Server (Microsoft, Inc., 1999).

TK-350 Camera

The TK-350 Camera was developed to collect panchromatic (510 – 760 nm) stereoscopic photography with up to 80 percent endlap between frames for the extraction of topographic

(elevation) data. The TK-350 Camera has a 350-mm focal-length lens. Images obtained by this camera have a scale of 1:660,000. The film format is 30 x 45 cm, which encompasses a single image covering 200 x 300 km, as shown in Figure 7-40a. TK-350 images can be enlarged to 1:50,000 scale without significant loss of image quality. The area covered by a single TK-350 image is covered by seven nested KVR-1000 images (Figure 7-40b) (Lavrov, 1997). After digitization, TK-350 imagery has a spatial resolution of approximately 10 x 10 m.

KVR-1000 Panoramic Camera

The KVR-1000 Panoramic Camera has a 1000-mm focal-length lens. It records panchromatic (510 – 760 nm) photographs at approximately 1:220,000 scale. KVR-1000 images may be enlarged to 1:10,000 scale without loss of detail. The geographic coverage of individual frames is 40 x 160 km, as shown in Figure 7-40b. Note the minimal amount of endlap obtained. After digitization, the imagery is provided at 2 x 2 m spatial resolution. Ground control for rectification is achieved through an onboard system employing GPS and laser altitude control systems. This enables the imagery to be rectified even when conventional ground control is not available (Aerial Images, Inc., 1999).

U.S. Space Shuttle Photography

NASA astronauts routinely document Earth processes during Space Transportation System (STS) missions using both analog and digital camera systems. These efforts have resulted in an impressive database of over 300,000 Earth images. Photographic documentation of Earth processes during manned spaceflights remains the cornerstone of the Space Shuttle Observations program, as it was with the earlier *Mercury*, *Gemini*, *Apollo*, and *Skylab* Earth observations programs (Lulla et al., 1994; 1996). During the Space Shuttle era, more than 200 selected sites of interest to geoscientists have been identified. Data from these sites are routinely acquired during Space Shuttle missions and cataloged into a publicly accessible electronic database according to the specific mission (e.g., STS-74) or by thematic topic.

Space Shuttle Analog Cameras

The primary analog cameras used during Space Shuttle missions are the Hasselblad and Linhof systems. NASA-modified Hasselblad 500 EL/M 70-mm cameras are used with large film magazines, holding 100 – 130 exposures. Standard lenses include a Zeiss 50-mm CF Planar f3.5, and a Zeiss 250-mm CD Sonnar f5.6. The Aero-Technika Linhof system can be fitted with 90- and 250-mm f5.6 lenses. This system uses large-format (100 x 127 mm) film.

The four windows in the aft part of the Space Shuttle are typically used to obtain photography of the Earth. The windows only allow 0.4 – 0.8 μm light to pass through. This results in the use of two primary film bases in the Hasselblad and Aero-Technika Linhof camera systems, including visible color (Kodak 5017/6017 Professional Ektachrome) and color-infrared (Kodak Aerochrome 2443) films.

Space Shuttle photographs are obtained at a variety of Sun angles, ranging from 1 to 80°, with the majority of pictures having sun angles of approximately 30°. Very low sun angle photography often provides unique topographic views of

remote mountainous areas otherwise poorly mapped. Sequential photographs with different look angles can, in certain instances, provide stereoscopic coverage. Seventy-five percent of the photographs in the archive cover the regions between 28° N and 28° S latitude, providing coverage for many little-known tropical areas. Twenty-five percent of the images cover regions between 30° to 60° N and S latitude.

Space Shuttle Electronic Still Camera (ESC)

Space Shuttle astronauts also acquire digital images of the Earth using an Electronic Still Camera (ESC), consisting of a specially modified Kodak DCS 460 camera with Nikon N90 camera body and lenses. Both color and monochrome digital images may be obtained. The camera uses a 3,000 x 2,000 astronomical grade CCD. Digital images are transmitted directly to the ground while in orbit.

The Space Shuttle Earth Observations Project (SSEOP) photography database of the Earth Science Branch at the NASA Johnson Space Center contains the records of more than 300,000 photographs of the Earth made from space during the last three decades. A select set of these photographs have been digitized for downloading using the File Transfer Protocol (http://images.jsc.nasa.gov).

Eventually, the International Space Station (ISS) will be launched. It will continue the NASA tradition of Earth observation from human-tended spacecraft. The ISS U.S. Laboratory Module will have a specially designed optical window with a clear aperture 50.8 cm in diameter that will be perpendicular to the Earth's surface most of the time (Lulla, 1997). Hasselblad, Linhof, Nikon, IMAX, digital cameras, and video recorders will be used to acquire images of the Earth.

 ### Digital Image Data Storage

Most digital remotely sensed data may be purchased on nine-track tape (6250 bpi), 4- or 8-mm tape, or on optical disks. The nine-track and 4- or 8-mm tapes must be read serially while it is possible to randomly select areas of interest from within the optical disk. This may result in significant savings of time when unloading remote sensor data. The 4- and 8-mm tape and compact disks are very efficient storage mediums, as opposed to the large number of nine-track tapes required to store most images. The companion volume on digital image processing (Jensen, 1996) describes the various types of digital formats in which the remote sensor data may be provided, including band-sequential, band-interleaved-by-line, and band-interleaved-by-pixel.

For a price, the commercial firms will provide radiometrically corrected data in a customer-specified format, map projection, Earth ellipsoid, pixel size, and level of geometric precision. Map-oriented products are usually available in three levels of geometric correction: (1) system-corrected, (2) precision-corrected (using ground-control points to adjust the satellite's predicted position to its actual geodetic position), or (3) terrain-corrected (using digital elevation data to adjust for relief displacement).

References

Abuelgasim, A., S. Gopal and A. H. Strahler, 1998, "Forward and Inverse Modeling of Canopy Directional Reflectance Using a Neural Network," *International Journal of Remote Sensing*, 19(3):453–471.

Aerial Images, Inc., 1999, *The SPIN-2 Story*, Raleigh: Aerial Images, Inc., http://AerialImages.com.

Analytical Imaging and Geophysics, 1999, *Hyperspectral Data Analysis and Image Processing Workshop*, AIG: Boulder, CO, March 15–19, 310 pp.

Asker, J. R., 1992, "Congress Considers Landsat 'Decommercialization' Move," *Aviation Week & Space Technology*, May 11, 18–19.

Asrar, G. and D. J. Dokken, 1993, *EOS Reference Handbook*. Washington, DC: Earth Science Support Office Document Resource Facility, 88–89.

Eidenshink, J. C., 1992, "The 1990 Conterminous U.S. AVHRR Data Set," *Photogrammetric Engineering & Remote Sensing*, 58(6):809–813.

Eidenshink, J. C. and J. L. Faundeen, 1999, *The 1-KM AVHRR Global Land Data Set: First Stages in Implementation*, Washington, DC: U.S. Geological Survey, http://edcwww.cr.usgs.gov/landdaac/1KM/paper.html.

Emerge Spatial, Inc., 1999, *Executive Summary — Emerge Specifications*, Billerica, MA: Litton Emerge Spatial, Inc., http://www.espatialweb.com.

England, G., 1994, *Airborne Multispectral Scanner*. Ann Arbor, MI: Daedalus Enterprises, Inc., 4 pp.

EOSAT, 1992a, *Landsat Technical Notes*. Lanham, MD: EOSAT, Inc., 4 pp.

EOSAT, 1992b, "The Remote Sensing Policy Act of 1992," *Landsat Data Users Notes*, 7(4):8.

Fegas, R. G., J. L. Cascio, and R. A. Lazar, 1992, "An Overview of FIPS 173, the Spatial Data Transfer Standard," *Cartography and Geographic Information Systems*, 19(5):278–293.

Filippi, A., 1999, "Hyperspectral Image Classification Using a Batch Descending Fuzzy Learning Vector Quantization Artificial Neural Network: Vegetation Mapping at the John F. Kennedy Space Center," unpublished Masters Thesis, Columbia: University of South Carolina, 276 pp.

Fisher, L. T., 1991, "Aircraft Multispectral Scanning with Accurate Geographic Control," *Geodetical Info Magazine*, 5(2):59–62.

Goetz, A., G. Vane, J. E. Solomon, and B. N. Rock, 1985, "Imaging Spectrometry for Earth Remote Sensing," *Science*, 228(4704):1147–1153.

Green, R. O., 1994, *AVIRIS Operational Characteristics*. Pasadena, CA: Jet Propulsion Lab, 10 pp.

Henderson, F. B., 1995, "Remote Sensing for GIS," *GIS World*, 8(2):42–44.

ITRES, Inc., 1999, *Calibrated Airborne Spectrographic Imager: CASI-2*, Canada: ITRES Research Ltd., http://www.itres.com.

Jensen, J. R., 1996, *Introductory Digital Image Processing: A Remote Sensing Perspective*, Upper Saddle River: Prentice-Hall, Inc., 318 pp.

Jensen, J. R., D. J. Cowen, J. Halls, N. Schmidt, and B. A. Davis, 1994, "Improved Urban Infrastructure Mapping and Forecasting for BellSouth Using Remote Sensing and GIS Technology," *Photogrammetric Engineering & Remote Sensing*, 60(3): 339–346.

Jensen, J. R., E. W. Ramsey, H. E. Mackey, E. J. Christensen, and R. R. Sharitz, 1987, "Inland Wetland Change Detection using Aircraft MSS Data," *Photogrammetric Engineering & Remote Sensing*, 53(5):521–529.

Jensen, J. R., S. Narumalani, and H. E. Mackey, 1994b, "Monitoring Commercial Forestry and Water Management Practices on a Cypress–Tupelo Swamp Forest in South Carolina," *14th Biennial Workshop on Color Photography and Videography in Resource Monitoring*. Bethesda, MD: American Society for Photogrammetry & Remote Sensing, 125–134.

a.

Additive Color Theory

Subtractive Color Theory

b.

c.

Plate 4-1   a) Sir Isaac Newton discovered that white light could be dispersed into its spectral components by passing white light through a prism (© David Parker, Photo Researchers, Inc.). b) *Additive color theory* — equal proportions of blue, green, and red light superimposed on top of one another creates white light, i.e., white light is composed of blue, green, and red light. The complementary colors yellow, magenta, and cyan are created by selectively adding together red and green, blue and red, and blue and green light, respectively. c) *Subtractive color theory* — equal proportions of blue, green, and red pigments yield a black surface. A yellow filter effectively absorbs all blue light, a magenta filter absorbs all green light, and a cyan filter absorbs all red light.

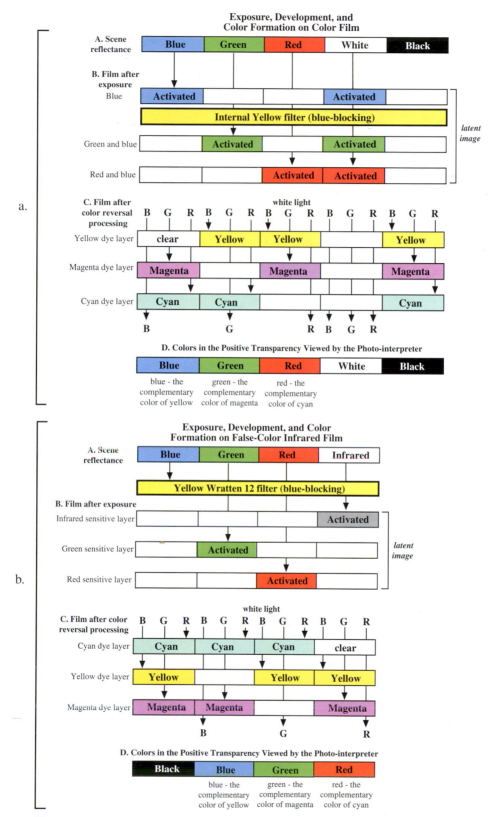

Plate 4-2   The process of exposing, developing, and forming color on a) normal color film, and b) false-color infrared film (courtesy Eastman Kodak Company).

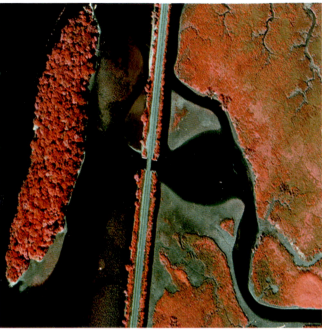

a.   b.

Plate 4-3   a) A normal (natural) color photograph of Tivoli North Bay along the Hudson River in New York. Color aerial photography records blue, green, and red light reflected from the scene. A haze filter (HF3) is usually placed in front of the lens to prevent scattered ultraviolet light from reaching the film plane. Vegetation shows up in green hues because the plants absorb more blue and red incident light than green light. b) A color-infrared aerial photograph records green, red, and near-infrared light reflected from the scene. A yellow (Wratten 12) filter is usually placed in front of the lens to prevent blue light from reaching the film plane. The land-water interface is especially well-delineated because water absorbs most of the incident near-infrared radiant energy causing it to appear dark. Vegetation appears in magenta hues because healthy vegetation reflects substantial amounts of near-infrared radiant flux while absorbing much of the green and red incident energy (after Berglund, 1999).

a.   b.

Plate 4-4   a) Normal (natural) color terrestrial photograph of a building at the University of Nebraska. b) Color-infrared photograph acquired using a Wratten 12 minus-blue filter (after Rundquist and Sampson, 1988).

Plate 5-1   Calibrated Airborne Multispectral Scanner (CAMS) data of one of the island keys near Key West, FL. The spatial resolution is 2.5 x 2.5 m. The image is a color composite of green, red, and near-infrared bands. The dense stands of healthy red mangrove (*Rhizophora mangale*) are 6 to 12 m tall. The tallest trees are found around the periphery of the island where tidal flushing provides a continuous flow of nutrients. Two areas of bare sand are located in the interior of the island. A coral reef surrounds the island (after Davis and Jensen, 1998).

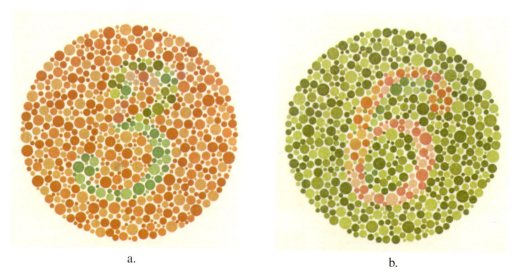

a.   b.

Plate 5-2   Some people do not perceive color like the majority of the population. This does not preclude them from becoming excellent image analysts. They only need to understand the nature of their color perception differences and adjust. Two color perception illustrations are provided. The viewer should see a "3" in (a) and a "6" in (b). If you see a "5", the next time you have an eye exam you might ask the optometrist to give you a color perception test to identify the nature of your color perception (courtesy American Society for Photogrammetry and Remote Sensing; Smith, J. T., 1968, *Manual of Color Aerial Photography*).

a. Landsat 4 Thematic Mapper Data (RGB = bands 4,3,2)

b. Landsat 4 Thematic Mapper Data (RGB = bands 7,4,2)

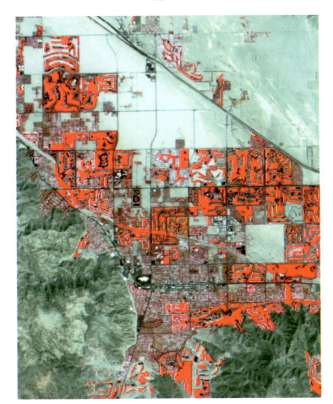

c. Landsat 7 Thematic Mapper Plus (RGB = bands 4,3,2)

d. Landsat 7 Thematic Mapper Plus (RGB = bands 7,4,2)

Plate 7-1   a,b) Landsat 4 Thematic Mapper color composite images of Charleston, SC, obtained on February 3, 1994. The data have a spatial resolution of 30 x 30 m (courtesy Space Imaging, Inc.). c,d) Landsat 7 Enhanced Thematic Mapper Plus (ETM[+]) color composite images of Palm Springs, CA, acquired in 1999. The numerous golf courses contrast sharply with the arid terrain (courtesy National Aeronautics and Space Administration and U.S. Geological Survey).

a. AVHRR Mosaic of the United States

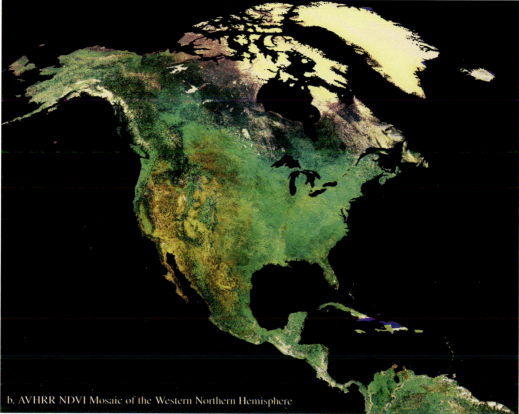

b. AVHRR NDVI Mosaic of the Western Northern Hemisphere

Plate 7-2   a) AVHRR mosaic of the conterminous United States derived from sixteen 1 x 1 km NOAA-8 and NOAA-9 AVHRR images obtained from May 24, 1984, to May 14, 1986, using channels 1 and 2. b) Normalized Difference Vegetation Index (NDVI) image of the Earth derived from a time series of AVHRR data (courtesy NOAA and U.S. Geological Survey).

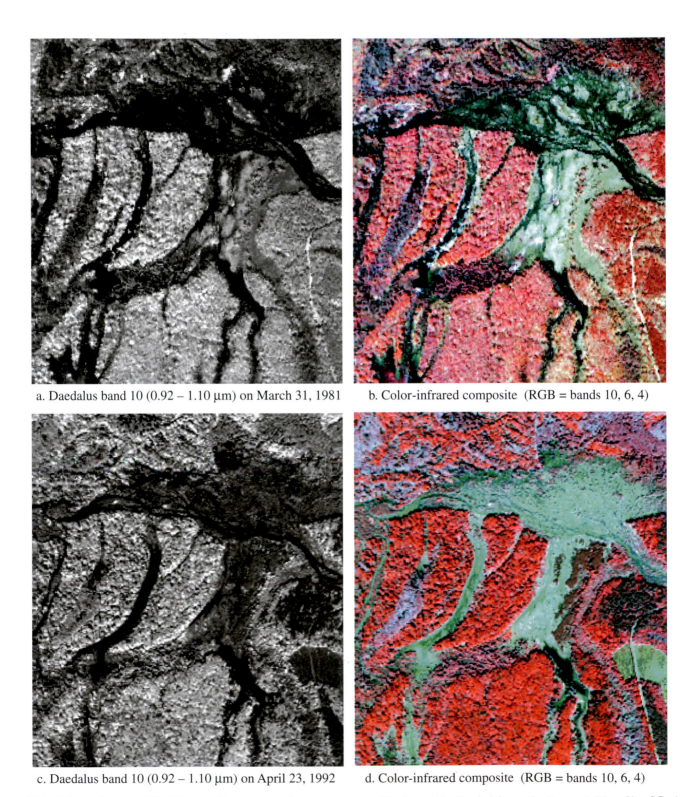

a. Daedalus band 10 (0.92 – 1.10 μm) on March 31, 1981   b. Color-infrared composite (RGB = bands 10, 6, 4)

c. Daedalus band 10 (0.92 – 1.10 μm) on April 23, 1992   d. Color-infrared composite (RGB = bands 10, 6, 4)

Plate 7-3   Daedalus DS-1260 aircraft multispectral scanner images of the Four Mile Creek delta on the Savannah River Site, SC. a) Band 10 near-infrared image obtained on March 31, 1981. b) Color-infrared composite. c) Band 10 image obtained on April 23, 1992. d) Color-infrared composite. Note the dramatic differences in the wetland vegetation distribution in the sloughs between 1981 and 1992. The Daedalus DS-1260 bands are different than the Daedalus AMS bands.

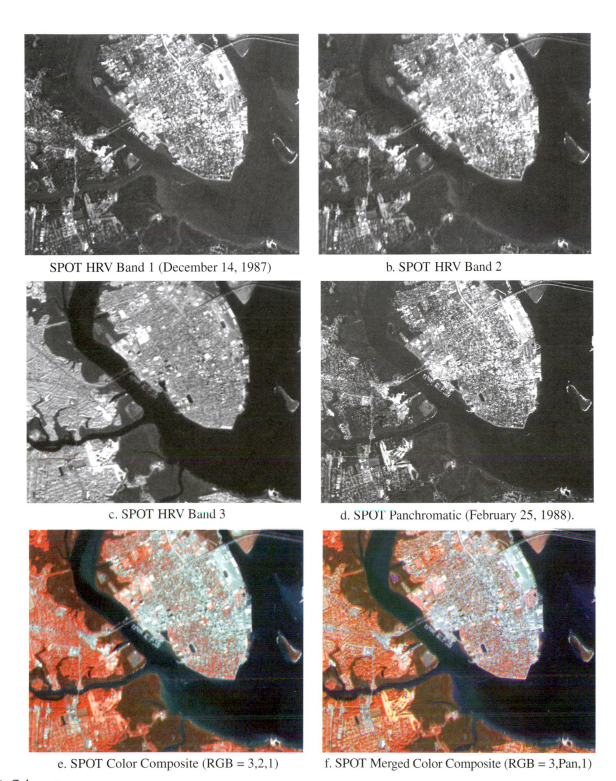

SPOT HRV Band 1 (December 14, 1987)

b. SPOT HRV Band 2

c. SPOT HRV Band 3

d. SPOT Panchromatic (February 25, 1988).

e. SPOT Color Composite (RGB = 3,2,1)

f. SPOT Merged Color Composite (RGB = 3,Pan,1)

Plate 7-4   a

to c) SPOT HRV multispectral 20 x 20 m data of Charleston, SC, obtained on December 14, 1987. d) Panchromatic 10 x 10 m image obtained on February 25, 1988. e) Color-infrared composite of the three 1987 multispectral bands. f) Panchromatic 10 x 10 m data merged with the 20 x 20 m multispectral data to improve the interpretability of the color-infrared composite (© SPOT Image, Inc.).

Portion of the First Global Ten-Day Synthesis
Image Produced Using the SPOT Vegetation Sensor

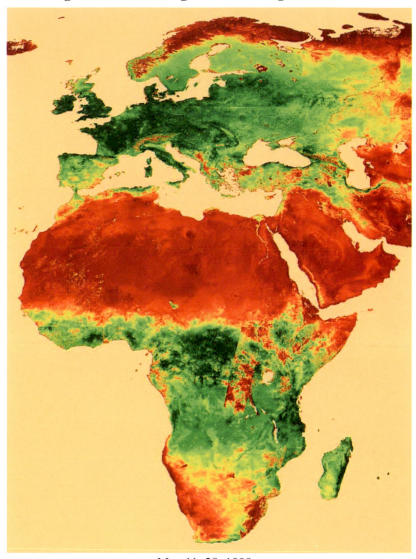

May 11–20, 1998

Plate 7-5   Portion of a Global Ten-Day Synthesis product generated using the SPOT 4 Vegetation sensor system launched on March 24, 1998. The daily product provides vegetation index information at a spatial resolution of 1 x 1 km over the whole Earth. The Ten-Day Global Synthesis is derived from analysis of a Daily Synthesis database (© SPOT Image, Inc.).

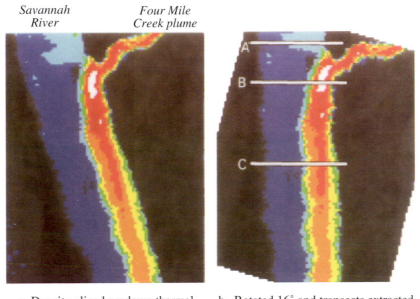

Savannah River   *Four Mile Creek plume*

a. Density sliced predawn thermal infrared (8 - 14 μm) data.

b. Rotated 16° and transects extracted.

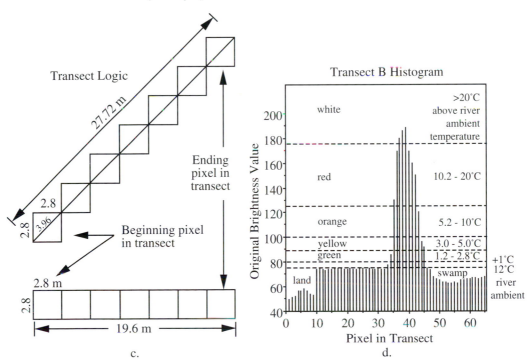

Transect Logic

27.72 m

2.8

2.8

3.96

Ending pixel in transect

Beginning pixel in transect

2.8 m

2.8

19.6 m

c.

Transect B Histogram

Original Brightness Value

200

180

160

140

120

100

80

60

40

0   10   20   30   40   50   60

Pixel in Transect

white   >20°C above river ambient temperature

red   10.2 - 20°C

orange   5.2 - 10°C

yellow   3.0 - 5.0°C
green   1.2 - 2.8°C   +1°C
                                12°C
land   swamp   river ambient

d.

Plate 8-1   a) Color-coded "density sliced" display of predawn thermal infrared data of the Four Mile Creek thermal plume in the Savannah River on March 28, 1981 (2.8 x 2.8 m spatial resolution). The data are colored according to the class intervals summarized in Table 8-7. b). Three transects are passed through the plume after rotating it 16°. c) Why it is important to rotate the image. d) Temperature information in Transect B displayed in a histogram format.

AVIRIS Image of Moffett Field, CA

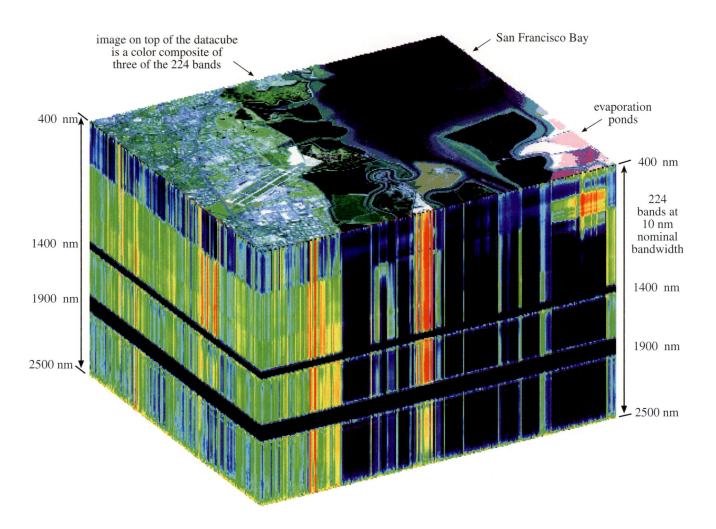

Plate 7-6   An Airborne Visible Infrared Imaging Spectrometer (AVIRIS) dataset of Moffett Field, CA, adjacent to San Francisco Bay. The data were acquired on August 20, 1992, by the NASA ER-2 aircraft (a U-2 modified for increased performance) at an altitude of 20,000 m (65,000 ft.) resulting in 20 x 20 m nominal spatial resolution. AVIRIS acquires images in 224 bands each 10 nm wide, in the 400 to 2,500 nm region (Green, 1994). Three of the 224 spectral bands of data were used to produce the color composite on top of the "hyperspectral datacube" to accentuate the structure in the water and in the evaporation ponds on the right. The sides of the cube depict just the edge pixels for all 224 of the AVIRIS spectral channels. The top of the sides are in the visible part of the spectrum (400 nm) and the bottom is in the middle-infrared (2,500 nm). The black areas in the datacube represent atmospheric absorption bands, especially at 1400 and 1900 nm. The sides are displayed in a pseudo-color ranging from black and blue (very low response) to red (high response). Of particular interest in the datacube is the small region of high response in the upper right corner of the larger side. This response is in the red part of the visible spectrum (about 700 nm) and is due to the presence of 1 cm (0.5 in.) long red brine shrimp in the evaporation pond (courtesy of R. O. Green, NASA Jet Propulsion Laboratory).

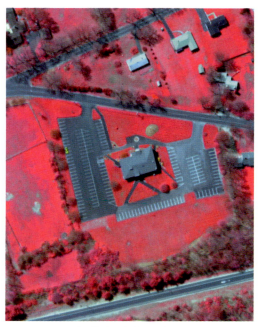

a. Natural Color   b. Color-infrared

Plate 7-7   Positive Systems, Inc. ADAR 5500 System digital camera imagery of a public building and residential neighborhood in Hartford, CN. The data were obtained at a spatial resolution of 0.5 x 0.5 m. a) Natural color image (RGB = red, green, blue). b) Color-infrared image (RGB = near-infrared, red, and green) (courtesy Positive Systems, Inc.).

a. Color-infrared   b. Natural Color

Plate 7-8   Litton Emerge Spatial, Inc. digital camera data. a) Color-infrared image (RGB = near-infrared, red, and green) of Dunkirk, NY, at 1 x 1 m obtained on December 12, 1998. The tennis court and track are covered with astroturf. b) Natural color image (RGB = red, green, and blue) of an area adjacent to a N.Y. Power Authority lake. The 1 x 1 ft data were acquired on October 13, 1997. The deciduous hardwood trees were captured senescing in this beautiful image (courtesy Litton Emerge Spatial, Inc.).

a.

b.

Plate 9-1   a) Space Shuttle photograph of the Nile River, Sudan. b) SIR-C/X-SAR color-composite image of C-band with HV polarization, L-band with HV polarization, and L-band with HH polarization. The data were acquired by the Space Shuttle Endeavor in April, 1994 (courtesy NASA Jet Propulsion Laboratory).

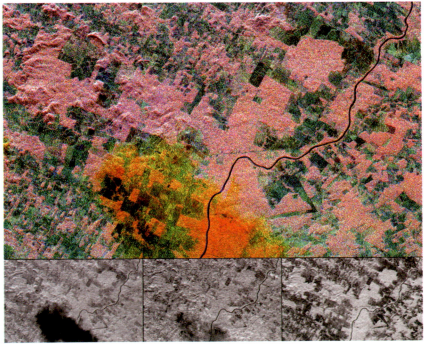

X-band VV   C-band HV   L-band HV

Plate 9-2   SIR-C/X-SAR image of a portion of Rondonia, Brazil, obtained on April 10, 1994. It is a color composite of X-band VV polarization, C-band HV polarization, and L-band HV polarization. A heavy rain appears as a black cloud in the X-band image, more faintly in the C-band image, and is relatively invisible in the L-band image. The bright pink is rainforest (courtesy NASA Jet Propulsion Laboratory).

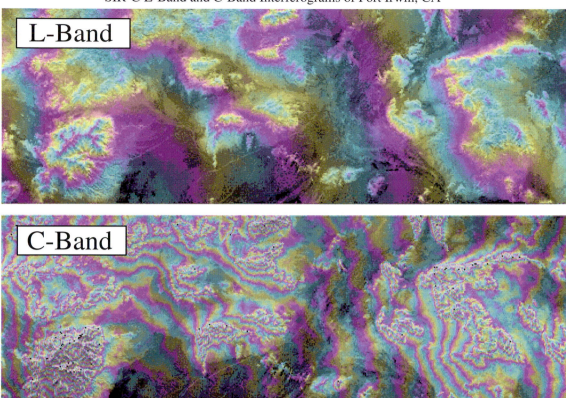

Plate 9-3   SIR-C/X-SAR interferograms of Fort Irwin, CA in the Mojave Desert. The colored bands provide detailed, quantitative elevation information that can be used to construct a digital elevation model of the terrain (courtesy NASA Jet Propulsion Laboratory).

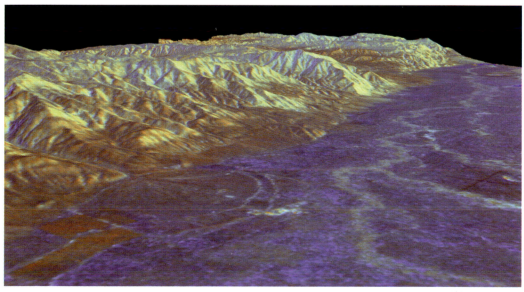

Plate 9-4   Three-dimensional perspective view of Owens Valley, near Bishop, CA, created by combining two spaceborne SIR-C/X-SAR images obtained in October, 1994, using interferometric techniques. The White Mountains are in the center of the image and rise to 3,000 m (10,000 ft). The Owens River and its tributaries are seen (courtesy NASA Jet Propulsion Laboratory).

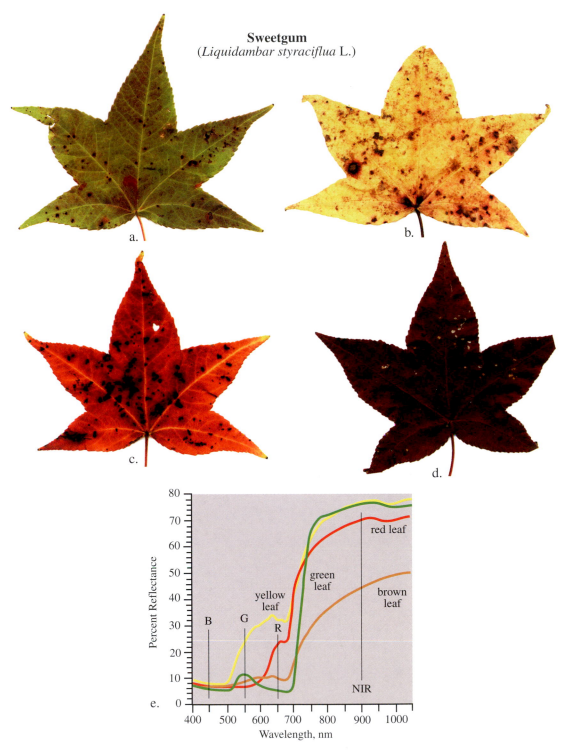

Sweetgum
(*Liquidambar styraciflua* L.)

Plate 10-1   a) Photosynthesizing green Sweetgum leaf (*Liquidambar styraciflua* L.) obtained from a tree on November 11, 1998. b-c) Senescing yellow and red Sweetgum leaves obtained from the tree. d) Senesced Sweetgum leaf that was on the ground. e) Geophysical & Environmental Research, Inc. (GER) 1500 spectroradiometer percent reflectance measurements over the wavelength interval 400 – 1050 nm.

Ground Reference Information Overlaid on A Single Channel of AVIRIS Imagery San Luis Valley, Colorado

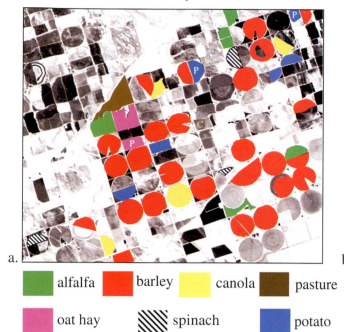

a.

🟩 alfalfa	🟥 barley	🟨 canola	🟫 pasture	
🟪 oat hay	▨ spinach	🟦 potato		

Vegetation Species Classification Map September 3, 1993

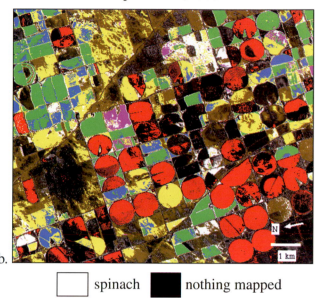

b.

☐ spinach	■ nothing mapped

Vegetation Senescence/Stress Map

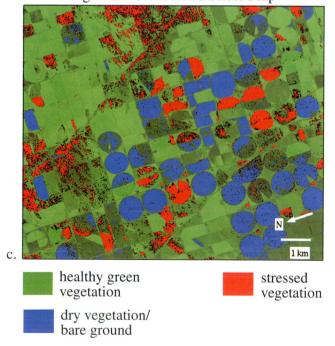

c.

🟩 healthy green vegetation	🟥 stressed vegetation
🟦 dry vegetation/ bare ground	

Plate 10-2   a) Ground reference information overlaid on a single channel of 20 x 20 m AVIRIS data obtained on September 3, 1993. b) Vegetation species classification map derived from analysis of AVIRIS data using USGS Tricorder hyperspectral data-analysis software. c) Vegetation senescence/stress map (after Clark et al., 1995).

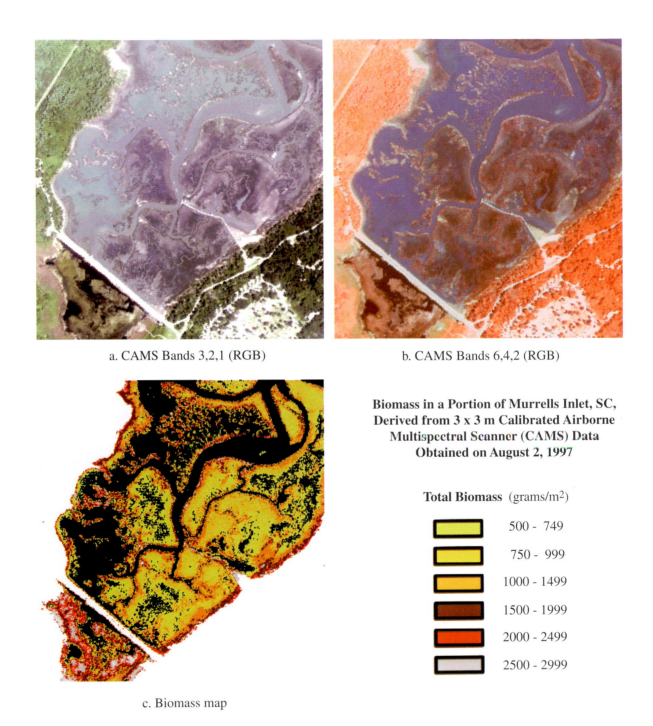

a. CAMS Bands 3,2,1 (RGB)

b. CAMS Bands 6,4,2 (RGB)

Biomass in a Portion of Murrells Inlet, SC, Derived from 3 x 3 m Calibrated Airborne Multispectral Scanner (CAMS) Data Obtained on August 2, 1997

Total Biomass (grams/m^2)

	500 - 749
	750 - 999
	1000 - 1499
	1500 - 1999
	2000 - 2499
	2500 - 2999

c. Biomass map

Plate 10-3   a) Natural color composite of a small portion of Murrells Inlet, SC, recorded by the NASA Calibrated Airborne Multispectral Scanner (CAMS) on August 2, 1997. b) Color-infrared color composite. c) Biomass (g/m^2) information extracted from the CAMS data (Jensen et al., 1998).

Phenological Cycles of San Joaquin and Imperial Valley, California, Crops and Landsat Multispectral Scanner Images of One Field During a Growing Season

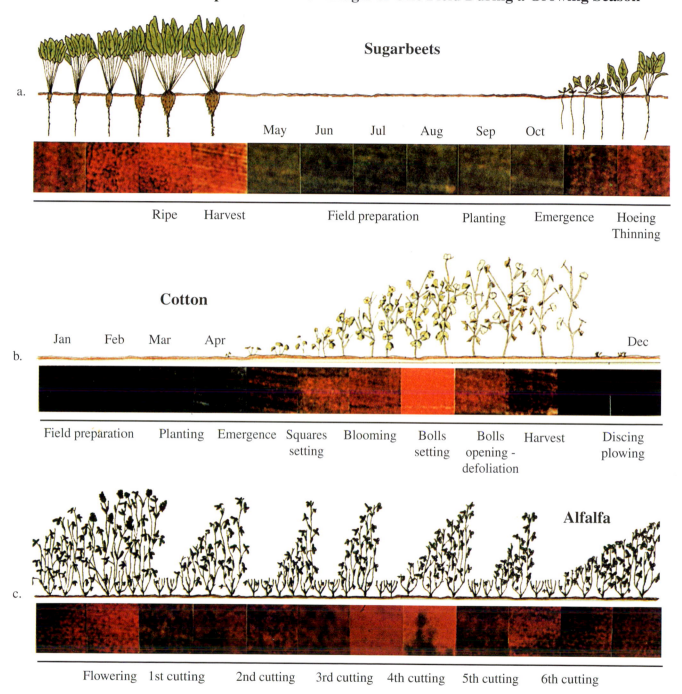

Plate 10-4   Phenological cycles of a) sugarbeets, b) cotton, and c) alfalfa grown in the San Joaquin and Imperial Valleys of Southern California. Landsat MSS images were obtained over a 12-month period in the San Joaquin Valley. The color composite images (RGB = bands 4,2,1) of three fields are extracted and placed below the crop calendar information (courtesy J. Estes, J. Sepan, T. Hardoin, and author).

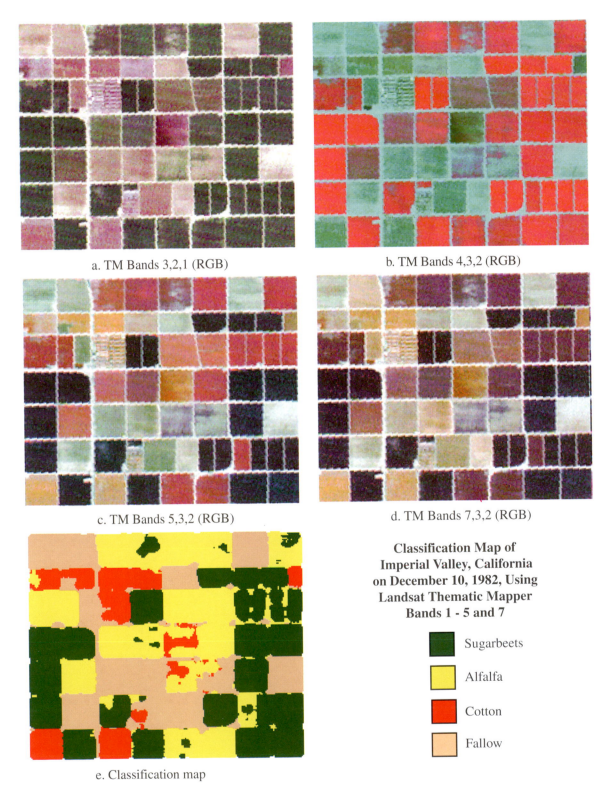

a. TM Bands 3,2,1 (RGB)

b. TM Bands 4,3,2 (RGB)

c. TM Bands 5,3,2 (RGB)

d. TM Bands 7,3,2 (RGB)

e. Classification map

**Classification Map of
Imperial Valley, California
on December 10, 1982, Using
Landsat Thematic Mapper
Bands 1 - 5 and 7**

Sugarbeets

Alfalfa

Cotton

Fallow

Plate 10-5   a) Landsat Thematic Mapper natural color composite image of a portion of the Imperial Valley, CA, obtained on December 10, 1982. b – d) Several other false-color composites. e) Classification map derived by digital image processing.

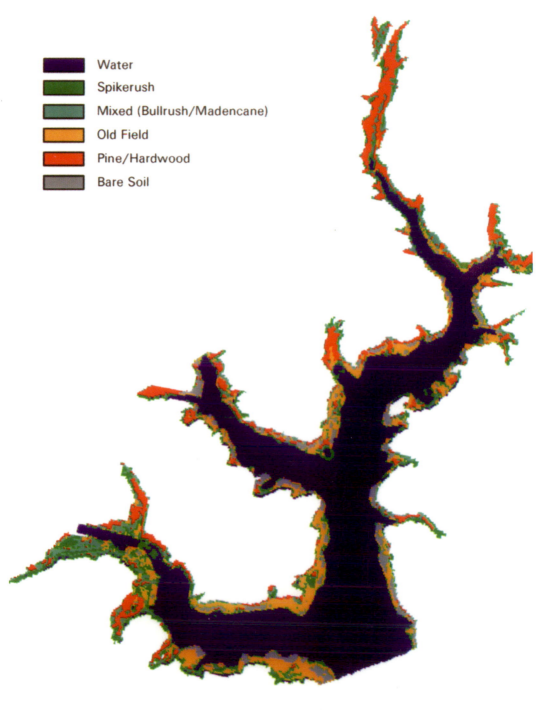

Water

Spikerush

Mixed (Bullrush/Madencane)

Old Field

Pine/Hardwood

Bare Soil

Plate 10-6   Wetland classification map of Par Pond reservoir located on the Savannah River Site near Aiken, SC. The map was derived from a digital multispectral analysis of SPOT HRV data obtained on October 25, 1994. Seven maps such as this were obtained from SPOT XS data from 1992 through 1994 and used to characterize the wetland succession caused by the drawdown of Par Pond reservoir (Jensen et al., 1997).

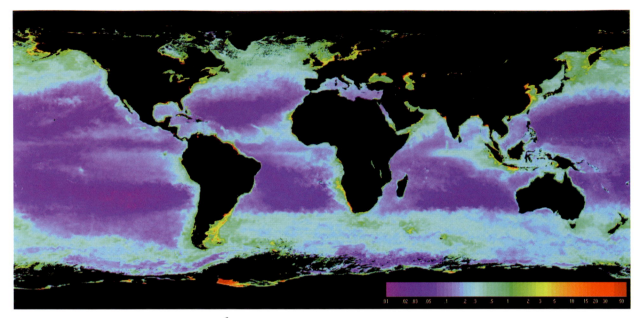

a. Global chlorophyll *a* (g/m³) derived from SeaWiFS imagery obtained from September 3, 1997, through December 31, 1997. The warmer the color, the greater the chlorophyll concentration.

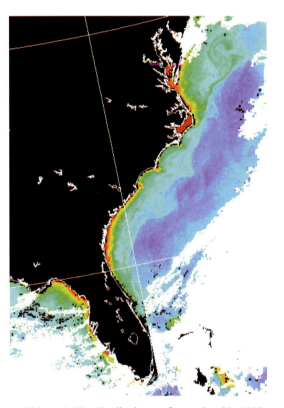

b. True-color SeaWiFS image of the Eastern Seaboard of the United States obtained on September 30, 1997.

c. Chlorophyll *a* distribution on September 30, 1997 derived from SeaWiFS data.

Plate 11-1   Examples of Sea-viewing Wide Field-of-View (SeaWiFS) remote sensor data (courtesy NASA Goddard Space Flight Center and Orbital Imaging Corporation (ORBIMAGE) used with permission).

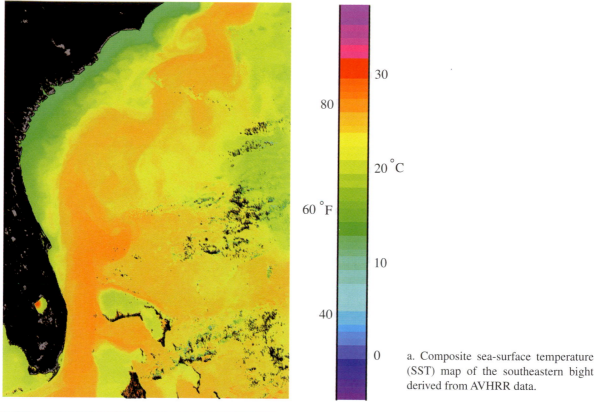

a. Composite sea-surface temperature (SST) map of the southeastern bight derived from AVHRR data.

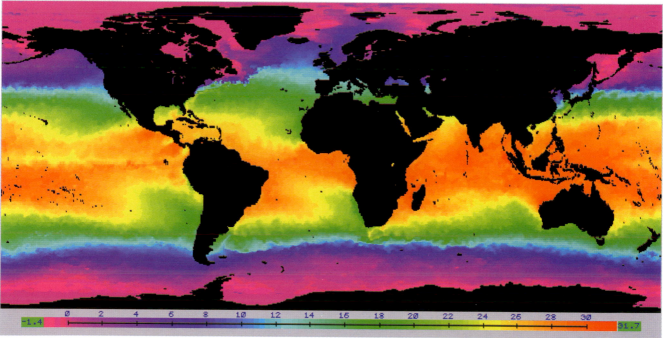

b. Worldwide sea-surface temperature (SST) map derived from NOAA-14 AVHRR data.

Plate 11-2   a) Sea-surface temperature (SST) map derived from a three-day composite of NOAA AVHRR thermal infrared data centered on March 4, 1999. Each pixel was allocated the highest surface temperature that occurred during the three days (courtesy of NOAA Coastal Services Center). b) Global ocean 50 x 50 km SST (°C) derived from March 9, 1999 through March 13, 1999 (95 hours) NOAA-14 AVHRR data (courtesy of NOAA/NESDIS).

Monthly Sea-Surface Temperature (°C)

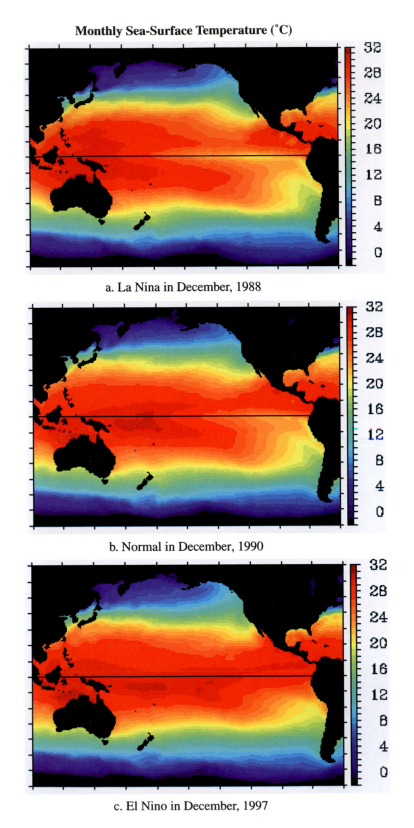

a. La Nina in December, 1988

b. Normal in December, 1990

c. El Nino in December, 1997

Plate 11-3   Reynolds monthly sea-surface temperature (SST) maps derived from *in situ* buoy data and remotely sensed data (courtesy NOAA/TAO/National Center for Environmental Prediction).

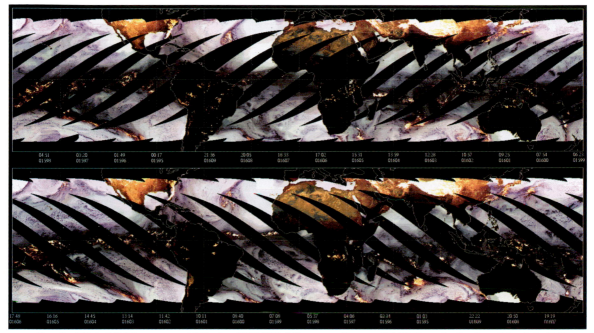

a. Tropical Rainfall Measurement Mission (TRMM) Microwave Imager (TMI) data obtained on March 9, 1998.

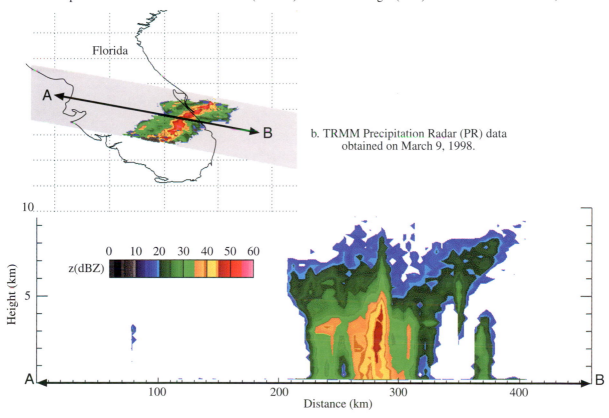

b. TRMM Precipitation Radar (PR) data obtained on March 9, 1998.

c. Along-track cross-section of TRMM Precipitation Radar data obtained on March 9, 1998.

Plate 11-4 a) Worldwide TRMM Microwave Imager data collected on March 9, 1998. b) Precipitation Radar data of southern Florida. c) Cross-section of Precipitation Radar data (courtesy NASA Goddard TRMM Office and Japan National Space Development Agency).

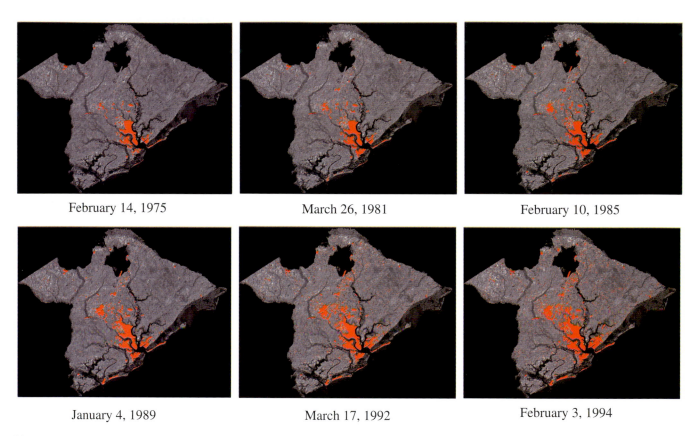

February 14, 1975   March 26, 1981   February 10, 1985

January 4, 1989   March 17, 1992   February 3, 1994

Plate 12-1   Level I Urban/non-urban land cover for Berkeley, Dorchester, and Charleston counties centered on Charleston, SC, derived from Landsat MSS 79 x 79 m data. The land cover information is draped over a near-infrared image (band 4).

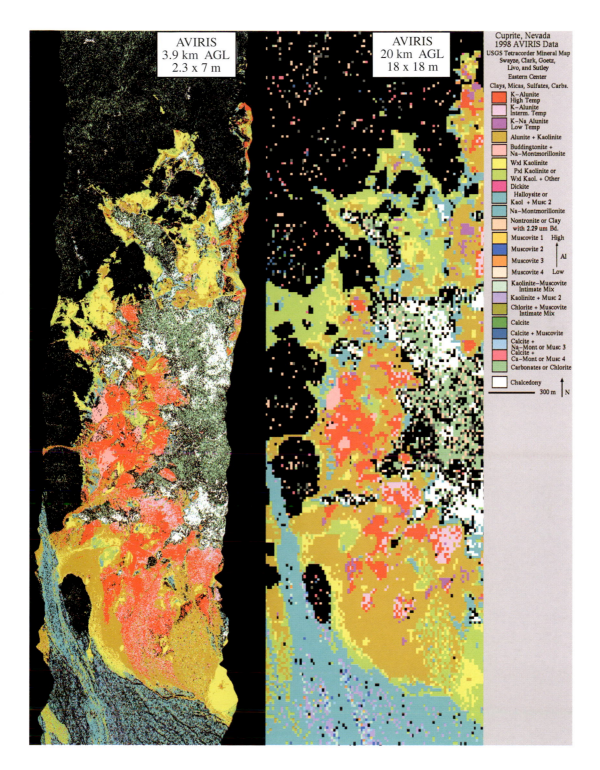

Plate 13-1   Mineral maps of Cuprite, NV, derived from low-altitude (3.9 km AGL) and high-altitude (20 km AGL) AVIRIS data obtained on October 11, 1998, and June 18, 1998, respectively. The hyperspectral data were analyzed using the USGS Tetracorder program (courtesy Swayze et al., 1999).

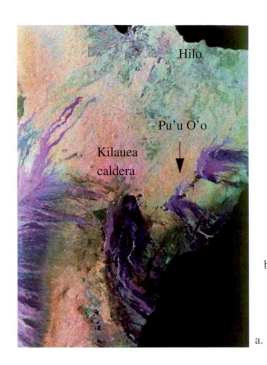

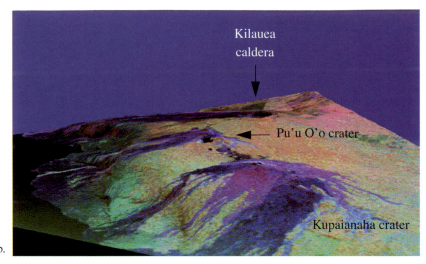

b.

a.

Plate 13-2   a) Composite Shuttle SIR-C/X-SAR image of Kilauea volcano, HI, (bands C, X, L) obtained on April 12, 1994. b) The image overlaid on a digital elevation model. The overland flow of lava on this shield volcano is very evident (courtesy NASA Jet Propulsion Lab).

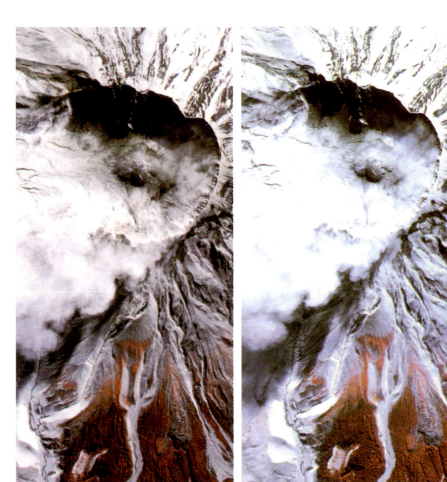

Plate 13-3   U.S. Geological Survey High Altitude Photography (HAP) color-infrared stereopair of Mount St. Helens, WA, on August 6, 1981 (photos 109-84, 85). The active lava dome in the center of the cone is visible. A sediment choked radial drainage pattern has developed. North is to the left (courtesy U.S. Geological Survey).

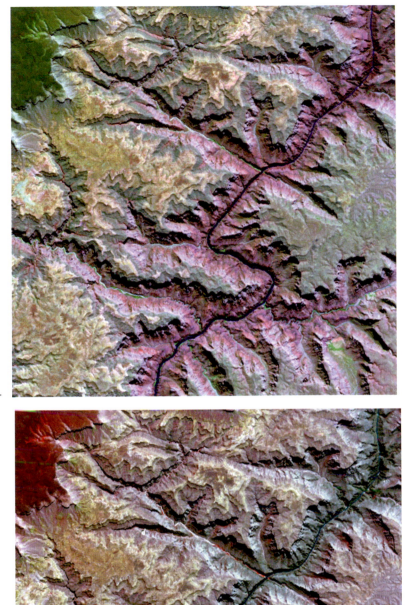

a.

b.

Plate 13-4   Landsat Thematic Mapper color composites of a portion of the Grand Canyon. a) TM Bands 7,4,2 = RGB. b)
TM Bands 4,3,2 = RGB (courtesy Space Imaging, Inc.).

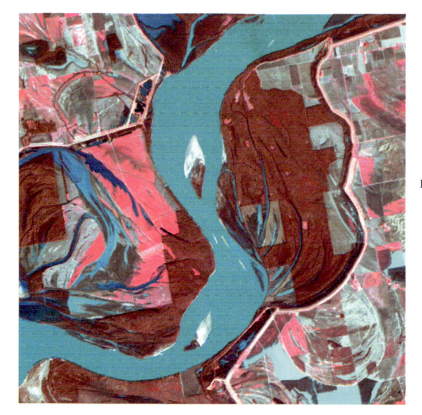

Plate 13-5   Landsat Thematic Mapper color posite (bands 4,3,2 = RGB) of th sissippi River (courtesy Space Im Inc.).

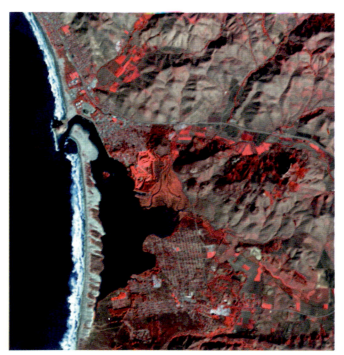

a. Landsat TM image of Morro Bay, CA (bands 4,3,2 = RGB).

b. Landsat TM image (bands 7,4,3 = RGB).

Plate 13-6   Landsat Thematic Mapper color composites of Morro Bay, CA (courtesy Space Imaging, Inc.).

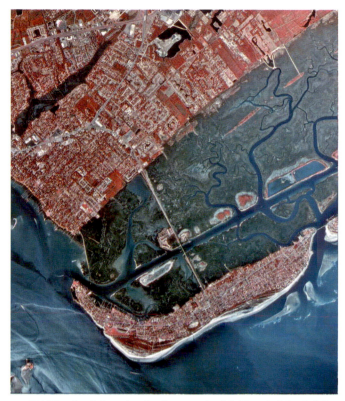

a. NAPP color-infrared orthophotograph of Sullivan's Island, SC.

b. ATLAS multispectral scanner data of the tidal flat behind Isle of Palms, SC.

Plate 13-7   a) Portion of a USGS NAPP color-infrared orthophoto quarter quadrangle of Sullivan's Island, SC (courtesy U.S. Geological Survey). b) ATLAS multispectral scanner data (2.5 x 2.5 m) of the tidal flats behind Isle of Palms, SC (bands 6,4,2 = RGB) (courtesy NASA Stennis Space Center).

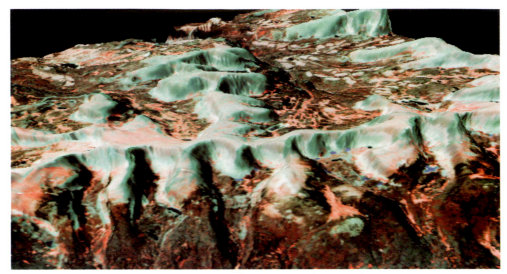

a. Uinta Mountain Range in Utah

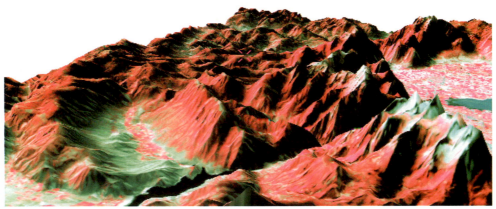

b. Wasatch Range in the Rocky Mountains of Utah

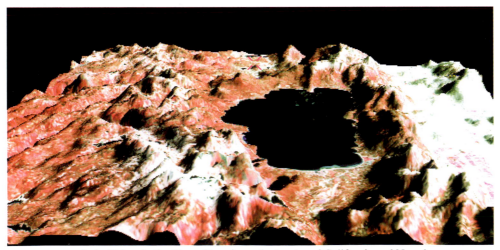

c. Lake Tahoe in the Sierra Nevada on the border of California and Nevada

Plate 13-8   Landsat color-composites of glaciated landscapes in portions of a) the Uinta Mountains in Utah, b) the Wasatch Range in Utah, and c) the Lake Tahoe region in the Sierra Nevada, CA. Please refer to the black-and-white illustrations in Chapter 13 for annotation (images courtesy U.S. Geological Survey and Space Imaging, Inc.).

Rubin, T. D., 1993, "Spectral Mapping with Imaging Spectrometers," *Photogrammetric Engineering & Remote Sensing*, 59(2):215–220.

Sampson, S. A., 1993, "Two Indices to Characterize Temporal Patterns in the Spectral Response to Vegetation," *Photogrammetric Engineering & Remote Sensing*, 59(4):511–517.

SBRC, 1994, *Space Sensors*. Goleta, CA: Santa Barbara Research Center, 33 pp.

Silvestrini, A., 1993, *Status of Landsat 6 Letter*. Lanham, MD: EO-SAT, Inc. 1 p.

Solomonson, V., 1984, "Landsat 4 and 5 Status and Results from Thematic Mapper Data Analyses," *Proceedings*, Machine Processing of Remotely Sensed Data. W. Lafayette, IN: Laboratory for the Applications of Remote Sensing, 13–18.

Solomonson, V. and D. L. Toll, 1991, "The Moderate Resolution Imaging Spectrometer—Nadir (MODIS-N) Facility Instrument," *Advances in Space Research*, 11(3):231–236.

Space Imaging, 1999a, *The New Era in the Information Age Begins*, Thornton, CO: Space Imaging, Inc., 5 pp.

Space Imaging, 1999b, *Space Imaging Catalog of Products and Services*, Thornton, CO: Space Imaging, Inc., Vol. 1, 50 pp.

Spiering, B. A., 1998, *Airborne Terrestrial Applications Sensor (ATLAS) Performance Verification Test Report*, SSC, MS: John C. Stennis Space Center, 35 pp.

SPOT, 1988, *SPOT User's Handbook*, 2 volumes. Reston, VA: SPOT Image Co., 200 pp.

SPOT, 1998a, *Spot Magazine*, Reston: Spot Image, Inc., 29: 36 pp.

SPOT, 1998b, *SPOT 4 Launch Press Brief*, Reston, VA: Spot Image, Inc., 50 pp.

SPOT, 1999a, *Satellite Imagery: An Objective Guide*, Reston: Spot Image, Inc., 31 pp.

SPOT, 1999b, *Spot home page*, http://spotimage.com.

Tan, H. L. and S. J. Hook, 1999, *ASTER home page*, Pasadena, CA: Jet Propulsion Laboratory, http://asterweb.jpl.nasa. gov/aster-home/subsystem.html.

Teillet, P. M. and B. N. Holben, 1993, "Towards Operational Radiometric Calibration of NOAA AVHRR Imagery in the Visible and Infrared Channels," *Canadian Journal of Remote Sensing*, 20(1):1–10.

Theodossiou, E. I. and I. J. Dowman, 1990, "Heighting Accuracy of SPOT," *Photogrammetric Engineering & Remote Sensing*, 56(11):1643–1649.

Toutin, T., and M. Beaudoin, 1995, "Real-Time Extraction of Planimetric and Altimetric Features from Digital Stereo SPOT Data Using a Digital Video Plotter," *Photogrammetric Engineering and Remote Sensing*, 61(1):63–68.

Townshend, J. R. G., C. O. Justice, and V. Kalb, 1987, "Characterization and Classification of South American Land Cover Types," *International Journal of Remote Sensing*, 8(8):1189–1207.

USGS, 1994, *Requirements Analysis Results for Land Cover and Land Use Data*. Reston, VA: United States Geological Survey, 76 pp.

Vane, G. and A. F. H. Goetz, 1993, "Terrestrial Imaging Spectrometry: Current Status, Future Trends," *Remote Sensing of Environment*, 44:117–126.

Thermal Infrared Remote Sensing 8

All objects that have a temperature above absolute zero (0 K) emit electromagnetic energy. Therefore, all the features we encounter in the landscape on a typical day such as vegetation, soil, rock, water, and even people emit thermal infrared electromagnetic radiation in the $3.0 - 14$ μm portion of the spectrum. As humans, we experience this thermal energy primarily through our sense of touch. For example, we can feel the thermal radiant energy from the Sun or the radiant energy from a fire on our face. However, our eyes cannot detect subtle differences in thermal infrared energy emanating from real world objects because our eyes are primarily sensitive to short-wavelength visible light from $0.4 - 0.7$ μm. Our eyes are not sensitive to the reflective infrared $(0.7 - 3.0$ μm) or thermal infrared energy $(3.0 - 14$ μm). Fortunately, engineers have developed detectors that are sensitive to thermal infrared radiation. These thermal infrared sensors allow humans to sense a previously invisible world of information as they monitor the thermal characteristics of the landscape.

When we are ill, one of the first things we do is take our temperature to determine if it is significantly different from the normal 98.6 °F (37 °C). If our temperature is elevated, it usually means something is wrong. Similarly, the various components of the landscape such as vegetation, soil, rock, water, concrete, and asphalt shingles have predictable thermal characteristics based on how they selectively absorb solar short-wavelength energy and radiate thermal infrared energy. Thermal infrared remote sensing systems record thermal infrared images that can be used to 1) determine the type of material in certain instances based on its thermal emission characteristics, and/or 2) evaluate if significant changes have taken place in the thermal characteristics of these phenomena through time. In this manner, it is possible to identify the surface physical manifestation of certain diseases in humans (e.g., perhaps the existence of a tumor), stress in plants, thermal pollution in water bodies, or the loss of heat from buildings due to faulty insulation.

Aerial thermal infrared imagery has not been widely available to the general public due to 1) the relatively high cost of the thermal sensor, 2) the cost of mobilizing an aircraft to acquire the remotely sensed data, and 3) the difficulty of calibrating and correctly interpreting the imagery. The public knows about thermal infrared remote sensing largely from close-range terrestrial applications such as those shown in Figure 8-1, including: residential and commercial heat-loss insulation studies, the use of handheld thermal imaging units to locate hot spots in homes and to find other human beings in a fire or at night, thermal images of humans to detect a variety of medical problems (sometimes referred to as *thermography),* and nondestructive testing and evaluation of electronic components. The public also sees thermal infrared imagery being used to allocate firefighting resources during a forest fire (Lytle, 1996) or to target enemy facilities as in the Gulf War in 1991 (McDaid and Oliver, 1997). The public in general does not realize that the nighttime GOES images of weather fronts displayed on the nightly news are thermal infrared images.

Aerial thermal infrared remote sensing will become more important in the future as additional orbital sensors obtain thermal data and costs decrease. Also, various government agencies such as the local police, drug enforcement and immigration border patrol officers are beginning to routinely use handheld thermal infrared sensors and forward-looking infrared sensors (FLIR) mounted in helicopters and other aircraft to look for missing persons and criminal activity.

History of Thermal Infrared Remote Sensing

The astronomer Sir Frederick William Herschel (1738–1822) discovered the infrared portion of the electromagnetic spectrum in 1800 and described it in his famous paper "Investigations of the Powers of the Prismatic Colours to Heat and Illuminate Objects: with Remarks." In 1879, S. P. Langley began a research program to find a superior radiation detector. A year later he invented the bolometer that was able to measure temperature variations of 1/10,000 °C. In World War I, S. O. Hoffman was able to detect men at 120 m and eventually aircraft. In the 1930s, Germany developed the Kiel system for discriminating between bombers and night fighters. The British and the United States also developed infrared surveillance techniques in World War II. In fact, the single most important development in infrared technology was the invention of the detector element by warring nations during World War II.

Early infrared detectors consisted of lead salt photodetectors (Fischer, 1983). Now we have very fast detectors consisting of mercury-doped germanium (Ge:Hg), indium antimonide (In:Sb) and other substances that are very sensitive to infrared radiation. We also have computers to rapidly process the energy recorded by the sensors and to display the thermal characteristics of the scene.

Thus, it took about a century for governments to understand that remote sensing in the thermal infrared region could provide valuable tactical reconnaissance information, especially since the images could be recorded both in the daytime and at night. In the 1950s, the government contracted with civilian firms to improve thermal infrared technology (e.g., Texas

Instruments). In the 1960s, some of these contractors received permission from the government to use the classified sensors to produce thermal infrared images for a few select civilian clients (Estes, 1966). In 1968, the government declassified the production of thermal infrared remote sensing systems that did not exceed a certain spatial resolution and temperature sensitivity. Thermal infrared remote sensing systems developed by Texas Instruments, Inc., Daedalus Enterprises, Inc., Rockwell International, Inc., etc., were first carried aloft by aircraft. Oil companies conducting geological exploration requested much of the early thermal infrared data and continue to be major consumers. Thermal infrared remote sensing systems mounted onboard aircraft continue to collect much of the on-demand thermal infrared data for public agencies (e.g., Environmental Protection Agency, Department of Energy, state departments of natural resources) and foreign governments.

The first declassified satellite remote sensor data were collected by the U.S. Television IR Operational Satellite (TIROS) launched in 1960. The coarse resolution thermal infrared data were ideal for monitoring regional cloud patterns and frontal movement. NASA launched the Heat Capacity Mapping Mission (HCMM) on April 26, 1978. It obtained 600 x 600 m spatial resolution thermal infrared data (10.5 – 12.6 μm) both day (1:30 p.m.) and night (2:30 a.m.). This was one of the first scientifically oriented (geology) thermal infrared systems. NASA's *Nimbus 7,* launched on October 23, 1978, had a Coastal Zone Color Scanner (CZCS) that included a thermal infrared sensor for monitoring sea-surface temperature. In 1980, NASA and the Jet Propulsion Laboratory developed the six-channel Thermal Infrared Multispectral Scanner (TIMS) that acquires thermal infrared energy in six bands at wavelength intervals of ≤ 1.0 μm (Quattrochi and Ridd, 1994). Successful studies using TIMS resulted in the development of the 15-channel Airborne Terrestrial Applications Sensor (ATLAS) (Lo et al., 1997).

The NOAA Geostationary Operational Environmental Satellite (GOES) collects thermal infrared data at a spatial resolution of 8 x 8 km for weather prediction. Full-disk images of the Earth are obtained every 30 minutes both day and night by the thermal infrared sensor. Also, the NOAA Advanced Very High Resolution Radiometer (AVHRR) collects thermal infrared local area coverage (LAC) data at 1.1 x 1.1 km and global area coverage (GAC) at 4 x 4 km.

Landsat Thematic Mapper 4 and 5 sensors were launched on July 16, 1982, and March 1, 1984, respectively, and collected 120 x 120 m thermal infrared data (10.4 – 12.5 μm) along with two bands of middle-infrared data (1.55 – 1.75

Figure 8-1   a) Thermal infrared image of radiant energy leaving a residential house, especially through the windows. Note the cool, insulated roof and metal water downspout. b) Thermal image of a firefighter. Handheld thermal imagers can be used to locate other firefighters or victims even in smoke-filled rooms. c) Thermal infrared image of a fire in the upper story of a home. d) Thermal image of nighttime crime in progress. e) Black-and-white panchromatic photograph of a parrot. f) Thermal infrared image of the same parrot. g) Thermal image of a power transformer revealing several very hot wires. h) Nondestructive thermal testing of a printed circuit board. i) Thermal image of a boat (courtesy FLIR Systems, Inc.).

and 2.08 – 2.35 μm). Landsat 7 was launched on April 15, 1999 with a 60 x 60 m well-calibrated thermal infrared sensor (10.4 – 12.5 μm). The Advanced Spaceborne Thermal Emission and Reflection Radiometer (ASTER) onboard *Terra* has six channels from 1.60 – 2.43 μm and five channels from 8.125 – 11.65 μm (Herring, 1998). Chapter 7 provides detailed information about the spatial, spectral, temporal, and radiometric characteristics of these and other thermal infrared sensor systems.

Thermal Infrared Radiation Properties

An image analyst should not interpret a thermal infrared image as if it were an aerial photograph or a typical image produced by an optical-mechanical multispectral scanner or charge-coupled-device (CCD) sensor system. Rather, the analyst must think *thermally*. He or she must understand 1) how the short-wavelength energy radiated from the Sun interacts with the atmosphere, 2) how it interacts with Earth surface materials (i.e., some of the energy is transformed into longer-wavelength energy), 3) how the energy emitted by the terrain interacts with the atmosphere once again, and finally, 4) how a remote sensing detector records the *emitted* thermal infrared electromagnetic radiation. The analyst should also understand how both the sensor system itself and the terrain can introduce noise into the thermal infrared image that might make the data less useful or lead to incorrect image interpretation.

Kinetic Heat, Temperature, Radiant Energy, and Radiant Flux

All objects in the real world having a temperature above absolute zero (0 K; -273.16 °C; -459.69 °F) exhibit random motion. The energy of particles of molecular matter in random motion is called *kinetic heat* (also referred to as internal, real, or true heat). When these particles collide they change their energy state and emit electromagnetic radiation as discussed in Chapter 2. The amount of heat can be measured in *calories*. We can measure the true kinetic temperature (T_{kin}) or concentration of this heat using a thermometer. We perform *in situ* (in-place) temperature measurement by placing the thermometer in direct physical contact with a plant, soil, rock or water body.

Fortunately for us, an object's internal kinetic heat is also converted to *radiant energy* (often called external or apparent energy), which allows us to utilize remote sensing tech-

nology. The electromagnetic radiation exiting an object is called *radiant flux* (Φ) and is measured in Watts as discussed in Chapter 2. The concentration of the amount of radiant flux exiting (emitted from) an object is its *radiant temperature* (T_{rad}). For most real world objects (except those composed of glass and metal) there is usually a high positive correlation between the true kinetic temperature of the object (T_{kin}) and the amount of radiant flux radiated from the object (T_{rad}). Therefore, we can utilize radiometers placed some distance from the object to measure its radiant temperature, which hopefully correlates well with the object's true kinetic temperature. *This is the basis of thermal infrared remote sensing.* Unfortunately, the relationship is not perfect, with the remote measurement of the radiant temperature always being somewhat less than the true kinetic temperature of the object. This is due to a thermal property called *emissivity*, to be discussed shortly.

Methods of Transferring Heat

The heat generated by the random motion of particles may be transferred from one location to another by conduction, convection, and radiation, as discussed in Chapter 2. Thermonuclear fusion taking place on the Sun produces a plasma of radiant flux consisting primarily of short-wavelength visible light that travels 93 million miles through the vacuum of space at the speed of light (3×10^8 m sec^{-1}) (Trefil and Hazen, 1995). Some of this short-wavelength energy passes through the atmosphere and is absorbed by the Earth's surface materials and reradiated (emitted) at longer wavelengths. Some of this emitted longer-wavelength electromagnetic radiation passes through the atmosphere once again and can be recorded using airborne thermal infrared detectors. Hopefully, the longer-wavelength radiation recorded by the detectors provides valuable information about the temperature characteristics of the Earth's surface.

Thermal Infrared Atmospheric Windows

Beyond the *visible* region of the electromagnetic spectrum, we encounter the *reflective infrared* region from 0.7 – 3 μm and the *thermal infrared* region from 3 – 14 μm (Figure 8-2). The only reason we can use remote sensing instruments to detect infrared energy in these regions is because the atmosphere allows a portion of the infrared energy to be transmitted from the terrain to the detectors. We call regions that pass energy *atmospheric windows*. Conversely, black areas in Figure 8-2 denote regions of the spectrum where the atmo-

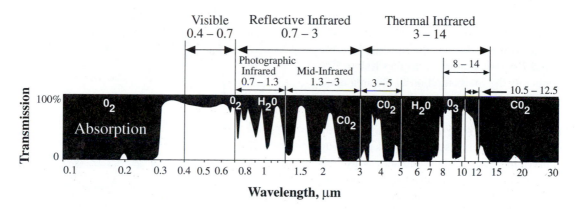

Figure 8-2   Atmospheric windows in the electromagnetic spectrum shown here are of significant value for remote sensing reflective and thermal infrared energy. Photographic films can be made sensitive to reflective energy from 0.7 – 3 μm. Electro-optical sensing systems can record infrared energy from 0.7 – 14 μm. The 3 – 5 μm region is especially useful for monitoring hot targets such as forest fires and geothermal activity. Vegetation, soil, and rock are best monitored using the 8 – 14 μm region for suborbital data collection. The 10.5 – 12.5 μm region is used when thermal imagery is acquired from orbital sensors above the Earth's ozone layer.

sphere absorbs most of the infrared energy present. We call these regions *absorption bands*. Water vapor (H_2O), carbon dioxide (CO_2), and ozone (O_3) are responsible for most of the absorption. The atmosphere "closes down," making it almost impossible to perform remote sensing of the environment in these regions. For example, atmospheric water vapor (H_2O) absorbs most of the energy exiting the terrain in the region from 5 – 7 μm, making it almost useless for thermal infrared remote sensing.

Remote sensing instruments can be engineered to be sensitive to the infrared energy present within just the atmospheric windows. For example, film emulsions can be made sensitive to reflected infrared energy in the window from 0.7 – 1.3 μm. Eastman Kodak's 2443 color infrared film works within this *photographic infrared* region and is ideal for monitoring vegetation and water (Chapter 4). Electro-optical detectors on Landsat Thematic Mapper 4 and 5 are sensitive to the reflective middle-infrared windows from 1.55 – 1.75 μm (TM band 5) and 2.08 – 2.35 μm (TM band 7).

Electronic detectors can also be made sensitive to photons of thermal infrared radiant energy exiting the terrain in the two primary thermal infrared windows: 3 – 5 μm and 8 – 14 μm. Suborbital thermal infrared remote sensing systems utilize these spectral bands. However, the Earth's ozone (O_3) layer absorbs much of the thermal energy exiting the terrain in an absorption band from approximately 9.2 – 10.2 μm. Therefore, satellite thermal infrared remote sensing systems often

only record data in the region from 10.5 – 12.5 μm (Figure 8-2) to avoid this absorption band. For example, ASTER band 12 is 8.925 – 9.275 μm and band 13 is 10.25 – 10.95 μm. The region from 9.276 – 10.24 μm is not sensed due to atmospheric absorption.

 Thermal Radiation Laws

A *blackbody* is a theoretical construct that absorbs all the radiation that falls on it and radiates energy at the maximum possible rate per unit area at each wavelength for any given temperature (Mulligan, 1980). No objects in nature are true blackbodies; however, we may think of the Sun as approximating a 6,000 K blackbody and the Earth as a 300 K blackbody. If we pointed a sensor at a blackbody, we would be able to record quantitative information about the total amount of radiant energy in specific wavelengths exiting the object and the dominant wavelength of the object. In order to do this, we utilize two important physical laws: the *Stefan-Boltzmann law* and *Wien's displacement law*.

Stefan-Boltzmann Law

The total spectral radiant exitance (M_b) measured in Watts m^{-2} leaving a blackbody (refer to Table 2-4) is proportional to the fourth power of its temperature (T). This is known as the *Stefan-Boltzmann law* and is expressed as (Mulligan, 1980):

$$M_b = \sigma T^4 \qquad (8\text{-}1)$$

where σ is the Stefan-Boltzmann constant equaling 5.6697 x 10^{-8} W m^{-2} K^{-4}, and T is temperature in degrees Kelvin. The total radiant exitance is the integration of all the area under the blackbody radiation curve (Figure 8-3). Notice how the Sun produces more spectral radiant exitance (M_b) at 6,000 K than the Earth at 300 K. As the temperature increases, the total amount of radiant energy measured in Watts per m^2 (the area under the curve) increases and the radiant energy peak shifts to shorter wavelengths. To determine the dominant wavelength for a blackbody at a specific temperature, we use Wien's displacement law.

Wien's Displacement Law

The relationship between the true temperature of a blackbody (T) in degrees Kelvin and its peak spectral exitance or dominant wavelength (λ_{max}) is described by *Wien's displacement law*:

$$\lambda_{max} = \frac{k}{T} \qquad (8\text{-}2)$$

where k is a constant equaling 2898 μm K. We can determine the dominant wavelength of any object by substituting its temperature into Equation 8-2. Remember from Chapter 2 and Figure 8-3 that the dominant wavelength of the 6,000 K Sun is 0.48 μm. The dominant wavelength for an 800 K red-hot object (Figure 8-3) is:

$$\lambda_{max} = \frac{2898 \mu m \cdot {}^\circ K}{800 \cdot {}^\circ K}$$

$$\lambda_{max} = 3.62 \mu m$$

We see a shift from longer to shorter wavelengths as the temperature of the blackbody increases. We can observe Wien's displacement law in real life. For example, when a poker is placed in the fire the tip progresses from dark red through orange and then to yellow. It never shifts into the green or blue portion of the spectrum because it is not that hot. Conversely, an acetylene torch has a hot flame and appears blue.

Why is knowing an object's dominant wavelength important to thermal infrared remote sensing? The dominant wavelength provides valuable information regarding the part of the thermal infrared spectrum in which we might want to

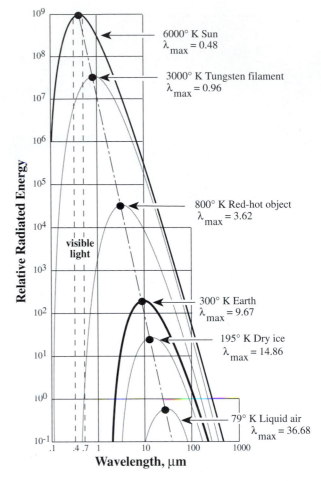

Figure 8-3 Blackbody radiation curves for several objects, including the Sun and the Earth, which approximate 6,000 K and 300 K blackbodies, respectively. The dominant wavelength, λ_{max}, shifts toward the short wavelength portion of the spectrum as the temperature of the object increases.

sense the object. For example, if we are looking for 800 K forest fires that have a dominant wavelength of approximately 3.62 μm, then the most appropriate remote sensing system might be a 3 – 5 μm thermal infrared detector. Conversely, if we are interested in soil, water, and rock ambient temperatures on the Earth's surface (300 K) with a dominant wavelength of 9.67 μm, then a thermal infrared detector operating in the 8 – 14 μm region might be most appropriate.

Emissivity

The world is not composed of radiating blackbodies. Rather, it is composed of *selectively radiating bodies* such as rock,

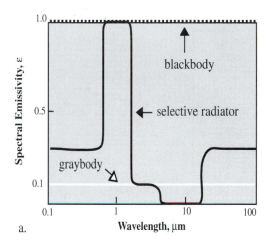

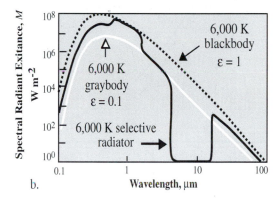

a.

b.

Figure 8-4   a) Spectral emissivity of a blackbody, a graybody, and a hypothetical selective radiator. b) Spectral radiant exitance distribution of the blackbody, graybody, and hypothetical selective radiator (after Slater, 1980).

soil, vegetation, and water that emit a certain proportion of the energy emitted from a blackbody at the same temperature. *Emissivity*, (ε) is the ratio between the radiant flux exiting a real world selective radiating body (M_r) and a blackbody at the same temperature (M_b):

$$\varepsilon = \frac{M_r}{M_b}. \tag{8-3}$$

All selectively radiating bodies have emissivities ranging from 0 to ≤ 1 that fluctuate depending upon the wavelengths of energy being considered. A *graybody* outputs a constant emissivity that is less than one at all wavelengths. Figure 8-4a depicts the emissivity of a blackbody, a graybody, and a hypothetical selective radiator over the wavelength interval 0.1 – 100 μm (Slater, 1980). Notice how the spectral emissivity of the hypothetical selective radiator fluctuates among levels of 0, 0.1, 0.3, and 1.0. The spectral radiant exitance of each of these 6,000 K bodies is shown in Figure 8-4b. Notice

how the different emissivities give rise to dramatic changes in the spectral radiant exitance distribution of the selective radiator. Where the selective radiator's emissivity is 1.0, it outputs the same amount of radiant energy as the blackbody. Where the selective radiator's emissivity is 0, it emits no spectral radiant exitance.

The spectral radiant exitance for several real world radiating bodies is shown in Figure 8-5. Note that the radiant energy exiting the substance is approximately the same as a blackbody at the same temperature for much of the spectral range, but that the curves depart in certain areas. If the area beneath each of the curves was summed (integrated) over the spectral wavelength interval of interest on the x-axis, we would find that the real world spectral radiant exitance was always less than the blackbody radiant exitance at the same temperature. Thus, the emissivity of the real world material would lie somewhere between 0 and 1 but would never be equal to 1. Some materials like distilled water have emissivities close to one (0.99) over the wavelength interval from 8 – 14 μm, as summarized in Table 8-1. Others such as polished aluminum (0.08) and stainless steel (0.16) have very low emissivities.

Why is it important to know about emissivity when conducting a thermal infrared remote sensing investigation? The reason is that two objects lying next to one another on the ground could have the same true kinetic temperature but have different apparent temperatures when sensed by a thermal radiometer simply because their emissivities are different. The emissivity of an object may be influenced by a number of factors, including:

• color — darker colored objects are better absorbers and emitters (i.e., they have a higher emissivity) than lighter colored objects, which tend to reflect more of the incident energy.

• surface roughness — the greater the surface roughness of an object relative to the size of the incident wavelength, the greater the surface area of the object and potential for absorption and reemission of energy.

• moisture content — the more moisture an object contains, the greater its ability to absorb energy and become a good emitter. Wet soil particles have a high emissivity similar to water.

• compaction — the degree of soil compaction can affect emissivity.

• field of view — the emissivity of a single leaf measured with a very high resolution thermal radiometer will have a

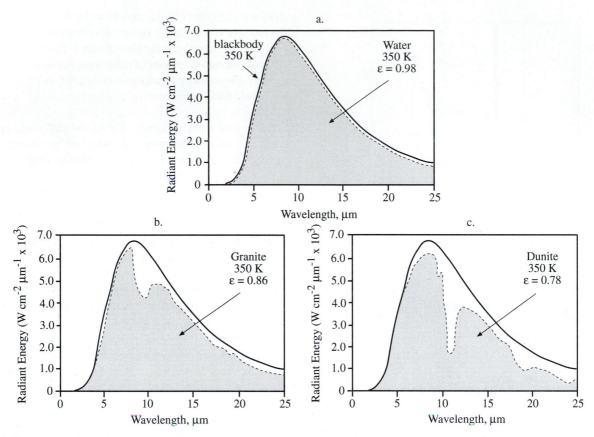

Figure 8-5   Radiant energy exiting a) water, b) granite, and c) dunite heated to 350 K compared with a blackbody at the same temperature.

different emissivity than an entire tree crown viewed using a more coarse spatial resolution radiometer.

- wavelength — the emissivity of an object is generally considered to be wavelength-dependent. For example, while the emissivity of an object is often considered to be constant throughout the 8 – 14 μm region, its emissivity in the 3 – 5 μm region may be different.

- viewing angle — the emissivity of an object can vary with sensor viewing angle.

Salisbury and D'Aria (1992) computed the emissivity for a variety of terrestrial materials in the 8 – 14 μm region, including igneous, metamorphic, and sedimentary rocks, desert varnish, soil, vegetation, water, and ice. Snyder et al. (1997) provided additional information on the bidirectional reflectance measurements of sands and soils in the 3 – 14 μm region. We must take into account an object's emissivity when we use our remote radiant temperature measurement to measure the object's true kinetic temperature. This is done by applying Kirchoff's radiation law.

Kirchoff's Radiation Law

In Chapter 2 we learned that the terrain intercepts incident (incoming) radiant flux (Φ_i). This incident energy interacts with terrain materials. The amount of radiant flux reflected from the surface (Φ_r), the amount of radiant flux absorbed by the surface (Φ_α), and the amount of radiant flux transmitted through the surface (Φ_τ) can be carefully measured as we apply the principle of conservation of energy and attempt to keep track of what happens to all the incident energy. The general equation for the interaction of *spectral (λ)* radiant flux with the terrain is:

$$\Phi_{i_\lambda} = \Phi_{r_\lambda} + \Phi_{\alpha_\lambda} + \Phi_{\tau_\lambda}. \tag{8-4}$$

Dividing each of the variables by the original incident radiant flux, Φ_{i_λ}:

$$\frac{\Phi_{i_\lambda}}{\Phi_{i_\lambda}} = \frac{\Phi_{r_\lambda}}{\Phi_{i_\lambda}} + \frac{\Phi_{\alpha_\lambda}}{\Phi_{i_\lambda}} + \frac{\Phi_{\tau_\lambda}}{\Phi_{i_\lambda}} \tag{8-5}$$

Table 8-1   Emissivity of Selected Materials from 8 – 14 µm
(Lyon, 1965; Marsh and Dozier, 1981; McDonnell
Douglas, 1982; Curran, 1985; Lillesand and Kiefer,
1994; Sabins, 1997).

Material	Emissivity, ε
water, distilled	0.99
water	0.92 – 0.98
water with petroleum film	0.972
concrete	0.71 – 0.90
asphalt	0.95
tar/stone	0.97
loamy soil, dry	0.92
loamy soil, wet	0.95
soil, sandy	0.90
brick, red and rough	0.93
vegetation, closed canopy	0.98
vegetation, open canopy	0.96
grass	0.97
wood, planed oak	0.90
deciduous forest	0.97 – 0.98
coniferous forest	0.97 – 0.99
stainless steel	0.16
aluminum, foil	0.05
aluminum, polished	0.08
aluminum, paint	0.55
polished metals	0.16 – 0.21
oxidized steel	0.70
granite	0.86
dunite	0.78
basalt, rough	0.95
snow	0.83 – 0.85
paint	0.90 – 0.96
human skin	0.98

allows us to rewrite Equation 8-4 as:

$$1 = r_\lambda + \alpha_\lambda + \tau_\lambda \tag{8-6}$$

where r_λ is spectral hemispherical reflectance by the terrain, α_λ is spectral hemispherical absorptance, and τ_λ is spectral hemispherical transmittance (Janza, 1975; Slater, 1980).

The Russian physicist *Kirchoff* found that in the infrared portion of the spectrum the spectral emissivity of an object generally equals its spectral absorptance, i.e., $\alpha_\lambda = \varepsilon_\lambda$. This is often phrased as *"good absorbers are good emitters and good reflectors are poor emitters"* (Kahle, 1980). Also, most real world materials are usually opaque to thermal infrared radiation, meaning that little radiant flux exits from the other side of the terrain element. Therefore, we may assume transmittance, $\tau_\lambda = 0$. Substituting emissivity for absorptance and removing transmittance from the equation yields (Schott, 1997):

$$1 = r_\lambda + \varepsilon_\lambda. \tag{8-7}$$

This relationship is important because it describes why objects appear as they do on thermal infrared imagery. *Because the terrain theoretically does not lose any incident energy to transmittance, all energy leaving the object must be accounted for by the relationship between reflectance (r_λ) and emissivity (ε_λ). If reflectivity increases, then emissivity must decrease. If emissivity increases, then reflectivity must decrease.* For example, water absorbs almost all incident energy and reflects very little. Therefore, water is a very good emitter and has a high emissivity close to 1. Conversely, a sheet-metal roof reflects most of the incident energy and absorbs very little, yielding an emissivity much less than 1. Therefore, metal objects such as cars, aircraft, and metal roofs usually appear very cold (dark) on thermal infrared imagery. For example, the metal hangar and aircraft in the nighttime thermal infrared imagery in Figure 8-6 appear cold. Note that several aircraft have their engines on, which appear bright, and that their jet blast is warming the tarmac.

The goal of thermal infrared remote sensing is to be able to point a radiometer at an object and have the recorded apparent radiant temperature (T_{rad}) equal the true kinetic temperature of the object (T_{kin}). Unfortunately, the radiant flux from a real world object at a given temperature is not the same as the radiant flux from a blackbody at the same temperature, largely due to the effects of emissivity. Knowing the emissivity of an object makes it possible to modify the Stefan-Boltzmann law originally applicable to blackbodies ($M_b = \sigma T^4$) so that it pertains to the total spectral radiant flux of real world materials (M_r):

$$M_r = \varepsilon \sigma T_{kin}^4. \tag{8-8}$$

The equation takes into account the temperature of the object and its emissivity to create a more accurate estimate of the radiant flux exiting an object and recorded by the thermal infrared sensor.

Figure 8-6   Nighttime thermal infrared imagery of an airport with several different types of jet aircraft. The metal hangar and the aircraft appear cool (dark) because of metal's low emissivity. The concrete tarmac has a relatively high emissivity and appears in lighter shades of gray. Seven of the aircraft jet engines are active, as evidenced by the bright bloom along their fuselages and their jet-wash heating the tarmac. One or two aircraft engines were recently turned off.

Thermal infrared remote sensing systems generally record the apparent radiant temperature, T_{rad} of the terrain rather than the true kinetic temperature, T_{kin}. If we assume that the incorporation of emissivity in Equation 8-1 has improved our measurement to the point that

$$M_r = \varepsilon \, \sigma \, T_{kin}^{\;4} \text{ and we assume that}$$

$$M_b = \sigma \, T_{rad}^{\;4} \text{ and}$$

$$M_r = M_b \text{ then,}$$

$$\sigma \, T_{rad}^{\;4} = \varepsilon \, \sigma \, T_{kin}^{\;4}. \qquad (8\text{-}9)$$

Therefore, the radiant temperature of an object recorded by a remote sensor is related to its true kinetic temperature and emissivity by the following relationship (Sabins, 1997):

$$T_{rad} = \varepsilon^{1/4} \, T_{kin} \qquad (8\text{-}10)$$

The relationship between true kinetic and radiant temperature for several different types of material are summarized in Table 8-2. It is clear from this table that if the effect of emissivity is not accounted for when analyzing remotely sensed apparent radiant temperature (T_{rad}), then the true temperature (T_{kin}) of the object will be underestimated (Curran, 1985). What if we wanted to determine the emissivity of the various objects within our study area? This can be done using a thermometer and a handheld thermal infrared radiometer. If we measure an object's true temperature with a thermometer (T_{kin}) and its apparent radiant temperature (T_{rad}) with a thermal radiometer, we can estimate its emissivity, ε, using the equation:

$$\varepsilon = \left(\frac{T_{rad}}{T_{kin}} \right)^4. \qquad (8\text{-}11)$$

Ideally we would collect both the *in situ* temperature measurement and the remote measurement simultaneously. Also, to get the best emissivity approximation, it would be good to

Table 8-2   Emissivity, True Kinetic, and Radiant Temperatures
of Selected Materials at 300 K and 27°C

Material	Emissivity	True Kinetic Temperature, T_{kin}		Radiant Temperature, T_{rad} $T_{rad} = \varepsilon^{1/4} T_{kin}$	
	ε	K	°C	K	°C
blackbody	1.00	300	27	300.0	27.0
distilled water	0.99	300	27	299.2	26.2
rough basalt	0.95	300	27	296.2	23.2
vegetation	0.98	300	27	298.5	25.5
dry loam soil	0.92	300	27	293.8	20.8

Table 8-3   Thermal Properties of Common Materials at 20 °C
(Janza, 1975)

	Thermal conductivity K	Thermal density p	Thermal capacity c	Thermal inertia P
Materials				
glass	0.0021	2.6	0.16	0.029
water	0.0013	1.0	1.0	0.036
wood	0.0050	0.5	0.327	0.009
Geologic Materials				
basalt	0.0050	2.8	0.20	0.053
dolomite	0.0120	2.6	0.18	0.075
granite	0.0075	2.6	0.16	0.056
gravel, sandy	0.0060	2.1	0.20	0.050
limestone	0.0048	2.5	0.17	0.045
obsidian	0.0030	2.4	0.17	0.035
sandstone	0.0120	2.5	0.19	0.075
shale	0.0042	2.3	0.17	0.041
slate	0.0050	2.8	0.17	0.049
soil, sandy	0.0014	1.8	0.24	0.024
soil, clay moist	0.0030	1.7	0.35	0.042

collect the remote measurement from a helicopter to try and
simulate 1) the size of the remote sensor IFOV of interest
(e.g., 20 x 20 m), and 2) the atmospheric effects that might
be encountered.

Thermal Properties of Terrain

Water, rock, soil, vegetation, the atmosphere, and human tissue all have the ability to conduct heat directly through them
(thermal conductivity) onto another surface and to store heat
(thermal capacity). Some materials respond to changes in
temperature more rapidly or slowly than others (thermal
inertia). It is useful to review these thermal properties, as
they have an impact on our ability to remotely sense thermal
information about various types of materials.

Heat or *Thermal capacity* (*c*) is a measure of the ability of a
material to absorb heat energy. It is the quantity of heat
required to raise the temperature of one gram of that material
by 1 °C (cal g^{-1} °C^{-1}) (Trefil and Hazen, 1995). Table 8-3
summarizes the thermal capacity of several materials at 20
°C. Water has the largest heat capacity of any common substance (1.00). The temperature of a lake usually varies very
little between night and day. Conversely, rocks do not store
heat well and exhibit significantly different temperatures in
the night and day.

Thermal conductivity (*K*) is a measure of the rate that a substance transfers heat through it (Campbell, 1996). It is measured as the number of calories that will pass through 1 cm^3
of material in 1 second when two opposite faces are maintained at 1°C difference in temperature (cal cm^{-1} sec^{-1} °C^{-1}).
The conductivity of a material can be variable due to the
amount of moisture present. The thermal conductivity of a
variety of materials is summarized in Table 8-3. Notice that
many rocks and soils are poor conductors of heat.

Thermal inertia (*P*) is a measurement of the thermal
response of a material to temperature changes and is measured in calories per cm^2 per second square root per 1°C (cal
cm^{-2} sec$^{-1/2}$ °C^{-1}). Thermal inertia is computed using the
equation

$$P = \sqrt{(K \cdot p \cdot c)} \qquad (8-12)$$

where *K* is thermal conductivity, *p* is density (g cm^{-3}), and *c*
is thermal capacity. Density is a very important biophysical
variable in this equation because thermal inertia generally
increases linearly with increasing material density. Table 8-3
summarizes the thermal inertia of a variety of materials.

It would be wonderful if we could remotely sense each of the aforementioned variables and then simply compute thermal inertia. Unfortunately, this is not the case, because conductivity, density, and thermal capacity must all be measured *in situ*. Nevertheless, it is possible to remotely sense and compute an *apparent thermal inertia* measurement per pixel in the following manner. A thermal infrared image is acquired over the same terrain in the nighttime and in the early daytime. The two images are geometrically and radiometrically registered to one another, and the change in temperature, ΔT, for a specific pixel is determined by subtracting the nighttime apparent temperature from the daytime apparent temperature. The apparent thermal inertia (ATI) per pixel is

$$ATI = \frac{1-A}{\Delta T} \qquad (8\text{-}13)$$

with A being the albedo (reflectance) measured in the visible spectrum during the daytime for the pixel of interest (Kahle et al., 1981; Sabins, 1997).

The best way to think about thermal inertia is to associate it with an inverse relationship with the measured temperature change, ΔT. Basically, a high ΔT value is usually associated with terrain materials that have a low thermal inertia value. Conversely, a low ΔT is usually associated with terrain materials that have a high thermal inertia value.

Geologists and other remote sensing scientists label areas in the image that have heterogeneous or homogeneous apparent thermal inertia characteristics to distinguish boundaries between bedrock and alluvial material, discriminate among rock units with similar spectral properties, and identify zones of hydrothermal alteration (Abrams et al., 1984; Kahle et al., 1984). However, Price (1985) cautions that apparent thermal inertia images should not be used in regions having variability in surface moisture (evaporation) like agricultural areas.

One of the first thermal infrared satellite remote sensing systems to collect both day (1:30 p.m.) and nighttime (2:30 a.m.) thermal infrared images of significant value for apparent thermal inertia mapping was the short-lived 1978 Heat Capacity Mapping Mission (HCMM) that acquired 600 x 600 m data in the region from 10.5 – 12.6 μm. The ASTER sensor onboard *Terra* collects five bands of day and nighttime thermal infrared data with 90 x 90 m spatial resolution and a 60 km swath width (Herring, 1998). Day and nighttime thermal infrared imagery is routinely collected using NASA TIMS and ATLAS sensors onboard suborbital aircraft (Quattrochi and Ridd, 1998).

 Thermal Infrared Data Collection

Thermal infrared remote sensor data may be collected by:

- across-track thermal scanners, and

- pushbroom linear and area-array charge-coupled-device (CCD) detectors.

It is useful to review the nature of thermal infrared sensor systems and their components and how various system parameters influence the type and quality of the thermal infrared data collected.

Thermal Infrared Multispectral Scanners

Chapter 7 introduced how multispectral scanners function. This section provides additional information about thermal infrared scanners.

Daedalus DS-1260, DS-1268, Airborne Multispectral Scanner (AMS), TIMS, and ATLAS

These scanners have provided much of the useful high spatial and spectral resolution thermal infrared data for monitoring the environment. The DS-1260 records data in 10 bands including a thermal infrared channel (8.5 – 13.5 μm). The DS-1268 incorporates the Landsat Thematic Mapper middle-infrared bands (1.55 – 1.75 μm and 2.08 – 2.35 μm). The AMS contains a hot-target, thermal infrared detector (3.0 – 5.5 μm) in addition to the standard thermal infrared detector (8.5 – 12.5 μm).

During the 1980s and early 1990s, many scientists utilized thermal infrared imagery acquired by the NASA Thermal Infrared Multispectral Scanner (TIMS) which had six bands ranging from 8.2 – 12.2 μm (Palluconi and Meeks, 1985). Many scientists now use the NASA Airborne Terrestrial Applications Sensor (ATLAS), which has six visible and near-infrared bands, two Thematic Mapper middle-infrared bands, and six thermal infrared bands from 8.2 – 12.2 μm. Specific bandwidths are summarized in Chapter 7. Both the TIMS and ATLAS sensors have a 2.0 milliradian instantaneous field of view (Wallace, 1999). Both sensor systems are operated by the NASA Stennis Space Center.

The basic principles of operation and components of the AMS, TIMS, and ATLAS are shown in Figure 8-7. The

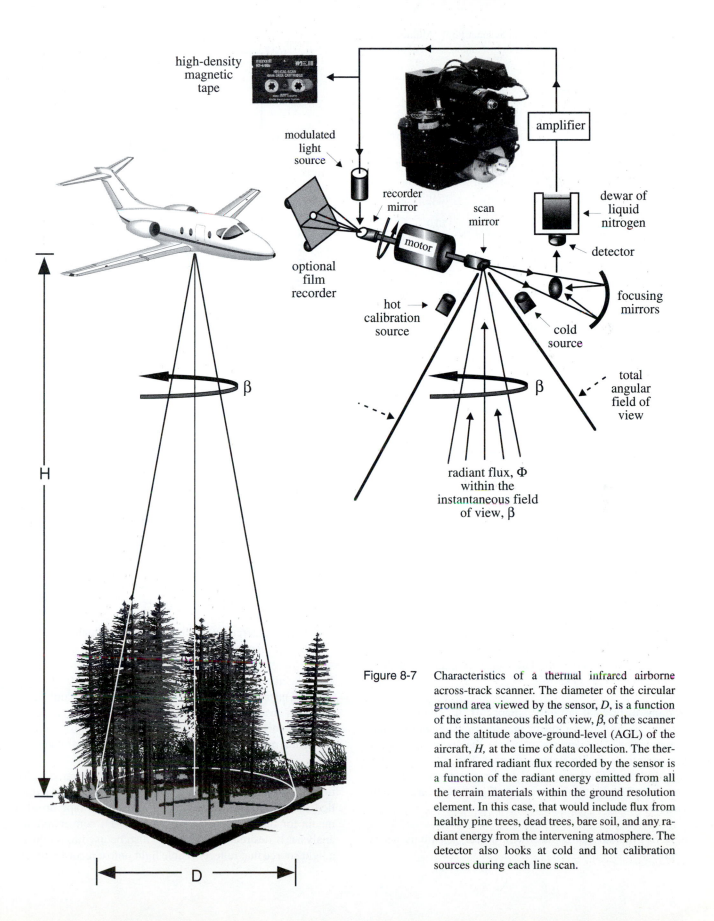

high-density magnetic tape

modulated light source

recorder mirror

optional film recorder

amplifier

scan mirror

dewar of liquid nitrogen

detector

motor

hot calibration source

cold source

focusing mirrors

total angular field of view

β

β

radiant flux, Φ within the instantaneous field of view, β

H

D

Figure 8-7 Characteristics of a thermal infrared airborne across-track scanner. The diameter of the circular ground area viewed by the sensor, *D*, is a function of the instantaneous field of view, β, of the scanner and the altitude above-ground-level (AGL) of the aircraft, *H*, at the time of data collection. The thermal infrared radiant flux recorded by the sensor is a function of the radiant energy emitted from all the terrain materials within the ground resolution element. In this case, that would include flux from healthy pine trees, dead trees, bare soil, and any radiant energy from the intervening atmosphere. The detector also looks at cold and hot calibration sources during each line scan.

Table 8-4   Aircraft Multispectral Scanner Flight Altitudes and Pixel Size Based on An Instantaneous Field of View of 2.5 milliradians

Flight Altitude AGL, m	Pixel Size, m
1,000	2.5
2,000	5.0
4,000	10.0
6,000	15.0
20,000	50.0
50,000	125.0

diameter of the circular ground area viewed by the sensor, D, is a function of the instantaneous field of view, β, of the scanner measured in milliradians (mrad) and the altitude of the scanner above-ground-level, H, where

$$D = H \times \beta. \qquad (8\text{-}14)$$

For example, if the IFOV of the scanner is 2.5 mrad, the ground size of the pixel in meters is a product of the IFOV (0.0025) and the altitude AGL in meters. Table 8-4 presents flight altitudes and corresponding pixel sizes at nadir (the point directly below the aircraft) for an IFOV of 2.5 mrad. IFOVs typically range from 0.5 – 5 milliradians.

When performing *across-track* scanning, an electric motor is oriented parallel with the aircraft fuselage and direction of flight and drives a 45° scanning mirror facet located at the end of the shaft which has a precise instantaneous field of view (e.g., 2.5 mrad). The mirror scans the terrain at a right angle (perpendicular) to the direction of flight. The mirror normally sweeps out a *total angular field of view* of 90 – 120° during each scan, depending on the sensor system. The mirror also views internal *hot* and *cold calibration sources* (targets) during each scan. The exact temperature of these calibration sources is known.

Photons of thermal infrared radiant flux, Φ, emitted by the terrain, are routed to a mirror that focuses the photons onto the *detector*. The detector converts the incoming radiant energy into an analog electrical signal. The greater the number of photons impacting the detector, the greater the signal strength. The infrared detectors (Figure 8-8) are usually composed of:

- *In:Sb* (indium antimonide) with a peak sensitivity near 5 μm;

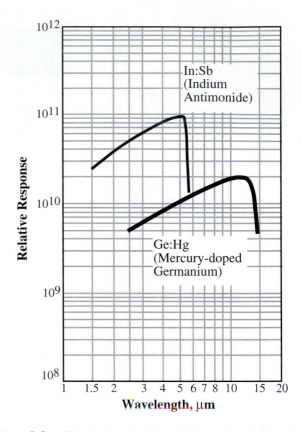

Figure 8-8   The peak spectral sensitivity of an indium antimonide (*In:Sb*) thermal infrared detector is approximately 5.0 μm, while the peak sensitivity of a mercury-doped germanium (*Ge:Hg*) detector is approximately 10 μm (after McDonnell Douglas, 1982).

- *Ge:Hg* (mercury-doped germanium) with a peak sensitivity near 10 μm, or

- *Hg:Cd:Te* (mercury-cadmium-telluride) which is sensitive over the range from 8 – 14 μm.

The detectors are cooled to low temperatures (-196°C; -243 °F; 73 K) using liquid helium or liquid nitrogen. Cooling the detectors ensures that the radiant energy (photons) recorded by the detectors comes from the terrain and not from the ambient temperature of objects within the scanner itself.

The Earth does not emit very much thermal infrared radiation; therefore, the relatively weak signal is usually amplified. The signal is then recorded on magnetic tape or other media for future analog-to-digital (A-to-D) conversion and analysis. If desired, the signal can also be used to modulate a light source that reflects visible light onto a recorder mirror located at the other end of the motorized shaft. Here the process is reversed and visible light radiant flux proportional to the amount of infrared energy received is used to expose photographic film pixel-by-pixel and line-by-line, creating a thermal infrared image of the terrain. To properly expose the photographic film, it must be advanced forward in relation to how fast the shaft is turning. The hard-copy thermal infrared image may be processed as a negative or positive print.

It is important to remember that the infrared radiant flux recorded by the sensor system is an integration of all the radiant flux emitted from the various materials within the IFOV and any radiant flux that the atmosphere might scatter into the IFOV of the sensor. For example, radiant flux emitted from the healthy pine trees, dead trees, bare soil, and the atmosphere would be integrated into a single measurement of the terrain shown in Figure 8-7.

The following factors should be considered when collecting aircraft MSS thermal infrared data:

- There is an inverse relationship between having high spatial resolution and high radiometric resolution when collecting thermal infrared data. The larger the radiometer instantaneous field of view, β, the longer the *dwell time* that an individual detector views the terrain within the IFOV during a single sweep of the mirror. A larger IFOV provides good *radiometric resolution,* which is the ability to discriminate between very small differences in radiant energy exiting the terrain element. In fact, the radiant energy *signal* measured may well be much stronger than any *noise* introduced from the sensor system components. When this takes place, we say that we have a good *signal-to-noise ratio.* Of course, the larger the IFOV, the poorer the ability to resolve fine spatial detail. Selecting a smaller IFOV will increase the spatial resolution, but the sensor will dwell a shorter time on each terrain element during a sweep of the mirror, resulting in poorer radiometric resolution and perhaps a poorer signal-to-noise ratio.

- Cutting in half the distance of a remote sensing detector from a point source quadruples the infrared energy received by that detector. The *inverse-square law* states that "the intensity of radiation emitted from a point source varies as the inverse square of the distance between source and receiver." Thus, we can obtain a more intense, strong thermal infrared signal if we can get the remote sensor detector as close to the ground as practical. For example, consider a blackbody point source, *S,* and two remote detectors (D_1 and D_2) of equal sensitive area, say, 1 cm^2. Detector D_1 is a distance *d* cm from *S,* and detector D_2 is at

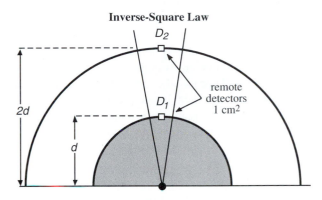

Inverse-Square Law

Blackbody Point Source, *S*

Figure 8-9   The inverse-square law reveals that the intensity of thermal radiation emitted from a blackbody point source, *S,* varies as the inverse square of the distance, *d,* between the source and remote detector receivers, D_1 and D_2.

a distance *2d* cm from *S* (Figure 8-9). From the Stefan-Boltzmann law (Equation 8-1), the total infrared energy radiated by *S* into a hemisphere is M_b Watts/cm^2 of the radiating surface. Thus, M_b is the total infrared energy radiated into a hemisphere of radius *d* on which sensor D_1 is located, that is, into a surface area of πd^2 cm^2. But M_b is also the total infrared energy radiated into a hemisphere of radius *2d* on which sensor D_2 is located, specifically, into a surface area of $4\pi d^2$ cm^2. Therefore,

$$\text{Infrared energy detected by } D_1 = \frac{M_b}{\pi d^2} = M_b' \quad (8\text{-}15)$$

$$\text{Infrared energy detected by } D_2 = \frac{M_b}{4\pi d^2} = \frac{M_b'}{4} \quad (8\text{-}16)$$

- Most thermal infrared remote sensing investigations try to maintain good radiometric and spatial resolution by 1) selecting a fairly large IFOV such as 2.5 mrad, and 2) flying at a relatively low altitude to obtain smaller pixel sizes. Unfortunately, at lower altitudes, the high spatial resolution may be outweighed by the fact that more flightlines are required to cover the area compared to more efficient coverage at higher altitudes with larger pixels. The pixel size and the geographic size of the survey are considered, objectives are weighed, and a compromise is reached. Multiple flightlines of aircraft MSS data are difficult to mosaic.

Geometric Correction of Across-Track Thermal Infrared Scanner Data

Thermal infrared scanning systems (actually all scanning systems) introduce numerous types of geometric error that must be understood because they impact 1) the quality of the imagery for visual or digital image processing and analysis, and 2) the creation of planimetric maps from the thermal infrared data. The most important considerations include:

- ground swath width,

- spatial ground resolution cell size,

- one-dimensional relief displacement, and

- tangential scale distortion.

Ground Swath Width (gsw): The ground swath width is the length of the terrain strip remotely sensed by the system during one complete across-track sweep of the scanning mirror. It is a function of the total angular field of view of the sensor system, θ, and the altitude of the sensor system above-ground-level, H (Figure 8-10). It is computed as

$$gsw = \tan\left(\frac{\theta}{2}\right) \times H \times 2. \tag{8-17}$$

For example, the ground swath width of an across-track scanning system with a 100° total field of view and an altitude above-ground-level of 6,000 m would be 14,301 m:

$$gsw = \tan\left(\frac{100}{2}\right) \times 6000 \times 2$$

$$gsw = 1.191753 \times 6000 \times 2$$

$$gsw = 14,301 \text{ m}.$$

If the total field of view were 90°, the ground swath width would be 12,000 m:

$$gsw = \tan\left(\frac{90}{2}\right) \times 6000 \times 2$$

$$gsw = 1 \times 6000 \times 2$$

$$gsw = 12,000 \text{ m}.$$

Most scientists utilizing across-track scanner data only use the central 70 percent of the swath width (35 percent on each

side of nadir) primarily because ground resolution elements have larger cell sizes the farther they are away from nadir.

Ground Resolution Cell Size (D): The diameter of the circular ground area viewed by the sensor, D, at nadir is a function of the instantaneous field of view, β, of the scanner measured in milliradians (mrad) and the altitude of the scanner above-ground-level, H, where $D = H \times \beta$. Interestingly, as the scanner's instantaneous field of view moves away from nadir on either side, the circle becomes an ellipsoid. One of the major reasons is that the distance from the aircraft to the resolution cell is increasing, as shown in Figure 8-10. In fact, the distance from the aircraft to the resolution cell on the ground, H_ϕ, is a function of the scan angle off-nadir, ϕ, at the time of data collection and the true altitude of the aircraft, H (Lillesand and Kiefer, 1994):

$$H_\phi = H \cdot \sec\phi. \tag{8-18}$$

Thus, the size of the ground-resolution cell increases as the angle increases away from nadir. The nominal (average) diameter of the elliptical resolution cell, D_ϕ, at this angular location from nadir has the dimension:

$$D_\phi = (H \cdot \sec\phi) \cdot \beta \tag{8-19}$$

in the direction of the line of flight, and

$$D_\phi = (H \cdot \sec^2\phi) \cdot \beta \tag{8-20}$$

in the orthogonal (perpendicular) scanning direction.

Scientists using thermal across-track scanner data usually only concern themselves with the spatial ground resolution of the cell at nadir, D. If it is necessary to perform precise quantitative work on pixels some angle ϕ off-nadir, then it may be important to remember that the radiant flux recorded is an integration of the radiant flux from all the surface materials in a ground-resolution cell with a constantly changing diameter. Using only the central 70 percent of the swath width reduces the impact of the larger pixels found at the extreme edges of the swath.

One-Dimensional Relief Displacement: Truly vertical aerial photographs have a single principal point directly beneath the aircraft at nadir at the instant of exposure. This perspective geometry causes all objects that rise above the local terrain elevation to be displaced from their proper planimetric position radially outward from the principal point (discussed in Chapter 6). For example, the four theoretical tanks in Figure 8-11a are each 50 ft high. The greater

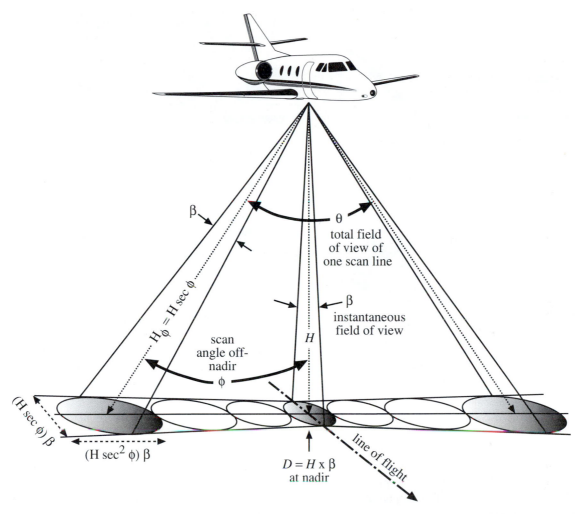

Figure 8-10 The ground resolution cell size along a single across-track scan is a function of: a) the distance from the aircraft to the observation, i.e., H is the altitude of the aircraft above-ground-level (AGL) at nadir and $H \sec \phi$ off-nadir; b) the instantaneous field of view of the sensor, β, measured in milliradians, and c) the scan angle off-nadir, ϕ. Thus, pixels off-nadir have semi-major and semi-minor axes (diameters) that define the resolution cell size. One-dimensional relief displacement and tangential scale distortion also occur in a direction perpendicular to the line of flight and parallel with a line scan.

the distance from the principal point, the greater the radial relief displacement of the top of the tank away from its base.

Thermal infrared images acquired using an across-track scanning system also contain relief displacement. However, instead of being radial from a single principal point, the displacement takes place in a direction that is perpendicular to the line of flight for each and every scan line as shown in Figure 8-11b. In effect, the ground-resolution element at nadir functions like a principal point for each scan line. At nadir, the scanning system looks directly down on the tank, and it appears as a perfect circle in Figure 8-11b. The greater the height of the object above the local terrain and the greater the distance of the top of the object from nadir (i.e., the line

of flight), the greater the amount of *one-dimensional relief displacement* present. One-dimensional relief displacement is introduced in both directions away from nadir for each sweep of the across-track mirror.

One-dimensional relief displacement can be beneficial or cause image-interpretation problems. For example, consider the aerial photograph and predawn thermal infrared image of the University of South Carolina campus shown in Figure 8-12. The science buildings exhibit radial relief displacement in the aerial photograph away from the principal point and one-dimensional relief displacement in the thermal infrared image perpendicular to the line of flight. Note how it is easy to view the side (façade) of the science buildings in the ther-

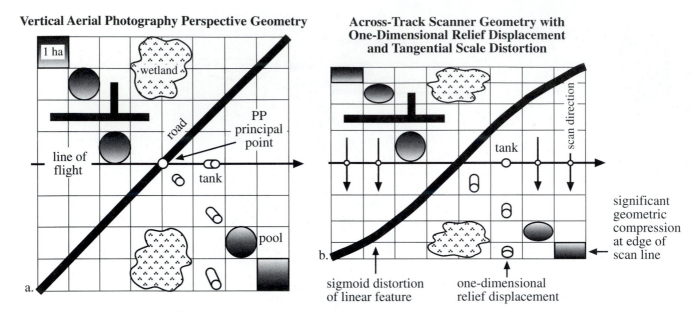

Figure 8-11   a) The hypothetical perspective geometry of a vertical aerial photograph obtained over level terrain. Four 50 ft tanks are distributed throughout the landscape and experience varying degrees of radial relief displacement the farther they are from the principal point (PP). b) Across-track scanning systems introduce one-dimensional relief displacement perpendicular to the line of flight and tangential scale distortion and compression the farther the object is away from nadir. Linear features trending across the terrain are often recorded with s-shaped or sigmoid curvature characteristics due to tangential scale distortion and compression.

mal infrared image. This is valuable if we want to study the temperature characteristics of the side of these buildings. However, if we wanted to evaluate the thermal characteristics of the road or objects immediately behind the buildings, they are obscured from view.

Aerial photography and predawn thermal infrared imagery of downtown New York City provide an even greater appreciation of one-dimensional relief displacement (Figure 8-13). In this case, the radial relief displacement in the aerial photograph makes it difficult to obtain information about the street pattern. In fact, the street pattern is almost completely obscured from view. Conversely, because the thermal imagery was obtained along a line of flight that was parallel with the street orientation, the one-dimensional relief displacement creates an excellent view of the temperature characteristics of the buildings and streets, especially those that lie perpendicular to the direction of flight. Notice the significant amount of thermal detail on the side of the Empire State Building that is visible because of the one-dimensional relief displacement. Also note the radiometrically cold metal on the top of the Empire State Building.

While some aspects of one-dimensional relief displacement may be of utility for visual thermal infrared image interpre-

tation, it seriously displaces the tops of objects projecting above the local terrain from their true planimetric position. Maps produced from such imagery contain serious planimetric errors. Thermal infrared imagery must be geometrically rectified in order to produce maps with some semblance of geometric accuracy. Methods of geometric rectification are summarized in Jensen (1996).

Tangential Scale Distortion: The mirror on a thermal infrared across-track scanning system rotates at a constant speed and typically views from 70° to 120° of terrain during a complete line scan. Of course, the amount depends on the specific sensor system. From Figure 8-10 it is clear that the terrain directly beneath the aircraft (at nadir) is closer to the aircraft than the terrain at the edges during a single sweep of the mirror. Therefore, because the mirror rotates at a constant rate, the sensor scans a shorter geographic distance at nadir than it does at the edge of the image. This relationship tends to *compress* features along an axis that is perpendicular to the line of flight. The greater the distance of the ground-resolution cell from nadir, the greater the image scale compression. This is called *tangential scale distortion*. Objects near nadir exhibit their proper shape. Objects near the edge of the flightline become compressed and their shape distorted. For example, consider the tangential geometric

Vertical Aerial Photograph

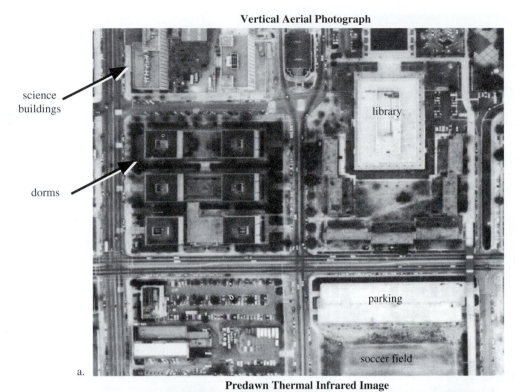

science
buildings

library

dorms

parking

soccer field

a.

Predawn Thermal Infrared Image

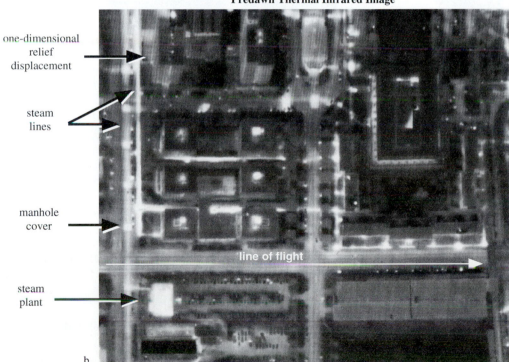

one-dimensional
relief
displacement

steam
lines

manhole
cover

line of flight

steam
plant

b.

Figure 8-12   a) Vertical panchromatic aerial photography of the University of South Carolina campus obtained on April 26, 1980. Note that the relief displacement of the science buildings is radial away from the principal point. b) Predawn 1 x 1 m thermal infrared imagery. Note the intricate underground steam-line network served by the steam plant. The science buildings exhibit one-dimensional relief displacement caused by the cross-track scanning system.

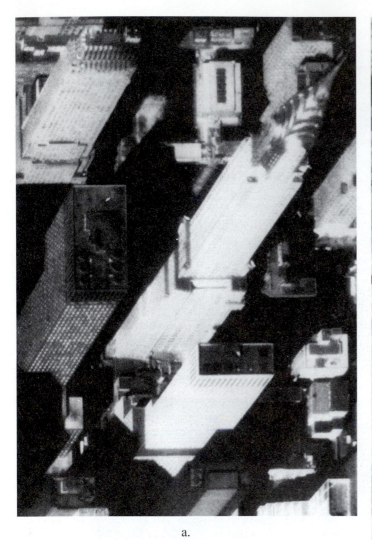

a.

line of flight ⟶

b.

Figure 8-13 a) Perspective aerial photograph of downtown New York City. Note the radial relief displacement of the Empire State Building away from the principal point (not shown). b) Predawn thermal infrared image of downtown New York City. Note the
one-dimensional relief displacement that is perpendicular to the line of flight. The metal on the top of the Empire State Building shows up as being very cold due to its low emissivity. The building on the left is very cool, while the Empire State Building radiates a significant amount of energy. The metal cars have a low emissivity, as do the circular metal evaporative coolers
on top of the nearest building.

distortion and compression of the circular swimming pools
and one hectare of land the farther they are from nadir in the
hypothetical diagram (Figure 8-11b).

This tangential scale distortion and compression in the far
range also causes linear features such as roads, railroads,
utility right of ways, etc., to have an *s-shape* or *sigmoid distortion* when recorded on thermal infrared imagery (Figure
8-11b). Interestingly, if the linear feature is parallel with or

perpendicular to the line of flight it does not experience sigmoid distortion.

Some sophisticated across-track scanning systems take tangential scale distortion into consideration and sweep the
exposing spot of light at the film recorder at a continuously
varying speed. Ideally, this is the type of thermal infrared
imagery available for analysis. When tangential scale distortion has not been removed, it is best to 1) use the central 70

percent of the field of view for analysis to minimize the effects of distortion, and 2) geometrically rectify the remote sensor data using ground-control point rectification discussed in Jensen (1996).

Even single flightlines of aircraft MSS data are difficult to rectify to standard map projection because of aircraft roll, pitch, and/or yaw during data collection (Jensen et al., 1988). Notches in the edge of a flightline of data are indicative of aircraft roll. Such data require significant human and machine resources to make the data planimetrically accurate. Several firms have placed GPS on the aircraft to obtain precise flightline coordinates, which are useful when rectifying the aircraft MSS data.

Radiometric Calibration of Thermal Scanner Data

To use the thermal infrared remote sensor data for practical purposes such as temperature mapping, it is necessary to calibrate the brightness values stored on the digital tape or hard disk to temperature values (Quattrochi and Goel, 1995). This radiometric calibration may be performed using 1) internal blackbody source referencing, or 2) external empirical referencing based on *in situ* data collection.

Internal Source Referencing: When an across-track scanning system is used to collect thermal infrared data, the detector first looks at a "cold" reference target, approximately 120° of terrain, and then a "hot" reference target during each line scan, as demonstrated in Figure 8-7. The true kinetic temperature of these cold and hot targets is constantly monitored by the remote sensing system and is recorded on disk, tape, or some other medium along with the image data for each line scan. If desired, all of the terrain brightness values collected during a scan can then be calibrated (converted) to *apparent* true temperature values based on their relationship to the cold and hot target information stored for each scan line. The radiometric resolution is usually accurate to within ± 0.2°C. This should be the ideal method of radiometric calibration because it involves no field work. Unfortunately, this method does not account for the intervening atmosphere 1) emitting spurious radiant energy into the IFOV of the sensor system, or 2) absorbing energy emitted from the ground before it reaches the detector optics.

External Referencing: To incorporate atmospheric effects, it may be necessary to perform external empirical referencing. This involves taking *in situ* measurements with 1) a thermometer that measures the true kinetic temperature of a material or water body, 2) a handheld *radiometer* that

measures the radiant temperature exiting the terrain in a specific instantaneous field of view, or 3) a *radiosonde* (a balloon carrying sensitive meteorological instruments) launched to obtain atmospheric profiles of temperature, barometric pressure, and water vapor.

Thermometers are straightforward *in situ* measurement devices. A radiometer is a handheld remote sensing instrument that you point at the terrain and an apparent radiant temperature reading is returned. Either type of temperature measurement should be obtained at the exact time that the remote sensor data are collected overhead. Ideally, more than 30 *in situ* samples are obtained. Unfortunately, this is not always practical due to the constraints of hiring people, boats, etc., and obtaining a sufficient number of high-quality thermometers and/or radiometers. The exact location of each of the *in situ* thermometer or radiometer measurements are obtained using GPS. The locations of the *in situ* data-collection points are then located in the rectified remote sensor data, and the brightness values at these locations are extracted. The accuracy of the geometric correction of the thermal infrared data becomes very important at this point.

The n thermometer or radiometer measurements collected *in situ* are then regressed with the corresponding n remote sensing brightness values obtained at the same geographic locations. For example, consider the eight *in situ* water-temperature measurements in Figure 8-14 and the corresponding uncalibrated thermal infrared remote sensing brightness values ($BV_{i,j}$) for these same eight locations. Linear and/or nonlinear curves are fit through the observations. The linear equation explained 86 percent of the variance while the 2nd-order polynomial explained 99 percent of the variance. Either of these equations may be used to relate the *in situ* temperature measurements to the remote sensing brightness values, in effect, radiometrically calibrating the remote sensor data. This method does not take into account the intervening atmosphere present at the time of data collection. Note that the linear equation is of the form $BV_{ij} = aT_{kin} + b$. If we want to take into account the emissivity (ε) of the terrain, we utilize the y-intercept (b) and slope of the relationship (a) and the equation:

$$BV_{ij} = a \cdot \varepsilon \cdot T_{kin}^{4} + b. \tag{8-21}$$

Rearranging the equation allows the true kinetic temperature T_{kin} of every pixel, BV_{ij} in the uncalibrated matrix of remote sensor data to be determined:

$$T_{kin} = \left(\frac{BV_{ij} - b}{a \cdot \varepsilon} \right)^{\frac{1}{4}}. \tag{8-22}$$

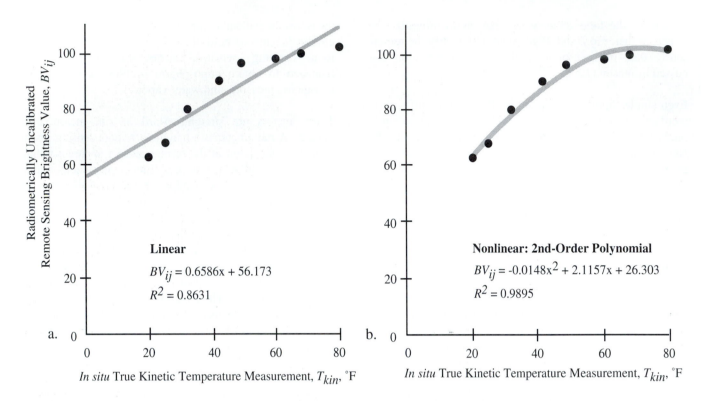

Figure 8-14   a) Linear regression line fit to eight *in situ* temperature measurements and uncalibrated remote sensing brightness values obtained for the same geographic location. b) Nonlinear 2nd-order polynomial equation fit to the same observations.

The radiometrically calibrated matrix of remote sensor data may then be used to make temperature maps. This method requires a costly and well-coordinated field data-collection program. Also, all the thermometers, radiometers, and GPS instruments must be carefully calibrated, and the *in situ* readings should be obtained at exactly the same time, if possible.

But what if it is not practical to collect *in situ* thermometer or radiometer measurements in the field over vast or intractable distances at the same time as the remote sensing overflight? How can we then take into account the deleterious effects of atmospheric absorption or emission on our thermal infrared measurements? The answer is the use of 1) radiosondes and atmospheric radiative transfer modeling, or 2) split-window atmospheric correction techniques.

As noted, the intervening atmosphere has a significant impact on the energy emitted by the terrain before it is actually recorded by the thermal infrared remote sensing system. In fact, the maximum atmospheric transmittance of thermal infrared radiation in the spectral region from 8.0 – 14 μm is only about 80 percent (Luvall, 1999). The amount of atmospheric absorption is primarily a function of the atmospheric

water vapor content, although there is an ozone absorption band around 9.5 μm. To obtain accurate thermal surface radiance values, radiosonde launches need to be made concurrently with daytime and nighttime overflights. Atmospheric profiles of temperature, barometric pressure, and humidity obtained using the radiosonde are transmitted to the ground. These data are then incorporated into an atmospheric transmission model (e.g., LOWTRAN or MODTRAN) to calculate atmospheric transmission characteristics (Quattrochi and Goel, 1995). The output from the LOWTRAN or MODTRAN model is combined with calibrated spectral response curves for the individual bands of the thermal infrared detectors being used and internal blackbody source referencing data recorded during the flight. All these data are then modeled and used to produce a lookup table for converting each pixel's brightness value into true kinetic temperature measurements (Anderson, 1992; Luvall, 1999). This is the most rigorous method of calibrating thermal infrared imagery. Unfortunately, very few persons have access to radiosonde technology. Sometimes atmospheric profile meteorological information from a nearby National Weather Service station (or airport) can be used instead of the radiosonde data as input to the atmospheric transmission model (Quattrochi and Goel, 1995).

Table 8-5   Spectral Sensitivity and Operating Temperatures for Selected Infrared Electro-Optical Focal Plane Arrays

Wavelength (μm)	Detector Material	Operating Temperature (K)
0.3 – 5.5	In:Sb	<90
1.0 – 3.0	PV Hg:Cd:Te	>150
3.0 – 5.0	PV Hg:Cd:Te	≤120
8.0 – 12.0	PV Hg:Cd:Te	50 – 80
12.0 – 25.0	Si:As	10

Scientists attempting to remove atmospheric effects and determine surface temperature (T_s) from AVHRR data have found that a *split-window* approach works well in certain instances (Czajkowski et al., 1997b). Basically, the brightness temperatures observed in one AVHRR thermal channel (T_4) are corrected for atmospheric effects by the linear difference between the brightness temperatures in AVHRR channels 4 and 5. An example of a split-window equation is:

$$T_s = a + T_4 + b(T_4 - T_5) \qquad (8\text{-}23)$$

where a and b are constants that can be estimated from model simulations (Becker and Li, 1990) or correlation with ground observations. The split-window method works best when estimating sea-surface temperatures. It is not as accurate for land-surface measurements, due to factors described in Kalluri and Dubayah (1995).

Pushbroom Linear and Area-Array Charge-Coupled-Device (CCD) Detectors

Until recently, it was difficult to make commercially available CCDs that had spectral sensitivity to mid- and long-wavelength infrared radiation. Therefore, most thermal infrared remote sensor data were collected with a single detector and a scanning mirror as previously discussed. This has changed (Wimmers and Smith, 1994; West, 1996). It is now possible to make both linear and area-arrays (sometimes referred to as "staring" focal-plane arrays) that are sensitive to mid- and thermal infrared radiation such as those summarized in Table 8-5. For example, a 640 x 480 Hg:Cd:Te long-wavelength infrared area array with its readout assembly attached is shown in Figure 8-15. Detector arrays greater than 1024 x 1024 elements are now available

Figure 8-15   A 640 x 480 Hg:Cd:Te (mercury-cadmium-telluride) long-wavelength infrared (LWIR) area array and readout assembly. This detector array is cooled to 50 – 80 K. New arrays have greater than 1000 x 1000 detectors.

and offer very low noise and state-of-the-art sensitivity for extremely low-background applications. Staring array detectors made of platinum silicide (Pt:Si) are also very popular (Silverman et al., 1992; West, 1996).

Linear and area staring arrays allow improved thermal infrared remote sensing to take place because (West, 1996):

• the solid-state micro-electronic detectors are smaller in size (e.g., 20 x 20 μm) and weight, require less power to operate, have fewer moving parts, and are more reliable;

• each detector in the array can view the ground resolution element for a longer time (i.e., it has a longer dwell time), allowing more photons of energy from within the IFOV to be recorded by the individual detector, resulting in improved radiometric resolution (the ability to resolve smaller temperature differences);

• each detector element in the linear or area array is fixed relative to all other elements, therefore, the geometry of the thermal infrared image is much improved relative to that produced by an across-track scanning system; and

• some linear and area thermal detectors now use a miniature Sterling closed-cycle cooling system that does not require the compressed gas-cooling apparatus (argon or liquid nitrogen) previously discussed (Finney, 1996).

Figure 8-16   A helicopter with a forward-looking infrared (FLIR) sensor system located under the nose (courtesy FLIR Systems, Inc.).

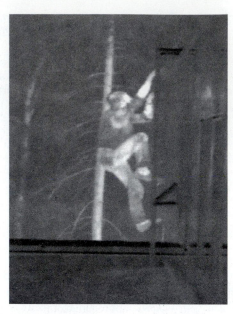

Figure 8-17   Nighttime thermal infrared image of a break-in (courtesy FLIR Systems, Inc.).

Forward-Looking Infrared (FLIR) Systems

During the 1991 Gulf War, the public saw day and nighttime *forward-looking infrared (FLIR)* images of the terrain and various targets. For decades, military organizations throughout the world have funded the development of FLIR-type systems that look obliquely ahead of the aircraft and acquire high-quality thermal infrared imagery, especially at night. In fact, their goal is usually "to own the night." Some FLIR systems collect the infrared energy based on the same principles as an across-track scanner (previously discussed), except that the mirror points *forward* about 45° and projects terrain energy during a single sweep of the mirror onto a linear array of thermal infrared detectors. Some systems use staring focal-plane array technology (Finney, 1996). An example of a FLIR system housed under the nose of an aircraft is shown in Figure 8-16.

FLIR and other thermal infrared systems are routinely used by law-enforcement agencies to see nighttime criminal activity and find lost people (Kruse, 1995). For example, a criminal is in the process of breaking into a second-story building in Figure 8-17. The person is approximately 98 °F, while the terrain is 70 – 80 °F. The moist rooftop is relatively cool (dark), the tree trunk and canopy relatively warm (bright), and the metal components of the ladder are cool (dark). Foresters and firefighters use FLIR to locate the perimeter of forest fires and effectively allocate firefighting resources. Security personnel at Gatwick Airport in London use pole-mounted FLIR surveillance systems to monitor the perimeter fences and make sure no unauthorized people gain access to the airport. Several other FLIR images are presented in Figure 8-18.

Thermal Infrared Environmental Considerations

When interpreting a thermal infrared image, it is useful to understand the diurnal cycle and how it relates to the temperature of objects on the Earth's surface.

Diurnal Temperature Cycle of Typical Materials

The diurnal cycle encompasses 24 hours. Beginning at sunrise, the Earth begins intercepting mainly short-wavelength energy (0.4 – 0.7 µm) from the Sun (Figure 8-19a). From dawn to dusk, the terrain intercepts the incoming short wavelength energy and reflects much of it back into the atmosphere, where we can use optical remote sensors to measure the reflected energy. However, some of the incident short-wavelength energy is absorbed by the terrain and then re-radiated back into the atmosphere as thermal infrared wavelength radiation (3 – 14 µm). The outgoing long-wave radiation peak usually lags two to four hours after the midday peak of incoming shortwave radiation, due to the time taken to heat the soil. The contribution of reflected short-wavelength energy and emitted long-wavelength energy causes an energy surplus to take place during the day, as shown in Figure 8-19a. Both incoming and outgoing shortwave radiation

Figure 8-18   Examples of airborne FLIR (forward-looking infrared) imagery. a) Nighttime image of people in a boat being rescued. b) Nighttime image of a person on a metal roof being pursued by another person on the ground. Note the warm transformers on the telephone pole and the low emissivity of the metal telephone lines. c) Helicopter view of policemen making an armed arrest. d) Nighttime image of a freighter and tugboat. e) Storage tank with a relatively cool lower layer of contents. f) Nighttime image of an urban area with water, boat, automobiles, and pedestrians (courtesy FLIR Systems, Inc.).

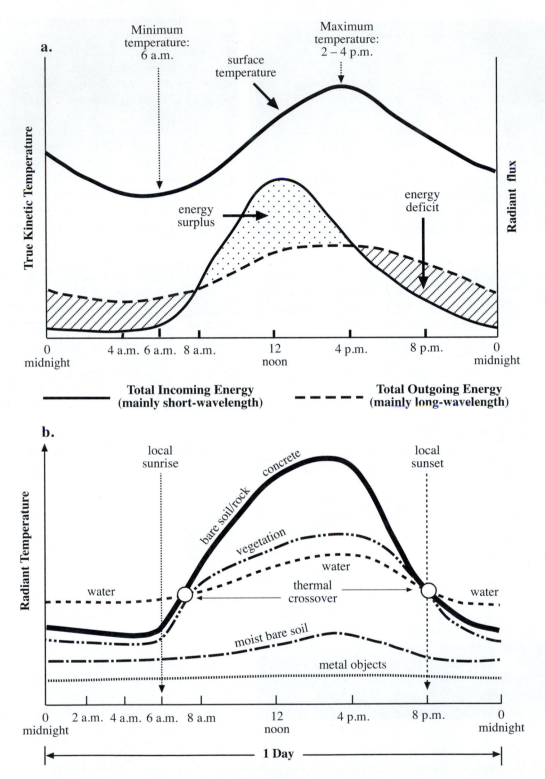

Figure 8-19   a) The diurnal cycle of reflected short-wavelength and emitted long-wavelength energy. Note the peak period of daily outgoing long-wavelength radiation and the general daily maximum temperature (after Marsh and Dozier, 1981). b) The diurnal radiant temperature of bare soil and rock, concrete, vegetation, water, moist bare soil and metal objects.

become zero after sunset (except for light from the moon and stars), but outgoing long-wave radiation exiting the terrain continues all night.

The typical diurnal temperature variations for soils and rock, water, vegetation, moist soil, and metal objects are shown in Figure 8-19b. If all of these curves lie exactly on top of one another, then remote sensing in the thermal infrared portion of the spectrum would be of no value because all the phenomena would have the same apparent radiant temperature. There would be no contrast in the imagery between the different phenomena. Fortunately, there are only two times during the day (after sunrise and near sunset) when some materials like soil, rock, and water have exactly the same radiant temperature. During this *crossover* time period it is generally not wise to acquire thermal infrared remotely sensed data.

Fortunately, some materials store heat more efficiently than others, i.e., they have a higher *thermal capacity*. For example, water has a much higher thermal capacity than soil and rock (Table 8-3). Its diurnal temperature range fluctuates very little when compared with the dramatic temperature fluctuation of soil and rock during a 24-hr period.

If we were interested in performing temperature mapping of terrain consisting of just soil, rock, and water, we could predict what the image would look like if we acquired thermal infrared imagery at about 2:00 p.m. and at 4:00 a.m. in the morning. The soil and rock would appear brighter than water in the daytime thermal imagery due to their higher apparent temperature. Rock and soil continue to radiate energy into the atmosphere during the night. During the early evening rock and soil (and concrete) are still warmer than much of the surrounding terrain. By midnight these surfaces have radiated most of the heat energy they absorbed during the day and gradually they recede in brightness. Conversely, water with its high thermal capacity, maintains a relatively stable surface temperature that may well be higher than the soil and rock (and concrete) by about 4:00 a.m. This results in water being brighter on the nighttime thermal infrared imagery than the soil and rock, and even the vegetation in many instances. Because vegetation contains water, it is usually cooler (darker) than soil and rock on 2:00 p.m. daytime imagery and warmer (brighter) than soil and rock on predawn imagery.

Vegetation tends to be slightly warmer than water throughout the day and cooler than water during predawn hours. Very moist soil tends to have a stable diurnal temperature cycle, as shown in Figure 8-19b, because the more water in the soil, the higher its thermal capacity. Metal objects such as cars and aluminum rooftops appear cool (dark) on both day and nighttime thermal infrared imagery due to their low emissivity (remember, good metal reflectors are poor absorbers, and therefore poor emitters). In fact, they are often the darkest objects in the imagery.

Several of the previously mentioned diurnal temperature cycle relationships are demonstrated in Figure 8-20. An ATLAS mission was flown over a large sandbar in the Mississippi River at 5:00 a.m. and 10:30 a.m. The data have a spatial resolution of 2.5 x 2.5 m. Daytime thermal bands 10, 11, and 12 reveal a dramatic difference in the temperature properties of sand and gravel on the sandbar. The different materials absorb the incident energy from the Sun differently, resulting in substantial differences in exitance from the sand and gravel surfaces in the three thermal infrared bands. During the day, the water is cooler than much of the surrounding countryside. The vegetation (V) on the bank of the river is also relatively cool. As expected, water is much warmer than the surrounding countryside at night and the vegetation is cooler than the water. There is still a noticeable difference between the large gravel and sand areas in the nighttime imagery, although not to the degree present in the daytime imagery. The goal of this project was to see if thermal infrared imagery was useful for discriminating between sand and gravel. It was practically impossible to discriminate between sand and gravel areas on 1:8,000-scale natural color aerial photography obtained at the same time as the 10:30 a.m. overflight.

Unless scientists are specifically trying to compute thermal inertia which requires both day and nighttime imagery, they often prefer to collect predawn thermal infrared imagery because:

- short-wavelength reflected energy from the Sun can create annoying shadows in daytime imagery;

- by 4:00 a.m., most of the materials in the terrain have relatively stable equilibrium temperatures, as shown in Figure 8-19b, i.e., the slopes are near zero; and

- convective wind currents usually settle down by the early morning, resulting in more accurate flightlines (less crabbing of the aircraft into the wind) and no wind smear or wind streaks on the imagery.

Pilots are now able to obtain very accurate flightlines of thermal infrared imagery at night using onboard GPS and inertial navigation systems.

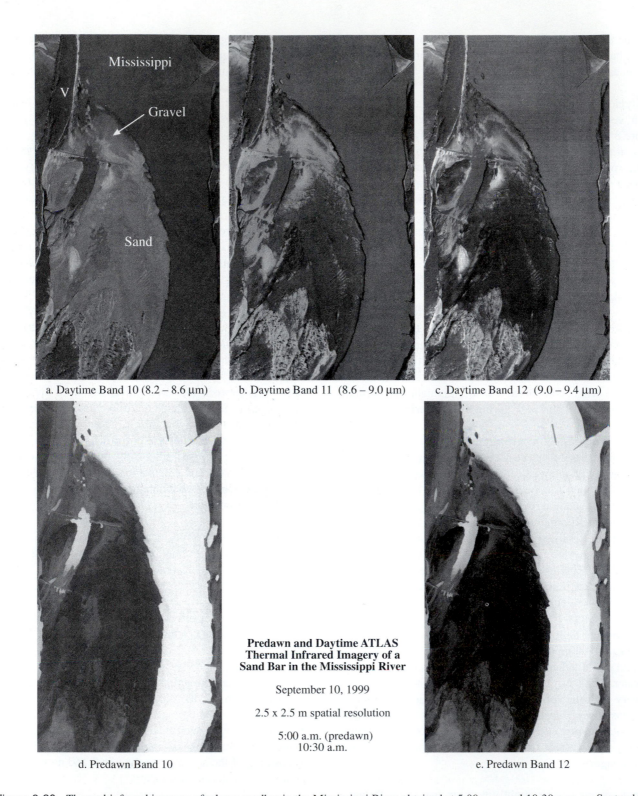

a. Daytime Band 10 (8.2 – 8.6 μm)   b. Daytime Band 11 (8.6 – 9.0 μm)   c. Daytime Band 12 (9.0 – 9.4 μm)

**Predawn and Daytime ATLAS
Thermal Infrared Imagery of a
Sand Bar in the Mississippi River**

September 10, 1999

2.5 x 2.5 m spatial resolution

5:00 a.m. (predawn)
10:30 a.m.

d. Predawn Band 10   e. Predawn Band 12

Figure 8-20   Thermal infrared imagery of a large sandbar in the Mississippi River obtained at 5:00 a.m. and 10:30 a.m. on September 10, 1999. The ATLAS Band 11 detector did not function properly during predawn data collection (Luders et al., 1999).

Savannah River Site

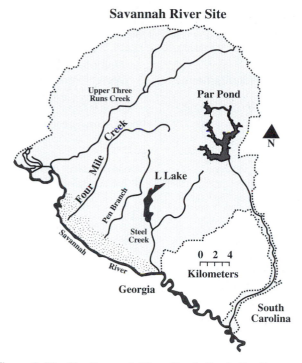

Figure 8-21 The Savannah River Site in South Carolina at one time extracted water from the Savannah River and used it to cool nuclear-reactor operations. The thermal effluent returned to the Savannah River swamp system via Four Mile Creek, where it was to dissipate. However, the thermal effluent on occasion reached the Savannah River, resulting in a thermal plume.

Examples of Thermal Infrared Remote Sensing

Nonpoint Source Pollution Monitoring

The Savannah River Site in South Carolina until recently extracted cooling water from the Savannah River and used it to dissipate heat in the reactors. The hot cooling water was then returned to the Savannah River swamp system via Four Mile Creek (Figure 8-21). Sometimes the thermal effluent reached the Savannah River and created a thermal plume as shown in the flightline of predawn thermal infrared imagery in Figure 8-22. A thermal infrared view of the plume at full spatial resolution is shown in Figure 8-23. A number of state and federal laws govern the characteristics of the plume in the Savannah River. For example, the South Carolina

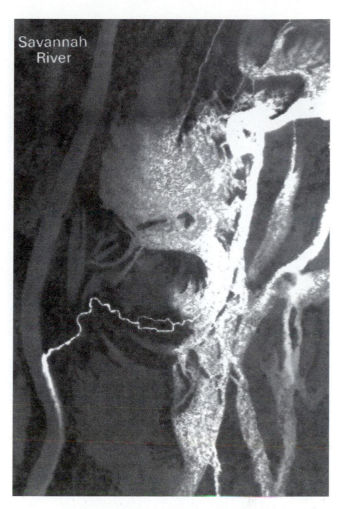

Figure 8-22 A predawn thermal infrared image of thermal effluent entering the Savannah River swamp system from Four Mile Creek on March 31, 1981, at 4:28 a.m. The water migrated through the swamp system and eventually entered the Savannah River, producing a thermal plume. The temperature and spatial distribution of this plume were governed by a number of state and federal statutes. This is a 2× reduction image of a portion of the flightline and consists of 1,065 rows and 632 columns. It provides a regional overview of the spatial distribution of the thermal effluent.

Department of Health and Environmental Control (DHEC) required that the plume be less than one-third the width of the river at temperatures >2.8°C above river ambient temperature. The Savannah River after a heavy rain is dangerous, fast moving, and contains substantial debris. It is difficult to obtain accurate temperature measurements by placing people in boats with thermometers that are at the mercy of the current. Furthermore, it is not practical to place *in situ* ther-

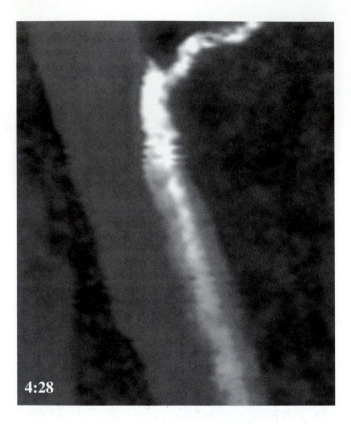

4:28

Figure 8-23   Predawn thermal infrared image (8.0 – 14 μm) of thermal effluent flowing into the Savannah River acquired at 4:28 a.m. on March 31, 1981, using a Daedalus DS-1260 multispectral scanning system.

Table 8-6   User Requirements for Monitoring the Temperature Distribution of a Thermal Plume in the Savannah River and Sensor System Parameters.

User Requirement	Remote Sensing System Parameter
The apparent temperature of each pixel must be accurate to within ± 0.2°C of its true temperature. Temperatures in the scene ≥ the ambient river temperature are of primary importance.	Sense in the 8 – 14 μm region using mercury-cadmium-telluride (Hg:Cd:Te) detector with an analog-to-digital conversion of 8 bits (0 – 255). Radiometrically calibrate to ±0.2°C with pixel values ranging from -3 to 48°C.
At least 30 pixels should fall within the 90-m width of the river. Therefore, a spatial resolution element of ≤ 3 x 3 m is required. Rectify all pixels to be within ±10 percent of their planimetric position.	Use a sensor with an IFOV of 2.5 milliradians flown at 1200 m (4,000 ft) AGL. Remove geometric sensor system distortion and resample to 2.8 x 2.8 m picture elements.
Acquire predawn imagery in the spring and analyze results rapidly.	Predawn data acquisition with stable atmospheric conditions. Calibrate and analyze.
Produce isotherm maps and extract quantitative temperature information for selected class intervals.	Apply image processing and cartographic principles to obtain quantitative information from the isothermal map and associated transects.

mometers in the river because of debris flowing at high rates of speed. Therefore, a remote sensing approach was used to obtain the required spatial temperature information.

Table 8-6 summarizes the user requirements and the sensor system parameters that were incorporated to obtain the necessary temperature data. An across-track multispectral scanning system equipped with a thermal infrared detector (Hg:Cd:Te) was used to record emitted radiant energy in the 8 – 14 μm region of the spectrum. This region was selected because the terrain was relatively cool, with a maximum emittance of thermal energy peaking at about 9.7 μm.

The width of the Savannah River was approximately 90 m (300 ft) in the vicinity of the plume. For scientists to obtain a minimum representative sample of 30 pixels within the river, each pixel had to record the surface temperature for an area ≤ 3 x 3 m. To achieve this size of pixel, a sensor with a 2.5 milliradian IFOV was used to produce a nominal spatial resolution at nadir of approximately 3.05 x 3.05 m when the sensor was flown at 1,220 m (4,000 ft) above-ground-level.

Because an across-track scanning system was used to record the radiant temperature of objects, tangential scale distortion was present. If not removed, the quality of the cartographic products would be diminished. Therefore, the image was geometrically corrected by modeling each pixel as a function of scan angle from nadir. This resulted in each pixel in a scan line having an effective spatial resolution of 2.8 x 2.8 m. The data were then geometrically corrected by selecting ground-control points and rectifying the data using nearest-neighbor resampling, as discussed in Jensen (1996).

For investigators to identify a thermal plume and map its temperature and spatial distribution, the plume, river, and surrounding land all had to be distinguishable on the remotely sensed imagery. Each day the maximum thermal contrast between Savannah River water and other land surfaces is reached just before dawn, as previously discussed when the river is warmer than the land surface. It was assumed that the hot plume would be distinguishable from the ambient Savannah River water and land surface day or night. An additional factor in favor of a predawn overflight was the fact that at night there are no tree shadows along the bank of the river that can cause spurious temperature measurements of the river or plume. These considerations resulted in the data being collected on March 31, 1981, at 4:28 a.m. under the influence of a high pressure weather system with an absolute humidity of 27%.

To comply with the ± 0.2°C requirement, the thermal infrared scanner data were calibrated using internal blackbody source referencing. While scanning a single line of data, the detector looked at a cold plate with a known temperature, the terrain, and then a hot plate with a known temperature. By knowing the relationship between the temperature of these plates, the analyst could calculate the apparent temperature of the terrain for each pixel, from -3 to 48°C. The major drawback of this calibration method was that it did not incorporate the effect of the intervening atmosphere on signal response. However, it appeared that the calibration was fairly successful. When the calibrated data were compared with two *in situ* surface-temperature measurements made in the Four Mile Creek plume during the time of the overflight, both remote sensing measurements were within ± 0.2°C of the *in situ* measurements.

Data Analysis:

Environmental scientists and ichthyologists are interested in the temperature and spatial distribution of thermal plumes and how they relate to the ambient river temperature. If a plume exists, it is important to determine where the plume is greater than a specified number of degrees above river ambient temperature. Depending on the time of year, thermal plumes may attract certain species of aquatic organisms and exclude others. Unfortunately, a hot plume extending across the river at some depth will inhibit certain species from swimming upstream during the spawning season.

Two fundamental types of data analysis were performed to extract quantitative temperature information. First, the spatial distribution of selected temperature class intervals > 2.8°C above river ambient temperature were mapped. Next, transects were passed through the isotherm map to document the cross-section surface area of the plume at specific temperature class intervals.

Mapping the spatial distribution of the temperature made it necessary to identify:

• the land (soil and vegetation) to make sure it was not confused with the thermal plume;

• the ambient river temperature; and

• the spatial distribution of the plume temperature > 2.8°C above river ambient temperature.

At 4:28 a.m. the river is much warmer than the surrounding land and easily distinguishable. Two methods were used to determine ambient river temperature. First, *in situ* temperature measurements of the Savannah River were available which placed the temperature above the plume at 12°C. Second, a sample of 200 pixels just above the Four Mile Creek plume yielded a mean of 12°C and a standard deviation of ±0.2°C. Thus, the ambient river temperature was determined to be 12°C.

Seven unequal class intervals were then selected. Three of the class intervals were for temperatures < 2.8°C above river ambient temperature and included the land and thermally unaffected river. Four other class intervals were used to highlight special ranges of temperature within the thermal plume that were > 2.8°C above river ambient temperature. Each of the seven class intervals were assigned a unique color with yellow, orange, red, and white corresponding to warmer temperatures in the plume (Table 8-7). Analysis of Color Plate 8-1a reveals that the hottest part of the plume was just offshore from the mouth of the creek (34.6°C; 94.28°F). The temperature of the creek itself was just as hot as the warmest part of the plume; however, the overhanging shrubbery along the banks of the creek dampened the remotely sensed apparent temperature of the creek along its main channel. Also, the northern eddy of Four Mile Creek warmed the water slightly just above the plume as it entered the Savannah River.

The color-coded isotherm map provided valuable information about the spatial distribution of the temperature of the plume as it entered the Savannah River and progressed downstream. However, more quantitative data were obtained by passing three transects (A, B, and C) through the plume, as shown in Color Plate 8-1b. The brightness values encountered along each transect are summarized in Table 8-7. These values were obtained only after the original image was geometrically rotated 16° clockwise so that the end points of each transect fell on the same scan line. This ensured that the number of meters in each temperature class along each transect was accurately measured. If the analyst extracts transects where the end points do not fall on the same scan line (or column), the hypotenuse of stair-stepped pixels must be considered instead of the simple horizontal pixel distance. These relationships are clearly demonstrated in Color Plate 8-1c.

Table 8-7   Savannah River Thermal Plume Apparent Temperature Transects

Transect[a]	Average Width of River[b]	Relationship of Class to Ambient River Temperature, 12°C						
		Class 1 Dark blue ambient	Class 2 Light blue +1°C	Class 3 Green 1.2°–2.8°C	Class 4 Yellow 3.0°–5.0°C	Class 5 Orange 5.2°–10°C	Class 6 Red 10.2°–20°C	Class 7 White >20°C
		Brightness Value Range for Each Class Interval						
		74–76	77–80	81–89	90–100	101–125	126–176	177–255
A	32 pixels = 89.6 m	15/42[c]	17/47.6	—	—	—	—	—
B	38 pixels = 106.4 m	25/70	1/2.8	1/2.8	2/5.6	1/2.8	5/14	3/8.4
C	34 pixels = 95.2 m	19/53.2	2/5.6	2/5.6	2/5.6	6/16.8	3/8.4	—

[a] Each transect was approximately 285 m in length (66 pixels at 2.8 m/pixel). Transect measurements in the river were made only after the image was rotated, so that the beginning and ending pixels of the transect fell on the same scan line.

[b] Includes one mixed pixel of land and water on each side of the river.

[c] Notation represents pixels and meters; for example, 15 pixels represent 42 m.

A histogram of transect B is shown in Color Plate 8-1d. The relationship between the original brightness values and the class intervals of the density-sliced map is provided in Table 8-7. By counting the number of pixels along a transect in specific temperature class intervals within the plume and counting the total number of pixels of river, it is possible to determine the proportion of the thermal plume falling within specific temperature class intervals (Jensen et al., 1983 and 1986). For example, in 1981 in South Carolina a thermal plume could not be >2.8°C above river ambient temperature for more than one-third of the width of the river. Transect information extracted from thermal infrared imagery and summarized in Table 8-7 were used to determine if the plume was in compliance.

The fundamental black-and-white thermal infrared image of the plume contained valuable information. However, it was also possible to digitally enhance the original thermal infrared image to visually appreciate the more subtle characteristics of the thermal plume and the surrounding phenomena. For example, Figure 8-24b is a low-frequency filtered image of the original contrast-stretched image. It emphasized the slowly varying components within the image. Conversely, the application of a high-frequency filter enhanced the high-frequency detail in the image. Note how several radiometric errors in the scan lines are more pronounced in the high frequency filtered image in Figure 8-24c. A minimum filter enhanced the core of the plume, while a maximum filter enhanced the entire plume (Figure 8-24d,e). An embossed

filter created a shaded-relief impression of the plume (Figure 8-24f). Three edge-enhancement algorithms (Roberts, Sobel, and Laplacian) highlighted different parts of the plume and the edges of the plume/land/water boundaries (Figure 8-24g,h,i). The logic and mathematics of these algorithms are summarized in Jensen (1996).

Thermal infrared surveys of the Four Mile Creek plume were routinely collected at least two times per year from 1981 through 1988 to ensure that the plume was in compliance with state and federal water-quality standards. Cooling towers were then built to receive the thermal effluent.

Residential Thermal Imagery Energy Surveys

Homes, office buildings, and industries are insulated using various materials so that in the winter, expensive heat does not easily exit. It is possible to utilize remote sensing to monitor the effectiveness of the insulation. However, the results are usually subjective rather than quantitative because many important parameters must be known to quantitatively document the exact insulation characteristics of every home or business in a neighborhood.

Table 8-8 summarizes the most important parameters to be taken into consideration when performing a residential thermal imagery energy survey. Colcord (1981) suggests the temperature (radiometric) resolution should be ± 0.8°C for

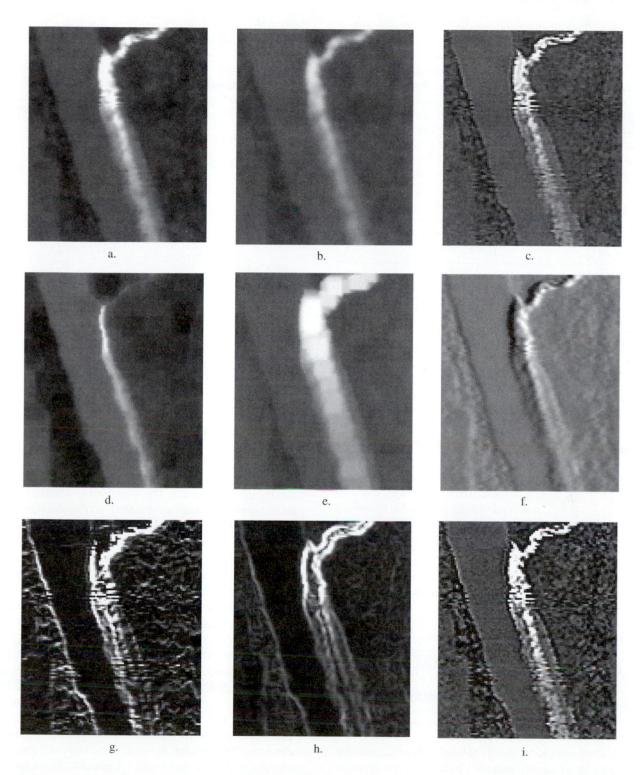

Figure 8-24   a) Original contrast-stretched image of the Four Mile Creek thermal plume in the Savannah River recorded on March 31, 1981, at 4:28 a.m. b) Low-frequency filtered image. c) High-frequency filtered image. d) Application of a minimum filter. e) Application of a maximum filter. f) An embossed filter gives the impression of shaded relief. g) A Roberts edge enhancement. h) A Sobel edge enhancement. i) A Laplacian edge enhancement.

Table 8-8   User Requirements for Conducting a Residential Thermal Imagery Energy Survey

User Requirement	Remote Sensing System Parameter
Temperature (radiometric) resolution should be $\pm$ 0.8°C for differentiation of R-10 and R-20 houses (Colcord, 1981).	Remotely sense in the 8–14 µm region using a mercury-cadmium-telluride (Hg:Cd:Te) detector. This region approximates a 300 K ambient house temperature condition. The true rooftop kinetic temperature of several houses is required if the data are to be calibrated to $\pm$0.8°C.
Obtain 15 x 15 cm spatial resolution cell to get many samples from each roof; however, 1 x 1 m spatial resolution may be sufficient. Rectify data to be within $\pm$1 pixel of their planimetric position.	Use a sensor with an IFOV of 2.5 milliradians flown at 400 m above-ground-level. This will result in a nominal ground spatial resolution of 1 x 1 m per pixel at nadir. Remove tangential scale distortion if possible and rectify to large-scale planimetric maps using ground-control points and nearest-neighbor resampling techniques.
Acquire predawn imagery in the spring and analyze results rapidly.	Acquire imagery when the sky is clear, wind is < 7 km/hr, there is low humidity and no moisture, snow, or ice on the roof. Pre-dawn imagery has no confusing shadows from surrounding buildings or trees.
Produce isotherm maps and extract qualitative and quantitative temperature information. Reduce the information to temperature class levels that the general population can understand (e.g., poor, fair, good, and excellent).	Obtain precise rooftop emissivity characteristics for the neighborhood of interest. Be aware of major differences in 1) rooftop materials, 2) roof pitch and orientation, and 3) whether the dwelling is occupied. Apply image processing and cartographic principles to obtain quantitative information from the thermal images.

Table 8-9   Emissivity of Building Materials from 8 – 14 µm (Colcord, 1981; Wolfe, 1985)

Material	Emissivity, ε
asphalt shingle (dry)	0.97
asphalt shingle (wet)	1.00
cedar-shake shingle (dry)	0.95
cedar-shake shingle (wet)	0.99
tar/stone	0.97
aluminum (sheet)	0.09
copper (oxidized)	0.78
iron (sheet-rusted)	0.69
tin (tin-plated sheet iron)	0.07
brick (red-common)	0.93
paint (average of 16 colors)	0.94
sand	0.90
wood (planed oak)	0.90
frost crystals	0.98

differentiation of an R-10 and R-20 house. This results in the discrimination of approximately four levels of apparent insulation (poor, fair, good, and excellent) based on visual examination. Most houses are approximately 80°F. This approximates a 300 K blackbody with a dominant wavelength of approximately 9.7 µm. Therefore, the scene should be inventoried using an 8 – 14 µm mercury-cadmium-tellu-

ride (Hg:Cd:Te) detector. The true rooftop kinetic temperature of several houses is required if the data are to be calibrated to $\pm$0.8°C.

The sensor system must acquire enough pixels on a rooftop to characterize its thermal environment. One pixel per rooftop is not sufficient. Some have recommended a 15 x 15 cm spatial resolution cell, but a 1 x 1 m spatial resolution appears to be sufficient for most homes (Colcord, 1981), excluding trailer parks. A sensor with an IFOV of 2.5 milliradians flown at 400 m above-ground-level would achieve a nominal ground spatial resolution of 1 x 1 m per pixel at nadir. This will result in problems in hilly terrain as local relief approaches 400 m. Tangential scale distortion should be removed and the data geometrically rectified using ground-control points and nearest-neighbor resampling techniques. The thermal infrared survey should take place under the following conditions, if possible (Anderson and Wilson, 1984):

• standing water, ice, or snow should not be present on the roof, as this dramatically changes the emissivity characteristics of the roof, as summarized in Table 8-9;

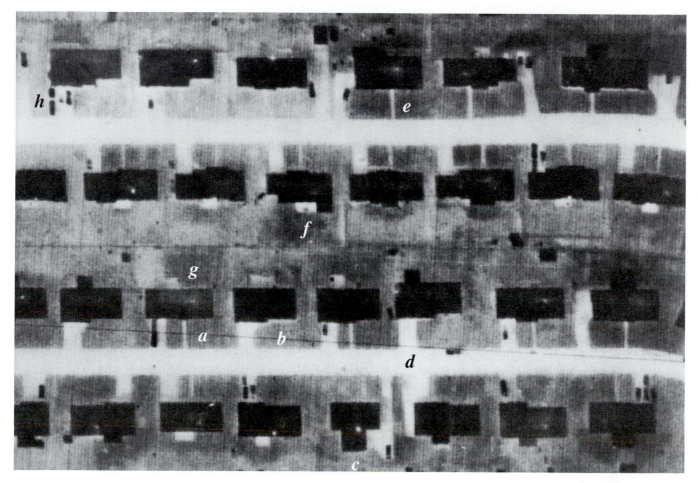

Figure 8-25 Predawn thermal infrared image of a residential subdivision in Fort Worth, TX. The image was acquired at an altitude of 250 m above-ground-level with an across-track scanning system and a 1 milliradian field of view at 6:45 a.m. on January 10, 1980. This yielded an effective ground spatial resolution at nadir of 0.25 x 0.25 m per pixel. The sky was clear.

- wind speed should be < 7 km/hr;

- humidity and ground fog should be minimal; and

- cloud cover should be minimal, as clouds radiate heat back onto the roof.

It is important to obtain precise rooftop-emissivity characteristics for the residential neighborhood or business districts of interest because different rooftop materials have different emissivities. For example, cedar shake shingles have an emissivity of 0.99 when wet and 0.95 when dry, while an aluminum rooftop has an emissivity of 0.09 (Table 8-9).

Ideally, all rooftops are flat. Unfortunately, roof pitch (how steep it is) and orientation toward or away from the sensor will impact the accuracy of the apparent temperature values

derived. Finally, people tend to lower their house temperature in the winter when they leave town for a few days, from approximately 72°F to approximately 55 – 60°F. Therefore, some rooftops that appear cold in a thermal infrared image (i.e., they appear to have good attic insulation) may in fact be unoccupied with the thermostat set low. Thermal infrared imagery of a residential area in Fort Worth, TX, is shown in Figure 8-25. It was acquired at an altitude of 250 m above-ground-level with an across-track scanning system with a 1 milliradian field of view at 6:45 a.m. on January 10, 1980. This yielded an effective ground spatial resolution at nadir of 0.25 x 0.25 m per pixel.

Theoretically, in Figure 8-25, the most poorly insulated home in the subdivision is at (a), while the house next door at (b) appears to be well insulated. There is a fire in the fireplace at home (c). The road network (d), concrete sidewalks *(e)*, and back-porch concrete pads *(f)* are all much warmer than the surrounding terrain. Some of the backyard terrain is more moist *(g)* than other areas. All metal vehicles *(h)* are dark (cold) because of their low emissivity. Most of the homes in this subdivision probably received about the same insulation; therefore, it is not surprising that their rooftop apparent temperatures appear similar.

Some utility companies subcontract for thermal infrared data and then attempt to convince homeowners that their insulation might not be as good as their neighbors'. The only problem with this logic is that 1) the neighbor's home might have a different roof material, 2) the orientation of the neighbor's roof toward the sensor might be different, and 3) the neighbor might keep their home much cooler. If these parameters are taken into consideration, then the rooftop temperature information extracted from the thermal infrared data may be of value to the homeowner.

Thermal infrared energy surveys are also performed routinely for public and industrial facilities. Figure 8-26 depicts an aerial photograph and predawn thermal infrared image of the Solomon Blatt fieldhouse on the University of South Carolina campus. The thermal imagery was acquired at 4:30 a.m. on March 10, 1983, at 500 m above-ground-level using a 2.5 milliradian IFOV sensor system. This resulted in a spatial resolution of approximately 1.25 x 1.25 m.

Analysis of the Urban Heat Island Effect

It is well known that an urban heat island exists over most urban areas compared to the relatively cooler nonurban surrounding countryside. Urban heat islands are caused by deforestation and the replacement of the land surface by nonevaporative and nonporous materials such as asphalt and concrete. In addition, air-conditioning systems introduce a significant amount of heat energy into the urban landscape. The result is reduced evapotranspiration and a general increase in urban landscape temperature.

Quattrochi and Ridd (1994) and Lo et al. (1997) evaluated several cities using high spatial resolution thermal infrared remote sensing systems to document the urban heat island effect. In general, they found that during the daytime hours commercial land cover exhibited the highest temperatures followed by services, transportation, and industrial land uses. The lowest daytime temperatures were found over water bodies, vegetation, and agricultural land use, in that order. Residential housing being composed of a heterogeneous mixture of buildings, grass, and tree cover exhibited an intermediate temperature, as expected.

At night, commercial, services, industrial, and transportation land cover types cooled relatively rapidly. Nevertheless, their temperatures even in the predawn early morning hours were still slightly higher than those for vegetation and agriculture. Water has a high thermal capacity; therefore, it is typically the warmest land cover during the predawn hours as previously discussed. Conversely, agriculture typically exhibits the lowest temperature at night.

Examples of daytime and nighttime thermal infrared images of Atlanta, GA, confirm these observations (Figure 8-27ab). The ATLAS channel 13 imagery (9.60 – 10.2 μm) at 10 x 10 m spatial resolution was obtained as part of the NASA EOS investigation Project ATLANTA (Quattrochi and Luvall, 1999). Shadows from tall buildings located in the Atlanta CBD are observed on the daytime imagery. The intense thermal exitance from buildings, pavement, and other surfaces typical of the urban landscape, as well as the heterogeneous distribution of these responses, stand in significant contrast to the relative "flatness" of the Atlanta thermal landscape at night. Also, the dampening effect that the urban forest has on upwelling thermal energy response is evident, particularly in the northeast portion of the daytime image where residential tree canopy is extensive. In the nighttime image there is still evidence, even in the very early morning, of the elevated thermal responses from buildings and other surfaces in the Atlanta CBD and from streets and freeways. Thermal energy responses for vegetation across the image are relatively uniform at night, regardless of vegetation type (e.g., grass, trees).

Such thermal information has been used to 1) model the relationship between Atlanta urban growth, land-cover change, and the development of the urban heat island phenomenon through time, 2) model the relationship between Atlanta urban growth and land-cover change and air quality through time, and 3) model the overall effects of urban development on surface-energy budget characteristics across the Atlanta urban landscape. Such data can also be used to recommend tree-planting programs that may be able to substantially decrease the urban heat island effect (Quattrochi and Ridd, 1998).

Use of Thermal Infrared Imagery for Forestry Applications

Quantitative information about forest canopy structure, biomass, age, and physiological condition have been extracted from thermal infrared data. Basically, a change in surface temperature can be measured by an airborne thermal infrared sensor (e.g., TIMS or ATLAS) by repeatedly flying over

Vertical Aerial Photograph

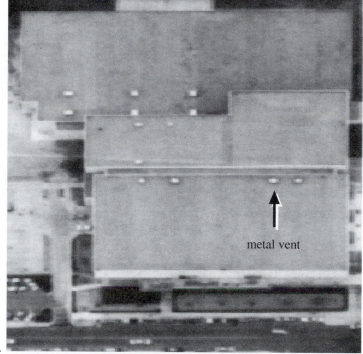

a.

Predawn Thermal Infrared Image

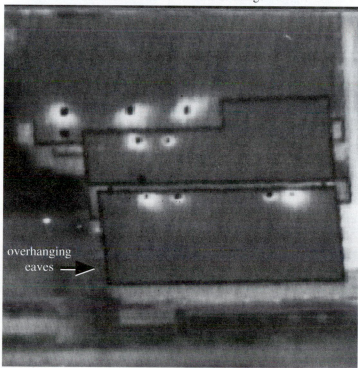

b.

Figure 8-26 a) Aerial photograph of the Solomon Blatt fieldhouse on the University of South Carolina campus, Columbia, SC. b) Thermal infrared image obtained at 4:30 a.m. on March 10, 1983, with a spatial resolution of 1.25 x 1.25 m. Note 11 cool metal vents, the hot air escaping from nine of the vents, and the cool overhanging eaves. The eaves are exposed on both sides to the cool early morning air.

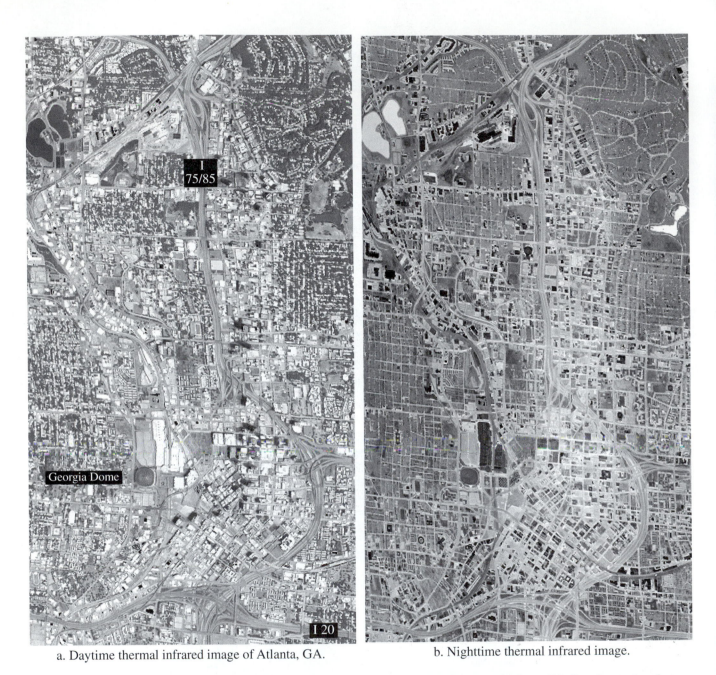

a. Daytime thermal infrared image of Atlanta, GA.   b. Nighttime thermal infrared image.

Figure 8-27   a) Daytime ATLAS airborne thermal infrared channel 13 (9.60 – 10.2 μm) image of Atlanta, GA. Prominent urban features are annotated for orientation. b) Nighttime, predawn thermal infrared image of Atlanta, GA (courtesy Dale Quattrochi and Jeff Luvall, Project ATLANTA – NASA Marshall Space Flight Center, Global Hydrology Climate Center, Huntsville, AL).

the same area a few times. Usually a separation of about 30 minutes results in a measurable change in surface temperature caused by the change in incoming solar radiation; however, a longer period of time may be required. Average surface net radiation (R_n) is measured *in situ* for the study area and is used to integrate the effects of the nonradiative

fluxes. The change in surface temperature (ΔT) from time period t_1 to t_2 (i.e., Δt) is the value that reveals how those nonradiative fluxes are reacting to radiant energy inputs. The ratio of these two parameters is used to compute a surface property defined by Luvall and Holbo (1989) and Luvall (1997) as a *Thermal Response Number* (TRN) (KJ m^{-2} °C^{-1}):

$$TRN = \frac{\sum\limits_{t_1}^{t_2}[(R_n)(\Delta t)]}{\Delta T} \quad . \qquad (8\text{-}24)$$

Terrain containing mostly soil and bare rock have the lowest TRN values, while forests have the highest. The TRN is a site-specific property that may be used to discriminate among various types of coniferous forest stands and some of their biophysical characteristics (Luvall, 1990; 1997).

References

Abrams, M. J., A. B. Kahle, F. D. Falluconi and J. P. Schieldge, 1984, "Geologic Mapping Using Thermal Images," *Remote Sensing of Environment*, 16(1):33.

Anderson, J. E., 1992, "Determination of Water Surface Temperatures Based on the Use of Thermal Infrared Multispectral Scanner Data," *Geocarto International*, 7(3):3–8.

Anderson, J. M. and S. B. Wilson, 1984, "The Physical Basis of Current Infrared Remote Sensing Techniques and the Interpretation of Data from Aerial Surveys," *Remote Sensing of Environment*, 5(1):1–18.

Avery, T. E. and G. L. Berlin, 1992, *Fundamentals of Remote Sensing and Airphoto Interpretation*, N.Y.: Macmillan Co., 472 pp.

Becker, F. and A. L. Li, 1990, "Towards a Local Split Window Method over Land Surfaces," *International Journal of Remote Sensing*, 11:369–393.

Becker, F. and A. L. Li, 1995, "Surface Temperature and Emissivity at Various Scales: Definition, Measurement and Related Problems," *Remote Sensing of Environment*, 12:225–253.

Campbell, J. B., 1996, *Introduction to Remote Sensing*, 2nd Ed., New York: Guilford Press.

Colcord, J. E., 1981, "Thermal Imagery Energy Surveys," *Photogrammetric Engineering & Remote Sensing*, 47(2):237–240.

Curran, P. J., 1985, *Principles of Remote Sensing*, New York: Longman, 282 pp.

Czajkowski, K. P., T. Mulhern, S. N. Goward, J. Cihlar, R. O. Dubayah, and S. D. Prince, 1997, "Biospheric Environmental Monitoring at BOREAS with AVHRR Observations," *Journal of Geophysical Research*, 102:651–663.

Czajkowski, K. P., S. N. Goward, T. Mulhern, S. J. Goetz, D. Shirey, S. Stadler, S. D. Prince and R. O. Dubayah, 1997b, "Recovery of Environmental Variables from Thermal Remote Sensing," *Scale in Remote Sensing and GIS*, (Eds.) D. A. Quattrochi and M. F. Goodchild, Chelsea: Lewis Publishers.

Estes, J. E., 1966, "Some Applications of Aerial Infrared Imagery," *Annals* of the Association of American Geographers, 56(4):673–682.

Finney, P., 1996, "IR Imaging with Uncooled Focal Plane Arrays," *Sensors*, (October), 36–43.

Fischer, W. A., 1983, "History of Remote Sensing," *Manual of Remote Sensing*, R. G. Reeves (Ed.), Falls Church, VA: American Society of Photogrammetry, 27–50.

FLIR Systems, 1999, *Infrared Imaging Systems*, Portland, OR: FLIR Systems, Inc., technical specifications and examples at http://www.FLIRsystems.com.

Herring, D., 1998, *NASA's Earth Observing System*, Washington, DC, NASA, 34 pp.

Janza, F. J., 1975, "Interaction Mechanisms," *Manual of Remote Sensing*, R. G. Reeves (Ed.), Falls Church, VA: American Society of Photogrammetry, 75–179.

Jensen, J. R., 1996, *Introductory Digital Image Processing: A Remote Sensing Perspective*, Upper Saddle River, N.J.: Prentice-Hall, Inc., 319 pp.

Jensen, J. R., M. E. Hodgson, E. J. Christensen, H. E. Mackey, L. R. Tinney and R. Sharitz, 1986, "Remote Sensing Inland Wetlands: A Multispectral Approach," *Photogrammetric Engineering & Remote Sensing*, 52(1)87–100.

Jensen, J. R., P. J. Pace and E. J. Christensen, 1983, "Remote Sensing Temperature Mapping: the Thermal Plume Example," *The American Cartographer*, 10:111–127.

Jensen, J. R., E. Ramsey, H. E. Mackey and M. E. Hodgson, 1988, "Thermal Modeling of Heat Dissipation in the Pen Branch Delta Using Thermal Infrared Imagery," *Geocarto International*, 4:17–28.

Kahle, A. B., 1980, "Surface Thermal Properties," *Remote Sensing in Geology*, B. Siegal and A. Gillespie (Ed.), New York: John Wiley, 257–273.

Kahle, A. B., J. P. Schieldge, M. J. Abrams, R. E. Alley and C. J. VeVine, 1981, *Geologic Applications of Thermal Inertia Imaging Using HCMM Data*, Pasadena: Jet Propulsion Lab Pub. #81–55.

Kahle, A. B., J. P. Schieldge and R. E. Alley, 1984, "Sensitivity of Thermal Inertia Calculations to Variations in Environmental Factors," *Remote Sensing of Environment*, 16:211–232.

Kalluri, S. N. and R. O. Dubayah, 1995, "Comparison of Atmospheric Correction Models for Thermal Bands of the AVHRR over FIFE," *Journal of Geophysical Research*, 100:411–418.

Kennedy, W. V, 1983, *Intelligence Warfare*, New York: Crescent Books, 208 pp.

Kruse, P. W., 1995, "Thermal Imagers Move from Military to Marketplace," *Photonics Spectra*, 29(3):103–108.

Lillesand, T. M. and R. W. Kiefer, 1994, *Remote Sensing and Image Interpretation*, New York: John Wiley & Sons, 750 pp.

Lo, C. P., D. A. Quattrochi and J. C. Luvall, 1997, "Application of High-resolution Thermal Infrared Remote Sensing and GIS to Assess the Urban Heat Island Effect," *International Journal of Remote Sensing*, 18(2):287–304.

Luders, J., S. Schill, J. R. Jensen, and G. Olson, 1999, *Sand and Gravel Particle Size Discrimination Using Airborne Terrestrial Applications Sensor (ATLAS)*, SSC, MS: NASA Stennis Space Center, Commercial Remote Sensing Report, 45 pp.

Luvall, 1997, "The Use of Remotely Sensed Surface Temperatures from an Aircraft Based Thermal Infrared Multispectral Scanner (TMS) to Estimate the Spatial Variability of Latent Heat Fluxes from a White Pine (*Pinus strobus* L.) Plantation," *Scale in Remote Sensing and GIS*, (Eds.), D. A. Quattrochi and M. F. Goodchild, Chelsea: Lewis Publishers, 169–185.

Luvall, J. C., 1999, personal correspondence, NASA Marshall Space Flight Center.

Luvall, J., C. D. Lieberman, M. Liberman, G. S. Hartshorn and R. Peralta, 1990, "Estimation of Tropical Forest Canopy Temperatures, Thermal Response Numbers, and Evapotranspiration Using an Aircraft-based Thermal Sensor," *Photogrammetric Engineering and Remote Sensing*, 56(10):1393–1401.

Luvall, J. C. and H. R. Holbo, 1989, "Measurement of Short-term Thermal Responses of Coniferous Forest Canopies Using Thermal Scanner Data," *Remote Sensing of Environment*, 27:1–10.

Lyon, R. J. P., 1965, "Analysis of Rocks by Spectral Infrared Emission," *Economic Geology*, 60:715-736.

Lytle, D., 1996, "Imager Finds 'Hot Spots' in Raging Blazes," *Photonics Spectra*, (October), 16–18.

Marsh, W. M. and J. Dozier, 1981, *Landscape: An Introduction to Physical Geography*, New York: Addison-Wesley, 637 pp.

McDaid, H. and D. Oliver, 1997, *Smart Weapons: Top Secret History of Remote Controlled Airborne Weapons*, New York: Barnes & Noble, 208 pp.

McDonnell Douglas, 1982, *Reconnaissance Handy Book for the Tactical Reconnaissance Specialist*, St. Louis: McDonnell Douglas, Inc., 183 pp.

Mulligan, J. F., 1980, *Practical Physics: The Production and Conservation of Energy*, New York: McGraw Hill, 526 pp.

Palluconi, F. D. and G. R. Meeks, 1985, *Thermal Infrared Multispectral Scanner (TIMS): An Investigator's Guide to TIMS Data*, Pasadena, CA: NASA Jet Propulsion Laboratory, JPL Publication #85–32, 26 pp.

Price, J. C., 1985, "On the Analysis of Thermal Infrared Imagery: The Limited Utility of Apparent Thermal Inertia," *Remote Sensing of Environment*, 18:59–73.

Quattrochi, D. A. and N. S. Goel, 1995, "Spatial and Temporal Scanning of Thermal Infrared Remote Sensing Data," *Remote Sensing Reviews*, 12:225–286.

Quattrochi, D. A. and J. C. Luvall, 1999, *High Spatial Resolution Airborne Multispectral Thermal Infrared Data to Support Analysis and Modeling Tasks in the EOS IDS Project ATLANTA*, http://wwwghcc.msfc.nasa.gov/atlanta/.

Quattrochi, D. A. and M. K. Ridd, 1994, "Measurement and Analysis of Thermal Energy Responses from Discrete Urban Surfaces Using Remote Sensing Data," *International Journal of Remote Sensing*, 15(10):1991–2022.

Quattrochi, D. A. and Merrill K. Ridd, 1998, "Analysis of Vegetation within a Semi-arid Urban Environment Using High Spatial Resolution Airborne Thermal Infrared Remote Sensing Data," *Atmospheric Environment*, 32(1):19–33.

Sabins, F. F., Jr., 1997, *Remote Sensing Principles and Interpretation*, New York: W. H. Freeman and Co., 494 pp.

Salisbury, J. W. and D. M. D'Aria, 1992, "Emissivity of Terrestrial Materials in the 8 – 14 μm Atmospheric Window," *Remote Sensing of Environment*, 42:83–106.

Sanders, J. S. and M. S. Currin, 1991, "Human Recognition of Infrared Images," *Proceedings of the Infrared Imaging Systems: Design, Analysis, Modeling, and Testing II*, Bellingham: SPIE - International Society for Photo-Optical Instrumentation Engineers,1488:144–155.

Schott, J. R., 1997, *Remote Sensing the Image Chain Approach*, New York: Oxford University Press, 394 pp.

Silverman, J., J. M. Mooney and F. D. Shepherd, 1992, "Infrared Video Cameras," *Scientific American*, (March):78–83.

Slater, P. N., 1980, *Remote Sensing: Optics and Optical Systems*, New York: Addison-Wesley, Inc., 575 pp.

Snyder, W. C., W. Zhengming, Y. Zhan and Y. Feng, 1997, "Thermal Infrared (3 – 14 μm) Bi-directional Reflectance Measurements of Sands and Soils," *Remote Sensing of Environment*, 60:101–109.

Trefil, J. and R. M. Hazen, 1995, *The Sciences: An Integrated Approach*, New York: John Wiley & Sons, 634 pp.

Wallace, K., 1999, *ATLAS - Airborne Terrestrial Applications Sensor Specifications*, SSC, MS: NASA Stennis Space Center, 20 pp.

West, L., 1996, "Innovations in IR Focal-Plane Array Cameras," *Lasers & Optronics*, 29–31.

Wimmers, J. T. and D. S. Smith, 1994, "Better, Smaller IR Imagers Lead the Way to New Applications," *Photonics Spectra*, 28(12):113–118.

Wolfe, W. L., 1985, *Infrared Handbook*, Ann Arbor, MI: ERIM.

Active and Passive Microwave, and LIDAR Remote Sensing

9

Passive remote sensing systems record electromagnetic energy that was reflected (e.g., blue, green, red, and near-infrared light) or emitted (e.g., thermal infrared energy) from the surface of the Earth. There are also *active* remote sensing systems that are not dependent on the Sun's electromagnetic energy or the thermal properties of the Earth. Active remote sensors create their own electromagnetic energy that 1) is transmitted from the sensor toward the terrain (and is largely unaffected by the atmosphere), 2) interacts with the terrain producing a backscatter of energy, and 3) is recorded by the remote sensor's receiver. The most widely used active remote sensing systems include:

- *active microwave (RADAR),* which is based on the transmission of long-wavelength microwaves (e.g., 3 – 25 cm) through the atmosphere and then recording the amount of energy backscattered from the terrain;

- *LIDAR,* which is based on the transmission of relatively short-wavelength laser light (e.g., 0.90 μm) and then recording the amount of light backscattered from the terrain; and

- *SONAR,* which is based on the transmission of sound waves through a water column and then recording the amount of energy backscattered from the bottom or from objects within the water column.

Of the three, RADAR remote sensing is the most widely used for Earth resource observations, although SONAR and LIDAR are quite useful for specific applications (e.g., bathymetric mapping and measurement of terrain elevation, respectively). It is also possible to record passive microwave energy that is naturally emitted from the surface of the Earth using a *passive microwave radiometer.* This chapter includes an overview of passive microwave remote sensing as well as the fundamental principles of LIDAR.

 ### History of Active Microwave (RADAR) Remote Sensing

James Clerk Maxwell (1831– 1879) provided the fundamental mathematical descriptions of the magnetic and electric fields associated with electromagnetic radiation. Then, Heinrich R. Hertz (1857 – 1894) increased our knowledge about the creation and propagation of electromagnetic energy in the microwave and radio portions of the spectrum. Hertz also studied the interaction of radio waves with metallic surfaces and thus initiated some of the early thinking that would eventually lead to radios and radars. Building on the fundamental physics principles discovered by Maxwell and Hertz, Guglielmo M. Marconi (1874 – 1937) constructed an antenna that transmitted and received radio signals. In 1901, he sent radio waves across the Atlantic and in 1909 shared the Nobel prize in physics for his work.

RADAR as we know it was investigated by A. H. Taylor and L. C. Young in 1922. These scientists positioned a high-frequency radio transmitter on one side of the Anacostia River near Washington, DC and a receiver on the opposite side. Ships passing up or down the river interrupted the very long wavelength radio signal (1 – 10 m) sent between the transmitter and receiver. Such systems provided the first clues that radio signals might be useful for detecting the distance to ships (i.e., the range) at sea. This had implications for ship navigation because the active radio transmission and reception could take place both at night and even in bad weather (Campbell, 1996). The military took an early interest in radar because in times of war it is important to know the location of all ships and planes (friendly and enemy) at all times. A phrase describing the process was "**ra**dio **d**etection **a**nd **r**anging" or *RADAR*. Although radar systems now use microwave wavelength energy almost exclusively instead of radiowaves, the acronym was never changed.

By 1935 Young and Taylor (and independently Sir Robert Watson-Watt in Great Britain) combined the antenna transmitter and receiver in the same instrument. Eventually, high-power transmissions in very specific bands of the electromagnetic spectrum were possible. These and other electronic advancements laid the groundwork for the development of RADAR during World War II for navigation and target location. By late 1936 experimental radars were working in the United States, Great Britain, Germany, and the Soviet Union. Of these nations, none was more vulnerable to air attack than Britain. London lay within 90 miles of foreign territory and within 275 miles of Germany. Under the direction of Air Chief Marshal Hugh Dowding, the British constructed 21 radars along the south and east coasts of England and the east coast of Scotland. The code name "Chain Home — CH" arose from a plan to build a "home chain" of radars in Britain to be followed by an overseas chain to defend threatened portions of the British empire. The CH radars were the primary British long-range early warning radars until well into World War II. The radars allowed incoming planes to be detected out to a maximum range of 50 miles. Without the radar, Britain would not have been able to effectively counter the German Luftwaffe bombers and fighter escorts especially during the Battle of Britain, which began in July 1940 (Fisher, 1988; Price,

1990). The circularly scanning Doppler radar that we watch every day during television weather updates to identify the geographic location of storms around cities is based on the same circularly scanning radar concept (plan-position indicator radar, PPI) found in the original World War II radars. PPI radars are also used for air-traffic control at airports.

RADAR images obtained from aircraft or spacecraft as we know them today were not available during World War II. The continuous-strip mapping capability of *side-looking airborne radar (SLAR)* was not developed until the 1950s. An important advantage of SLAR is its ability to obtain reconnaissance images over vast regions to the left or right of the aircraft (Sabins, 1997). This is called *long-range standoff* because the pilot can fly at the edge of friendly air-space while obtaining detailed RADAR imagery far into potentially unfriendly air space (Avery and Berlin, 1992). It also became possible to perform *radargrammetric* measurement — the science of extracting quantitative geometric object information from radar images (Leberl, 1990; Henderson and Lewis, 1998).

The military began using SLARs in the 1950s. By the mid-1960s some systems were declassified. There are two primary types of SLAR: *real aperture radar* (also known as brute-force radar) and *synthetic aperture radar (SAR)*, where the word "aperture" means *antenna*. Real aperture radars use an antenna of fixed length, e.g., 1 – 2 m. Synthetic aperture radars also use a 1 – 2 m antenna, but they are able to synthesize a much larger antenna (e.g., perhaps 600 m in length), which has improved resolving power. SARs achieve very fine resolution from great distances. For example, an 11-m SAR antenna on an orbital platform can be synthesized electronically to have a synthetic length of 15 km. Later it will be demonstrated how important this capability is to SAR azimuth resolution. Most military and civilian commercial RADARS are now synthetic aperture radars.

In the 1960s and 1970s, both real aperture and synthetic aperture SLARs were used extensively for rapid Earth resource reconnaissance mapping of vast, previously unmapped regions of the continents. Early studies focused on continents perennially shrouded in cloud cover. The first large-scale project for mapping terrain with side-looking airborne radar was a complete survey of the province of Darien, which connects Panama and South America. In 1968, Westinghouse Electric, Inc., in cooperation with Raytheon, Inc., employed a real aperture SLAR that Westinghouse developed for the U.S. Army and successfully made a 20,000 km^2 mosaic of this area (Jensen et al., 1977). Up to this time, the area had never been seen or mapped in its entirety because of almost perpetual cloud cover.

Table 9-1.   Characteristics of Selected Earth Orbiting Synthetic Aperture Radars (SAR)

Parameter	SEASAT	SIR-A	SIR-B	SIR-C/X-SAR	ALMAZ-1	ERS-1,2	JERS-1	RADARSAT
Launch date	June 26, 1978	November 12, 1981	October 5, 1984	April 1994 October 1994	March 31, 1991	1991 1995	February 11,1992	November 1995
Nationality	USA	USA	USA	USA	Soviet Union	Europe	Japan	Canada
Wavelength, cm	L - (23.5)	L - (23.5)	L - (23.5)	X - (3.0) C - (5.8) L - (23.5)	S - (9.6)	C - (5.6)	L - (23.5)	C - (5.6)
Depression angle, ° (near to far-range) [incident angle]	73 − 67° [23°]	43 − 37° [50°]	75 − 35° [15 − 64°]	variable [15 − 55°]	59 − 40° [30 − 60°]	67° [23°]	51° [39°]	70 − 30° [10° − 60°]
Polarization	HH	HH	HH	HH, HV, VV,VH	HH	VV	HH	HH
Azimuth resolution, m	25	40	17 − 58	30	15	30	18	8 − 100
Range resolution, m	25	40	25	10 − 30	15 − 30	26	18	8 − 100
Swath width, km	100	50	10 − 60	15 − 90	20 − 45	100	75	50 − 500
Altitude, km	800	260	225 and 350	225	300	785	568	798
Latitude coverage, °	10° - 75° N	41° N - 36° S	60° N - 60° S	57° N - 57° S	73° N - 73° S	near-polar orbit	near-polar orbit	near-polar orbit
Mission duration	105 days	2.5 days	8 days	10 days	18 months	--	6.5 years	--

Goodyear Aerospace, Inc., and the Aero Service Division of Western Geophysical, Inc., adapted a synthetic aperture radar built by Goodyear and installed it in an Aero Service Caravelle jet for civilian surveys. In 1971 they initiated project RADAM (Radar of the Amazon) to map the Amazon Basin in Venezuela and Brazil. Approximately 4 million km^2 of land area (about half the size of the United States) were recorded and assembled into radar mosaics. Goodyear and Aero Service eventually surveyed the Amazon rain forest in Brazil, Venezuela, eastern Colombia, Peru, and Bolivia. Subsequent radar investigations mapped Guatemala, Nigeria, and Togo, and portions of Indonesia, the Philippines, Peru, and other regions that were previously unmapped (Leberl, 1990).

NASA has launched two successful SARs, SEASAT and the ongoing Space Shuttle Imaging Radar experiments. *SEASAT* (for "sea satellite") was launched in 1978 to obtain L-band (23.5 cm) 25 x 25 m spatial resolution oceanographic information (Table 9-1). This was the first orbital SAR that provided public-domain data. It also provided valuable land information, but functioned for only 105 days. The *Shuttle Imaging Radar* experiment A *(SIR-A)* with its L-band (23.5

cm) 40 x 40 m SAR was launched in 1981, SIR-B was launched in 1984 with its 17 x 25 m resolution, and SIR-C in 1994 with its multifrequency and multipolarization capability. SIR-A, SIR-B, and SIR-C missions lasted only 2.5, 8, and 10 days, respectively.

The former Soviet Union launched the ALMAZ-1 S-band (9.6 cm) radar in 1991. The European Space Agency (ESA) launched their *European Remote Sensing Satellite ERS-1* with its C-band (5.6 cm) imaging radar in 1991 and *ERS-2* in 1995. Japan launched the L-band (23.5 cm) *Japanese Earth Resources Satellite JERS-1* in 1992. The Canadian government placed the C-band (5.6 cm) *RADARSAT-1* in orbit in 1995. RADARSAT-2 is scheduled for launch early in the twenty-first century. The system characteristics of these orbital SAR systems are summarized in Table 9-1 and will be discussed later in this chapter.

Active and passive microwave remote sensing will continue to be very important research areas in the coming decades. Such sensors provide the only viable information for the tropical portions of the world where extensive, fragile ecosystems are at risk and under perennial cloud cover. For a

a. Intermap LearJet 36 Star 3*i*

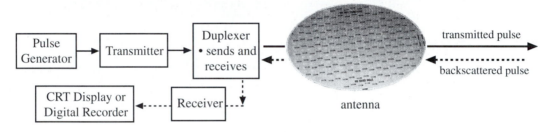

b. Typical active microwave system components

Figure 9-1   a) A side-looking airborne radar (SLAR) called the Star 3*i* IFSAR (interferometric synthetic aperture radar) is mounted underneath a LearJet 36 aircraft (courtesy Intermap Technologies, Inc.). b) The active microwave system components and flow of energy through a typical SLAR. It consists of a pulse-generating device, transmitter, duplexer, antenna, receiver, and a digital recorder. Some systems have a quick-look CRT display to make sure data is being collected. The 1-m synthetic aperture antenna shown can synthesize an antenna hundreds of meters long.

more detailed history of radar development, see Leberl (1990) and Henderson and Lewis (1998). The primary and secondary advantages of RADAR remote sensing are summarized in Table 9-2.

Active Microwave System Components

Active microwave imagery is obtained using instruments and principles that are different from those used when acquiring data in the visible, near-, mid-, and thermal infrared portions of the spectrum using passive remote sensing techniques. Therefore, it is necessary to understand the active microwave system components and how electromagnetic energy is sent, received, and recorded by the sensor system before an analyst can accurately interpret a radar image. The discussion is based initially on the system components and functions of a *real aperture side-looking airborne radar* (SLAR). The discussion then expands to

include *synthetic aperture radars* (SAR) that have improved capabilities.

Sending and Receiving a Pulse of Microwave Electromagnetic Energy - System Components

A typical active microwave SLAR is shown mounted on an aircraft in Figure 9-1a. It consists of a pulse-generating device, a transmitter, a duplexer that carefully coordinates when the active microwave energy is transmitted and received, an antenna, a receiver, a recording device such as a high-density digital tape recorder or hard disk, and typically a CRT monitor so that the technician in the plane can make sure radar imagery is actually being collected (Figure 9-1b). The radar can also be mounted onboard a satellite. The pulse-generating device sends a pulse of electromagnetic energy at a specific wavelength (frequency) to the transmitter. When in sending mode, the duplexer then sends the polarized pulse of energy through the antenna toward the Earth's surface.

Table 9-2.   Advantages of RADAR Remote Sensing of the Environment (after Leberl, 1990)

Advantages
Primary
• Active microwave energy penetrates clouds and serves as an all-weather remote sensing system.
• Synoptic views of large areas, for mapping at 1:25,000 to 1:400,000. Satellite coverage of entire cloud-shrouded countries is possible.
• Coverage can be obtained at user-specified times, even at night.
• Permits imaging at shallow look angles, resulting in different perspectives that cannot always be obtained using aerial photography.
• Senses in wavelengths outside the visible and infrared regions of the electromagnetic spectrum, providing information on surface roughness, dielectric properties, and moisture content.
Secondary
• May penetrate vegetation, sand, and surface layers of snow.
• Has its own illumination, and the angle of illumination can be controlled.
• Enables resolution to be independent of distance to the object, with the size of a resolution cell being as small as 1 x 1 m.
• Images can be produced from different types of polarized energy (HH, HV, VV, VH).
• May operate simultaneously in several wavelengths (frequencies) and thus has multifrequency potential.
• Can measure ocean wave properties, even from orbital altitudes.
• Can produce overlapping images suitable for stereoscopic viewing and radargrammetry.
• Supports interferometric operation using two antennas for 3-D mapping, and analysis of incident-angle signatures of objects.

Wavelength, Frequency and Pulse Length

The pulse of electromagnetic radiation sent out by the transmitter through the antenna is of a specific wavelength and duration (i.e., it has a *pulse length* measured in microseconds, μsec). The wavelengths of energy most commonly used in imaging radars are summarized in Table 9-3. The wavelengths are much longer than visible, near-infrared, mid-infrared, or thermal infrared energy used in other remote sensing systems (Figure 9-2). Therefore, microwave energy is usually measured in centimeters rather than micrometers (Carver, 1988). The unusual names associated with the radar wavelengths (e.g., K, K_a, K_u, X, C, S, L, and P) are an artifact of the original secret work on radar remote sensing when it was customary to use the alphabetic descriptor instead of the actual wavelength or frequency. These descriptors are still used today in much of the radar scientific

Table 9-3.   RADAR Wavelengths and Frequencies Used in Active Microwave Remote Sensing Investigations

RADAR Band Designations (common wavelengths shown in parentheses)	Wavelength (λ) in cm	Frequency (υ) in GHz
K_a (0.86 cm)	0.75 – 1.18	40.0 – 26.5
K	1.19 – 1.67	26.5 – 18.0
K_u	1.67 – 2.4	18.0 – 12.5
X (3.0 and 3.2 cm)	2.4 – 3.8	12.5 – 8.0
C (7.5, 6.0 cm)	3.9 – 7.5	8.0 – 4.0
S (8.0, 9.6, 12.6 cm)	7.5 – 15.0	4.0 – 2.0
L (23.5, 24.0, 25.0 cm)	15.0 – 30.0	2.0 – 1.0
P (68.0 cm)	30.0 – 100	1.0 – 0.3

literature. SIR-C and the NASA Jet Propulsion Lab suborbital AIRSAR radar operate in more than one frequency. These systems produce *multifrequency radar* imagery.

Table 9-3 and Figure 9-2 also provide the radar band designation in units of frequency measured in billions of cycles per second (Gigahertz or GHz, 10^9 cycles sec^{-1}). Earth resource image analysts seem to grasp the concept of wavelength more readily than frequency, so the convention is to describe a radar in terms of its wavelength. Conversely, engineers generally prefer to work in units of frequency because as radiation passes through materials of different densities, frequency remains constant while velocity and wavelength change. Since wavelength (λ) and frequency (υ) are inversely related to the speed of light (c), it really does not matter which unit of measurement is used as long as one remembers the following relationships:

$$c = \lambda \upsilon \tag{9-1}$$

$$\lambda = \frac{3 \times 10^8 \text{m sec}^{-1}}{\upsilon} \tag{9-2}$$

$$\upsilon = \frac{3 \times 10^8 \text{m sec}^{-1}}{\lambda}. \tag{9-3}$$

The following simple equation can be used to rapidly convert frequencies into units of radar wavelength:

$$\lambda \text{ in centimeters} = \frac{30}{\upsilon \text{ (in GHz)}}. \tag{9-4}$$

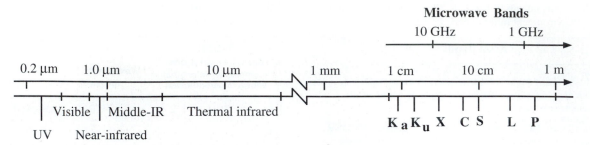

Figure 9-2	The wavelength and frequency of commonly used RADAR bands. RADAR antennas transmit and receive very long wavelength energy measured in centimeters, unlike the relatively short wavelength visible, near-infrared, middle-infrared, and thermal infrared regions measured in micrometers (after Carver, 1988).

Azimuth and Range Direction, Depression Angle, Incident Angle, and Polarization

There are additional parameters that must be known about the nature of the pulse of microwave energy that is sent by the radar antenna to the terrain, including aircraft *azimuth direction*, *radar range* or *look direction*, *depression angle*, *incident angle*, and *polarization*. The following discussion is based on the use of an airborne radar system, although the principles are equally applicable to satellite radar systems.

Azimuth Direction: In a typical SLAR configuration the antenna is mounted beneath and parallel to the aircraft fuselage (Figure 9-3). The aircraft travels in a straight line that is called the *azimuth flight direction*. Pulses of active microwave electromagnetic energy illuminate strips of the terrain at right angles (orthogonal) to the aircraft's direction of travel, which is called the *range* or *look direction*. Figure 9-3 depicts an aircraft equipped with a SLAR system that is illuminating the terrain on one side of the aircraft but not beneath it. The pulses of energy sent out in the range direction only illuminate a certain part of the terrain. The terrain illuminated nearest the aircraft in the line of sight is called the *near-range* (Figure 9-3). The farthest point of terrain illuminated by the pulse of energy is called the *far-range*.

Range Direction: The *range* or *look direction* for any *radar* image is the direction of the radar illumination that is at right angles to the direction the aircraft or spacecraft is traveling. Look direction usually has a significant impact on feature interpretation. The extent to which linear features are enhanced or suppressed on the imagery depends significantly on their orientation relative to a given look direction of radar illumination. Generally, objects that trend (or strike) in a direction that is orthogonal (perpendicular) to the range or look direction are enhanced much more than those objects in the terrain that lie parallel to the look direction. Consequently, linear features that appear dark or are imperceptible

in a radar image using one look direction may appear bright in another radar image with a different look direction. A good example of this is demonstrated in Figure 9-4 which shows radar imagery of an area in Nigeria, West Africa, that were obtained using two different look directions. Note how certain terrain features are emphasized and/or deemphasized in the two images.

Depression Angle: The *depression angle* (γ) is the angle between a horizontal plane extending out from the aircraft fuselage and the electromagnetic pulse of energy from the antenna to a specific point on the ground (Figure 9-3). The depression angle within a strip of illuminated terrain varies from the *near-range depression angle* to the *far-range depression angle*. The *average depression angle* of a radar image is computed by selecting a point midway between the near and far-range in the image strip. Summaries of radar systems often only report the average depression angle.

Incident Angle: The *incident angle* (θ) is the angle between the radar pulse of electromagnetic energy and a line perpendicular to the Earth's surface where it makes contact. When the terrain is flat, the incident angle (θ) is the complement ($\theta = 90 - \gamma$) of the depression angle (γ). However, if the terrain is sloped, there is no relationship between depression angle and incident angle. The incident angle best describes the relationship between the radar beam and surface slope. A schematic diagram of the relationship is shown in Figure 9-5. Many mathematical radar studies assume the terrain surface is flat (horizontal) therefore, the incident angle is assumed to be the complement of the depression angle.

Polarization: Unpolarized energy vibrates in *all* possible directions perpendicular to the direction of travel. Radar antennas send and receive *polarized* energy. This means that the pulse of energy is filtered so that its electrical wave vibrations are only in a single plane that is perpendicular to the direction of travel. The pulse of electromagnetic energy

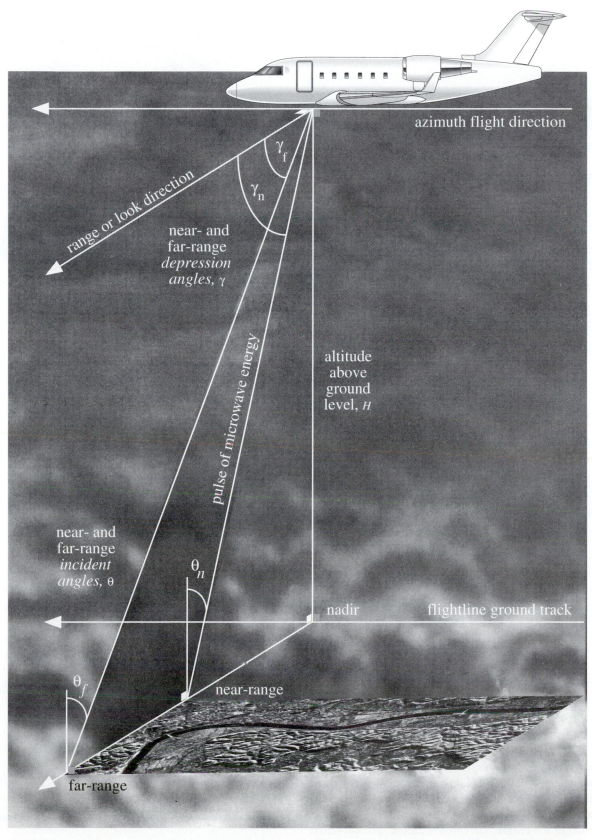

azimuth flight direction

range or look direction

γ_f

γ_n

near- and far-range *depression angles,* γ

pulse of microwave energy

altitude above ground level, H

near- and far-range *incident angles,* θ

θ_n

nadir

flightline ground track

θ_f

near-range

far-range

Figure 9-3   Geometric characteristics of radar imagery acquired by a side-looking airborne radar (SLAR) through cloud cover. All the nomenclature assumes that the terrain is flat.

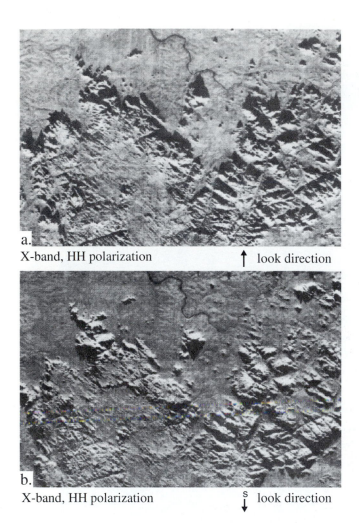

a.

X-band, HH polarization   ↑ look direction

b.

X-band, HH polarization   S↓ look direction

Figure 9-4   a) X-band image of the Kaduna State in Nigeria created by flying east-west and having the RADAR
look north. b) This image was obtained by flying the
aircraft east-west with the RADAR looking south.
Generally, it is good practice to always orient a radar image so that the look direction is toward the
viewer. This causes the shadows to fall toward the
analyst and keeps him or her from experiencing
pseudoscopic illusion (i.e., topographic inversion).

sent out by the antenna may be *vertically* or *horizontally
polarized* as shown in Figure 9-6. The transmitted pulse of
electromagnetic energy interacts with the terrain and some
of it is backscattered at the speed of light toward the aircraft
or spacecraft where it once again must pass through a filter.
If the antenna accepts the backscattered energy, it is
recorded. Various types of backscattered polarized energy
may be recorded by the radar. For example, it is possible to:

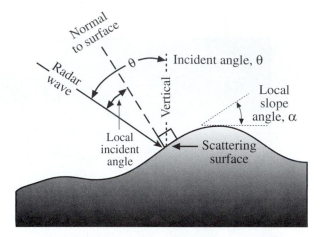

Figure 9-5   The relationship between radar incident angle (θ),
true vertical, and local slope angle (α) for non-level
terrain (after Henderson and Lewis, 1998).

- send vertically polarized energy and receive only
  vertically polarized energy (designated *VV*),

- send horizontal and receive horizontally polarized energy
  (*HH*),

- send horizontal and receive vertically polarized energy
  (*HV*), or

- send vertical and receive horizontally polarized energy
  (*VH*).

HH and *VV* configurations produce *like-polarized* radar
imagery. *HV* and *VH* configurations produce *cross-polarized*
imagery. The ability to record different types of polarized
energy from a resolution element in the terrain results in
valuable Earth resource information in certain instances. For
example, Figure 9-7 demonstrates how a northern Arizona
basalt lava flow is much easier to delineate in the HH polarization real aperture K_a-band radar image than in the HV
polarization image acquired at the same time. More will be
said about how the terrain interacts with a pulse of polarized
electromagnetic energy in the section on radar environmental considerations.

*Slant-Range versus Ground-Range RADAR Image
Geometry*

Radar imagery has a different geometry than that produced
by most conventional remote sensor systems, such as cam-

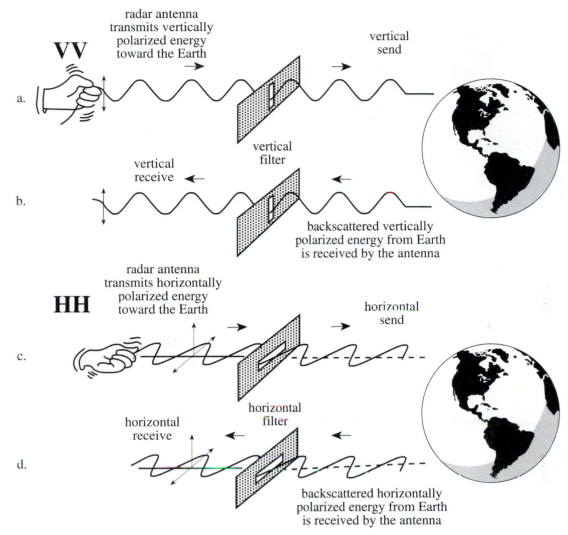

Figure 9-6   a) VV – Polarization. The radar antenna transmits vertically polarized energy toward the terrain. b) Some of the vertically polarized energy is *not* depolarized by the terrain. A vertical filter in the antenna allows only vertically polarized energy backscattered from the terrain to be recorded. c) HH – Polarization. The radar antenna transmits horizontally polarized energy toward the terrain. d) Some of the horizontally polarized energy is *not* depolarized by the terrain. A horizontal filter in the antenna allows only horizontally polarized energy backscattered from the terrain to be recorded.

eras, multispectral scanners or area-array detectors. Therefore, one must be very careful when attempting to make radargrammetric measurements. First, uncorrected radar imagery is displayed in what is called *slant-range geometry*, i.e., it is based on the actual distance from the radar to each of the respective features in the scene. For example, in Figure 9-8 we see two fields, A and B, that are the same size in the real world. One field is in the near-range close to the aircraft and one is in the far-range. Field A in the near-range is compressed much more than field B in the far-range in a slant-range display (Ford et al., 1980). It is possible to con-

vert the *slant-range display (S_{rd})* information into the true *ground-range display (G_{rd})* on the x-axis so that features in the scene are in their proper planimetric (x,y) position relative to one another in the final radar image. The following equation, based on the Pythagorean theorem applied to a right triangle, transforms the slant-range distance, S_{rd}, at the very beginning of field A to a corrected ground-range distance, G_{rd}, based on the trigonometric relationship between the altitude of the sensor above-ground datum (H) and the other two sides of the right triangle, S_{rd} and G_{rd}, shown in Figure 9-8:

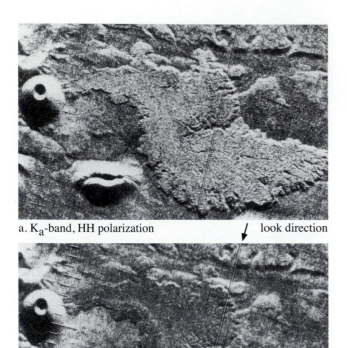

a. K$_a$-band, HH polarization   ↓ look direction

b. K$_a$-band, HV polarization   north ⟶

Figure 9-7   a) K$_a$-band, HH polarization real aperture radar image of a cinder cone and basalt lava flow in north-central Arizona. b) Simultaneously obtained HV polarization image with the same look direction. The strong response of the lava flow in the direct return image (HH) and the weak return on the cross-polarized image (HV) indicates that the blocky flow is highly polarized. This is due to the direct reflection of blocks that are large relative to the wavelength (courtesy of NASA; Carver, 1988).

$$S_{rd}^2 = H^2 + G_{rd}^2 \qquad (9\text{-}5)$$

$$G_{rd} = \sqrt{S_{rd}^2 - H^2} \; . \qquad (9\text{-}6)$$

It is also possible to transform the slant-range display to a ground-range display using the relationship between the height of the antenna above the local ground level, H, and the depression angle (γ) at the point of interest using the following equation (Ford et al., 1980):

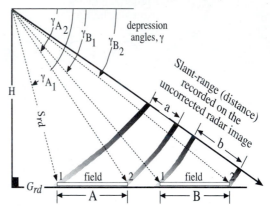

**Slant-Range Display
versus Ground-Range Display**

Ground-range (distance) on corrected radar image

Figure 9-8   Uncorrected radar images have slant-range geometry, where objects in the near-range are compressed more than objects in the far-range. In this example, field A and B are the same size and have no local relief. The slant-range image display is a plane connecting the antenna at altitude H, with the farthest point in the far-range. When the corner of field A is illuminated with a pulse of energy at a depression angle of γ_A, it is projected onto the slant-range display as shown. The same thing occurs at the end of field A and for both sides of field B. This causes field A to be compressed much more than field B in the slant-range image display.

$$G_{rd} = H \sqrt{\frac{1}{\sin^2\gamma} - 1} \qquad (9\text{-}7)$$

If we wanted to measure the true ground-range distance between two points in a radar image such as between points 1 and 2 in field A, we may use the following relationship (Henderson and Lewis, 1998):

$$G_{rd_{1-2}} = H(\cot\gamma_2 - \cot\gamma_1) \qquad (9\text{-}8)$$

which takes into account the altitude of the sensor, H, and the depression angle (γ) to points 1 and 2 in the image.

These equations assume that the terrain is flat. It does not correct for distortion due to radar layover (foreshortening) caused by topographic relief (to be discussed). Radar image analysts should always inquire as to whether they are viewing radar imagery that has been converted from slant-range to ground-range geometry. Most radar systems and data providers now provide the data in ground-range geometry.

Computing Range Resolution

To determine the spatial resolution at any point in a radar image, it is necessary to compute the resolution in two dimensions: the *range* and *azimuth resolutions*. Radar is in effect a ranging device that measures the distance to objects in the terrain by means of sending out and receiving pulses of active microwave energy (Figure 9-9). The *range resolution* in the across-track direction is proportional to the length of the microwave pulse. The shorter the pulse length, the finer the range resolution. *Pulse length* is a function of the speed of light (*c*) multiplied by the duration of the transmission (τ). The length of time that the microwave energy (e.g., L-band, 23.5 cm) is actually transmitted is measured in microseconds (10^{-6} sec) and typically ranges from 0.4 – 1.0 microsecond. This translates into a pulse length ranging from 8 – 210 m. The pulse length must travel to the target and back to the sensor. Therefore, it is necessary to divide by 2 to measure the slant-range resolution. To scale it to ground-range, it is multiplied by the cosine of the depression angle (γ). Thus, the equation for computing the *range resolution* becomes:

$$R_r = \frac{\tau \cdot c}{2\cos\gamma}. \qquad (9\text{-}9)$$

One might ask, Why not select an extremely short pulse length to obtain an extremely fine range resolution? The reason is that as the pulse length is shortened, so is the total amount of energy that illuminates the target of interest. Soon we would have such a weak backscattered return signal that it would be of no value. Therefore, there is a trade-off between shortening the pulse length to improve range resolution and having enough energy in the transmitted pulse to receive a strong signal from the terrain, which is the heart of microwave remote sensing.

The general rule is that signals reflected from two distinct objects in the terrain (e.g., two houses) can be resolved if their respective range distances are separated by at least half the pulse length. For example, consider Figure 9-9 in which the terrain is being illuminated with a single pulse of microwave energy that lasts 10^{-7} seconds. This translates to a pulse length of 30 m and therefore has a range resolution of 15 m. The fate of the single pulse of microwave energy is monitored for four brief time periods. At *time n* the pulse has not impacted any homes yet. At *time n+1* a portion of the pulse has been reflected back toward the antenna while the remaining part of the pulse continues across-track. By *time n+2* homes 2, 3, and 4 have reflected a part of the incident microwave energy back to the antenna. Because houses 1 and 2 were greater than 15 m apart, they will appear as distinct fea-

tures in the radar imagery. However, houses 3 and 4 were less than 15 m apart; therefore, their two returns will overlap and they will be perceived by the antenna as one broad object. They will probably be difficult to resolve as individual houses in the radar image.

While the pulse length remains constant through the near- and far-range, the range resolution varies linearly from the near- to the far-range. For example, consider Figure 9-10 which depicts towers 1 and 2 that are 30 m apart in the near-range and towers 3 and 4 that are 30 m apart in the far-range. If we use Equation 9-9 to compute the range resolution in the far-range with a depression angle of 40° and a pulse length duration of transmission of 0.1 μsec, the range resolution is:

$$R_r = \frac{(0.1\times 10^{-6}\,\text{sec})\cdot(3\times 10^{8}\,\text{m sec}^{-1})}{2\cos 40°}$$

$$R_r = \frac{\left(0.1\times\dfrac{1\,\text{sec}}{1,000,000}\right)\times\dfrac{300,000,000\,\text{m}}{\text{sec}}}{2\times 0.766}$$

$$R_r = \frac{\dfrac{0.1\,\text{sec}}{1,000,000}\times\dfrac{300,000,000\,\text{m}}{\text{sec}}}{2\times 0.766}$$

$$R_r = \frac{30\,\text{m}}{1.532}$$

$$R_r = 19.58\,\text{m}.$$

Therefore, towers 3 and 4 in the far-range must be separated by more than 19.58 m to resolve the individual towers on the radar image. Because towers 3 and 4 are separated by 30 m, it is possible to identify the individual towers. Conversely, towers 1 and 2, located at a depression angle of 65° in the near-range, would not be resolved because the range resolution in this area would be 35.5 m. The two towers would probably appear as a single bright return.

Computing Azimuth Resolution

Thus far we have only identified the length in meters of an active microwave resolution element at a specific depression angle and pulse length in the range (across-track) direction. To know both the length and width of the resolution element, we must also compute the width of the resolution element in the direction the craft is flying — the azimuth direction. *Azimuth resolution* (R_a) is determined by computing the width of the terrain strip that is illuminated by the radar beam. Real aperture active microwave radars produce a lobe-shaped beam similar to the one shown in Figure 9-11, which is nar-

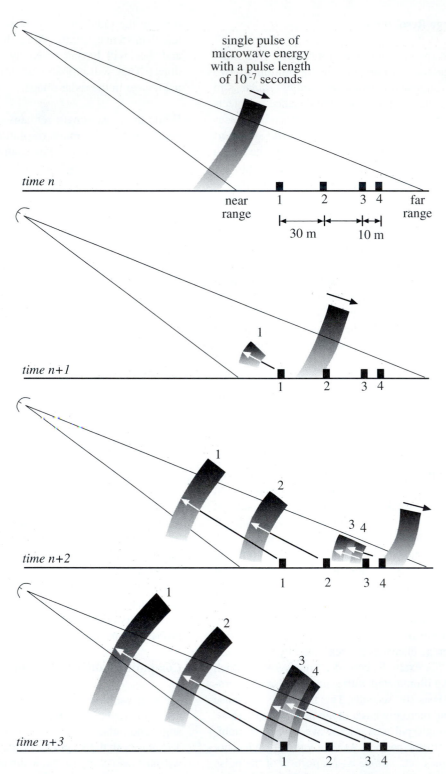

Figure 9-9   The distance of an object (its range) from the aircraft antenna is determined by the length of time required for the pulse of microwave energy traveling at the speed of light to reach the object and be reflected back to the antenna. Signals from houses 1 and 2 will arrive sooner than the signals returned from the two houses farther away (3 and 4). In this example, the 10^{-7} second pulse length equates to 30 m. The resolution across-track is equal to half the pulse length, in this case 15 m. Houses 1 and 2 will be resolved as distinct objects on the radar image. Two objects (houses 3 and 4) separated by less than half the radar pulse length will be perceived by the antenna as one broad object.

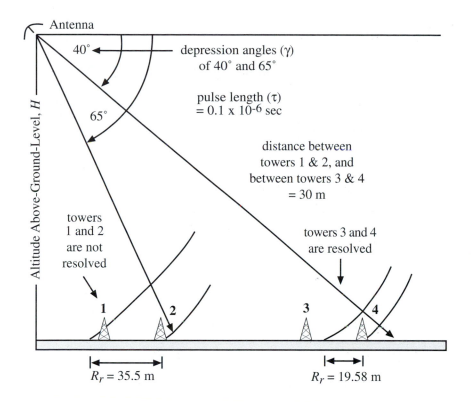

Antenna

40°

depression angles (γ) of 40° and 65°

65°

pulse length (τ) = 0.1 x 10⁻⁶ sec

distance between towers 1 & 2, and between towers 3 & 4 = 30 m

Altitude Above-Ground-Level, H

towers 1 and 2 are not resolved

towers 3 and 4 are resolved

1   2   3   4

$R_r = 35.5$ m $R_r = 19.58$ m

Figure 9-10 Computing the range resolution at two different depression angles (40° and 65°) for a real aperture radar with a pulse length of 0.1 x 10⁻⁶ sec. The towers can be resolved in the far-range but not in the near-range (after Sabins, 1997).

rower in the near-range and spreads out in the far-range. Basically, the angular beam width is directly proportional to the wavelength of the transmitted pulse of energy, i.e., the longer the wavelength, the wider the beam width, and the shorter the wavelength, the narrower the beam width. Therefore, in real aperture (brute force) radars a shorter wavelength pulse will result in improved azimuth resolution. Unfortunately, the shorter the wavelength, the poorer the atmospheric and vegetation penetration capability.

Fortunately, the beam width is also inversely proportional to antenna length (L). This means that the longer the radar antenna, the narrower the beam width and the higher the azimuth resolution. The relationship between wavelength (λ) and antenna length (L) is summarized in Equation 9-10, which can be used to compute the *azimuth resolution* (Henderson and Lewis, 1998):

$$R_a = \frac{S \times \lambda}{L} \qquad (9-10)$$

where S is the *slant-range distance* to the point of interest. The equation can be used to compute the azimuth resolution at any location between the near- and far-range. For example, consider the conditions shown in Figure 9-11 where the near slant-range is 20 km and the far slant-range is 40 km. Tanks 1 and 2 and tanks 3 and 4 are separated by 200 m. If

an X-band radar (3 cm) is used with a 500 cm antenna, then Equation 9-10 can be used to compute the near-range *azimuth resolution*:

$$R_a = \frac{20 \text{ km} \times 3 \text{ cm}}{500 \text{ cm}}$$

$$R_a = \frac{20,000 \text{ m} \times 0.03 \text{ m}}{5 \text{ m}}$$

$$R_a = \frac{600 \text{ m}}{5 \text{ m}}$$

$$R_a = 120 \text{ m}.$$

The far-range azimuth resolution at the 40 km distance is:

$$R_a = \frac{40 \text{ km} \times 3 \text{ cm}}{500 \text{ cm}}$$

$$R_a = \frac{40,000 \text{ m} \times 0.03 \text{ m}}{5 \text{ m}}$$

$$R_a = \frac{1200 \text{ m}}{5 \text{ m}}$$

$$R_a = 240 \text{ m}.$$

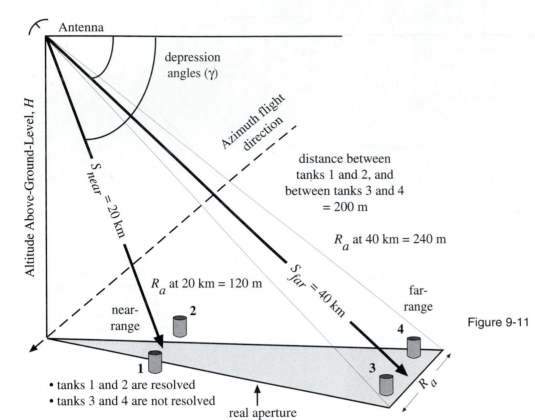

Figure 9-11 Computing the azimuth resolution at two different slant-range distances (20 and 40 km) for a real aperture radar with an X-band wavelength of 3 cm and a 500 cm antenna. The tanks can be resolved in the near range but not in the far-range.

Tanks 1 and 2 in the near-range would most likely be resolved because the azimuth resolution at this slant-range distance (120 m) is less than the distance between tanks 1 and 2 (200 m). Conversely, tanks 3 and 4 in the far-range would probably not be resolved, because at a slant-range distance of 40 km the azimuth resolution is 240 m, much greater than the 200 m separating the tanks.

There is a trigonometric relationship between the slant-range distance (S) and the depression angle (γ) and the height of the aircraft or spacecraft platform above the local datum, H:

$$S = \frac{H}{\sin \gamma} . \tag{9-11}$$

The equation for computing the azimuth resolution becomes:

$$R_a = \left(\frac{H}{\sin \gamma} \right) \cdot \frac{\lambda}{L} . \tag{9-12}$$

Ideally, one could build an extremely long antenna and place it on an aircraft to achieve very high azimuth resolution. Unfortunately, there is a practical limit (about 2 m) to the size of antenna that can be mounted on an aircraft (although

the size is virtually unlimited in outer space!). Fortunately, there are ways to electronically synthesize a longer antenna, which is the heart of the synthetic aperture radar concept to be discussed shortly.

While we have not yet defined how a synthetic aperture radar (SAR) works, it is instructive to point out at this time a significant difference between the computation of the azimuth resolution for a real aperture radar versus a synthetic aperture radar. The equation for the azimuth or along-track resolution for a point target in a synthetic aperture radar (SAR_a) is:

$$SAR_a = \frac{L}{2} \tag{9-13}$$

where L is the antenna length. As Henderson and Lewis (1998) point out,

This is one of the more incredible equations in the discipline of radar remote sensing. The omission of the slant-range distance (S) from the equation means that the azimuth resolution of a SAR system is independent of range distance or sensor altitude. Theoretically, the spatial resolution in the azimuth direction from a SAR imaging system will be the same from an aircraft platform as it is from a spacecraft. There is no other remote sensing system with this capability!

Equation 9-13, however, is not the only parameter that has an impact in SAR data. The coherent nature of the SAR signal produces speckle in the image. To remove the speckle, the image is usually processed using several *looks*, i.e., an averaging takes place. For example, four looks (N) might be averaged. This dramatically improves the interpretability of the SAR image data. However, the azimuth resolution must be adjusted by the equation:

$$ SAR_a = N\left(\frac{L}{2}\right) . \tag{9-14} $$

The SIR-C SAR had a 12 m antenna, which would yield a 6 m along-track resolution if Equation 9-13 were used. However, the speckle in the SIR-C SAR data were processed using $N = 4$ looks to improve the interpretability of the data. Thus, the adjusted azimuth resolution is 24 m.

This discussion first summarized how the range and azimuth resolution are computed for specific locations within the real aperture radar beam swath. Note that the resolution element might have different dimensions, i.e., the range resolution could be 10 m and the azimuth resolution 20 m. Therefore, because scale is constantly changing throughout a real aperture radar image, it is important to be careful when making radargrammetric measurements over large regions. When using such data, it is best to compute the range and azimuth resolution for the major region you are interested in and then make measurements. If a new area is selected for study, then new resolution measurements should be computed before area measurements are obtained.

RADAR Relief Displacement, Image Foreshortening, Layover, Shadows, and Speckle

Geometric distortions exist in almost all radar imagery, including (Ford et al., 1980): foreshortening, layover, and shadowing. When the terrain is flat, it is a straightforward matter to use Equation 9-6 to convert a slant-range radar image into a ground-range radar image that is planimetrically correct in x,y. However, when trees, tall buildings, or mountains are present in the scene, relief displacement in the radar image occurs. In radar relief displacement, the horizontal displacement of an object in the image caused by the object's elevation is in a direction toward the radar antenna (Figure 9-12a). Because the radar image is formed in the

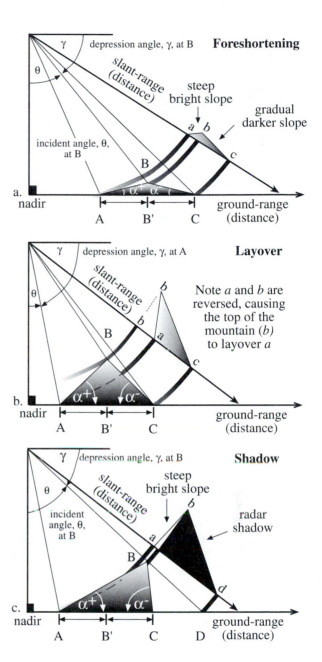

Figure 9-12 a) Radar foreshortening in the slant-range display. Slope *AB* in the ground-range display and *ab* in the slant-range display are supposed to be equal as are *BC* and *bc*. Instead, slope *ab* is shorter and steeper while slope *bc* is more gradual and longer than it should be. b) Image layover of foreslope *AB* in the slant-range display (*ba*) caused by the incident angle θ being smaller than the foreslope angle α^+. Slant-range *a* and *b* are reversed, causing layover. c) A substantial shadow is produced at *bd* because the backslope angle (α^-) is greater than the depression angle (γ) (after Ford et al., 1980).

range (cross-track) direction, the higher the object, the closer it is to the radar antenna, and therefore the sooner (in time) it is detected on the radar image. This contrasts sharply with relief displacement in optical aerial photography where the relief displacement is radially outward from the principal point (center) of a photograph. The elevation-induced distortions in radar imagery are referred to as *foreshortening* and *layover*.

Foreshortening and Layover

Basically, all terrain that has a slope inclined toward the radar will appear compressed or foreshortened relative to slopes inclined away from the radar. The effect is illustrated in Figure 9-12a. The foreshortening factor, F_f, is approximately:

$$F_f = \sin(\theta - \alpha) \qquad (9\text{-}15)$$

where the *incident angle* θ is the angle between the vertical plane at nadir and a line that links the imaging radar antenna to a feature on the ground, and α is the slope angle of the surface. Alpha is positive (α^+) where the slope is inclined toward the radar (foreslope), and negative (α^-) where the slope is inclined away from it (backslope).

For example, consider the relatively low relief mountain in Figure 9-12a. It is perfectly symmetrical with the distance from AB' and B'C being equal when plotted in the ground-range (i.e., on a quality planimetric map). A single pulse of microwave energy illuminating the terrain will first encounter the base of the mountain (A) and record it in the slant-range image at *a*. Because the top of the mountain (B) extends above the terrain and is relatively close to the antenna, it is recorded at *b*. The base of the mountain at *C* is recorded at *c* in the slant-range display. The higher the object above the terrain, the more foreshortening that will take place. In this case, the radar image will have a very bright short foreslope and a darker backslope. It would be difficult to make a map of mountainous features using this radar image because the top of the mountain, even though it was not very high, is displaced from its true ground-range planimetric position in the radar image, i.e., *ab* does not equal *bc*. Foreshortening is influenced by the following factors:

- *object height*: The greater the height of the object above local datum, the greater the foreshortening.

- *depression angle* (or *incident angle*): The greater the depression angle (γ) or smaller the incident angle (θ), the greater the foreshortening. A good example is found in Figure 9-13a,b where the ERS-1 sensor with its large 67°

depression angle and its 23° incident angle introduces more foreshortening than the JERS-1 radar with its 51° depression angle and 39° incident angle. Also consider the foreshortened cinder cone in Arizona when recorded on a radar image versus a conventional vertical aerial photograph (Figure 9-13c,d).

- *location of objects in the across-track range:* Features in the near-range portion of the swath are generally foreshortened more than identical features in the far-range. Foreshortening causes features to appear to have steeper slopes than they actually have in nature in the near-range of the radar image and to have shallower slopes than they actually have in the far-range of the image (Campbell, 1996).

Image *layover* is an extreme case of image foreshortening. It occurs when the incident angle (θ) is smaller than the foreslope (α^+), i.e., $\theta < \alpha^+$. This concept is illustrated in Figure 9-12b. In this case, the mountain has so much relief that the summit (B) backscatters energy toward the antenna before the pulse of energy even reaches the base of the mountain (A). Remember that in terms of planimetric distance from the nadir point directly beneath the aircraft, the base of the mountain (A) is much closer than the summit (B), as documented by the ground-range distance display. However, because the mountain summit (B) reflects the incident microwave energy sooner than the base of the mountain (A), the summit (b) in the slant-range radar image actually *lays over* (hence the terminology) the base of the mountain recorded on the radar image at *a*. Once again, the summit of the mountain (a) is significantly displaced from its true planimetric position. This distortion cannot be corrected even when the surface topography is known. Great care must be exercised when interpreting radar images of mountainous areas where the thresholds for image layover exist. The bright white ridges in Figure 9-14 represent severe SIR-C L-band (HH) radar layover in the San Gabriel Mountains just east of Los Angeles, CA.

Shadows

Shadows in radar images can enhance the geomorphology and texture of the terrain. Shadows can also obscure the most important features in a radar image, such as the information behind tall buildings or land use in deep valleys. If certain conditions are met, any feature protruding above the local datum can cause the incident pulse of microwave energy to reflect all of its energy on the foreslope of the object and produce a black shadow for the backslope. A backslope is in radar shadow when its angle (α^-) is steeper than the depression angle (γ), i.e., $\alpha^- > \gamma$ (Figure 9-12c). If the backslope

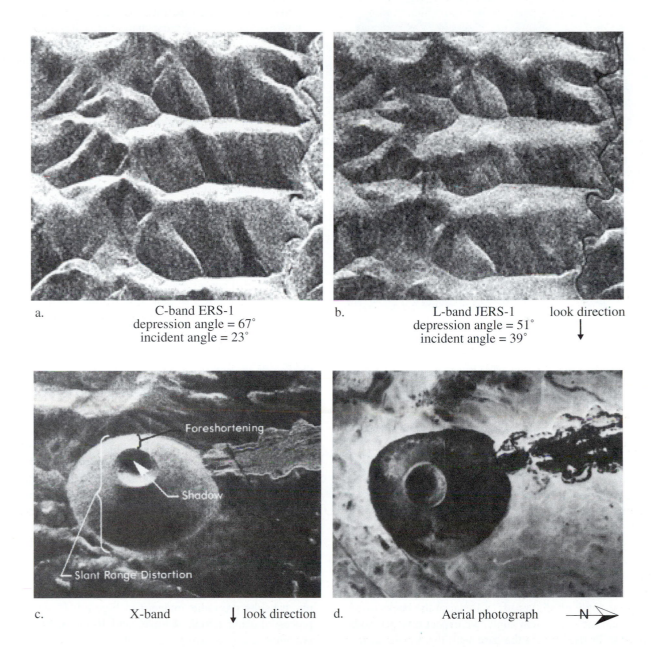

a.   C-band ERS-1
depression angle = 67°
incident angle = 23°

b.   L-band JERS-1   look direction
depression angle = 51°
incident angle = 39°

c.   X-band   ↓ look direction   d.   Aerial photograph   ⟶ N

Figure 9-13   a) The C-band ERS-1 image of the White Mountains exhibits substantial foreshortening due to the satellite SAR's small incident angle (courtesy Alaska SAR Facility; © 1992, European Space Agency). b) The L-band JERS-1 image with a larger incident angle has significantly less image foreshortening (courtesy Alaska SAR Facility; © NASDA). c) X-band aircraft synthetic aperture radar (SAR) image of a cinder cone in Arizona. The shape of the basically cylindrical cinder cone is distorted in the slant-range look direction, appearing more elliptical in shape than it does in the aerial photograph. Foreshortening (relief displacement toward the radar antenna) occurs in the near-range. There is also a dark shadow under the rim of the crater at the top of the cone. d) Vertical panchromatic aerial photograph of SP Mountain, AZ (courtesy of Eric Kasischke, ERIM International, Inc.).

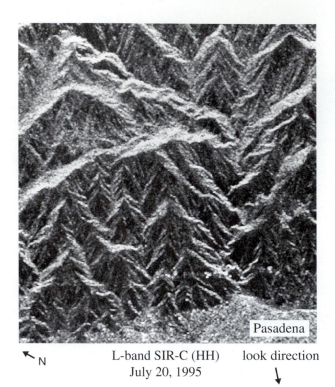

N   L-band SIR-C (HH)   look direction
July 20, 1995

Figure 9-14   L-band SIR-C (HH) image of the San Gabriel Mountains and a portion of the city of Pasadena. Extensive layover is found in the mountainous terrain, causing the ridge lines to be displaced toward the antenna (courtesy NASA Jet Propulsion Lab).

• Unlike aerial photography, where light may be scattered into the shadow area and then recorded on film, there is no information within the radar shadow area. It is black.

• Two terrain features (e.g., mountains) with identical heights and fore- and backslopes may be recorded with entirely different shadows, depending upon where they are in the across-track. A feature that casts an extensive shadow in the far-range might have its backslope completely illuminated in the near-range.

• Radar shadows occur only in the cross-track dimension. Therefore, the orientation of shadows in a radar image provides information about the look direction and the location of the near- and far-range.

Shadows in radar imagery are valuable when evaluating geomorphic characteristics. Radar shadows often function like low Sun-angle photography, enhancing lineaments and other geologic conditions. Terrain information such as topographic height, slope, etc., can be used to configure radar overflights or, purchase radar data that have the most appropriate depression angles and near- and far-range characteristics to emphasize terrain shadows.

Speckle

Speckle is a grainy salt-and-pepper pattern in radar imagery (Figure 9-15) that is present due to the coherent nature of the radar wave, which causes random constructive and destructive interference, and hence random bright and dark areas in a radar image. The speckle can be reduced by processing separate portions of an aperture and recombining these portions so that interference does not occur (Kasischke et al., 1984). As previously mentioned, this process, called multiple *looks* or noncoherent integration, produces a more pleasing appearance, and in some cases may aid in interpretation of the image but at a cost of degraded resolution. For example, consider the radar imagery in Figure 9-15 which was processed using 1-look, 4-looks, and 16-looks. Most interpreters would prefer working with the 4- or 16-look imagery.

Synthetic Aperture Radar Systems

A major advance in radar remote sensing has been the improvement in azimuth resolution through the development of *synthetic aperture radar* (SAR) systems. Remember, in a real aperture radar system that the size of the antenna (L) is inversely proportional to the size of the angular beam width (Equation 9-10). Therefore, great improvement in azimuth

equals the depression angle ($\alpha^- = \gamma$), then the backslope is just barely illuminated by the incident energy. This is called grazing illumination because the radar pulse just grazes the backslope. The backslope is fully illuminated when it is less than the depression angle ($\alpha^- < \gamma$).

Figure 9-12c demonstrates how a large shadow from a relatively steep backslope might be produced. In this case, we have a backslope of approximately 85° ($\alpha^- = 85°$) and a depression angle of 45° ($\gamma = 45°$). Because the backslope is greater than the depression angle, we expect this area to be in shadow. In fact, this is the case with the terrain surface *BCD* in the ground-range being in complete shadow in the slant-range radar image display (*bd*). In the ground-range display the distance from the summit at B to the back base of the mountain at C is relatively short. But in the slant-range radar image, *bd* is very long. This particular radar image would also experience image foreshortening (but not layover because *A* is recorded by the antenna before *B*) and have a very bright return from the foreslope.

Some important characteristics of radar shadows are:

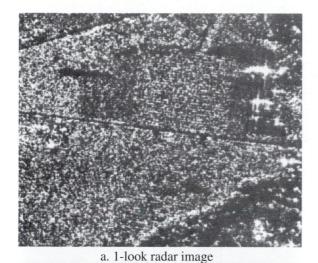

a. 1-look radar image

b. 4-look radar image

c. 16-look radar image

Figure 9-15   Radar speckle reduction using multiple-look techniques (courtesy of Eric Kasischke, ERIM International, Inc.).

resolution could be realized if a longer antenna were used. Engineers have developed procedures to *synthesize* a very long antenna electronically. Like a brute force or real aperture radar, a synthetic aperture radar also uses a relatively small antenna (e.g., 1 m; refer to Figure 9-1) that sends out a relatively broad beam perpendicular to the aircraft. The major difference is that a greater number of additional beams are sent toward the object. Doppler principles are then used to monitor the returns from all these additional microwave pulses to synthesize the azimuth resolution to become one very narrow beam.

The *Doppler principle* states that the frequency (pitch) of a sound changes if the listener and/or source are in motion relative to one another. For example, an approaching train whistle will have an increasingly higher frequency pitch as it approaches. This pitch will be highest when it is directly perpendicular to the listener (receiver). This is called the point of zero Doppler. As the train passes by, its pitch will decrease in frequency in proportion to the distance it is from the listener (receiver). This principle is applicable to all harmonic wave motion, including the microwaves used in radar systems.

Figure 9-16 depicts the Doppler frequency shift due to the relative motion of a terrain object at times *n*, *n+1*, *n+2*, *n+3*, and *n+4* through the radar beams due to the forward motion of the aircraft. The Doppler frequency diagram reveals that the frequency of the energy pulse returning from the target increases from a minimum at time *n* to a maximum at point *n+3*, normal (at a right angle) to the aircraft. Then, as the target recedes from *n+3* to *n+4*, the frequency decreases.

How is a synthetic aperture image actually produced? A long antenna can be synthesized using a short antenna by taking advantage of the aircraft's motion and the Doppler principle. It is assumed that the terrain is stable and *not* moving. It is also assumed that the object of interest remains a fixed distance away from the aircraft's flightline. As the aircraft flies along a straight line, a short antenna sends out a series of microwave pulses at regular intervals (Jensen et al., 1977). As an object (black dot) enters the antenna's beam (Figure 9-16a), it backscatters a portion of the pulse it receives toward the antenna. At some point in the aircraft's path the object will be an integral number of microwave wavelengths away; between those points it will not be. For example, in (a) we see that the object is first 9 wavelengths away, then 8 (b), then 7 (c), then 6.5 (d), at which point the object is at a right angle to the antenna, i.e., the shortest distance and area of zero Doppler shift. From then on the distance between the aircraft and the object will be increasing, i.e., perhaps 7 wavelengths away at location (e). The antenna receives the

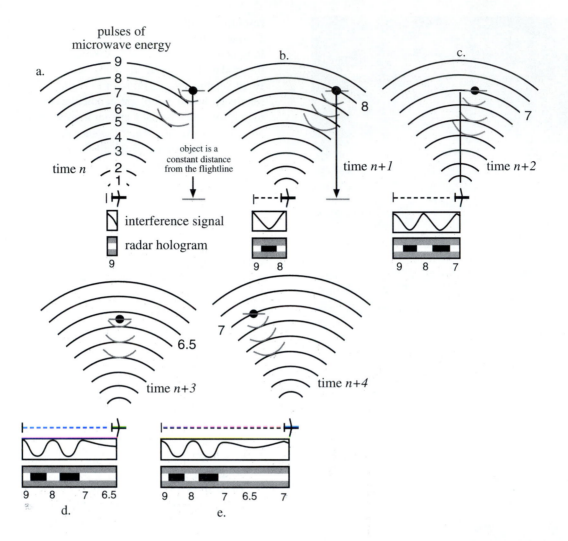

Figure 9-16   A long antenna can be synthesized by a short antenna (e.g., 1 m long) by taking advantage of the aircraft's motion and Doppler principles. As the craft flies along a straight line, a mounted short real antenna sends out a series of pulses at regular intervals. Each pulse consists of a train of coherent microwaves. Although the length of the pulses determines the resolution across the track, it is the wavelength of the microwave radiation that determines the resolution along the track. As an object (black dot) enters the antenna's beam (*a*), it reflects a portion of the pulses it receives back toward the antenna. At some point in the aircraft's path the object is an integral number of microwave wavelengths away; between these points it will not be. For example, in the illustration the object is first 9 wavelengths away at *a*, then 8 at *b*, then 7 at *c*, then 6.5 at *d*, where the object is at right angles to the antenna. From then on, the crest is increasing its distance from the object, e.g., point *e*. The antenna receives the series of reflected waves and electronically combines them with a train of reference wavelengths (not shown), causing the two series of oscillations to interfere. The interference signal emerges as a voltage that controls the brightness of a spot scanning across a cathode-ray tube. At the times that a returned pulse coincides with a reference pulse, the interference is constructive; the voltage will be high and the moving spot will be bright. At the times that the phase of the returned wavelength does not coincide with the phase of the reference frequency the interference is destructive; the voltage will be low and the moving spot will be dim. The moving spot thus traces out a series of light and dark dashes of unequal length that are recorded on a strip of data film moving at a velocity proportional to the velocity of the aircraft (after Jensen et al., 1977; Kasischke et al., 1984).

series of reflected waves (gray lines in illustrations a – e) and electronically combines them with a train of reference wavelengths (not shown), causing the two series of oscillations to interfere. The interference signal emerges as a voltage that controls the brightness of a spot scanning across the screen of a cathode-ray tube. When the returned pulse coincides with a reference pulse, the interference is constructive; the voltage will be high and the moving spot will be bright. When the phase of the returned wavelength does not coincide with the phase of the reference frequency, the interference is destructive; the voltage will be low and the moving spot will be dim or dark. The moving spot thus tracks out a series of light and dark dashes of unequal length that are recorded on a strip of data film moving at a velocity proportional to the velocity of the aircraft. The series of opaque and transparent dashes on the film are actually a one-dimensional interference pattern; the film on which they are recorded is a *radar hologram.*

When the developed hologram is illuminated by a source of coherent light (Figure 9-17a), each transparent dash functions as a separate source of coherent light. Below the hologram there will be a single point where the resulting light waves all constructively interfere. In this example, the 9th wavelength of light (thick curved lines) from the transparent dash created by the 9th microwave will meet the 8th wavelength of light (regular lines) from the transparent dash created by the 8th microwave, and both will meet the 7th wavelength of light (thin curves) from the transparent dash created by the 7th microwave and so on (only the 9th, 8th, and 7th patterns are shown in this example for clarity). At that one point, light from the entire length of the interference pattern is focused to form a miniature image of the original object. Figure 9-17b demonstrates how the holographic image is reconstructed and recorded on film. After processing the negative film to become a positive print, the radar image is ready for analysis.

The record of Doppler frequency enables the target to be resolved on the image film as though it was observed with an antenna of length L, as shown in Figure 9-18 (Leberl, 1990; Sabins, 1997). This synthetically lengthened antenna produces the effect of a very narrow beam with constant width in the azimuth direction, shown by the shaded area in Figure 9-18. For both real and synthetic aperture systems, range resolution is determined by pulse length and depression angle. Generally, synthetic aperture images are higher in azimuth resolution than real aperture radars.

The aforementioned method is often called synthetic aperture radar *optical correlation* because of all the precision optics utilized. It is also possible to use *SAR digital correlation* techniques to record and process the amplitude and phase history of the radar returns. Digital correlation does not produce an intermediate radar film. The digital process is computation intensive. Major advantages of digital correlation include the ability to 1) perform both radiometric and geometric corrections rapidly during onboard processing (good for emergencies such as oil spills, floods, fires, etc.), 2) telemeter the processed radar data directly to the ground to support real-time decision making, and/or 3) store the digitally processed SAR data on high-density digital tapes (HDDTs) for subsequent digital processing on the ground. Many of the commercial (e.g., Intermap Star 3*i*) and government SARs (e.g., RADARSAT) use digital SAR correlation techniques.

The Radar Equation

The following discussion provides additional quantitative information about the radar signal. A radar image is a two-dimensional representation of the power (or voltage) returned to the radar from a specific area on the ground presented as a picture element (pixel). This returned power is usually quantized to a radiometric scale of 256 values (8 bits) for presentation on image processing systems (Leberl, 1990). RADARSAT is quantized to 11 bits. To understand how to interpret radar images, we should understand the nature of the power scattered back toward the radar antenna. In the most simple case, this can be stated verbally as suggested by Moore (1983):

$$\begin{pmatrix} \text{Power} \\ \text{received} \end{pmatrix} = \begin{pmatrix} \text{Power per unit} \\ \text{area at target} \end{pmatrix} \times \begin{pmatrix} \text{Effective scattering} \\ \text{area of the target} \end{pmatrix}$$
$$\times \begin{pmatrix} \text{Spreading loss of} \\ \text{reradiated signal} \end{pmatrix} \times \begin{pmatrix} \text{Effective antenna} \\ \text{receiving area} \end{pmatrix}$$

where the power per unit area at the receiver is the energy scattered back from the terrain — *backscatter*. The spreading loss occurs because the signal starts from the backscattering point source on the ground (e.g., a large rock) and spreads out in all directions, so that the power per unit area is less at a greater distance than it would be near the scatterer. Therefore, the strength of the backscatter toward the receiver is a product of the power per unit area illuminating the target, times the effective scattering area of the target, and then the retransmission of this wave back toward the receiving antenna. The actual size of the receiving antenna also makes a difference.

The fundamental radar equation is derived by combining these word quantities to create the mathematical expression (Moore, 1983; Kasischke et al., 1984):

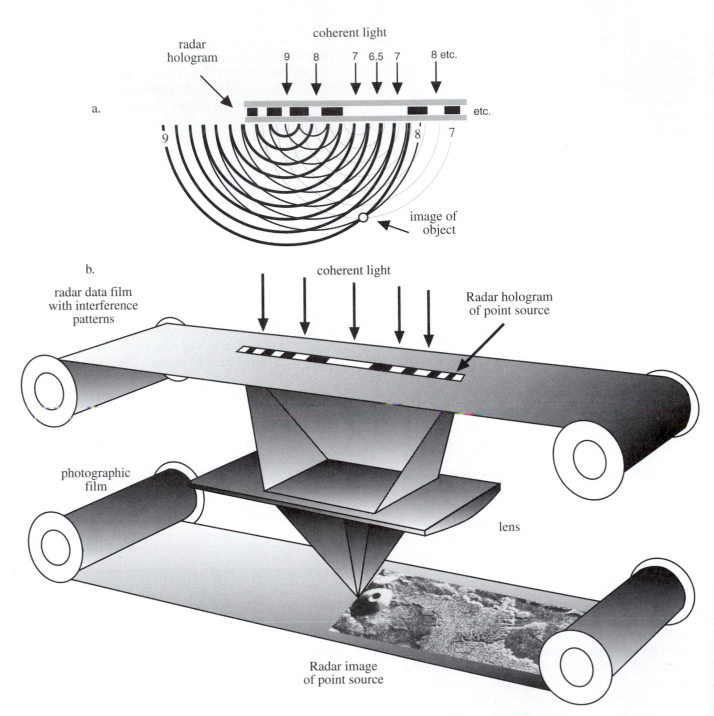

Figure 9-17   a) When the developed hologram is illuminated by laser coherent light, each transparent dash acts as a separate source of coherent light. Below the hologram there is a single point where the resulting light waves all constructively interfere. Here the 9th wavelength of light from the transparent dash created by the 9th microwave will meet the 8th wavelength of light from the transparent dash created by the 8th microwave, and both will meet with the 7th wavelength of light from the transparent dash created by the 7th microwave. At that one point light from the entire length of the interference pattern is focused to form a miniature image of the original object. b) As the data film is advanced through the beam of laser light, the reconstructed image is recorded on another moving strip of film. Because the data film is holographic only in the along-track coordinate, the images in the across-track coordinate must be focused with a cylindrical lens (after Jensen et al., 1977).

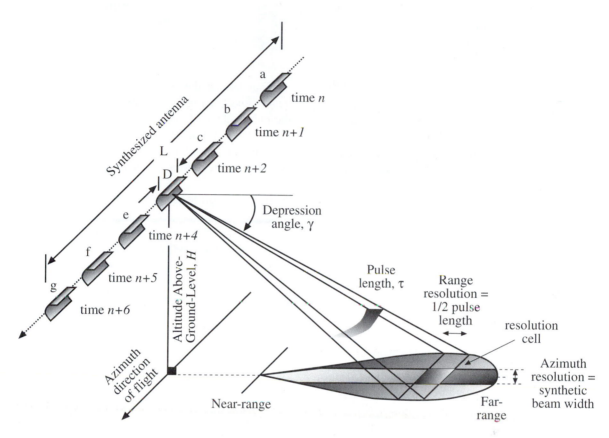

Figure 9-18 A synthetic antenna of length L is produced by optically or digitally processing the phase histories of microwave radar returns sent and received by a real antenna of length D sent at times n. The synthetic aperture radar has a range resolution that is approximately equal to half the pulse length. Note how the azimuth resolution remains constant from the near-range to the far-range (after Sabins, 1997).

$$P_r = \frac{P_t \cdot G_t \cdot \sigma \cdot A_r}{(4\pi)^2 \cdot R^4} \qquad (9\text{-}16)$$

where P_r is power received, P_t is the power transmitted toward the target, G_t is the gain of the antenna in the direction of the target, R is the range distance from the transmitter to the target, σ is the effective backscatter area of the target (often called the radar cross-section), and A_r is the area of the receiving antenna. This equation can be broken down into seven parts for even more clarity:

$$P_r = P_t \cdot G_t \cdot \frac{1}{4\pi R^2} \cdot \sigma \cdot \frac{1}{4\pi R^2} \cdot A_r \qquad (9\text{-}17)$$

$$1 = 2 \cdot 3 \cdot 4 \cdot 5 \cdot 6 \cdot 7$$

The power received [1] by the radar system, P_r, is a function of a pulse of electromagnetic energy, P_t [2], that has been focused down to an angular beam width by the antenna so that the flux becomes higher by a factor [3] of G_t over a spherically expanding wave [4]. The focused energy illuminates an *area* on the ground that has a cross-section of σ [5]. The *radar cross-section* is defined as the equivalent of a perfectly reflecting area that reflects isotropically (spherically). The energy backscattered from the ground once again spherically expands from the source [6]. Finally, the receiving antenna area (A_r) intercepts a portion of the reflected wave and records it [7].

Most radars use the same antennas for transmitting (A_t) and receiving (A_r). Consequently, the gain factors of the antennas may be combined using the relationship between gain and receiving aperture as (Moore, 1983):

$$G = G_t = G_r = \frac{4\pi A_r}{\lambda^2} \qquad (9\text{-}18)$$

where λ is the wavelength (or frequency) of the radar system. Substituting this value in Equation 9-16 or 9-17 results in a modified radar equation (Carver et al., 1985):

$$P_r = \frac{P_t \cdot G^2 \cdot \sigma \cdot \lambda^2}{(4\pi)^3 \cdot R^4}. \qquad (9\text{-}19)$$

Thus, we find that the radar equation can be viewed as a product of system parameters and other environmental terrain parameters that produce the backscatter cross-section, σ. Because the system parameters are well known, their effects are typically removed from the radar images, i.e., the system parameters may be set to unity (1).

It is the effects of terrain on the radar signal that we are most interested in, i.e., the amount of radar cross-section, σ, reflected back to the receiver, per unit area (A) on the ground. This is called the *radar backscatter coefficient* ($\sigma°$) and is computed as:

$$\sigma° = \frac{\sigma}{A} \qquad (9\text{-}20)$$

where σ is the radar cross-section. The radar backscatter coefficient determines the percentage of electromagnetic energy reflected back to the radar from within a resolution cell, e.g., 10 x 10 m. The actual $\sigma°$ for a surface depends on a number of terrain parameters, like geometry, surface roughness, moisture content, and the radar system parameters (wavelength, depression angle, polarization; Leberl, 1990). $\sigma°$ is a dimensionless quantity characteristic of the scattering behavior of all the elements contained in a given ground cell. Because $\sigma°$ can vary over several orders of magnitude, it is expressed as a logarithm with units of decibels (dB) that usually range from -5 to +40 dB.

The total radar cross-section of an area (A) on the ground therefore becomes $(\sigma°A)$, and the final form of the radar equation for an area-extensive target becomes (Henderson and Lewis, 1998):

$$P_r = P_t(\sigma°A)\left(\frac{G^2 \cdot \lambda^2}{(4\pi)^3 \cdot R^4}\right). \qquad (9\text{-}21)$$

A digital SAR image is created that consists of a two-dimensional array (matrix) of picture elements (pixels) with the intensity (called the brightness) of each pixel proportional to the power of the microwave pulse reflected back from the corresponding ground cell (Waring et al., 1995). The reflected radar signal is proportional to the backscattering coefficient ($\sigma°$) of a given ground cell.

RADAR Environmental Considerations

Having defined the radar backscatter coefficient, $\sigma°$, as a quantitative measure of the intensity of energy returned to the radar antenna from a specific area on the surface of the Earth, it is important to identify the environmental parameters within the resolution cell on the ground that are responsible for backscattering the incident energy.

Surface Roughness Characteristics

Surface roughness is the terrain property that strongly influences the strength of the radar backscatter. When interpreting aerial photography, we often use the terminology "rough" (coarse), "intermediate," or "smooth" (fine) to describe the surface texture characteristics (refer to Chapter 5). It is possible to extend this analogy to the interpretation of radar imagery if we keep in mind that the *surface roughness* we are talking about may be envisioned at the *microscale*, *mesoscale*, and/or *macroscale*.

Microscale surface roughness is usually measured in centimeters (i.e., the height of stones, size of leaves, or length of branches in a tree) and not in hundreds or thousands of meters as with topographic relief or mountains. The amount of microwave energy backscattered toward the sensor based on microscale components is a function of the relationship between the wavelength of the incident microwave energy (λ), the depression angle (γ), and the local height of objects (h in cm) found within the resolution cell being illuminated. We may use the *modified Rayleigh criteria* to predict what the Earth's surface will look like in a radar image if we know the microscale surface roughness characteristics and the radar system parameters (λ,γ,h) mentioned (Peake and Oliver, 1971). For example, one would expect that an area with *smooth surface roughness* would send back very little backscatter toward the antenna, i.e., it acts like a specular reflecting surface, where most of the energy bounces off the terrain away from the antenna (Figure 9-19a). The small amount of backscattered energy returned to the antenna is recorded and shows up as a dark area on the radar image. The quantitative expression of this *smooth criteria* is:

$$h < \frac{\lambda}{25\sin\gamma}. \qquad (9\text{-}22)$$

To illustrate we will compute what the local height (h) of objects (e.g., grass and rocks in this example) must be in

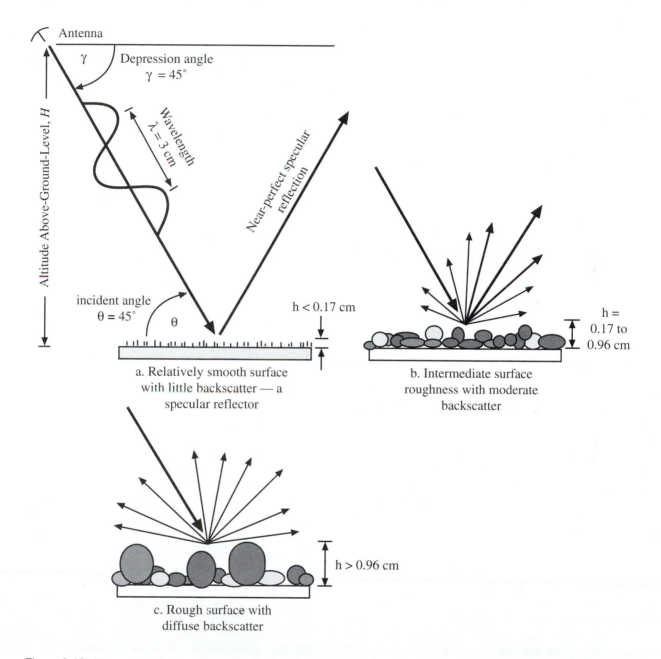

Figure 9-19 Expected surface roughness backscatter from terrain illuminated with 3 cm wavelength microwave energy. a) With a depression angle of 45° and a local relief of < 0.17 cm, the terrain represents a relatively smooth surface and should produce very little backscatter toward the sensor. This specular reflector will appear dark on the radar image. b) Local relief from 0.17 – 0.96 cm represents intermediate surface roughness and should produce a medium gray response on the radar image. c) When the local surface roughness relief is > 0.96 cm, a rough surface exists that will produce very diffuse backscatter. This diffuse reflector will produce a bright return on the radar image due to the large amount of microwave energy reflected back toward the antenna.

order to produce a smooth (dark) radar return when an X-band (λ = 3.0 cm) radar is used with a depression angle (γ) of 45° (Figure 9-19a). Substituting these values into Equation 9-22 yields:

$$h < \frac{3\,cm}{25\sin 45°}$$

$$h < \frac{3\,cm}{25 \times 0.70711}$$

$$h < 0.17\,cm\,.$$

This means that a resolution cell uniformly filled with dry grass that has a surface height of < 0.17 cm should produce relatively little radar backscatter and will therefore be recorded on the radar image in a dark tone.

A bright return is expected if the following modified Rayleigh *rough criteria* are used:

$$h > \frac{\lambda}{4.4\sin \gamma}\,. \qquad (9\text{-}23)$$

Substituting the same wavelength and depression angle into this equation yields:

$$h > \frac{3\,cm}{4.4\sin 45°}$$

$$h > \frac{3\,cm}{4.4 \times 0.70711}$$

$$h > 0.96\,cm\,.$$

Therefore, if the uniformly distributed rocks in the resolution cell had a local relief of > 0.96 cm (Figure 9-19c), then a strong return from the terrain would be expected and would be recorded as a bright tone in the radar image.

If the terrain were composed of rocks with local relief between 0.17 and 0.96 cm, it would be considered to have an *intermediate surface roughness* for this combination of wavelength and depression angle and would produce an intermediate tone of gray on the radar image.

It is important to remember that the radar backscatter is dependent on the wavelength and the depression angle. For

Table 9-4. Modified Rayleigh Surface Roughness Criteria for Three Different Radar Systems at Two Different Depression Angles

Surface Roughness Category	Aircraft K_a-band $\lambda = 0.86$ cm $\gamma = 45°$	Aircraft X-band $\lambda = 3$ cm $\gamma = 45°$	Seasat L-band $\lambda = 23.5$ cm $\gamma = 70°$
Smooth, cm	h < 0.048	h < 0.17	h < 1.00
Intermediate, cm	h = 0.048 to 0.276	h = 0.17 to 0.96	h = 1.00 to 5.68
Rough, cm	h > 0.276	h > 0.96	h > 5.68

example, consider Table 9-4 where the modified Rayleigh criteria are computed for three different radar wavelengths (λ = 0.86, 3.0, and 23.5 cm) at two different depression angles (γ = 45° and 70°). Terrain with a local relief of 0.5 cm would appear very bright on the K_a-band imagery and as an intermediate shade of gray on the X-band imagery. In this case the only parameter changed was the wavelength. If we changed both the wavelength and the depression angle to be that of the Seasat SAR (Table 9-4), we see that the same 0.5 cm local relief would produce a smooth (dark) return in this L-band imagery. The significance of this relationship is that the same terrain will appear differently in radar imagery as a function of the sensor's depression angle and wavelength (Henderson and Xia, 1998). Therefore, it is difficult to create "radar image interpretation keys" of selected phenomena. An analyst must constantly keep this in mind when interpreting radar imagery. Later we will see that aircraft or spacecraft look-direction also impacts the backscattered energy.

The incident radar energy may also be scattered according to both *mesoscale* and *macroscale surface roughness* criteria (Henderson and Lewis, 1998). As mentioned, the microscale roughness is a function of the size of leaves, twigs, and perhaps branches within an individual resolution cell. Conversely, mesoscale surface roughness would be a function of the characteristics within numerous resolution cells, perhaps covering the entire forest canopy. Based on this logic, we would expect forest canopies to generally show up with a more coarse texture than grasslands when recorded in the same image at the same depression angle and wavelength. Finally, *macroscale* roughness would be influenced greatly by how the entire forest canopy is situated on the hillside. Macroscale roughness is significantly influenced by the topographic slope and aspect of the terrain with the existence or absence of shadows playing a very important role in creating image surface roughness.

Electrical Characteristics (Complex Dielectric Constant) and the Relationship with Moisture Content

A radar sends out a pulse of microwave energy that interacts with the Earth's terrain. Different types of terrain conduct this electricity better than others. One measure of a material's electrical characteristics is the *complex dielectric constant,* defined as a measure of the ability of a material (vegetation, soil, rock, water, ice) to conduct electrical energy. Dry surface materials such as soil, rock, and even vegetation have dielectric constants from 3 to 8 in the microwave portion of the spectrum. Conversely, water has a dielectric constant of approximately 80. The most significant parameter influencing a material's dielectric constant is its *moisture content.* Therefore, the amount of moisture in a soil, on a rock surface, or within vegetative tissue may have a significant impact on the amount of backscattered radar energy.

Moist soils reflect more radar energy than dry soils, which absorb more of the radar wave, depending on the dielectric constant of the soil material. Radar images may be used to estimate *bare* ground soil moisture content when the terrain is devoid of most other material such as plants and rocks and has a uniform surface roughness. The amount of soil moisture influences how deep the incident electromagnetic energy penetrates into the material. If the soil has a high surface soil moisture content, then the incident energy will only penetrate a few centimeters into the soil column and be scattered more at the surface producing a stronger, brighter return.

The general rule of thumb for how far microwave energy will penetrate into a dry substance is that the penetration should be equal to the wavelength of the radar system. However, active microwave energy may penetrate extremely dry soil several meters. For example, Figure 9-20 depicts four views of a part of the Nile River, near the Fourth Cataract in the Sudan. The top image is a photograph taken by the crew of the Space Shuttle *Columbia* in November 1995. The three radar images were acquired by the Shuttle Imaging Radar C/ X-band Synthetic Aperture Radar (SIR-C/X-SAR) onboard Space Shuttle *Endeavor* in April 1994; C-band HV, L-band HV, and L-band HH. Each radar image provides some unique information about the geomorphology of the area. The thick white band in the top right of the radar image is an ancient channel of the Nile that is now buried under layers of sand. This channel cannot be seen in the photograph, and its existence was not known before the radar imagery was processed. The area to the left in both images shows how the Nile is forced to flow through a set of fractures that causes

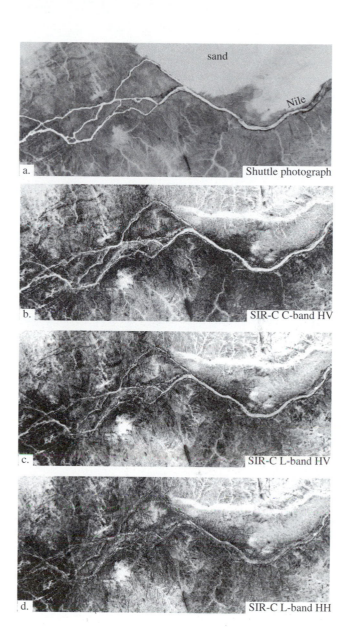

Figure 9-20 Example of radar penetration of dry soil along the Nile River, Sudan. a) Black-and-white version of a color-infrared photograph acquired by Space Shuttle *Columbia* crew in November 1995; b-d) SIR-C/ X-SAR images acquired by the Space Shuttle *Endeavor* in April 1994. Subtle, different information is recorded in each of the three radar images. Each reveals an ancient, previously unknown channel of the Nile. Radar brightness values are inverted in these examples (courtesy NASA Jet Propulsion Laboratory).

the river to break up into smaller channels, suggesting that the Nile has only recently established this course. The radar images have allowed scientists to develop new theories to explain the origin of the "Great Bend" of the Nile in the Sudan. Color Plate 9-1 is a color composite of the three radar images. It provides more revealing visual information about the regional geomorphology.

Ocean surfaces have a very high dielectric constant with most of the radar energy being reflected at the water's surface. The dielectric constant of snow depends upon how much liquid water is in the snow. Thus, microwave remote sensing can be used to determine snow water content. Similarly, there are differences in the dielectric constants of ice, depending upon age, degree of compaction, and type of ice. Microwave remote sensing has been useful for the extraction of these biophysical variables. Finally, healthy agricultural crops and forest canopy leaves often have relatively large surface areas and high moisture content (high relative turgidity). Therefore, it is not surprising that dense canopies of moist vegetation reflect radar energy well. In effect, a fully turgid vegetated forest canopy acts like a cloud of water droplets hovering above the surface of the Earth. Also, the steeper the depression angle, the greater the sensitivity of the radar to soil and vegetation moisture content.

Vegetation Response to Microwave Energy

Emphasis is being placed on quantitatively monitoring the spatial distribution, biomass, gross and net primary productivity, and condition of global vegetation communities, especially forests (approximately 33 percent of the Earth's land surface), semi-arid ecosystems of grassland/steppe/desert, agricultural land (10 percent), and wetlands. Scientists are interested in how energy and matter move within these vegetated ecosystems. Many of these vegetated areas are shrouded in perennial cloud cover. Synthetic aperture radar imagery may provide some of the following vegetation biophysical parameters (Carver, 1988):

- canopy water content,

- vegetation type,

- biomass by component (foliage, higher-order stems and main stem), and

- canopy structure (including green leaf area index), leaf orientation, main stem (trunk) geometry and spatial distribution, stem, branch size, and angle distributions.

Any plant canopy (forest, agriculture, grassland, etc.) may be thought of as a seasonally dynamic three-dimensional water-bearing structure consisting of foliage components (leaves) and woody components (stems, trunk, stalks, and branches). Remote sensing systems such as the Landsat Thematic Mapper, SPOT, or aerial photography sense reflected optical wavelength energy measured in micrometers that is reflected, scattered, transmitted, and/or absorbed by the first few layers of leaves and stems of a healthy vegetation canopy. We typically get little information about the internal characteristics of the canopy, much less information about the surface soil characteristics lying below the canopy. Conversely, active microwave energy can penetrate the canopy to varying depths depending upon the frequency, polarization, and incident angle of the radar system. Microwave energy responds to objects in the plant's structure that are measured in centimeters and even decimeters. It is useful to identify the relationship between the canopy components and how they influence the radar backscattering.

If a radar sends a pulse of vertically or horizontally polarized microwave energy toward a stand of trees, it interacts with the components present and scatters some of the energy back toward the sensor. The amount of energy received is proportional to the nature of the energy sent (its frequency and polarization) and is dependent upon whether or not the canopy components depolarize the signal, how far the signal penetrates into the canopy, and whether it eventually interacts with the ground soil surface.

Penetration Depth and Polarization

Like-polarization backscatter toward the sensor results from single reflections from canopy components such as the leaves, stems, branches, and trunk. These returns are generally very strong and are recorded as bright signals in like-polarized radar imagery (HH or VV). This is often called canopy *surface scattering*. Conversely, if the energy is scattered multiple times within a diffuse volume such as a stand of pine trees (i.e., from a needle, to a stem, to the trunk, to a needle), the energy may become depolarized. This is often called *volume scattering*. A radar can measure the amount of depolarized volume scattering that takes place. For example, it is possible to configure a radar to send a vertically polarized pulse of energy. Some of this energy becomes depolarized in the canopy and exits toward the sensor in the horizontal domain. The depolarized energy may then be recorded by the sensor in VH mode – vertical send and horizontal receive.

Kasischke and Bourgeau-Chavez (1997) provide insight as to how the radar backscattering coefficient, $\sigma°$, from both woody (forested) and non-woody (e.g., brush, scrub, crops) environments is produced when the terrain is impacted by microwave energy.

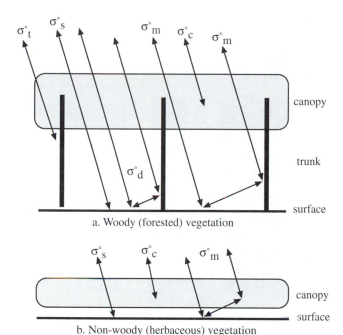

Figure 9-21   The major sources of scattering from a) woody and b) herbaceous vegetation canopies where σ°_c is the backscatter coefficient of the crown layer of smaller woody branches and foliage (i.e., surface scattering), σ°_m is the multiple-path scattering between the ground and canopy layer, σ°_t is direct scattering from the tree trunks, σ°_s is direct surface backscatter from the ground, and σ°_d is the double-bounce scattering between the tree trunks and ground (courtesy of American Society for Photogrammetry & Remote Sensing; Kasischke and Bourgeau-Chavez, 1997).

To understand radar-scattering from complex vegetation covers, it is necessary to think in terms of the different canopy layers affecting the radar signature. For wetlands containing shrubs and trees, there are three distinct layers to consider, as shown in Figure 9-21a: 1) a canopy layer that consists of small branches and foliage (leaves), 2) a trunk layer that consists of large branches and trunks or boles, and 3) a surface layer that may or may not be covered by water if wetland is present. For wetlands that do not contain woody plants, a simple two-layer model can be used (Figure 9-21b): 1) a canopy layer consisting of herbaceous vegetation, and 2) a surface layer that may be covered by water or soil. The backscattering coefficient exiting a woody vegetation canopy toward the radar system is σ°_w and can be expressed as (Wang et al., 1995; Dobson et al., 1995; Kasischke and Bourgeau-Chavez; 1997):

$$\sigma^\circ_w = \sigma^\circ_c + \tau_c^2 \tau_t^2 (\sigma^\circ_m + \sigma^\circ_t \; \sigma^\circ_s + \sigma^\circ_d) \qquad (9\text{-}24)$$

where

- σ°_c is the backscatter coefficient of the canopy layer of smaller woody branches and foliage (i.e., surface scattering),

- τ_c is the transmission coefficient of the vegetation canopy,

- τ_t is the transmission coefficient of the trunk layer,

- σ°_m is the multiple-path scattering between the ground and canopy layer,

- σ°_t is direct scattering from the tree trunks,

- σ°_s is direct surface backscatter from the ground, and

- σ°_d is the double-bounce scattering between the trunks and the ground.

By eliminating all terms associated with the trunk layer, it is possible to determine the total radar-scattering coefficient from terrain with non-woody, herbaceous vegetation, σ°_h:

$$\sigma^\circ_h = \sigma^\circ_c + \tau_c^2 (\sigma^\circ_s + \sigma^\circ_m) . \qquad (9\text{-}25)$$

The terms in Equations 9-24 and 9-25 are dependent on 1) the type of vegetation present (which has an impact on surface roughness), 2) the wavelength and polarization of the incident microwave energy, 3) the dielectric constant of the vegetation, and 4) the dielectric constant of the ground surface. The scattering and attenuation in the equations are all directly proportional to the dielectric constant. Live vegetation, with a higher water content (turgidity) has a higher dielectric constant than drier or dead vegetation. The presence of dew or moisture acts to increase the dielectric constant of vegetated surfaces (Kasischke and Bourgeau-Chavez, 1997). Often, the primary quantity governing the attenuation coefficient of a vegetation canopy is the water content per unit volume, not necessarily the actual structure and geometry of the leaves, stems, and trunk of the plants.

The condition of the ground layer is also very important in microwave scattering from vegetation surfaces. There are two properties of this layer that are important, including: 1) the micro- and mesoscale surface roughness (relative to the radar wavelength previously discussed), and 2) the reflection coefficient. In general, a greater surface roughness 1) increases the amount of microwave energy backscattered (increasing σ°_s), and 2) decreases the amount of energy scattered in the forward direction (decreasing σ°_m and σ°_d). The reflection coefficient is dependent on the dielectric constant (or conductivity) of the ground layer. A dry ground layer has a low dielectric constant and therefore has a low reflection coefficient. As soil moisture increases, so does the dielectric constant and, hence, the reflection coefficient. Given a constant surface roughness, as the soil dielectric constant increases, so does both the amount of backscattered and forward scattered microwave energy (resulting in increases in σ°_m, σ°_s, and σ°_d).

If there is a layer of water over the ground surface of a vegetated landscape such as in wetland environments, two things happen: 1) it eliminates any surface roughness, and 2) it significantly increases the reflection coefficient. In terms of microwave scattering, the elimination of any surface roughness means that all the energy is forward scattered, eliminating the surface backscattering term (σ°_s) in the equations; and, the increased forward scattering and higher reflection coefficient lead to significant increases in the ground-trunk and ground-canopy interaction terms σ°_d and σ°_m respectively (Kasischke and Bourgeau-Chavez, 1997).

Penetration Depth and Frequency

The longer the microwave wavelength, the greater the penetration into the plant canopy. For example, Figure 9-22 depicts the response of a hypothetical pine forest to microwave energy. Surface scattering takes place at the top of the canopy as the energy interacts with the leaves (or needles) and stems. Volume scattering by the leaves, stems, branches, and trunk takes place throughout the stand, and surface scattering can occur again at the soil surface. A comparison of the response of X-, C-, and L-band microwave energy incident to the same canopy is presented in Figure 9-23a–c. The shorter wavelength X-band (3 cm) energy is attenuated most by surface scattering at the top of the canopy by foliage and small branches. The C-band (5.8 cm) energy experiences surface scattering at the top of the canopy as well as some volume scattering in the heart of the stand. Little energy reaches the ground. L-band (23.5 cm) microwave energy penetrates farther into the canopy, where volume scattering among the leaves, stems, branches, and trunk cause the beam to become depolarized. Also, numerous pulses may be transmitted to the ground, where surface scattering from the soil-vegetation boundary layer may take place. Longer P-band radar (not shown) would afford the greatest penetration through the vegetation and mainly reflect off large stems and the soil surface (Waring et al., 1995).

Radar Backscatter and Biomass

Radar backscatter increases approximately linearly with increasing biomass until it saturates at a biomass level that

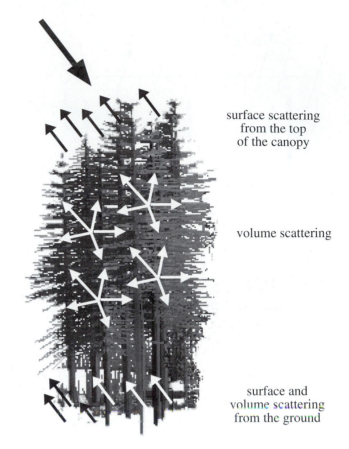

surface scattering
from the top
of the canopy

volume scattering

surface and
volume scattering
from the ground

Figure 9-22   The types of active microwave surface and volume scattering that might take place in a hypothetical pine forest stand (after Carver, 1988).

depends on the radar frequency. For example, Dobson et al. (1992) found that the biomass saturation level was about 200 tons/ha of Loblolly pine using P-band and 100 tons/ha at L-band, and that the C-band backscattering coefficient showed much less sensitivity to total aboveground biomass. Wang et al. (1994) evaluated Loblolly pine using ERS-1 SAR backscatter data. They also found that the C-band functioned poorly due to its high sensitivity to soil moisture and the steep local incident angle of the sensor (23°). Generally, backscatter at lower frequencies (P- and L-bands) is dominated by scattering processes involving the major woody biomass components (trunks and branches), while scattering at high frequencies (C- and X-bands) is dominated by scattering processes in the top crown layer of branches and foliage. Radar canopy measurements have also been found to be correlated with leaf-area-index (LAI) measurements (Franklin et al., 1994).

Some other general observations about SAR vegetation interpretation include:

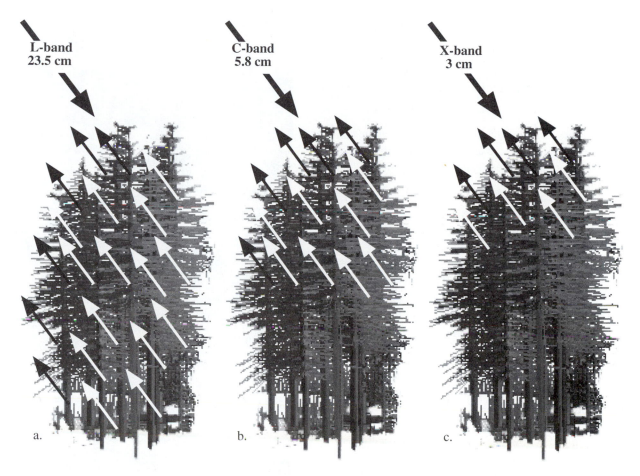

Figure 9-23   Theoretical response of a pine forest stand to X-, C-, and L-band microwave energy. The shorter the wavelength, the greater the contribution from *surface scattering*. The longer the wavelength, the greater the penetration into the material and the greater the *volume scattering*.

- Vertically polarized energy is highly attenuated by the vertically oriented canopy components (leaves, stems, branches, and trunk) while horizontally polarized energy is not.

- The brighter the return on like-polarized radar images (HH or VV), the greater the contribution from surface scattering.

- The brighter the return on cross-polarized images (HV or VH), the greater the contribution from volume (internal canopy) scattering (Avery and Berlin, 1992).

- When the radar wavelength is approximately the same size as the canopy components (e.g., C- or X-band), substantial surface and volume scattering will take place and little energy may reach the ground. Consequently, shorter wavelength radars (2 – 6 cm) may be preferred when

monitoring crop canopies and tree leaves. Longer wavelength radars (9 – 30 cm) exhibit substantial volume scattering as incident energy interacts with larger trunk and branch components. Considerable surface scattering from the underlying soil may also occur which can cause confusion.

- Cross-polarized images (HV or VH) are less sensitive to slope variations. This suggests vegetation monitoring in mountainous areas may best be performed using cross-polarization techniques. Also, the same row crop planted in different directions can produce like-polarized images that are difficult to interpret. This ambiguity may be reduced when cross-polarized images are available in addition to like-polarized images.

- One of the key variables necessary to model the hydrological cycle is how much water is being stored in vegetation canopies and the rate of evapotranspiration taking place. Generally, the more moisture in the structure, the greater the dielectric constant and the higher the radar backscatter return. Active microwave remote sensing is capable of sensing canopy (or leaf) water content in certain instances.

- Radar imagery can provide some information on landscape-ecology patch size and canopy gaps that are of value when monitoring ecosystem fragmentation and health (Sun and Ranson, 1995).

Shuttle Imaging Radar C- and X-Band SAR (SIR-C/X-SAR) images of the Amazon rain forest obtained on April 10, 1994, are displayed in black-and-white in Figure 9-24 and as a color composite in Color Plate 9-2. These images may be used to demonstrate several of the concepts mentioned. First, the multifrequency radar data reveals rapidly changing land-use patterns, and it also demonstrates the capability of the different radar frequencies to detect and penetrate heavy rainstorms. The top X-band image has VV polarization, the center C-band image is HV, and the bottom L-band image is HV. A heavy downpour in the lower center of the image appears as a black "cloud" in the X-band image, more faintly in the C-band image, and is invisible in the L-band image. When combined in the color composite, the rain cell appears red and yellow. Although radar can usually penetrate through clouds, short radar wavelength (high frequency), such as X- and C-band, can be changed by unusually heavy rain cells. L-band, at 23.5 cm (9 in.) wavelength, is relatively unaffected by such rain cells. Such information has been used to estimate rainfall rates (NASA JPL, 1996). Of course, there is very little backscatter from the river in all three radar images, causing it to appear black.

The area shown is in the state of Rondonia, in western Brazil. The pink areas in the color composite are pristine tropical rain forests, and the blue and green patches are areas where the forest has been cleared for agriculture. Radar imaging can be used to monitor not only the rain forest modification but also the rates of recovery of abandoned fields. Inspection of the black-and-white images reveals that as the wavelength of the radar progresses from X- to C- to L-band, it becomes easier to discriminate cleared land from rain forest. In the L-band black-and-white image, the cleared land appears relatively dark while the rainforest appears very bright. Evidently, the cleared land contains sufficient scatterers to cause it to appear bright at X- and C-band frequencies, while at L-band frequencies the incident energy to the cleared fields is reflected away from the radar receiver. The rain forest appears relatively bright in all images because of

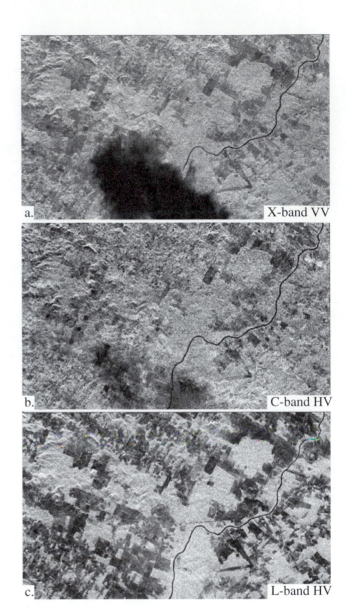

Figure 9-24   SIR-C/X-SAR images of a portion of Rondonia, Brazil, obtained on April 10, 1994. a) X-band image with VV polarization. b) C-band image is HV. c) L-band image is HV. A heavy rain in the lower center of the image appears as a black "cloud" in the X-band image, more faintly in the C-band image, and is invisible in the L-band image. L-band, at 24 cm (9 in) wavelength, is relatively unaffected by such rain cells. Also, the L-band image differentiates the pristine rain forest (bright) from the darker clear-cut areas because it penetrates farther into the canopy, experiencing greater volume scattering. Water, of course, is dark in all three bands. A color composite of the three images is found in Color Plate 9-2 (courtesy NASA Jet Propulsion Lab).

1) the tremendous number of leaves, branches, stems, and trunks that can contribute to canopy surface scattering and volume scattering within the canopy (L-band penetrates the greatest distance into the canopy and is brighter), and 2) there is a high moisture content in the rain forest canopy, further increasing the amount of backscatter present in all three bands (NASA JPL, 1996).

Water Response to Microwave Energy

Flooding occurs periodically in many areas, and cloud cover obscures the collection of data using optical and thermal sensors. Fortunately, the smooth surface of standing water reflects almost all the incident microwave radiation away from the sensor (unless there is a strong wind with lots of chop, which can cause confusion), resulting in lower backscatter than a dry surface. When standing water is present under vegetation, a unique corner-reflection backscatter interaction between surface water and tree stems results in an extremely high backscatter and allows inundation to be clearly mapped (Hess et al., 1990; Waring et al., 1995). Ford et al. (1986) demonstrated how wetlands along the Savannah River appeared especially bright compared to standing water and other upland land cover such as Loblolly pine, due to this corner-reflector condition between the water and cypress-tupelo tree trunks.

Assessing Soil Moisture

L-band radar penetrates into bare, damp, smooth soil to a maximal depth of approximately 10 cm. Shorter wavelengths penetrate to only 1 – 3 cm. In agricultural fields that have smooth soil surfaces and biomass of less than 1 mg/ha, moisture content of surface layers can be fairly accurately determined (Wang et al., 1994). Once vegetation biomass exceeds a certain limit, the ability of radar to sense surface soil-water conditions decreases rapidly (Waring et al., 1995). Under a dense forest canopy, the amount of moisture held in the leaves is so large that it interferes with any direct assessment of soil-water status (Jackson and Schmugge, 1991).

Thus, the presence of vegetation over the soil surface can add considerable complexity to measuring soil moisture using microwave remote sensing techniques. Because the vegetation's transmissivity decreases with increasing microwave frequency, it is best to use the longest wavelength available for soil moisture mapping (Dobson and Ulaby, 1998). This would be P-band ($\lambda = 68$ cm) for aircraft SARs and L-band ($\lambda = 23$ cm) for satellite SARs. If the vegetation biomass is < 0.5 kg/m^2, the effect of the vegetation backscatter may be ignored for like-polarized L-band data, i.e., σ°_{VV}

and σ°_{HH}. Unfortunately, when the biomass is > 0.5 kg/m^2, it is not currently possible to disentangle the separate soil moisture and vegetation backscatter contributions.

Urban Structure Response to Microwave Energy

Urban buildings, cars, fences, bridges, etc., act as corner reflectors that send much of the incident energy back toward the antenna. This generally results in bright signatures for urban phenomena on the radar image. Unfortunately, this plethora of bright returns is often confusing in radar images of urban areas especially when 1) the cardinal effect takes place, and/or 2) the imagery is of relatively low spatial resolution.

A SIR-C/X-SAR image of Los Angeles, CA, is found in Figure 9-25. It has approximately 30 x 30 m resolution. The radar look-direction is from the top to the bottom of the image. Only the major terrain features can be resolved, including the freeways, major urban development, the Pacific Ocean, and the mountainous terrain. In general, it is not possible to distinguish between commercial and residential communities. Interestingly, there are some very bright polygonal areas in the image. These are produced by the radar *cardinal effect*. Early in radar remote sensing research it was noted that reflections from urban areas, often laid out according to the cardinal points of the compass, caused significantly larger returns when the linear features were illuminated by the radar energy at an angle orthogonal (perpendicular) to their orientation (Raney, 1998). For example, the residential communities of San Fernando and Santa Monica are for all practical purposes similar in nature to other residential communities in their vicinity (Figure 9-25). However, due to the cardinal effect caused by their unique orientation relative to the SIR-C sensor system at the time of data collection, they appear in the radar image to contain land-cover that is dramatically different than the other residential communities found throughout the Los Angeles basin.

It has been demonstrated repeatedly that two identical tracts of urban land (e.g., two single-family housing tracts) built at the same time, with the same lot size, and using the same materials will appear dramatically different from one another on radar imagery if one is laid out in a different orientation than the other, e.g., one is laid out with streets trending northeast and one is laid out with streets trending due north. Similarly, the regular spacing of agricultural row crops can produce a similar effect. Also, the same parcel of urban land may appear quite different on radar imagery acquired on two different dates if practically any of the sys-

SIR-C/X-SAR Image of Greater Los Angeles, California

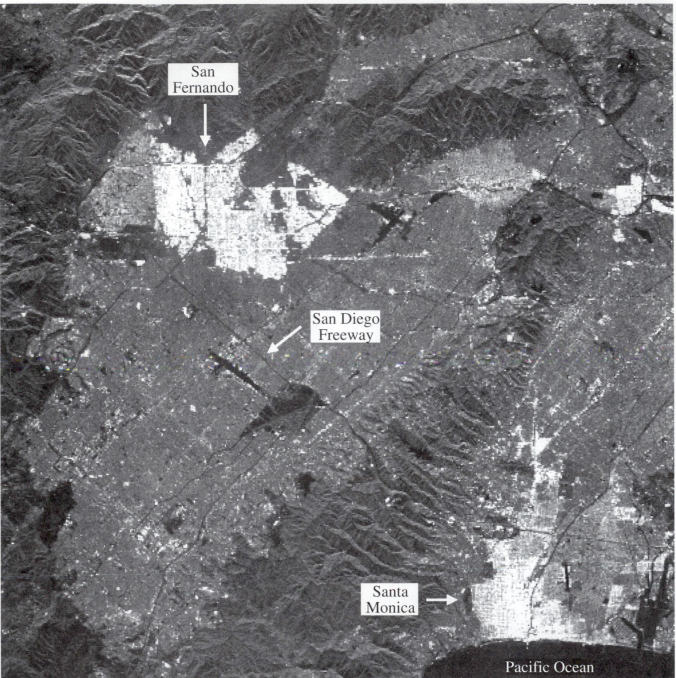

Figure 9-25 The cardinal effect is responsible for the pronounced bright signature of portions of Santa Monica and San Fernando in this SIR-C/X-SAR image of the Los Angeles basin obtained on October 3, 1994. The look direction is from the top to the bottom of the image (courtesy NASA Jet Propulsion Laboratory).

tem parameters are changed, especially look direction. This plays havoc when attempting to perform digital change detection using radar imagery.

RADARSAT, JERS-1, and ERS-1,2, with their relatively coarse spatial resolutions (Table 9-1) may be of value for obtaining general Level I land-cover and land-use information. When attempting to extract Level II and III urban land-cover and land-use information, optical remote sensor data is usually superior. However, if high spatial resolution radar is available, then detailed urban information may be extracted. For example, a photograph and a radar image of the Pentagon are found in Figure 9-26. Henderson and Xia (1998) provide examples of how settlement patterns and socioeconomic information such as population estimates may be extracted from radar data.

SAR Remote Sensing from Space

The following sections summarize the characteristics of the major satellite SAR sensor systems.

Seasat

Seasat was launched by NASA on June 26, 1978, and functioned for 105 days. It carried an L-band (23.5 cm) active microwave SAR at an altitude of 800 km. The antenna was 10.7 x 2.16 m in dimension. It collected HH-polarized data at an incident angle of 23°. It had a range resolution of 25 m and an azimuth resolution of 25 m. The swath width was 100 km. The data were processed with "4 looks." Seasat had an orbital repeat cycle of 17 days. The data were processed originally optically and then digitally.

Shuttle Imaging Radar SIR-A, SIR-B, SIR-C

Several very important scientific radar instruments have been carried aboard NASA's Space Shuttle and operated for a number of days before returning to Earth. SIR-A and SIR-B were launched on November 12, 1981, and October 5, 1984, respectively, and were in orbit for 2.5 and 8 days. Both payloads consisted of an L-band (23.5 cm) SAR.

SIR-A had a 9.4 x 2.16 m antenna with HH polarization. The incident angle was 50°. The sensor had a range and azimuth resolution of 40 m with 6 looks. The swath width was 50 km. The Shuttle was in orbit at 260 km above the Earth. The data were processed optically.

SIR-B had a 10.7 x 2.16 m antenna with HH polarization. It had an incident angle of 15 – 64°. Its azimuth resolution was 17 – 58 m and its range resolution was 25 m with 4 looks. The swath width was 10 – 60 km. It operated at 225 and 350 km above the Earth. The data were processed both optically and digitally.

SIR-C was a significant breakthrough in radar remote sensing. It was a joint project between the United States (NASA JPL) and a consortium of European groups. SIR-C carried aloft a three frequency SAR: X-band (3 cm), C-band (5.8 cm), and L-band (23.5 cm). The three antennas were placed on a common platform in the Shuttle bay. The L- and C-bands had quad polarization (i.e., HH, HV, VV, and VH) while the X-band had VV polarization. The incident angle was from 15 – 55°. The range resolution was 10 - 30 m, and the azimuth resolution was 30 m with approximately 4 looks. The swath width was 15 – 90 km. It was flown at 225 km above the Earth. The data were processed digitally. *This was the first true multifrequency, multipolarization spaceborne SAR.* Data from all three bands are excellent and widely available through JPL and other locations. Several examples of SIR-C data are found in this chapter and in the color plate section.

RADARSAT

RADARSAT was launched by the Canadian government on November 4, 1995, into a near-polar, Sun-synchronous orbit 798 km above the Earth. It has a dawn-to-dusk orbit, meaning it crosses the equator at dawn and dusk (approximately 6:00 a.m. and p.m. local time) and is rarely in eclipse or darkness. The orbital inclination is 98.6° with a period of 100.7 minutes and 14 orbits per day. It has a single C-band (5.6 cm) active microwave sensor that transmits at 5.3 GHz frequency at a pulse length of 42.0 μs. The antenna size is 15 x 1.5 m. Its polarization is horizontal-send and horizontal-receive (HH) (Raney, 1991).

Unlike many of the other sensors to be discussed, RADARSAT provides a range of spatial resolutions and geographic coverages. There are seven image sizes, termed *beam modes,* summarized in Figure 9-27 and Table 9-5. Each beam mode is defined by the geographic area it covers and the spatial resolution. The beam modes range from *Fine,* which covers a 50 x 50 km area and has a 10 x 10 m spatial resolution, to *ScanSAR Wide,* which covers a 500 x 500 km area at 100 x 100 m spatial resolution.

RADARSAT obtains data using a range of incident angles from less than 20° (steep angle) to almost 60° (shallow

a. Oblique Photograph of the Pentagon

b. Radar Image of the Pentagon

Figure 9-26   a) Low-oblique aerial photograph of the Pentagon in Washington, DC. b) Synthetic aperture radar image of the Pentagon (courtesy of Federation of American Scientists).

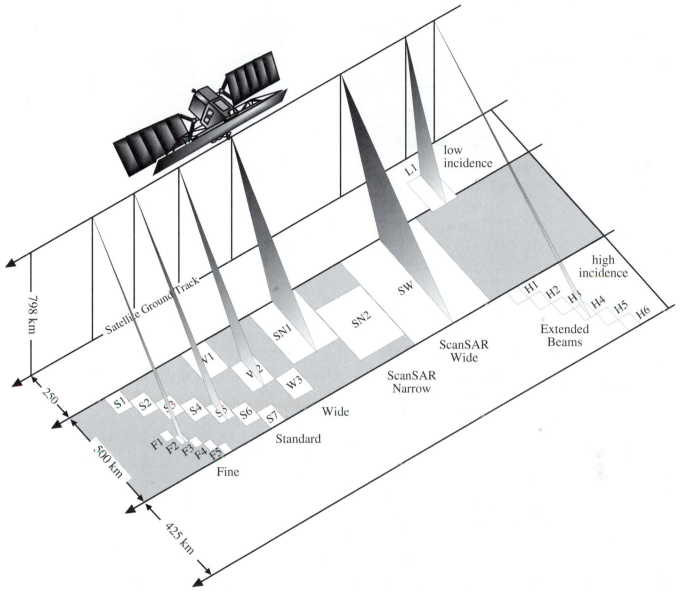

Figure 9-27 The beam modes and incident angle options for acquiring RADARSAT data. The size of the footprints provide information on the spatial resolution. Refer to Table 9-5 for specifications (after RADARSAT, International, Inc.).

angle). Within each beam mode, a number of incident angle positions are available (Figure 9-27 and Table 9-5). These are called *beam positions*. For example, Standard beam mode, which covers a 100 x 100 km area, has seven beam positions. By specifying a beam position, one of seven 100 x 100 km images within a 500 km accessible swath will be collected. Factors influencing the choice of beam mode include the sensitivity of the application to incident angle, type of terrain, stereo requirements, spatial resolution desired, and how often coverage of the area is required.

RADARSAT's orbit has a 24-day cycle, meaning it returns to the same location every 24 days. However, it can be pointed to provide a more frequent revisit cycle. The user also has the option of collecting imagery based on two different look directions. As RADARSAT descends from the North Pole (a descending orbital pass), it views the Earth in a westerly direction. As it ascends from the South Pole (an ascending orbital pass) it views the Earth in an easterly direction. This can be very useful when working in areas with high relief, when we are interested in highlighting fea-

Table 9-5.   RADARSAT Beam Position Characteristics (RADARSAT, 1998)

Operational Beam Mode	Beam Position	Incident Angle Positions (Degrees)	Nominal Spatial Resolution (m)	Nominal area (km)	Number of Processing Looks
Fine (5 positions)	F1	37 – 40	10	50 x 50	1 x 1
	F2	39 – 42			
	F3	41 – 44			
	F4	43 – 46			
	F5	45 – 48			
Standard (7 positions)	S1	20 – 27	30	100 x 100	1 x 4
	S2	24 – 31			
	S3	30 – 37			
	S4	34 – 40			
	S5	36 – 42			
	S6	41 – 46			
	S7	45 – 49			
Wide (3 positions)	W1	20 – 31	30	165 x 165	1 x 4
	W2	31 – 39		150 x 150	
	W3	39 – 45		130 x 130	
ScanSAR Narrow (2 positions)	SN1	20 – 40	50	300 x 300	2 x 2
	SN2	31 – 46			
ScanSAR Wide	SW1	20 – 49	100	500 x 500	2 x 4
Extended High (6 positions)	H1	49 – 52	25	75 x 75	1 x 4
	H2	50 – 53			
	H3	52 – 55			
	H4	54 – 57			
	H5	56 – 58			
	H6	57 – 59			
Extended Low	L1	10-23	35	170 x 170	1 x 4

tures with a particular orientation, and/or when the study requires imagery acquired in the early morning or early evening. RADARSAT-2 will be launched in the early twenty-first century.

European Space Agency ERS-1

The European Space Agency launched the ERS-1 on July 16, 1991. It has a C-band (5.6 cm) SAR with a 10 x 1 m antenna and VV polarization. It functions with an incident angle of 23°. The range resolution is 26 m and the azimuth resolution is 30 m with 6 looks. It has a swath width of 100 km. It is in orbit approximately 785 km above the Earth. The data are processed digitally. An identical ERS-2 was launched in 1995. ERS-1 and ERS-2 have on occasion been operated in tandem to provide image pairs for SAR interferometry research.

JERS-1

The National Space Development Agency (NASDA) of Japan launched the Japanese Earth Resource Satellite (JERS-1) on February 11,1992. It is very similar to the original Seasat. It has an L-band (23.5 cm) SAR with an 11.9 x 2.4 m antenna and HH polarization. The incident angle is 39°. Range and azimuth resolution are both 18 m with 3 looks. The swath width is 75 km. The SAR orbits at approximately 568 km above the Earth. The data are processed digitally. JERS-1 operation was terminated on October 12, 1998.

Almaz-1

This SAR was placed in orbit by the former Soviet Union on March 31, 1991 and functioned for 1.5 years. It consisted of an S-band (9.6 cm) SAR with a 1.5 x 15 m antenna and HH polarization. The incident angle ranged from 30 – 60°. Range resolution was 15 – 30 m and azimuth resolution was 15 m with greater than 4 looks. The swath width was 20 – 45 km. It was placed in a 300 km orbit above the Earth. The data were processed digitally.

 ### Radar Interferometry

The previous discussion of synthetic aperture radar systems was primarily concerned with collecting a single image of the terrain. It is possible to acquire multiple SAR images of the terrain from aircraft or spacecraft to extract valuable three-dimensional and velocity information. *Imaging radar interferometry* is the process whereby radar images of the same location on the ground are recorded by antennas at 1) different locations, or 2) different times (Madsen and Zebker, 1998). Analysis of the resulting two interferograms allow very precise measurements of the range to any specific x,y,z point found in each image of the interferometric pair. The precision may be at the sub-wavelength scale.

Interferometric Topographic Mapping

Topographic mapping based on SAR interferometry relies on acquiring data from two different look angles and assumes that the scene imaged did not move between data acquisitions. The two measurements can be from two radars placed on the same platform separated by a few meters. This is called *single-pass interferometry*. The first single-pass interferometric SAR is the Shuttle Radar Topography Mission (SRTM) to be launched early in the twenty-first century. It has a C-band and an X-band antenna separated by 60 m. Interferometry may also be conducted using a single radar that obtains two measurements on different orbital tracks that are closely spaced but a day or so longer apart. This is the methodology used for the SIR-C and ERS-1,2 interferometry and is called *multiple-pass* or repeat pass interferometry.

Interferometric SAR data can provide extremely high precision topographic information (x,y,z) that is just as accurate as digital elevation models derived using traditional optical photogrammetric techniques. However, interferometry can

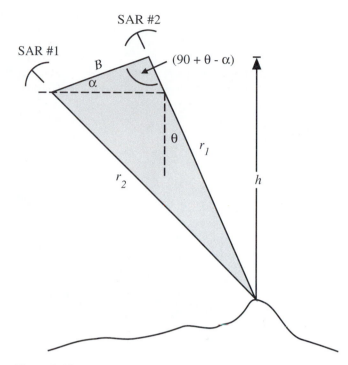

Figure 9-28 The geometric relationship between two satellite SAR systems used for interferometry to extract topographic information (after NASA JPL, 1999b).

operate through clouds, day or night. This is important for cloud-shrouded tropical or arctic environments or when disasters strike and one cannot wait for a window to obtain optical (photographic) data. Digital elevation information derived from SIR-C interferometric data are displayed in Color Plates 9-3 and 9-4. The cartographic community is especially interested in interferometric SAR.

Interferometric SAR obtains digital elevation information in the following manner (JPL, 1999b). First, the two radar images must be precisely registered. Then their geometry is such that the two radars and any object on the ground form a triangle (Figure 9-28). If we know the distance (range) from each radar to the object on the ground (r_1 and r_2), the distance between the two radars (the baseline B) and the angle of that baseline, α (with respect to the horizontal), we can use trigonometry (the cosine rule) to calculate the height, h, of one of the radars above the position of the object on the ground. From Figure 9-28 we know that:

$$h = r_1 \cos\theta \qquad (9\text{-}26)$$

and by the cosine rule,

$$(r_2)^2 = (r_1)^2 + B^2 - 2r_1 B \cos(90 + \theta - \alpha) \qquad (9\text{-}27)$$

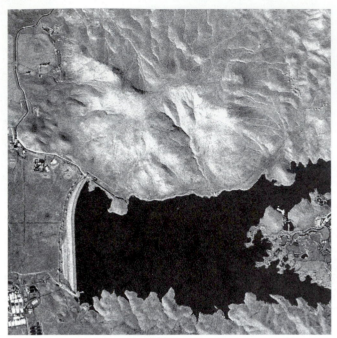

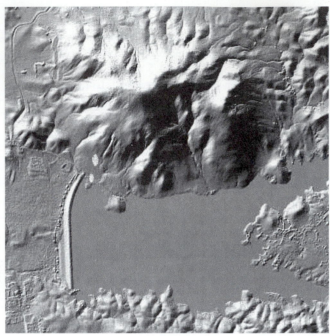

a. Intermap Star 3*i* X-band image

b. Intermap digital terrain model derived from IFSAR data

Figure 9-29   a) Intermap X-band Star 3*i* orthorectified image of Bachelor Mountain, CA. b) Digital elevation model of the same area derived using interferometric synthetic aperture radar (IFSAR) techniques (orthorectified imagery and Global Terrain Digital Elevation Model provided by Intermap Technologies, Inc.).

which is the same as:

$$(r_2)^2 = (r_1)^2 + B^2 - 2r_1 B \sin(\theta - \alpha). \qquad (9\text{-}28)$$

We solve for θ, then for h (using $h = r_1 \cos \theta$). This calculation is repeated for every point on the ground within the image. If we determine the precise height of one of the radars above sea level, we can produce a map of the surface heights. It turns out that we can only precisely measure the relative distance $(r_2 - r_1)$ from the phase difference between each pair of radar measurements. This relative distance can be related to the height, h, after some additional algebra (SIR-CED home page; JPL, 1999b).

The Shuttle Radar Topography Mission (SRTM) will use C-band and X-band interferometric synthetic aperture radars (IFSARS) to acquire topographic data over 80 percent of the Earth's land mass (between 60° N and 56° S) during an 11 day mission. The topographic maps produced will meet Interferometric Terrain Height Data (ITHD-2) specifications. Thus, the first worldwide collection of digital elevation data will be acquired using interferometry, not photogrammetry. The interferometer pairs will also be of significant value for many Earth resource applications.

Private commercial companies also provide interferometric synthetic aperture radar (IFSAR) data. One of the most widely used is the Intermap X-band Star 3*i* system which generates high quality 3 x 3 m X-band microwave imagery plus a detailed digital elevation model of the terrain. A good example is shown in Figure 9-29 with the X-band image of Bachelor Mountain, CA, on the left and the digital elevation model on the right. Of course, the radar image and the DEM are registered. Such data are very valuable for land-use and land-cover analysis as well as watershed hydrologic studies.

Interferometric Velocity Mapping

If the look angles of multiple data acquisitions are held constant, there is no sensitivity to topography and the interferometry can be used to extract information about things that have changed in the scene. Quantitative information about the velocity of objects that moved between the two observations may be made. Interferometry has been successfully applied to measuring movement along fault lines, measuring seismic displacement due to earthquakes, mapping glacier velocity, monitoring ocean currents, and measuring wave spectra. In addition, interferometry can be used to determine if man-made objects in the scene have moved. This is very powerful for change detection purposes.

Passive Microwave Remote Sensing

There is a growing interest in the measurement of passive microwave energy to monitor some of the more important global hydrologic variables such as soil moisture, precipitation, ice water content, and sea-surface temperature. In fact, several sensors onboard Earth Observing System PM-1 will include specialized passive microwave radiometers.

As discussed in Chapter 2, the Earth approximates a 300 K blackbody with a dominant wavelength of approximately 9.7 μm. While the dominant wavelength may be 9.7 μm, a continuum of energy is emitted from the Earth and the atmosphere. In fact, the Earth passively emits a steady stream of microwave energy. The only difference is that 1) the Earth's materials do not emit a tremendous amount of passive microwave energy, and 2) what energy it does emit is relatively weak in intensity due to its long wavelength. Nevertheless, a suite of radiometers have been developed that can record subtle, passive microwave energy. The instruments measure the *brightness temperature* of the terrain or the atmosphere and its constituents (Engman and Gurney, 1991).

Passive Microwave Radiometers

Passive microwave remote sensing devices may be 1) profiling radiometers, or 2) scanning radiometers. A profiling radiometer simply stares at the terrain directly beneath the craft at nadir (or off-nadir if desired) and records the radiance from within the instantaneous field of view of the sensor. The output is a profile of the microwave brightness temperature recorded as the aircraft or spacecraft moves forward. A scanning microwave radiometer is much like the scanning thermal infrared radiometer discussed in Chapter 8. It collects data across-track as the craft moves forward. The result is a matrix of brightness temperature values that can be used to construct a passive microwave image.

Passive microwave radiometers generally record energy in the region between 0.15 and 30 cm (between 1 and 200 GHz), well beyond the thermal infrared region (3 – 14 μm). The microwave frequencies most commonly used are centered at 1, 4, 6, 10, 18, 21, 37, 50, 85, 157, and 183 GHz. This means that it is theoretically possible to acquire multi-

Figure 9-30 SSM/I passive microwave radiometer image of the Amazon Basin obtained at a frequency of 85 GHz with vertical polarization (courtesy DMSP).

spectral passive microwave imagery. The actual bandwidths (range of frequencies) recorded are usually fairly broad so that enough passive microwave energy is available to be recorded by the antenna. Similarly, the spatial resolution of passive microwave radiometers is usually large so that sufficient energy is collected within the instantaneous field of view to be recorded by the antenna. Aircraft sensors flying closer to the ground may have spatial resolutions measured in meters while most satellite passive microwave scanning radiometers have a spatial resolution measured in kilometers. The sensor is actually a large antenna sensitive to passive microwave energy.

Special Sensor Microwave/Imager (SSM/I)

One of the first passive microwave sensors was the Special Sensor Microwave/Imager (SSM/I) onboard the Defense Meteorological Satellite Program (DMSP) satellites since 1987. Fortunately, the Department of Defense releases the data to the scientific community (Figure 9-30). The SSM/I is a four-frequency, linearly polarized passive microwave radiometric system that measures atmospheric, ocean and terrain microwave brightness temperatures at 19.35, 22.23, 37.0, and 85.5 GHz. The SSM/I rotates continuously about an axis parallel to the local spacecraft vertical and measures the upwelling scene brightness temperatures. It is calibrated using cold sky radiation and a hot reference absorber. The swath is approximately 1400 km. The data are converted into sensor counts and transmitted to the National Environmental Satellite, Data, and Information Service (NESDIS). The SSM/I is an excellent sensor for measuring the brightness temperature of very large regions. For example, a SSM/I passive microwave image of almost the entire Amazon Basin is shown in Figure 9-30.

NOAA has developed a SSM/I rainfall algorithm that utilizes the 85.5 GHz channel to detect the scattering of upwelling radiation by precipitation-sized ice particles within the rain layer. The scattering technique is applicable over the land and the ocean. Rain rates can be derived indirectly, based on the relationship between the amount of ice in the rain layer and the actual rainfall on the surface. In addition, a scattering-based global land rainfall algorithm has been developed. Monthly rainfall at 100 x 100 km and 250 x 250 km grids have been produced for the period from July, 1987 to the present (Ferraro, 1997; Li et al., 1998).

TRMM Microwave Imager (TMI)

The Tropical Rainfall Measuring Mission (TRMM) is sponsored by NASA and the National Space Development Agency (NASDA) of Japan to study the tropical rainfall and the associated release of energy that helps to power global atmospheric circulation. It carries five instruments and was launched on November 27, 1997. The TRMM Microwave Imager (TMI) is a passive microwave sensor designed to provide quantitative rainfall information over a 487-mile (780 km) swath. It is based on the design of the SSM/I. It measures the intensity of radiation at five frequencies: 10.7 (45 km spatial resolution), 19.4, 21.3, 37, and 85.5 GHz (5 km spatial resolution). Dual polarization at four frequencies provides nine channels. The new 10.7 GHz frequency provides a more linear response for the high rainfall rates common in tropical rainfall.

Calculating the rainfall rates from both the SSM/I and TMI sensors requires complicated calculations because water bodies such as oceans and lakes emit only about one-half the energy specified by Planck's radiation law at microwave frequencies. Therefore, they appear to have only about half their actual temperature at the surface and appear very "cold" to a passive microwave radiometer. Fortunately, raindrops appear to have a temperature that equals their real temperature and appear "warm" or bright to a passive microwave radiometer. The more raindrops, the warmer the whole scene appears. Research over the last three decades has made it possible to obtain relatively accurate rainfall rates based on the temperature of the passive microwave scene.

Land is very different from oceans in that it emits about 90 percent of its real temperature at microwave frequencies. This reduces the contrast between the rain droplets and the land. Fortunately, the high-frequency microwaves (85.5 GHz) are strongly scattered by ice present in many raining clouds. This reduces the microwave signal of the rain at the satellite and provides a contrast with the warm land background, allowing accurate rainfall rates to be computed over land as well.

Advanced Microwave Scanning Radiometer (AMSR)

The AMSR is to be flown as an EOS PM-1 sensor in polar, Sun-synchronous orbit. It will be an eight frequency passive microwave radiometer that measures frequencies at 6.9, 10.7, 18.7, 23.8, 36.5, and 89 (HV polarization) and 50.3 and 52.8 (VV polarization). It will have a 7 km field of view at 89 GHz and 60 km field of view at 6.9 GHz. It will have a 1600 km swath width. The AMSR will measure total water-vapor content, total liquid-water content, precipitation, snow-water equivalent, soil moisture (using the 6.9 and 10.7 GHz frequencies), sea-surface temperature (SST), sea-surface wind speed, and sea-ice extent. The specifications of the sensors are subject to change.

From this discussion it should be clear that passive microwave remote sensing is making a significant contribution to our Earth science knowledge.

 ## Light Detection and Ranging (LIDAR)

Elevation information is a critical component of most geographic databases. Several methods have been derived to measure the elevation of features, with varying degrees of accuracy and cost (Flood and Gutelius, 1997), including *in situ* field surveying, photogrammetry, SAR interferometry, and LIDAR data collection. Field surveys generally require a team of people to measure distances and angles across the landscape. Methods for such surveys are well developed and can result in very accurate information, and may include the use of GPS instruments. However, the field methods are time-consuming and expensive on a per-point basis. In rugged or heavily vegetated areas, the ground-based surveys can be difficult. Due to these obstacles, sometimes the density of x,y,z observations is not very high, making it necessary to interpolate some distance between points to produce a digital elevation model of the area.

Photogrammetric techniques are also routinely used to collect x,y,z topographic information (refer to Chapter 6). This method traditionally covers a larger area with a more dense collection of x,y,z points. In fact, Chapter 6 summarizes how digital elevation models are extracted directly from stereoscopic aerial photography. An advantage of photogrammetric techniques is that it allows the analyst to select critical features such as ridgelines or breaklines, where a greater density of x,y,z observations can be extracted. LIDAR does not allow the analyst to control the placement of individual x,y,z measurements.

LIDAR is a relatively new technology that offers an alternative to *in situ* field surveys and photogrammetric techniques for the collection of elevation data. Although technologically complex, LIDAR provides a methodology that is accurate, timely, capable of operating in difficult terrain, and increasingly affordable (Flood and Gutelius, 1997).

LIDAR Sensor System

The first optical laser was developed in 1960 by Hughes Aircraft, Inc. Soon laser-based instruments were computing distance by measuring the travel time of light from a laser transmitter to a target and then back to a laser receiver (Ritchie, 1996). Early uses of LIDAR employed a profiling laser, which only collected measurements directly underneath the aircraft, creating a single transect of measurements across the landscape (Jensen et al., 1987). Modern LIDAR systems offer several improvements over these earlier profiling systems. The integrative use of kinematic GPS and inertial navigation systems on airborne LIDAR platforms has allowed the technology to mature into commercially available systems whose price and cost of operation are generally similar to that of photogrammetric equipment used for similar purposes (Flood and Gutelius, 1997).

The essence of LIDAR technology is the measurement of laser pulse travel time from the transmitter to the target and back to the receiver. Since the laser pulse travels at the speed of light (3×10^8 m s^{-1}), very accurate timing is necessary to obtain fine vertical resolutions. A one nanosecond timing resolution, for example, allows a range measurement accuracy of about 30 cm. Timing technology allows for measurements of < 5 cm (Ritchie, 1996).

As the aircraft moves forward, a scanning mirror directs the laser pulses back and forth across-track (Figure 9-31a). The maximum off-nadir scan angle for the instrument is generally tunable to the needs of a particular data-collection mission. This results in a series of data points arranged across the flightline, as shown in Figure 9-31b. This example displays hits on bare ground, a power transmission line and tower, and the effects of the laser interacting with a tree canopy. Multiple flightlines can be combined to cover the desired area. Data-point density is dependent on the number of pulses transmitted per unit time, the scan angle of the instrument (Figure 9-31c), the elevation of the aircraft

above-ground-level, and the forward speed of the aircraft. Note that the greater the scan angle off-nadir, the more vegetation that will have to be penetrated to receive a pulse from the ground assuming a uniform canopy.

LIDAR data avoids the problems of aerial triangulation and orthorectification because each LIDAR measurement is individually georeferenced (Flood and Gutelius, 1997). This requires measurement of several important factors, including:

- the laser pulse travel time from the LIDAR instrument to the target (ground) and back,

- the scan angle of the LIDAR at the time of the laser pulse,

- the effect of atmospheric refraction on the speed of light,

- the attitude (pitch, roll, and heading) of the aircraft at the time of the laser pulse, and

- the position of the LIDAR instrument in three-dimensional space at the time of the laser pulse.

The LIDAR records the scan angle of the mirror and the travel time of the laser pulse to and from the target. Atmospheric refraction has a slight effect on the speed of light that should be taken into account in the conversion from laser pulse travel time to distance. Aircraft attitude is recorded by an inertial navigation unit that takes careful measurements of the aircraft's pitch, roll, and heading. Combined with the information recorded from the scanning mirror, the attitude measurements allow precise determination of where the LIDAR instrument was pointed at the time of an individual laser pulse. Aircraft position is determined from GPS equipment located on the aircraft.

During the LIDAR overflight, one or more GPS base stations must be recording positional data on the ground in the area of the overflight. These base stations are located at known points, and the data they collect is used to remove any GPS error resulting from atmospheric effects and selective availability. This differential correction allows the aircraft location to be known within 5 – 10 cm accuracy in all three axes (Vaughn et al., 1996).

The combination of all these factors allows three-dimensional georeferenced coordinates to be determined for each laser pulse. There are several other factors and calibrations required for accurate LIDAR georeferencing, including timing accuracy, range dispersion within the laser footprint, "range walk" due to backscatter strength variations, and syn-

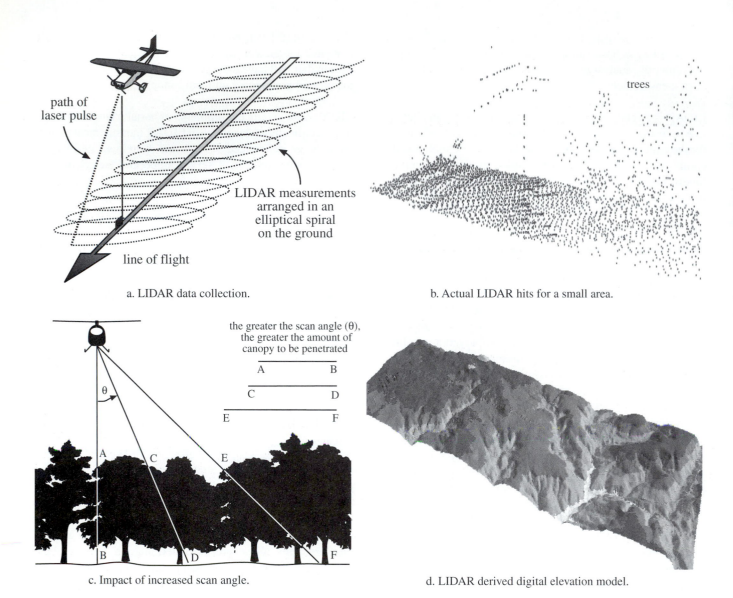

a. LIDAR data collection.

b. Actual LIDAR hits for a small area.

c. Impact of increased scan angle.

d. LIDAR derived digital elevation model.

Figure 9-31   a) Logic of LIDAR data collection showing the line of flight and the elliptical spiral of points that are backscattered from the terrain as the aircraft moves forward. b) LIDAR-derived location (x,y) and elevation (z) data for a small portion of the study area. The LIDAR does well on the unvegetated left portion of the area. It is so sensitive that it even hits the power transmission line in the center. It begins to interact with the tree canopy on the right. c) Assuming that the tree canopy has a uniform height and canopy density, then the greater the scan angle from nadir the greater the amount of canopy that will have to be penetrated to reach the ground to acquire the information necessary to create a digital elevation model. d) A reduced image of a LIDAR-derived digital elevation model for an area near Aiken, SC.

chronization errors among the data streams (Vaughn et al., 1996; Ridgway et al., 1997; Krabill et al., 1995).

Accuracy of LIDAR Measurements

Using test flights over Lake Crowley, CA, Vaughn et al. (1996) reported vertical accuracies within ± 2-13 cm of a

local USGS tide gauge with a mean error of ± 6 cm. Another project at Lake Crowley found the absolute height difference to be 1 – 4 cm (Ridgway et al., 1997). During their work on the Greenland ice sheet, Krabill et al. (1995) used a runway for testing the instrument's accuracy from the air and found the measurements to be accurate within approximately ± 10 cm of the runway's surveyed elevation. Recent research indicates that LIDAR can provide measurements with vertical accuracy of less than 10 cm. Vaughn et al. (1996) believe that if care is taken in the numerous calibrations and corrections, then consistent accuracies of ± 5 cm can be obtained.

Canopy Penetration Capability

One of the unique properties of LIDAR is the ability to penetrate vegetation canopy and to map the surface below. Although some of the laser energy will be backscattered by vegetation above the ground surface, only a portion of the laser needs to reach the ground to produce a surface measurement. Both of these partial returns (vegetation and ground) can be recorded by the LIDAR instrument, allowing measurements of both vegetation canopy height and ground-surface elevation. Some LIDAR instruments are capable of recording up to five laser returns from a single pulse. Figure 9-31b shows a section of LIDAR data where both ground measurements and canopy measurements were obtained.

The ability to see both vegetation and the ground surface has useful applications in measuring vegetation attributes. Nilsson (1996) examined the usefulness of LIDAR for estimating timber volume as a part of Sweden's National Forest Inventory. Weltz et al. (1994) used a profiling LIDAR system to map vegetation height and canopy cover at the Walnut Gulch experimental watershed in Arizona. This study demonstrated the ability of LIDAR to distinguish between different vegetation communities and ground-cover types. Nelson et al. (1988) successfully used LIDAR to estimate forest biomass and volume to within 2.6 percent and 2.0 percent, respectively, on 38 test plots in southwestern Georgia.

If the purpose of a LIDAR overflight is to collect data for creating digital elevation models (DEMs), the presence of vegetation can be a nuisance. In areas covered by dense vegetation, the majority of the LIDAR measurements will be in the canopy, with only a few reaching the ground. Hendrix (1999) found that up to 93% of LIDAR pulses never reached the ground in mixed bottomland hardwoods near Aiken, SC. Separating the ground measurements from measurements of vegetation, buildings, and other structures can be difficult, especially if only a few measurements of the ground are made through the canopy. A single LIDAR overflight can produce millions of measurements, and current research is seeking to identify the best methods for automated identification and extraction of ground measurements.

References

Aero Service Corp., 1978, *SAR Synthetic Aperture Radar*, Houston, TX: Aero Service Corporation, 14 pp.

Avery, T. E. and G. L. Berlin, 1992, *Fundamentals of Remote Sensing and Airphoto Interpretation*, 5th Ed., Upper Saddle River: Prentice-Hall, Inc., 472 pp.

Blom, R. G., L. R. Schenck and R. E. Alley, 1987, "What are the Best Radar Wavelengths, Incident Angles, and Polarizations for Discrimination Among Lava Flows and Sedimentary Rocks? A Statistical Analysis," *IEEE Transactions on Geoscience and Remote Sensing*, GE-25(2):209–212.

Campbell, J. B.,1996, *Introduction to Remote Sensing*. New York: The Guilford Press.

Carver, K. R., 1988, SAR *Synthetic Aperture Radar – Earth Observing System*, Vol. IIf, Washington: NASA Instrument Panel Report, 233 pp.

Carver, K. R., E. Elachi and F. T. Ulaby, 1985, "Microwave Remote Sensing from Space," *Proceedings of the IEEE*, 7(6):970–996.

Committee on Earth Sciences, 1990, *Our Changing Planet: The FY 1991 Research Plan - U.S. Global Change Research Program*. Washington: Federal Coordinating Council for Science, Engineering, & Technology, 60 pp.

Curran, R. J., 1989, "NASA's Plans to Observe the Earth's Atmosphere with LIDAR," *IEEE Transactions on Geoscience and Remote Sensing*, 27:154–163.

Dobson, M. C. and F. T. Ulaby, 1998, "Mapping Soil Moisture Distribution with Imaging Radar," Lewis and Henderson (Eds.), *Principles and Applications of Imaging Radar*, New York: John Wiley & Sons, 407–433.

Dobson, M. C., F. T. Ulaby, T. LeToan, A. Beaudoin, E. S. Kasischke and N. Christensen, 1992, "Dependence of Radar Backscatter on Coniferous Forest Biomass," *IEEE Transactions on Geoscience and Remote Sensing*, 30(2):412–415.

Dobson, M. C., F. T. Ulaby, L. E. Pierce, T. L. Sharik, K. M. Bergen, J. Kellndorfer, J. R. Kendra, E. Li and Y. C. Lin, 1995, "Estimation of Forest Biomass Characteristics in Northern Michigan with SIR-C/S-SAR Data," *IEEE Trans. Geosci. Remote Sensing*, 33:877–894.

Elachi, C., et al. 1982, "Shuttle Imaging Radar Experiment," *Science*, 218, 996–1004.

Engman, E. T. and R. J. Gurney, 1991, *Remote Sensing in Hydrology*, New York: Chapman and Hall, 225 pp.

Evans, D. L., E. R. Stofan, T. D. Jones and L. M. Godwin, 1994, "Earth from Sky," *Scientific American*, (December): 69–75.

Ferraro, R. R., 1997, "SSM/I Derived Global Rainfall Estimates for Climatological Applications," *Journal of Geophysical Research*, 102(16):715–716.

Fisher, D. E., 1988, *A Race on the Edge of Time: RADAR — The Decisive Weapon of World War II*, N.Y.: Paragon House Publishing, 371 pp.

Fitch, J. P., 1988, *Synthetic Aperture Radar*, New York: Springer-Verlag, 170 pp.

Flood, M., and B. Gutelius, 1997, "Commercial Implications of Topographic Terrain Mapping Using Scanning Airborne Laser Radar," *Photogrammetric Engineering and Remote Sensing*, 63:327.

Ford, J. P., R. G. Blom, M. L. Bryan, M. K. Daily, T. H. Dixon, C. Elachi and E. C. Xenos, 1980, *Seasat Views North America the Caribbean and Western Europe with Imaging Radar*, JPL Publication #80 - 67, Los Angeles, CA: NASA Jet Propulsion Laboratory, 1–6 pp.

Ford, J. P., J. B. Cimino and C. Elachi, 1983, *Space Shuttle Columbia View the World with Imaging Radar: the SIR-A Experiment*, Los Angeles, CA: NASA Jet Propulsion Laboratory, 179 pp.

Ford, J. P., J. B. Cimino, B. Holt and M. R. Ruzek, 1986, *Shuttle Imaging Radar Views the Earth from Challenger: SIR-B Experiment*, LA: NASA Jet Propulsion Laboratory, 135 pp.

Franklin, S. E., M. B. Lavigne, B. A. Wilson and E. R. Hunt, 1994, "Empirical Relations Between Balsam Fir Stand Conditions and ERS-1 SAR data in Western Newfoundland," *Canadian Journal of Remote Sensing*, 20:124–130.

Henderson, F. M. and Z Xia, 1998, "Radar Applications in Urban Analysis, Settlement Detection and Population Estimation," in Henderson and Lewis (Eds.) *Principles and Applications of Imaging Radar*, New York: John Wiley & Sons, 733–768.

Henderson, F. M. and A. J. Lewis, 1998, "Radar Fundamentals: The Geoscience Perspective," in Henderson and Lewis (Eds.) *Principles and Applications of Imaging Radar*, New York: John Wiley & Sons, 131–181.

Hendrix, C., 1999, *Parameterization of LIDAR Interaction with Vegetation Canopy in a Forested Environment*, Columbia: University of South Carolina, Masters Thesis, 150 pp.

Hess, L. L., J. M. Melack and D. S. Simonett, 1990, "Radar Detection of Flooding Beneath the Forest Canopy: A Review," *International Journal of Remote Sensing*, 5:1313–1325.

Jackson, T. J. and T. J. Schmugge, 1991, "Vegetation Effects on the Microwave Emission of Soils," *Remote Sensing of Environment*, 36(3):203–212.

Jensen, H., L. C. Graham, L. J. Porcello and E. N. Leith, 1977, "Side Looking Airborne Radar," *Scientific American*, 23: 84–95.

Jensen, J. R., 1983, "Biophysical Remote Sensing," *Annals of the Association of American Geographers*, 73:111–132.

Jensen, J. R., 1996, *Introductory Digital Image Processing: a Remote Sensing Perspective*. Englewood Cliffs, NJ: Prentice-Hall, 318 pp.

Jensen, J. R. et al., 1989, "Remote Sensing," *Geography in America*, G. Gaile and C. Wilmott (Eds.), New York: Merrill Co.

Jensen, J. R., M. E. Hodgson, H. E. Mackey, Jr., and W. Krabill, 1987, "Correlation Between Aircraft MSS and LIDAR Remotely Sensed Data on a Forested Wetland, *Geocarto International*, 4:39-54.

JPL, 1999a, *Shuttle Radar Topography Mission*, Pasadena: Jet Propulsion Laboratory, http://www-radar.jpl.nasa.gov/srtm/tech_factsheet.html.

JPL, 1999b, *Module 2: How Radar Imaging Works: Interferometry*, Pasadena, CA: Jet Propulsion Lab, SIR-C educational Web site.

Kasischke, E. S. and L. L. Bourgeau-Chavez, 1997, "Monitoring South Florida Wetlands Using ERS-1 SAR Imagery," *Photogrammetric Engineering & Remote Sensing*, 63(3):281–291.

Kasischke, E. S., G. A. Meadows and P. L. Jackson, 1984, *The Use of Synthetic Aperture Radar to Detect Hazards to Navigation*, Ann Arbor, MI: Environmental Research Institute of Michigan, Publication # 169200-2-F., 194 pp.

Krabill, W. B., R. H. Thomas, C. F. Martin, R. N. Swift, and E. B. Frederick, 1995, "Accuracy of Airborne Laser Altimetry over the Greenland Ice-sheet," *International Journal of Remote Sensing*, 16(7):1211–1222.

Leberl, F. W., 1990, *Radargrammetric Image Processing*, Norwood: Artech House, 595 pp.

Li, Q., R. Ferraro and N. C. Grody, 1998, "Detailed Analysis of the Error Associated with the Rainfall Retrieved by the NOAA/NESDIS SSM/I Rainfall Algorithm: Part 1," *Journal Geophysical Research*, 103(11):419–427.

Lillesand, T. M. and R. W. Kiefer, 1994, *Remote Sensing and Image Interpretation*. 2nd Ed., New York, NY: John Wiley & Sons, 750 pp.

Madsen, S. N. and H. A. Zebker, 1998, "Imaging Radar Interferometry," in Henderson and Lewis (Eds.) *Principles and Applications of Imaging Radar*, New York: John Wiley & Sons, 359–380.

Moore, R. K., 1983, "Imaging Radar Systems," *Manual of Remote Sensing*, 2nd Ed., R. Colwell, (Ed.), Falls Church, VA: American Society for Photogrammetry & Remote Sensing, 429–474.

NASA, 1990, *EOS — A Mission to Planet Earth*. Washington: National Aeronautics and Space Administration, 38 pp.

NASA JPL, 1996, "Brazil Rainforest," at *Imaging Radar Home Page* (http:/www.jpl.gov).

Nelson, R., W. Krabill, and J. Tonelli, 1988, "Estimating Forest Biomass and Volume Using Airborne Laser Data," *Remote Sensing of Environment*, 24:247–267.

Nilsson, M., 1996, "Estimation of Tree Heights and Stand Volume using an Airborne LIDAR system," *Remote Sensing of Environment*, 56(1):1–7.

Olmsted, C., 1993, *Alaska SAR Facility Scientific SAR User's Guide*, Anchorage: Alaska SAR Facility.

Peake, W. H. and T. L. Oliver, 1971, *The Response of Terrestrial Surfaces at Microwave Frequencies,* Electroscience Laboratory Report Air Force Avionics Laboratory #TR-70-301, Columbus: Ohio State University.

Price, A., 1990, "The Eyes of England," *Air & Space,* 5(4): 82–93.

RADARSAT International, 1998, *Radarsat Illuminated*, British Columbia: RADARSAT International, 110 pp.

Ramsey, E. W., 1995, "Monitoring Flooding in Coastal Wetlands Using Radar Imagery and Ground-based Measurements," *International Journal of Remote Sensing*, 16(13):2495–2502.

Raney, K., 1991, "RADARSAT," *Proceedings of IEEE*, 79(6).

Raney, K., 1998, "Radar Fundamentals: Technical Perspective," Henderson and Lewis (Eds.), *Principles and Applications of Imaging Radar*, New York: John Wiley & Sons, 42–43.

Ridgway, J. R., J. B. Minster, N. Williams, J. L. Bufton, and W. B. Krabill, 1997, "Airborne Laser Altimeter Survey of Long Valley, California," *Geophysical Journal Intl.*, 131(2):267–280.

Ritchie, J. C., 1996, "Remote Sensing Applications to Hydrology: Airborne Laser Altimeters," *Hydrological Sciences Journal*, 41(4):625–636.

Sabins, F. F., 1997, *Remote Sensing: Principles and Interpretation*. 3nd Ed., New York: W. H. Freeman & Co., 494 pp.

Sun, G. and K. J. Ranson, 1995, "A Three-dimensional Radar Backscatter Model for Forest Canopies," *IEEE Transactions on Geoscience and Remote Sensing*, 33:372–382.

Vaughn, C. R., J. L. Bufton, and D. Rabine, 1996, "Georeferencing of Airborne Laser Altimeter Measurements," *International Journal of Remote Sensing*, 17(11):2185–2200.

Wang, Y., E. S. Kasischke, J. M. Melack, F. W. Davis and N. L. Christensen, 1994, "The Effects of Changes in Loblolly Pine Biomass and Soil Moisture on ERS-1 SAR Backscatter," *Remote Sensing of Environment*, 49:25–31.

Wang, Y., E. S. Kasischke, F. W. Davis, J. M. Melack and N. L. Christensen, 1995, "The Effects of Changes in Forest Biomass on Radar Backscatter from Tree Canopies," *International Journal of Remote Sensing*, 16:503–513.

Waring, R. H., J. Way, E. R. Hunt, L. Morrissey, K. J. Ranson, J. F. Weishampel, R. Oren and S. E. Franklin, 1995, "Imaging Radar for Ecosystem Studies," *BioScience*, 45(10):715–723.

Weller, G., 1983, *Science Program for an Imaging Radar Receiving Station in Alaska*, Los Angeles, CA: Jet Propulsion Laboratory, 45 pp.

Weltz, M. A., J. C. Ritchie, and H. D. Fox, 1994, "Comparison of Laser and Field Measurements of Vegetation Height and Canopy Cover," *Water Resources Research*, 30(5):1311–1319.

Wu, Y., and A. H. Strahler, 1994, "Remote Estimation of Crown Size, Stand Density and Biomass on the Oregon Transect," *Ecological Applications*, 4:299–312.

Remote Sensing of Vegetation 10

Approximately 70 percent of the Earth's land surface is covered with vegetation. Furthermore, vegetation is one of the most important components of ecosystems. Knowledge about variations in vegetation species and community distribution patterns, alterations in vegetation phenological (growth) cycles, and modifications in the plant physiology and morphology provide valuable insight into the climatic, edaphic, geologic, and physiographic characteristics of an area (Jones et al., 1998). Scientists have devoted a significant amount of effort to develop sensors and visual and digital image processing algorithms to extract important vegetation biophysical information from remotely sensed data (e.g., Frohn, 1998; Huete and Justice, 1999). Many of the remote sensing techniques are generic in nature and may be applied to a variety of vegetated landscapes, including (Danson, 1998; Lyon et al., 1998):

- agriculture,

- forests,

- rangeland,

- wetland, and

- manicured urban vegetation.

This chapter introduces the fundamental concepts associated with vegetation biophysical characteristics and how remotely sensed data can be processed to provide unique information about these parameters. It then summarizes some of the vegetation indices developed to extract biophysical vegetation information from digital remote sensor data. Several of the metrics used by landscape ecologists to extract meaningful parameters about vegetation patch shape, size, etc., using remote sensor data are then summarized. Case studies are provided throughout the chapter on remote sensing of agriculture, inland wetland, coastal wetland, and biodiversity.

 Photosynthesis Fundamentals

Oil and coal today provide more than 90 percent of the energy needed to power automobiles, trains, trucks, ships, airplanes, factories, and a myriad of electrically energized appliances, computers, and communication systems. The energy within oil and coal was originally "captured" from the Sun by plants growing millions of years ago that were transformed into fossil fuels by geological forces. Therefore, photosynthesis, at least indirectly, is not only the principal means of enabling a civilized society to function normally but also the sole means of sustaining life — except for a few bacteria that derive their energy from sulfur salts and other inorganic compounds. This unique photosynthetic manufacturing process of green plants furnishes raw material, energy, and oxygen. In photosynthesis, energy from the Sun is harnessed and packed into simple sugar molecules made from water and carbon dioxide (CO_2) with the aid of chlorophyll. Oxygen (O_2) is given off as a by-product of the process.

The naturalist Joseph Priestly discovered in 1772 that when he placed a candle in an inverted jar it would burn out quickly, long before it ran out of wax. He also found that a mouse would die if placed under the same jar. He believed that the air was *injured* by the candle and the mouse but that it could be restored by placing a plant under the jar. Jan Ingen-Housz built on Priestly's experiments and discovered in 1788 that the influence of sunlight on the plant could cause it to rescue the mouse in a few hours. In 1796 the French pastor Jean Senebier discovered that it was the carbon dioxide, CO_2, in the jar that was the injured air and that it was actually taken up by the plants. Finally, Theodore de Saussure demonstrated that the increase in the mass of the plant as it grows was due not only to the uptake of CO_2 but also to the uptake of water, H_2O. In the early twentieth century, scientists found that the oxygen for photosynthesis was derived from the water. In effect, light energy entering the plant splits the water into oxygen and hydrogen. The photosynthetic process is described by the equation:

$$6CO_2 + 6H_2O + \text{light energy} \rightarrow C_6H_{12}O_6 + 6O_2 \quad (10\text{-}1)$$

Photosynthesis is an energy-storing process that takes place in leaves and other green parts of plants in the presence of light. The light energy is stored in a simple sugar molecule (glucose) that is produced from carbon dioxide (CO_2) present in the air and water (H_2O) absorbed by the plant primarily through the root system. When the carbon dioxide and the water are combined and form a sugar molecule ($C_6H_{12}O_6$) in a chloroplast, oxygen gas (O_2) is released as a by-product. The oxygen diffuses out into the atmosphere. The photosynthetic process begins when sunlight strikes *chloroplasts*, small bodies in the leaf that contain a green substance called chlorophyll.

Plants have adapted their internal and external structure to perform photosynthesis. This structure and its interaction with electromagnetic energy has a direct impact on how leaves and canopies appear spectrally when recorded using remote sensing instruments.

Spectral Characteristics of Vegetation

A healthy green leaf intercepts incident radiant flux (Φ_i) directly from the Sun or from diffuse skylight scattered onto the leaf. This incident electromagnetic energy interacts with the pigments, water, and intercellular air spaces within the plant leaf. The amount of radiant flux reflected from the leaf (Φ_r), the amount of radiant flux absorbed by the leaf (Φ_α), and the amount of radiant flux transmitted through the leaf (Φ_τ) can be carefully measured as we apply the energy balance equation and attempt to keep track of what happens to all the incident energy. The general equation for the interaction of *spectral* (λ) radiant flux on and within the leaf is

$$\Phi_{i_\lambda} = \Phi_{r_\lambda} + \Phi_{\alpha_\lambda} + \Phi_{\tau_\lambda}. \quad (10\text{-}2)$$

Dividing each of the variables by the original incident radiant flux, Φ_{i_λ}:

$$\frac{\Phi_{i_\lambda}}{\Phi_{i_\lambda}} = \frac{\Phi_{r_\lambda}}{\Phi_{i_\lambda}} + \frac{\Phi_{\alpha_\lambda}}{\Phi_{i_\lambda}} + \frac{\Phi_{\tau_\lambda}}{\Phi_{i_\lambda}} \quad (10\text{-}3)$$

yields

$$i_\lambda = r_\lambda + \alpha_\lambda + \tau_\lambda \quad (10\text{-}4)$$

where r_λ is spectral hemispherical reflectance of the leaf, α_λ is spectral hemispherical absorptance, and τ_λ is spectral hemispherical transmittance by the leaf. Most remote sensing systems function in the $0.35 - 3.0$ μm region measuring primarily reflected energy. Therefore, it is useful to think of this relationship as:

$$r_\lambda = i_\lambda - (\alpha_\lambda + \tau_\lambda) \quad (10\text{-}5)$$

where the energy reflected from the plant leaf surface is equal to the incident energy minus the energy absorbed directly by the plant for photosynthetic or other purposes and the amount of energy transmitted directly through the leaf onto other leaves or the ground beneath the canopy.

Dominant Factors Controlling Leaf Reflectance

Pioneering work by Gates et al. (1965), Gausmann et al. (1969), Myers (1970) and others demonstrated the importance of understanding how leaf pigments, internal scattering, and leaf water content affect the reflectance and transmittance properties of leaves (Peterson and Running,

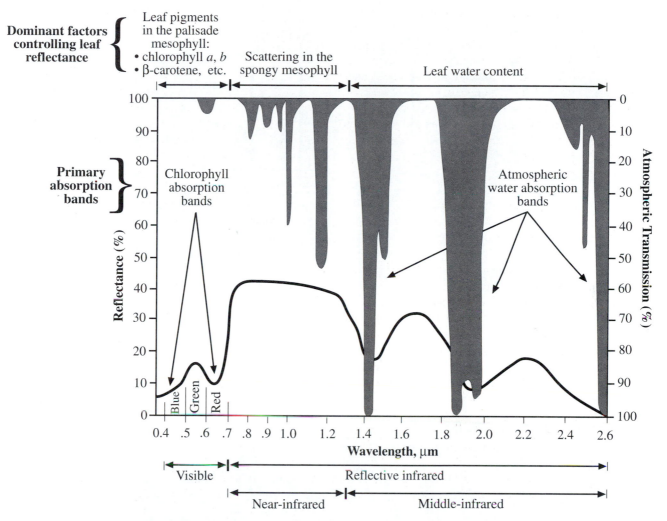

Figure 10-1 Spectral reflectance characteristics of healthy, green vegetation for the wavelength interval 0.4 – 2.6 μm. The dominant factors controlling leaf reflectance are the various leaf pigments in the palisade mesophyll (e.g., chlorophyll *a* and *b*, and β-carotene), the scattering of near-infrared energy in the spongy mesophyll, and the amount of water in the plant. The primary chlorophyll absorption bands occur at 0.43 – 0.45 μm and 0.65 – 0.66 μm in the visible region. The primary water absorption bands occur at 0.97, 1.19, 1.45, 1.94, and 2.7 μm.

1989). Dominant factors controlling leaf reflectance in the region from 0.35 – 2.6 μm are summarized in Figure 10-1.

Visible Light Interaction with Pigments in the Palisade Mesophyll Cells

The process of food-making via photosynthesis determines how a leaf and the associated plant canopy actually appear radiometrically on remotely sensed images. A healthy leaf needs three things to make food:

- carbon dioxide (CO_2),

- water (H_2O), and

- irradiance (E_λ) measured in $W\,m^{-2}$.

The carbon dioxide from the air and the water provided by the root and stem system represent the fundamental raw materials of photosynthesis. Sunlight provides the irradiance (E_λ) that powers photosynthesis.

The leaf is the primary photosynthesizing organ. A cross-section of a typical green leaf is shown in Figure 10-2. The cell structure of leaves is highly variable depending upon species and environmental conditions during growth. Carbon dioxide enters the leaf from the atmosphere through tiny pores called *stomata* or *stoma*, located primarily on the underside of the leaf on the *lower epidermis*. Each stomata is

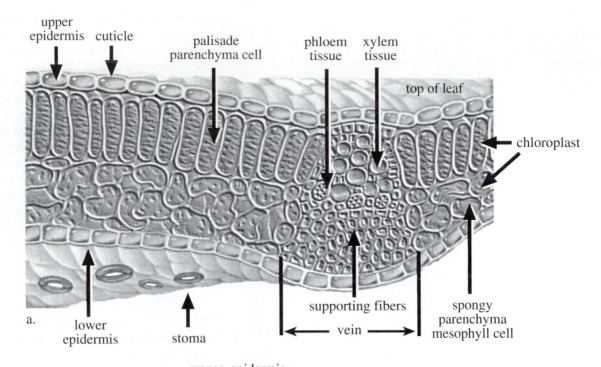

a.

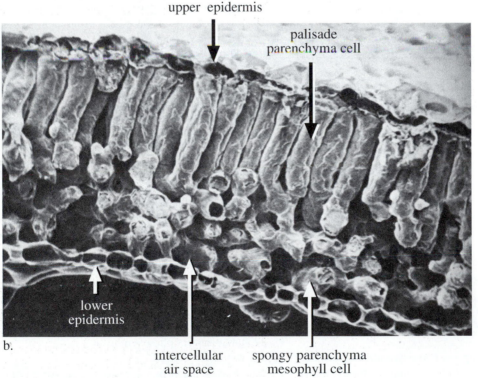

b.

Figure 10-2   a) Hypothetical cross-section of a typical healthy green leaf, showing both the top and underside of the leaf. The chlorophyll pigments in the palisade parenchyma cells have a significant impact on the absorption and reflectance of visible light (blue, green, and red), while the spongy parenchyma mesophyll cells have a significant impact on the absorption and reflectance of near-infrared incident energy. b) Electron microscope image of a green leaf.

surrounded by *guard cells* that swell or contract. When they swell, the stomata pore opens and allows carbon dioxide to enter the leaf. A typical sunflower leaf might have 2 million stomata, but they make up only about 1 percent of the leaf's surface area. Usually, there are more stomata on the bottom of a leaf; however, on some leaves the stomata are evenly distributed on both the upper and lower epidermis.

The top layer of leaf *upper epidermis* cells has a *cuticular* surface that diffuses but reflects very little light (Philpott, 1971). It is variable in thickness but is often only 3 – 5 μm thick with cell dimensions of approximately 18 x 15 x 20 μm. It is useful to think of it as a waxy, translucent material similar to the cuticle at the top of your fingernail. Leaves of many plants that grow in bright sunlight have a thick cuticle that can filter out some light and guard against excessive plant water loss. Conversely, some plants such as ferns and some shrubs on the forest floor must survive in shaded conditions. The leaves of many of these plants have a thin cuticle so that the plant can collect as much of the dim sunlight as possible for photosynthesis.

Many leaves in direct sunlight have hairs growing out of the upper (and lower) epidermis, causing them to feel fuzzy. These hairs can be beneficial, as they reduce the intensity of the incident sunlight to the plant. Nevertheless, much of the visible and near-infrared wavelength energy is transmitted through the cuticle and upper epidermis to the palisade parenchyma mesophyll cells and spongy parenchyma mesophyll cells below.

Photosynthesis occurs inside the typical green leaf in two kinds of food-making cells — *palisade parenchyma* and *spongy parenchyma* mesophyll cells. Most leaves have a distinct layer of long palisade parenchyma cells in the upper part of the mesophyll and more irregularly shaped, loosely arranged spongy parenchyma cells in the lower part of the mesophyll. The palisade cells tend to form in the portion of the mesophyll toward the side from which the light enters the leaf. In most horizontal (planophile) leaves the palisade cells will be toward the upper surface, but in leaves that grow nearly vertical (erectophile), the palisade cells may form from both sides. In some leaves the elongated palisade cells are entirely absent and only spongy parenchyma cells will exist within the mesophyll.

The cellular structure of the leaf is large compared to the wavelengths of light that interact with it. Palisade cells are typically 15 x 15 x 60 μm, while spongy mesophyll parenchyma cells are smaller. The palisade parenchyma mesophyll plant cells contain chloroplasts with chlorophyll pigments.

The chloroplasts are generally 5 – 8 μm in diameter and about 1 μm in width. As many as 50 chloroplasts may be present in each parenchyma cell. Within the chloroplasts are long slender *grana* strands (not shown) within which the chlorophyll is actually located (approximately 0.5 μm in length and 0.05 μm in diameter). The chloroplasts are generally more abundant toward the upper side of the leaf in the palisade cells and hence account for the darker green appearance of the upper leaf surface compared with the bottom lighter surface.

A molecule, when struck by a wave or photon of light, reflects some of the energy or it can absorb the energy and thus enter into a higher energy or excited state (refer to Chapter 2). Each molecule absorbs or reflects its own characteristic wavelengths of light. Molecules in a typical green plant have evolved to absorb wavelengths of light in the visible region of the spectrum (0.35 – 0.70 μm) very well and are called *pigments*. An *absorption spectrum* for a particular pigment describes the wavelengths at which it can absorb light and enter into an excited state. Figure 10-3a presents the absorption spectrum of pure chlorophyll pigments in solution. Chlorophyll *a* and *b* are the most important plant pigments absorbing blue and red light: chlorophyll *a* at wavelengths of 0.43 and 0.66 μm and chlorophyll *b* at wavelengths of 0.45 and 0.65 μm (Curran, 1983; Farabee, 1997). A relative lack of absorption in the wavelengths between the two chlorophyll absorption bands produces a trough in the absorption efficiency at approximately 0.54 μm in the green portion of the electromagnetic spectrum (Figure 10-3a). Thus, it is the relatively lower absorption of green wavelength light (compared to blue and red light) by the leaf that causes healthy green foliage to appear green to our eyes.

There are other pigments present in the palisade mesophyll cells that are usually masked by the abundance of chlorophyll pigments. For example, there are yellow *carotenes* and pale yellow *xanthophyll* pigments, with strong absorption primarily in the blue wavelength region. The β-carotene absorption spectra is shown in Figure 10-3b with its strong absorption band centered at about 0.45 μm. *Phycoerythrin* pigments may also be present in the leaf which absorb predominantly in the green region centered at about 0.55 μm, allowing blue and red light to be reflected. *Phycocyanin* pigments absorb primarily in the green and red regions centered at about 0.62 μm, allowing much of the blue and some of the green light (i.e., the combination produces cyan) to be reflected (Figure 10-3b). Because chlorophyll *a* and *b* chloroplasts are also present and have a similar absorption band in this blue region, they tend to dominate and mask the effect of the other pigments present. When a plant undergoes senescence in the fall or encounters stress, the chlorophyll

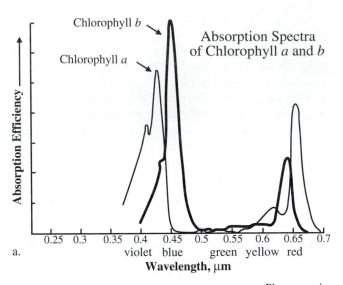

a.

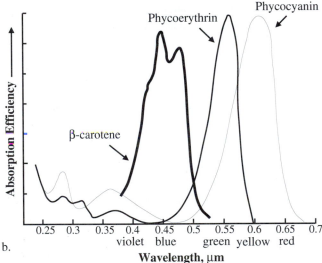

b.

Figure 10-3   a) Absorption spectra of chlorophyll *a* and *b* pigments. Chlorophyll *a* and *b* pigments in a leaf absorb much of the incident blue and red wavelength energy. b) Absorption spectra of β-carotene, which absorbs primarily in the blue. Other pigments that might be found in a leaf include Phycoerythrin which absorbs primarily green light, and Phycocyanin which absorbs primarily green and red light (after Farabee, 1997).

pigment may disappear, allowing the carotenes and other pigments to become dominant. For example, in the fall, chlorophyll production ceases, causing the yellow coloration of the carotenes and other specific pigments in the tree foliage to become more visible to our eyes. In addition, some trees produce great quantities of *anthocyanin* in the fall, causing the leaves to appear bright red.

The two optimum spectral regions for sensing the chlorophyll absorption characteristics of a leaf are believed to be 0.45 – 0.52 μm and 0.63 – 0.69 μm (Figure 10-3a). The former region is characterized by strong absorption by carotenoids and chlorophylls, whereas the latter is characterized by strong chlorophyll absorption. Remote sensing of chlorophyll absorption within a canopy represents a fundamental biophysical variable useful for many biogeographical investigations. The absorption characteristics of plant canopies may be coupled with other remotely sensed data to identify vegetation stress, yield, and other hybrid variables. Thus, many remote sensing studies are concerned with monitoring what happens to the *photosynthetically active radiation* (PAR) as it interacts with individual leaves and/or the plant canopy. The use of high spectral resolution imaging spectrometers are particularly useful for measuring the absorption and reflectance characteristics of the photosynthetically active radiation.

To demonstrate these principles, consider the spectral reflectance characteristics of four different leaves obtained from a single healthy Sweetgum tree (*Liquidambar styraciflua* L.) on November 11, 1998, in Columbia, SC (Color Plate 10-1). The green leaf (a), yellow leaf (b), and red leaf (c) were still on the Sweetgum tree at the time of data collection. The dark brown leaf (d) was collected on the ground beneath the tree.

A GER 1500 (Geophysical & Environmental Research, Inc.) handheld spectroradiometer was used to obtain spectral reflectance measurements from each of the leaves. The spectroradiometer obtained spectral reflectance measurements in 512 bands in the ultraviolet, blue, green, red, and near-infrared spectral regions from 350 – 1050 nm. Percent reflectance measurements were obtained in the laboratory by measuring the amount of energy reflected from the surface of the leaf (the target) divided by the amount of energy reflected from a Spectralon® reflectance reference (percent reflectance = target/reference * 100). The reflectance measurements for each of the leaves from 400 – 1050 nm were plotted in a percent reflectance graph (Color Plate 10-1e).

The green leaf (Color Plate 10-1a) was still photosynthesizing and yielded a typical healthy green reflectance spectra with strong chlorophyll absorption bands in the blue and red regions (approximately 6 percent reflectance at 450 nm and 5 percent at 650 nm, respectively), and a peak in reflectance in the green region of the visible spectrum (11 percent at 550 nm). Approximately 76 percent of the incident near-infrared radiant flux was reflected from the leaf at 900 nm.

The yellow leaf (Color Plate 10-1b) was undergoing senescence. As the influence of the chlorophyll pigments diminished, relatively greater amounts of green (24 percent at 550 nm) and red (32 percent at 650 nm) light were reflected from the leaf, resulting in a yellow appearance. At 750 nm the yellow leaf reflected less near-infrared radiant flux than the healthy green leaf. However, near-infrared reflectance at 900 nm was about 76 percent, very similar to the healthy green leaf.

The red leaf (Color Plate 10-1c) reflected 7 percent of the blue at 450 nm, 6 percent of the green energy at 550 nm, and approximately 23 percent of the incident red energy at 650 nm. Near-infrared reflectance at 900 nm dropped to 70 percent.

The dark brown leaf (Color Plate 10-1d) produced a spectral reflectance curve with low blue (7 percent at 450 nm), green (9 percent at 550 nm), and red reflectance (10 percent at 650 nm). This combination produced a dark brown appearance. Near-infrared reflectance dropped to 44 percent at 900 nm.

It is important to understand the physiology of the plants under investigation and especially their pigmentation characteristics so that we can appreciate how a typical plant will appear when chlorophyll absorption starts to decrease, either due to seasonal senescence or environmental stress. As demonstrated, when a plant is under stress and/or chlorophyll production decreases, the lack of chlorophyll pigmentation typically causes the plant to absorb less in the chlorophyll absorption bands. Such plants will have a much higher reflectance, particularly in the green and red portion of the spectrum, and therefore may appear yellowish or *chlorotic*. In fact, Carter (1993) suggests that increased reflectance in the visible spectrum is the most consistent leaf reflectance response to plant stress. Infrared reflectance responds consistently only when stress has developed sufficiently to cause severe leaf dehydration (to be demonstrated shortly).

Leaf spectral reflectance is most likely to indicate plant stress first in the sensitive 535 – 640 and 685 – 700 nm visible light wavelength ranges. Increased reflectance near 700 nm represents the often reported "blue shift of the red edge," i.e., the shift toward shorter wavelengths of the red-infrared transition curve that occurs in stressed plants when reflectance is plotted versus wavelength (Cibula and Carter, 1992). The shift toward shorter wavelengths in the region from 650 – 700 nm is particularly evident for the yellow and red reflectance curves shown in Color Plate 10-1e. Remote sensing within these spectrally narrow ranges may provide improved capability to detect plant stress not only in individual leaves but for whole plants and perhaps for densely vegetated canopies (Carter, 1993; Carter et al., 1996).

Normal color film is sensitive to blue, green, and red wavelength energy. Color-infrared film is sensitive to green, red, and near-infrared energy after minus-blue (yellow) filtration (Rundquist and Sampson, 1988; refer to Chapter 4). Therefore, even the most simple camera with color or color-infrared film and appropriate band-pass filtration (i.e., only certain wavelengths of light are allowed to pass) can be used to remotely sense differences in spectral reflectance caused by the pigments present in the palisade parenchyma layer of cells in a typical leaf. However, to detect very subtle spectral reflectance differences in the relatively narrow bands suggested by Cibula and Carter (1992) and Carter et al. (1996), it may be necessary to use a high spectral resolution imaging spectroradiometer that has very narrow bandwidths.

Near-Infrared Energy Interaction Within the Spongy Mesophyll Cells

In a typical healthy green leaf, the near-infrared reflectance increases dramatically in the region from 700 - 1200 nm. For example, the healthy green leaf in the previous example reflected approximately 76 percent of the incident near-infrared energy at 900 nm. Healthy green leaves absorb radiant energy very efficiently in the blue and red portions of the spectrum where incident light is required for photosynthesis. But immediately to the long wavelength side of the red chlorophyll absorption band, why does the reflectance and transmittance of plant leaves increase so dramatically, causing the absorptance to fall to low values (Figure 10-1)? This condition occurs throughout the near-infrared wavelength range where the direct sunlight incident on plants has the bulk of its energy. If plants absorbed this energy with the same efficiency as they do in the visible region, they could become much too warm and the proteins would be irreversibly denatured. As a result, plants have adapted so they do not use this massive amount of near-infrared energy and simply reflect it or transmit it through to underlying leaves or the ground.

The spongy mesophyll layer in a green leaf controls the amount of near-infrared energy that is reflected. The spongy mesophyll layer typically lies below the palisade mesophyll layer and is composed of many cells and intercellular air spaces as shown in Figure 10-2. It is here that the oxygen and carbon dioxide exchange takes place for photosynthesis and respiration. In the near-infrared region, healthy green vegetation is generally characterized by high reflectance (40 – 60 percent), high transmittance (40 – 60 percent) through the leaf onto underlying leaves, and relatively low absorptance (5 – 10 percent). Notice that a healthy green leaf's reflectance and transmittance spectra throughout the visible and near-infrared portion of the spectrum are almost mirror

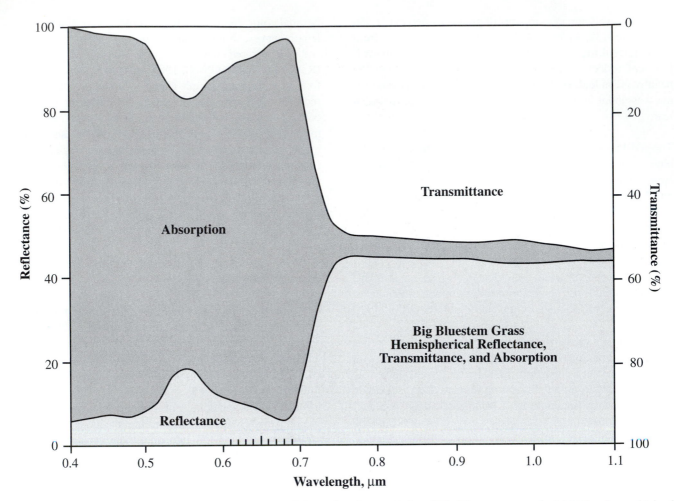

Figure 10-4   Hemispherical reflectance, transmittance, and absorption characteristics of Big Bluestem grass adaxial leaf surfaces obtained using a laboratory spectroradiometer. The reflectance and transmittance curves are almost mirror images of one another throughout the visible and near-infrared portions of the electromagnetic spectrum. The blue and red chlorophyll in plants absorb much of the incident energy in the visible portion of the spectrum (0.4 – 0.7 μm) (after Walter-Shea and Biehl, 1990). Imaging spectrometers such as AVIRIS are capable of identifying small changes in the absorption and reflection characteristics of plants because the sensors often have channels that are only 10 nm apart, i.e., we could have 10 channels in the region from 0.6 – 0.7 μm (600 to 700 nm).

images of one another, as shown in Figure 10-4 (Walter-Shea and Biehl, 1990).

The high diffuse reflectance of the near-infrared (0.7 – 1.2 μm) energy from plant leaves is due to the internal scattering at the cell wall-air interfaces within the leaf (Gausmann et al., 1969; Peterson and Running, 1989). A water vapor absorption band exists at 0.92 – 0.98 μm; consequently, the optimum spectral region for sensing in the near-infrared region is believed to be 0.74 – 0.90 μm (Tucker, 1978).

The main reasons that healthy plant canopies reflect so much near-infrared energy are:

• the leaf already reflects 40 – 60 percent of the incident near-infrared energy from the spongy mesophyll (Figure 10-4), and

• the remaining 45 – 50 percent of the energy penetrates (i.e., is transmitted) through the leaf and can be reflected once again by leaves below it.

This is called *leaf additive reflectance*. For example, consider the reflectance and transmission characteristics of the hypothetical two-layer plant canopy shown in Figure 10-5. Assume that leaf 1 reflects 50 percent of the incident near-infrared energy back into the atmosphere and that the

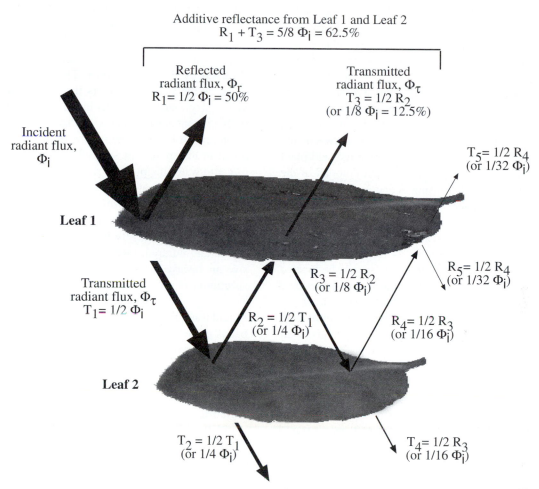

Figure 10-5 A hypothetical example of additive reflectance from a canopy with two leaf layers. Fifty percent of the incident radiant flux, Φ_i, to leaf 1 is reflected (R_1) and the other 50 percent is transmitted onto leaf 2 (T_1). Fifty percent of the radiant flux incident to leaf 2 is transmitted through leaf 2 (T_2), the other 50 percent is reflected toward the base of leaf 1 (R_2). Fifty percent of the energy incident at the base of leaf 1 is transmitted through it (T_3) while the remaining 50 percent (R_3) is reflected toward leaf 2 once again. At this point, an additional 12.5 percent (1/8) reflectance has been contributed by leaf 2, bringing the total reflected radiant flux to 62.5 percent. However, to be even more accurate, one would have to also take into account the amount of energy reflected from the base of leaf 1 (R_3) onto leaf 2, the amount reflected from leaf 2 (R_4), and eventually transmitted through leaf 1 once again (T_5). This process would continue.

remaining 50 percent of the near-infrared energy is transmitted through leaf 1 onto leaf 2. The transmitted energy then falls on leaf 2 where 50 percent again is transmitted (25 percent of the original) and 50 percent is reflected. The reflected energy then passes back through leaf 1 which allows half of that energy (or 12.5 percent of the original) to be transmitted and half reflected. The resulting total energy exiting leaf 1 in this two-layer example is 62.5 percent of the incident energy. Therefore, the greater the number of leaf layers in a healthy, mature canopy, theoretically the greater the infrared reflectance. Conversely, if the canopy is only composed of a single, sparse leaf layer then the near-infrared reflectance will

not be as great because the energy that is transmitted through the leaf layer may be absorbed by the ground cover beneath.

It is important to point out that changes in the near-infrared spectral properties of healthy green vegetation may provide information about plant senescence and/or stress. For example, consider the four leaves and their spectral reflectance characteristics shown in Color Plate 10-1. The photosynthesizing green leaf (a) exhibited strong chlorophyll absorption in the blue and red wavelength regions, an understandable increase in green reflectance, and approximately 76 percent reflectance in the near-infrared region. After a certain point, near-infrared reflectance decreased as the leaves senesced (b – d). However, if the leaves were to dry out significantly during senescence, we would expect to see much higher reflectance in the near-infrared region (to be discussed shortly).

Scientists have known since the 1960s that a *direct* relationship exists between response in the near-infrared region and various biomass measurements. Conversely, it has been shown that an *inverse* relationship exists between the response in the visible region, particularly the red, and plant biomass. The best way to appreciate this is to plot all of the pixels in a typical remote sensing scene in red and near-infrared reflectance space. For example, Figure 10-6a depicts where approximately 10,000 pixels in a typical agricultural scene are located in red and near-infrared multispectral feature space (i.e., within the gray area). Dry bare soil fields and wet bare soil fields in the scene would be located at opposite ends of the soil line. This means that a wet bare soil would have very low red and near-infrared reflectance. Conversely, a dry bare soil area would probably have high red and high near-infrared reflectance. As a vegetation canopy matures, it reflects more near-infrared energy while at the same time absorbing more red radiant flux for photosynthetic purposes. This causes the spectral reflectance of the pixel to move in a perpendicular direction away from the soil line. As biomass increases and as the plant canopy cover increases, the field's location in the red and near-infrared spectral space moves farther away from the soil line.

Figure 10-6b demonstrates how just one agricultural pixel might move about in the red and near-infrared spectral space during a typical growing season. If the field was prepared properly, it would probably be located in the moist bare soil region of the soil line with low red and low near-infrared reflectance at the beginning of the growing season. After the crop emerges, it would depart from the soil line, eventually reaching complete canopy closure. At this point the reflected near-infrared radiant flux would be high and the red reflectance would be low. After harvesting, the pixel would probably be found once again on the soil line but perhaps in a drier condition.

The relationship between red and near-infrared canopy reflectance has resulted in the development of numerous remote sensing vegetation indices and biomass-estimating techniques that utilize multiple measurements in the visible and near-infrared region (e.g., Richardson and Everitt, 1992; Lyon et al., 1998). The result is a linear combination that may be more highly correlated with biomass than either red or near-infrared measurement alone. Several of these algorithms are summarized in the section on Vegetation Indices in this chapter.

Middle-Infrared Energy Interaction with Water in the Spongy Mesophyll

Plants require water to grow. A leaf obtains water that was absorbed by the plant's roots. The water travels from the roots, up the stem, and enters the leaf through the *petiole*. Veins carry water to the cells in the leaf. If a plant is watered so that it contains as much water as it can possibly hold at a given time, it is said to be fully *turgid*. Much of the water is found in the spongy mesophyll portion of the plant. If we forget to water the plant or rainfall decreases, the plant will contain an amount of water that is less than it can potentially hold. This is called its *relative turgidity*. It would be useful to have a remote sensing instrument that was sensitive to how much water was actually in a plant leaf. Remote sensing in the middle-infrared, thermal infrared (Chapter 8), and passive microwave (Chapter 9) portion of the electromagnetic spectrum can provide such information to a limited extent.

Liquid water in the atmosphere creates five major absorption bands in the near-infrared through middle-infrared portions of the electromagnetic spectrum at 0.97, 1.19, 1.45, 1.94, and 2.7 μm (Figure 10-1). The fundamental vibrational water-absorption band at 2.7 μm is the strongest in this part of the spectrum (there is also one in the thermal infrared region at 6.27 μm). However, there is also a strong relationship between the reflectance in the middle-infrared region from 1.3 – 2.5 μm and the amount of water present in the leaves of a plant canopy. Water in plants absorb incident energy between the absorption bands with increasing strength at longer wavelengths. In these middle-infrared wavelengths, vegetation reflectance peaks occur at about 1.6 and 2.2 μm, between the major atmospheric water absorption bands (Figure 10-1).

Water is a good absorber of middle-infrared energy, so the greater the turgidity of the leaves, the lower the middle-infrared reflectance. Conversely, as the moisture content of leaves decreases, reflectance in the middle-infrared region increases substantially. As the amount of plant water in the intercellular air spaces decreases, this causes the incident middle-infrared energy to be more intensely scattered at the interface of the intercellular walls resulting in greater middle-infrared reflectance from the leaf. For example, consider the spectral reflectance of Magnolia leaf samples at five different moisture conditions displayed over the wavelength interval from 0.4 – 2.5 μm (Figure 10-7). The middle-infrared wavelength intervals from about 1.5 – 1.8 μm and from 2.1 – 2.3 μm appear to be more sensitive to changes in the moisture content of the plants than the visible or near-infrared portions of the spectrum (i.e., the y-axis distance between the spectral reflectance curves is greater as the

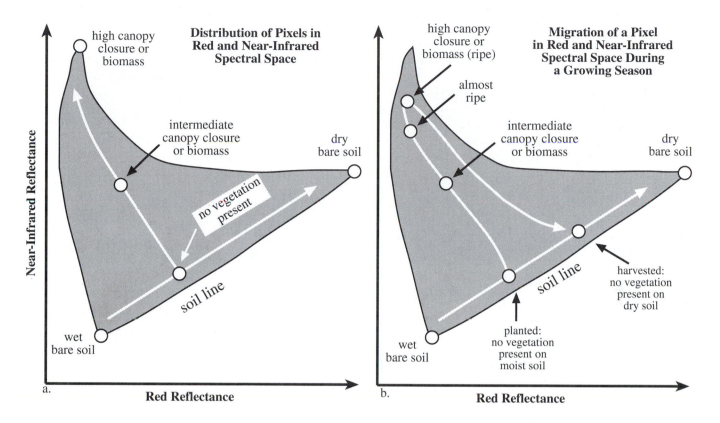

Figure 10-6 a) The distribution of all the pixels in a scene in red and near-infrared multispectral space is found in the gray shaded area. Wet and moist bare soil fields are located along the soil line. The greater the biomass and/or crop canopy closure, the greater the near-infrared reflectance and the lower the red reflectance. This condition moves the pixel's spectral location a perpendicular direction away from the soil line. b) The migration of a single vegetated agricultural pixel in red and near-infrared multispectral space during a growing season is shown. After the crop emerges, it departs from the soil line, eventually reaching complete canopy closure. After harvesting, the pixel will be found on the soil line, but perhaps in a drier soil condition.

moisture content decreases). Also note that substantive changes in the visible reflectance curves (0.4 – 0.7 µm) did not begin to appear until the plant moisture in the leaves decreased to about 50 percent. When the relative water content of the plant decreases to 50 percent, almost any portion of the visible, near- and middle-infrared regions might provide some valuable spectral reflectance information.

Leaf reflectance in the middle-infrared region is inversely related to the absorptance of a layer of water approximately 1 mm in depth (Carter, 1991). The degree to which incident solar energy in the middle-infrared region is absorbed by vegetation is a function of the total amount of water present in the leaf and the leaf thickness. If proper choices of sensors and spectral bands are made, it is possible to monitor the relative turgidity in plants.

Most optical remote sensing systems (except radar) are generally constrained to function in the wavelength intervals from 0.3 – 1.3, 1.5 – 1.8, and 2.0 – 2.6 µm due to the strong atmospheric water absorption bands at 1.45, 1.94, and 2.7 µm. Fortunately, as demonstrated in Figure 10-7, there is a strong "carry over" sensitivity to water content in the 1.5 – 1.8 and 2.0 – 2.6 µm regions adjacent to the major water absorption bands. This is the reason that the Landsat Thematic Mapper (4 and 5) and Landsat 7 Enhanced Thematic Mapper Plus (ETM+) were made sensitive to two bands in this region: band 5 (1.55 – 1.75 µm) and band 7 (2.08 – 2.35 µm). The 1.55 – 1.75 µm middle-infrared band has consistently demonstrated a sensitivity to canopy moisture content. For example, Pierce et al. (1990) found that this band and vegetation indices produced using it were correlated with canopy water stress in coniferous forests.

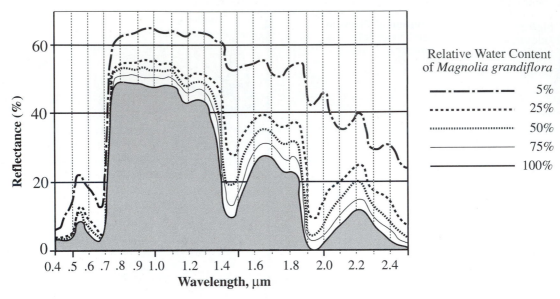

Figure 10-7   Reflectance response of a single Magnolia leaf (*Magnolia grandiflora*) to decreased relative water content. As moisture content decreased, reflectance increased throughout the 0.4 – 2.5 μm region. However, the greatest increase occurred in the middle-infrared region from 1.3 – 2.5 μm (after Carter, 1991).

Much of the water in a plant is lost via transpiration. *Transpiration* occurs as the Sun warms the water inside the leaf, causing some of the water to change its state to water vapor that escapes through the stomata. The following are several important functions that transpiration performs:

- It cools the inside of the leaf because the escaping water vapor contains heat.

- It keeps water flowing up from the roots, through the stem, to the leaves.

- It ensures a steady supply of dissolved minerals from the soil.

As molecules of water vapor at the top of the leaf in the tree are lost to transpiration, the entire column of water is pulled upward. Plants lose a considerable amount of water through transpiration each day. For example, a single corn plant can lose about 4 quarts (3.8 liters) of water on a very hot day. If the roots of the plant cannot replace this water, the leaves wilt, photosynthesis stops, and the plant dies. Thus, monitoring the moisture content of plant canopies, which is correlated with rates of transpiration, can provide valuable information on the health of a crop or stand of vegetation. Thermal infrared and passive microwave remote sensing have provided valuable plant canopy evapotranspiration information.

The most practical application of plant moisture information is the regional assessment of crop water conditions for irrigation scheduling, stress assessment, and yield modeling for agriculture, rangeland, and forestry management.

Advanced Vegetation Spectral Reflectance Characteristics: the Bidirectional Reflectance Distribution Function (BRDF)

It would be wonderful if a vegetated canopy such as a mature corn crop or Loblolly pine plantation reflected the same amount of radiant flux toward the sensor irrespective of 1) the solar incidence and azimuth angles, or 2) the sensor viewing geometry. This would mean that the vegetated canopy was a true Lambertian surface, i.e., it reflects incident energy equally well in all directions. Unfortunately, considerable research has demonstrated that this is not the case (e.g., Kimes, 1983). In fact, the spectral radiant flux leaving a vegetation canopy is significantly impacted by a number of factors, many of which are listed in Table 10-1. We briefly discuss the primary vegetation parameters and then delve more deeply into Sun illumination and sensor system geometric characteristics.

The amount and spectral distribution of radiant flux leaving a vegetated canopy is influenced by the type of vegetation present. For example, grasses hopefully reflect energy differently than a mature stand of trees. Some vegetated canopies

Table 10-1.   Major Variables Affecting the Bidirectional Reflectance Distribution Function (BRDF) of a Vegetated Canopy

Variables Affecting Bidirectional Reflectance from a Vegetated Canopy

Illumination	• Geometry
	   - angle-of-incidence of Sun (or Radar)
	   - azimuth
	• Spectral characteristics (λ)

Sensor	• Geometry
	   - angle of view (e.g., 0° nadir)
	   - azimuth look direction (0 – 360°)
	• Spectral sensitivity (λ)
	• IFOV (milliradians)

Vegetation	• Canopy
	   - type (plant or tree nominal class)
	   - closure (%)
	   - orientation
	       - systematic (e.g., rows 0 – 360°)
	       - unsystematic (random)
	• Crown
	   - shape (e.g., circular, conical)
	   - diameter (m)
	• Trunk or Stem
	   - density (units per m^2)
	   - tree diameter-at-breast-height (DBH)
	• Leaf
	   - leaf-area-index (LAI)
	   - leaf-angle-distribution (LAD) (planophile, erectophile)

Understory	same as vegetation
Soil	• Texture • Color • Moisture content

have 100 percent *canopy closure* meaning that the understory and soil beneath are not visible through the canopy. Conversely, some canopies have < 100 percent canopy closure, allowing portions of the understory and/or soil to reflect energy into the instantaneous field of view (IFOV) of the sensor, creating a hybrid or mixed pixel. It is possible to disentangle the spectral contribution from the individual land covers, but it requires considerable image processing expertise.

Some vegetation is oriented randomly while other vegetation is often systematically arranged in rows in a cardinal direction (0 – 360°). Individual tree crowns often have unique shapes (e.g., a conical Ponderosa pine crown or circular Blackjack oak crown) with unique tree crown diameters that

may be measured. Tree trunks or plant stems have a certain density (e.g., number of trunks per unit area) with unique *diameter-at-breast-height* (DBH) values.

Leaf-area-index (LAI) is the total one-sided (or one half of the total all-sided) green leaf area per unit ground-surface area. It is an important biological parameter because 1) it defines the area that interacts with solar radiation and provides much of the remote sensing signal, and 2) it is the surface that is responsible for carbon absorption and exchange within the atmosphere (Chen and Black, 1992). Some canopies have substantially higher leaf-area-indices than others.

The *leaf-angle-distribution* (LAD) may change throughout the day as the leaves orient themselves toward or away from the incident radiation. Some leaves lie predominantly in the horizontal plane (e.g., planophile leaves in many broadleaf trees), while others are oriented vertically (e.g., erectophile leaves of coastal cordgrass marsh).

All of these vegetation factors can have a significant impact on the reflection of incident light toward the sensor system. Therefore, it is best to hold as many of them as constant as possible when attempting to extract biophysical information such as primary productivity or biomass using multiple dates of remote sensor data. Even if we held these variables as constant as possible, the zenith and azimuth angles of the incident solar radiation and the azimuth and viewing angle of the sensor system can introduce such dramatic effects that we may not be able to compare our spectral reflectance measurements obtained at one time with those of another. Fortunately, if we know a great deal about the vegetation characteristics of the canopy (just discussed) and obtain some information about the bidirectional reflectance characteristics of the solar angle-of-incidence and the sensor viewing geometry, it is possible to calibrate the remote sensor data to extract useful biophysical information.

Bidirectional Reflectance Distribution Function (BRDF): Early measurements of bidirectional reflectance obtained over various vegetation and soil surfaces demonstrated that most terrestrial surfaces exhibit non-Lambertian (*anisotropic*) reflectance characteristics (Kimes, 1983). In the past, bidirectional reflectance effects of the land surface did not play a major role in global change and ecological analysis, although they are assumed to be crucial for multi-temporal studies with varying Sun incidence angles. This was mainly due to a lack of bidirectional data available from remote sensors or acquired in field campaigns. The impact of the bidirectional reflectance distribution function is still not well understood despite the fact that we know it exists in much of the remotely sensed data commonly used for Earth

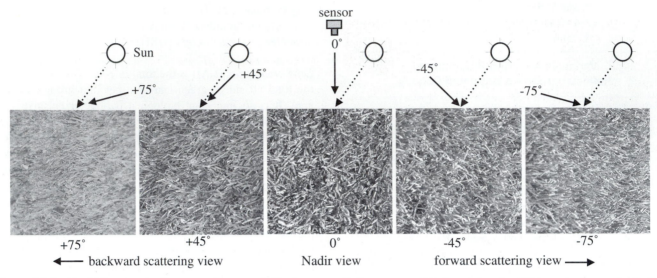

Figure 10-8   The bidirectional reflectance effect on a field of ryegrass (*Lolium perenne L.*) observed under different viewing angles in the solar principal plane from a FIGOS mounted camera. Solar zenith angle was at 35° as indicated by the dashed arrows. The sensor viewing angles are shown as black lines at nadir (0°) and at ±45° and ±75° off-nadir (after Sandmeier and Itten, 1999).

observation, especially for sensors with large fields of view such as the NOAA AVHRR 1.1 x 1.1 km data. A great amount of data from sensors with off-nadir viewing capability is becoming available, especially from commercial satellites such as Space Imaging's IKONOS, and NASA's *Terra* MODIS and MISR (Chapter 7). To properly apply these data in land-use change and ecologically relevant studies, the BRDF should be well understood to calibrate the data (Walter-Shea and Biehl, 1990; Abuelgasim et al., 1998).

Most terrain surfaces (soil, vegetation, and even water) reveal a relationship between the amount of reflected radiance, and 1) the geometric characteristics of the Sun's irradiance, and 2) the sensor viewing geometry. Hence the term *bidirectional*. Depending upon the specific irradiance and sensor viewing angles, most surfaces appear brighter or darker. We have all witnessed this effect. If we walk around a patch of grass or a pool of deep water, it may appear brighter or darker depending upon our viewing angle in relation to the constant incident angle of the Sun. For example, Figure 10-8 depicts a portion of a ryegrass lawn (*Lolium perenne L.*) illuminated by sunlight with a constant solar zenith angle of 35°. Individual photographs of the same grass canopy are obtained at nadir (0°) and at ±45° and ±75° off-nadir (Sandmeier and Itten, 1999). The terrain is generally brighter when the sensor records back-scattered energy as opposed to forward-scattered energy as the diagram reveals (i.e., the image on the left produced by the +75° back-scattered energy appears to be the brightest). Note,

however, that the +75° and -75° images are both much brighter than the image acquired at nadir, 0°.

The bidirectional effect in remote sensing data is most obvious when the angle of illumination and sensor viewing angle are nearly identical and in the same plane. This can produce a *hot spot* (i.e., a shadow is not visible). The solar principal plane is formed when the source, target, and the sensor are in the same plane (in the solar azimuth angle). This is where the BRDF effects are most pronounced. An example of a hot spot in aerial photography is shown in Figure 10-9.

Fortunately, scientists have developed instruments to measure the bidirectional reflectance distribution function of various surfaces. One of the most useful instruments is the *goniometer*. Here we briefly review the field goniometer system (FIGOS) designed by S. Sandmeier. FIGOS consists of a computer-controlled GER 3700 spectroradiometer sensitive to the wavelength interval from 300 – 2,450 nm in 704 bands with a spectral resolution of 1.5 nm for the interval 300 – 1,050 nm and 8.4 nm for the interval 1,050 – 2,450 nm. It consists of three major parts: a zenith arc rail, an azimuth rail, and a motorized sled where the radiometer is mounted (Figure 10-10a,b). The aluminum goniometer weighs approximately 230 kg and may be transported to field sites as shown. Mounted on the goniometer, the sensor views an area in the very center of the circle approximately 10.5 cm in diameter at nadir and about 41 cm (major axis) in the most extreme view zenith angle position of 75°.

$$f_r(\theta_i, \varphi_i;\theta_r, \varphi_r;\lambda) = \frac{dL_r(\theta_i, \varphi_i;\theta_r, \varphi_r;\lambda)}{dE_i(\theta_i, \varphi_i;\lambda)}. \qquad (10\text{-}6)$$

Note that dL_r and dE_i are directional quantities measured in solid angle steradians (sr^{-1}) as shown in Figure 10-10c.

Normally, a *bidirectional reflectance factor* (*BRF*, also referred to as *R*) is computed which is the radiance dL_r reflected from a surface in a specific direction divided by the radiance dL_{ref}, reflected from a loss-less Lambertian reference panel measured under identical illumination geometry (Sandmeier and Itten, 1999):

$$R(\theta_i, \varphi_i;\theta_r, \varphi_r;\lambda) = \qquad (10\text{-}7)$$

$$\left[\frac{dL_r(\theta_i, \varphi_i;\theta_r, \varphi_r;\lambda)}{dL_{ref}(\theta_i, \varphi_i;\theta_r, \varphi_r;\lambda)}\right] \times R_{ref}(\theta_i, \varphi_i;\theta_r, \varphi_r;\lambda)$$

where R_{ref} is a calibration coefficient determined for the spectral reflectance panel. Bidirectional reflectance factors (BRF) are dimensionless. Both BRF and BRDF take on values from zero to infinity. Values > 1 for BRF and $> 1/\pi$ for BRDF are obtained in peak reflectance directions, such as the hot spot, where the reflected flux from a target surface is higher than the flux from a Lambertian surface.

It is also possible to develop an *anisotropy factor* which is used to analyze the spectral variability in BRDF data. Anisotropy factors (ANIF) allow separation of spectral BRDF effects from the spectral signature of a target. They are calculated by normalizing bidirectional reflectance data R to nadir reflectance, R_o (Sandmeier et al., 1998a; Sandmeier and Itten, 1999):

$$ANIF(\theta_i, \varphi_i;\theta_r, \varphi_r;\lambda) = \frac{R(\theta_i, \varphi_i;\theta_r, \varphi_r;\lambda)}{R_o(\theta_i, \varphi_i;\lambda)}. \qquad (10\text{-}8)$$

So what do these measurements tell us about the BRDF of a typical surface? To answer this question, consider the nadir-normalized BRDF data (i.e., the ANIF data) of perennial ryegrass (*Lolium perenne L.*) shown in Figure 10-10d. During goniometer data collection the Sun zenith angle was 35°. Spectral results from four of the 704 possible spectroradiometer bands are presented using just the viewing zenith angle in the solar principal plane. It reveals that BRDF effects were very pronounced in the blue (480 nm) and red (675 nm) chlorophyll absorption bands previously discussed, whereas in the green and particularly in the low absorbing near-infrared range, relatively low BRDF effects were observed. Thus, persons using remote sensor data might consider radiometrically adjusting the brightness values associated with the blue and red bands, but not necessarily the near-infrared bands

Figure 10-9 A hot spot near the left fiducial mark on a vertical aerial photograph of the Savannah River swamp in South Carolina. The Cypress-Tupelo forested wetland has relatively uniform percent canopy closure in this area. Its tone and texture should appear relatively homogeneous but they do not. A hot spot is produced when the angle of illumination and sensor viewing angle are nearly identical and in the same plane.

Thus, it is possible in a very short period of time while the Sun is approximately in the same zenith arc (θ_i) and azimuth (φ_i), to vary the position of the radiometer to determine if the amount of radiant flux in very specific wavelength regions (e.g., blue, green, red, near-infrared, middle-infrared) leaving the target is influenced by the sensor angle-of-view (θ_r) and azimuth (φ_r). If it is not, then we do not have to worry about BRDF effects. If the reflectance results are not uniform as will be demonstrated, then we may need to concern ourselves with adjusting for BRDF effects if we want to compare remote sensor data obtained on multiple dates (with varying Sun azimuth and zenith angle) or by sensors with multiple look angles (e.g., SPOT data acquired at 0° nadir on day 1 and 20° off-nadir on day 2).

The *bidirectional reflectance distribution function* (BRDF), f_r (sr^{-1}), is formally defined as the ratio of the radiance dL_r ($\text{W m}^{-2}\,\text{sr}^{-1}\,\text{nm}^{-1}$) reflected in one direction (θ_r, φ_r) to the Sun's incident irradiance dE_i ($\text{W m}^{-2}\,\text{nm}^{-1}$) from direction ($\theta_i, \varphi_i$) (Sandmeier, 1999; Sandmeier and Itten, 1999):

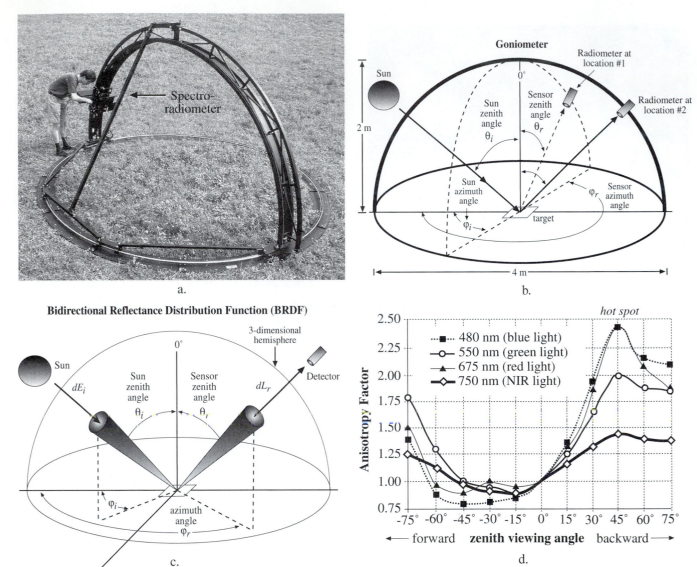

Figure 10-10 a) The field goniometer system (FIGOS) developed by S. Sandmeier. b) In this hypothetical example, the spectroradiometer is located at two locations. Spectral measurements are made at location #1 with a Sun zenith angle of θ_i and Sun azimuth angle of φ_i and a sensor angle of view of θ_r and azimuth angle of φ_r. The GER spectroradiometer records the amount of radiance leaving the target in 704 spectral bands. The radiometer is then moved to location #2 (i.e., perhaps only varying the sensor azimuth angle as shown) where it acquires radiance data again in 704 bands. c) The concept and parameters of the BRDF. A target is bathed in irradiance (dE_i) from a specific zenith and azimuth angle, and the sensor records the radiance (dL_r) exiting the target of interest at a specific azimuth and zenith angle. d) Anisotropy factor for ryegrass (*Lolium perenne L.*) at multiple zenith viewing angles in the solar principal plane for four of the 704 bands. The Sun zenith angle was 35° (after Sandmeier and Itten, 1999). e) Second generation Sandmeier Field Goniometer (SFG) obtaining smooth cordgrass (*Spartina alterniflora*) readings at North Inlet, SC, on October 21, 1999 (Sandmeier, 1999; Schill et al., 2000).

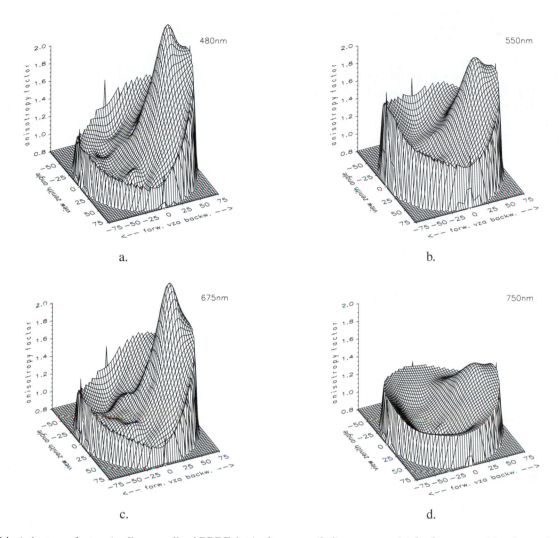

Figure 10-11 Anisotropy factors (nadir-normalized BRDF data) of ryegrass (*Lolium perenne L.*) for four spectral bands acquired with the FIGOS instrument with a Sun zenith angle of 35° (after Sandmeier and Itten, 1999).

under investigation. Typical for vegetated surfaces, all four bands exhibit a bowl shape, hot spot, and forward-scattering component.

It is actually more interesting to view the ryegrass anisotropy factors for the four wavelengths of interest according to not only the viewing zenith angle of 0° but in a range from ±75° (Figure 10-11). Ideally, the entire three-dimensional surface should be relatively flat as with the 750 nm near-infrared example, meaning that measurements in this specific band are relatively free of BRDF effects. Conversely, the 480 nm and 675 nm bands exhibit significant anisotropy factors. In the high-absorbing (i.e., low-reflecting) wavelength ranges, multiple scattering effects are reduced due to the relatively low amount of radiation in the canopy. Thus, the contrast

between shadowed and illuminated canopy components are enhanced, which then enhances the BRDF effects. BRDF effects are rather small in the low-absorbing (i.e., high-reflecting) green and near-infrared wavelength ranges where multiple scattering effects are strong and diminish the contrast in the canopy (Sandmeier et al., 1998b).

Future field research using goniometers coupled with off-nadir pointing aircraft and satellite remote sensing systems such as *Terra* MISR (Chapter 7) will eventually document when we should be concerned with BRDF effects. Such information may help us 1) identify and select the bands that are impacted the least by BRDF effects, 2) recognize that there are certain Sun and/or sensor angle-of-view relationships that should be avoided when collecting remotely sensed data, and/or 3) provide us with methods to radiometrically adjust the remote sensor data to minimize BRDF effects.

One should be careful, however, not to infer that BRDF effects are all bad. Strahler (1994) and Qi et al. (1995) point out that multidirectional remote sensing measurements of the terrain (e.g., at viewing angles off-nadir) can provide complementary information to that provided by the nadir measurement. In fact, a single nadir view remote sensing measurement obtains information about the surface as if the surface had no vertical structures, which is usually not the case in practice, while off-nadir view measurements reveal different aspects of the vertical structures such as vegetation height. *Consequently, to objectively characterize vegetation biophysical parameters, a single viewing geometry (e.g., at nadir) may be insufficient.*

Modeling Canopy Reflectance

Scientists have attempted to predict exactly how much radiant energy in specific wavelengths should be exiting a given leaf or vegetated canopy based on a number of factors, including (Danson, 1998): the leaf-area-index (LAI), soil reflectance properties below the canopy, amount of direct and/or diffuse skylight onto the canopy, the leaf-angle-distribution (ranging from erectophile canopies with vertical leaves at 90° inclination to planophile canopies dominated by horizontal leaves with 0° inclination), and the BRDF influenced by the Sun angle and sensor viewing angle geometry just discussed. For example, the SAIL model (Scattering by Arbitrarily Inclined Leaves) has been widely used in remote sensing research for investigating the spectral and directional reflectance properties of vegetation canopies. It uses radiative transfer equations to model energy interacting with a vegetation canopy in three distinct streams, including a downward flux of direct radiation and a downward and upward flux of diffuse radiation (Verhoef, 1984; Goel, 1988). It assumes that the canopy may be represented by small absorbing and scattering elements (e.g., leaves) with known optical properties, distributed randomly in horizontal layers and with known angular distribution. The model has been used to simulate the effects of off-nadir viewing, to simulate spectral shifts of the red-edge, and to estimate canopy properties directly from remotely sensed data (Jacquemoud et al., 1995).

Similarly, the Li-Strahler model (1985; 1992) yields estimates of the size and density of trees from remotely sensed images. The signal received by the sensor is modeled as consisting of reflected light from tree crowns, their shadows, and the background within the field of view of the sensor.

The reflected signal is modeled as a linear combination of four components and their areal proportions (Woodcock et al., 1997):

$$S = K_g G + K_c C + K_t T + K_z A \qquad (10\text{-}9)$$

where S is the brightness value of a pixel, K_g, K_c, K_t, and K_z stand for the areal proportions of sunlit background, sunlit crown, shadowed crown, and shadowed background, respectively, and G, C, T, and A are the spectral signatures of the respective components. This is called a mixture model, or a scene component model.

The SAIL and Li-Strahler models represent important attempts to explicitly model the energy-matter interactions taking place above, in, and below the vegetative canopy. Scientists strive to invert the models and use the reflectance characteristics recorded by the remote sensor system to predict the characteristics of specific types of structure within the canopy, such as tree height, density, leaf-area-index, etc. Unfortunately, it is often difficult to calibrate such models because so much information must be known about the leaf and canopy characteristics, atmospheric conditions, Sun angle and viewing geometry, and terrain slope and aspect. This is an important area of remote sensing research.

Imaging Spectrometry of Vegetation

As discussed in Chapter 7, an imaging spectrometer collects a continuous reflectance spectrum of the Earth's surface. But instead of having just four (e.g., SPOT) or eight (e.g., Landsat 7 Enhanced Thematic Mapper Plus) spectral channels to characterize the vegetation spectral characteristics, an imaging spectrometer records information in hundreds of spectral channels.

Imaging spectrometry has great potential for monitoring vegetation type and biophysical characteristics (Goetz, 1995). Vegetation reflectance spectra are often quite informative, containing information on the vegetation chlorophyll absorption bands in the visible region, the sustained high reflectance in the near-infrared region, and the effects of plant water absorption in the middle-infrared region.

Geologists have been using hyperspectral imagery for years to discriminate between different rock types based on their spectral reflectance and absorption characteristics. They have prepared an exhaustive spectral library of the most important minerals, soils, and rock types (e.g., Clark et al., 1993). When calibrated airborne hyperspectral imagery are acquired, they often compare the spectra obtained from the airborne data to the spectra stored in the database to deter-

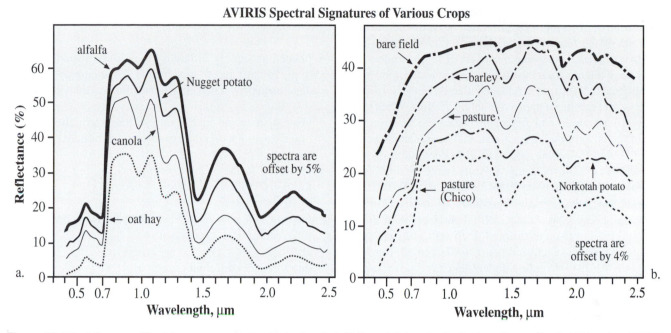

Figure 10-12 a) Spectra of healthy green vegetation in the San Luis Valley of Colorado obtained on September 3, 1993, using AVIRIS; 224 channels at 20 x 20 m pixels. b) Most of these other vegetation types exhibited reduced chlorophyll absorption effects. Note that the spectra are offset for clarity (after Clark et al., 1995). Differences in the reflectance and/or absorption characteristics of the vegetation make it possible in certain instances to distinguish one vegetation type from another or to identify if the vegetation is undergoing stress by some agent.

mine mineral type and other characteristics. Several examples are provided in Chapter 13 (Remote Sensing of Soils, Minerals and Geomorphology).

Ideally, there would be a vegetation spectral databank available that contains information on: 1) the spectral reflectance and emittance characteristics of every type of vegetation in the region from $0.35 - 14$ μm, 2) how these data appear on different dates throughout the pertinent growing seasons, and 3) what these spectra should look like in the event of stress or insect infestation. Unfortunately, such a database does not yet exist. Instead, it is still usually necessary to collect *in situ* spectroradiometer data of the vegetation of interest at the time of the overflight and then use this data to calibrate the spectral reflectance information derived from the airborne spectroradiometer. When this is performed carefully, it is possible to extract vegetation type and condition information from the hyperspectral data.

For example, consider the AVIRIS data collected over the San Luis Valley of Colorado on September 3, 1993, shown in Color Plate 10-2a (Clark et al., 1995). The 224 spectral channels of 20 x 20 m AVIRIS data were radiometrically corrected to remove the effects of atmospheric water vapor and geometrically rectified to a standard map projection. Land

cover of some of the fields at the time of the overflight are annotated in color on top of the single AVIRIS channel in Color Plate 10-2a. AVIRIS reflectance spectra for some of these fields, including potato, alfalfa, barley, oat hay, canola, and pasture are portrayed in Figure 10-12a,b.

The alfalfa, canola, oat hay, and Nugget potato spectra show the plants to be green and healthy (Figure 10-12a). Barley had lost all of its chlorophyll response at the time of the overflight (Figure 10-12b). Norkotah potatoes were not being irrigated as they were about to be harvested, and consequently they show weak chlorophyll and cellulose absorptions, with soil (clay) absorption from exposed soil. Field investigation revealed that the potatoes were also being sprayed with a defoliant. Thus, they exhibit decreased chlorophyll absorption along with a shift of the red-edge to shorter wavelengths as we would expect. The Chico and pasture spectra exhibit combinations of chlorophyll and cellulose (dry vegetation) absorption. The spectra from the known fields was used in a special computer program called *Tricorder*, developed at the U.S. Geological Survey, to identify other pixels in the study area with similar spectral response (Clark et al., 1995). The result was a classification map that was approximately 96 percent accurate (Color Plate 10-2b).

Clark et al. (1995) point out that the long-wavelength side of the chlorophyll absorption (approximately 0.68 – 0.73 μm) forms one of the most extreme slopes found in spectra of naturally occurring common materials. The absorption is usually very intense, ranging from a low reflectance of < 5 percent (near 0.68 μm) to a near-infrared reflectance maximum of ≥ 50 percent (at approximately 0.73 μm). When the chlorophyll absorption in the plant decreases, the overall *width* of the absorption band decreases. These factors cause a shift to shorter wavelengths as the chlorophyll absorption decreases. This is known as the "red-edge shift" or the "blue shift of the red edge" previously discussed and can be caused by natural senescence, water deprivation, or toxic materials (Rock et al., 1986). Clark et al. (1995) found that the AVIRIS data could detect red-edge shifts of < 0.1 nm. The red-edge shift information was analyzed using the Tricorder program and used to identify vegetation undergoing any type of stress in the study area (Color Plate 10-2c).

Temporal Characteristics of Vegetation

Timing is very important when attempting to identify different vegetation types or to extract useful vegetation biophysical information (e.g., biomass, chlorophyll characteristics) from remotely sensed data. Selecting the most appropriate date(s) for data collection requires an intimate knowledge of the plants' temporal *phenological* (growth) *cycle*. Plants whose leaves drop seasonally are *deciduous*. Plants whose leaves remain from season to season are *evergreen*.

A noted remote sensing scientist, Dave Simonett, often said "Green is green is green." By this he meant that if two different crops (e.g., corn and cotton) were planted at the same time and had complete canopy closure on the same date as the remotely sensed data were collected, the spectral reflectance characteristics of the two crops would most likely appear to be very similar throughout the visible and near-infrared portion of the spectrum. His comment was based on the use of the relatively broad-band sensors available in the 1970s and 1980s. We now have hyperspectral sensors that allow us to sample in relatively narrow portions of the spectrum, hopefully identifying unique absorption features that will allow us to discriminate between one vegetation type and another or to extract biophysical information.

Nevertheless, Simonett's observation is still correct in many instances. Often our only hope of discriminating between two crops using relatively coarse spectral resolution remote sensor data (e.g., Landsat Thematic Mapper, SPOT, IKONOS) is if the crops were:

- planted at slightly different times in the growing season (e.g., 10 days apart), which caused one canopy to be less developed (lower percent canopy closure) than the other;

- one crop received significantly different irrigation than the other, causing it to produce more or less biomass;

- one crop matured more rapidly than the other (e.g., through fertilization or careful weeding);

- the row spacing or field orientation was dramatically different for the two crops; or

- one crop has a different canopy structure.

The difference in crop percent canopy closure, soil moisture, biomass, or the difference in row spacing or orientation might cause one crop to have dramatically different reflectance properties due to the influence of the background soil or understory materials present. Therefore, it is often the proportion of background material present within the instantaneous field of view of the sensor system (e.g., perhaps 1 x 1 ft for high resolution color-infrared aerial photography; 30 x 30 m for Landsat Thematic Mapper imagery) that allows us to discriminate between the two vegetation types. The amount of understory background material present is largely a function of the stage of the plant in its phenological cycle. Therefore, if a scientist is trying to differentiate (classify) between several crops, wetland, or forest types using remote sensor data, it is essential to know the phenological cycle characteristics of all of them. This information is then used to determine the optimum time of year to collect the remotely sensed data to discriminate one land cover or vegetation type from another.

Another important temporal factor cannot be overlooked. Plants require water to grow. Therefore, their most productive growth period is often associated with the most intense periods of precipitation and associated cloud cover. For example, Figure 10-13 identifies the annual percent cloud cover statistics for several areas in the United States. Except for California, with its low humidity and low cloud cover, most of the areas have their greatest cloud cover at exactly the time when scientists may want to collect the maximum amount of remote sensor data during the growing season. Consequently, scientists must juggle the identification of the optimum date of remote sensor data using the phenological calendar with the likely spectre of considerable cloud cover

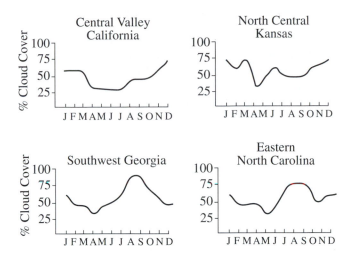

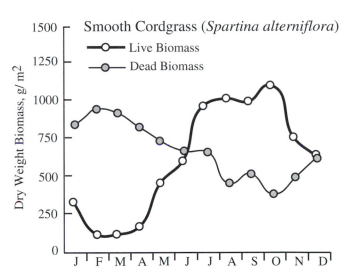

Figure 10-13 Predicted percent cloud cover in four areas in the United States. The greatest amount of cloud cover is often associated with the height of the growing season, except for California. This can significantly complicate the collection of remote sensing data for vegetation analysis.

Figure 10-14 Phenological cycle of Smooth Cordgrass (*Spartina alterniflora*) biomass in South Carolina. The greatest amount of live biomass is present from July through October. The greatest amount of dead biomass is present from January through March. The marsh contains approximately equal proportions of live and dead biomass in December and June (after Dame and Kenny, 1986; Jensen et al., 1998).

being present. This is why we need higher temporal resolution sensors.

The development of forests, grassland, and wetland canopies follow relatively predictable cycles each year except when they are under stress from a pathogen or when unseasonably high or low temperatures occur. Similarly, all managed agricultural crops have relatively well-known phenological cycles within a given region, but these may be modified by individual farmers. Therefore, it is useful to review the phenological cycles of both natural vegetation systems and managed agricultural systems to gain insight into how important the cycles are when using remote sensing to extract vegetation biophysical information. Several case studies are provided.

Natural Phenological Cycles

Many plant species found in forests, wetland, rangeland, etc., have relatively unique phenological growth cycles (Yool et al., 1997). They leaf-out, grow to maturity, and senesce at approximately the same time each year. However, unseasonably cold or warm temperatures in the spring or fall can shift their annual phenological cycles by as much as 30

days. Therefore, the analyst must be aware of whether the remotely sensed data were collected in a typical or atypical year. If a scientist is attempting to classify vegetation using remote sensor data, it is often wise to collect the data early in the growing season when the vegetation are developing at different rates, which yield different percent canopy closures, hopefully creating differences in their spectral signatures. For example, if one is attempting to discriminate among various inland wetland species (e.g., cattail, willow, buttonbush) in the southeastern United States, it is best to collect imagery from February 15 through May 1. After May 1, most of the wetland vegetation has achieved complete canopy closure, and the spectral signatures of the various canopies become similar, i.e., "Green is green is green."

Conversely, if the goal is to monitor the biomass of the vegetation, it is useful to collect remote sensor data at the height of the growing season when the maximum biomass exists. Monitoring biomass through time can provide important information about the stability of the natural ecosystem and whether significant change is taking place. The following section demonstrates this logic applied to monitoring coastal Smooth Cordgrass wetland (Figure 10-14).

Phenological Cycle of Coastal Smooth Cordgrass in South Carolina

The annual phenological cycle of evergreen Smooth Cordgrass (*Spartina alterniflora*), which grows extensively along the eastern coast of the United States is shown in Figure 10-14. The decomposed *Spartina* organic matter (detritus) is carried out to sea and represents the foundation of the food chain that supports the Eastern Seaboard marine fishery industry. The time of greatest *Spartina* growth and dry weight biomass occurs from July through October. This is the optimum time for remote sensing data collection. It serves little purpose to obtain imagery of the *Spartina* in January through March, when it is in a senesced (dormant) state. Still, one often encounters persons who attempt to use winter imagery collected for some other purpose to monitor the *Spartina* wetland.

Figure 10-15 displays nine bands of 3 x 3 m NASA Calibrated Airborne Multispectral Scanner (CAMS) data of Murrells Inlet, SC, obtained on August 2, 1997, during the high biomass time of year. *In situ* Smooth Cordgrass total dry biomass (g/m^2) measurements were obtained at 27 locations on August 2 and 3, 1997. The *in situ* data were then correlated with individual band brightness values and vegetation transforms of the original CAMS data (Jensen et al., 1998). One of the most impressive relationships was between the fundamental near-infrared band 6 CAMS data and the *in situ* measurements, which yielded a correlation coefficient of 0.88 and an r^2-value of 0.774 meaning that approximately 77 percent of the variance was accounted for (Figure 10-16).

Color Plate 10-3 is a map of the spatial distribution of total dry biomass for a small portion of Murrells Inlet on August 2, 1997. It was produced using the regression equation:

$$y = 3.4891x - 23.9 \qquad (10\text{-}10)$$

where y is total biomass (g/m^2) and x is the CAMS band 6 brightness value. Studies such as this provide important baseline biophysical information that can be compared against future studies to determine if deleterious changes in biomass are occurring in the inlet. Comparative biomass studies would require that most of the system and environmental variables remain constant (e.g., approximately the same sensor system configuration and viewing geometry, altitude above-ground-level, atmospheric conditions, anniversary date, time of day, and tidal cycle). The limitations of using a regression approach when predicting biophysical variables using remotely sensed data are summarized in Qi et al. (1995).

Managed Phenological Cycles

Crops associated with mechanized forestry and agriculture are established, grown, and harvested according to relatively predictable phenological cycles (Rundquist and Sampson, 1988; Thenkabail et al., 1994). It would be wonderful if all farmers planted the same crop on exactly the same date in a given growing season. This would make the analysis of remotely sensed data straightforward. Fortunately for us, who eat the bread of their labors, they do not do this, but rather plant the crop based on current meteorological conditions, the availability of equipment, and heuristic rules-of-thumb. Therefore, not all crops of the same species are planted during the same month or harvested during the same month. If two identical corn crops are planted 20 days apart, their spectral signatures will likely be very different throughout their respective phenological cycles. This condition may cause problems when attempting to perform crop identification using remote sensor data.

County agricultural extension agents and land-grant university scientists know the local and regional crop phenological cycles very well. They often represent the most important sources for valuable crop phenology and soils information and any idiosyncrasies associated with local farming practices. It is instructive to provide examples of several agricultural phenological cycles that span the geography of the conterminous United States to demonstrate the importance of vegetation phenological information. This will include case studies concerning hard red winter wheat in the Midwest United States, several major crops in South Carolina, and remote sensing of sugar beets, cotton, and alfalfa in the Imperial Valley, CA. Each example includes detailed phenological information displayed in different formats.

Phenological Cycle of Hard Red Winter Wheat in the Midwest United States

The detailed phenological cycle of hard red winter wheat (*triticale*) grown on the prairie soils in the Midwest United States is shown in Figure 10-17. The crop is established in October and November. It lies dormant under snow cover (if present) until March, when growth resumes. Hard red winter wheat greens up in April and produces heads of grain in May. It matures by mid-June. It is usually ripe and harvested by early July. Remotely sensed data acquired in October and November provide information on the amount of land prepared during the crop establishment period. Imagery acquired during the green-up phase in April and May can be used to extract information on standing-crop biomass and perhaps predict the harvested wheat yield.

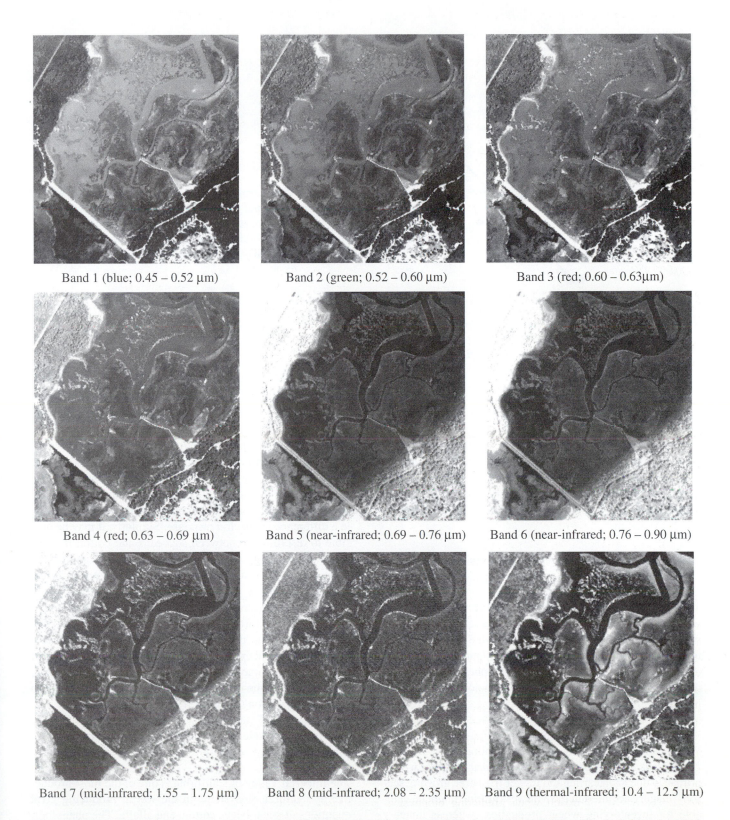

Band 1 (blue; 0.45 – 0.52 μm)   Band 2 (green; 0.52 – 0.60 μm)   Band 3 (red; 0.60 – 0.63μm)

Band 4 (red; 0.63 – 0.69 μm)   Band 5 (near-infrared; 0.69 – 0.76 μm)   Band 6 (near-infrared; 0.76 – 0.90 μm)

Band 7 (mid-infrared; 1.55 – 1.75 μm)   Band 8 (mid-infrared; 2.08 – 2.35 μm)   Band 9 (thermal-infrared; 10.4 – 12.5 μm)

Figure 10-15 Nine bands of 3 x 3 m Calibrated Airborne Multispectral Scanner (CAMS) data of Murrells Inlet, SC, obtained on August 2, 1997. The data were obtained at 4,000 ft above-ground-level.

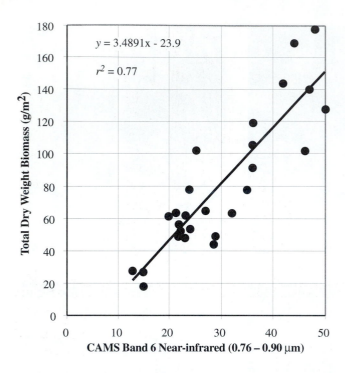

$y = 3.4891x - 23.9$

$r^2 = 0.77$

Figure 10-16 Relationship between Calibrated Airborne Multi-spectral Scanner (CAMS) band 6 brightness values and *in situ* measurement of *Spartina alterniflora* total dry biomass (g/m^2) at 27 locations in Murrells Inlet, SC, obtained on August 2 and 3, 1997 (Jensen et al., 1998).

Detailed crop calendar information such as this is used by governments, individuals, and private companies to acquire remote sensing data at optimum dates in the phenological cycle in order to model and predict the production of specific food for sale on the world market. Such monitoring may in certain instances also be used to predict future agricultural disasters due to severe drought, hopefully alleviating suffering due to famine.

Phenological Cycle of Agricultural Crops in South Carolina

The phenological cycles of natural and cultivated vegetation varies greatly by region. For example, if a scientist wanted to monitor the condition of the wheat crop in South Carolina, it is *not* possible to use the Midwest wheat phenological information presented in Figure 10-17. The phenological cycles for wheat and South Carolina's other major cash crops are presented in Figures 10-18 and 10-19.

Wheat heads in late May and early June in the Midwest. Wheat heads in early May in South Carolina, approximately 15 days ahead of the Midwest phenological cycle for wheat, primarily due to an earlier and longer frost-free growing season. Imagery acquired in the Midwest in June might be valuable for inventorying wheat production, whereas imagery obtained in June in South Carolina might well reveal that the wheat crop has already been harvested.

The crop calendars also reveal that it should be straightforward to identify wheat production in South Carolina by acquiring a single image in March or early April when wheat is the only crop in existence with a complete canopy closure. Conversely, it may be difficult to differentiate between tobacco and corn unless we obtain a mid-June image and hope that the tobacco is leafed out more than corn, resulting in a difference in spectral response. Similarly, it may be possible to discriminate between soybeans and cotton if a late July or early August date is selected and soybeans have a greater canopy closure than cotton. If all the phenological cycles of the major crops were exactly the same for South Carolina, with complete canopy closure at the same time of year, it would be difficult to use remotely sensed data to discriminate between the vegetation types. However, because their phenological cycles are staggered throughout the growing season to some degree, it is possible to discriminate between them if the imagery is collected at optimum times in the growing season (Savitsky, 1986).

Phenological Cycle of Agricultural Crops in the Imperial Valley, California

The Imperial Valley of California is one of the most productive agricultural ecosystems in the world, producing great quantities of sugar beets, cotton, and alfalfa. The phenological cycles of several of its major crops are summarized in Color Plate 10-4 (Byrd, 1998; GRSU, 1999). Each calendar

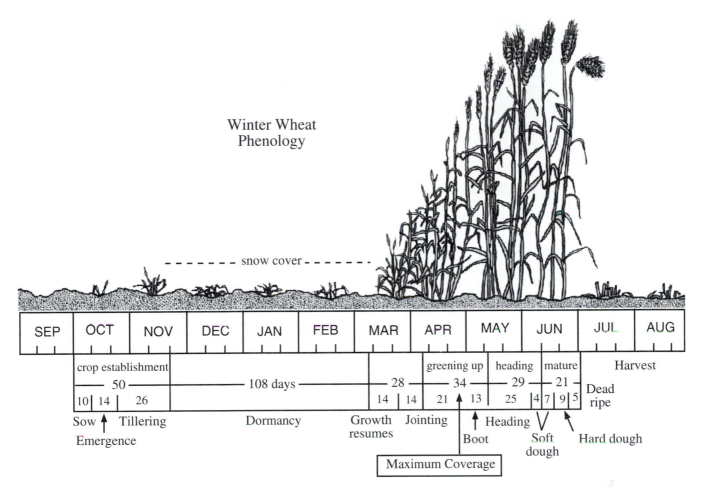

Figure 10-17 The phenological cycle of hard red winter wheat in the Great Plains of the United States. The crop is established in October and November. It lies dormant under snow cover until March, when growth resumes. The plants green up in April, produce heads of grain in May, and mature by mid-June. The wheat is usually dead ripe and harvested by early July. Remotely sensed data acquired in October and November provide information on the amount of land prepared during the crop establishment period. Imagery acquired during the green-up phase in April and May can be used to extract information on standing-crop biomass and perhaps predict the harvested wheat yield.

depicts a single field monitored throughout a 12-month period using 79 x 79 m Landsat Multispectral Scanner (MSS) data (RGB = bands 4,2,1). The brighter red (magenta) the signature, the greater the amount of biomass and crop canopy closure. Sugarbeets are established in early September, emerge in November, and are harvested in April and early May. By law, cotton must be planted in March and harvested and plowed under in November of each year to control Bole Weevil. In 1982, the plow-down date was January 1. Alfalfa is planted year-round and may be harvested five or six times per year as depicted in the crop calendar.

It is possible to identify the type of crop in each field if imagery is obtained at times of the year that maximize the spectral contrast between crops. For example, Figure 10-20

depicts seven bands of Landsat Thematic Mapper imagery obtained on December 10, 1982. Six of the bands are at 30 x 30 m, while the thermal infrared band (6) is at 120 x 120 m spatial resolution. A ground reference crop map provided by the Imperial Valley Irrigation Board is also included (Haack and Jampoler, 1995).

The four color composites shown in Color Plate 10-5 demonstrate the importance of the middle-infrared bands when attempting to discriminate among crop types. The fundamental visible (RGB = bands 3,2,1) and near-infrared color composites (RGB = bands 4,3,2) are not nearly as effective for visual determination of the crop types as the visible and middle-infrared composites (RGB = bands 5,3,2 and RGB = bands 7,3,2). Note that all the sugarbeet fields show up as

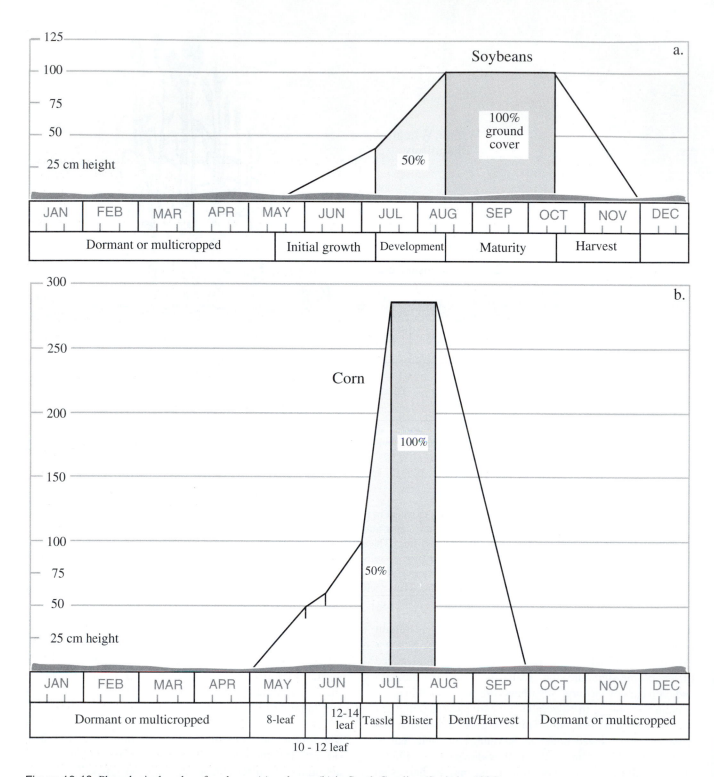

Figure 10-18 Phenological cycles of soybeans (a) and corn (b) in South Carolina (Savitsky, 1986).

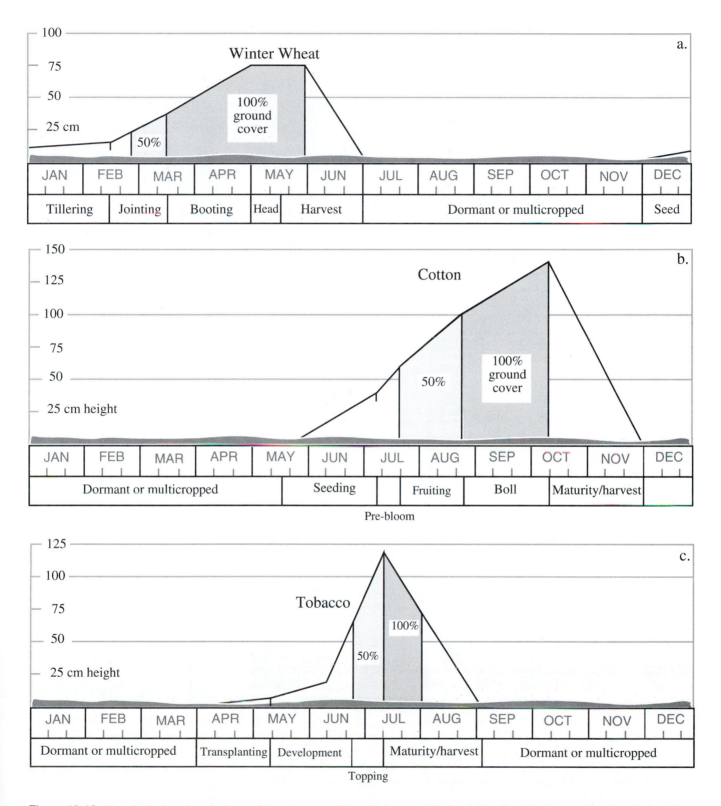

Figure 10-19 Phenological cycles of winter wheat (a), cotton (b), and tobacco (c) in South Carolina. The information was obtained from county agricultural extension agents, Clemson land-grant university extension personnel, and field work (Savitsky, 1986).

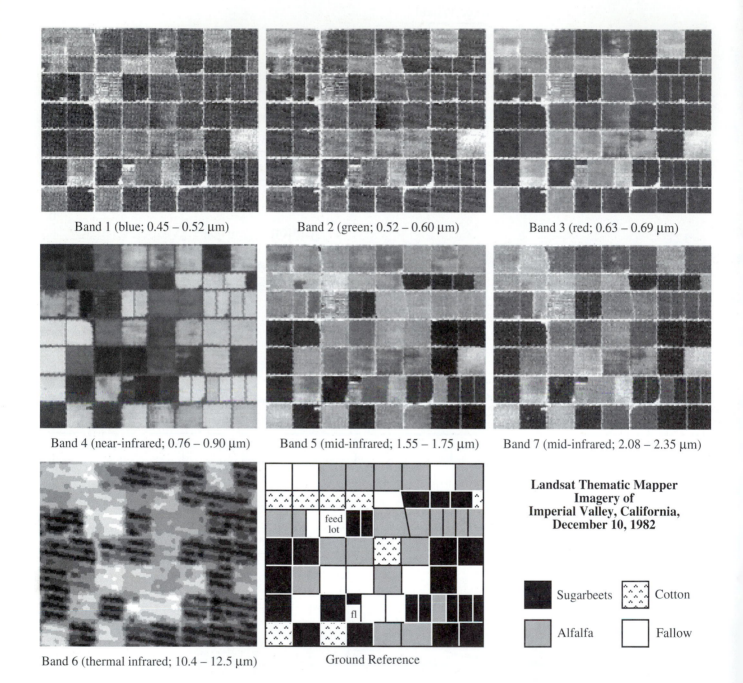

Band 1 (blue; 0.45 – 0.52 μm)   Band 2 (green; 0.52 – 0.60 μm)   Band 3 (red; 0.63 – 0.69 μm)

Band 4 (near-infrared; 0.76 – 0.90 μm)   Band 5 (mid-infrared; 1.55 – 1.75 μm)   Band 7 (mid-infrared; 2.08 – 2.35 μm)

Band 6 (thermal infrared; 10.4 – 12.5 μm)   Ground Reference

**Landsat Thematic Mapper
Imagery of
Imperial Valley, California,
December 10, 1982**

Sugarbeets   Cotton

Alfalfa   Fallow

feed lot

fl

Figure 10-20 Individual bands of Landsat Thematic Mapper imagery of a portion of the Imperial Valley, CA, obtained on December 10, 1982. Bands 1 – 5 and 7 have a spatial resolution of 30 x 30 m. Band 6 has a spatial resolution of 120 x 120 m. Sugarbeets are bright in the near-infrared band 4 data and dark in both the middle-infrared bands (5 and 7). Fallow fields are gray in all bands except band 4 (near-infrared), which is dark (images courtesy of Space Imaging, Inc.).

Table 10-2.   Tonal Characteristics of Individual Crop Types in Black-and-White Landsat Thematic Mapper Images of Imperial Valley, CA (Figure 10-20) obtained on December 10, 1982 (D = dark; G = mid-gray; Br = brighter)

Crop Type	TM1	TM2	TM3	TM4	TM5	TM7
Sugarbeets	D	D	D	Br	D	D
Alfalfa	D	D	D	Br	G	G
Cotton	G	G	G	D	Br	Br
Fallow land	G	G	G	D	G	G

black in the middle-infrared color composite (Color Plate 10-5c; RGB = bands 5,3,2) and in the individual band 5 image (Figure 10-20).

Because of the significant spectral contrast among the crops in this December 10, 1982, Landsat TM image, it is possible to perform a visual classification of the crops based on an examination of the individual bands in Figure 10-20. For example, after careful examination we see that the heuristic rules summarized in Table 10-2 could be used to classify the scene.

It is also possible to program a computer to take into account these same spectral characteristics and perform a classification. Color Plate 10-5e displays a remote sensing derived digital classification map of the study area. When compared with the ground reference map in Figure 10-20 it is about 85 percent accurate. It was produced using TM bands 1 – 5 and 7 and a supervised maximum likelihood classification algorithm [please refer to Jensen (1996) for information about digital image classification]. Haack and Jampoler (1994; 1995) evaluated this same Landsat TM scene and achieved an overall accuracy of 90 percent using a different set of bands.

The accuracy of a remote sensing derived crop classification map is always dependent upon there being a significant difference in the spectral response between the various crop types. The only way to identify when this maximum contrast among spectral response should take place is to evaluate the phenological crop calendars and select the appropriate dates of imagery for analysis. Then classification maps, such as the one shown in Color Plate 10-5, can be used to predict the amount of acreage in specific crops at a given instant in time. Such information coupled with agricultural-meteorological models can be used to predict crop yield.

Vegetation Indices

Since the 1960s, scientists have extracted and modeled various vegetation biophysical variables using remotely sensed data. Much of the effort has gone into the development of *vegetation indices* — defined as dimensionless, radiometric measures that function as indicators of relative abundance and activity of green vegetation, often including leaf-area-index (LAI), percentage green cover, chlorophyll content, green biomass, and absorbed photosynthetically active radiation (APAR). A vegetation index should (Running et al., 1994; Huete and Justice, 1999):

- maximize sensitivity to plant biophysical parameters, preferably with a linear response in order that sensitivity be available for a wide range of vegetation conditions, and to facilitate validation and calibration of the index;

- normalize or model external effects such as Sun angle, viewing angle, and the atmosphere for consistent spatial and temporal comparisons;

- normalize internal effects such as canopy background variations, including topography (slope and aspect), soil variations, and differences in senesced or woody vegetation (nonphotosynthetic canopy components); and

- be coupled to some specific measurable biophysical parameter such as biomass, LAI, or APAR as part of the validation effort and quality control.

There are more than 20 vegetation indices in use. A select few are summarized in Table 10-3. Many are functionally equivalent (redundant) in information content (Perry and Lautenschlager, 1984), while some provide unique biophysical information (Qi et al., 1995). It is useful to review the historical development of the main indices and provide information about recent advances in index development. More detailed summaries are found in Richardson and Everitt (1992), Running et al. (1994), and Lyon et al. (1998).

Cohen (1991) suggests that the first true vegetation index was the *Simple Ratio* (SR), which is the near-infrared (NIR) to red reflectance ratio described in Birth and McVey (1968):

$$SR = \frac{NIR}{Red}.$$

(10-11)

Rouse et al. (1974) developed what is now called the generic *Normalized Difference Vegetation Index* (NDVI):

$$\text{NDVI} = \frac{NIR - red}{NIR + red}. \quad (10\text{-}12)$$

The NDVI index was widely adopted and applied to the original Landsat MSS digital remote sensor data. Deering et al. (1975) added 0.5 to the NDVI to avoid negative values and took the square root of the result to stabilize the variance. This index is referred to as the *Transformed Vegetation Index* (TVI). These three indices respond to changes in the amount of green biomass and chlorophyll content.

Kauth and Thomas (1976) produced an orthogonal transformation of the original Landsat MSS data space to a new four-dimensional feature space. It was called the *Tasseled Cap* or *Kauth-Thomas* transformation. It created four new axes: the soil brightness index (B), greenness vegetation index (G), yellow stuff index (Y), and non-such (N). The names attached to the new axes indicate the characteristics the indices were intended to measure. The coefficients in the following formula are from Kauth et al. (1979):

$$B = 0.332_{mss1} + 0.603_{mss2} + 0.675_{mss3} + 0.262_{mss4} \quad (10\text{-}13)$$

$$G = -0.283_{mss1} - 0.66_{mss2} + 0.577_{mss3} + 0.388_{mss4} \quad (10\text{-}14)$$

$$Y = -0.899_{mss1} + 0.428_{mss2} + 0.076_{mss3} - 0.041_{mss4} \quad (10\text{-}15)$$

$$N = -0.016_{mss1} + 0.131_{mss2} - 0.452_{mss3} + 0.882_{mss4} \quad (10\text{-}16)$$

Crist (1985) derived the visible, near-infrared, and middle-infrared coefficients for transforming Landsat Thematic Mapper imagery into brightness, greenness, and wetness variables:

$$B = 0.0243_{tm1} + 0.4158_{tm2} + 0.5524_{tm3} + 0.5741_{tm4} + 0.3124_{tm5} + 0.2303_{tm7} \quad (10\text{-}17)$$

$$G = -0.1603_{tm1} - 0.2819_{tm2} - 0.4939_{tm3} + 0.794_{tm4} - 0.0002_{tm5} - 0.1446_{tm7} \quad (10\text{-}18)$$

$$W = 0.0315_{tm1} + 0.2021_{tm2} + 0.3102_{tm3} + 0.1594_{tm4} - 0.6806_{tm5} - 0.6109_{tm7} \quad (10\text{-}19)$$

The derived wetness image makes special use of the Thematic Mapper middle-infrared bands (tm_5 and tm_7) and has been shown to be sensitive to plant moisture content. The thermal infrared band (tm_6) is not used in the calculation of the vegetation indices.

The Tasseled Cap transformation is a global vegetation index. Theoretically, it may be used anywhere in the world to disaggregate the amount of soil brightness, vegetation, and moisture content in individual pixels in a Landsat MSS or Thematic Mapper image. Practically, however, it is better to compute the coefficients based on local conditions. Jackson (1983) provides a computer program for this purpose.

Hardisky et al. (1983) found that an *Infrared Index* (II):

$$\text{II} = \frac{NIR_{TM4} - MidIR_{TM5}}{NIR_{TM4} + MidIR_{TM5}} \quad (10\text{-}20)$$

was more sensitive to changes in plant biomass and water stress than NDVI for wetland studies.

Richardson and Wiegand (1977) used the perpendicular distance to the "soil line" as an indicator of plant development. The "soil line," a two-dimensional analog of the Kauth-Thomas soil brightness index, was estimated by linear regression. The *Perpendicular Vegetation Index* (PVI) based on MSS band 4 data was:

$$PVI = \sqrt{(0.355_{mss4} - 0.149_{mss2})^2 + (0.355_{mss2} - 0.852_{mss4})^2} \quad (10\text{-}21)$$

Hay et al. (1979) proposed a vegetation index called *Greenness Above Bare Soil* (GRABS):

$$GRABS = G - 0.09178\,B + 5.58959 \quad . \quad (10\text{-}22)$$

The calculations were made using the Kauth-Thomas transformation greenness (*G*) and soil brightness index (*B*) applied to Sun-angle and haze-corrected Landsat MSS data. The resulting index is similar to the Kauth-Thomas greenness vegetation index since the contribution of soil brightness index (*B*) is less than 10 percent of GVI.

Rock et al. (1986) utilized a *Moisture Stress Index* (MSI):

$$\text{MSI} = \frac{MidIR_{TM5}}{NIR_{TM4}} \quad (10\text{-}23)$$

based on the Landsat Thematic Mapper near-infrared and middle-infrared bands.

Hunt et al. (1987) developed the *Leaf Water Content Index* (LWCI) to assess water stress in leaves:

$$\text{LWCI} = \frac{-\log[1 - (NIR_{TM4} - MidIR_{TM5_{ft}})]}{-\log[1 - NIR_{TM4} - MidIR_{TM5_{ft}}]} \quad (10\text{-}24)$$

where *ft* represents reflectance in the specified bands when leaves are at their maximum relative water content (RWC) defined as:

$$\text{RWC} = \frac{\text{field weight} - \text{oven dry weight}}{\text{turgid weight} - \text{oven dry weight}} \times 100 \quad (10\text{-}25)$$

Musick and Pelletier (1988) demonstrated a strong correlation between the *MidIR Index* and soil moisture:

Table 10-3.   Selected Remote Sensing Vegetation Indices.

Vegetation Index	Equation	References
Simple Ratio (SR)	$$SR = \frac{NIR}{Red}$$	Birth and McVey, 1968
Normalized Difference Vegetation Index (NDVI)	$$NDVI = \frac{NIR - red}{NIR + red}$$	Rouse et al., 1974; Deering et al., 1975
Kauth-Thomas Transformation Brightness Greenness Yellow stuff Non-such Brightness Greenness Wetness	Landsat Multispectral Scanner (MSS) $B = 0.332_{mss1} + 0.603_{mss2} + 0.675_{mss3} + 0.262_{mss4}$ $G = -0.283_{mss1} - 0.66_{mss2} + 0.577_{mss3} + 0.388_{mss4}$ $Y = -0.899_{mss1} + 0.428_{mss2} + 0.076_{mss3} - 0.041_{mss4}$ $N = -0.016_{mss1} + 0.131_{mss2} - 0.452_{mss3} + 0.882_{mss4}$ Landsat Thematic Mapper (TM) $B = 0.0243_{tm1} + 0.4158_{tm2} + 0.5524_{tm3} + 0.5741_{tm4} + 0.3124_{tm5} + 0.2303_{tm7}$ $G = -0.1603_{tm1} - 0.2819_{tm2} - 0.4939_{tm3} + 0.794_{tm4} - 0.0002_{tm5} - 0.1446_{tm7}$ $W = 0.0315_{tm1} + 0.2021_{tm2} + 0.3102_{tm3} + 0.1594_{tm4} - 0.6806_{tm5} - 0.6109_{tm7}$	Kauth and Thomas, 1976; Kauth et al., 1979 Crist, 1985
Infrared Index (II)	$$II = \frac{NIR_{TM4} - MidIR_{TM5}}{NIR_{TM4} + MidIR_{TM5}}$$	Hardisky et al., 1983
Perpendicular Vegetation Index (PVI)	$$PVI = \sqrt{(0.355_{mss4} - 0.149_{mss2})^2 + (0.355_{mss2} - 0.852_{mss4})^2}$$	Richardson and Wiegand, 1977
Greenness Above Bare Soil (GRABS)	$$GRABS = G - 0.09178B + 5.58959$$	Hay et al., 1979;
Moisture Stress Index (MSI)	$$MSI = \frac{MidIR_{TM5}}{NIR_{TM4}}$$	Rock et al., 1986
Leaf Relative Water Content Index (LWCI)	$$LWCI = \frac{-\log[1 - (NIR_{TM4} - MidIR_{TM5_{ft}})]}{-\log[1 - NIR_{TM4} - MidIR_{TM5_{ft}}]}$$	Hunt et al., 1987
MidIR Index	$$MidIR = \frac{MidIR_{TM5}}{NIR_{TM7}}$$	Musick and Pelletier, 1988
Soil Adjusted Vegetation Index (SAVI) and Modified SAVI (MSAVI)	$$SAVI = \frac{(1 + L)(NIR - red)}{NIR + red + L}$$	Huete, 1988; Huete and Liu, 1994; Running et al., 1994; Qi et al., 1995
Atmospherically Resistant Vegetation Index (ARVI)	$$ARVI = \left(\frac{p^*_{nir} - p^*_{rb}}{p^*_{nir} + p^*_{rb}}\right)$$	Kaufman and Tanre, 1992; Huete and Liu, 1994
Soil and Atmospherically Resistant Vegetation Index (SARVI)	$$SARVI = \frac{p^*_{nir} - p^*_{rb}}{p^*_{nir} + p^*_{rb} + L}$$	Huete and Liu, 1994; Running et al., 1994
Enhanced Vegetation Index (EVI)	$$EVI = \frac{p^*_{nir} - p^*_{red}}{p^*_{nir} + C_1 p^*_{red} - C_2 p^*_{blue} + L}(1 + L)$$	Huete and Justice, 1999

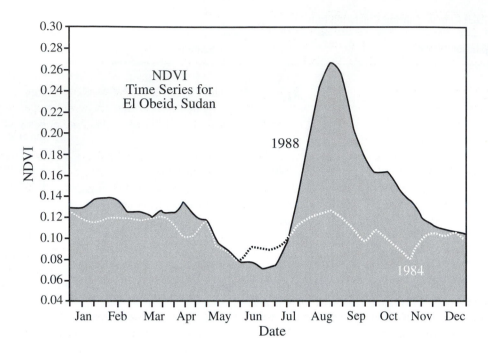

Figure 10-21 A time series of 1984 and 1988 NDVI measurements for the region around El Obeid, Sudan, in sub-Saharan Africa. The NDVI measurements were derived by analyzing AVHRR global area coverage (GAC) data (after Hutchinson, 1991).

$$MidIR = \frac{MidIR_{TM5}}{NIR_{TM7}}. \qquad (10\text{-}26)$$

The utility of the normalized difference vegetation index (NDVI) and related indices for satellite and airborne assessment of the Earth's vegetation cover has been demonstrated for almost three decades. The time series analysis of seasonal NDVI data have provided a method of estimating net primary production over varying biome types, of monitoring phenological patterns of the Earth's vegetated surface, and of assessing the length of the growing season and dry-down periods (Huete and Liu, 1994; Ramsey et al., 1995). For example, Figure 10-21 documents the NDVI for the sub-Saharan African region around El Obeid, Sudan, for 1984 and 1988 (Hutchinson, 1991). The NDVI were computed using the global area coverage (approximately 4 x 4 km) Advanced Very High Resolution Radiometer (AVHRR) data. The 1988 growing season was the best agricultural season in this region for 20 years.

Global vegetation analysis was initially based on linearly regressing NDVI values (derived from AVHRR, Landsat MSS, Landsat TM, and SPOT HRV data) with *in situ* measurements of leaf-area-index (LAI), absorbed photosynthetically active radiation (APAR), percent cover, and/or biomass. This empirical approach revolutionized global sci-

ence land-cover biophysical analysis in just one decade (Running et al., 1994). Unfortunately, studies have found that the empirically derived NDVI products can be unstable, varying with soil color and moisture conditions, bidirectional reflectance distribution function (BRDF) effects previously discussed, atmospheric conditions, and the presence of dead material in the canopy itself (Qi et al., 1995). For example, Goward et al. (1991) found errors of ±50 percent in NDVI images used for global vegetation studies derived from the NOAA Global Vegetation Index product. What is needed are globally accurate NDVI related products that do not need to be calibrated by *in situ* measurement within each geographic area yet will remain constant under changing atmospheric and soil background conditions (Huete and Justice, 1999).

Therefore, it is not surprising that emphasis has been given to the development of improved vegetation indices that will take advantage of calibrated hyperspectral sensor systems such as the Moderate Resolution Imaging Spectrometer (Running et al., 1994). Although the NDVI has been shown to be useful in estimating vegetation properties, many important external and internal influences restrict its global utility. The improved indices incorporate a soil adjustment factor and/or a blue band for atmospheric normalization.

The *Soil Adjusted Vegetation Index* (SAVI) introduces a soil calibration factor, *L*, to the NDVI equation to minimize soil background influences resulting from first-order soil-plant spectral interactions (Huete, 1988; Huete et al., 1992):

$$SAVI = \frac{(1+L)(NIR - red)}{NIR + red + L}.$$ (10-27)

An *L* value of 0.5 in reflectance space was found to minimize soil brightness variations and eliminate the need for additional calibration for different soils (Huete and Liu, 1994). The utility of SAVI for minimizing the soil "noise" inherent in the NDVI has been corroborated in several studies (Bausch, 1993). Qi et al. (1995) developed a modified SAVI called MSAVI that used an iterative, continuous *L* function to optimize soil-adjustment and increase the dynamic range of SAVI.

SAVI was made less sensitive to atmospheric effects by normalizing the radiance in the blue, red, and near-infrared bands. This became the *Atmospherically Resistant Vegetation Index* (ARVI):

$$ARVI = \left(\frac{p^*_{nir} - p^*_{rb}}{p^*_{nir} + p^*_{rb}} \right)$$ (10-28)

where

$$p^*_{rb} = p^*_{red} - \gamma(p^*_{blue} - p^*_{red}).$$ (10-29)

The technique requires prior correction for molecular scattering and ozone absorption of the blue, red, and near-infrared remote sensor data, hence the term p^*. ARVI uses the difference in the radiance between the blue channel and the red channel to correct the radiance in the red channel and thus reduce atmospheric effects. Unless the aerosol model is known *a priori*, gamma (γ) is normally equal to 1.0 to minimize atmospheric effects. Kaufman and Tanre (1992) provide guidelines where different gammas might be used over continental, maritime, desert (e.g., the Sahel), or heavily vegetated areas.

Huete and Liu (1994) integrated the *L* function from SAVI and the blue-band normalization in ARVI to derive a *Soil and Atmospherically Resistant Vegetation Index* (SARVI) that corrects for both soil and atmospheric noise, as would a modified SARVI (MSARVI):

$$SARVI = \frac{p^*_{nir} - p^*_{rb}}{p^*_{nir} + p^*_{rb} + L}$$ (10-30)

and

$$MSARVI = \frac{2\,p^*_{nir} + 1 - \sqrt{\left[(2\,p^*_{nir} + 1)^2 - \gamma(p^*_{nir} - p^*_{rb}) \right]}}{2}$$ (10-31)

Huete and Liu (1994) performed a sensitivity analysis on the original NDVI and improved vegetation indices (SAVI, ARVI, SARVI, MSARVI) and concluded that:

- If there were a total atmospheric correction then there would mainly be "soil noise" and the SAVI and MSARVI would be the best equation to use and the NDVI and ARVI would be the worst;

- If there were a partial atmospheric correction to remove the Rayleigh and ozone components, then the best vegetation indices would be the SARVI and MSARVI, with the NDVI and ARVI being the worst; and

- If there were no atmospheric correction at all, i.e., no Rayleigh, ozone, or aerosol correction, the SARVI would become slightly worse but still would have the least overall noise. The NDVI and ARVI would have the most noise and error.

The MODIS Land Discipline Group recently proposed the *Enhanced Vegetation Index* (EVI) for use with MODIS data:

$$EVI = \frac{p^*_{nir} - p^*_{red}}{p^*_{nir} + C_1 p^*_{red} - C_2 p^*_{blue} + L}(1 + L)$$ (10-32)

The EVI is a modified NDVI with a soil adjustment factor, *L*, and two coefficients, C_1 and C_2, which describe the use of the blue band in correction of the red band for atmospheric aerosol scattering. The coefficients, C_1, C_2, and *L*, are empirically determined as 6.0, 7.5, and 1.0, respectively. This algorithm has improved sensitivity to high biomass regions and improved vegetation monitoring through a de-coupling of the canopy background signal and a reduction in atmospheric influences (Huete and Justice, 1999).

Scientists throughout the world are studying the role of terrestrial vegetation in large-scale global processes. This is necessary in order to understand how the Earth functions as a system. The more rigorous remote sensing derived vegetation indices may eventually be used to accurately inventory the global distribution of vegetation types as well as their biophysical (e.g., LAI, biomass, APAR) and structural properties (e.g., percent canopy closure). Monitoring these characteristics through space and time will provide valuable information for understanding the Earth as a system.

Landscape Ecology Metrics

Vegetation indices are useful for monitoring the condition and health of vegetated pixels. However, such per-pixel analysis does not provide any information about the nature of surrounding pixels, including their site and association characteristics. Landscape ecology principles have been developed that increasingly incorporate remote sensor data to assess the health and diversity of vegetation and other variables within entire ecosystems. This has resulted in the development of numerous *landscape metrics* or *indicators* that are of significant value when analyzing rangeland, grassland, forests, and wetland. Numerous government agencies, such as the Environmental Protection Agency, base much of their environmental modeling and landscape characterization on these metrics and indicators (EPA, 1994; 1995; Jones et al., 1998). Therefore, it is useful to provide a brief review of their origin and summarize several of the more important landscape ecology metric variables that can be extracted from remotely sensed data.

The term *landscape ecology* was first introduced by the German geographer Carl Troll (1939) who made widespread use of the then new technique of aerial photography. Troll intended for the term landscape ecology to distinguish his approach for using such imagery to interpret the interaction of water, land surfaces, soil, vegetation, and land use from that of conventional photographic interpretation and cartography (Golley, 1993). Landscape ecology has been intensively practiced in Europe for many decades and became generally recognized in the United States in about 1980. Since then, landscape ecology has rapidly evolved as a discipline, spurred by the synergistic interactions between remote sensing and GIS techniques and advances in ecological theory (Golley, 1993; EPA, 1994).

Landscape ecology is the study of the structure, function, and changes in heterogeneous land areas composed of interacting organisms (Bourgeron and Jensen, 1993). It is the study of the interaction between landscape patterns and ecological processes, especially the influence of landscape pattern on the flows of water, energy, nutrients, and biota (Turner, 1989). What distinguishes landscape ecology from the many separate disciplines that it embraces (e.g., geography, biology, ecology, hydrology) is that it provides a hierarchical framework for interpreting ecological structure, function, change, and resiliency at multiple scales of inquiry.

Traditional measures to protect the environment, such as preventing water pollution or protecting biodiversity, often

focused on specific effluent discharges or fine-scale habitat requirements. This method has been described as the fine-filter approach. In contrast, the coarse-filter approach to resource conservation states that "by managing aggregates (e.g., communities, ecosystems, landscapes), the components of these aggregates will be managed as well" (Bourgeron and Jensen, 1993). In other words, the most cost-effective strategy to maintain the resiliency and productivity of ecological systems is to conserve (or restore) the diversity of species, ecosystem processes, and landscape patterns that create the systems (Jensen and Everett, 1993; Bourgeron and Jensen, 1993). Applying this coarse-filter management method requires that landscape patterns be evaluated at multiple spatial and temporal scales rather than simply at the traditional scales of stream reach or forest stand.

Hierarchy theory allows landscape ecologists to integrate multiple scales of information to determine whether landscape patterns are sufficient to allow ecological processes to operate at the necessary scales. The objective is to investigate changes in the distribution, dominance, and connectivity of ecosystem components and the effect of these changes on ecological and biological resources (Forman, 1995; Golley, 1989). For example, ecosystem fragmentation has been implicated in the decline of biological diversity and ecosystem sustainability at a number of spatial scales (Forman, 1995; Jones et al., 1998). Determining status and trends in the pattern of landscapes is critical to understanding the overall condition of ecological resources. Landscape patterns thus provide a set of indicators (e.g., pattern shape, dominance, connectivity, configuration) that can be used to assess ecological status and trends at a variety of scales.

A hierarchical framework also permits two important types of comparisons: 1) to compare conditions within and across landscapes, and 2) to compare conditions across different types of ecological risks. Such ecological risks include the risk of erosion, loss of soil productivity, loss of hydrologic function, and loss of biodiversity.

Scalable units are needed to address landscape ecology issues at multiple scales within a hierarchical framework. Examples of scalable units include patches, patterns, and landscapes. A *patch unit* is a set of contiguous measurement units (e.g., pixels) that have the same numerical value. A *pattern unit* is a collection of measurement units and/or patch units that have the property of being the minimum unit descriptor of a larger spatial area. The scales of assessment questions and indicators suggest two types of landscape units: watersheds and landscape pattern types (LPT) (Wickham and Norton, 1994). Watersheds and LPTs capture or bound four important flow processes operating within and among landscapes: flows of energy, water, nutrients, and biota. Scales of watersheds and LPTs range from approximately 10^3 to 10^6 units in extent, and from 1 to 100 ha.

Landscape Indicators and Patch Metrics

Jones et al. (1996; 1998) and the Environmental Protection Agency suggest that landscape integrity can be monitored by carefully watching the status of the following indicators:

- land-cover composition and pattern,

- riparian extent and distribution,

- ground water,

- greenness pattern,

- degree of biophysical constraints, and

- erosion potential.

Monitoring these landscape indicators requires precise, repeatable measurements of terrain *patches* such as individual forest stands, rangeland, wetland, and/or agricultural fields (Table 10-4). It is also important to identify patches of pure urban structure such as residential and commercial land use. These measurements of terrain patches are routinely referred to as *landscape pattern* and *structure metrics*.

Numerous landscape structure metrics have been developed (e.g., Turner, 1989; Ritters et al., 1995; Schuft et al., 1999). O'Neill et al. (1997) suggest that the health of an ecosystem could be monitored if the following three metrics (indices) were monitored through time: dominance, contagion, and fractal dimension.

Dominance, D, is the information theoretic index that identifies the extent to which the landscape is dominated by a single land-cover type. The metric, $0 < D < 1$, is computed as:

$$D = 1 - \left[\sum_k \frac{(-P_k \cdot \ln P_k)}{\ln(n)} \right] \qquad (10\text{-}33)$$

where $0 < P_k < 1$ is the proportion of land-cover type k, and n is the total number of land-cover types present in the landscape.

Contagion, C, expresses the probability that land cover is more "clumped" than the random expectation. The index, $0 < C < 1$, is:

$$C = 1 - \left[\sum_i \sum_j \frac{(-P_{ij} \cdot \ln P_{ij})}{2\ln(n)} \right] \qquad (10\text{-}34)$$

where P_{ij} is the probability that a pixel of cover type i is adjacent to type j.

The *fractal dimension, F,* of patches indicates the extent of human reshaping of landscape structure (O'Neill et al., 1988; 1997). Humans create simple landscape patterns; nature creates complex patterns. The fractal dimension index is calculated by regressing the log of the patch perimeter against the log of the patch area for each patch on the landscape. The index equals twice the slope of the regression line. Patches of four or fewer pixels are excluded because resolution problems distort their true shape.

O'Neill et al. (1997) suggest that this set of three indices may capture fundamental aspects of landscape pattern that influence ecological processes. Significant changes in these indices for an ecosystem might indicate that perhaps deleterious processes are at work in the environment. For example, consider a small ecosystem that exhibits a less modified fractal dimension, is highly clumped, and has relatively few land-cover types within it. It might appear in a three-dimensional landscape metric space at location *a* in Figure 10-22. If this small ecosystem were subdivided with several new roads and fragmented, its location might move in three-space to *b* with many land-cover types being introduced (dominance change), be less clumped (contagion), with a more modified fractal dimension. This could be good or bad. In fact, the relationships between the metric values and how they actually relate to ecological principles are still being determined. *Remote sensing of vegetation within these patches is very important and constitutes one of the major factors responsible for whether or not the metrics are robust and useful for ecological modeling.*

Ritters et al. (1995) reviewed 55 patch metrics and concluded that the following metrics accounted for most of the variance, including: the number of attribute cover types in a region (*n*); contagion (previously discussed); average perimeter-area ratio; perimeter-area scaling; average large-patch density-area scaling; standardized patch shape; and patch perimeter-area scaling. The algorithms for all 55 metrics are provided in the paper.

Applying the principles of landscape ecology requires an understanding of the natural variability of landscape patterns and processes across both space and time. Estimates of this variability are essential to determining whether the current condition of landscape is sustainable, given its historic patterns and processes (Jensen and Everett, 1993). Moreover,

Remote Sensing of Vegetation Change

Analyzing an individual date of remote sensor data to extract meaningful vegetation biophysical information is often of value. However, to appreciate the dynamics of the ecosystem, it is necessary to monitor the vegetation through time and determine what changes in succession are taking place. Relatively high temporal resolution satellite data is often of value when conducting such successional studies. The following case study documents the usefulness of SPOT remote sensor data and the methods used to monitor wetland change in a freshwater reservoir.

Remote Sensing Inland Wetland Successional Changes

Inland wetlands assimilate pollutants, control floods, and serve as breeding, nursery, and feeding grounds for fish and wildlife. Unfortunately, the conterminous United States lost 53 percent of its wetlands to agricultural, residential, and/or commercial land use from the 1780s to 1980s (Dahl, 1990). Accurate and timely information on wetland distribution is essential for effective protection and management. This case study documents the use of multiple-season and multiple-year SPOT satellite multispectral data for monitoring inland wetland successional changes in Par Pond on the Savannah River Site, SC, caused by the draw-down of the reservoir's water level from June 1991 through October 1994 (Jensen et al., 1997). The study brings to bear many of the considerations discussed in this chapter.

Par Pond Reservoir Study Area

Par Pond is a 1012-ha reservoir originally created in 1958 to receive reactor cooling water, but now receives flow-through water from the Savannah River. Par Pond has a stable, predictable cattail and waterlily macrophyte community when water levels fluctuate < 1 m per year.

Approximately 30 macrophyte plant species are found in Par Pond, but only a few dominate:

- Cattails (*Typha* spp.) are persistent emergent macrophytes that exist year-round in Par Pond. They begin to "green up" in early to mid-April and often form a dense, green canopy by late May, as documented in the phenological calendar shown in Figure 10-23. Cattails senesce in late September or early October, remaining brown throughout

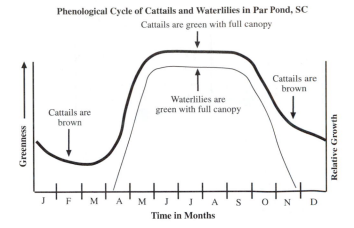

Phenological Cycle of Cattails and Waterlilies in Par Pond, SC

Figure 10-23 Annual phenological cycle of inland wetland cattails and waterlilies in Par Pond on the Savannah River Site in South Carolina. Cattails are persistent emergent marsh vegetation that green up earlier than waterlilies in late March and early April. They begin senescing in late October and lie dormant over the winter. Waterlilies disappear completely by November each year and reappear in late April. Both cattail and waterlily have a full, dense canopy throughout the summer months.

the winter months. Cattails are normally found in shallow water < 1 m in depth adjacent to the reservoir shoreline.

- Waterlilies (*Nymphaea odorata*) are dominant floating macrophytes that do not persist through the winter (Figure 10-23). They appear at the outer edge of the cattails by late April or early May and reach full growth 6 – 8 weeks later. Waterlily beds expand into the open-water areas of the reservoir through September and typically extend to depths of 2 – 5 m. The beds persist until mid-October to mid-November and then disappear.

Before June, 1991, the water level of Par Pond was relatively stable at 200 ft above mean sea level (msl), fluctuating < 0.15 m in most years. Extensive beds of cattails and waterlilies from 20 – 150 m in width extended out into the lake. From June 23, 1991, through September 15, 1991, the water level of Par Pond was lowered to 181 ft above msl for reservoir repair. The 19 ft (approximately 6 m) drop in water level turned a submerged zone along the shoreline about 50 – 250 m in width into upland. This exposed the emergent and non-emergent aquatic macrophyte beds of the Par Pond shoreline to drying conditions for about four years, until the water level was raised to 200 ft once again in 1995.

Table 10-5. SPOT HRV Images Used to Document the Vegetation Succession of Par Pond

Date	SPOT Data Type	Reference Color-Infrared Aerial Photography
March 14, 1992	HRV-XS	January 31, 1992
May 5, 1992	HRV-XS	May 2, 1992
October 13, 1992	HRV-XS	August 10, 1992
May 10, 1993	HRV-XS	April 23, 1993
October 28, 1993	HRV-XS	January 22, 1994
April 2, 1994	HRV-XS	April 18, 1994
October 25, 1994	HRV-XS	

Remote Sensor Data

Seven SPOT multispectral (XS) images acquired in the spring and fall of 1992, 1993, and 1994 were analyzed (Table 10-5). Near-anniversary dates were selected whenever possible to minimize changes in reflectance caused by Sun angle and seasonal soil moisture (Jensen et al., 1995). Color-infrared aerial photography was also available for this experiment. SPOT image-to-map rectification was based on the use of approximately 50 ground-control points per date and nearest-neighbor resampling to a Universal Transverse Mercator projection. Total geometric root-mean-square-error (RMSE) was $\pm$ 0.5 pixel ($\pm$ 20 m) for each date of imagery. All terrain > 200 ft above msl was masked out of the study area.

Image Classification and Change Detection

Fifty spectrally unique clusters were extracted from each date of imagery using an Interactive Self Organizing Data Analysis Technique — ISODATA (refer to Jensen, 1996). The clusters were labeled to produce seven land-cover maps. A large-scale version of the land-cover classification of the October 25, 1994, SPOT image is shown in Color Plate 10-6. The overall accuracy for two of the classification maps (October 28, 1993, and April 2, 1994) was 75.6 percent and 77.9 percent, respectively (Huang, 1996). This level of accuracy is about what is expected from simple three-band 20 x 20 m SPOT multispectral data (Jensen et al., 1995).

The goal of this study was to document the ecological change in vegetation cover in the draw-down areas using multiple dates of remote sensor data. This required the use of a change-detection algorithm that provided "from-to" map information, i.e., a 20 x 20 m pixel of Dead Vegetation in 1992 changed into a pixel of Spikerush by 1993. Twenty-one change detection maps were produced and may be seen in Jensen at al. (1997).

Results

The percentage of land-cover in each category on each date from March 14, 1992, through October 25, 1994, for the Par Pond study area are graphed in Figure 10-24a. This information documents the four year vegetation succession trends on the Par Pond exposed shoreline. On March 14, 1992, Dead Vegetation occupied approximately 35.48 percent of the study area. The cattail and waterlily aquatic macrophytes did not survive the draw-down. Also present were 7.46 percent Bare Soil and 8.74 percent early invasion by Spikerush. As time passed, the Dead Vegetation areas were colonized by other species. By October 25, 1994, no Dead Vegetation was present.

Two major types of successional development took place on the Dead Vegetation and Bare Soil draw-down areas. First, there was a significant increase in Spikerush in 1992. By May 5, 1992, Spikerush occupied 20.28 percent of the study area. As the soil dried out in subsequent years, Spikerush coverage declined to about 10.75 percent by October 25, 1994.

Spikerush and Bare Soil were succeeded primarily by Old Field species as expected, and at a very rapid rate. The most dramatic increase was from May 10, 1993 to October 28, 1993, which saw Old Field land cover increase from almost nothing to 25.62 percent of the study area. Pine/Hardwood seedling coverage also increased steadily from almost no cover in 1992 to 10.25 percent by October 25, 1994. If the draw-down areas were unmolested for approximately 75 years, oak/hickory climax vegetation would be present in all the draw-down areas.

Using the land-cover statistics derived from multiple-date remote sensor data, it was possible to develop predictive successional models for each of the land-cover vegetation types in the draw-down area. For example, Figure 10-24b depicts the actual and predicted percent of the scene occupied by Spikerush at different dates since the initiation of draw-down on June 23, 1991. The Spikerush succession was modeled using a fifth-order polynomial where

$$\%Spikerush = -0.0129x^5 + 0.4018x^4 - 4.0989x^3 + 15.383x^2 - 14.425x,$$

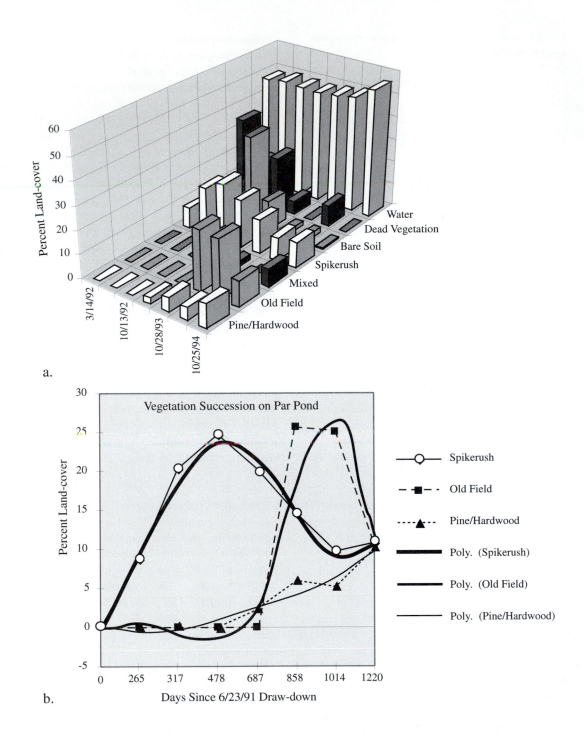

Figure 10-24 a) Perspective view of the percentage of land cover in Par Pond from March 14, 1992, through October 25, 1994, derived from analysis of seven dates of SPOT HRV XS data. Extensive areas of dead vegetation were present in early 1992. Spikerush dominated in late 1992 and early 1993. Old field natural succession dominated in late 1993 and throughout 1994. Pine and hardwood seedlings were colonizing extensively by October 25, 1994. b) The percent land-cover of spikerush, old field, and pine/hardwood vegetation on draw-down areas of Par Pond from June 23, 1991, through October 25, 1994, are presented as raw data and modeled with the use of fifth-order polynomial equations (Jensen et al., 1997).

with $R^2 = 0.992$. Using this equation it was possible to predict the percent of Spikerush that would be present at user-specified dates after a serious reservoir draw-down. Similarly, Old Field and Pine/Hardwood land-cover succession were modeled using the equations

%Old Field = $-0.0416x^5 + 0.6421x^4 - 3.0604x^3 + 5.1495x^2 - 2.3275x$, and

%Pine/Hardwood = $0.00762x^5 - 0.1543x^4 + 1.0971x^3 - 2.9539x^2 + 2.4909x$.

The polynomial model for Old Field ($R^2 = 0.89$) did not explain the succession as well as the Pine/Hardwood predictive model ($R^2 = 0.94$). The first derivative of these models can be used to predict the rate-of-change (gradient) in percent of land cover on a specific day since draw-down.

The analysis of the land-cover and change-detection maps derived from satellite remote sensor data quantitatively documented 1) the spatial distribution of the successional changes in land cover, and 2) the rate of successional change in Par Pond caused by the draw-down.

Numerous state and government agencies recognize the importance of utilizing remote sensor data for monitoring their sensitive vegetation resources. For example, the first comprehensive inventory of the vegetation at the plant community level of the Everglades in Florida was recently completed using a combination of satellite imagery, aerial photography, GPS, and GIS technologies (Doren et al., 1999; Welch et al., 1999).

References

Abuelgasim, A. A., S. Gopal and A. H. Strahler, 1998, "Forward and Reverse Modeling of Canopy Directional Reflectance Using a Neural Network," *Remote Sensing of Environment*, 59:453–471.

Bausch, W. C., 1993, "Soil Background Effects on Reflectance-based Crop Coefficients for Corn," *Remote Sensing of Environment*, 46: 1–10.

Birth, G. S. and G. McVey, 1968, "Measuring the Color of Growing Turf with a Reflectance Spectrophotometer," *Agronomy Journal*, 60:640–643.

Bourgeron, P. S. and M. E. Jensen, 1993, "An Overview of Ecological Principles for Ecosystem Management," Jensen, M. E., and R. Everett, 1993, "An Overview of Ecosystem Management Principles, in Jensen, M. E., and P. S. Bourgeron (eds.), *Eastside Forest Ecosystem Health Assessment*, Vol. II, *Ecosystem Management: Principles and Applications*, Missoula, MN: USDA Forest Service, 410–60.

Byrd, D., 1998, Correspondence with water superintendent and agricultural biologist at the Imperial County Agricultural Commissioners Office, Imperial Valley, CA.

Carter, G. A., 1991, "Primary and Secondary Effects of the Water Content on the Spectral Reflectance of Leaves," *American Journal of Botany,* 78(7);916–924.

Carter, G. A., 1993, "Responses of Leaf Spectral Reflectance to Plant Stress," *American Journal of Botany*, 80(3):2310–243.

Carter, G. A., W. G. Cibula and R. L. Miller, 1996, "Narrow-band Reflectance Imagery Compared with Thermal Imagery for Early Detection of Plant Stress," *Journal of Plant Physiology*, 148:515–522.

Chen, T. M. and T. A. Black, 1992, "Defining Leaf Area Index for Non-flat Leaves," *Plant, Cell and Environment*, 15:421–429.

Cibula, W. G. and G. A. Carter, 1992, "Identification of a Far-Red Reflectance Response to Ectomycorrhizae in Slash Pine," *International Journal of Remote Sensing*, 13(5):925–932.

Clark, R. N., T. V. King, C. Ager and G. A. Swayze, 1995, "Initial Vegetation Species and Senescence/Stress Indicator Mapping in the San Luis Valley, Colorado, Using Imaging Spectrometer Data," *Proceedings*, Summitville Forum 1995, H. H. Posey, J. A Pendelton, and D. Van Zyl, (Eds.), Colorado Geological Survey Special Publication 38, 64-69.; http://speclab.cr.usgs.gov.

Clark, R. N., G. A. Swayze, A. Gallagher, T. V. King and W. M. Calvin, 1993, *The U.S. Geological Survey, Digital Spectral Library 0.2 to 3.0 μm*; U.S. Geological Survey Open File Report 93-592, 1340 pp.

Cohen, W. B., 1991, "Response of Vegetation Indices to Changes in Three Measures of Leaf Water Stress," *Photogrammetric Engineering & Remote Sensing*, 57(2):195–202.

Crist, E. P., 1985, "A Thematic Mapper Tasseled Cap Equivalent for Reflectance Factor Data," *Remote Sensing of Environment*, 17:301–306.

Curran, P. J., 1983, "Estimating Green LAI from Multispectral Aerial Photography," *Photogrammetric Engineering and Remote Sensing*, 49:1709–1720.

Dahl, T. W., 1990, *Wetlands Losses in the United States 1780s to 1980s*. Washington:U.S. Department of Interior and U.S. Fish and Wildlife Service.

Dame, R. F. and P. D. Kenny, 1986, "Variability of *Spartina Alterniflora* Primary Production in the Euhaline North Inlet Estuary," *Marine Ecology*, 32:71–80.

Danson, F. M., 1998, "Teaching the Physical Principles of Vegetation Canopy Reflectance Using the SAIL Model," *Photogrammetric Engineering & Remote Sensing*, 64(8):8010–812.

Deering, D. W., J. W. Rouse, R. H. Haas and J. A. Schell, 1975, "Measuring Forage Production of Grazing Units from Landsat MSS Data," *Proceedings*, Tenth International Symposium on Remote Sensing of Environment, Ann Arbor: ERIM, 2:1169–1178.

Doren, R. F., R. Rutchey and R. Welch, 1999, "The Everglades: A Perspective on the Requirements and Applications for Vegetation Map and Database Products," *Photogrammetric Engineering and Remote Sensing*, 65(2):155–161.

Edwards, T. C., C. G. Homer, S. C. Bassett, A. Falconer, R. D. Ramsey and D. W. Wight, 1995, *Utah Gap Analysis: An Environmental Information System*, Logan: Utah Cooperative Fish & Wildlife Research Unit, 68 pp.

EPA, 1994, *Landscape Monitoring and Assessment Research Plan*, Washington, DC: EPA, EPA/620/R-94/009, 53 pp.

EPA, 1995, *Mid-Atlantic Landscape Indicators Project Plan*, Washington, DC: EPA, EPA/620/R-95/003, 37 pp.

Farabee, M. J., 1997, Photosynthesis, http://gened.emc.maricopa.edu/bio/bio181/BIOBK/BioBookPS.html

Forman, R. T. T., 1995, "Some General Principles of Landscape and Regional Ecology," *Landscape Ecology*, 10(3):133–142.

Frohn, R. C., 1998, *Remote Sensing for Landscape Ecology*, Boca Raton, FL: Lewis Publishers, 99 pp.

Gates, D. M., J. J. Keegan, J. C. Schleter and V. R. Weidner, 1965, "Spectral Properties of Plants," *Applied Optics*, 4(1):11–20.

Gausmann, H. W., W. A. Allen and R. Cardenas, 1969, "Reflectance of Cotton Leaves and their Structure," *Remote Sensing of Environment*, 1:110–22.

Goel, N. S., 1988, "Models of Vegetation Canopy Reflectance and their Use in Estimation of Biophysical Parameters from Reflectance Data," *Remote Sensing Reviews*, 4:1–212.

Goetz, A. F. H., 1995, "Imaging Spectrometry for Remote Sensing: Vision to Reality in 15 Years," *Proceedings*, SPIE International Society for Optical Engineers, 2480:12 pp.

Golley, F. B., 1989, Landscape Ecology and Biological Conservation, *Landscape Ecology*, 2:201–202.

Golley, F. B., 1993, "Development of Landscape Ecology and its Relationship to Environmental Management," Jensen, M. E., and P. S. Bourgeron (eds.), *Eastside Forest Ecosystem Health Assessment*, Vol. II, Missoula: USDA Forest Service, 37–44.

Goward, S. N., B. Markham, D. G. Dye, W. Dulaney and J. Yang, 1991, "Normalized Difference Vegetation Index Measurements from the Advanced Very High Resolution Radiometer," *Remote Sensing of Environment*, 35:257–277.

GRSU, 1999, *Phenological Calendars of Selected Crops*, Santa Barbara: Geography Remote Sensing Unit.

Haack, B. and S. Jampoler, 1994, "Agricultural Classification Comparisons Using Landsat Thematic Mapper Data," *ITC Journal*, 1994(2):113–118.

Haack, B. and S. Jampoler, 1995, "Colour Composite Comparisons for Agricultural Assessments," *International Journal of Remote Sensing*, 16(9):15810–1598.

Hardisky, M. A., V. Klemas and M. Smart, 1983, "The Influence of Soil Salinity, Growth Form, and Leaf Moisture on the Spectral Radiance of Spartina Alterniflora Canopies," *Photogrammetric Engineering & Remote Sensing*, 49:77–83.

Huang, X., 1996, *A Machine Learning Approach to Automated Construction of Knowledge Bases for Expert Systems for Remote Sensing Image Analysis with GIS Data*, unpublished Ph.D. dissertation, Columbia: University of South Carolina.

Huete, A. R., 1988, "A Soil-adjusted Vegetation Index (SAVI)," *Remote Sensing of Environment*, 25:295–309.

Huete, A. R., G. Hua, J. Qi, A. Chehbouni and W. J. Van Leeuwem, 1992, "Normalization of Multidirectional Red and Near-Infrared Reflectances with the SAVI," *Remote Sensing of Environment*, 40:1–20.

Huete, A. and C. Justice, 1999, *MODIS Vegetation Index (MOD 13) Algorithm Theoretical Basis Document*, Greenbelt: NASA Goddard Space Flight Center, http://modarch.gsfc.nasa.gov/MODIS/LAND/#vegetation-indices, 129 pp.

Huete, A. F. and H. Q. Liu, 1994, "An Error and Sensitivity Analysis of the Atmospheric- and Soil-Correcting Variants of the Normalized Difference Vegetation Index for the MODIS-EOS," *IEEE Transactions on Geoscience and Remote Sensing*, 32(4):897–905.

Hunt, E. R., B. N. Rock and P. S. Nobel, 1987, "Measurement of Leaf Relative Water Content by Infrared Reflectance," *Remote Sensing of Environment*, 22:4210–435.

Hutchinson, C. F., 1991, "Uses of Satellite Data for Famine Early Warning in sub-Saharan Africa," *International Journal of Remote Sensing*, 12(6):1405–1421.

Jackson, R. D., 1983, "Spectral Indices in *n*-Space," *Remote Sensing of Environment*, 13:409–421.

Jacquemoud, S., F. Baret, B. Andrieu, F. M. Danson and K. W. Jaggard, 1995, "Extraction of Vegetation Biophysical Parameters by Inversion of PROSPECT + SAIL Models on Sugar Beet Canopy Reflectance Data: Application to TM and AVIRIS Sensors," *Remote Sensing of Environment*, 52:163–172.

Jensen, J. R., 1983, "Biophysical Remote Sensing," *Annals*, Association of American Geographers, 73(1):111–132.

Jensen, J. R., 1996, *Introductory Digital Image Processing: A Remote Sensing Perspective*, Upper Saddle River, NJ: Prentice-Hall, Inc., 319 pp.

Jensen, J. R. and M. E. Hodgson, 1985, "Remote Sensing Forest Biomass: An Evaluation Using High Resolution Remote Sensor Data and Loblolly Pine Plots," *Professional Geographer*, 37(1):46–56.

Jensen, J. R., C. Coombs, D. Porter, B. Jones, S. Schill and D. White, 1998, "Extraction of Smooth Cordgrass (Spartina alterniflora) Biomass and Leaf Area Index Parameters from High Resolution Imagery," *Geocarto International*, 13(4):25–46.

Jensen, J. R., X. Huang and H. E. Mackey, 1997, "Remote Sensing of Sucessional Changes in Wetland Vegetation as Monitored During a Four-Year Drawdown of a Former Cooling Lake," *Applied Geographic Studies*, 1(1):31–44.

Jensen, J. R., D. J. Cowen, J. D. Althausen, S. Narumalani and O. Weatherbee, 1993, "An Evaluation of the CoastWatch Change Detection Protocol in South Carolina," *Photogrammetric Engineering and Remote Sensing*, 59(6): 1039–1046.

Jensen, J. R., K. Rutchey, M. S. Koch and S. Narumalani, 1995, "Inland Wetland Change Detection in the Everglades Water Conservation Area 2A Using a Time Series of Normalized Remotely Sensed Data," *Photogrammetric Engineering & Remote Sensing*, 61(2): 199–209.

Jensen, M. E. and P. S. Bourgeron (eds.), 1993, *Eastside Forest Ecosystem Health Assessment,* Vol. II, *Ecosystem Management: Principles and Applications*, Missoula, MT: USDA Forest Service, PNW-GTR-318.

Jensen, M. E. and R. Everett, 1993, "An Overview of Ecosystem Management Principles," in Jensen, M. E., and P. S. Bourgeron (eds.), *Eastside Forest Ecosystem Health Assessment,* Vol. II, *Ecosystem Management: Principles and Applications*, Missoula: USDA Forest Service, 7–16.

Jones, K. B., K. H. Ritters, J. D. Wickham, R. D. Tankersley, R. V. O'Neill, D. J. Chaloud, E. R. Smith and A. C. Neale, 1998, *An Ecological Assessment of the United States: Mid-Atlantic Region,* Washington: EPA, 103 pp.

Jones, B., J. Walker, K. H. Ritters, J. D. Wickham and C. Nicoll, 1996, "Indicators of Landscape Integrity," J. Walker and D. J. Reuter (Eds.), *Indicators of Catchment Health: A Technical Perspective,* Melbourne: CSIRO, 155–168.

Kaufman, Y. J. and D. Tanre, 1992, "Atmospherically Resistant Vegetation Index (ARVI) for EOS-MODIS," *IEEE Transactions on Geoscience and Remote Sensing*, 30(2):261–270.

Kauth, R. J. and G. S. Thomas, 1976, "The Tasseled Cap — A Graphic Description of the Spectral-Temporal Development of Agricultural Crops as Seen by Landsat," *Proceedings*, Machine Processing of Remotely Sensed Data, West Lafayette, IN: Laboratory for the Applications of Remote Sensing, 41–51.

Kauth, R. J., P. F. Lambeck, W. Richardson, G. S. Thomas, and A. P. Pentland, 1979, "Feature Extraction Applied to Agricultural Crops as Seen by Landsat," *Proceedings*, LACIE Symposium, Houston: NASA, 705–721.

Kimes, D. S., 1983, "Dynamics of Directional Reflectance Factor Distributions for Vegetation Canopies," *Applied Optics*, 22(9):1364–1372.

Li, X. and A. H. Strahler, 1985, "Geometric-optical Modeling of a Conifer Forest Canopy," *IEEE Trans. Geosci. Remote Sensing*, GE-23(5):705–721.

Li, X. and A. H. Strahler, 1992, "Geometric-optical bidirectional Reflectance modeling of the Discrete-Crown Vegetation Canopy:

Effect of Crown Shape and Mutual Shadowing," *IEEE Trans. Geosci. Remote Sensing*, 30:276–292.

Lyon, J. G., D. Yuan, R. S. Lunetta and C. D. Elvidge, 1998, "A Change Detection Experiment Using Vegetation Indices," *Photogrammetric Engineering & Remote Sensing*, 64(2):143–150.

Moran, M. S., R. D. Jackson and G. F. Hart, 1990, "Obtaining Surface Reflectance Factors from Atmospheric and View Angle Corrected SPOT-1 HRV Data," *Remote Sensing of Environment*, 32:203–214.

Myers, V. I., 1970, "Soil, Water and Plant Relations," *Remote Sensing with Special Reference to Agriculture and Forestry*, Washington, DC: National Academy of Sciences, 253–297.

O'Neill, R. V., J. R. Krummel, R. H. Gardner, G. Sugihara, B. Jackson, D. L. DeAngelis, B. T. Milne, M. G. Turner, B. Zygmunt, S. W. Christensen, V. H. Dale and R. L. Graham, 1988, "Indices of Landscape Pattern," *Landscape Ecology*, 1:153–162.

O'Neill, T., 1993, "New Sensors Eye the Rain Forest," *National Geographic*, September:118–130.

O'Neill, R. V., C. T. Hunsaker, K. B. Jones, K. H. Ritters, J. D. Wickham, P. Schwarz, I. A. Goodman, B. Jackson and W. S. Baillargeon, 1997, "Monitoring Environmental Quality at the Landscape Scale," *BioScience*, 47(8):513–519.

Pace, S., K. O'Connell and B. Lachman, 1997, *Using Intelligence Data for Environmental Needs*, Washington, DC: Rand Corp., 75 pp.

Perry, C. R. and L. F. Lautenschlager, 1984, "Functional Equivalence of Spectral Vegetation Indices," *Remote Sensing of Environment*, 14:169–182.

Peterson, D. L. and S. W. Running, 1989, "Applications in Forest Science and Management," *Theory and Applications of Optical Remote Sensing*, Doghouses Asrar, New York: John Wiley & Sons, 4210–473.

Pierce, L. L. and S. W. Running, 1988, "Rapid Estimation of Coniferous Forest Leaf Area Index Using a Portable Integrating Radiometer," *Ecology*, 69(6):1762–1767.

Pierce, L. L., S. W. Running and G. A. Riggs, 1990, "Remote Detection of Canopy Water Stress in Coniferous Forests Using NS001 Thematic Mapper Simulator and the Thermal Infrared Multispectral Scanner," *Photogrammetric Engineering & Remote Sensing*, 56(5):5710–586.

Qi, J., F. Cabot, M. S. Moran and G. Dedieu, 1995, "Biophysical Parameter Estimations Using Multidirectional Spectral Measurements," *Remote Sensing of Environment*, 54:71–83.

Ramsey, R. D., A. Falconer and J. R. Jensen, 1995, "The Relationship between NOAA-AVHRR NDVI and Ecoregions in Utah," *Remote Sensing of Environment*, 53:188–198.

Richardson, A. J. and C. L. Wiegand, 1977, "Distinguishing Vegetation from Soil Background Information," *Remote Sensing of Environment*, 8:307–312.

Richardson, A. J. and J. H. Everitt, 1992, "Using Spectral Vegetation Indices to Estimate Rangeland Productivity," *Geocarto International*, 1:63–77.

Ritters, K. H., R. V. O'Neill, C. T. Hunsaker, J. D. Wickham, D. H. Yankee, S. P. Timmins, K. B. Jones and B. L. Jackson, 1995, "A Factor Analysis of Landscape Pattern and Structure Metrics," *Landscape Ecology*, 10(1):23–39.

Rock, B. N., J. E. Vogelmann, D. L. Williams, A. F. Voglemann and T. Hoshizaki, 1986, "Remote Detection of Forest Damage," *Bio Science*, 36: 439 pp.

Rouse, J. W., R. H. Haas, J. A. Schell and D. W. Deering, 1974, "Monitoring Vegetation Systems in the Great Plains with ERTS," *Proceedings*, Third Earth Resources Technology Satellite-1 Symposium, Greenbelt: NASA SP-351, 3010–317.

Rundquist, D. C. and S. A. Sampson, 1988, *A Guide to the Practical Use of Aerial Color-infrared Photography in Agriculture*, Lincoln: Conservation and Survey Division, 27 pp.

Running, S. W., C. O. Justice, V. Solomonson, D. Hall, J. Barker, Y. J. Kaufmann, A. H. Strahler, A. R. Huete, J. P. Muller, V. Vanderbilt, Z. M. Wan, P. Teillet and D. Carneggie, 1994, "Terrestrial Remote Sensing Science and Algorithms Planned for EOS/MODIS," *Intl. Journal of Remote Sensing*, 15(17):3587–3620.

Sandmeier, S. R., 1999, *Guidelines and Recommendations for the Use of the Sandmeier Field Goniometer for the NASA Stennis Space Center*, Washington, DC, NASA, 47 pp.

Sandmeier, S. R. and K. I. Itten, 1999, "Field Goniometer System (FIGOS) for Acquisition of Hyperspectral BRDF Data," *IEEE Transactions on Geoscience & Remote Sensing*, 37(2):978–986.

Sandmeier, S., C. Muller, B. Hosgood and G. Andreoli, 1998a, "Physical Mechanisms in Hyperspectral BRDF Data of Grass and Watercress," *Remote Sensing of Environment*, 66:222-233.

Sandmeier, S., C. Muller, B. Hosgood and G. Andreoli, 1998b, "Sensitivity Analysis and Quality Assessment of Laboratory BRDF Data," *Remote Sensing of Environment*, 64(2):176–191.

Savitsky, B. G., 1986, *Agricultural Remote Sensing in South Carolina: A Study of Crop Identification Capabilities Utilizing Landsat Multispectral Scanner Data*, unpublished masters thesis, Columbia: University of South Carolina Geography Department, 78 pp.

Savitsky, B. G., 1998, "Overview of Gap Analysis" *GIS Methodologies for Developing Conservation Strategies*, B. G. Savitsky and T. E. Lacher, Jr., (Eds.), New York: Columbia University Press, 151–158.

Schill, S., J. R. Jensen and D. Porter, 2000, "Bidirectional Reflectance Distribution Function (BRDF) Modeling of Smooth Cordgrass (*Spartina alterniflora*) Using the Sandmeier Field Goniometer," *Proceedings*, American Society for Photogrammetry & Remote Sensing, Washington, DC, May 22-26, in press.

Schuft, M. J., T. J. Moser, P. J. Wiginton, Jr., D. L. Stevens, Jr., L. S. McAllister, S. S. Chapman and T. L. Ernst, 1999, "Development of Landscape Metrics for Characterizing Riparian-Stream Networks," *Photogrammetric Engineering & Remote Sensing*, 65(10):1157-1167.

Scott, J. M., J. J. Jacobi and J. E. Estes, 1987, "Species Richness: A Geographic Approach to Protecting Future Biological Diversity," *BioScience*, 37:782–788.

Scott, J. M., F. Davis, B. Csuti, R. Noss, B. Butterfield, C. Groves, H. Anderson, S. Caicco, F. D'Erchia, R. Edwards, J. Ulliman and R. G. Wright, 1993, "Gap Analysis: A Geographic Approach to Protection of Biological Diversity," *Wildlife Monographs,* 123, 1–41.

Strahler, A., 1994, "Measuring and Modeling the Bidirectional Reflectance of Plant Canopies," in *Sixth International Symposium on Physical Measurements and Signatures in Remote Sensing,* Vald'Isere, France, 17–24, January.

Swain, P. H. and S. M. Davis, 1978, *Remote Sensing: The Quantitative Approach.* New York: McGraw-Hill, 333 pp.

Thenkabail, P. S., D. A. Ward and J. G. Lyon, 1994, "Impacts of Agricultural Management Practices on Soybean and Corn Crops Evident in Ground Truth Data and Thematic Mapper Vegetation Indices," *Transactions of the American Society of Agricultural Engineers*, 37(3):9810–995.

Troll, C., 1939, Luftbildplan and okologische Bodenforschung, *A. Ges. Erdkunde*, Berlin: 241–298.

Tucker, C, 1978. "A Comparison of Satellite Sensor Bands for vegetation Monitoring, "Photogrammetric *Engineering & Remote Sensing*, 44:13610–1380.

Turner, M. G., 1989, "Landscape Ecology: The Effect of Pattern on Process," *Annual Review Ecological Systems*, 20:171–197.

Verhoef, W., 1984, "Light Scattering by Leaf Layers with Application to Canopy Reflectance Modeling: The SAIL Model," *Remote Sensing of Environment*, 16:125–141.

Walter-Shea, E. A. and L. L. Biehl, 1990, "Measuring Vegetation Spectral Properties," *Remote Sensing Reviews*, 5(1):179–205.

Welch, R., M. Madden and R. F. Doren, 1999, "Mapping the Everglades," *Photogrammetric Engineering and Remote Sensing*, 65(2):163–170.

Wickham, J. D. and D. J. Norton, 1994, "Mapping and Analyzing Landscape Patterns," *Landscape Ecology*, 9(1):7–23.

Woodcock, C. E., J. B. Collins, V. D. Jakabhazy, X. Li, S. A. Macomber and Y. Wu, 1997, "Inversion of the Li-Strahler Canopy Reflectance Model for Mapping Forest Structure," *IEEE Transactions on Geoscience and Remote Sensing*, GE-35(2):405–414.

Yool, S. R., M. J. Makaio, and J. M. Watts, 1997, "Techniques for Computer-Assisted Mapping of Rangeland Change," *Journal of Range Management*, 50(3):3007–314.

Remote Sensing of Water

Water is life. Fortunately for mankind, it covers approximately 74 percent of the Earth's surface. Nowhere else in the known universe is such an abundance of liquid water found. Almost 97 percent of the Earth's volume of water is in the great saline oceans. Only about 0.02 percent of the Earth's water is found in freshwater streams, rivers, lakes, and reservoirs. The remaining water is contained in underground aquifers (0.6 percent), the Earth's atmosphere in the form of water vapor (0.001 percent), and the permanent icecap (approximately 2.2 percent).

Water exists in various states on Earth, including: freshwater, saltwater, water vapor, rain, snow, and ice. Meteorologists, oceanographers, hydrologists, some geographers, and others devote their lives to measuring, monitoring, and predicting the spatial distribution, volume, and movement of water as it progresses through the hydrologic cycle.

It is possible to obtain *in situ* measurements of various hydrologic (water) parameters such as precipitation, water depth, temperature, salinity, velocity, volume, etc., at very specific locations on the Earth. For example, the U.S. Geological Survey maintains a dense network of *in situ* river-flow gauges on major streams and rivers that provide continuous records of river stage (height) and velocity. Major cities and airports collect *in situ* precipitation (rain and snow) information. Departments of health and environmental control are often mandated to collect water-quality samples from rivers, lakes, reservoirs, and estuaries. These point measurements are very important. If enough of the point observations are collected throughout a region, it is possible to interpolate between the point observations and infer regional geographic patterns. Unfortunately, there are usually not enough point observations to create a statistically significant distribution map. In fact, it is often difficult to obtain regional spatial information using *in situ* point observations for a number of the most important hydrologic variables, including:

- water-surface area (streams, rivers, ponds, lakes, reservoirs, and seas),

- water constituents (organic and inorganic),

- water depth (bathymetry),

- water-surface temperature,

- snow-surface area,

- snow-water equivalent,

- ice-surface area,

- ice-water equivalent,

- cloud cover,

- precipitation, and

- water vapor.

Therefore, it is not surprising that a significant amount of research has been undertaken to develop remote sensing methods that can obtain quantitative, spatial measurements of these important hydrologic variables (e.g., Asrar and Dozier, 1994; Ikeda and Dobson, 1995). This chapter introduces some of the fundamental principles associated with remote sensing surface water and its constituents, clouds, water vapor, precipitation, and snow. It ends with a nonpoint source pollution water quality study.

Remote Sensing Surface Water Biophysical Characteristics

This section reviews how remote sensing can be used to inventory and monitor the spatial extent, organic/inorganic constituents, depth, and temperature of water in rivers, lakes, reservoirs, seas, and oceans. It is important to first obtain an appreciation for the energy-matter interactions that may impact our ability to perform an accurate aquatic remote sensing investigation.

Water Surface, Subsurface Volumetric, and Bottom Radiance

The total radiance, (L_t) recorded by the sensor onboard the aircraft or satellite is a function of the electromagnetic energy from the four sources identified in Figure 11-1 (Bukata et al., 1995):

$$L_t = L_p + L_s + L_v + L_b \qquad (11-1)$$

where

- L_p is the portion of the radiance recorded by a remote sensing instrument resulting from the downwelling solar (E_{Sun}) and sky (E_{sky}) radiation that never actually reaches the water surface. This is atmospheric noise and may be considered to be the unwanted *path radiance* discussed in Chapter 2;

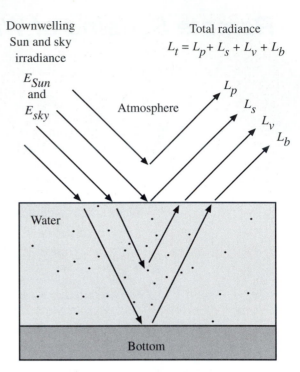

Figure 11-1   Water bodies receive irradiance from the Sun (E_{Sun}) and atmosphere (E_{sky}). The total radiance upwelling (exiting) from a water body toward the remote sensor is a function of the radiance from atmospheric scattering (L_p), water-surface radiance (L_s), subsurface volumetric radiance (L_v), and radiance from the bottom (L_b) of the water body (after Bukata et al., 1995).

- L_s is the radiance from the downwelling solar and sky radiation that reaches the air-water interface (sometimes called the *free-surface layer* or *boundary layer*) but only penetrates it a millimeter or so and is then essentially *reflected* from the water surface. This reflected energy contains valuable spectral information about the near-surface characteristics of the water body (Figure 11-2a). Unfortunately, if the solar zenith angle and sensor viewing angle are almost identical, then we may get a purely specular reflection from the surface of the water body, which provides very little useful spectral information (Figure 11-2b). Such *sunglint* is to be avoided whenever possible. This is why we rarely collect remotely sensed data at nadir within one or two hours of local noon.

- L_v is the radiance from the downwelling solar and sky radiation that actually penetrates the air-water interface, interacts with the water and organic/inorganic constituents and then exits the water column without encountering the bottom (called *subsurface volumetric radiance*). This

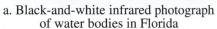

a. Black-and-white infrared photograph
of water bodies in Florida

b. Black-and-white infrared photograph with sunglint

Figure 11-2   a) A black-and-white infrared photograph of an area in Florida containing natural vegetation, a golf course, and fairly deep
nonturbid water bodies. b) In this example, specularly reflected sunglint is present that seriously impedes our ability to extract
useful information. The only data of value are the wind streak patterns on the water surface that provide some information on
wind direction at the time of data collection.

radiance provides valuable information about the internal
bulk characteristics of the water column.

- L_b is that portion of the recorded radiance resulting from
the downwelling solar and sky radiation that penetrates the
air-water interface, reaches the bottom of the waterbody, is
propagated back through the water column, and then exits
the water column. If we want to obtain information about
the bottom, such as when performing bathymetric (depth)
mapping or coral reef mapping, then this radiance from the
bottom may be of significant value (Mumby et al., 1997;
Pasqualini et al., 1997). However, radiance from the
bottom makes it very difficult to properly characterize the
water column above it. Basically, it is difficult to
disentangle or disaggregate L_v and L_b.

The goal of most aquatic remote sensing is to extract the
radiance of interest from all the other radiance components
being recorded by the sensor system. For example, the scientist interested in identifying the organic and inorganic constituents in the water column (e.g., suspended sediment or
chlorophyll a) is most concerned with isolating the subsurface volumetric radiance (L_v) computed as:

$$L_v = L_t - (L_p + L_s + L_b). \qquad (11\text{-}2)$$

This usually involves careful radiometric correction of the
remote sensor data to remove atmospheric attenuation (L_p),
surface sun-glint and other surface reflection (L_s), and bottom reflectance (L_b).

Scientists interested in monitoring water depth (bathymetry)
or bottom characteristics such as reef structure are most
interested in precise measurements of bottom radiance (L_b),
so they must attempt to remove atmospheric, surface, and
subsurface volume radiance.

Spectral Response of Water as a Function of Wavelength

When conducting a remote sensing investigation on water
bodies, it is first useful to understand how *pure* water selectively absorbs and/or scatters the incident, downwelling sunlight in the water column. Later we will consider how the
incident light is affected when the water column is not pure,
but contains organic and inorganic materials.

Pure water is free from organic and inorganic matter. Bukata
et al. (1995) summarized the absorption coefficient $\alpha(\lambda)$, the
scattering coefficient $b(\lambda)$, and the total attenuation coeffi-

Table 11-1. Optical Properties of Pure Water (derived from various sources by Bukata et al., 1995)

Wavelength (nm)	Absorption $\alpha(\lambda)$ (m⁻¹)	Scattering $b(\lambda)$ (m⁻¹)	Total Attenuation $c(\lambda)$ (m⁻¹)
250 – ultraviolet	0.190	0.032	0.2200
300 – ultraviolet	0.040	0.015	0.0550
320 – ultraviolet	0.020	0.012	0.0320
350 – ultraviolet	0.012	0.0082	0.0202
400 – violet	0.006	0.0048	0.0108
420 – violet	0.005	0.0040	0.0090
440 – violet	0.004	0.0032	0.0072
460 – dark blue	0.002	0.0027	0.0047
480 – dark blue	0.003	0.0022	0.0052
500 – light blue	0.006	0.0019	0.0079
520 – green	0.014	0.0016	0.0156
540 – green	0.029	0.0014	0.0304
560 – green	0.039	0.0012	0.0402
580 – yellow	0.074	0.0011	0.0751
600 – orange	0.20	0.00093	0.2009
620 – orange	0.24	0.0082	0.2408
640 – red	0.27	0.00072	0.2707
660 – red	0.310	0.00064	0.3106
680 – red	0.38	0.00056	0.3806
700 – red	0.60	0.0005	0.6005
740 – near-infrared	2.25	0.0004	2.2504
760 – near-infrared	2.56	0.00035	2.5604
800 – near-infrared	2.02	0.00029	2.0203

cient $c(\lambda)$ of pure water molecules at wavelengths from 250 – 800 nm from a number of studies (Table 11-1). Several important relationships are observed when the absorption and scattering data are graphed, as shown in Figure 11-3.

The most noticeable characteristic is that the *least* amount of absorption *and* scattering of incident light in the water column (therefore the best transmission) takes place in the blue wavelength region from approximately 400 – 500 nm, with the minimum located at approximately 460 – 480 nm. These

wavelengths of violet to light blue light penetrate further than any other type of light into the water column (Clark et al., 1997).

Incident green and yellow light from 520 – 580 nm is absorbed very well by the water column with relatively little scattering taking place. Similarly, scattering of orange and red wavelength energy (580 – 740 nm) by water molecules becomes insignificant when compared to absorption by water molecules. Almost all of the incident near- and middle-infrared (740 – 3,000 nm) radiant flux entering a deep, pure water body is absorbed with negligible scattering taking place (Figure 11-3).

Thus, molecular scattering of violet and blue light (< 520 nm) in a water column and significant absorption of green, yellow, orange, and red wavelength light (520 – 700 nm) in the same water column, cause pure water to appear blue to our eyes. The blue color is especially evident in pure mid-ocean water and deep nonturbid inland water bodies.

Monitoring the Surface Extent of Water Bodies

The best wavelength region for discriminating land from pure water is the near-infrared and middle-infrared regions at wavelengths between 740 – 2,500 nm. In the near- and middle-infrared regions, water bodies appear very dark, even black, because they absorb almost all of the incident radiant flux, especially when the water is deep and pure with no suspended sediment or organic matter in it (Figure 11-2a). Conversely, land surfaces are typically composed of vegetation and bare soil that reflect significant amounts of near- and middle-infrared energy, as summarized in previous chapters. This causes the land surfaces to appear relatively bright in near- and middle-infrared bands of imagery.

For example, consider the green (band 1), red (band 2), and near-infrared (band 3) SPOT HRV XS images of a portion of the Palancar Reef just offshore the island of Cozumel, Quintana Roo, Mexico (Figure 11-4). Underwater visibility in this portion of the Caribbean Sea is usually 80 – 120 ft. Incident green and red wavelength energy from the Sun penetrate through the water column and are reflected by bottom sand, hard-bottom, and coral reef. Note how the details of the reef are more clearly seen in the green (band 1) and red (band 2) images than in the near-infrared band 3 image (Figure 11-4).

While the subsurface information available in the green and red SPOT images (bands 1 and 2) is useful for many reasons, it is not of much value when trying to identify the land-water interface, i.e., the actual edge of the island of Cozumel. For-

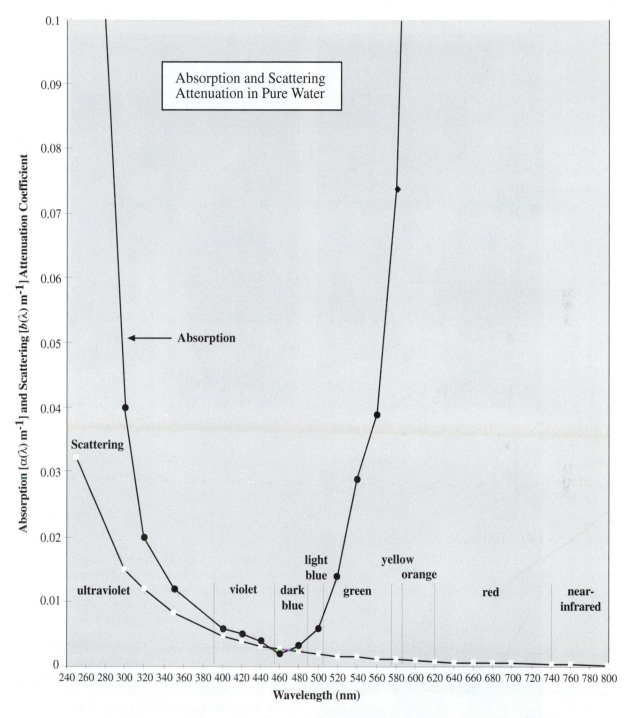

Figure 11-3 Absorption and scattering of light in pure water. Molecular water absorption dominates in the ultraviolet (< 400 nm) and in the yellow through the near-infrared portion of the spectrum (> 580 nm). Almost all of the incident near- and middle-infrared (740 – 2,500 nm) radiant flux entering a pure water body is absorbed with negligible scattering taking place. This is why water is so dark on black-and-white infrared or color-infrared film. Scattering in the water column is especially important in the violet, dark blue, and light blue portions of the spectrum (400 – 500 nm). This is the reason water appears blue to our eyes. These data were derived from a variety of sources by Bukata et al. (1995). The graph truncates the absorption attenuation information in the ultraviolet and in the yellow through near-infrared regions because the attenuation is so great. Refer to Table 11-1 for absorption attenuation information in these regions.

Palancar Reef on Cozumel Island, Quintana Roo, Mexico

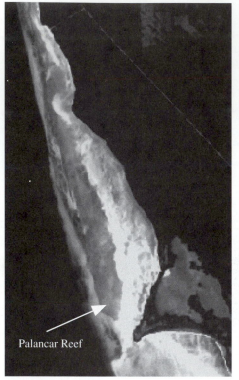

Palancar Reef

a. SPOT Band 1 (0.50 – 0. 59 μm; green)

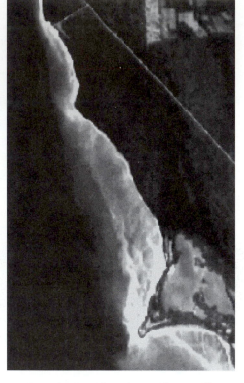

b. SPOT Band 2 (0.61 – 0. 68 μm; red)

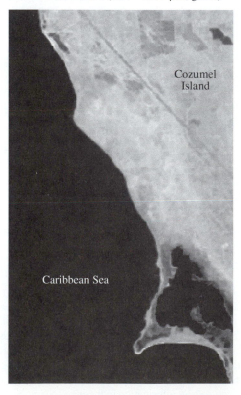

Cozumel
Island

Caribbean Sea

c. SPOT Band 3 (0.791 – 0.89 μm; near-infrared)

Figure 11-4   Individual bands of SPOT XS data of Cozumel, Mexico. Incident green and red radiant flux (bands 1 and 2) penetrate through the water column and are reflected by the sand, hard-bottom, and Palancar Reef. Conversely, much of the green and red energy is absorbed by terrestrial vegetation, causing it to appear dark. It is difficult to identify the land-water interface (boundary) using these visible bands. Water absorbs almost all of the incident near-infrared radiant flux, while the terrestrial vegetation reflects most of the incident near-infrared energy. The result is that water appears almost black on the near-infrared (band 3) data while the upland vegetated terrain appears bright. This makes it very easy to identify the land-water boundary (© SPOT Image, Inc.).

tunately, the water absorbs almost all the incident near-infra-red radiant flux (as shown in Figure 11-4c), causing the SPOT HRV band 3 sensor to record ocean water as black, while the upland vegetation reflects much of the near-infra-red radiant flux, causing it to appear bright. Therefore, it is a straightforward task to delineate the land-water interface using the SPOT near-infrared band 3 image.

Care must be exercised, however, when there are organic and inorganic constituents in the water column (especially those near the surface) because these materials will cause near-infrared surface reflection and subsurface volumetric scattering to take place, dramatically increasing the amount of near-infrared radiant flux leaving the surface of the water body. For example, consider the Space Shuttle photograph of the Mississippi River in Figure 11-5 where the suspended sediment causes the water body to appear relatively bright. The suspended sediment in the water column causes signifi-cant scattering and reflection of radiant flux back toward the sensor system, causing the water to appear almost as bright as the few land features in this bird's foot delta.

Spectral Response of Water as a Function of Organic and Inorganic Constituents — Monitoring Suspended Minerals (Turbidity), Chlorophyll, and Dissolved Organic Matter

---

Thus far we have mainly considered the spectral response of pure water. Most natural water bodies, however, contain a variety of organic (e.g., phytoplankton chlorophyll *a*) and inorganic (e.g., suspended minerals) constituents. When nat-ural waters contain a mixture of these materials, one of the most difficult remote sensing problems is to disentangle (extract) quantitative information about these specific con-stituents from the remotely sensed data (Goodin et al., 1993).

When conducting water-quality studies or trying to predict water productivity using remotely sensed data, we are usu-ally most interested in the *subsurface volumetric radiance*, L_v (Figure 11-1), which is the radiance from the down-welling solar and sky radiation that actually penetrates the air-water interface, interacts with the water and organic/inor-ganic constituents, and then exits the water column toward the sensor without encountering the bottom. The subsurface volumetric radiance exiting the water column toward the sensor (L_v) is a function of the concentration of pure water (*w*), inorganic suspended minerals (*SM*), organic chlorophyll *a* (*Chl*), dissolved organic material (*DOM*), and the *total* amount of absorption and scattering attenuation that takes

place in the water column due to each of these constituents, $c(\lambda)$, i.e.,

$$L_v = f[w_{c(\lambda)}, SM_{c(\lambda)}, Chl_{c(\lambda)}, DOM_{c(\lambda)}]. \qquad (11\text{-}3)$$

It is instructive to look at the effect that each of these constit-uents has on the spectral reflectance characteristics of a water column.

Suspended Minerals

Minerals such as silicon, aluminum, and iron oxides are found in suspension in most natural water bodies. The parti-cles range from fine clay particles ($3 - 4$ μm in diameter), to silt ($5 - 40$ μm), to fine-grain sand ($41 - 130$ μm), and coarse grain sand ($131 - 250$ μm). The sediment comes from a vari-ety of sources including upland agricultural cropland ero-sion, weathering of mountainous terrain, shoreline erosion caused by natural waves or boat traffic, and volcanic erup-tion (ash). Most of the suspended mineral sediment is con-centrated in the inland and nearshore water bodies (Bukata et al., 1995). Clear, deep ocean (often referred to as Case I water) far from shore rarely contains suspended minerals greater than 1 μm in diameter. Thus, suspended mineral con-centration is usually of no significance to deep ocean remote sensing studies. This is important since the contributions from suspended minerals can often be ruled out when con-ducting a deep ocean remote sensing investigation. Con-versely, inland and nearshore water bodies may carry a significant load of suspended sediment that can dramatically impact the spectral reflectance characteristics of the water bodies (Nanu, 1993).

Monitoring the type, amount, and spatial distribution of sus-pended minerals in inland and nearshore water bodies is very important. For example, soil erosion in a watershed contrib-utes sediment loads to surface waters, which results in faster filling of major rivers, reservoirs, farm ponds, flood-control impoundments, and estuaries. This can shorten the useful life of reservoirs, ponds, and flood-control devices and require dredging of rivers and estuaries. For example, the reduction in storage capacity in reservoirs in the United States caused by the infusion of suspended sediment results in a loss of $100 million annually (Julien, 1995). Sediment also affects water quality and its suitability for drinking, rec-reation, and industrial purposes. It serves as a carrier and storage agent of pesticides, absorbed phosphorus, nitrogen, and organic compounds and can be an indicator of pollution. Suspended sediments can impede the transmission of solar radiation and reduce photosynthesis in submerged aquatic vegetation and near-bottom phytoplankton. The aquatic veg-

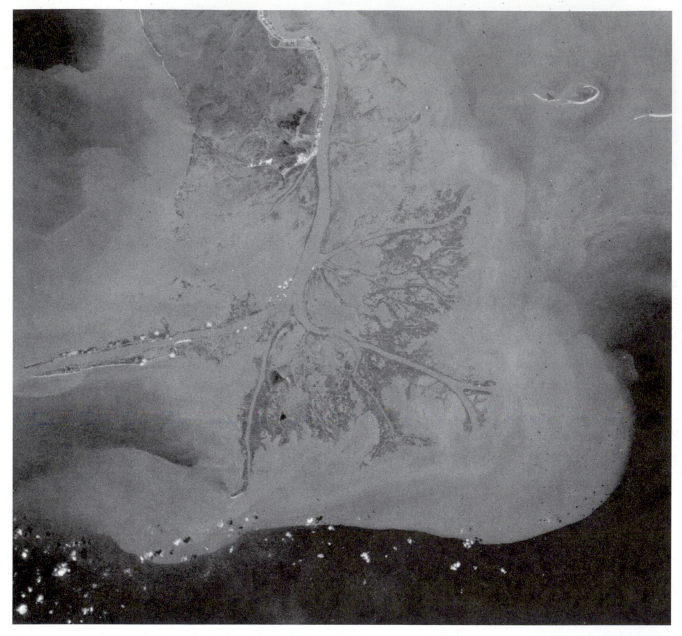

Figure 11-5 Photograph of the Mississippi River delta just below New Orleans, LA, obtained by Space Shuttle astronauts during Mission STS 51. The suspended sediment in the water is reflecting much of the incident radiant flux back into the atmosphere. Conversely, the more pure water, farther off-shore, is absorbing most of the incident radiant flux, causing it to appear dark. The boundary (front) between the sediment-laden water and the nonturbid is quite evident (courtesy K. Lulla, NASA Johnson Space Center).

etation and phytoplankton play a vital role in the food chain of the aquatic ecosystem.

Fortunately, remote sensing can be used in certain instances to monitor the suspended mineral concentrations in water bodies. This usually requires obtaining *in situ* measurements

of suspended mineral concentrations and relating it to the remote sensor data to derive a quantitative relationship. It is good practice to collect both the remote sensor data and the *in situ* suspended sediment measurements on days that have little wind. Wind-roughened surface water creates specular reflections, which can be deleterious to remote sensing of

Figure 11-6 A secchi disk is used to measure suspended sediment in water bodies by lowering it into the water column and determining the depth at which it disappears. The accuracy of the measurement is a function of the visual acuity of the observer, which can vary dramatically.

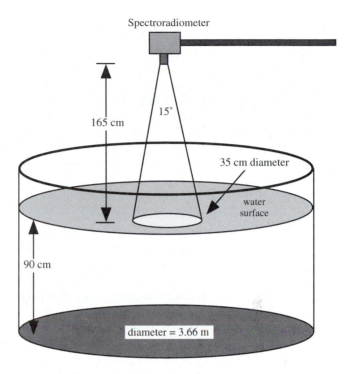

Figure 11-7 A spectroradiometer may be used to measure *in situ* spectral reflectance characteristics of pure water and water with various suspended sediment and chlorophyll *a* concentrations (after Lodhi et al., 1997).

suspended sediment concentrations (Han and Rundquist, 1998).

The spectral reflectance of suspended sediment in surface waters is a function of both the quantity and characteristics (particle size, absorption) of the material in the water. Suspended sediment concentration is measured *in situ* using either a secchi disk or a nephelometric turbidity unit (NTU) detector. The secchi disk (Figure 11-6) is a circular plate that is lowered into the water until it cannot be seen. This secchi depth is inversely correlated with the amount of suspended material in the water. The greater the amount of suspended material, the less the secchi depth, i.e., it will disappear from view relatively quickly. This method relies on human visual perception, which is not constant among scientists. The more rigorous nephelometer detector passes light through a sample of water obtained at various depths to determine its transmission characteristics that are related to suspended material in the water.

Research has documented the general impact of increased suspended mineral concentration in water bodies in the visible and near-infrared portion of the spectrum. For example, consider the experiment conducted by Lodhi et al. (1997), which determined the spectral reflectance characteristics of two Nebraska soil types (clayey and silty) at various suspended sediment concentrations in water.

Figure 11-7 depicts the experimental design where a spectroradiometer was used to collect spectral reflectance data from a height of 165 cm directly above a controlled water surface. The spectral measurements were made in 252 discrete spectral bands between 368 and 1,114 nm. The spectroradiometer was calibrated with a gray card (18 percent reflectance) that was cross-referenced to a $BaSO_4$ calibration panel. Reflectance was calculated as a simple ratio using the equation (Markham and Barker, 1986):

$$R(\lambda) = \frac{L(\lambda)}{S(\lambda)} \times Cal(\lambda) \times 100 \qquad (11\text{-}4)$$

where $L(\lambda)$ is the radiance measured from the water surface), $S(\lambda)$ is the radiance from the $BaSO_4$ panel, and $Cal(\lambda)$ is the calibration factor for the $BaSO_4$ panel.

Figures 11-8a,b depict the spectral reflectance of clear water and water with varying suspended sediment concentrations of clayey and silty soil. Spectral reflectance of the clear water drops continuously after about 580 nm due to absorption in the water column, as previously described (refer to Figure 11-3). As the suspended sediment concentration is increased, reflectance increases at all wavelengths for both clayey and silty soils. The clayey soil (Figure 11-8a) had more organic matter and was darker in color, which resulted in approximately 10 percent lower volume reflectance at all wavelengths than the light-colored silty soil (Figure 11-8b). Reflectance increased in the 580 - 690 nm region and in the near-infrared region as more minerals were suspended in the water bodies. *Thus, the peak reflectance shifts toward longer wavelengths in the visible region as more suspended sediments are added.* More green, red, and near-infrared radiant flux is reflected from the water body and recorded by the remote sensing system. A water body with suspended sediment in it will generally appear brighter in imagery than a nearby waterbody without any suspended sediment.

Correlation coefficients (r) were computed to describe the relationship between suspended sediment concentration and the reflectance at each of the 252 spectral sampling points. For the clayey soil, the values ranged from 0.28 – 0.97, and for the silty soil, the range was 0.78 – 0.98. For both soils the values of $r > 0.90$ occurred in the near-infrared region between 714 and 880 nm, as expected (Bhargava and Mariam, 1990; Han and Rundquist, 1994).

These results suggest that:

- the visible wavelength range of 580 – 690 nm may provide information on the *type* of suspended sediments (soil) in surface waters; and

- the near-infrared wavelength range of 714 – 880 nm may be a useful wavelength range for determining the *amount* of suspended minerals in surface waters where suspended minerals are the predominant constituent.

These relationships may not be universally applicable. Scientists attempting to remotely sense suspended sediment concentrations should investigate the spectral absorption and scattering characteristics of the water and soils in their geographic area of interest.

Chlorophyll

Plankton is the generic term used to describe all the living organisms (plant and animal) present in a water body that cannot resist the current (unlike fish). Plankton may be sub-divided further into algal plant organisms (*phytoplankton*), animal organisms (*zooplankton*), bacteria (*bacterioplankton*), and lower plant forms such as algal fungi. *Phytoplankton* are small single-celled plants smaller than the size of a pinhead. Micrographs of the blue radiant energy reflected from a diatom and a single cell of green algae are shown in Figure 11-9a,b. In both instances, the chlorophyll *a* pigments in the plants absorb most of the incident blue radiant flux, causing the photosynthetic portion of each to appear dark.

Phytoplankton, like plants on land, are composed of substances that contain carbon. Phytoplankton sink to the ocean or water-body floor when they die. Zooplankton migrate to the surface at night to feed on live phytoplankton and then sink to greater depths during the day. When zooplankton die, they also sink to the bottom, carrying their carbon with them. The carbon in the dead phytoplankton and zooplankton is soon covered by other sediments. Also, phytoplankton use carbon dioxide and produce oxygen during the photosynthetic process. In this way, the water bodies and oceans act as a *carbon sink*, a place that disposes of global carbon, which otherwise can accumulate in the atmosphere as carbon dioxide. Other global sinks include land vegetation and soil. However, the carbon in these sinks frequently is returned to the atmosphere as carbon dioxide by burning or decomposition. No one knows exactly how much carbon the inland water bodies and ocean accumulate. As such, the characteristics of phytoplankton and zooplankton are very important to our knowledge of the global carbon cycle as it represents a significant carbon sink for increased levels of carbon dioxide in the atmosphere (Bukata et al., 1995).

All phytoplankton in water bodies contain the photosynthetically active pigment chlorophyll *a*. However, chlorophyll *c*, *d*, and even *e* may also be present and at various depths. There are two other phytoplankton photosynthesizing agents: *carotenoids* and *phycobilins*. Bukata et al. (1995) suggest, however, that chlorophyll *a* may be considered a reasonable surrogate for the organic component of optically complex natural waters. Fortunately, because different types of phytoplankton have different concentrations of chlorophyll, they appear as different colors to sensitive remote sensors. Thus, recording the color of an area of the ocean or other water body allows us to estimate the amount and general type of phytoplankton in that area and tells us about the health and chemistry of the water body. Comparing images taken at different times tells us about changes that occur over time and the processes at work.

Chlorophyll *a* introduced to pure water changes its spectral reflectance characteristics, i.e., its color. For example, Figure 11-10a depicts the spectral reflectance characteristics of

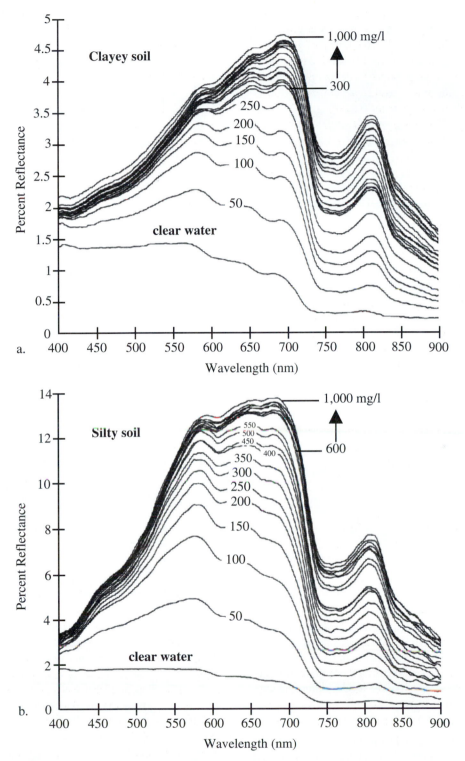

Figure 11-8   a) *In situ* spectral reflectance measurements of clear water and clear water with various levels of *clayey* soil suspended sediment concentrations from 0 – 1,000 mg/l. b) *In situ* spectral reflectance measurements of clear water and clear water with various levels of *silty* soil suspended sediment concentrations. Note that the silty soil had approximately 10 percent more volume reflectance at all wavelengths when compared with the clayey soil. In both cases, the peak reflectance shifted toward longer wavelengths in the visible region as more suspended sediments were added (after Lodhi et al., 1997).

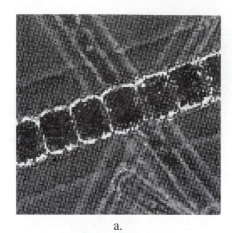

a.

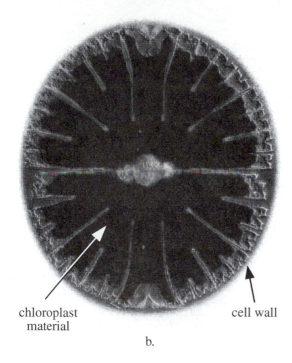

chloroplast
material

cell wall

b.

Figure 11-9   a) A micrograph of blue reflected light from a photosynthesizing diatom. b) A micrograph of blue reflected light from a green algae cell (*Micrasterias
sp.*). The chloroplast material in the center shows
up in dark tones because the chlorophyll *a* pigments absorb much of the incident blue light during photosynthesis.

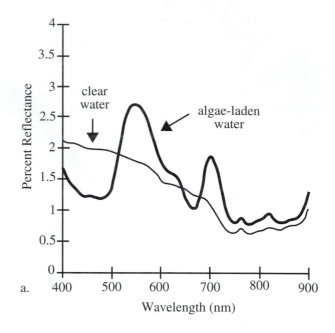

a.

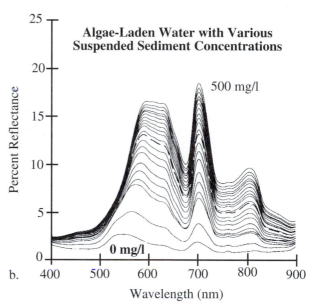

b.

Figure 11-10   a) Percent reflectance of clear and algae-laden water based on *in situ* spectroradiometer measure-
ment. b) Percent reflectance of algae-laden water
with various concentrations of suspended sediment
ranging from 0 – 500 mg/l (after Han, 1997).

clear water and the same water laden with algae consisting
primarily of chlorophyll *a* (Han, 1997). Clear water reflected
approximately 2 percent between 400 and 500 nm and
dropped gradually to less than 1 percent at wavelengths
beyond 710 nm, as expected. Conversely, four pronounced
scattering/absorption features of chlorophyll are evident in
the algae-laden water:

- strong chlorophyll *a* absorption of blue light between 400
and 500 nm;

- strong chlorophyll *a* absorption of red light at
approximately 675 nm (Gitelson, 1992);

Table 11-2. Spectral characteristics of the two most important ocean-color remote sensing systems: the Nimbus-7 Coastal Zone Color Scanner (CZCS) and the Sea-viewing Wide Field of View Sensor (SeaWiFS)

Sensor	Band	Bandcenter (nm)	Bandwidth (nm)
CZCS	1	443	423 – 463
	2	520	500 – 520
	3	550	520 – 570
	4	670	650 – 690
	5	750	650 – 850
	6	11,500	9,500 – 13,500
SeaWiFS	1	412	402 – 422
	2	443	433 – 453
	3	490	480 – 500
	4	510	500 – 520
	5	555	545 – 565
	6	670	660 – 680
	7	765	745 – 785
	8	865	845 – 885

- reflectance maximum around 550 nm (green peak) caused by *relatively* lower absorption of green light by algae (Gitelson, 1992); and

- prominent reflectance peak around 690 – 700 nm caused by an interaction of algal-cell scattering and a minimum combined effect of pigment and water absorption. The height of this peak above the baseline (absorption trough) can be used to accurately measure chlorophyll amount (Rundquist et al., 1995).

Basically, as chlorophyll concentration increases in the water column, there is a significant decrease in the relative amount of energy reflected in the blue and red wavelengths but an increase in green wavelength reflectance.

When both suspended mineral sediment and chlorophyll are present in the water body at the same time, a dramatically different spectral response is produced. For example, Figure 11-10b shows what happens to the spectral response of water as red loam sediment concentrations from 0 – 500 mg/l are added to water that contains algae (chlorophyll *a*). For algae-

laden water, the peak reflectance in the visible region shifted from 547 nm (green light) at 0 mg/l to 596 nm (orange) at 500 mg/l.

Rundquist et al. (1996) and Han and Rundquist (1997) obtained accurate estimates of algal chlorophyll pigment amount in surface water using a simple near-infrared (705 nm) / red (670 nm) ratio when the concentration of chlorophyll was relatively low. Computing the first derivative of reflectance around 690 nm produced the best results when algal chlorophyll concentration was relatively high.

Chlorophyll in Ocean Water: It is generally assumed that a remote estimate of near-surface chlorophyll concentration constitutes a remote estimate of near-surface biomass (or primary productivity) for deep ocean (Case I) water where there is little danger of suspended mineral sediment contamination. Since the launch of Landsat 1 in 1972 and especially the Coastal Zone Color Scanner (1978 – 1986) (Table 11-2), numerous studies (e.g., Gordon et al., 1983) have documented a relationship between selected spectral bands and ocean chlorophyll concentration (*Chl*) using the equation:

$$Chl = x\left[\frac{L(\lambda_1)}{L(\lambda_2)}\right]^y \qquad (11\text{-}5)$$

where $L(\lambda_1)$ and $L(\lambda_2)$ are the upwelling radiances at selected wavelengths recorded by the remote sensing system and x and y are empirically derived constants. Bukata et al. (1995) point out that this approach does not take into account the absorption and scattering characteristics of the inorganic and organic constituents of the study area. The only reason it is successful is that in the mid-ocean, chlorophyll pigments are the principal colorant of Case I waters.

The Sea-viewing Wide Field of View Sensor (SeaWiFS) launched in 1997 was designed by NASA and private industry (ORBIMAGE, Inc.) to build on all that was learned using the Coastal Zone Color Scanner. SeaWiFS has additional bands at 412 nm (to identify yellow substances through their blue wavelength absorption), at 490 nm (to increase sensitivity to chlorophyll concentration), and in the 765 and 865 nm near-infrared (to assist in the removal of atmospheric attenuation). Interestingly, there is no SeaWiFS band at 700 nm, which could be quite useful.

Because of the role of phytoplankton in the global carbon cycle, SeaWiFS data are being used to assess the ocean's role in the global carbon cycle and to examine oceanic factors that affect global climate change (Hooker et al., 1992). SeaWiFS data are being used to identify the magnitude and variability of the annual cycle of primary production by marine phytoplankton and to determine the distribution and timing of spring blooms. The observations help scientists understand the dynamics of ocean and coastal currents, the physics of mixing, and the relationships between ocean physics and large-scale patterns of productivity. The data are filling the gap in ocean biological observations between those of the test-bed CZCS and the Moderate Resolution Imaging Spectrometer (MODIS).

Recent SeaWiFS' research incorporates atmospheric correction and improved algorithms applied to ocean imagery. The most important SeaWiFS algorithms involve the use of band ratios of 443/555 nm and 490/555 nm (Aiken et al., 1995; Falkowski et al., 1998). For example, consider several examples of the oceanic applications of SeaWiFS data in Color Plate 11-1. The first image is a global map of chlorophyll *a* derived from a composite of images obtained from September 3, 1997, through December 31, 1997 (Color Plate 11-1a). The next image is a true-color composite of a portion of the eastern United States at 1.13 x 1.13 km (Color Plate 11-1b). The true-color image is a color composite of bands 670, 555, and 412 nm (RGB) and highlights vegetation associated with land morphology. White areas are clouds and dense aerosols. Color Plate 11-1c presents the chlorophyll *a* concentrations on September 30, 1997. The red colors reveal high concentrations of chlorophyll in the water, the yellows/greens indicate intermediate concentrations of chlorophyll, and the blues/purples reveal low concentrations of chlorophyll. The Japanese Ocean Colour and Temperature Scanner (OCTS) was launched on August 17, 1996, and also provided chlorophyll *a* distribution information.

The primary instrument for assessing ocean productivity on the EOS *Terra* spacecraft is MODIS. MODIS bands 8 – 16, ranging from 405 – 877 nm at 1000 x 1000 m spatial resolution, are particularly well suited to the collection of information on ocean color, phytoplankton concentration, and biogeochemistry. However, due to sunglint over a portion of the MODIS swath as the satellite passes over the equator, some imagery will be of poor quality. This gap in the ocean-color data will be partially filled by the Multi-angle Imaging SpectroRadiometer (MISR) also on EOS *Terra* (Martonchik, 1999).

Chlorophyll in Coastal and Inland Water: For inland and near-coastal water bodies, it is often difficult to disentangle the information about the phytoplankton pigments in the remote sensor data from the effects of suspended inorganic materials or dissolved organic matter (DOM). This normally requires the use of 1) sophisticated atmospheric correction techniques applied to the remote

sensor data (e.g., Ramsey et al., 1992) and a complex multiple-component extraction methodology (e.g., Bukata et al., 1995) or 2) the use of a derivative spectra technique described in Goodin et al. (1993).

Many of the studies referred to thus far relied on the correlation of local *in situ* measurements of suspended sediment, chlorophyll *a*, etc., with the remote sensor data. These *local algorithms* are only good for the particular location and cannot usually be transported through space or time. What is needed are transportable algorithms that are spatially and temporally invariant, meaning that they will work most anywhere, anytime. Such algorithms may be applied to Sea-WiFS and/or the EOS *Terra* MODIS data to produce maps of phytoplankton pigment concentrations (e.g., chlorophyll *a*) as a routine product.

Dissolved Organic Material

Sunlight penetrates into the water column a certain *photic depth* (the vertical distance from the water surface to the 1 percent subsurface irradiance level). Phytoplankton within the photic depth of the water column consume nutrients and convert them into organic matter via photosynthesis. This is called *primary production*. Zooplankton eat the phytoplankton and create organic matter. Bacterioplankton decompose this organic matter. All this conversion introduces *dissolved organic matter* (DOM) into oceanic, nearshore, and inland water bodies. In certain instances, there may be sufficient dissolved organic matter in the water to reduce the penetration of light in the water column (Bukata et al., 1995).

The decomposition of phytoplankton cells yields carbon dioxide, inorganic nitrogen, sulfur, and phosphorous compounds. The more productive the phytoplankton, the greater the release of dissolved organic matter. In addition, *humic substances* may be produced. These often have a yellow appearance and represent an important colorant agent in the water column, which may need to be taken into consideration. These dissolved humic substances are called *yellow substance* or *Gelbstoffe* and can 1) impact the absorption and scattering of light in the water column, and 2) change the color of the water.

There are sources of dissolved organic matter other than phytoplankton. For example, the brownish-yellow color of the water in many rivers in the northern United States is due to the high concentrations of tannin from the eastern hemlock (*Tsuga canadensis*) and various other species of trees and plants grown in bogs in these areas (Hoffer, 1978). These tannins can create problems when remote sensing inland water bodies.

Water Penetration (Bathymetry)

Based on Figure 11-3, it is clear that the optimum spectral wavelength to obtain bathymetric (depth) information is approximately 0.48 μm. This is exactly what was discovered by Kodak scientists when they devised a special water penetration film in the 1970s that was made sensitive to the wavelength interval 0.44 – 0.54 μm after filtration. This information was also used when creating the blue-sensitive band (0.45 – 0.52 μm) on the Landsat Thematic Mapper sensor system, which is often called the water penetration band. If the water column is exceptionally clear, it is possible to see subsurface features to a depth of 10 – 30 m.

Bathymetric mapping in the 0.44 – 0.54 μm portion of the spectrum requires that the water be almost free from organic and inorganic constituents such as chlorophyll and suspended sediments that would cause scattering and/or absorption to take place and obscure the bottom topography. However, as demonstrated in the Cozumel, Mexico example (Figure 11-4), even the green and red wavelength region are of value for water penetration when the water is relatively clear.

In addition, one must also take into account the fact that the light from the Sun is bent from its true course in *both* the atmosphere and in the water column, causing bathymetric (z) information in the imagery to *not* be in its proper planimetric (x,y) position. Therefore, the index-of-refraction previously discussed in Chapter 2 must be taken into consideration whenever remote sensing methods are used to perform bathymetric mapping. Most bathymetric charting is now performed using active sonar instruments that bounce sound waves off the bottom and record the depth directly beneath the vessel. However, ships cannot navigate in shallow bays, estuaries, rivers, etc., so remote sensing water-penetration and bathymetric studies will continue to be important.

Water Surface Temperature

We know from Chapter 8 (Thermal Infrared Remote Sensing) that we may obtain the temperature of inland water bodies and sea-surface temperature (SST) during daylight hours and at night using thermal infrared remote sensing techniques. Unlike land surfaces, however, water bodies transfer energy primarily through convection. Therefore, any heat energy introduced into the system may be transferred hundreds of meters into the water body. This mixing is responsible for the relatively uniform surface temperature of a water body both day and night. Thus, water bodies generally have high thermal inertia, meaning that there may only be a few degrees difference between the daytime and nighttime water-surface temperature. This results in water bodies appearing relatively cooler than the land during daylight hours and relatively warmer than surrounding land during predawn hours, as demonstrated in the thermal infrared daytime and nighttime images of Atlanta, GA, in Chapter 8.

Water has an emissivity (ε) very close to 1. Therefore, it is possible to obtain relatively accurate water-surface temperature measurements because the remote sensor radiant temperature measurement (T_{rad}) is approximately equal to the true kinetic temperature (T_{kin}), assuming the effects of the intervening atmosphere are accounted for. It is important to remember, however, that it is only a surface measurement. If there is a significant change in temperature just a few meters below the surface (i.e., a thermocline), it may not be detected by the thermal infrared detectors.

Two of the most important thermal infrared remote sensing instruments used to measure sea-surface temperature are found on the NOAA Advanced Very High Resolution Radiometer and the NOAA Geostationary Operational Environmental Satellite (GOES). Their spatial, spectral, and temporal characteristics are summarized in Chapter 7. Data from these sensors is in the public domain, available at a reasonable price, and are used heavily to obtain sea-surface temperature (SST) measurements over large oceanic surfaces usually at a spatial resolution of > 1 x 1 km. For example, sea-surface temperature measurements off the southeastern coast of the United States obtained on March 4, 1999, by the NOAA AVHRR are shown in Figure 11-11 and in Color Plate 11-2a. The sea-surface temperature images depict very well the cool water adjacent to the Georgia and South Carolina shoreline caused by the influx of cool water from the major rivers. Also note the location of the warm Gulf Stream winding its way northward from the tip of Florida along the Eastern Seaboard.

El Nino and La Nina

A global sea-surface temperature (SST) map is shown in Color Plate 11-2b. It was created using day/night NOAA-14 thermal data collected from March 9–13, 1999. SST maps such as this have been indispensable in mapping the distribution and movement of the El Nino Southern Oscillation (ENSO). In non-El Nino conditions, the trade winds blow toward the west across the tropical Pacific. These winds pile up warm surface water in the western Pacific, so that the sea surface is about 0.5 m higher near Indonesia than near Ecuador, South America. The sea-surface temperature is also

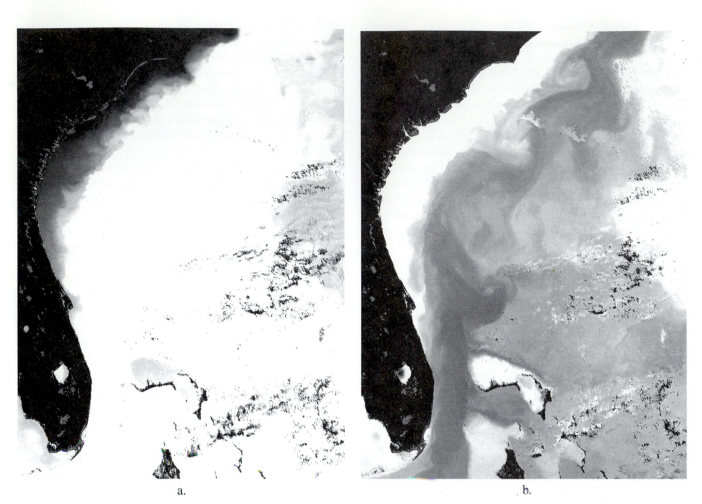

a.   b.

Figure 11-11 Sea-surface temperature (SST) maps derived from a three-day composite of NOAA AVHRR thermal infrared data centered on March 4, 1999. Each pixel was allocated the highest surface temperature that occurred during the three days. a) This temperature map has been adjusted to highlight the cool water adjacent to the Georgia and South Carolina coastline. b) This map has been adjusted to highlight the circulation patterns associated with the warm offshore Gulf Stream. A color composite of these images that provides additional information is found in Color Plate 11-2a (courtesy NOAA).

about 8°C higher in the west. Cooler temperatures exist off South America due to an upwelling of cold water from deeper levels. This cold water is nutrient-rich, supporting high levels of primary productivity, diverse marine ecosystems, and major fisheries. Rainfall is found in rising air over the warmest water, and the eastern Pacific is relatively dry.

Every two to seven years the trade winds relax in the central and western Pacific Ocean. When this occurs, the ocean currents and winds off the western coast of South America shift, bringing warm water eastward, displacing the nutrient-rich cold water that normally wells up from deep in the ocean. This invasion of warm water disrupts the marine food chain and the economies of coastal communities. Fishermen named the phenomenon El Nino (the Christ Child) because

it occurred during the Christmas season. Thus, El Nino is characterized by unusually warm ocean temperatures, especially in the eastern equatorial Pacific Ocean. La Nina is characterized by unusually cool ocean temperatures in the equatorial Pacific Ocean. Tree ring analysis has documented that these southern oscillations are not new, but have been occurring systematically for more than a hundred years (NOAA, 1999).

El Nino can be seen in measurements of the sea-surface temperature. For example, Color Plate 11-3 depicts three different sea-surface temperature conditions for selected Decembers: a La Nina in 1988, a normal month in 1990, and an El Nino in 1997. These are Reynolds Monthly SST (°C) maps derived by analyzing satellite thermal infrared data in conjunction with daily temperature measurements telemetered from an array of NOAA-sponsored buoys strategically placed in the equatorial Pacific Ocean (Reynolds, 1988; Reynolds and Smith, 1994). This is a good example of the utility of using both *in situ* and remote sensing to improve the accuracy of a product. Unfortunately, despite our ability to accurately map the sea-surface temperature of the El Nino Southern Oscillation, a substantial amount of life and property is still lost when they occur, mainly due to poor preparation by local authorities (Suplee, 1999).

When it is necessary to map the temperature of relatively small reservoirs, lakes, ponds, etc., then thermal infrared sensors flown on suborbital aircraft may be required. An example of using an aircraft-mounted thermal infrared multispectral scanner to map the temperature of water in the Savannah River is found in Chapter 8 (Thermal Infrared Remote Sensing).

 Precipitation

Precipitation may be measured operationally on a local or regional basis using active microwave sensors. For example, Nexrad weather radar systems operated on the ground by the National Weather Service in the United States provide excellent real-time precipitation information. Similar systems function in many other countries. However, the majority of the world's land surfaces do not enjoy such coverage by weather radars or even rain gauge networks, and will not for the foreseeable future. Meteorological satellites are the only systems capable of acquiring rainfall data over vast distances. Fortunately, it is now possible to monitor entire continents on a continuous basis and to do so for a small fraction of the cost of a surface-based observing network of equivalent spatial density (Petty, 1995).

Unfortunately, instead of directly measuring rainfall using a rain gauge, remote sensing methods must rely on using *indirect* remotely sensed data on cloud reflectance, cloud-top temperature, and/or the presence of frozen precipitation aloft to estimate the surface rain rate. Petty (1995) reviews the various satellite-based precipitation estimation techniques.

Visible – Infrared Techniques

The earliest precipitation estimation methods were based on the assumption that the brightness of reflected sunlight in the visible and near-infrared portion of the spectrum from clouds might give an indication of their thickness and thus of their likelihood of bearing rain. Unfortunately, not all bright clouds produce precipitation. Similarly, the temperature of the top of clouds was used to estimate precipitation. Generally, the colder the cloud top, the greater the likelihood of precipitation. Unfortunately, not all cold cloud tops produce precipitation. Nevertheless, some very useful precipitation estimation techniques were developed based on visible and infrared data, including 1) twice-daily estimates of outgoing longwave radiation from polar-orbiting infrared sensors onboard AVHRR to estimate monthly rainfall, and 2) a GOES Precipitation Index (GPI) based on an analysis of thermal infrared data, where pixels were classified as "raining" at the rate of 3 mm hr^{-1} if the cloud-top temperature was < 235 K (Arkin and Meisner, 1987). Until recently, the GPI was the standard climatological rainfall product that provided a climate-scale precipitation record throughout the tropics and subtropics. Algorithms have also been developed that utilize multiple images obtained through time to refine the rate of rainfall estimates.

Active and Passive Microwave Techniques

A major milestone in satellite precipitation-rate estimation was the development of microwave remote sensing instruments that respond in a more physically *direct* way to the presence of precipitation-size water and/or ice particles within the clouds, although remaining insensitive to non-precipitating clouds (Petty, 1995). The first Special Sensor Microwave Imager (SSM/1) was launched in 1987 and included a high-frequency channel at 85.5 GHz. Basically, precipitation-size ice particles and large raindrops significantly reduce the emissivity of the cloud and thus depress its brightness temperature to below a nominal background level. This allows a rainfall estimate to be computed. This sensor could distinguish rainfall over land with reasonable consistency at 15 x 15 km spatial resolution. The SSM/I has provided worldwide precipitation estimates since 1987.

The single most important event in rainfall estimation (especially for tropical and subtropical regions) was the launch of the Tropical Rainfall Measuring Mission (TRMM) on November 27, 1997, by NASA and the National Space Development Agency (NASDA) of Japan (TRMM, 1999). The satellite is in a 350 km orbit at an inclination of 35°. TRMM carries five instruments onboard: the Precipitation Radar (PR), the TRMM Microwave Imager (TMI), Visible Infrared Scanner (VIRS), Lightning Imaging Sensor (LIS), and the Clouds and Earth's Radiant Energy System (CERES).

The TRMM Microwave Imager (TMI) is a passive microwave sensor designed to provide quantitative rainfall information on the integrated column precipitation content over a 487 mile (780 km) swath. It is best suited for rainfall estimates over oceans where data are needed most for climate model verification. It measures the intensity of radiation at five frequencies: 10.7 (45 km spatial resolution), 19.4, 21.3, 37, and 85.5 GHz (5 km spatial resolution). It has dual polarization at four of the frequencies. The 10.7 GHz frequency provides a more linear response for the high rainfall rates common in tropical rainfall. A composite image of all the TRMM Microwave Imager orbital paths on March 9, 1998, are presented in Figure 11-12 and Color Plate 11-4a. Note the intense line of thunderstorms passing over southern Florida.

The TRMM Precipitation Radar (PR), the first in space, measures the three-dimensional rainfall distribution over both land and oceans. It defines the layer depth of the precipitation and provides information about the rainfall actually reaching the surface, which is used to determine the latent heat of the atmosphere (TRMM, 1999). Passive microwave sensors have difficulty measuring rain over land areas (discussed in Chapter 9), whereas the scanning radar operating at 13.8 GHz (HH) provides accurate precipitation data over land at a spatial resolution of 4.3 km at nadir and a swath width of 220 km. TRMM Precipitation Radar data over Florida obtained on March 9, 1998 is shown as an oblique image in Color Plate 11- 4b and as an along-track cross-section in Color Plate 11-4c. The cross-section provides a significant amount of information about cloud height and precipitation.

The Visible Infrared Scanner (VIRS) provides high resolution information on cloud coverage, type, and cloud-top temperatures. It is a five channel cross-track scanning radiometer operating at 0.63, 1.6, 3.75, 10.80, and 12 µm. It obtains 2.1 km data at nadir with a swath width of approximately 720 km.

The Lightning Imaging Sensor (LIS) inventories the global incidence of lightning using an optical remote sensing system operating at 0.7774 µm with a spatial resolution of 5 x 5 km at nadir and a 590 km swath width. The Clouds and Earth's Radiant Energy System (CERES) is a visible and infrared sensor designed especially to measure emitted and reflected radiative energy from the surface of the Earth, and from the atmosphere and its constituents (e.g., clouds, aerosols). It is a broad-band scanning radiometer with a total channel spectral range of 0.3 – 50 mm.

TRMM's unique combination of sensor wavelengths, coverages, and resolving capabilities with its low-altitude, non-

Sun-synchronous orbit provide monthly precipitation amounts to a 500 x 500 km grid.

 Aerosols and Clouds

NASA Earth Science Enterprise scientists state that more information is needed on the following variables if we are to be able to model and understand the forces modifying the Earth's global climate system (Kahn, 1999):

- the amount and type of atmospheric particles (aerosols), including those formed by nature and by human activities;

- the amount, type, and height of clouds; and

- the distribution of land-surface cover, including vegetation canopy structure.

Aerosols

Aerosol particles may be solid or liquid and range in size from 0.01 micrometers to several tens of micrometers. Cigarette smoke particles are in the middle of this size range; typical cloud drops are 10 or more micrometers in diameter. Aerosols tend to cool the surface below them because most aerosols are bright particles that reflect Sunlight back to space reducing the amount of solar radiation that can be absorbed by the surface below (Kahn, 1999). The magnitude of this effect depends on the size and composition of the aerosols, and on the reflecting properties of the underlying surface. Some believe that the cooling effect due to increased amounts of aerosols may actually reduce or delay some of the expected warming due to increases in the amount of atmospheric carbon dioxide. Perhaps more smog is good! Others disagree.

Although remote sensing aerosol retrieval science has been taking place for more than 20 years, until recently there was no method of obtaining information on aerosol amount and distribution (called optical depth) and particle properties. The EOS *Terra* Multi-angle Imaging SpectroRadiometer (MISR) will collect such information using four spectral bands (blue, green, red, and near-infrared) and nine look angles (refer to Chapter 7 for a detailed description). Four of the linear-array sensors look forward at an oblique angle (26.1˚, 45.6˚, 60˚, and 70.5˚, respectively), one looks directly down (nadir), and four look aft at the same oblique angles.

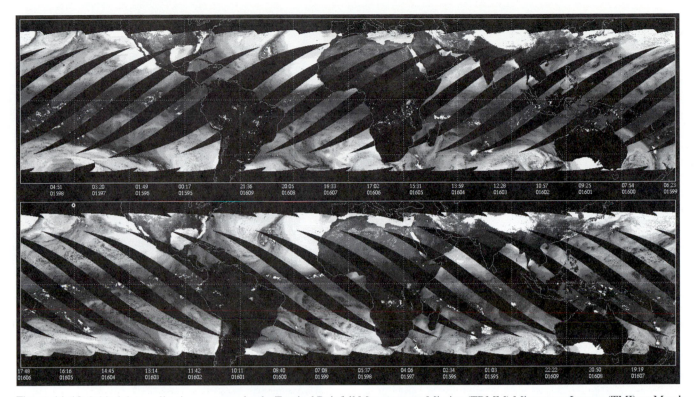

Figure 11-12 Orbital data-collection coverage by the Tropical Rainfall Measurement Mission (TRMM) Microwave Imager (TMI) on March 9, 1998. Notice that the 35° angle of inclination obtains good coverage of the tropical and subtropical portions of the Earth. An intense frontal system is over southern Florida. Also please refer to Color Plate 11-4a (courtesy NASA Goddard TRMM Office and Japan National Space Development Agency).

This provides multiple looks at exactly the same part of the atmosphere (and the surface of the Earth), allowing quantitative information to be derived about aerosol location and content. The aerosol characteristics of the entire Earth may be obtained every nine days.

Clouds

Clouds play a major role in controlling the Earth's climate. More than any other component of the climate system, clouds affect the flow of energy within the Earth's atmosphere. A cloud may warm or cool the Earth, depending upon its thickness and height above the surface. Low, thick clouds reflect incoming solar radiation back to space, which causes cooling. High clouds trap outgoing infrared radiation and produce greenhouse warming. Because cloud type, height, moisture content, and location are so variable, their effect on global climate is very difficult to measure. Many believe that the largest uncertainty in climate prediction models is how to characterize and parameterize the radiative and physical properties of clouds.

Clouds in Visible Imagery

The first meteorological satellites only measured visible energy reflected from clouds (0.4 – 0.7 μm). New GOES sensors provide data in both the visible and thermal infrared portion of the spectrum, as summarized in Chapter 7. In the daylight hours, visible imagery provides detailed views of the cloud patterns that closely match our visual senses, i.e., clouds usually appear bright while land and water appear darker on the images. Consider typical visible images of both the western and eastern United States obtained by GOES-east and GOES-west on April 17, 1998 (Figure 11-13a,b). Note the large low-pressure frontal system approaching Washington state from the Bay of Alaska. There is also a significant low-pressure system located along the Eastern Seaboard and throughout the southeast. GOES was first launched on October 16, 1975. Since that time many new GOES satellites have been parked at 35,790 km in geostationary orbit to obtain visible and infrared imagery. They are used to monitor frontal systems, intense thunderstorm activity associated with tornadoes, and hurricanes. Europe has a similar system, called METEOSAT.

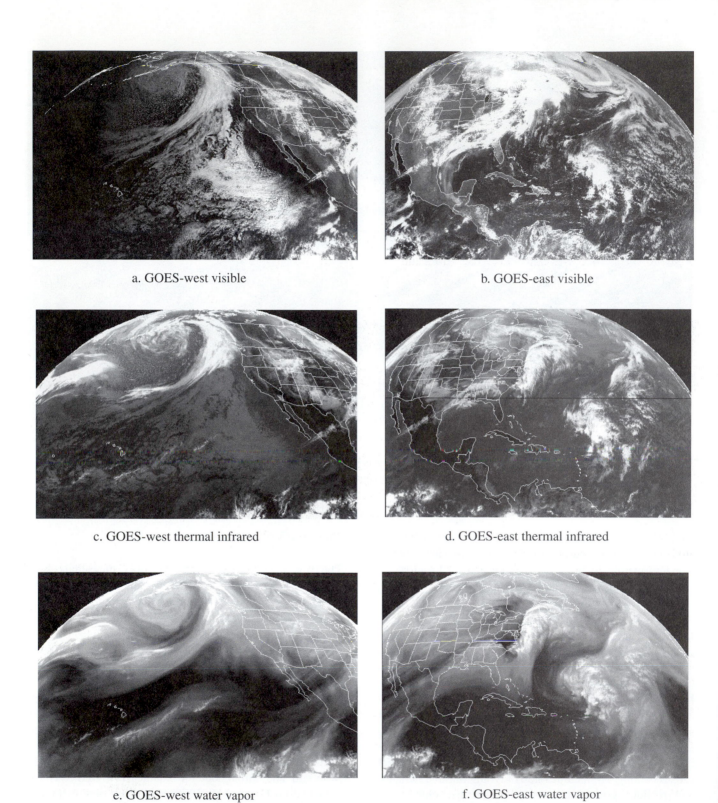

a. GOES-west visible

b. GOES-east visible

c. GOES-west thermal infrared

d. GOES-east thermal infrared

e. GOES-west water vapor

f. GOES-east water vapor

Figure 11-13 Examples of GOES-east and GOES-west images of the United States and portions of Central America obtained on April 17, 1998 (courtesy NOAA).

Visible imagery can only be obtained during the daytime. However, a light-sensitive instrument onboard the Defense Meteorological Satellite Program (DMSP) can obtain visible images at night. This is done by recording the features illuminated at night by moonlight.

In general, clouds do not reflect solar radiation equally well in all directions. Therefore, a single measurement of reflectivity from a single direction (e.g., at nadir) makes it difficult to determine the total amount of light reflected by the cloud (its *albedo*) relative to the incident energy. One solution is to obtain multiple images of a cloud from different vantage points. This introduces stereoscopic parallax. The EOS *Terra* MISR collects stereoscopic cloud information by viewing each cloud from nine angles previously discussed. The stereoscopic data can be analyzed to yield three-dimensional quantitative information about cloud height, structure, thickness, shape, and roughness of cloud tops. Accurate albedo information can also be computed.

Clouds in Thermal Infrared Imagery

The most common thermal infrared band used in meteorological investigations is 10 – 12.5 µm. The atmosphere is relatively transparent to this wavelength energy upwelling from the Earth's surface and clouds. Also, thermal infrared images can be obtained at night, so we can have a continuous 24-hr record of cloud patterns and do not miss important meteorological events taking place at night.

In satellite meteorology, infrared images are normally inverted, i.e. the greater the radiance of a pixel, the darker the pixel. This way, clouds, which are usually colder than the surface of the Earth, appear white, and the warmer ground or ocean surfaces appear darker than clouds, as in visible images (Kidder and VanderHaar, 1995). Thermal infrared GOES-west and GOES-east images of the United States are shown in Figure 11-13c,d. It is easier to identify the edge of frontal systems using the thermal infrared imagery. Also, the land-water boundaries are more distinct.

It has been known for some time that it is possible to extract information on the type of clouds and their height using multispectral remote sensing. One of the earliest relationships is demonstrated in Figure 11-14 where visible and infrared data can be used to differentiate between the sea, land, cumuliform clouds, semitransparent high clouds, and convective clouds (like thunderstorms). Tall convective cumulonimbus clouds are cold and bright. The sea and land surface are warm and dark (Desbois et al., 1982). The analyst extracts the pixel value in the visible and thermal infra-

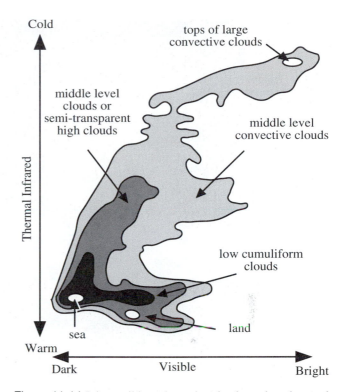

Figure 11-14 It is possible to determine cloud type based on multispectral measurements in the visible and thermal infrared portion of the electromagnetic spectrum (after Desbois et al., 1982).

red bands, locates it in the diagram, and identifies the nature of the cloud under investigation.

Numerous scientists have attempted to detect the presence of clouds using a thermal infrared cloud temperature threshold value. The goal is to identify low-temperature pixels that correspond to medium- or high-altitude clouds. The main problem is usually the specification of the exact temperature threshold value. For example, Derrien et al. (1993) developed an automated cloud-detection procedure based on thresholding NOAA AVHRR 11 µm band data. Franca and Cracknell (1995) developed five different cloud-detection algorithms based on visible and infrared spectral information using NOAA AVHRR daytime data for tropical areas.

Cloud-height information extracted from thermal infrared data can be used to generate pseudo three-dimensional oblique images of major storm events. For example, the AVHRR visible and infrared images in Figure 11-15a,b capture Hurricane Andrew moving toward Louisiana in 1992. In these examples, the visible (band 1; 0.58 – 0.68 µm) and near-infrared images (band 2; 0.725 – 1.10 µm) are draped

a.

b.

Figure 11-15 a) AVHRR visible band 1 (0.58 – 0.68 µm) imagery of Hurricane Andrew bearing down on Louisiana obtained on August 25, 1992, at 20:20 UT. The apparent height of the perspective rendering is inversely proportional to the cloud temperatures observed in band 5 (11.5 – 12.5 µm). b) AVHRR band 2 (0.725 – 1.10 µm) near-infrared data (courtesy of F. Hasler, K. Palaniappan, M. Manyin and H. Pierce, NASA Goddard Space Flight Center).

over the cloud-height information derived from thermal infrared data (band 5; 11.5 – 12.5 μm).

The TRMM Visible Infrared Scanner (VIRS) launched in 1997 provides high-resolution information on cloud coverage, type, and cloud-top temperatures using a five channel cross-track scanning radiometer. The *Terra* MISR sensor collects information in only the visible and near-infrared portions of the spectrum, while the *Terra* Clouds and Earth's Radiant Energy System (CERES) sensor collects data from just one look angle, but across the entire solar spectrum. CERES measures both solar-reflected and Earth-emitted radiation from the top of the atmosphere to the surface. It also determines cloud properties including amount, height, thickness, and particle size. Thus, the VIRS, MISR, and CERES instruments complement one another in the collection of cloud information. MODIS obtains cloud-top information from bands 33 – 36 in the thermal infrared region from 13.185 – 14.385 μm.

 Water Vapor

Water vapor is essential for precipitation. It is possible to detect and map water vapor by sensing in water-vapor absorption bands. Several wavelengths can be used, but the most common is centered around 6.7 μm. METEOSAT 1, launched in 1978 by the European Space Agency, was the first geostationary satellite to obtain images of mid- to upper-troposphere water vapor in the 6.7 μm region in addition to visible and 10 – 12 μm infrared images (Kidder and VanderHaar, 1995).

GOES sensors routinely provide water-vapor images obtained in the 6.7 μm region. At this wavelength, most of the radiation sensed by the satellite comes from the atmospheric layer between 300 and 600 km, i.e., from the middle layers of the troposphere. Figures 11-13e,f depict water-vapor images of the western and eastern United States. Note the tremendous amount of water vapor in the approaching storm in the western United States and the departing storm in the eastern United States. Also note the lack of water vapor on this particular date around Hawaii and in the lower portion of the Gulf of Mexico near Cozumel, Mexico. The weather was very comfortable on this day in these areas. The relative humidity is likely to be higher in bright areas than in dark areas on a water-vapor image. Bright and dark areas may also indicate rising and sinking motions, respectively. MODIS has several bands that are sensitive to atmospheric

water vapor, including band 17 (890 – 920 nm), 18 (931 – 941 nm), and 19 (915 – 965 nm).

 Snow

Snow runoff can be both beneficial and disastrous. Spring runoff is the lifeblood for most agricultural activities in semiarid climates and provides much of the potable drinking water for mankind (in addition to groundwater). Unfortunately, a tremendous buildup of snow or an untimely thaw can cause severe flooding and consequently loss of life and property. Therefore, meteorologists, hydrologists, and glaciologists work diligently to identify 1) the geographic extent of the snowpack on a seasonal basis, and 2) its water equivalent.

Snow in the Visible Spectrum

When clouds are not present in a visible and/or near-infrared image, it is a relatively straightforward task to identify the spatial distribution of snow because it is generally much brighter than the vegetation, soil, or water nearby that is not covered with snow (Engman, 1995). Also, one of the major diagnostic clues that can be used to distinguish between clouds and snow is to obtain numerous images of the terrain in a relatively short time period using AVHRR or GOES data. Clouds move; snow does not.

Snow in the Middle-Infrared and Microwave Regions

However, if one can only obtain a single image of the terrain and it contains both clouds and snow, then it is possible to utilize the middle-infrared portion of the spectrum to differentiate between snow and cloud cover. The reason is made clear in Figure 11-16. Throughout the visible and near-infrared portions of the spectrum, clouds and snow reflect approximately equal amounts of radiant flux. In the middle-infrared portion of the spectrum (especially 1.5 – 2.5 μm) clouds continue to reflect or emit a substantial amount of energy, while snow reflectance and/or emission reduces dramatically. This relationship was first documented by Skylab sensors operating in the region 1.55 – 1.75 μm. In this band, clouds have a very high reflectance and appear white in the imagery, while the snow has a very low reflectance and appears black. Thus, the Landsat Thematic Mapper band 5 (1.55 – 1.75 μm) and the SPOT 4 sensor with its middle-

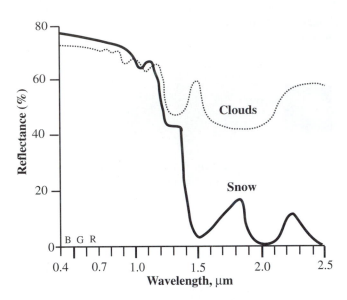

Figure 11-16 Reflectance of clouds and snow in the wavelength interval 0.4 to 2.5 μm (Bowker et al., 1985; after Avery and Berlin, 1992).

infrared band can be used to discriminate between clouds and snow cover. The decrease in reflectance of snow from 80 – 90 percent at wavelengths < 1 μm to < 10 percent at approximately 1.5 μm is extraordinary when compared with the reflectance characteristics of other Earth surface materials. As snow ages, the reflectance generally decreases in the infrared region but not in the visible.

The microwave region of the spectrum offers promise for retrieving important snow information (Engman, 1995). Depending upon the wavelength, estimates of the depth, water content, and the amount/presence of liquid water in the snow pack are possible. In addition, clouds are transparent to many of the microwave frequencies so that mapping of the snow area and properties is possible even in regions where clouds are common. Aircraft and Space Shuttle Synthetic Aperture Radar (SAR) measurements have shown that SAR can be used to discriminate between snow and glaciers and between wet and dry snow (Shi et al., 1994).

Water Quality Modeling Using Remote Sensing and GIS

State and county agencies are required to reduce point source discharges to surface waters. Point sources are relatively easy to control and regulate because of their known location. A phenomenon more common and more difficult to detect is *nonpoint source pollution* (NPS). NPS is defined as pollution originating from urban runoff, construction, hydrologic modification, silviculture, mining, agriculture, irrigation return flows, solid-waste disposal, atmospheric deposition, stream bank erosion, and sewage disposal. Impacts of NPS include, decreased recreational water use; reduction of water storage capacity in streams, lakes and estuaries; clogging of drainage ditches and irrigation canals; excessive enrichment and sedimentation of surface waters that contribute to the loss of fish, shellfish and wildlife habitat; and a reduction in the aesthetic qualities of the aquatic environment (Corbitt, 1990).

Traditional *in situ* measurement techniques have had a limited effect on identifying and modeling nonpoint source pollution. Scientists have found that an integration of *in situ* and remote sensing data collection with GIS modeling techniques can provide some very useful water quality information. In fact, remote sensing data collection combined with GIS modeling offers a means of identifying and ranking NPS potential in surface waters with output that is relatively easy for the general public to understand. Schultz (1988) provides examples of how remotely sensed biophysical variables such as slope, soils and land-cover characteristics can be used as input to hydrological models.

An Integrated Remote Sensing and GIS Water Quality Model

This brief case study evaluates water quality degradation and impacts from urban runoff within the Withers Swash watershed in Myrtle Beach, SC. Withers Swash is an urbanized watershed (34% impervious) that covers approximately 4.2 mi^2 and includes industrial, residential, commercial, agricultural, recreational, and undeveloped land uses. The watershed is divided into two subbasins (Figure 11-17a), which drain into a tidal pool designed to capture sediment before emptying into the Atlantic Ocean. A study was conducted using remotely sensed data to derive land-cover information which was then integrated with other spatial information such as hydrology, topography, and soil data, into a GIS database. The data were then analyzed using the *Agricultural NonPoint Source* (AGNPS) pollution model to generate water quality predictions (USDA, 1998; Schill and Jensen, 1998).

The AGNPS water-quality model uses eight directional flow attributes determined by topography and channel characteristics to route stormwater throughout the watershed to a single outlet cell (Cronshey and Theurer, 1998). Approximately

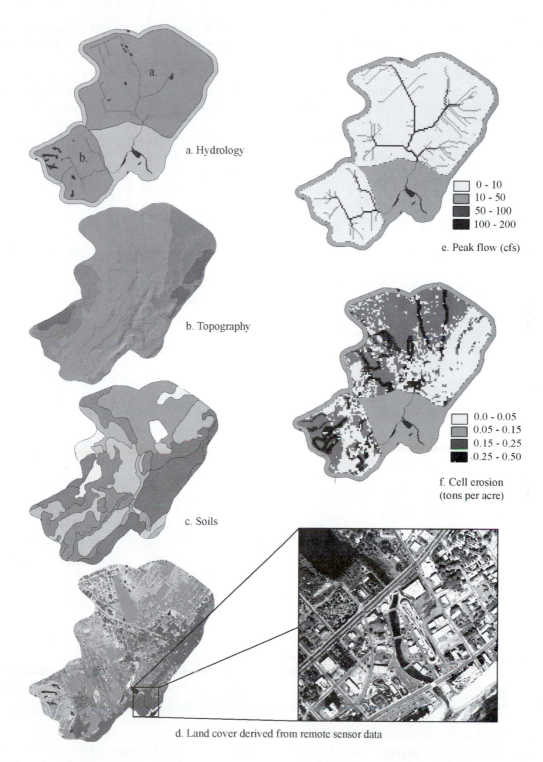

a. Hydrology

e. Peak flow (cfs)

- 0 - 10
- 10 - 50
- 50 - 100
- 100 - 200

b. Topography

f. Cell erosion
(tons per acre)

- 0.0 - 0.05
- 0.05 - 0.15
- 0.15 - 0.25
- 0.25 - 0.50

c. Soils

d. Land cover derived from remote sensor data

Figure 11-17 Nonpoint source pollution modeling using the Agricultural NonPoint Source (AGNPS) pollution water quality model applied to two subbasins (a) in the Withers Swash watershed in Myrtle Beach, SC. Modeling was based on 30 x 30 m cells and a 2.68 in. rain in 24 hours. b) Topography was derived photogrammetrically and used to extract slope information. c) Soils data were sampled in the field. d) Land cover characteristics were derived from analysis of aerial photography and ATLAS multispectral scanner data. Two examples of model output include concentrated peak flow in ft^3 per second (e), and total cell erosion in tons per acre (f) (Schill and Jensen, 1998).

one third of the model input parameters are derived from topography (Figure 11-17b). These parameters include terrain slope, the Universal Soil Loss Equation (USLE) slope length, and slope shape. Soil attributes including erodability (USLE K factor) and texture, were verified through field sampling and sieve tests (Figure 11-17c). Land-cover information was derived from 1:10,000-scale color-infrared aerial photography and from NASA's Airborne Terrestrial Applications Sensor (ATLAS) 3 x 3 m multispectral data (Figure 11-17d). From these land-cover attributes, the SCS curve number (assigned by soil hydrologic group), Manning roughness coefficient, USLE C (cropping) and P (practice) factors, soil condition constant, and chemical oxygen demand (COD) were derived based on lookup tables in Young et al. (1994).

Before executing the model, all spatial data were assigned to a 30 x 30 m cell (0.2 x 0.2 acre) for each of the two sub-basins. A single storm event was simulated using the following parameters: storm energy intensity (*r* factor) of 95.39, storm duration of 24 hours, total storm rainfall equal to 2.68 in., and rainfall nitrogen concentrations of 0.80 ppm. An example of two model output parameters include concentrated peak flow in ft^3 per second (cfs) (Figure 11-17e) and total cell erosion in tons per acre (Figure 11-17f). Model results have been compared with *in situ* measurements and used to locate areas of increased sediment and nutrient loss and to indicate the spatial extent of nutrient and sediment transport, deposition and accumulation. This information has been used to spatially allocate vegetated buffer strips and other best management practices (BMPs).

 References

Aiken, J., G. F. Moore, D. K. Clark and C. C. Trees, 1995, *The Sea-WiFS CZCS-Type Pigment Algorithm*, NASA Technical Memorandum 104566, Vol. 29, Greenbelt: NASA Goddard Space Flight Center, 34 pp.

Arkin, P. A. and B. N. Meisner, 1987, "Spatial and Annual Variation in the Diurnal Cycle of Large Scale Tropical Convective Cloudiness and Precipitation," *Monthly Weather Review*, 115:1009–1032.

Asrar, G. and J. Dozier, 1994, *EOS: Science Strategy for the Earth Observing System*, Woodbury, New York: American Institute of Physics.

Avery, T. E. and G. L. Berlin, 1992, *Fundamentals of Remote Sensing and Airphoto Interpretation*, New York: Macmillan.

Bhargava, D.S. and D. W. Mariam, 1990, "Spectral Reflectance Relationships to Turbidity Generated by Different Clay Materials," *Photogrammetric Engineering & Remote Sensing*, 56:225-229.

Bowker, D. E., R. E. Davis, D. L. Myrick, K. Stacy and W. T. Jones,1985, *Spectral Reflectances of Natural Targets for Use in Remote Sensing Studies*, Washington: National Aeronautics & Space Administration, Publication #1139.

Bukata, R. P., J. H. Jerome, K. Y. Kondratyev and D. V. Pozdnyakov, 1995, *Optical Properties and Remote Sensing of Inland and Coastal Waters*, New York: CRC Press, 362 pp.

Campbell, J. B., 1996, *Introduction to Remote Sensing*, New York: The Guilford Press.

Cary, T., 1994, "A World of Possibilities: Remote Sensing Data for Your GIS," *Geo Info Systems*, September, 38–42.

Clark, C. D., H. T. Ripley, E. P. Green, A. J. Edwards and P. J. Mumby, 1997, "Mapping and Measurement of Tropical Coastal Environments with Hyperspectral and High Spatial Resolution Data," *International Journal of Remote Sensing*, 18(2):237–242.

Corbitt, R. A., 1990, *Standard Handbook of Environment Engineering*, New York: McGraw Hill, Inc., pp. 7.50–7.57.

Cronshey, R. G. and F. D. Theurer, 1998, "AnnAGNPS — Nonpoint Pollutant Loading Model," *Proceedings,* First Federal Interagency Hydrologic Modeling Conference, Las Vegas, NV: 9–16.

Derrien, M., B. Farki, L. Harang, H. LeGleau, A. Noyalet, D. Pochic and A. Sairouni, 1993, "Automatic Cloud Detection Applied to NOAA-11 AVHRR Imagery," *Remote Sensing of Environment*, 46:246-267.

Desbois, M., G. Seze and G. Szejwach, 1982, "Automatic Classification of Clouds on METEOSAT Satellite Imagery: Applications to High-level Clouds," *Journal of Applied Meteorology*, 21:401–412.

Engman, E. T. and R. J. Gurney, 1991, *Remote Sensing in Hydrology*, New York: Chapman and Hall, 225 pp.

Engman, E. T., 1995, "The Use of Remote Sensing Data in Watershed Research," *Journal of Soil and Water Conservation*, 50(5):438–440.

Falkowski, P. G., M. J. Behrenfeld, W. E. Esaias, W. Balch, J. W. Campbell, R. L. Iverson, D. A. Kiefer, A. Morel and J. A. Yoder, 1998, *Satellite Primary Productivity Data and Algorithm Development: A Science Plan for Mission to Planet Earth,* NASA Technical Memo 104566, Vol. 42, Greenbelt, MD: NASA Goddard Space Flight Center, 36 pp.

Franca, G. B. and A. P. Cracknell, 1995, "A Simple Cloud Masking Approach Using NOAA AVHRR Daytime Data for Tropical Areas," *International Journal of Remote Sensing,* 16(9):1697-1705.

Gitelson, A. A., 1992, "The Peak Near 700 nm on Radiance Spectra of Algae and Water: Relationships of its Magnitude and Position with Chlorophyll Concentration, *International Journal of Remote Sensing,* 13:3367–3373.

Goodin, D. G., L. Han, R. N. Fraser, D. C. Rundquist and W. A. Stebbins, 1993, "Analysis of Suspended Solids in Water Using Remotely Sensed High Resolution Derivative Spectra," *Photogrammetric Engineering and Remote Sensing,* 59(4):505–510.

Gordon, H. R., D. K. Clark, J. W. Brown, O. B. Brown, R. H. Evans and W. W. Broenkow, 1983, "Phytoplankton Pigment Concentrations in the Middle Atlantic Bight: Comparison of Ship Determinations and CZCS Estimates," *Applied Optics,* 22:20–36.

Han, L., 1997, "Spectral Reflectance with Varying Suspended Sediment Concentrations in Clear and Algae-Laden Waters," *Photogrammetric Engineering and Remote Sensing,* 63(6):701–705.

Han, L. and D. C. Rundquist, 1994, "The Response of Both Surface Reflectance and the Underwater Light Field to Various Levels of Suspended Sediments: Preliminary Results," *Photogrammetric Engineering & Remote Sensing,* 60:1463–1471.

Han, L. and D. C. Rundquist, 1997, "Comparison of NIR/RED Ratio and First Derivative of Reflectance in Estimating Algal-Chlorophyll Concentration: A Case Study in a Turbid Reservoir," *Remote Sensing of Environment,* 62:253–261.

Han, L. and D. C. Rundquist, 1998, "The Impact of a Wind-roughened Water Surface on Remote Measurements of Turbidity," *International Journal of Remote Sensing,* 19(1):195–201.

Hoffer, R., 1978, "Spectral Characteristics of Water and Snow," *Remote Sensing: The Quantitative Approach,* P. Swain and S. Davis (Eds.), New York: McGraw Hill, Inc.

Hooker, S. B., W. E. Esaias, G. C. Feldman, W. W. Gregg and C. R. McClain, 1992, *An Overview of SeaWiFS and Ocean Color,* NASA Technical Memorandum 104566, Vol. 1, Greenbelt, MD: Goddard Space Flight Center, 24 pp.

Ikeda, M. and F. W. Dobson, 1995, *Oceanographic Applications of Remote Sensing,* New York: CRC Press, 492 pp.

Ishizaka, J., H. Fukushima and M. Kishino, 1997, "Ocean Colour and Temperature Scanner Update," *Backscatter,* 8(1):11–16

Jensen, J. R., 1996, *Introductory Digital Image Processing: A Remote Sensing Perspective.* Saddle River: Prentice-Hall, 318 pp.

Julien, R. Y., 1995, *Erosion and Sedimentation,* New York: Cambridge University Press, 279 pp.

Kahn, R., 1999, *A General Introduction to MISR,* Pasadena: NASA Jet Propulsion Lab, http://www-misr.jpl.nasa.gov/miintro.html.

Kidder, S. Q. and T. H. VanderHaar, 1995, *Satellite Meteorology: An Introduction,* New York: Academic Press, 466 p.

Lodhi, M. A., D. C. Rundquist, L. Han and M. S. Juzila, 1997, "The Potential for Remote Sensing of Loess Soils Suspended in Surface Waters," *Journal of the American Water Resources Association,* 33(1):111–127.

Markham, B. L. and J. C. Barker, 1986, "Landsat MSS and TM Post-Calibration Dynamic Ranges, Exoatmospheric Reflectance and at Satellite Temperature," *Landsat Technical Notes,* 1:3–8.

Martonchik, J., 1999, *MISR's Study of the Earth's Surface,* MISR home page, http://www-misr.jpl.nasa.gov/misci.html.

Mumby, P. J., E. P. Green, A. J. Edwards and C. D. Clark, 1997, "Coral Reef Habitat-mapping: How Much Detail can Remote Sensing Provide?" *Marine Biology,* 130:193–202.

Nanu, L., 1993, "Effect of Suspended Sediment Depth Distribution on Coastal Water Spectral Reflectance: Theoretical Simulation," *International Journal of Remote Sensing,* 14(2):225–239.

NOAA, 1999, *What is an El Nino?,* Washington, DC: NOAA.

Pasqualini, V., C. Pergent-Martine, C. Fernandez and G. Pergent, 1997, "The Use of Airborne Remote Sensing for Benthic Cartography: Advantages and Reliability," *International Journal of Remote Sensing,* 18(5):1167–1177.

Petty, G. W., 1995, "The Status of Satellite-Based Rainfall Estimation over Land," *Remote Sensing of Environment,* 51:125–137.

Ramsey, E. W., J. R. Jensen, H. E. Mackey and J. Gladden, 1992, "Remote Sensing of Water Quality in Active to Inactive Cooling Water Reservoirs," *International Journal of Remote Sensing,* 13(18):3465–3488.

Reynolds, R. W., 1988, "A Real-time Global sea-surface Temperature Analysis" *Journal of Climate*, 1:75–86.

Reynolds, R. W. and T. M. Smith, 1994, "Improved Global sea-surface Temperature Analyses Using Optimum Interpolation," *Journal of Climate*, 7:929–948.

Rundquist, D. C., J. F. Schalles and J. S. Peake, 1995, "The Response of Volume Reflectance to Manipulated Algal Concentrations Above Bright and Dark Bottoms at Various Depths in an Experimental Pool," *Geocarto International*, 10:5–14.

Rundquist, D. C., L. Han, J. F. Schalles and J. S. Peake, 1996, "Remote Measurement of Algal Chlorophyll in Surface Waters: The Case for the First Derivative of Reflectance Near 690 nm," *Photogrammetric Engineering & Remote Sensing*, 62(2):195–200.

Schill, S. R., 1997, *Modeling Impacts of Coastal Development on Water Quality in Beaufort County, South Carolina Using a Remote Sensing/GIS Approach,* masters thesis, Columbia, SC: Department of Geography, University of South Carolina, 152 pp.

Schill, S. R. and J. Jensen, 1998, "Modeling to Predict Stormwater Runoff in Withers Swash Myrtle Beach, SC," *Proceedings*, Urban and Regional Information Systems, Charlotte, NC: 83–93.

Schultz, G. A., 1988, "Remote Sensing in Hydrology," *Journal of Hydrology*, 100:239–265.

Shi, J., J. Dozier and H. Rott, 1994, "Snow Mapping in Alpine Regions with Synthetic Aperture Radar," *IEEE Transactions on Geoscience and Remote Sensing*, 32(1).

Suplee, C., 1999, "El Nino and La Nina: Nature's Vicious Cycle," *National Geographic*, 195(3):73–95.

TRMM, 1999, *Tropical Rainfall Measurement Mission (TRMM) Home Page*, Greenbelt, MD: National Aeronautics and Space Administration, http//trmm.gsfc.nasa.gov/trmm_office.

U.S. Department of Agriculture, 1998, *AGNPS 98 Agricultural Non-Point Source Pollution Model (Version 5.0)*, USDA (Agricultural Research Service), Morris, MN, http://www.ftc.nrcs.usda.gov/huwq/agnps.htm.

Young, R. A., C. A. Onstad, D. D. Bosch, and W. P. Anderson, 1994, *Agricultural Non-Point Source Pollution Model (AGNPS)*, Version 4.0, U.S Department of Department of Agriculture, St. Joseph, MI.

Remote Sensing the Urban Landscape 12

U rban landscapes are composed of a diverse assemblage of materials (concrete, asphalt, metal, plastic, shingles, glass, water, grass, shrubbery, trees, and soil) arranged by humans in complex ways to build housing, transportation systems, utilities, commercial and industrial facilities, and recreational landscapes. The goal of this construction is hopefully to improve the quality of life. In many instances, *urbanization* is taking place at a dramatic rate, with and without planned development (Hagerstrand, 1995). A significant number of businessmen and women and public organizations constantly require up-to-date information about the city and suburban infrastructure (Gottmann, 1994). For example, detailed urban information is required by:

- city, county, and regional councils of governments which legislate zoning regulations to hopefully improve the quality of life in the urbanized areas;

- city and state departments of commerce which are mandated to stimulate development, often to increase the tax base;

- tax assessor offices which maintain legal geographic descriptions of every parcel of land, assess its value, and levy a tax-millage rate;

- county and state departments of transportation that maintain existing facilities, hopefully build new facilities without damaging the environment, and prepare for future transportation demand;

- public and private utility companies (water, sewer, gas, electricity, telephone, cable) which attempt to predict where new demand will occur and plan for the most efficient and cost-effective method of delivering the services;

- public service commissions which are mandated to make the utility services available economically to the consumer;

- departments of parks, recreation, and tourism which attempt to improve our public recreation facilities;

- departments of emergency management and preparedness which are responsible for mitigating destruction and allocating resources in the event of a disaster;

- private real estate companies which are paid to find the ideal location for industrial, commercial, and residential development; and

• developers who continually build residential, commercial, and industrial facilities to stay in business.

These are educated, professional users of urban information (Cullingworth, 1997). The urban/suburban land they manage or develop is of significant monetary value. Therefore, it is not surprising that city, county, and state agencies as well as private companies invest hundreds of millions of dollars each year obtaining aerial photography and other forms of remotely sensed data to extract the required urban information. Much of the required information simply cannot be obtained by driving around and conducting a windshield survey or on-site (*in situ*) investigation.

 Urban/Suburban Resolution Considerations

Many of the detailed urban/suburban attributes that businesses and agencies require are summarized in Table 12-1 and Figure 12-1. This chapter reviews how remotely sensed data may be of value for the collection of these attributes. To remotely sense urban phenomena, it is necessary to appreciate the urban attribute's temporal, spectral, and spatial resolution characteristics.

Urban/Suburban Temporal Resolution Considerations

Three types of temporal resolutions should be considered when monitoring urban environments using remote sensor data. First, urban/suburban phenomena often progress through an identifiable *developmental cycle* much like vegetation progresses through a *phenological cycle* (Chapter 10). For example, Jensen and Toll (1983) documented a 10-stage single-family residential housing development cycle at work in suburban Denver, CO, that progressed from stage-1 (undeveloped rangeland), to stage-10 (fully landscaped residential housing), often within just six months (Figure 12-2a). The stages of residential development were identified based on the presence or absence of five major factors, including:

• partial or complete clearing of the parcel of land (i.e., the amount of original vegetation that was allowed to remain),

• land subdivision (i.e., parcels surveyed and graded),

• roads (dirt or paved),

• buildings (presence or absence), and

• landscaping (partial or complete).

Single-family residential housing developments in San Diego, CA, undergo a similar cycle (Figures 12-2bc). It is imperative that the image analyst understand the temporal development cycle of the urban phenomena being analyzed. If it is not understood, embarrassing and costly interpretation mistakes can be made.

The second type of temporal resolution is how often it is possible for the remote sensor system to collect data of the urban landscape, e.g., every 8 days, 16 days, or on demand. Up-to-date information is critical for most urban applications. Generally, sensors that can be pointed off-nadir (e.g., SPOT HRV) have higher temporal resolution than sensors that only sense the terrain at nadir (e.g., Landsat Thematic Mapper). Urban applications are usually not as time-sensitive as those dealing with highly dynamic phenomena such as vegetation where a life cycle might take place during a single season. For these reasons, most urban applications (except traffic count transportation studies and emergencies) only require that imagery be collected every year or so. Orbital characteristics of the satellite platform and the latitude of the study area also impact the revisit schedule. Remote sensor data may be collected on demand from suborbital aircraft (airplanes, helicopters) if weather conditions permit.

Finally, temporal resolution may refer to how often land managers require this type of information. For example, local planning agencies may need precise population estimates every 5 – 7 years in addition to the estimates provided by the decennial census. The temporal resolution requirements for many important urban applications are summarized in Table 12-1 and shown graphically in Figure 12-1.

Urban/Suburban Spectral Resolution Considerations

Most image analysts would agree that when extracting urban/suburban information from remotely sensed data, it is more important to have high spatial resolution (often ≤ 5 x 5 m) than high spectral resolution (i.e., a large number of bands). For example, local population estimates based on building unit counts usually require a minimum spatial resolution of 0.25 – 5 m (0.82 ft – 16.4 ft) to detect, distinguish between, and/or identify the type of individual buildings. Practically any visible band (e.g., green or red) or near-infrared spectral band at this spatial resolution will do. Of course, there must be sufficient spectral contrast between the object

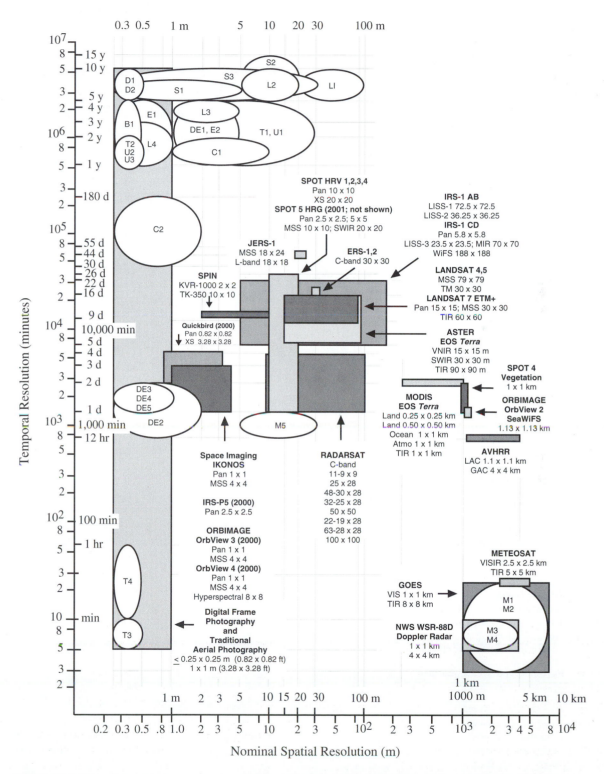

Figure 12-1 The clear polygons represent the spatial and temporal requirements for selected urban attributes listed in Table 12-1. Gray boxes depict the spatial and temporal characteristics of the major remote sensing systems that may be used to extract the required urban information (Jensen and Cowen, 1999).

Table 12-1.   Relationship Between Urban/Suburban Attributes and the Minimum Remote Sensing Resolutions Required to Provide Such Data (Jensen and Cowen, 1999)

Attributes	Minimum Resolution Requirements		
	Temporal	Spatial	Spectral
Land-Use/Land-Cover			
L1 – USGS Level I	5 – 10 years	20 – 100 m	V - NIR - MIR - Radar
L2 – USGS Level II	5 – 10 years	5 – 20 m	V - NIR - MIR - Radar
L3 – USGS Level III	3 – 5 years	1 – 5 m	Pan - V - NIR - MIR
L4 – USGS Level IV	1 – 3 years	0.25 – 1 m	Panchromatic (Pan)
Building and Property-Line Infrastructure			
B1 – building perimeter, area, volume, height and cadastral information (property lines)	1 – 5 years	0.25 – 0.5 m	Pan - Visible
Transportation Infrastructure			
T1 – general road centerline	1 – 5 years	1 – 30 m	Pan -V - NIR
T2 – precise road width	1 – 2 years	0.25 – 0.5 m	Pan - Visible
T3 – traffic count studies (cars, airplanes, etc.)	5 – 10 min	0.25 – 0.5 m	Pan - Visible
T4 – parking studies	10 – 60 min	0.25 – 0.5 m	Pan - Visible
Utility Infrastructure			
U1 – general utility line mapping and routing	1 – 5 years	1 – 30 m	Pan -V - NIR
U2 – precise utility line width, right-of-way	1 – 2 years	0.25 – 0.6 m	Pan - Visible
U3 – location of poles, manholes, substations	1 – 2 years	0.25 – 0.6 m	Panchromatic
Digital Elevation Model (DEM) Creation			
D1 – large-scale DEM	5 – 10 years	0.25 – 0.5 m	Pan -Visible
D2 – large-scale slope map	5 – 10 years	0.25 – 0.5 m	Pan -Visible
Socioeconomic Characteristics			
S1 – local population estimation	5 – 7 years	0.25 – 5 m	Pan -V - NIR
S2 – regional/national population estimation	5 – 15 years	5 – 20 m	Pan -V - NIR
S3 – quality of life indicators	5 – 10 years	0.25 – 30 m	Pan -V - NIR
Energy Demand and Conservation			
E1 – energy demand and production potential	1 – 5 years	0.25 – 1 m	Pan -V - NIR
E2 – building-insulation surveys	1 – 5 years	1 – 5 m	TIR
Meteorological Data			
M1 – weather prediction	3 – 25 min	1 – 8 km	V - NIR - TIR
M2 – current temperature	3 – 25 min	1 – 8 km	TIR
M3 – clear air and precipitation mode	6 – 10 min	1 km	WSR-88D Radar
M4 – severe weather mode	5 min	1 km	WSR-88D Radar
M5 – monitoring urban heat island effect	12 – 24 hr	5 – 30 m	TIR
Critical Environmental Area Assessment			
C1 – stable sensitive environments	1 – 2 years	1 – 10 m	V - NIR - MIR
C2 – dynamic sensitive environments	1 – 6 months	0.25 – 2 m	V - NIR - MIR - TIR
Disaster Emergency Response			
DE1 – pre-emergency imagery	1 – 5 years	1 – 5 m	Pan -V - NIR
DE2 – post-emergency imagery	12 hr – 2 days	0.25 – 2 m	Pan - NIR - Radar
DE3 – damaged housing stock	1 – 2 days	0.25 – 1 m	Pan - NIR
DE4 – damaged transportation	1 – 2 days	0.25 – 1 m	Pan - NIR
DE5 – damaged utilities, services	1 – 2 days	0.25 – 1 m	Pan - NIR

of interest (e.g., a building) and its background (e.g., the surrounding landscape) in order to detect, distinguish between, and identify the object from its background.

While hyperspectral data may not be required for most urban applications, there are still optimum portions of the electromagnetic spectrum that are especially useful for extracting certain types of urban/suburban information (Table 12-1). For example, USGS Level III land-cover is best acquired using the visible (0.4 – 0.7 μm; V), near-infrared (0.7 – 1.1 μm; NIR), middle-infrared (1.5 – 2.5 μm; MIR), and/or panchromatic (0.5 – 0.7 μm) portions of the spectrum. Building perimeter, area, and height information is best acquired using black-and-white panchromatic (0.5 – 0.7 μm) or color imagery (0.4 – 0.7 μm). The thermal infrared portion of the spectrum (3 – 12 μm; TIR) may be used to obtain urban temperature measurements. Active microwave sensors may obtain imagery of cloud-shrouded or tropical urban areas (e.g., Japanese JERS-1 L-band, Canadian RADARSAT C-band, and European Space Agency ERS-1,2 C-band).

Urban/Suburban Spatial Resolution Considerations

Trained image analysts do rely on black-and-white tone or color in aerial photography or other types of imagery to extract useful urban information. However, in many instances the *geometric* spatial elements of image interpretation such as object shape, size, texture, orientation, pattern, and shadow silhouette are often just as useful if not more useful (all elements are discussed in Chapter 5). Generally, the higher the spatial resolution of the remote sensor data, the more detailed information that can be extracted in the urban environment. But how do we know what spatial resolution imagery to use for a specific urban application? Fortunately, there are several methods available for assessing the interpretability of imagery for urban applications based largely on the spatial resolution of the remote sensor data. Some are more widely adopted than others.

NIIRS Criteria

One solution might be to use the military and/or civilian versions of the National Image Interpretation Rating Scales (NIIRS) developed by the Image Resolution Assessment and Reporting Standards Committee (IRARS). NIIRS is the metric used by the intelligence community to characterize the usefulness of imagery for intelligence purposes (Leachtenauer, 1996, et al., 1998; Logicon, 1995, 1997; Pike, 1998). Interestingly, many urban image-interpretation tasks are not unlike those required for intelligence applications. The NIIRS criteria consist of 10 rating levels (0 – 9) for a given

type of imagery arrived at through evaluation by trained image analysts. The IRARS committee makes it clear that spatial resolution (ground-resolved distance) is only one of the measures of the interpretability of an image, although it is a very important one (Figure 12-3). Other factors such as film quality, atmospheric haze, contrast, angle of obliquity, and noise can reduce the ability of a well-trained analyst to detect, distinguish between, and identify military and civilian objects in an image. While it would be useful to use the NIIRS criteria, it is not optimum for this discussion because 1) the civil NIIRS criteria were only recently made available (Hothem et al., 1996; Leachtenauer et al., 1998), 2) there has not been sufficient time for the civilian community to familiarize itself with the concept, and consequently 3) the civilian community has never reported their collective experiences in urban/suburban information extraction using the NIIRS system.

Area Weighted Average Resolution

A figure of merit for measuring the resolvability of a film camera system often used by photogrammetrists is called the *area weighted average resolution* (AWAR), measured in line-pairs-per-millimeter (lp/mm) (Light, 1993). A line pair is the width of one black bar and one white space as contained on resolution targets recorded in an aerial photograph. Together, they form a pair and serve as a measure of image quality for the aerial film camera industry. The five essential elements that make up the system AWAR are the lens, film, image blur (smear) on the film due to aircraft forward velocity, angular motion, and the resolution of the duplicating film. Also, scene contrast of the Earth and atmosphere play a role in system resolution.

Scientists have studied the general relationship between aerial photography scale and AWAR. For example, Light (1993, 1996) documented that if we assume that the Earth is a low-contrast scene, the 1:40,000-scale National Aerial Photography Program (NAPP) photography exhibits approximately 39 lp/mm and yields approximately 25 μm for the size of 1 lp in the image. At 1:40,000 scale, 25 μm equate to a ground resolution of 1 x 1 m for low-contrast scenes. Therefore, a minimum of 1 m ground resolution can be expected throughout the photographic mission. In fact, the USGS digital orthophoto quarter-quad files produced from 1:40,000-scale NAPP photography are provided at 1 x 1 m (3.28 x 3.28 ft) spatial resolution by scanning the photography with a pixel size of 11 μm. Light (1998) suggests that there is a general linear relationship for larger scale aerial photography obtained using metric cameras, i.e., 1:20,000-scale photography equates to approximately 0.5 x 0.5 m (1.64 x 1.64 ft); 1:10,000-scale photography to 0.25 x

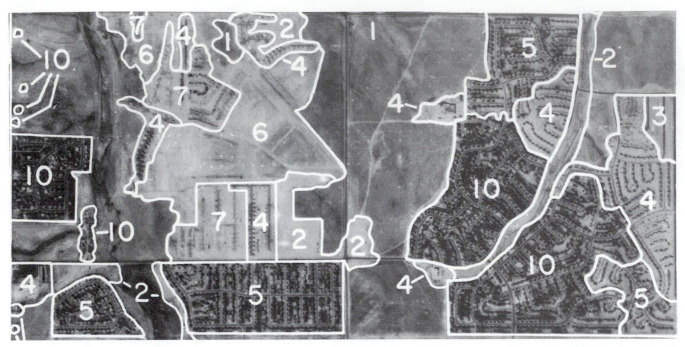

a. Single-family residential development near Denver, CO. A parcel of land may progress from rangeland (stage 1) to fully developed, land-scaped, single-family residential housing (stage 10).The stages of residential development were identified based on the presence or absence of five factors, including clearing, subdivision, roads, buildings, and degree of landscaping (Jensen and Toll, 1983).

b. Cleared, subdivided, terraced lots with dirt roads on steep slopes being developed for single-family residential housing in San Diego, CA.

c. Single-family residential housing in various stages of construc-tion. Rows one and two in the top left have foundation footings. Rows three and four have sub-flooring and are being framed.

Figure 12-2   Examples of stages in the development cycle of single-family residential housing in Denver, CO, and San Diego, CA. Remote sensor data may capture urban land use in one of many stages of development.

0.25 m (0.82 x 0.82 ft); and 1:5,000-scale photography to 0.125 x 0.125 m (0.41 x 0.41 ft). Unfortunately, most persons using remote sensor data for practical urban applications do not report their findings in terms of AWAR criteria.

Nominal Spatial Resolution (Ground-Resolved Distance)

The civilian user community usually reports the utility of a given type of imagery for extracting urban information based on the comparatively easy-to-understand concept of *nominal spatial resolution* (also called *ground-resolved distance*). For example, the Landsat 7 Enhanced Thematic Mapper Plus has six multispectral bands at 30 x 30 m nominal spatial resolution and a single 60 x 60 m thermal infrared band. The SPOT 3 HRV sensor has three multispectral bands with 20 x 20 m nominal spatial resolution and a single panchromatic band at 10 x 10 m.

Another general, nominal spatial resolution rule is that there needs to be a minimum of four spatial observations (e.g., pixels) within an urban object to identify it. Stated another way, the sensor spatial resolution should be one-half the width of the smallest object of interest. For example, to identify mobile homes that are 5 m wide, the minimum spatial resolution of imagery needed without haze or other problems is ≤ 2.5 x 2.5 m pixels (Cowen et al., 1995).

The temporal, spectral, and spatial resolution requirements for selected urban attributes summarized in Table 12-1 and Figure 12-1 were synthesized from subjective, practical experience reported in journal articles, symposia, chapters in books, and government and society manuals (Chisnell and Cole, 1958; Stone, 1964; Branch, 1971; Ford, 1979; Jensen et al., 1983; Avery and Berlin, 1992; Light (1993, 1996); Greve, 1996; Jensen, 1996; Philipson, 1997; Haack et al., 1997; Keister, 1997; Cowen and Jensen, 1998; Pike, 1998; Ridley et al., 1998; Jensen and Cowen, 1999; and others in the individual sections). Ideally, there would always be a remote sensing system that could obtain images of the terrain that satisfy the urban attributes' resolution requirements (Table 12-1). Unfortunately, this is not always the case.

 Remote Sensing Land Use and Land Cover

The term *land use* refers to how the land is being used by human beings. *Land cover* refers to the biophysical materials found on the land. For example, a state park may be used for recreation but have a coniferous forest cover. Urban land-use or land-cover information is required for a great variety of applications including residential-industrial-commercial site selection, population estimation, tax assessment, development of zoning regulations, etc. (Green et al., 1994; Cullingworth, 1997). This means that urban information collected for one application might be useful in another.

Land-Use/Land-Cover Classification Schemes

The best way to insure that urban information derived from remote sensor data is useful in many applications is to organize it according to a standardized *land-use* or *land-cover classification scheme*.

American Planning Association "Land-Based Classification Standard"

The most comprehensive hierarchical classification system for urban/suburban land use is the *Land-Based Classification Standard* (LBCS) under development by the American Planning Association (1999). This standard updates the 1965 *Standard Land Use Coding Manual* (Urban Renewal Administration, 1965), which is cross-referenced with the *1987 Standard Industrial Classification* (SIC) *Manual* (Bureau of the Budget, 1987) and the updated *North American Industrial Classification Standard* (NAICS). The LBCS requires extensive input from *in situ* site surveys, aerial photography, and satellite remote sensor data to obtain information at the parcel level on the following five characteristics: activity, function, site development, structure, and ownership (American Planning Association, 1999). The system provides a unique code and description for almost every commercial and industrial land-use activity. The LBCS does not provide information on land-cover or vegetation characteristics in the urban environment, as it relies on the Federal Geographic Data Committee standards on this topic.

The LBCS is still under development. Users are encouraged to keep abreast of the LBCS and to utilize it for intensive urban studies that require detailed commercial/industrial classification codes.

U.S. Geological Survey "Land-Use/Land-Cover Classification System for Use with Remotely Sensed Data"

The U.S. Geological Survey *Land-Use/Land-Cover Classification System* (Anderson et al., 1976; USGS, 1992) was originally designed to be resource-oriented (land cover) in

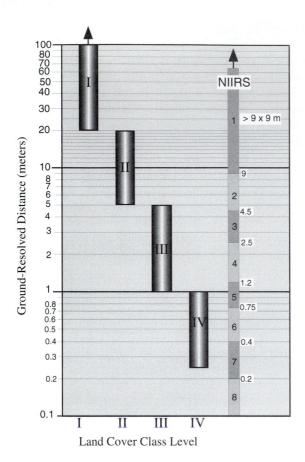

Figure 12-3 The general relationship between the USGS Land-use/Land-cover Classification System and the spatial resolution of the remote sensing system (often referred to as ground-resolved distance in meters). The National Image Interpretability Rating Scale (NIIRS) is also provided for comparison. A NIIRS zero rating suggests that the interpretability of the image is precluded by obscuration, degradation, or very poor resolution.

Table 12-2. Four Levels of the USGS Land-Use/Land-Cover Classification System and Representative Types of Remotely Sensed Data Typically Used to Provide the Information. Some of the systems are *proposed*.

Classification Level	Remote Sensor Data Characteristics
I	Landsat MSS (79 x 79 m), Thematic Mapper (30 x 30 m), Indian LISS (72.5 x 72.5 m; 36.25 x 36.25 m; 23.5 x 23.5 m), SPOT HRV XS (20 x 20 m), RADARSAT (100 x 100 m), aerial photography 1:120,000 to 1:240,000 scale.
II	Landsat 7 Enhanced Thematic Mapper Plus pan (15 x 15 m), SPOT HRV pan (10 x 10 m), SPOT HRV XS (20 x 20 m), Indian IRS pan (5.8 x 5.8 m), Space Imaging IKONOS (1 x 1 m panchromatic and 4 x 4 m multispectral), RADARSAT (11 x 9 m), aerial photography 1:60,000 to 1:120,000 scale.
III	Indian IRS pan (5.8 x 5.8 m), Space Imaging IKONOS (1 x 1 m panchromatic), aerial analog or digital photography 1:20,000 to 1:60,000 (1 – 3 m).
IV	Space Imaging IKONOS pan (1 x 1 m), EarthWatch *Quickbird* pan (0.8 x 0.8 m), aerial analog or digital photography 1:6,000 to 1:20,000 (0.25 – 1 m).

contrast with various people or activity (land use) oriented systems, such as the *Standard Land-Use Coding Manual* (Jensen et al., 1983). The USGS rationale was that "although there was an obvious need for an urban-oriented land-use classification system, there was also a need for a resource-oriented classification system whose primary emphasis would be the remaining 95 percent of the United States land area." The USGS classification system is designed to be driven primarily by the interpretation of remote sensor data obtained at various scales and resolutions (Figure 12-3; Table 12-2) and not data collected *in situ*. The USGS system addresses this need with eight of the nine level I categories treating land area that is not in urban or built-up categories (Tables 12-3 to 12-6). The classification system was initially

developed to include land-use data that was visually photo-interpreted, although it has been widely used for digital multispectral classification studies as well.

While the USGS *Land-use/Land-cover Classification System* was not originally designed exclusively to incorporate detailed urban attribute information, it nevertheless has been used extensively for more than two decades for urban land-use studies. This is usually performed by embellishing the classification system with detailed Level III, IV, and V urban class definitions such as those shown in Tables 12-3 to 12-5. The utility of the modified system is that it may be adapted to include as many levels as desired yet is upwardly compatible with all the USGS Level I and II land-use and land-cover data compiled by neighboring cities, counties, states, or nations.

Urban Land-Use/Land-Cover Classification (Levels I to IV) Using Remote Sensor Data

The general relationship between USGS land-cover classification system levels (I – IV) and the nominal spatial resolution of the sensor system (ground-resolved distance in

Table 12-3. Urban Residential, Commercial and Services – Level I and II are from the "U.S. Geological Survey Land-Use/Land-Cover Classification System for Use with Remotely Sensed Data." Levels III, IV, and V are logical extensions.

Classification Level

1 Urban or Built-up Land

11 Residential
  111 Single-Family Residential
    1111 House, houseboat, hut, tent
    1112 Mobile home
  112 Multiple-Family Residential
    1121 Duplex
    1122 Triplex
    1123 Apartment complex or Condominium ($\geq$ 4 attached units)
    1124 Mobile home (trailer) park

12 Commercial and Services
  121 Commercial
    1211 Automotive (cars, trucks, motorcycles)
      12111 Dealership
      12112 Gas station
      12113 Junkyard
      12114 Service or repair
    1212 Boat
      12121 Dealership
      12122 Marina
      12123 Service or repair
    1213 Department Store
      12131 Isolated department store (e.g., WalMart, K-Mart)
      12132 Mall complex
    1214 Financial and Construction
      12141 Banking, Financial Services
      12142 Insurance
      12143 Real estate
      12144 Construction
    1215 Food and Drug
      12151 Drugstore
      12152 Market
      12153 Restaurant
    1216 Funeral
      12161 Mortuary
      12162 Cemetery-mausoleum
    1217 Housing (temporary)
      12171 Hotel
      12172 Motel
      12173 Campground
    1218 House & Garden
    1219 Recreation (commercial and public)
      12191 Amusement park (e.g., Go-carts, miniature golf)
      12192 Baseball
      12193 Football/soccer/track
      12194 Golf
      12195 Swimming
      12196 Stadiums (multipurpose)
      12197 Tennis
      12198 Theater (includes multiplex and drive-in)
    1220 Utility
      12201 Gas – electric
      12202 Water – sewer
      12203 Waste disposal [solid (e.g., landfill) or liquid]
    1221a Warehousing/shipping
  122 Services (Public and Private)
    1221 Public Buildings and Facilities
      12211 Administration
      12212 Education
        122121 Elementary
        122122 Middle
        122123 High School
        122124 University or college
      12213 Fire, Police, Rescue
      12214 Health and fitness centers
      12215 Library
      12216 Post Office
      12217 Parks
    1222 Medical
      12221 Hospital
      12222 Offices
    1223 Religion (churches)

Table 12-4. Urban Industrial Land-Use Classification – Level I and II are from the "U.S. Geological Survey Land-Use/Land-Cover Classification System for Use with Remotely Sensed Data." Levels III, IV, and V are logical extensions.

Classification Level

13 Industrial
  131 Extraction
    1311 Dredging
      13111 Sand and gravel
      13112 Placer mining for gold, platinum, and diamonds
    1312 Open pit mine
      13121 Copper
      13122 Gravel pit – rock quarry
      13123 Iron ore
      13124 Salt
    1313 Shaft mine
      13131 Bauxite
      13132 Coal
      13133 Lead
      13134 Limestone
    1314 Strip mining
      13141 Coal
      13142 Oil shale
    1315 Well
      13151 Petroleum
      13152 Natural gas
      13153 Water
    1316 Cutting field
      13161 Forest clear-cutting or high-grading
    1317 Dredging or mining spoil piles
  132 Processing
    1321 Mechanical
      13211 Agriculture grain mill
      13212 Aluminum
      13213 Hydroelectric power plant
      13214 Ore concentration
      13215 Sewage treatment
      13216 Sugarbeet refining
      13217 Textile
      13218 Timber sawmill
      13219 Water treatment and purification
    1322 Chemical
      13221 Chlorine and caustic soda
      13222 Fertilizer
      13223 Hazardous-waste disposal
      13224 Petroleum refinery
      13225 Pharmaceutical
      13226 Magnesia – magnesium
      13227 Textile (synthetic)
      13228 Wood treatment
    1323 Heat
      13231 Calcium carbide
      13232 Cement
      13233 Clay products
      13234 Copper smelting
      13235 Glass
      13236 Iron and steel
      13237 Lead
      13238 Lime
      13239 Nuclear facilities
      13240 Thermal electric power plant
      13241 Zinc
  133 Fabrication
    1331 Heavy
      13311 Heavy machinery (bulldozers, etc.)
      13312 Railroad machinery manufacture and repair
      13313 Shipbuilding
      13314 Structural steel fabrication
    1332 Light
      13321 Aircraft assembly
      13322 Automotive assembly (e.g., cars, trucks, motorcycles)
      13323 Boat building and repair
      13324 Canneries
      13325 Electronics
      13326 Housing (prefabricated)
      13327 Meat packing
      13328 Plastic products

Table 12-5.   Urban Transportation, Communications, Utilities, and Industrial and Commercial Complexes – Level I and II classes are from the "U.S. Geological Survey Land-Use/Land-Cover Classification System for Use with Remotely Sensed Data." Levels III, IV, and V are logical extensions.

Classification Level

14 Transportation, Communications, and Utilities
    141 Transportation
        1411 Roads and Highways
            14111 Dirt
            14112 Paved
            14113 Limited access (freeway, toll)
            14114 Interchange
            14115 Parking
            14116 Bridge
        1412 Railroad
            14121 Track
            14122 Marshalling (classification) yard
            14123 Terminal
            14124 Bridge
        1413 Airport
            14131 Runway, tarmac
            14132 Hangar
            14133 Terminal
        1414 Boats/Ships
    142 Communications
        1421 Sign, billboard
        1422 Radio/Television/Cable
    143 Utility Facilities
        1431 Electricity
        1432 Natural gas
        1433 Petroleum
        1434 Water

15 Industrial and Commercial Complexes
    151 Industrial complex (park)
    152 Commercial complex (mall)

16 Mixed Urban or Built-up Land

17 Other Urban or Built-up Land

Table 12-6.   Rural Land-Use and Land-Cover including Agriculture, Rangeland, Forest, Water, Wetland, Barren Land, Tundra, and Perennial Ice and Snow. Levels I and II are from the "USGS Land-Use/Land-Cover Classification System for Use with Remotely Sensed Data" (Anderson et al., 1976).

Classification Level

2 Agriculture
    21 Cropland and pasture
    22 Orchards, groves, vineyards, nurseries, and
        ornamental horticultural areas
    23 Confined feeding operations
    24 Other agricultural land

3 Rangeland
    31 Herbaceous Rangeland
    32 Shrub–Brush and Rangeland
    33 Mixed Rangeland

4 Forest Land
    41 Deciduous Forest Land
    42 Evergreen Forest Land
    43 Mixed Forest Land

5 Water
    51 Streams and Canals
    52 Lakes
    53 Reservoirs
    54 Bays and Estuaries

6 Wetland
    61 Forested Wetland
    62 Nonforested Wetland

7 Barren Land
    71 Dry Salt Flats
    72 Beaches
    73 Sandy Areas Other Than Beaches
    74 Bare Exposed Rock
    75 Strip Mines, Quarries, and Gravel Pits
    76 Transitional Areas
    77 Mixed Barren Land

8 Tundra
    81 Shrub and Brush Tundra
    82 Herbaceous Tundra
    83 Bare Ground Tundra
    84 Wet Tundra
    85 Mixed Tundra

9 Perennial Snow or Ice
    91 Perennial Snowfields
    92 Glaciers

meters) is presented in Figure 12-3. Generally, USGS Level I classes may be inventoried effectively using sensors with a nominal spatial resolution of 20 – 100 m such as the Landsat Multispectral Scanner (MSS) with 79 x 79 m nominal spatial resolution, the Thematic Mapper (TM) at 30 x 30 m, SPOT HRV XS at 20 x 20 m, and Indian LISS 1-3 (72.5 x 72.5 m; 36.25 x 36.25 m; 23.5 x 23.5 m, respectively). Of course, any sensor system with higher spatial resolution can be used to acquire Level I data. Color Plate 12-1 presents Level I urban vs. non-urban land-cover information for three counties centered on Charleston, SC, extracted from Landsat MSS data acquired on six dates from February 14, 1975, to February 3, 1994. For global projects, Level I land-cover may be extracted from more coarse spatial resolution imagery (e.g., AVHRR at 1.1 x 1.1 km) (Belward et al., 1999).

Typical urban spectral reflectance signatures are summarized in Figure 12-4. Urban land cover typically appears "steel-gray" on color-infrared color-composite images because the urban terrain, consisting primarily of concrete and asphalt roads, parking areas, shingles, and bare soil, typ-

ically reflects high proportions of the incident green, red, and near-infrared radiant flux (Rundquist and Sampson, 1988). The steel-gray urban signature contrasts sharply with vegetated surfaces that appear bright red or magenta on color-infrared images because the vegetation reflects substantial amounts of near-infrared energy while absorbing much of the incident green and red wavelength energy, as discussed in Chapter 4. Water absorbs most of the incident radiant flux, causing it to appear dark on color-infrared imagery, easily distinguishable from the urban landscape. Color composites of Charleston, SC, produced using many of the major sensor systems, are provided in Chapter 7.

Laboratory Spectroradiometer Reflectance
Characteristics of Common Urban Materials

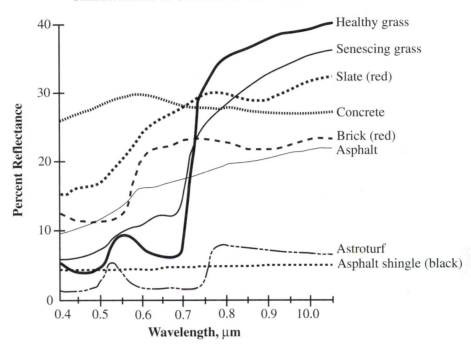

Figure 12-4   Percent reflectance curves for common materials found in urban environments. The reflectance spectra were obtained in a controlled laboratory environment using a GER 1500 spectroradiometer.

Sensors with a minimum spatial resolution of 5 – 20 m are generally required to obtain Level II information. The SPOT HRV and the Russian SPIN-2 TK-350 are the only operational satellite sensor systems providing 10 x 10 m panchromatic data. RADARSAT provides 11 x 9 m spatial resolution data for Level I and II land-cover inventories even in cloud-shrouded tropical landscapes. Landsat 7 with its 15 x 15 m panchromatic band was launched on April 15, 1999.

More detailed Level III classes may be inventoried using a sensor with a spatial resolution of approximately 1 – 5 m (Welch, 1982; Forster, 1985) such as IRS-1CD pan (5.8 x 5.8 m data resampled to 5 x 5 m) or large-scale aerial photography. Space Imaging, Inc., IKONOS provides 1 x 1 m panchromatic and 4 x 4 m multispectral data. Future sensors that may provide such data include EarthWatch, Inc., Quickbird (0.82 x 0.82 m pan; 3.28 x 3.28 m multispectral), ORBIMAGE, Inc., OrbView 3 (1 x 1 m pan; 4 x 4 m multispectral), and IRS P5 (2.5 x 2.5 m pan). The synergistic use of high spatial resolution panchromatic data (e.g., 1 x 1 m), merged with lower spatial resolution multispectral data (e.g., 4 x 4 m), provides an image-interpretation environment that is superior to using panchromatic data alone (Jensen, 1996).

Level IV classes and cadastral (property-line) information is best monitored using high spatial resolution panchromatic sensors, including aerial photography ($\leq 0.25 – 1$ m), digital frame imagery such as that provided by Positive Systems, Inc., or Litton Emerge Spatial, Inc. ($\leq 0.25 – 0.5$ m), and Space Imaging IKONOS 1 x 1 m panchromatic data.

Urban land-use and land-cover classes in Levels I through IV have temporal attribute requirements ranging from 1 – 10 years (Table 12-1 and Figure 12-1). All of the sensors mentioned have temporal resolutions of less than 55 days, so the temporal resolution of the land-use/land-cover attributes is satisfied by the current and proposed sensor systems.

Urban landscapes are notoriously complex. While simple binary urban/non-urban information is valuable for assessing growth trends (Color Plate 12-1), it does not provide the level of land-use/land-cover detail necessary to make wise decisions. Therefore, it is useful to provide additional detail about the extraction of the following urban land-use information from remote sensor data:

• residential housing,

- commercial and services,

- industrial,

- transportation infrastructure, and

- communications and utilities.

 Residential Land Use

The *home* in almost all cultures is the single most important possession a person acquires in a lifetime. It shelters the family and usually represents the household's greatest economic investment. Inventorying the location, type, condition, and number of residences is one of the most important tasks performed using remote sensor data. Discriminating between Level II residential, commercial, industrial, and transportation land-uses requires a sensor system with a spatial resolution from 5 – 20 m. Identifying Level III residential land-use classes requires a spatial resolution from 1 – 5 m (Table 12-1; Figure 12-3).

Prior to investigating how residential land use appears on remotely sensed data, it is important to introduce the concept of *form* and *function*. Basically, the function of a building often dictates its form. In fact, we often say "form follows function" or "form is dictated by function." For example, the need to house a single family often results in the creation of a detached dwelling composed of several attached rooms (this is not true in the central business district). Similarly, the preparation of finished steel typically requires a very large, linear building (form) that can house the various steel processing activities (function). An analyst must be careful, however, because a building's use may change over time. Thus, a knowledge of the sequence-of-occupance of buildings in a region becomes important.

Single-Family versus Multi-Family Residential

A *single-family residence* is detached from any other housing structure. In developed countries it usually has a single driveway, one front sidewalk leading to the front door, a front yard, a backyard (often fenced), and a garage or carport. It is usually $\leq$ 3 stories in height. Several single-family homes in Richmond, CA are in the process of being framed in Figure 12-5a. Completely landscaped single-family residences in Boca Raton, FL, and San Diego, CA, are shown in

Figure 12-5b,c. Mobile homes (trailers) in developed countries are usually much smaller than other single-family homes (approximately 5 m single-wide; 10 m double-wide), rarely have a garage (but may have a carport), and may or may not have a paved driveway or sidewalk (Figure 12-5d).

In developing countries, single-family residences may consist of a building, hut, tent, lean-to, or igloo, depending upon the culture. It may be permanent or temporary (seasonal). It is often difficult to identify such structures in developing countries because they are made of the same materials found in the surrounding countryside, resulting in low object-to-background contrast (Welch, 1982; Jensen et al., 1983).

In developed countries, *multiple-family residences* usually have more than one sidewalk, may have large collective above-ground or below-ground parking garages or parking lots, may be $\geq$ 2 stories in height, and share front and back yards (Figure 12-6a). There may be a community pool and/or tennis court. *Duplexes* (two attached housing units) and *triplexes* (three attached housing units) usually have two or three walkways and driveways, respectively. Sometimes a common parking area is provided. It is often difficult to determine how many families occupy single residences in developing countries.

When there are more than three attached housing units, it is usually called an apartment or condominium complex (Figure 12-6b,c). Such complexes often occupy an extensive area of land, rise many stories above the ground, and in large cities may be high-rise buildings. It is usually possible to determine if a building is a single or multiple-family residence using imagery with a spatial resolution from 1 – 5 m. Sometimes it is necessary to merge multiple types of imagery such as Landsat Thematic Mapper and radar imagery to identify villages in tropical parts of the world (Haack and Slonecker, 1994).

The reader should take note that on numerous occasions visual, heads-up on-screen image interpretation of urban structure often outperforms computer-assisted digital image processing (Jensen, 1996). Visual image interpreters are able to take into account the *site*, *situation*, and *association* of structures. These elements of image interpretation have been very difficult to incorporate into analytical methods of digital image classification. Only now are analytical neural network-based image interpretation systems beginning to include the fundamental elements of visual image interpretation into the image-analysis process (Jensen and Qiu, 1998). For example, Figure 12-7 depicts the user interface associated with a neural network-based digital image processing system that assists an image interpreter to arrive at the cor-

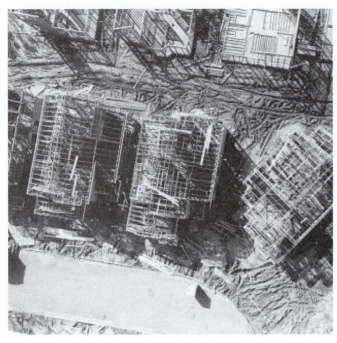

a. Several single-family residences being framed in Richmond, CA (level 1111) (courtesy Cris C. Benton).

b. A single-family residence in Boca Raton, FL, with one sidewalk, one driveway, and a fenced backyard (level 1111).

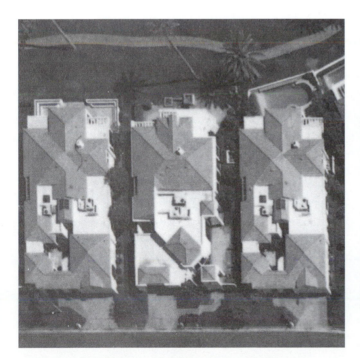

c. Closely spaced single-family residences in San Diego, CA, with one sidewalk, one driveway, and occasionally a small swimming pool (level 1111).

d. Single-family trailers (mobile homes) in a trailer park near Myrtle Beach, SC (level 1112).

Figure 12-5 Examples of single-family residential housing (level 111).

a. Condominium residential housing in Colorado (level 1123). The number of driveways, sidewalks, and roof structures indicates how many housing units are present. The top-left condo is a four-plex. The top right is a seven-plex. The bottom two condos are five-plex.

b. Two-story apartment complex in Columbia, SC (level 1123), with numerous sidewalks and parking around the perimeter.

c. Twelve-story student-housing apartment complex at the University of South Carolina (level 1123).

Figure 12-6 Examples of multiple-family residential housing (level 112).

a.

Figure 12-7 a) The user-interface of a Neural Network Image Interpretation System that assists an image analyst in the classification of difficult urban land-use. In this example, the analyst used visual on-screen digitizing to outline an unknown object-of-interest. b) The interpreter then used a point-and-click dialog box to identify primitive attributes within the unknown object-of-interest, such as the existence of parking lot(s), sidewalks, etc. The empirical information is passed to the neural network, which then compares this pattern with previous patterns (actually weights) and attempts to identify the unknown object as a multiple-family residential housing complex (Jensen and Qiu, 1998).

b.

rect land-cover or land-use classification for a particular parcel of land. In the past, the interpreter had to rely on complex branching dichotomous keys to assist in the interpretation process. Now, the image analyst can draw a polygon around a feature such as the one shown in Figure 12-7a and then enter in the characteristics observed in the polygon of interest using a simple point-and-click menu such as the one shown in Figure 12-7b. The characteristics might include the number of sidewalks, the presence of a driveway and/or garage, parking lots, multiple stories, and general size of the phenomena. The neural network then processes this information and reports back to the image analyst what it believes is the most appropriate land-use or land-cover code. The neural network is trained using real-world empirical examples, and in effect it *learns*. The neural network classification system can also be used to photointerpret other land uses, including commercial, services, and industrial (Jensen and Qiu, 1998). Neural network or expert system-assisted interpretation systems will most likely be the means in the future by which most detailed urban information is eventually extracted from digital remote sensor data.

Building and Cadastral (Property-Line) Infrastructure

In addition to fundamental nominal scale land-use and land-cover information (i.e., identifying whether an object is a single-family residence or a commercial building), transportation planners, utility companies, tax assessors, and others require more detailed information on building footprint perimeter, area, and height, driveways, patios, fences, pools, storage buildings, and the distribution of landscaping every one - five years (Table 12-1).

These building and property-line parameters are best obtained using stereoscopic (overlapping) panchromatic aerial photography or other remote sensor data with a spatial resolution of $\leq 0.25 - 0.5$ m (Jensen, 1995; Warner et al., 1996). For example, panchromatic stereoscopic aerial photography with 0.25 x 0.25 m (1 ft) spatial resolution was used to extract the exact dimensions of individual houses and outbuildings, trees, pools, driveways, fences, and contours for the single-family residential area in Figure 12-8a,b. In many instances, the fence lines are the cadastral property lines. If the fence lines are not visible or are not truly on the property line, the property lines can be identified by a surveyor in the field and the information overlaid onto the orthophotograph or planimetric map database to represent the legal cadastral (property) map. Many municipalities in the United States use high spatial resolution imagery as the source for some of the cadastral information and as an image

backdrop upon which surveyed cadastral and tax information are portrayed.

Detailed building height and volume data can also be extracted from high spatial resolution (0.25 – 0.5 m) stereoscopic imagery (Jensen et al., 1996). For example, consider the wire-frame three-dimensional outline of the United States Capitol shown in Figure 12-8c. The perimeter, area, volume, and square footage of this building are now available. Lo (1999) suggests that such detailed building information may allow the exact type of roof to be identified (e.g., ridge, mansard, hipped, lean-to, flat, etc.), which may be used to infer the date of building construction.

Socioeconomic Characteristics Derived from Single- and Multi-Family Residential-Housing Information

Selected socioeconomic attributes may be extracted directly from remote sensor data or via surrogate information derived from the imagery. Two of the most important are *population estimation* and *quality of life* indicators, derived primarily from the single- and multiple-family residential-housing stock information.

Population Estimation

Knowing how many people live within a specific geographic area or administrative unit (e.g., city, county, state, country) is very powerful information. In fact, some have suggested that the global effects of increased population density and ecosystem land-cover conversion may be much more significant than those arising from climate change (Skole, 1994). Population estimation can be performed at the local, regional, and national level based on:

- counts of individual dwelling units (requires 0.25 – 5 m nominal spatial resolution),

- measurement of urbanized land areas (often referred to as settlement size), and

- estimates derived from land-use/land-cover classification (Lo, 1995; Sutton et al., 1997).

Remote sensing techniques may provide population estimates that approach the accuracy of traditional census methods if sufficiently accurate *in situ* data are available to calibrate the remote sensing model. Unfortunately, the ground-based population estimations are often woefully inaccurate (Clayton and Estes, 1979). In many instances, the

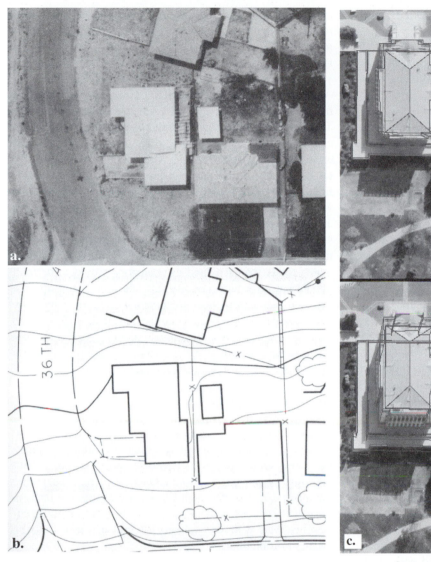

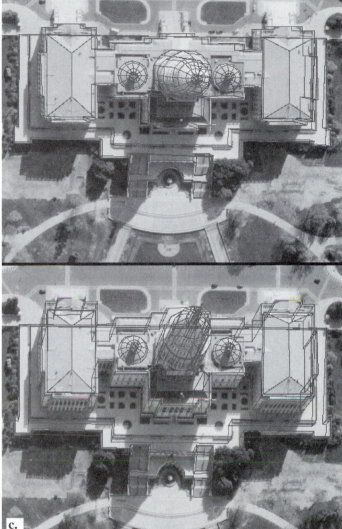

a. Vertical panchromatic 1:2,400-scale aerial photography.

b. Planimetric (x,y) cadastral information photogrammetrically extracted from the stereoscopic aerial photography, including: 2 ft contours, building footprints, fences, retaining walls, and large trees (Jensen and Cowen, 1999).

c. Wire-frame information of the U.S. Capitol extracted from large-scale vertical aerial photography, including precise building footprint, height, area, perimeter, and volume. The two photographs may be viewed stereoscopically (courtesy LH Systems, LLC, San Diego, CA, and Marconi Integrated Systems, Inc.; © SOCET Set).

Figure 12-8   Extraction of cadastral information from stereoscopic vertical aerial photography using a) traditional, and b,c) soft-copy photogrammetric techniques. Such information is vital to the accurate inventory of residential, commercial, and industrial land-use. It can be acquired using traditional surveying or photogrammetric techniques.

remote sensing methods may be superior to the ground-based methods.

The most accurate remote sensing method of estimating the population of a local area is to *count individual dwelling units* based on the following assumptions (Forster, 1985; Lindgren, 1985; Lo, 1995; Holz, 1988; Haack et al., 1997):

- The imagery must have sufficient spatial resolution to allow identification of individual structures even through sparse tree cover, and to determine whether the structures are residential, commercial, or industrial.

- Some estimation of the average number of persons per dwelling unit must be available.

- Some estimate of the number of homeless, seasonal, and migratory workers is required.

- It is assumed all dwelling units are occupied, and only *n* families live in each unit (calibrated using *in situ* information).

This is usually performed every five to seven years and requires high spatial resolution remotely sensed data (0.25 – 5 m). For example, individual dwelling units in Irmo, SC, were extracted from 2.5 x 2.5 m aircraft multispectral data (Cowen et al., 1995). Correlation of the remote sensing derived dwelling unit data with Bureau of the Census dwelling unit data for the 32 census block area yielded an $r^2 = 0.81$ which accounted for 81 percent of the variance. These findings suggest that the new high spatial resolution panchromatic sensors may provide a good source of information for monitoring the housing stock of a community on a routine basis. This will enable local governments to anticipate and plan for schools and other services with data that has a much more frequent temporal resolution than the decennial census. This data will also be of value for real estate, marketing, and other business applications (Lo, 1995). Unfortunately, the dwelling-unit approach is not suitable for a regional/national census of population because it is too time-consuming and costly (Sutton et al., 1997). In fact, Broome (1998) suggests that this method requires so much *in situ* data to calibrate the remote sensor data that it can become operationally impractical. Research is required to document the utility of the method in a variety of cultures and population densities. Therefore, other methods have been developed.

Scientists have known for some time that there is a relationship between the simple urbanized built-up area (settlement size) extracted from a remotely sensed image and settlement population (Tobler, 1969; Olorunfemi, 1984):

$$r = a \times P^b \qquad (12-1)$$

where *r* is the radius of the populated area circle, *a* is an empirically derived constant of proportionality, *P* is the population, and *b* is an empirically derived exponent. Estimates of these parameters are fairly consistent at regional scales, but the estimate of the *a* parameter varies between regions. For example, Sutton et al. (1997) used Defense Meteorological Satellite Program Operational Linescan System (DMSP-OLS) visible near-infrared nighttime 1 x 1 km imagery to inventory urban extent for the entire United States (Figure 12-9). When the data were aggregated to the state or county level, spatial analysis of the clusters of the saturated pixels predicted population with an $r^2 = 0.81$ once

again. Unfortunately, the use of "DMSP imagery underestimated the population density of urban centers and overestimated the population density of suburban areas" (Sutton et al., 1997).

Another widely adopted population estimation technique is based on the use of the Level I – III land-use information. This approach assumes that land use in an urban area is closely correlated with population density. Researchers establish an empirical value for the population density for each land use by field survey or census data (e.g., multiple-family residential housing may contain five persons per pixel when using 30 x 30 m Thematic Mapper data; rural forested areas might have only 0.20 persons per pixel). Then, by measuring the total area for each land-use category, they estimate the total population for that category. Summing the estimated totals for each land-use category provides the total population projection (Lo, 1995). The urban built-up area and land-use data method can be based on more coarse spatial resolution multispectral remote sensor data (5 – 20 m) every 5 – 15 years. Henderson and Xia (1998) discuss how radar imagery can be of value when conducting population estimates using these same techniques.

Quality of Life Indicators

Lo and Faber (1998) suggest that adequate income, decent housing, good education and health services, and good physical environment (e.g., no hazards, refuse) are important indicators of social well-being and *quality of life*. Evaluating the quality of life of a population on a continuing basis is important because it helps planners and government agencies involved with the delivery of human services to be aware of problem areas.

In the past, most quality of life studies were based almost exclusively on the analysis of socioeconomic information derived from *in situ* enumeration. Remotely sensed images have rarely been used in social studies. Only recently have factor analysis studies documented how quality of life indicators (such as house value, median family income, average number of rooms, average rent, and education) can be estimated by extracting the urban attributes found in Table 12-7 from relatively high spatial resolution (0.25 – 30 m) imagery (Monier and Green, 1953; Green, 1957; McCoy and Metivier, 1973; Tuyahov et al., 1973; Henderson and Utano, 1975; Jensen et al., 1983; Lindgren, 1985; Holz, 1988; Avery and Berlin, 1992; Haack et al., 1997; Lo and Faber, 1998). Note that the attributes are arranged by *site* (building and lot) and *situation*. The site may be situated in positive and negative surroundings.

Figure 12-9 Defense Meteorological Satellite Program Operational Linescan System (DMSP-OLS) visible near-infrared nighttime 1 x 1 km imagery of the conterminous United States. When the data were aggregated to the state or county level, spatial analysis of the clusters of the saturated pixels predicted population with an $r^2 = 0.81$ (Sutton et al., 1997; courtesy American Society for Photogrammetry & Remote Sensing).

These remote sensing-derived attributes must be correlated with *in situ* census observations to compute the quality of life indicators. Lo and Faber (1998) suggest that in many cities green vegetation is very precious and costly, and that only the wealthy can afford to include grass and trees in their living environment. They found that greenness information derived from TM data was highly correlated with quality of life. Quality of life indicators are usually collected every 5 – 10 years.

Energy Demand and Conservation

Local urban/suburban energy demand (sometimes referred to as heat-load density) may be estimated using remotely sensed data. First, the square footage (or m^2) of individual buildings is determined from high spatial resolution imagery. Local ground-reference information about energy consumption is then obtained for a representative sample of homes in the area. Regression relationships are then derived

to predict the anticipated energy consumption for the region. It is also possible to predict how much solar photovoltaic energy potential a geographic region has by modeling the individual rooftop square footage and orientation with known photovoltaic generation constraints. Both these applications require high spatial resolution imagery (0.25 – 0.5 m) (Clayton and Estes, 1979; Angelici et al., 1980; Curran and Hobson, 1987).

Regional and national energy consumption may also be predicted using DMSP imagery (Welch, 1980). Unfortunately, DMSP imagery of urbanized areas are recorded at 6-bit (values from just 0 – 63) radiometric resolution causing most of the urban, energy-consuming areas to saturate at a brightness value of 63 (Elvidge et al., 1997; Sutton et al., 1997).

Numerous studies have documented how high spatial resolution (1 – 5 m) predawn thermal infrared imagery (8 – 14 µm) can be used to inventory the relative quality of housing insulation if 1) the rooftop material is known (e.g., asphalt versus

Table 12-7.   Urban/Suburban Attributes that May Be Extracted from Remote Sensor Data and Used to Assess Housing Quality and/or Quality of Life

Factor	Attributes
Site	**Building** • single or multiple-family • size (m²) • height (m) • age (derived by convergence of evidence) • garage (attached, detached) **Lot** • size (m²) • front yard (m²) • backyard (m²) • street frontage (m) • driveway (paved, unpaved) • fenced • pool (in-ground, above-ground) • patio, deck • outbuildings (sheds) • density of buildings per lot • percent landscaped • health of vegetation (e.g., NDVI greenness) • fronts paved or unpaved road • abandoned autos • refuse
Situation	**Adjacency to Community Amenities** • schools • shopping • churches • hospitals • fire station • open space, parks, golf courses • marsh **Adjacency to Nuisances or Hazards** • heavy street traffic • railroad or switchyard • airports and/or flightpath • freeway • located on a floodplain • sewage-treatment plant • industrial area • power plant or substation • overhead utility lines • steep terrain

wood shingles), 2) no moisture is present on the roof, and 3) the orientation and slope of the roof are known (Colcord, 1981; Eliasson, 1992). If energy conservation or the generation of solar photovoltaic power were important in countries, these variables would probably be collected every one to five years.

 Commercial and Services Land Use

People engage in commerce (business) to provide food, clothing, and shelter for themselves and their extended families. Many work in the service industries providing education, medical, sanitation, fire and police protection, etc., to improve the quality of life for the entire community. Interestingly, many of these activities are visible in the landscape. Commercial and service activities often have a unique *cultural signature* defined as the assemblage of specific materials and structures that characterize certain activities in a particular culture. Each culture has commerce and services that are unique to that culture. For example, there are many types of commerce found in Asiatic urban environments that simply are not found in Western urban landscapes. Therefore, an interpreter must understand well the cultural milieu of the country of interest before he or she attempts to inventory and map the commerce and service activities using remote sensor data. In fact, if an interpreter has not actually seen the phenomena on the ground and does not understand completely the cultural significance and purpose of the commerce or activity, then it is unlikely that a correct interpretation will be made.

The author has extended the *USGS Land-Use/Land-Cover Classification System* in Tables 12-3 to 12-5 to include Level IV and sometimes Level V classes. This extension is primarily for U. S. and Western European landscapes but may be generally applicable to other developed nations. It represents a common-sense hierarchical system that can be aggregated back to Levels I and II. It is not complete in content; rather, it reflects an attempt to categorize the land use that accounts for approximately 95 percent of the urban land use encountered in most developed cities. It uses common terminology whenever possible. A detailed definition of each class is not provided here. However, it is instructive to provide numerous examples of the various types of land use and comment on unique characteristics that allow one type of commerce or service to be distinguished from others.

The Central Business District

The *central business district* (CBD) of many modern cities is the most distinctive urban landscape with its high-rise buildings and the highest density of land use. Many CBDs have changed considerably in recent years, particularly in the United States. Urban geographers use the term *deindustrial-*

ization to describe the loss of manufacturing industries in cities today, and the rise of *tertiary* and *quaternary* commercial activities, notably financial services, insurance, and real estate (Lo, 1999). There are also many other professional consulting services that characterize the modern cities and the phenomenon of *suburbanization*. Downtown Seattle, WA, is an excellent example of a central business district (Figure 12-10). Numerous high-rise office buildings house the financial services, insurance, and real estate firms. However, some of the towers also contain single- and multiple-family residences.

The CBDs of some cities in the United States have been 1) transformed into poverty areas, 2) gentrified, meaning that only older people live in them and that very few young families are present, and/or 3) reconstructed with new buildings to attract professional middle-class people and their small families (usually ≤ 2 children) to return and stay. These and other conditions make it difficult to identify the exact land use of a particular parcel of land in the central business district. *In situ* observation is usually required in addition to the analysis of remotely sensed data.

Commercial Land Use

Automotive and Boat

The developed world and especially the United States is enamored with the automobile and other forms of personal transportation, including trucks, sport utility vehicles, recreation vehicles, and boats. Much of our disposable income is spent acquiring and maintaining these vehicles. Therefore, the urban landscape is littered with dealerships to sell them, service stations to fill them up with petroleum distillate products, and businesses that repair them. Finally, their remains can often be seen in automobile junkyards.

New and used car, truck, and boat dealerships are relatively easy to identify due to the myriad of vehicles or boats parked systematically in large parking lots, the presence of large showrooms, and attached service centers (Figure 12-11a). Isolated service and repair businesses are more difficult to identify but usually have smaller parking lots with many cars present in the lot in various states of repair.

Gas service stations often have a unique separate island structure with its own canopy where the fuel is dispensed, a detached building that is usually a mini-market, and occasionally another building where automobiles are washed. Gas stations are often located on corner lots (Figure 12-11b).

Numerous cars are often located around the periphery of the service stations.

Junkyards have a large number of vehicles arranged in a haphazard fashion, many missing hoods and trunks. Also, grass may be growing between the cars, making it straightforward to distinguish it from a dealership.

Department Stores

A department store (e.g., WalMart, K-Mart, Sears-Roebuck) contains many types of businesses located within a single building (e.g., sporting goods, garden, clothing, market, toys). The building usually is one of the largest (m^2) in the region (Figure 12-12 location *a*). A very large parking lot is adjacent to it. Isolated department stores often have storage locations in the parking lot where carts can be returned.

Conversely, a mall is a collection of multiple department stores (Figure 12-12 locations *bcd*). It is usually the largest building complex in the region (m^2). The department stores within the mall are usually connected by a central walkway that is often enclosed. Skylights are often located above the central walkway. The mall is often completely encircled with single or multiple levels of parking, with perhaps some underground parking. Department stores and malls are often located at the junction of major highways. They are usually straightforward to identify because of their extreme size and situation.

Finance and Construction

There are often a tremendous number of banks, brokerage firms, insurance agencies, construction firms, and real estate offices in the central business district of a major city (e.g., refer to the image of Seattle, WA, in Figure 12-10). Parking may be visible above-ground, in multiple-story above-ground parking structures, or underground. For example, consider the Affinity Office Building located in downtown Columbia, SC (Figure 12-13a). It is the tallest building in the city and houses banking, brokerage firms, insurance agencies, and an airline ticket office. Parking for this building is primarily above-ground. It is impossible to identify the specific activities taking place within this office building. Therefore, it is often necessary to simply label the building or parcel as being *commercial* land use. Field work or census enumeration can provide some of the information.

Large multiple-story buildings like this are commonplace in major cities of the world. On large-scale aerial photography, relief displacement of the top of the building is a function of 1) how high the building is above the local datum, 2) the alti-

Figure 12-10 A portion of the *central business district* (CBD) of downtown Seattle, WA. The region has been deforested. It is primarily composed of impervious high-rise buildings that contain financial services, insurance, and real estate companies. Numerous multiple-story apartment complexes are also present. A rectangular, systematically surveyed transportation system serves the area. A new office building is under construction in the top-left portion of the image. This is a 1 x 1 m ADAR 5500 digital frame camera image (courtesy of Positive Systems, Inc.).

a. Automobile dealership in Columbia, SC (level 12111). Cars for sale are arranged systematically for buyer inspection and ease of exit and return.

b. Typical automotive service station with multiple islands for dispensing gasoline or diesel fuel and a main building consisting of service bays or mini-market (level 12112). The circular tops of the underground petroleum storage tanks are visible in this example.

Figure 12-11 Examples of commercial automobile establishments.

tude of the aircraft at the instant of exposure, and 3) how far the building is located away from the principal point of the photograph. Also, the buildings tend to cast long shadows. The relief displacement and the shadows can both hinder and help in building identification. In this example, relief displacement aids in the identification, as the sides (façade) of buildings are actually visible providing a good impression of building vertical relief. Shadows help to set the buildings off from the remainder of the scene. When using such aerial photography, it is necessary to use photogrammetric techniques to compensate for the relief displacement if one desires to map the exact perimeter of the building in its proper planimetric (x,y) position. One cannot simply draw a polygon around the top of the building and assume that it is in its proper planimetric position. One might think that relief displacement would be absent from high spatial resolution imagery obtained from sensors placed in orbit. Unfortunately, this is not the case, as demonstrated in the Space Imaging, Inc., IKONOS 1 x 1 m image of downtown Washington, DC, where building displacement is still apparent (Figure 7-33 in Chapter 7).

Isolated, detached banking facilities typically have a drive-through portico structure where financial transactions take place (Figure 12-13b). One parking area is usually associated with walk-in banking, while the other portion of the parking area services the drive-through traffic. One must be careful not to confuse the drive-through bank building with a gas station that also has a drive-through portico.

Construction companies often have heavy equipment (backhoes, dump trucks, etc.) located on the property as well as stacks or piles of raw materials (e.g., lumber, cement, palettes of bricks).

Food and Drug

The urban landscape is replete with many food and drug business establishments. Chainstore markets and drugstores occupy large buildings with extensive parking facilities. They may be isolated, but are often found attached. For example, consider the attached drugstore (1) and supermarket (2) shown in Figure 12-14a. The food and drug establishments share a large common parking lot. The exact land-use code of such buildings is often difficult to determine without *in situ* inspection because the large extended building and parking facilities may appear similar to other commercial enterprises.

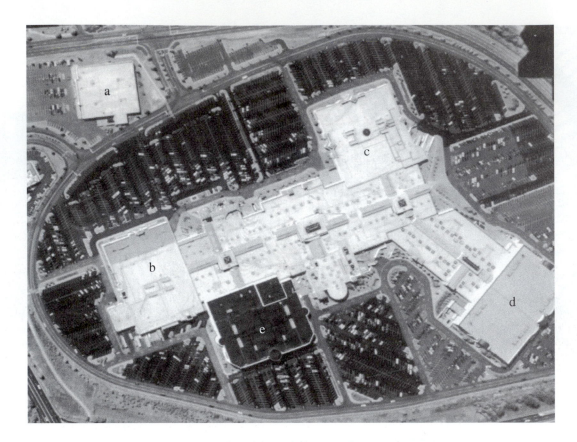

Figure 12-12 Example of department store and mall commercial land use in some developed countries. An iso-lated Toys-R-Us, Inc., department store (a) is independent of the mall (level 12131). The mall commercial complex (level 12132) is anchored by three major chains: Sears-Roebuck, Inc. (b), Belk, Inc. (c), and Dillards, Inc. (d). The central walkway of the mall is enclosed with occasional skylights. The mall is completely surrounded by parking, some of which is asphalt (dark) and some concrete (light). The roof of Whites department store (e) is asphalt, barely distinguishable from the parking lot. Note the numerous air-conditioning units on each building.

Fast-food restaurants generally occupy relatively small, sin-gle-story buildings. They are located on both major and minor roads, have modest but adequate parking facilities, and often exhibit a single automobile drive-through window lane that encircles the building (Figure 12-14bc). This is important because drive-through banks typically have multi-ple drive-through lanes. There are usually numerous entrances and exits to the lot. The lot is usually modestly landscaped. Sometimes there are elaborate playgrounds with gymnasium equipment present in front of the restaurants.

More expensive, upscale restaurants are larger in size (m²), have well-manicured lawn and tree landscaping, occupy a much larger lot, and have larger parking areas than fast-food restaurants. Both fast-food and upscale restaurants have a large number of vents on their rooftops. Sometimes the smoke discharged from the vents discolors the rooftops.

Funeral and Cemetery

Having traveled the equivalent of 100 times around the Earth in our automobiles during a lifetime (the Earth is only 25,000 miles in circumference) and eaten at far too many fast-food restaurants, American and Western Europeans finally reside in a cemetery or mausoleum. Interestingly, cemeteries often confound image interpreters, yet they have unique characteristics (Figure 12-15). In developed nations they generally occupy an expensive and extensive tract of land, often adjacent to churches. The landscaping is meticu-lous. The road network is intricate, with many narrow roads. Often, the roads follow the contour of the land. But most important, there are hundreds of systematically spaced small white dots on the landscape. There is often a shed on the property with heavy equipment (backhoe) and vaults located in the shed or on the ground out of sight from the public.

a. Commercial financial services, including banking (level 12141), insurance (level 12142), and real estate transactions (level 12143), take place in the Affinity Office Building in downtown Columbia, SC. Building relief displacement in this original 1:6,000-scale vertical aerial photography provides detailed information about the façade (side) of the building and aids in building image interpretation.

b. An isolated small banking establishment (level 12141), complete with the main building and attached drive-through teller system. In this example, one parking area is associated with walk-in banking while the other portion of the parking area services the drive-through customers.

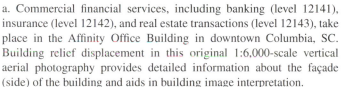

Figure 12-13 Examples of commercial financial services (level 1214).

A mortuary or funeral home may be located on the property. It is often a single large building with ample parking.

Housing (Temporary)

People throughout the world love to go on overnight and extended vacations. This has resulted in a plethora of small motels that cater to the motorist and larger hotels that are used for both business and pleasure. One of the major differences between a motel and a hotel is size. Motels tend to be smaller and have fewer stories (Figure 12-16a). Hotels are often imposing structures rising ≥ 4 stories (Figure 12- 16b). Both motels and hotels usually have large parking areas and swimming pool(s). Motel parking is often directly in front of the rental unit, whereas hotels often have large underground parking areas that are difficult to detect. Hotels often have tennis courts, whereas motels usually do not. Large hotels tend to be located near major thoroughfares, airports, and limited-access highways while motels may be located just

about anywhere. Hotels usually have more luxuriant landscaping.

There are also specialized campgrounds that cater specifically to the motorist. These have unique attributes, including individual driveways where the car is parked and concrete pads where a tent or recreational vehicle may be placed. There is often a centrally located swimming pool and other recreational amenities. The campground is often quite large and located near a major highway. Campgrounds are usually well-manicured.

House and Garden

This type of business (e.g., Lowes, Home Depot) usually has a single large building, attached greenhouse(s), and rows of shrubbery outside or adjacent to the building. Also visible are palettes of fertilizer and other materials, yard machinery such as lawn tractors and trailers sitting outside the building,

a. The Heart of Carolina Motel (level 12172). Many of the cars are parked directly in front of the patron's rooms.

b. Temporary housing at the Columbia Plaza Hotel (level 12171). Note the high-rise hotel complex, the large parking area, and attached convention hall with its own elevated ramp for loading and unloading convention exhibits. The pool is inside the building.

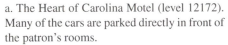

Figure 12-16 Commercial motel and hotel temporary housing (level 1217).

structure at the apex of the fan (the screen). It typically casts a long shadow, depending on the time of day. A fan-shaped road network is found within the site, and there are poles placed systematically about every 5 m on the ground that hold the detachable speaker. The concession stand is almost always located in the center of the site. There is usually no landscaping, the terrain being composed of asphalt or gravel.

Warehousing/Shipping

Warehouses receive goods and then reorganize the materials according to a bill of laden for shipment via trucks, railroad, or ship to an interim or final destination. They are usually very large buildings with modest landscaping. In the case of truck transport, there is usually a large parking lot with an elevated (5 ft high) docking system on one or multiple sides of the building. Forklifts are often present in the yard. Occasionally, some nonperishable commodities or container cargo boxes may be seen outside in the parking area. The

most diagnostic features are the tractor trailers that are situated with their back doors against the building or arranged systematically throughout the site out of harm's way.

Shipping and warehousing along harbor waterfronts usually involves extremely large warehouses situated on the docks or immediately inland. For example, consider the aerial photograph of a portion of the Seattle, WA, harbor in Figure 12-23. Here, piles of minerals are being loaded onto the ship from the conical storage structures using a sophisticated overhead conveyor-belt system. Extensive large warehouses are also located nearby. The minerals were brought to the shipping facility most likely by the adjacent railroad.

Other Commercial

Finally, there is the matter of what to do about all the other myriad of commercial land-uses that are found in almost every urban landscape (hobby shops, Fotomats, clothing

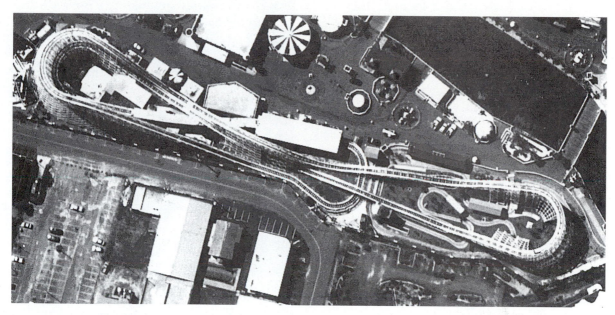

a. The roller-coaster recreational facility at the Myrtle Beach, SC, Pavilion amusement park (level 12191).

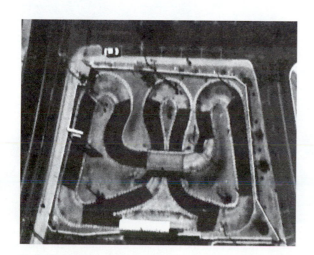

b. A typical go-cart track. The overpass and most of the corners are concrete, while the remainder of the track is asphalt.

c. A water-slide ride that empties into a pool.

Figure 12-17 Commercial recreation amusement-park facilities (level 12191).

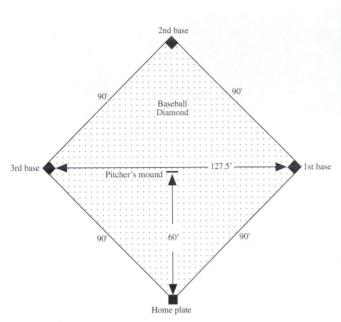

a. Dimensions of a baseball diamond.

b. Capitol City Bombers professional minor league baseball stadium, Columbia, SC (originally 1:6,000 scale).

c. High school baseball diamond.

d. Atlanta Fulton County stadium, home of the professional Atlanta Braves baseball team for many years (originally 2 x 2 m).

Figure 12-18 Examples of high school, minor league, and professional baseball diamonds (level 12192). The baseball-diamond dimensions remain the same while the facilities for the team and spectators change dramatically.

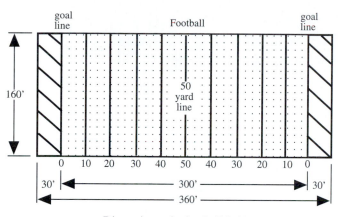

a. Dimensions of a football field.

b. University of South Carolina football stadium (level 12193).

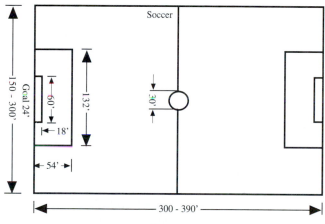

c. Dimensions of a soccer field.

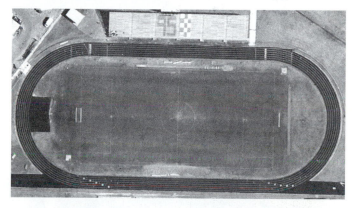

d. High school soccer field and track (level 12193). The goal posts reveal that the field is also used for football.

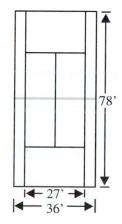

e. Dimensions of a tennis court.

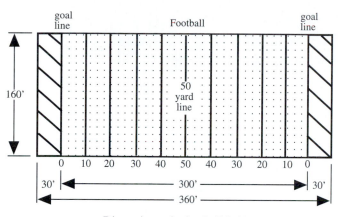

f. Multiple tennis courts at a tennis club (level 12197).

Figure 12-19 Examples of a university football stadium, high school soccer field and track, and community tennis club.

Figure 12-20 Low-oblique aerial photography of a golf course near Atlanta, GA, with multiple long, curvilinear fairways, bright sandtraps, and manicured greens (level 12194).

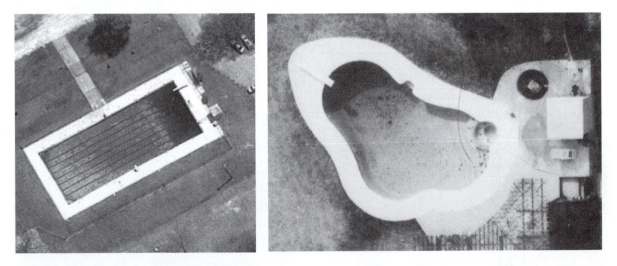

Figure 12-21 Large-scale vertical aerial photography of swimming pools (level 12195) and associated decking. The pool on the left has swimming lanes. The pool on the right has debris in it. Both pools have shallow and deep ends.

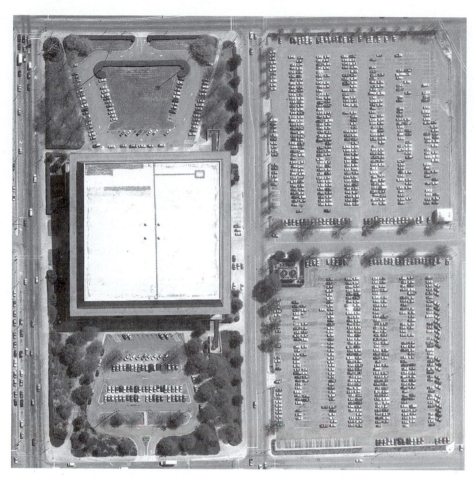

Figure 12-22 The multipurpose coliseum (stadium) associated with the University of South Carolina (level 12196). A recreational or sporting event is in full production on this date.

stores, bowling alleys, etc.). It is impossible to determine what type of business is being conducted from their remote sensing attributes. Only *in situ* investigation will allow them to escape from land-use class *17, Other Urban or Built-up Land* category in Table 12-5.

Services (Public and Private)

In developed countries, mankind has organized a sophisticated infrastructure of public services to improve the quality of life. Some of the buildings and grounds associated with these services have relatively diagnostic characteristics that allow them to be identified in remote sensor data.

Public Buildings and Facilities (Administration, Fire, Police, Rescue, Postal, Libraries, Prisons)

Preeminent public facilities such as state capitols (Figure 12-24a) and major administrative buildings (Figure 12-24b) are relatively easy to identify in remote sensor data. They usually have striking architecture because they are supposed to last and inspire us for many generations. They typically have well-landscaped grounds and ample parking, which may be located in multiple-story parking garages. National monuments such as the Statue of Liberty in New York and L'Arc de Triomphe in Paris, France, often have very diagnostic architecture (Figure 12-24cd, respectively).

Conversely, the only way to identify fire-and-rescue services is if the engines, rescue vehicles, and/or firefighting or rescue equipment are temporarily visible outside the building. Also, some fire stations have a two or three-story tower nearby where they practice firefighting. It is possible to identify a police station if the cars parked outside have numbers on their roofs.

Regional post offices are large buildings with extensive parking and a significant number of postal trucks and smaller postal vehicles present. Local post offices are diffi-

Figure 12-23 Shipping and warehousing of grain and other commodities in the Seattle, WA, harbor. Note the conical storage structures, the elaborate conveyor-belt system, the large rectangular warehouses, and railroad spurs that transport raw materials to and from the facility.

cult to identify because they are small, well-landscaped buildings, with ample parking nearby. In fact, they may be indistinguishable from local libraries.

Prisons are usually easy to identify because they have large buildings, ample parking, guard towers, multiple rows of very high fencing, and restricted personal and automobile access. Exercise fields are usually present within the fenced perimeter.

Education

Universities are relatively easy to identify in high spatial resolution remote sensor data. Except in the downtown areas of the largest cities (e.g., New York, Chicago, Atlanta), universities usually exhibit a geographically extensive collection of buildings, open spaces, recreational facilities (swimming pools, baseball, football, soccer, track, tennis), gymnasiums, coliseums, and stadiums. Many of the education buildings are multiple-story. Also, large dormitories are often present. The problem is usually deciding where the formal university begins and ends. Two different examples of universities are provided in Figure 12-25. The University of South Carolina was founded in 1801. All of the original administration buildings, student dormitories, and teaching facilities were systematically oriented to look out upon a landscaped park called the "horseshoe." Conversely, the oblique aerial photo-

graph of a portion of the University of California at Santa Barbara reveals the unsystematic placement of facilities interspersed with a complex network of sidewalks.

Most high schools and middle schools have large buildings, ample parking, and outdoor fields for baseball, football, soccer, track, and/or tennis. High schools usually have a football stadium and a pool, whereas middle schools do not. In cold climates, the pool may be indoors. High schools and middle schools are often fenced.

Elementary schools have relatively smaller buildings, ample parking, and a few outdoor fields. They rarely have a pool or a football stadium. Elementary schools may be fenced.

Medical

It is difficult to identify medical business offices or health and fitness clubs. They appear similar to many other nondescript commercial activities that can take place in a building. However, it is relatively easy to identify hospitals. They are usually very large multistory buildings with many additions. Ample parking is necessary. Walkways, tramways, etc., often connect one building to another. Sometimes it is possible to detect the emergency-room automotive entrance. Some hospitals have a helicopter landing pad on the ground or on top of a building.

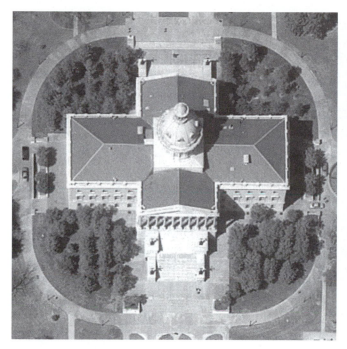

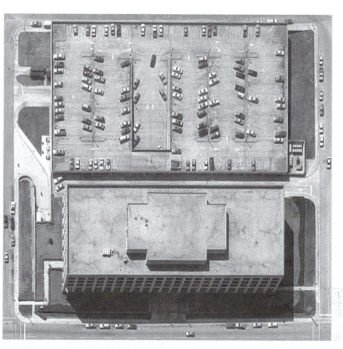

a. State Capitol of South Carolina in Columbia (level 12211).

b. Public office building with multiple-story parking garage.

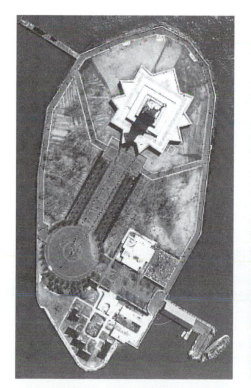

c. Ellis Island and the Statue of Liberty National Park in New York, NY (12217).

d. Low-oblique aerial photograph of L'Arc de Triomphe in Paris (12217).

Figure 12-24 Public administration buildings, national parks, and monuments.

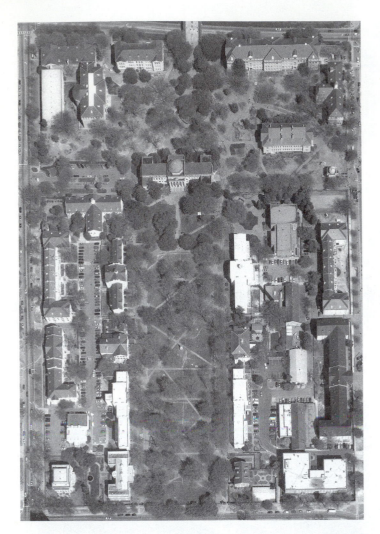

a. Vertical aerial photograph of the manicured "horseshoe" of the University of South Carolina, founded in 1801. At one time most of the buildings surrounding the horseshoe were for administrators, classes, and students and their attendants. Now most of the buildings facing the horseshoe are administrative. This type of architectural design is common on other southern campuses such as the University of Georgia.

b. Low-oblique photograph of the University of California at Santa Barbara in Goleta, CA, with its varied architectural forms, well-manicured landscape, and intricate network of walkways between buildings.

Figure 12-25 Examples of university public education facilities (level 122124).

Industrial Land Use

This section introduces the concept of image interpretation for the analysis of aerial photography and other types of high spatial resolution remote sensor data for extracting industrial land-use information. The heuristic rules discussed are designed to enable the interpreter who is not a specialist in industrial image interpretation to identify those industries primarily from their images alone. If the image analyst cannot specifically identify an industry, however, he or she should still be able to place it in a category that limits the kind of industry it may be. The term industry includes those businesses engaged in the *extraction* of raw materials, the *processing* of these same materials, and the *fabrication* of intermediate and finished products (NPIC, 1961).

First, however, it should come as no surprise that military intelligence gathering organizations throughout the world are very interested in monitoring the industrial land-use of foreign countries. They have developed excellent industrial target image analysis manuals that may also be used for civilian applications. Several of the more important manuals focus on 1) general industrial target analysis logic (DoD, 1976), 2) the coke, iron and steel industries (DoD, 1978a), 3) power and water facilities (DoD, 1977), and 4) the petroleum industry (DoD, 1978b). Copies of these manuals may be viewed at the Federation of American Scientists home page (Appendix A).

Industrial Land-Use Classification Logic

Industries often have unique assemblages of raw materials, equipment, final products and waste, as well as buildings that characterize the industry. Sometimes these *industry-related components* are visible on the photography or imagery; at other times the equipment or material may be sheltered in buildings or otherwise obscured from view. In this case, images of other components, not significant in themselves but closely associated with the hidden essential components, may still be recorded on the imagery. Moreover, from their distinctive forms, patterns, and relationships, one frequently can infer the kind of material or equipment obscured from view. For instance, unusually shaped buildings may be specially designed to house specific kinds of equipment. Similarly, chimneys and stacks by their number and arrangement may indicate the specific kind of furnace or oven from which smoke or noxious gases are being vented.

Identification of the various industry-related components in an image may allow the industry to be placed into one of three major industrial categories (Chisnell and Cole, 1958; NPIC, 1961; Stone, 1964; Avery and Berlin, 1992):

- extraction,

- processing, and

- fabrication.

When attempting to categorize an industry using remote sensing data, it is recommended that the analyst 1) first decide whether it is an extraction, processing, or fabrication industry; 2) if it is a processing industry, decide whether it is chemical, heat, or mechanical processing, in that order; and 3) if it is a fabrication industry, decide whether it is light or heavy fabrication. There are a tremendous variety of industries. It will only be possible to provide a few representative examples of each major category of industry.

Extraction Industries

Extractive industries extract the natural resources of the Earth with the minimum handling required to accumulate raw materials in a form suitable for transportation or processing. The extraction usually takes place from a shaft mine, open-pit mine, well, or cutting field, e.g., forestry clear-cutting or high-grading (Table 12-4). Extractive industries typically have several of the following diagnostic recognition features:

- the presence of open-pit excavations, ponds, mine openings or derricks;

- normal and oversized handling equipment including bulldozers, dump trucks, cranes, power shovels, dredges, tree-cutters, and mine cars;

- an elaborate transportation system to move the raw material about the site, including conveyors, pipelines, railroads, drag lines, and/or road network;

- extracted bulk materials stored in piles, ponds, hoppers, or tanks; and/or

- piles or ponds of waste that have been separated from the valuable material of interest.

All the extractive industry features are rarely seen at any one location. Often the surrounding countryside shows evidence of reclamation, reforestation, etc., in an attempt to repair the effects of the extractive industry.

A 1:19,200-scale black-and-white panchromatic stereopair of the Kennecott Ridgeway Mining Company located near Ridgeway, SC, obtained on March 16, 1992, is shown in Figure 12-26. Two very large open-pit mines are visible in the photographs. They are surrounded primarily by loblolly pine plantations. The raw ore bearing material is extracted from the earth using hugh shovels within the open-pit mines. It is transported to the central facility by dump trucks using a complex network of roads. The gold is separated from the ore-bearing rock using a chemical process at the mine central processing facility. Two elevated piles of material that have already been processed (often referred to as tailings piles) are visible. Note that a substantial amount of the processed rock material also has been used to construct a catchment basin. The mine was decommissioned in 1999. The process of restoration is well underway.

The largest *open-pit mine* in the world is the Bingham Canyon Copper Mine near Salt Lake City, UT, operated by Kennecott Copper Corporation. It is 4 km in length and has a depth of 800 m. It is one of the few features that man has created that can be seen from space (another is the Great Wall of China). More than 3 billion tons of copper have been extracted, more than five times the material excavated for the Panama Canal. Figure 12-27a is a low-oblique aerial photograph of the mine. Figure 12-27b is a terrestrial photograph taken within the mine. The terrestrial photograph provides information on just how large the mine is when we discover that the dark linear features located along the farthest wall are actually four long trains that are used to remove material from the mine. The railroad tracks spiral down to the lowest level.

An open-pit gravel mine in South Carolina is shown in Figure 12-28a. Material is dynamited from the walls of the excavation and carried by truck to a collection point at the base of the mine. A conveyor belt then carries the unsorted material up 100 m to the surface, where it is sorted into various grades of gravel and sand. The sorted material is then transported by trucks to final destinations. Precipitation cannot escape the pit, resulting in an internal drainage pond at the base of the mine.

Figure 12-28b depicts a small portion of a petroleum extraction well field situated in the mountainous terrain behind Ventura, CA. Each clearing contains one or more oil wells that are connected via an intricate system of dirt roads. The oil is actually pumped by pipeline from the wells to collecting points not present in the photograph.

Processing Industries

Processing industries subject the accumulated raw materials to mechanical, chemical, or heat treatment to render them suitable for further processing, or to produce materials from which goods and equipment can be made.

The processing industries are characterized by the presence of facilities for the storage and handling of large quantities of bulk materials. These materials may be stored in the open in piles, ponds or reservoirs, or in storage containers such as silos, bins, hoppers, bunkers, and open or closed tanks. The materials are handled by pipelines, conveyors, and fixed cranes, as well as railroad cars and other mobile equipment. The conveyor-belt system used to sort the gravel in the open-pit mine previously discussed is a good example. In addition, other outdoor equipment are commonly present, including blast furnaces, kilns, chemical-processing towers, large chimneys or many stacks, indicative of the kind of processes being carried out.

The processes involved usually require large quantities of power, which is indicated by the presence of coal piles, fuel tanks, boiler houses, or transformer yards if electric power is employed or produced. The buildings that house the processing equipment frequently are large or at least complex in outline and roof structure. Since the processing of raw materials usually involves refinement, piles and ponds of waste are common, and care is required to distinguish between waste versus useful raw materials stored in a similar manner.

The processing industries can be subdivided on the basis of image components into three subcategories:

* mechanical,

* chemical, and

* heat processing.

As the names suggest, these subcategories have functional significance.

Mechanical-Processing Industries

Mechanical-processing industries are engaged in sizing, sorting, separating, or otherwise changing the physical form or appearance of the raw materials. The image components that characterize the mechanical-processing industries are the bulk materials stored in piles, ponds, or reservoirs, or such outdoor equipment as silos, bins, bunkers or open

Kennecott Ridgeway Mining Company
Ridgeway, South Carolina

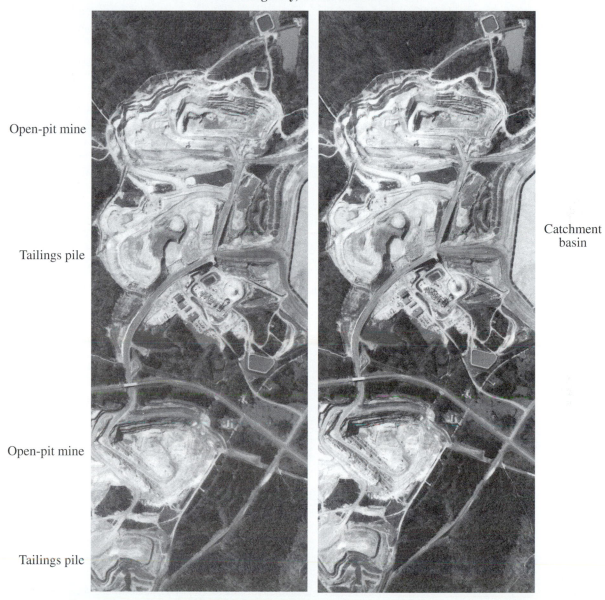

Figure 12-26 Panchromatic stereopair of the Kennecott Ridgeway Mining Company located near Ridgeway, SC, on March 16, 1992 (level 1312) (courtesy of Kennecott Ridgeway Mining Company).

a. Low-oblique aerial photograph of the Bingham Canyon Copper Mine near Salt Lake City, UT, the largest open-pit mine in the world.

b. Terrestrial photograph of the mine reveals ore trains, huge shovels, and a dense network of power-line poles.

Figure 12-27 The Bingham Canyon Copper Mine near Salt Lake City, UT, is an extractive industry (level 13121).

a. An open-pit gravel mine in South Carolina (level 13122). The gravel is transported from the base of the mine via a single conveyor belt to several sorters at the top of the mine. Trucks transport the sorted material to final destinations.

b. Petroleum extraction from oil wells in the mountains behind Ventura, CA (level 13151). Each clearing contains one or more oil wells that are connected via an intricate system of dirt roads. The original scale of the photography was 1:125,000.

Figure 12-28 Examples of open-pit mining and petroleum extraction industries.

a. Low-oblique photograph of mechanical processing at an agricultural mill where wheat is processed into baking flour for human consumption and animal feed. The grain is stored in the silos and the processing takes place in the buildings. Flour and feed are transported to market by truck (level 13211).

b. Low-oblique photograph of mechanical processing taking place at a sewage-treatment plant in Newport Beach, CA (level 13215). Solids are mechanically separated from liquids, hence the term mechanical processing. Note the settling and aeration ponds. Purified effluent is pumped 500 m to a submerged outfall in the Pacific Ocean (level 13215).

Figure 12-29 Examples of mechanical-processing industries (level 1321).

tanks, as well as an abundance of handling equipment such as conveyors, launders, cranes, rail cars, and other mobile equipment. The processing involved may require large quantities of power, as indicated by the presence of boiler houses with their fuel supply, or transformer yards when electric power is employed. Many of the buildings may be large or at least complex in outline and roof structure. Piles or ponds of waste are quite common. The industries in the mechanical-processing category differ from the other processing industries in that they have few pipelines, closed or tall tanks, or stacks other than on boiler houses. Furthermore, there is an absence of the kilns associated with the heat-processing industries.

Mechanical processing takes place at the Adluh Flour Meal and Feed Company (level 13211) in Columbia, SC (Figure 12-29a). Grain arrives at the mill by way of a railroad spur behind the buildings or by truck. Three large concrete silos store the unprocessed wheat. Wheat milling takes place in the large, complex building to the right. Tractor trailers backed up to the building transport the finished flour to local and regional market.

Figure 12-29b depicts a low-oblique photograph of mechanical processing taking place at a sewage-treatment plant (level 13215) in Newport Beach, CA. The sewage is routed from the nearby communities to the plant via underground pipelines. Mechanical processing at the plant separates the solids from the liquid. Various settling ponds, enclosed tanks, aerators, and anerobic bacteria facilities purify the sewage so that it can eventually be pumped 500 m to a submerged outfall in the Pacific Ocean. Water purification installations are also mechanical-processing industries.

A hydroelectric power plant utilizes the "hydrostatic head" of water stored in the lake to spin turbines that create electricity. For example, consider the hydroelectric generating facilities at the Lake Murray Dam near Columbia, SC (Figure 12-30). Four intake towers transport water down 200 ft to the turbine house where the water pressure spins the turbines generating electricity. The water then enters the Saluda River. The electricity is transported to the regional power grid via a large substation complex. There is also a thermal-electric power plant at this location which will be discussed in the heat-processing industry section.

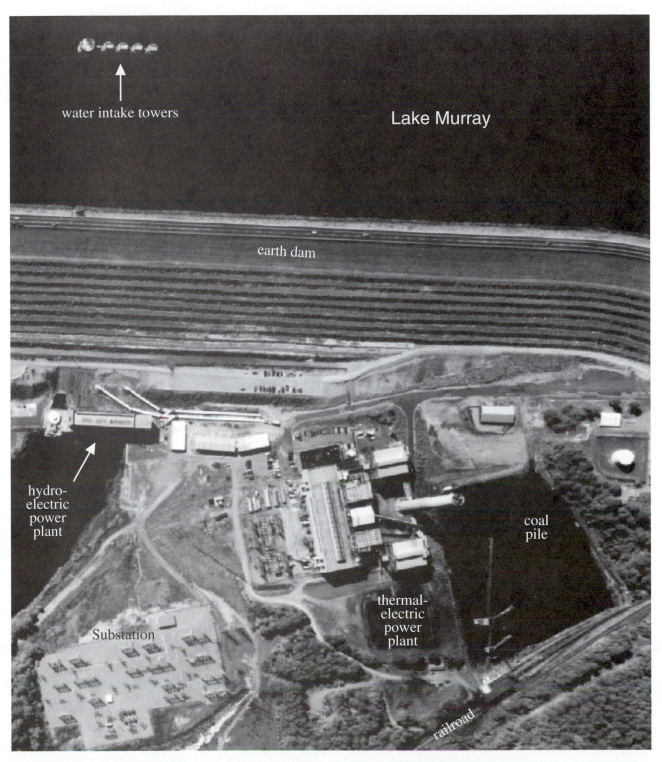

Figure 12-30 Aerial photograph of the Lake Murray Dam, SC, hydroelectric power plant (level 13213) and the Saluda thermal-electric power plant. Four intake towers provide the high-pressure water that drives the turbines that produce the electricity. The thermal-electric power plant burns coal which produces steam to drive the separate turbines that produce electricity. The coal is delivered by railroad and transported from the railroad cars to the pile via conveyor belts. The electricity is distributed from the site onto the regional power grid via the substation.

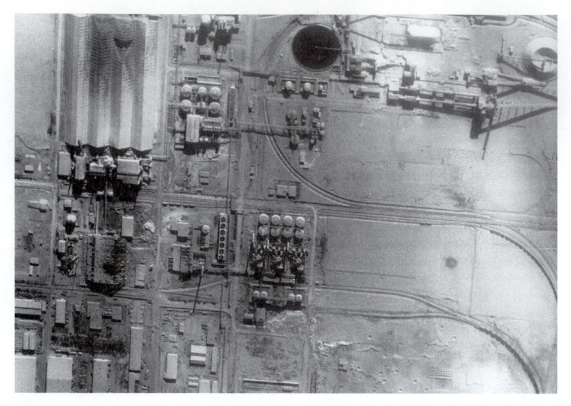

Figure 12-31 Imagery of the Al Qaim superphosphate fertilizer plant in Iraq, obtained during the Gulf War in 1991 (U.S. Navy TARPS image released under the Freedom of Information Act to William M. Arkin).

Chemical-Processing Industries

Chemical-processing industries utilize chemicals or chemical processes to separate or rearrange the chemical constituents of the raw materials. Petroleum refineries and fertilizer plants are good examples. Pressure, heat and catalysts, or other chemicals may be employed. Closed vessels for holding or handling fluids, gases, or suspensions of solids in liquids, and the use of fluid flow in processing the raw materials are typical of these industries. Consequently, the image components that characterize the chemical-processing industries are an abundance of closed tanks, pipelines, and such large processing equipment as towers for cracking or distillation. As with all processing industries, those in this category have facilities for the storage and handling of bulk materials and require large quantities of power. The buildings frequently are complex, and extensive waste piles or ponds are common.

The Al Qaim superphosphate fertilizer plant in Iraq is shown in Figure 12-31. Note the existence of numerous enclosed storage tanks, a large conical pile of raw materials, a circular pond, and large processing buildings (several with recent bomb damage). Phosphates may be used to produce explo-

sives such as those used in the Oklahoma federal building bombing. That is why they are targeted during war.

Industry generates a significant amount of chemical hazardous wastes. Some companies use remote sensing data to document the initial conditions under which the materials are stored and then monitor the surface expression after they are stored. For example, Figure 12-32a is a low-oblique aerial photograph of the Low-Level Waste Management site at the Department of Energy's (DOE) Nevada Test Site. This location provides efficient disposal of radioactive chemical wastes generated in the nuclear weapons program. Waste is trucked to the site in Department of Transportation approved containers (in this example drums) and buried in the large trench. As it fills, the pit is covered with soil (Pike, 1998).

Figure 12-32b is a low-oblique photograph of radioactive waste being stored in a tank farm at the Hanford, WA, DOE facility. The tank farms store by-product materials left over from plutonium extraction operations prior to permanent disposal. This by-product material has no useful purpose and is stored in approximately 177 underground storage tanks with a cumulative total of 55 million gallons capacity, with some individual tanks ranging up to 1,000,000 gallons. The

a. Radioactive hazardous waste being stored in drums at the Department of Energy's Nevada Test Site Low-Level Waste Management facility. The waste is eventually covered with sand (courtesy Department of Energy, ID WB212; John Pike, FAS).

b. Hazardous-waste storage tanks at the DOE Hanford, WA facility. At one time approximately 2,300 personnel supported the total tank farm activities at Hanford (courtesy Department of Energy, ID 303-058-001; John Pike, Federation of American Scientists).

Figure 12-32 Examples of chemical hazardous-waste-site monitoring using large-scale low-oblique aerial photography (level 13223).

trailers and cars in the lower right of the photograph provide some idea of the size of the tanks (Pike, 1998).

Petroleum chemical-processing industries convert raw crude oil into gasoline, diesel fuel, synthetic textiles, and plastic. The plants usually have an extensive, often dense network of above or underground pipelines such as those found in a petroleum refinery in Texas (Figure 12-33a). Storage tanks are often completely surrounded by Earth dike revetments that can contain the chemical if a spill occurs. For example, Figure 12-33b depicts a tank farm at the petroleum refinery at Al Basrah, Iraq, in 1991. Several of the tanks are on fire and leaking. The Earth dikes are containing much of the spill. Figure 12-33c depicts the Atlantic Richfield petroleum storage and transport facilities in Seattle, WA.

Heat-Processing Industries

The heat-processing industries utilize primarily heat to refine, separate, or re-form the raw materials, or to derive energy from them. Iron and steel production and thermal-electric power plants that burn oil or coal are good examples. The image components that distinguish this category are the large quantities of coal or other fuel, large chimneys, large numbers of stacks, flues and kilns of various kinds. Although pipelines and tanks frequently are employed in these industries, they are usually not abundant, and their presence is outweighed by the image components evidencing the use of heat. As with the other processing industries, these have extensive facilities for handling and storing bulk materials and require large quantities of fuel or power. Large outdoor equipment, such as blast furnaces and kilns, often are employed. The buildings frequently are complex, and piles of waste are common.

Iron and steel heat-processing industries are usually straightforward to identify on remote sensor data. For example, Figure 12-34 is a historical aerial photograph of a small portion of the Bethlehem Steel Company at Sparrows Point, MD in 1952. The photograph records numerous long buildings that house the steel rolling mills and blast furnaces, many support buildings, and considerable smoke from large stacks.

Nuclear facilities are also relatively easy to identify. The Fast Flux Test Facility at the Hanford site in eastern Washington is a 400-megawatt thermal reactor cooled by liquid sodium (Figure 12-35a). It was built in 1978 to test plant equipment and fuel for the U.S. Government's liquid metal reactor development program. This program demonstrated the technology of commercial breeder reactors. The containment dome is easily identified in the image. Above-ground testing of nuclear materials has been banned for decades. However, underground testing continues. Figure 12-35b depicts the results of one test at the Nevada Test Site where

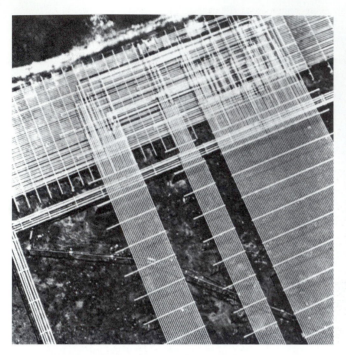

a. Large-scale vertical aerial photograph of a dense network of pipes at a petroleum refinery in Texas.

b. Petroleum refinery at Al Basrah, Iraq. Note the large tanks and Earth dikes that separate them. This photograph was acquired in 1991 during the Gulf War, and some tanks are on fire and leaking (U.S. Navy TARPS image released under the Freedom of Information Act to William M. Arkin).

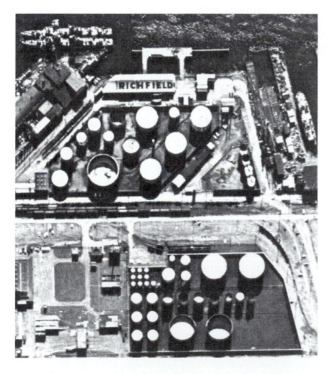

Figure 12-33 Examples of chemical-processing industries (level 1322).

c. Chemical processing, storage, and transportation of petroleum products at the Atlantic Richfield facilities in Seattle, WA. Note the petroleum tank farms, tanker ships, and railroad tank cars.

Figure 12-34 Bethlehem Steel Company at Sparrows Point, MD on April 23, 1952.

a 100 kiloton explosive was buried under 635 feet of desert alluvium and detonated on July 6, 1962, displacing 12 million tons of earth. Underground tests conducted by nations throughout the world often leave craters such as these as evidence of what has taken place.

Below are some additional fundamental image-recognition features that may be used to distinguish among the three types of processing industries:

Mechanical-Processing Industries:
- few pipelines or closed tanks,
- little fuel,
- conveyor-belt systems often present,
- few stacks,
- no kilns.

Chemical-Processing Industries:
- many closed or tall tanks, including gasholders,
- many pipelines,
- much large outdoor processing equipment.

Heat-Processing Industries:
- a few pipelines or tanks,
- large active chimneys and/or stacks,
- large quantities of fuel, and
- kilns.

Fabrication Industries

Fabrication industries assemble the mechanical and chemical subcomponents into finished products such as automobiles, trucks, boats and ships, trailers, heavy equipment (e.g., bulldozers), plastic products, and electronic devices. The fabrication industries may be subdivided into heavy and light fabrication. *Heavy-fabrication* industries often have tall heavy steel frame one-story buildings, storage yards with heavy lifting equipment and cranes, and occasionally rail lines entering buildings. *Light-fabrication* industries often make use of light steel and wood-frame buildings (perhaps multistory), and typically lack heavy-lifting equipment.

a.

b.

Figure 12-35 a) The Department of Energy Fast Flux Test Facility at Hanford, WA, has a 400-megawatt thermal reactor. The dome is the containment structure. b) The Sedan Crater was formed when a 100 kiloton explosive buried 635 ft beneath the desert was detonated on July 6, 1962 at the Nevada Test Site, displacing 12 million tons of earth. The massive crater is 320 ft deep and 1,280 ft in diameter. Two craters formed from other detonations are seen in the distance (courtesy Department of Energy).

a. Heavy fabrication taking place at the Kline Iron and Steel Company in Columbia, SC. Iron and steel stock is sorted by size and type in the two central open-air storage yards. Fabrication takes place inside the various buildings. Raw materials and finished products are transported to and from the facility via trucks or the railroad spur (level 13314).

b. A detailed vertical aerial photograph of a power plant under construction. The project contains significant quantities of prefabricated structural iron and steel (courtesy Eastman Kodak Company).

Figure 12-36 Examples of iron and steel fabrication and the use of such materials in the building of a power plant.

Heavy Fabrication

Below are some of the fundamental image-recognition features that may be used to distinguish among the major heavy-fabrication industries. Heavy fabrication is almost universally associated with large buildings.

Heavy Machinery Fabrication:
- storage yard for heavy metal shapes,
- piles of structural steel and other materials,
- finished heavy machinery.

Railroad Machinery Manufacture and Repair:
- railroad locomotives and/or cars under construction,
- railroad track network.

Shipbuilding:
- storage yards for steel plate,
- extensive dry-dock and ramp facilities,
- ships under construction,
- large cranes.

Structural Steel Fabrication:
- storage yards of structural steel,
- structural steel fabrication yard.

A structural steel fabrication industry is shown in Figure 12-36a. The actual fabrication of the iron and steel girders used in building and bridge construction takes place inside the large elongated buildings. The iron and steel stock materials are sorted by size and type in the central open-air storage yard. Two overhead cranes (not visible) move the iron and steel raw materials back and forth (left to right) into the buildings for fabrication. Raw and finished materials are transported to and from the business either by truck or via the railroad spur.

A vertical perspective of how such structural steel is used in the building of a power plant is presented in Figure 12-36b.

Aerial images such as this can document the progress on specific dates in the construction process.

Light Fabrication

Below are some of the fundamental image-recognition features that can be used to distinguish among the major light-fabrication industries.

Aircraft Assembly:
- aircraft or seaplane ramps or taxiways adjacent to or connected to the fabrication building,
- hangars,
- finished aircraft stored in the open.

Automobile Assembly (cars, trucks, motorcycles):
- single- or multiple-story buildings,
- finished parked vehicles arranged in an orderly system,
- test track(s),
- tractor-trailer shipping facilities with cars on trailers,
- rail shipping container cars.

Boat Building and Repair:
- boat components stored in the open,
- fiberglass forms,
- building where fiberglass application takes place,
- boat trailers,
- finished boats in open storage.

Meat Packing:
- holding pens,
- a few tanks,
- railroad siding,
- tractor-trailer parking.

Textiles:
- tractor-trailer shipping and receiving,
- bales of cotton or other raw material, and
- pipelines and tanks when creating synthetic fabrics.

An automotive component manufacturing company in Columbia, SC, is shown in Figure 12-37a. Raw materials that are not susceptible to damage by the Sun or precipitation are sorted and stored by type in the open air. Fabrication takes place inside the building. The finished products are shipped from the shipping bays located at the back of the building.

A fish-canning plant located in San Pedro Harbor, CA, is found in Figure 12-37b. The ships moor at the dock next to the plant. The manufacturing process (canning) takes place within the large building. Numerous tanks hold chemicals

used during the canning process. The plant is discharging effluent into the harbor using an underground pipe.

 Transportation Infrastructure

Transportation planners often use remote sensor data to 1) update transportation network maps, 2) evaluate road and railroad conditions, 3) study urban traffic patterns at choke points such as tunnels, bridges, shopping malls, and airports, and 4) conduct parking studies (Mintzer, 1983; Haack et al., 1997).

Roads and Highways

Roads may be unpaved or paved. Paved highways may be accessible to motorists at almost any location along their route or they may be limited-access (freeways, toll). The general updating of a road or highway network centerline map is a fundamental task. This is often done every one to five years and, in areas with minimum tree density, can be accomplished using imagery with a spatial resolution of 1 – 30 m (Lacy, 1992). If more precise road dimensions are required such as the exact width of the road and sidewalks, then a spatial resolution of 0.25 – 0.5 m is required (Jensen et al., 1994). Currently, only large-scale aerial photography such as that previously shown in Figure 12-8 can provide such planimetric information. In addition, consider the aerial photograph of the intersection shown in Figure 12-38a. The stoplights suspended across the street, individual cars and trucks, and even the right-of-way white road markings are visible. Future unmanned aerial vehicles (UAVs) carrying lightweight digital cameras might well provide similar high-resolution information at a reasonable cost.

Highway interchanges are used to route traffic onto and off of limited-access highways. They come in an endless variety of shapes and sizes and are relatively easy to distinguish on remote sensor data. For example, consider the classic limited access interchange shown in Figure 12-38b. It is gently sloping and has very long entrance and exit ramps that allow traffic to merge carefully with other traffic.

Next to meteorological investigations, traffic count studies of automobiles, airplanes, boats, pedestrians, and people in groups require the highest temporal resolution data, often ranging from 5 to 10 minutes. It is difficult to resolve the type of car or boat using even 1 x 1 m data. This task requires high spatial resolution imagery from 0.25 – 0.5 m. Such

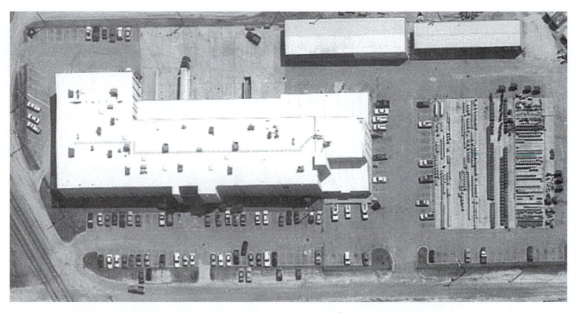

a. Light fabrication company. Some of the raw materials are sorted by type in the open air. Fabrication takes place in the building.

b. A fish-canning plant (level 13324) in San Pedro Harbor, CA, and its associated pollution.

Figure 12-37 Examples of light fabrication (manufacturing).

a. High spatial resolution image of an intersection acquired using kite aerial photography.

b. A limited-access highway interchange in Texas.

c. Large scale aerial photography of a parking lot.

d. Aerial photography of highway and railroad bridges.

Figure 12-38 Examples of roads, limited-access highways, interchanges, parking, and bridges recorded using various scales of aerial photography (level 1411).

information can only be acquired using aerial photography or video sensors that are 1) located on the top edges of buildings looking obliquely at the terrain, or 2) placed in aircraft or helicopters and flown repetitively over the study areas. Figure 12-38c captures the parking characteristics of a parking lot at a single instant in time. When such information is collected at an optimum time of day, future parking and traffic movement decisions can be made. Parking studies require the same high spatial resolution (0.25 – 0.5 m) but slightly lower temporal resolution (10 – 60 minutes).

Road, railroad, and bridge conditions (cracks, potholes, etc.) are routinely monitored both *in situ* and occasionally using high spatial resolution remote sensor data. For example, Figure 12-38d presents a panchromatic image of a highway and railroad bridge. Careful inspection by a trained analyst provides significant information about the condition of the road and bridge. Road and bridge condition can be documented using high spatial resolution aerial photography (< 0.25 x 0.25 m).

Railroads

Railroad tracks, locomotives, cars, and terminal facilities are relatively easy to distinguish on high spatial resolution imagery because:

- the tracks rarely move from year to year,

- the tracks have no right angles, only gradual curvilinear turns,

- railroad bridges become narrow as opposed to highway bridges, which generally stay the same width as the road,

- railroad crossing signposts and their attendant shadows are located at almost every major intersection with a road,

- railroad tracks rarely cross other railroad tracks, and

- locomotives and railroad cars are large, long, and linear, making them fairly easy to identify.

Much of the actual work of a railroad takes place in the railroad marshalling (classification) yard. It is here that railroad cars going the same direction down the line or railroad cars going to the identical destination are placed in a specific train. Figure 12-39 depicts an eleven track railroad marshalling yard in Montana. One part of the yard has eight tracks and the other has three. Note that the eight-track marshalling yard terminates at the right. There are 175 railroad cars visible in the photograph.

Airport

Civilian and military airports usually have:

- long runway(s),

- flight-traffic control tower(s),

- terminal(s) [civilian],

- hangars,

- an extensive tarmac road network connecting the hangars and terminals with the runways,

- large parking areas, and

- visible aircraft.

Rural airports often have a single runway, a windsock, visible planes, and a hangar or two.

Figure 12-40a depicts a biplane recorded at approximately 6 x 6 in. spatial resolution. The biplane photograph was acquired using kite aerial photography. A portion of the El Toro Marine Air Station, CA (1 x 1 ft) recorded using a conventional camera is found in Figure 12-40b. An IKONOS (1 x 1 m) image of the Ronald Reagan National Airport is presented in Figure 12-40c. The civilian airport has an extensive terminal system to move people to and from the plane. Conversely, some military airports do not have terminals, as the crews are ferried to and from the aircraft by vehicle.

Boats and Ships

Small boats and large ships are relatively easy to identify using high spatial resolution remote sensor data for the following reasons:

- They are often seen as isolated man-made reflective objects on a relatively uniform water surface which can increase their object-to-background contrast.

- Boats and ships have relatively unique shapes (i.e., the length is almost always at least two times greater than the width, with a relatively pointed bow) because they need to hydrodynamically slice through the water with a minimum amount of friction.

Boats and ships range in size from the small family rowboats, power boats, and sailboats from 3 – 10 m in length (Figure 12-41a) to extremely large cruise ships, passenger

Figure 12-39 An eleven track railroad marshalling (classification) yard (level 14122) in Montana. One part of the yard has eight tracks and the other has three.

liners, oil tankers, battleships, submarines, and aircraft carriers. Figure 12-41b depicts twelve commercial barges on the Mississippi River being pushed by two tugboats. Numerous large military ships are in storage in National City, San Diego, CA, in Figure 12-41c.

 Communications and Utilities

Urban/suburban environments are enormous consumers of electrical power, natural gas, telephone service, and potable water (Haack et al., 1997). In addition, they create great quantities of refuse, wastewater, and sewage. The removal of storm water from urban impervious surfaces is also a serious problem (Schultz, 1988). Automated mapping/facilities management (AM/FM) and geographic information systems (GIS) have been developed to manage extensive right-of-way corridors for various utilities, especially pipelines (Jad-

kowski et al., 1994). The most fundamental task is to update maps to show a general centerline of the utility of interest such as a power line right-of-way. This is relatively straightforward if the utility is not buried and 1 – 30 m spatial resolution remote sensor data is available. It is also often necessary to identify prototype utility (e.g., pipeline) routes (Feldman et al., 1995). Such studies require more geographically extensive imagery, such as Landsat Thematic Mapper data. Therefore, the majority of the actual and proposed right-of-way may be observed well on imagery with 1 – 30 m spatial resolution obtained once every one to five years.

When it is necessary to inventory the exact location of the utility footpads, transmission towers, utility poles, manhole covers, the true centerline of the utility, the width of the utility right-of-way, and the dimensions of buildings, pumphouses, and substations, then it is necessary to have a spatial resolution of from 0.25 – 0.6 m (Jadkowski et al, 1994). For example, Figure 12-42a,b depicts a water-storage tower and the satellite transmission facilities associated with a television station.

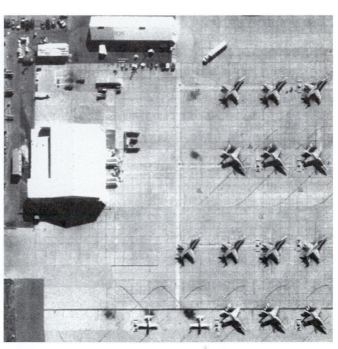

a. A vintage biplane recorded at approximately 6 x 6 in. spatial resolution using kite aerial photography (courtesy Cris C. Benton).

b. Jet aircraft at the El Toro Marine Air Station, CA, recorded at approximately 1 x 1 ft spatial resolution.

c. Panchromatic 1 x 1 m image of Ronald Reagan National Airport in Washington, DC, recorded by the IKONOS satellite in 1999 (courtesy Space Imaging, Inc.).

Figure 12-40 Examples of aircraft, tarmac, hangars, and military and civilian terminals (level 1413).

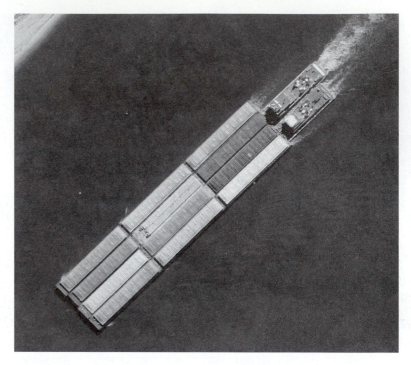

a. Sailboats at anchor in a harbor near San Francisco, CA (courtesy Cris C. Benton).

b. High spatial resolution vertical aerial photograph of twelve barges on the Mississippi River being pushed by two tugboats.

c. Four battleships, two troop transports, and an aircraft carrier in storage at National City, San Diego, CA.

Figure 12-41 Examples of boats and ships recorded on remotely sensed imagery (level 1414).

a. Fenced water tower with adjacent maintenance building.

b. WIS TV-10 television station is in the building on the right. It has a tall antenna on the roof and a large circular satellite dish in the parking lot. A government office building is on the left.

Figure 12-42 Examples of utility transmission facilities (level 143).

 ## Digital Elevation Model (DEM) Creation

Most GIS used for socioeconomic or environmental planning in the urban environment include a digital elevation model (Cowen et al., 1995). Analysts often forget that the digital elevation models are derived almost exclusively from analysis of stereoscopic remote sensor data (Jensen, 1995). It is possible to extract relatively coarse z-elevation information using SPOT 10 x 10 m data, SPIN-2 data (Lavrov, 1997), and Landsat Thematic Mapper 30 x 30 m data (Gugan and Dowman, 1988). Any DEM to be used in an urban/suburban application should have z-elevation and x,y coordinates that meet geospatial positioning accuracy standards (FGDC, 1997). A sensor that can provide such information at the present time is stereoscopic large-scale metric aerial photography with a spatial resolution of $\leq 0.25 - 0.5$ m.

Ridley et al. (1998) found that simulated 1 x 1 m stereoscopic satellite data when processed using standard off-the-shelf DEM generation packages yielded a z-elevation RMSE ranging from $1.5 - 2$ m after editing and showed considerable potential for creating DEMs for use in national mapping. Digital desktop soft-copy photogrammetry is revolutionizing the creation and availability of DEMs (Jensen, 1995) and should be of significant value when applied to the commercial high spatial resolution remote sensor data.

Terrain elevation in urban environments does not change very rapidly. Therefore, a DEM of an urbanized area need only be acquired once every $5 - 10$ years unless there is significant development and the analyst desires to compare two different date DEMs to determine change in terrain elevation, identify unpermitted additions onto buildings, or identify changes in building heights. Figure 12-43 depicts 1) a large-scale aerial photograph of downtown Columbia, SC, 2) a digital elevation model of the same area extracted from the stereoscopic photography depicting the height of every building, 3) the orthophotograph draped over the DEM, creating a virtual reality representation of a major street, and 4) use of the DEM for modeling the optimum location for locating a cellular phone transceiver (Petrie and Kennie, 1990; Jensen, 1995; Cowen and Jensen, 1998). Architects, planners, engineers, and real estate personnel are beginning to use such information for a variety of purposes.

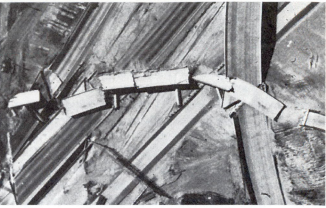

a. Overturned tractor-trailer tanker in Alaska (Jensen and Cowen, 1999).

b. Aerial photography of collapsed spans of a raised highway near Saugus, CA, immediately after an earthquake in 1971. This span was supported by pillars that could not withstand the quake.

c. Tornado damage near Mechanicsville, NY. This is 1 x 1 ft spatial resolution imagery obtained on June 1, 1998, using a digital frame camera. Fire and medical rescue workers are on the scene (courtesy Litton Emerge, Inc.).

Figure 12-44 Examples of high spatial resolution imagery acquired immediately after disasters using traditional metric aerial photography (a,b) and a digital frame camera (c).

 Observations

Table 12-1 and Figure 12-1 reveal that there are a number of remote sensing systems that currently provide some of the desired urban/socioeconomic information when the required spatial resolution is poorer than 4 x 4 m and the temporal resolution is between 1 and 55 days. However, very high spatial resolution data (< 1 x 1 m) is required to satisfy many of the socioeconomic data requirements. In fact, as shown in Figure 12-1, the only sensor that currently provides such data on demand is analog or digital aerial photography (0.25 – 0.5 m). Space Imaging, Inc., IKONOS, with its 1 x 1 m panchromatic data; ORBIMAGE, Inc., OrbView 3 (2000), with its 1 x 1 m panchromatic data; and EarthWatch Inc.,

Quickbird (2000), with its 0.8 x 0.8 m panchromatic data, may still not satisfy all of the data requirements. None of the sensors can provide the 5- to 60-minute temporal resolution necessary for traffic and parking studies except for 1) repetitive aerial photography (very costly), or 2) the placement of digital or video cameras on the edge of tall buildings to obtain an oblique view. The GOES satellite constellation (east and west) and the European Meteosat provide sufficient national and regional weather information at reasonable temporal (3 – 25 minutes) and spatial resolutions (1 – 8 km and 2.5 – 5 km, respectively). Ground-based National Weather Service Weather Surveillance Radar provides sufficient spatial resolution (1 x 1°) and temporal resolution (5 – 10 minutes) for precipitation and intense storm tracking in urban environments.

References

American Planning Association, 1999, *Land-Based Classification Standards: Draft Classification*, American Planning Association Research Department, http://www.planning.org/lbcs.

Anderson, J. R., E. Hardy, J. Roach, and R. Witmer, 1976, *A land-use and land-cover Classification System for Use with Remote Sensor Data*, Washington, DC: U.S. Geological Survey Profession Paper 964, 28 pp.

Angelici, G. L., N. A. Bryant, R. K. Fretz and S. Z. Friedman, 1980, *Urban Solar Photovoltaic Potential: An Inventory and Modeling Study Applied to the San Fernando Valley Region of Los Angeles*. Pasadena, JPL Report. 80–43.

Atkinson, P. M. and P. J. Curran, 1997, "Choosing an Appropriate Spatial Resolution for Remote Sensing Investigations," *Photogrammetric Engineering*, 63(12):1345–1351.

Avery, T. E. and G. L. Berlin, 1992, *Fundamentals of Remote Sensing and Airphoto Interpretation*, New York: Macmillan, 377–404.

Belward, A. S., J. E. Estes and K. D. Kline, 1999, "The IGBP-DIS Global 1-km Land-Cover Data Set DISCover: A Project Overview," *Photogrammetric Engineering & Remote Sensing*, 65(9):1013-1020.

Branch, M. C., 1971, *City Planning and Aerial Information*. Cambridge, Harvard University Press, 283 pp.

Broome, F., 1998, correspondence, Washington: Bureau of Census.

Bureau of the Budget, 1987, *Standard Industrial Classification Manual*, Washington, DC: Government Printing Office.

Chisnell, T. C. and G. E. Cole, 1958, "Industrial Components - A Photo Interpretation Key on Industry," *Photogrammetric Engineering*, 24:590–602.

Cinti, F. A., 1994, "South California Shakes Again: The Earthquake of January 17, 1994," *Systema Terra – Remote Sensing and the Earth*, 3(2):27–30

Clayton, C. and J. E. Estes, 1979, "Distributed Parameter Modeling of Urban Residential Energy Demand," *Photogrammetric Engineering & Remote Sensing*, 45:106–115.

Colcord, J. E., 1981, "Thermal Imagery Energy Surveys," *Photogrammetric Engineering & Remote Sensing*, 47:237–240.

Cowen, D. J. and J. R. Jensen, 1998, "Extraction and Modeling of Urban Attributes Using Remote Sensing Technology," *People and Pixels: Linking Remote Sensing and Social Science*, Washington, DC: National Academy Press, 164–188.

Cowen, D., J. R. Jensen, P. Bresnahan, D. Ehler, D. Traves, X. Huang, C. Weisner and H. E. Mackey, 1995, "The Design and Implementation of an Integrated GIS for Environmental Applications," *Photogrammetric Engineering and Remote Sensing*, 61(11):1393–1404.

Crum, T. D. and R. L. Alberty, 1993, "The WSR-88D and the WSR-88D Operational Support Facility," *Bulletin of the American Meteorological Society*, 74(9):1669–1687.

Cullingworth, B., 1997, *Planning in the USA: Policies, Issues and Processes*, London: Routledge, 280 pp.

Curran, P. J., and T. A. Hobson, 1987, "Landsat MSS Imagery to Estimate Residential Heat-Load Density," *Environment and Planning*, 19:1597–1610.

Davis, B. A., 1993, "Mission Accomplished," *NASA Tech Briefs*, 17(1):14–16.

DoD, 1976, *Industrial Target Analysis Supplemental Reading*, Washington, DC: U.S. Army Intelligence Center School, SupR 62810-11.

DoD, 1977, *Power/Water Facility Clue Sheet*, Fort Belvoir, VA: Defense Mapping School, DMS No. 527.

DoD, 1978a, *Coke, Iron and Steel Industries*, Fort Belvoir, VA: Defense Mapping School, DMS No. 555.

DoD, 1978b, *Petroleum Industries*, Fort Belvoir, VA: Defense Mapping School, DMS No. 553.

Eliasson, I., 1992, "Infrared Thermography and Urban Temperature Patterns," *International Journal of Remote Sensing*, 13(5):869–879.

Elvidge, C. D., K. E. Baugh, E. A. Kihn, H. W. Kroeh and E. R. Davis, 1997, "Mapping City Lights with Nighttime Data from the DMSP Operational Linescan System," *Photogrammetric Engineering & Remote Sensing*, 63(6):727–734.

Feldman, S. C., R. E. Pelletier, E. Walser, J. R. Smoot and D. Ahl, 1995, "A Prototype for Pipeline Routing Using Remotely Sensed Data and Geographic Information System Analysis," *Remote Sensing of Environment*, 53:123–131.

FGDC, 1997, *Draft Geospatial Positioning Accuracy Standards.* Washington, Federal Geographic Data Committee, 32 pp.

Ford, K., 1979, R*emote Sensing for Planners.* Rutgers, State Univ. of New Jersey, 219 pp.

Forster, B. C., 1985, "An Examination of Some Problems and Solutions in Monitoring Urban Areas from Satellite Platforms," *International Journal of Remote Sensing*, 6(1):139–151.

Gottmann, J., 1994, "Towards a Global Urbanization - The Post-Industrial City," Systema *Terra - Remote Sensing and the Earth*, 3(3):4–7.

Green, K., D. Kempka and L. Lackey, 1994, "Using Remote Sensing to Detect and Monitor Land-Cover and Land-Use Change," *Photogrammetric Engineering & Remote Sensing*, 60:331–337.

Green, N. E., 1957, "Aerial Photographic Interpretation and Social Structure of the City," *Photogrammetric Engineering*, 23:89–96.

Greve, C. W., 1996, *Digital Photogrammetry: An Addendum to the Manual of Photogrammetry*, Bethesda, MD: American Society for Photogrammetry & Remote Sensing, 247 pp.

Gugan, D. J. and I. J. Dowman, 1988, "Topographic Mapping from SPOT Imagery," *Photogrammetric Engineering & Remote Sensing*, 54(10):1409–1404.

Haack, B. K. and E. T. Slonecker, 1994, "Merged Spaceborne Radar and Thematic Mapper Digital Data for Locating Villages in Sudan," *Photogrammetric Engineering & Remote Sensing*, 60(10):1253–1257.

Haack, B. K., S Guptill, R. Holz, S. Jampoler, J. R. Jensen and R. Welch, 1997, "Chapter 15: Urban Analysis and Planning," *Manual of Photographic Interpretation*, Bethesda, American Society for Photogrammetry & Remote Sensing, 517–553.

Hagerstrand, T., 1995, "Remote Sensing, GIS and the Landscape Mantle," *Systema Terra — Remote Sensing and the Earth*, 4(2):7–10.

Henderson, F. M. and J. J. Utano, 1975, "Assessing General Urban Socioeconomic Conditions with Conventional Air Photography," *Photogrammetria*, 31:81–89.

Henderson, F. M. and Z. Xia, 1998, "Radar Applications in Urban Analysis, Settlement Detection and Population Estimation," *Principles and Applications of Imaging Radar*, 3rd Ed., *Manual of Remote Sensing*, New York: John Wiley & Sons, 733–768.

Holz, R. K., 1988, "Population Estimation of Colonias in the Lower Rio Grande Valley Using Remote Sensing Techniques," Paper presented at the *Annual Meeting of the Association of American Geographers*, Phoenix, AZ.

Hothem, D., J. M. Irvine, E. Mohr and K. B. Buckely, 1996, "Quantifying Image Interpretability for Civil Users, *Proceedings*, AS-PRS/ACSM Annual Convention, Bethesda: ASPRS, 292–298.

Jadkowski, M. A., P. Convery, R. J. Birk and S. Kuo, 1994, "Aerial Image Databases for Pipeline Rights-of-Way Management," *Photogrammetric Engineering & Remote Sensing*, 60(3):347–353.

Jensen, J. R., et al., 1983, "Urban/Suburban land-use Analysis," *Manual of Remote Sensing*, 2nd ed., R. N. Colwell, ed., Falls Church, VA, American Society of Photogrammetry, 1571–1666.

Jensen, J. R., 1995, "Issues Involving the Creation of Digital Elevation Models and Terrain Corrected Orthoimagery Using Soft-Copy Photogrammetry," *Geocarto International: A Multidisciplinary Journal of Remote Sensing*, 10(1): 1–17.

Jensen, J. R., 1996, *Introductory Digital Image Processing: A Remote Sensing Perspective.* Upper Saddle River, Prentice-Hall, 318 pp.

Jensen, J. R. and D. C. Cowen, 1999, "Remote Sensing of Urban/Suburban Infrastructure and Socio-Economic Attributes," *Photogrammetric Engineering & Remote Sensing*, 65:611–622.

Jensen, J. R. and F. Qiu, 1998, "A Neural Network Based System for Visual Landscape Interpretation Using High Resolution Remotely Sensed Imagery," *Proceedings*, Annual Meeting of the American Society for Photogrammetry & Remote Sensing, Tampa, FL; 15 pp. on compact disk.

Jensen, J. R. and D. L. Toll, 1983, "Detecting Residential Land-Use Development at the Urban Fringe," *Photogrammetric Engineering & Remote Sensing,* 48:629–643.

Jensen, J. R., D. C. Cowen, J. Halls, S. Narumalani, N. Schmidt, B. A. Davis, and B. Burgess, 1994, "Improved Urban Infrastructure Mapping and Forecasting for BellSouth Using Remote Sensing and GIS Technology," *Photogrammetric Engineering & Remote Sensing*, 60(3):339–346.

Jensen, J. R., J. Halls and J. Michel, 1998, "A Systems Approach to Environmental Sensitivity Index (ESI) Mapping for Oil Spill Contingency Planning and Response," *Photogrammetric Engineering & Remote Sensing*, 64(10):1003-1014.

Jensen, J. R., X. Huang, D. Graves and R. Hanning, 1996, "Cellular Phone Transceiver Site Selection," *Raster Imagery in Geographic Information Systems*, S. Morain and S. Baros, (Eds.), Santa Fe, OnWard Press, 117–125.

Keister, M. D. (Ed.), 1997, *Multispectral Imagery Reference Guide*, Fairfax: Logicon Geodynamics, 210 pp.

Kidder, S. Q. and T. H. V. Haar, 1995, *Satellite Meteorology*, New York: Academic Press, 87–144.

Lacy, R., 1992, "South Carolina Finds Economical Way to Update Digital Road Data," *GIS World,* 5(10):58–60.

Lavrov, V. N., 1997, "Space Survey Photocameras for Cartographic Purposes," *Proceedings* of the Fourth International Conference on Remote Sensing for Marine and Coastal Environments, Ann Arbor: ERIM, 7 pp.

Leachtenauer, J. C., 1996, "National Imagery Interpretability Rating Scales Overview and Product Description," *Proceedings*, ASPRS Annual Convention, Bethesda: ASPRS, 1:262–272.

Leachtenauer, J. C., K. Daniel and T. Vogl, 1998, "Digitizing Satellite Imagery: Quality and Cost Considerations," *Photogrammetric Engineering & Remote Sensing,* 64:29–34.

Light, D. L., 1993, "The National Aerial Photography Program as a Geographic Information System Resource," *Photogrammetric Engineering & Remote Sensing*, 59(1):61–65.

Light, D. L., 1996, "Film Cameras or Digital Sensors? The Challenge Ahead for Aerial Imaging," *Photogrammetric Engineering & Remote Sensing*, 62(3):285–291.

Light, D. L., 1998, personal communication, Emerge, Inc.

Lindgren, D. T., 1985, *Land-use Planning and Remote Sensing.* Boston, Martinus Nijhhoff Inc.

Lo, C. P., 1986, "The Human Population," *Applied Remote Sensing.* New York: Longman, 40–70.

Lo, C. P., 1995, "Automated Population and Dwelling Unit Estimation from High-Resolution Satellite Images: A Geographic Information System Approach," *International Journal of Remote Sensing*, 16(1):17–34.

Lo, C. P., 1999, Personal Correspondence, Dept. of Geography, University of Georgia, Athens, GA.

Lo, C. P. and B. J. Faber, 1998, "Integration of Landsat Thematic Mapper and Census Data for Quality of Life Assessment," *Remote Sensing of Environment,* 62(2):143-157.

Lo, C. P., D. A. Quattrochi and J. C. Luvall, 1997, "Application of High-Resolution Thermal Infrared Remote Sensing and GIS to Assess the Urban Heat Island Effect," *International Journal of Remote Sensing,* 18(2):287–304.

Logicon, 1995, *Multispectral Users Guide*, Fairfax: Logicon Geodynamics, 102 pp.

Logicon, 1997, *Multispectral Imagery Reference Guide*, Fairfax: Logicon Geodynamics, 100 pp.

McCoy, R. M. and E. D. Metivier, 1973, "House Density vs. Socioeconomic Conditions," *Photogrammetric Engineering*, 39:43–49.

Milazzo, V. A., 1980, *A Review and Evaluation of Alternatives for Updating U.S. Geological Survey land-use and land-cover Maps*, Washington: U.S.G.S., Circular #826, 19 pp.

Mintzer, O. W., Ed., 1983, "Engineering Applications," *Manual of Remote Sensing*, 2nd Ed., R. N. Colwell, ed., Falls Church, American Society for Photogrammetry, 1955–2109.

Monier, R. B. and N. E. Green, 1953, "Preliminary Findings on the Development of Criteria for the Identification of Urban Structures from Aerial Photographs," *Annals of the Association of American Geographers*, special issue.

Naval Photographic Interpretation Center, 1961, *Photographic Interpretation Keys: Major Industries,* Washington: NPIC, 90 pp.

Nichol, J. E., 1994, "A GIS-Based Approach to Microclimate monitoring in Singapore's High-Rise Housing Estates," *Photogrammetric Engineering & Remote Sensing*, 60(10):1225–1232.

Olorunfemi, J. F., 1984, "land-use and Population: a Linking Model," *Photogrammetric Engineering & Remote Sensing*, 50:221–227.

Petrie, G. and T. J. M. Kennie, 1990, *Terrain Modeling in Surveying and Civil Engineering.* London, Whittles Publishing, 351 pp.

Philipson, W., 1997, *Manual of Photographic Interpretation.* Bethesda, American Society for Photogrammetry & Remote Sensing, 830 pp.

Pike, J., 1998, *Space Policy Project of the Federation of American Scientists,* Washington: Federation of American Scientists, http://www.fas.org/irp/imint/niirs.htm. Source of the spatial resolutions associated with the National Image Interpretability Rating System (NIIRS) in Figure 12-3.

Ridley, H. M., P. M. Atkinson, P. Alpin, J. Muller and I. Dowman, 1998, "Evaluating the Potential of the Forthcoming Commercial U.S. High-Resolution Satellite Sensor Imagery at the Ordnance Survey," *Photogrammetric Engineering & Remote Sensing*, 63(8):997–1005.

Roth, M., T. R. Oke and W. J. Emery, 1989, "Satellite-derived Urban Heat Islands from Three Coastal Cities and the Utilization of Such Data in Urban Climatology," *International Journal of Remote Sensing*, 10(11):1699–1720.

Rundquist, D. C. and S. A. Sampson, 1988, *A guide to the Practical Use of Aerial Color-infrared Photography in Agriculture*, Lincoln: Conservation and Survey Division, Institute of Agriculture and Natural Resources, 27 pp.

Schultz, G. A., 1988, "Remote Sensing in Hydrology," *Journal of Hydrology*, 100(1988):239–265.

Schweitzer, B. and B. McLeod, 1997, "Marketing Technology that Is Changing at the Speed of Light," *Earth Observation Magazine*. (6):7, 22–24.

Skole, D. L., 1994, "Data on Global Land-Cover Change: Acquisition, Assessment, and Analysis," *Changes in land-use and land-cover: A Global Perspective* (W. B. Meyer and B. L. Turner, Eds.), Cambridge: Cambridge University Press, 437–472.

Slonecker, E. T., D. M. Shaw and T. M. Lillesand, 1998, "Emerging Legal and Ethical Issues in Advanced Remote Sensing Technology," *Photogrammetric Engineering & Remote Sensing*, 64(6):589–595.

Stone, K. H., 1964, "A Guide to the Interpretation and Analysis of Aerial Photos," *Annals of the Association of American Geographers*, 54:318–328.

Sutton, P., D. Roberts, C. Elvidge and H. Meij, 1997, "A Comparison of Nighttime Satellite Imagery and Population Density for the Continental United States," *Photogrammetric Engineering & Remote Sensing*, 63(11):1303–1313.

Tobler, W., 1969, Satellite Confirmation of Settlement Size Coefficients, *Area*, 1:30–34.

Tuyahov, A. J., C. S. Davies and R. K. Holz, 1973, "Detection of Urban Blight Using Remote Sensing Techniques," *Remote Sensing of Earth Resources*, 2:213–226.

Tyler, W. and I. W. Ginsberg, 1995, "Monitoring Urban Growth with Historical Landsat Data and Computer Visualization," *Systema Terra - Remote Sensing and the Earth*, 4(1):47–52.

Urban Renewal Administration Department of Commerce, 1965, *Standard land-use Coding Manual: A. Standard System for Identifying and Coding land-use Activities*, Washington, DC: U.S. Government Printing Office.

Vogelmann, J. E., T. Sohl and S. M. Howard, 1998, "Regional Characterization of Land-cover Using Multiple Sources of Data," *Photogrammetric Engineering & Remote Sensing*, 64(1):45–57.

Wagman, D., 1997, Fires, Hurricanes Prove No Match for GIS. *Earth Observation Magazine*, 6(2):27–29.

Wagner, J. J., 1994, "Flood 1993 in Western Europe – ERS-1 on the Job," *Systema Terra – Remote Sensing and the Earth*, 3(2):42–25.

Warner, W. S., R. W. Graham and R. E. Read, 1996, "Chapter 15: Urban Survey," *Small Format Aerial Photography,* Scotland, Wittles Publishing, 253–256.

Welch, R., 1980, "Monitoring Urban Population and Energy Utilization Patterns from Satellite Data," *Remote Sensing of Environment,* 9(1):1–9.

Welch, R., 1982, "Spatial Resolution Requirements for Urban studies," *International Journal of Remote Sensing*, 3:139–146

Remote Sensing of Soils, Minerals, and Geomorphology

13

Only 26 percent of the Earth's surface is exposed land. The remaining 74 percent is covered by water (including inland seas, lakes, reservoirs, and rivers). Very few people actually live on boats or on structures located in water. Almost all of humanity lives on the terrestrial, solid Earth comprised of bedrock and the weathered bedrock we call *soil*. Humankind is able to obtain a relatively abundant harvest in certain parts of the world from this soil. They are also able to extract important minerals from the bedrock and derivative materials that we use in industrial/commercial processes, hopefully to improve the quality of life on Earth. It is important to have accurate information about the location, quality, and abundance of soils, minerals, and rocks in order to conserve these often irreplaceable natural resources.

Bedrock is continually weathered and eroded by the combined effects of water, wind, and/or ice. The materials are then moved to other locations via mass transport. These unconsolidated sedimentary materials are called surficial deposits. Remote sensing can play a limited role in the identification, inventory, and mapping of soils that are on the surface of the Earth, especially when *surficial* soils are not covered with dense vegetation. This chapter introduces the fundamental issues associated with remote sensing the spectral characteristics of soils. The impact of soil grain size, organic matter, and water content on soil spectral reflectance are identified. Remote sensing may also assist in the modeling of soil erosion, providing biophysical information for the Universal Soil Loss Equation and other hydrologic models (Pickup and Chewings, 1988).

In addition to soils, remote sensing can provide information about the chemical composition of rocks and minerals that are on the Earth's surface, and not completely covered by dense vegetation. Emphasis is placed on understanding unique *absorption bands* associated with specific types of rocks and minerals as recorded using imaging spectroscopy techniques (Clark, 1999). In certain instances, remote sensing can even be used to identify geobotanical relationships that can be used to identify soil geochemistry or rock types (Morrissey et al., 1984; Schwaller, 1987).

The chapter concludes with an overview of how general geologic information may be extracted from remotely sensed data, including information on lithology, structure, drainage patterns, and geomorphology (landforms) (Butler and Walsh, 1997; Walsh et al., 1997). Remote sensing data are generally of limited value for detecting deep, subsurface geologic features unless they have a surface surrogate expression that can be extrapolated to depth.

CHAPTER 13

Soil Characteristics and Taxonomy

Soil is unconsolidated material at the surface of the Earth that serves as a natural medium for growing plants. Plant roots reside within this material and extract water and nutrients. Soil is the weathered material between the atmosphere at the Earth's surface and the bedrock below the surface to a maximum depth of approximately 200 cm (USDA, 1998). A mature, fertile soil is the product of centuries of physical and chemical weathering of rock, combined with the addition of decaying plants and other organic matter (Loynachan et al., 1999). Soil is essential to the Earth's life-support system on the land. Agronomists refer to this as the *solum*.

Soil is a mixture of inorganic mineral particles and organic matter of varying size and composition. The particles make up about 50 percent of the soil's volume. Pores containing air and/or water occupy the remaining volume. The weathered, unconsolidated organic and inorganic mineral material that lies on top of the bedrock shown in Figure 13-1 varies greatly in composition and thickness throughout the Earth. In the heartland of continents such as North America, it may be 25 – 300 m deep. On steep mountain slopes or in deserts it may be almost completely absent. Permafrost soils may exist in arctic climates.

Soil Horizons

Biological, chemical, and physical processes create vertical zonation within the upper 200 cm or so of soils in which there is comparatively free movement of gravity water and groundwater capillary moisture. This results in the creation of relatively horizontal layers, or *soil horizons*. There are several standard horizons in a typical soil profile situated above the bedrock, including (Figure 13-1): *O, A, E, B, C, R,* and *W* that may be distinguishable from one another based on their color (hue, value, chroma), texture, and chemical properties (USDA, 1998).

The humus-rich topsoil, or *O horizon,* contains more than 20 percent partially decayed organic matter. Thus, it is a complex mixture of inorganic soil particles and decaying organic matter. *O* horizon soils typically have a dark brown or even black surface layer ranging in thickness from a few centimeters to several meters in areas where dense plant cover exists. This horizon is created by the interaction of water, other

chemicals, heat, organic material, and air among the soil particles. Plant root systems extract much of their water and nutrients from within this "zone of life" (Marsh and Dozier, 1981).

The *A horizon* is a *zone of eluviation* or *leaching* formed at the surface or below an *O* horizon where water moving up and down in the soil column leaches out minerals in solution (ions) and clay colloids from within the soil and relocates them to other horizons below. *A* horizons exhibit obliteration of all or much of the original rock structure. In a humid (wet) climate, clay minerals, iron oxides, and dissolved calcite are usually moved downward. This leaching may cause the *A* horizon to be pale and sandy, but the uppermost part is often darkened by humus (decomposed plant material) that collects at the top of the soil. The *A* horizon soils may be influenced by cultivation or animal pasturing in agricultural environments.

Sometimes an *E horizon* exists between the zone of eluviation (*A* horizon) and the zone of illuviation (*B* horizon). *E* horizons have mineral layers in which the main feature is the loss of silicate clay, iron, aluminum, or some combination of these materials, leaving a concentration of sand and silt particles. All or much of the original rock structure is usually obliterated in an *E* horizon.

The downward movement of ions and colloids in the soil often terminates at a certain location in the soil column where the upward pressure of capillary groundwater offsets that of the percolating gravity water from surface precipitation or irrigation. Here the materials may be deposited in the *B* horizon, the *zone of illuviation*. This layer is commonly rich in clay and is colored red or yellow by iron oxides (Loynachan et al., 1999). Over time, the zone of illuviation can collect such a mass of colloids and minerals that the interparticle air spaces become clogged, cementing the particles together. This can lead to the development of an impervious *hardpan*.

The *C horizon* is simply weathered parent material, lying below the *B* horizon. Most are mineral layers. The parent material is commonly subjected to physical and chemical weathering from frost action, roots, plant acids, and other agents. In some cases, the *C* horizon is transitional between unweathered bedrock (*R*) below and the developing soil above.

Sometimes there is a water layer within or beneath the soil. The water may be permanently frozen (*Wf*) or not permanently frozen (*W*).

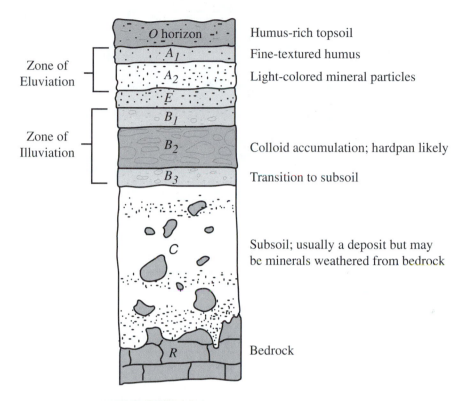

Figure 13-1 A standard soil profile as defined by the U.S. Department of Agriculture. The major soil horizons are O, A, E, B, C, situated on top of bedrock, R. The O horizon contains partially decayed organic matter. Subhorizons are transitional to and between the horizons (U.S. Department of Agriculture, 1998).

Soil Grain Size and Texture

The average diameter of grains of soil in a soil horizon is one of the major variables used to identify the taxonomy of a soil. There are three universally recognized soil grain size classes: *sand*, *silt*, and *clay*. Figure 13-2 identifies three different scales used to classify soil particles based on their diameters. The U.S. Department of Agriculture scale is (Loynachan et al., 1999):

- Sand: (a) a soil particle between 0.05 and 2.0 mm in diameter; (b) a soil composed of a large fraction of sand-size particles.

- Silt: (a) a soil particle between 0.002 and 0.05 mm in diameter; (b) a soil composed of a large fraction of silt-size particles.

- Clay: (a) a soil particle < 0.002 mm in equivalent diameter; (b) a soil that has properties dominated by clay-size particles.

Sand, silt, and clay-size particles play different roles in the soil formation process. Soil particles with sand-size diameters enhance soil drainage because water can percolate freely in the large air spaces between the large soil particles. Conversely, silt and clay-size soil particles enhance the movement and retention of soil capillary water. The very small clay-size soil particles carry electrical charges that attract and hold minute particles of dissolved minerals (ions) such as potassium and calcium. Because the ions are attached to the clay particles, they are not readily washed away. Thus, some clay in a soil horizon helps to maintain soil fertility by retaining nutrient-rich potassium, calcium, and other dissolved minerals (Marsh and Dozier, 1981).

Soil horizons contain various proportions of sand, silt, and clay-size particles. *Soil texture* is the relative proportion of sand, silt, and clay in a soil. A typical soil's *texture* is the percentage in *weight* of particles in various size classes. The USDA soil-texture triangle shown in Figure 13-3 identifies the percentages of sand, silt, and clay, comprising standard soil types. For example, a loam soil found in the lower center of the diagram consists of 40 percent sand, 40 percent silt,

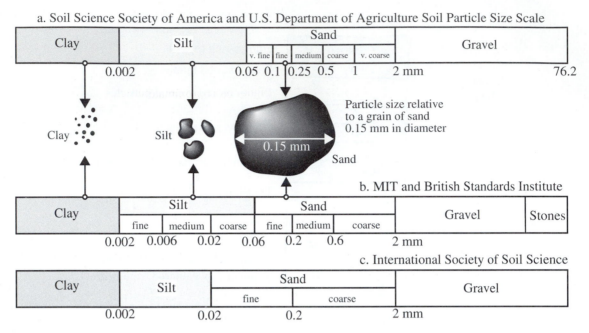

Figure 13-2   Three soil particle size scales: a) Soil Science Society of America, b) Massachusetts Institute of Technology and British Standards Institute, and c) International Society of Soil Science scale (after Marsh and Dozier, 1981; Loynachan et al., 1999).

and 20 percent clay. If a soil has a greater concentration of sand – say, 60 percent – and 10 percent clay and 30 percent silt, it is called a sandy loam soil, etc.

Soil Taxonomy

We no longer identify a "soil type." Rather, soil scientists determine the *soil taxonomy* (Petersen, 1999). *Keys to Soil Taxonomy* have been used by the USDA Natural Resources Conservation Service (USDA, 1998) since 1975 to qualitatively and quantitatively differentiate between soil taxa. The highest category of the U.S. Soil Taxonomy is Soil Order. Each order reflects the dominant soil-forming processes and the degree of soil formation. The 12 dominant U.S. Soil Orders are: Alfisols (high-nutrient soils), Andisols (volcanic soils), Aridisols (desert soils), Entisols (new soils), Gelisols (tundra soils), Histosols (organic soils), Inceptisols (young soils), Mollisols (prairie soils), Oxisols (tropical soils), Spodosols (forest soils), Ultisols (low-nutrient soils), and Vertisols (swelling-clay soils). Additional dichotomous keys allow the scientist to classify the soil into Suborders, Great Groups, Subgroups, Family Level, and Soil Series.

The Soil Taxonomy is a relatively complex system based on the use of elimination dichotomous keys that allow the soil scientist to evaluate various characteristics in a soil profile, including soil color (hue, value, chroma), soil-texture class,

moisture content, bulk density, porosity, and chemistry. For example, the first entry into the Key to Soil Orders is, Does the soil have 1) permafrost within 100 cm of the soil surface, *or* 2) gelic materials within 100 cm of the soil surface and permafrost within 200 cm of the soil surface? If it does, it is a Gelisol. If it does not, the analyst then progresses through the dichotomous key until he or she finds that the soil meets all the conditions identified.

 Remote Sensing of Soil Properties

From the previous discussion it is clear that most of the information used by soil scientists to map a soil series in a county is obtained by direct observation in the field. It is essential that subsurface soil profiles be examined and careful biological, chemical, and physical measurements be obtained within each soil horizon. Thus, it is not realistic to expect remote sensing to be a universal panacea that can be used to perform soil taxa mapping without *in situ* data collection. Rather, we should understand that under certain conditions it is possible for remotely sensed data to be of value to the soil scientist as he or she collects all the pertinent material necessary to perform accurate soil classification. In fact, many soil scientists find that remotely sensed images of

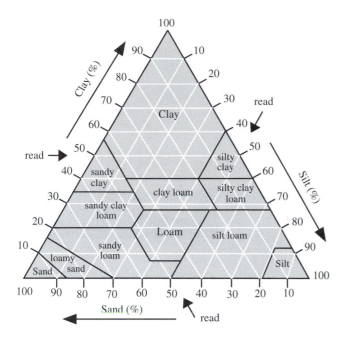

Figure 13-3 Percentages of clay (< 0.002 mm), silt (0.002 – 0.05 mm), and sand (0.05 – 2.0 mm) in the basic soil-textural classes (USDA, 1998).

the terrain under investigation are essential to the soil mapping process (Petersen, 1999). Fortunately, some soil property characteristics may be measured remotely under ideal conditions.

For example, optical remote sensing instruments such as aerial photography, multispectral scanners, and hyperspectral remote sensors can record the spectral reflectance characteristics of the surface properties of soils if they are not totally obscured by dense multiple-story shrubs or trees with their coalescing canopies. Theoretically, the total upwelling radiance from an exposed soil recorded by the sensor onboard the aircraft or satellite, L_t, is a function of the electromagnetic energy from the sources identified in Figure 13-4 and summarized as:

$$L_t = L_p + L_s + L_v \qquad (13-1)$$

where,

- L_p: This is the portion of the recorded radiance resulting from the downwelling solar (E_{sun}) and sky (E_{sky}) radiation that never actually reaches the soil surface. This is unwanted atmospheric path radiance noise and should ideally be removed from the imagery prior to trying to extract information about surficial soils or minerals.

- L_s: Some of the downwelling solar and sky radiation reaches the air-soil interface (*boundary layer*) and penetrates it approximately 1/2 wavelength (λ) deep into the soil. If the major wavelength of light being investigated is green light, the depth of penetration into the soil column would be approximately 0.275 μm (i.e., 1/2 of 0.55 μm). The amount of radiant flux exiting the soil column based on the reflection and scattering taking place at this depth is L_s. The characteristics of the soil organic matter (decomposed vegetation or animal material) and inorganic (mineral) constituents and the amount of soil moisture present have a significant impact on the amount of energy absorbed, scattered, and/or reflected from this surficial portion of the soil/rock matrix. Figure 13-4a depicts a soil with well-developed *O* and *A* horizons. It is likely that most of the energy reflected from this soil will be representative of the constituents of the *O* horizon. As the amount of humus (organic matter) diminishes, the surface reflectance will be more representative of the characteristics of the *A* horizon (Figure 13-4b). If both the *O* and *A* horizons are almost nonexistent, as shown in Figure 13-4c, the surface reflectance may be a function of the weathered subsoil (regolith) or even the bedrock if it is completely exposed, as on a steep mountain slope.

- L_v: Some of the incident downwelling solar and sky radiation may be able to penetrate perhaps a few millimeters or even a centimeter or two into the soil column. This may be referred to as *volume scattering*, L_v. Unlike water (refer to Chapter 11), however, there may be very little volumetric visible, near- and middle-infrared radiant flux exiting the soil profile from any appreciable depth. In fact, almost all the specular boundary layer reflectance (L_s) and volumetric scattering (L_v) may take place in the first few millimeters of the soil profile. The amount of volumetric radiant flux scattered or reflected back into the atmosphere is a function of the wavelength of incident energy (i.e., longer wavelength active microwave energy may penetrate farther into the soil, as has been demonstrated in the Sahara), the type and amount of organic/inorganic constituents, the shape and density of the minerals, the degree of mineral compaction, and the amount of soil moisture present. Figures 13-4b,c suggest that as the *O* horizon becomes less well-developed, perhaps more incident energy interacts with the soil particles in the *A* horizon (Figure 13-4b) or even the subsoil and bedrock (Figure 13-4c).

The goal of most soil and mineral remote sensing is to extract the radiance of interest from all the other radiance components being recorded by the sensor system. For example, the scientist interested in identifying the organic and

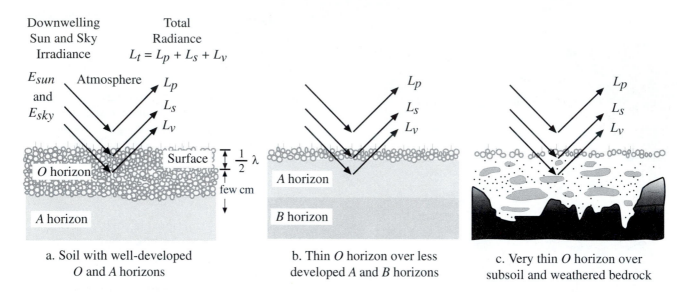

Figure 13-4   Soils and rocks receive irradiance from the Sun (E_{Sun}) and atmosphere (E_{sky}). The total radiance upwelling from a soil/rock matrix toward the remote sensor is a function of radiance from unwanted atmospheric scattering (L_p), a substantial amount of radiance (L_s) reflected or scattered from the upper surface of the soil/rock matrix (approximately 1/2 wavelength deep), a small amount of subsurface volumetric radiance (L_v), and a very small amount of radiance from subsequent soil/rock substrate.

inorganic (mineral) constituents in the very top layers of the soil profile is most concerned with measuring the integrated spectral response of the surface and subsurface radiance, i.e., L_s and L_v:

$$L_s + L_v = L_t - L_p. \qquad (13-2)$$

This involves careful radiometric correction of the remote sensor data to remove atmospheric attenuation (L_p). Ideally we could disentangle the individual contribution of L_s and L_v to the reflected radiant flux. Unfortunately, this is difficult, and usually we must be content analyzing an integration (summation) of these two radiance constituents. Nevertheless, it is possible to make some general observations about how surficial soils appear in remote sensor data based on their spectral reflectance properties.

The spectral reflectance characteristics of soils are a function of several important characteristics, including:

- soil texture (percentage of sand, silt, and clay),
- soil moisture content (e.g., dry, moist, saturated)
- organic matter content,
- iron-oxide content, and
- surface roughness.

Generally, a dry soil that contains almost no organic matter exhibits a relatively simple, less complex spectral reflectance curve than those associated with terrestrial vegetation

(Chapter 10) or algae-laden water bodies (Chapter 11). For example, Figure 13-5 depicts laboratory spectroradiometer reflectance characteristics of dry silt and dry sand soils. This demonstrates one of the most consistent characteristics of dry soils: *increasing reflectance with increasing wavelength, especially in the visible, near- and middle-infrared portions of the spectrum.* However, as a soil gains moisture, and/or additional organic content or iron oxide, its spectral response may depart from the simple curve. Therefore, it is useful to review how these parameters influence the spectral response of surficial soils. A goal of remote sensing is to disentangle the spectral response recorded from a surficial soil and be able to identify the proportions and/or influence of the characteristics within the instantaneous field of view of the sensor system.

Soil Texture and Moisture Content

There is a relationship between the size of the soil particles found in a mass of soil (e.g., m^3) and the amount of moisture that the soil can store. Figure 13-6a depicts several theoretical grains of sand. Incident radiant flux may be reflected from the surface of the sand grains yielding specular reflectance, or the incident energy may penetrate some distance into the sand particle. The energy may then be absorbed by the particle and turned into heat energy or exit the particle and be scattered or absorbed by other particles. The void

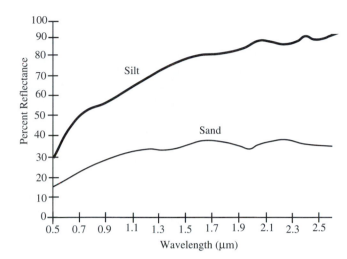

Figure 13-5 *In situ* spectroradiometer reflectance curves for dry silt and sand soils. Reflectance generally increases with increasing wavelength throughout the visible, near- and middle-infrared portions of the spectrum.

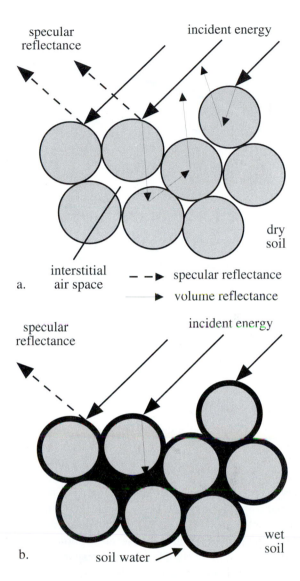

Figure 13-6 a) Incident radiant energy may be reflected from the surface of the dry soil, or it penetrates into the soil particles, where it may be absorbed or scattered. The total reflectance leaving the soil is a function of the specular reflectance and the internal volume reflectance. b) As soil moisture increases, each particle may be encapsulated with a thin membrane of capillary water. The interstitial spaces may also fill with water. The greater the amount of water in the soil, the greater the absorption of incident electromagnetic energy and the lower the soil reflectance.

between the soil particles is called the *interstitial air space*. The total reflectance per wavelength (e.g., blue, green, red, near-infrared light) for a given dry soil with no organic content and no iron oxides is a function of the average of both the *soil specular reflected energy* from the soil particle surfaces and the *soil volume reflectance* taking place due to internal scattering (Vincent, 1997).

The finer clay soils have particles that are packed very closely to one another. The interstitial air spaces between the soil particles are very small. Conversely, sand particles are very large and contain relatively large interstitial air spaces. When precipitation occurs or groundwater rises into the soil horizon, the individual particles may become surrounded by a thin membrane of capillary water. Water may also occupy the interstitial air spaces (Figure 13-6b). The densely packed clayey soil, with each of its small particles holding a membrane of water around it, can hold a tremendous amount of water. Conversely, sandy soils with their significantly larger soil particles and large air spaces 1) drain much more rapidly than clayey soils, and 2) are dried out much more rapidly by evaporation than the clayey soils. So what does this have to do with the spectral reflectance characteristics of soils? Basically, the amount of moisture held in the surficial soil layer is a function of the soil texture. The finer the soil texture, the greater the soil's ability to maintain a high moisture content in the presence of precipitation. The greater the soil moisture, the more incident radiant energy absorbed and the less reflected energy.

This is demonstrated for sandy soils with little surface vegetation in Figure 13-7a. The dry sandy soil has a relatively simple, increasing spectral response throughout the region from 0.5 – 2.6 µm. However, as soil moisture increases, the

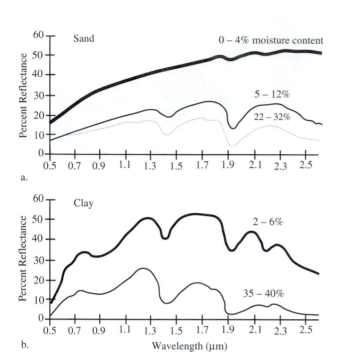

Figure 13-7 Higher moisture content in sandy (a) and clayey soil (b) results in decreased reflectance throughout the visible and near-infrared region, especially in the water-absorption bands at 1.4, 1.9, and 2.7 μm (Hoffer, 1978).

water in the upper few centimeters of the soil begins to selectively absorb significant amounts of incident energy in the water-absorption bands at 1.4, 1.9, and 2.7 μm. The soil moisture also absorbs more incident radiant energy in the spectral regions adjacent to these absorption bands. The result is a much more complex spectral response, with characteristic dips in the reflectance curve at the 1.4, 1.9, and 2.7 μm atmospheric water-absorption bands. Also note that the amount of reflected green, red, near- and middle-infrared radiant energy is dramatically reduced as the moisture content increases.

This same relationship holds for clayey soils as demonstrated in Figure 13-7b, where increased moisture in the upper few centimeters of the soil dampens the entire spectral response throughout the wavelength interval from 0.5 – 2.6 μm and deepens the absorption around the water-absorption bands. Notice, however, that the water-absorption bands appear to be much more active. This is because clayey soils with their fine soil texture manage to hold more moisture in the upper portion of the soil horizon, which allows the moisture to absorb some of the incident radiant flux, creating significant dips in and around the water-absorption bands.

Experienced image analysts know that remote sensor data of exposed soil surfaces obtained after a major precipitation event such as a thunderstorm or prolonged frontal activity will be noticeably *darker* than if the imagery were acquired after many days without precipitation. This is because the water in the surficial soil absorbs much of the incident radiant energy, especially in the visible and near-infrared portions of the spectrum, resulting in less radiance exiting toward the sensor system.

If high spectral and radiometric resolution sensors are available, it may be possible to differentiate between soils with different soil textures. This is because almost all soils that have a moderate to large proportion of clay-size particles exhibit strong *hydroxyl absorption bands* at approximately 1.4 and 2.2 μm. The spectral response of the sandy soil in Figure 13-7a was not influenced by the hydroxyl absorption band at 2.2 μm whereas the spectral response of the clayey soil in Figure 13-7b exhibited significant absorption at both 1.4 and 2.2 μm. Of course, it is only possible to differentiate between the clayey and silt/sandy soil texture characteristics if the soils are almost dry and contain very little organic matter that could mask the relationship, and the sensor is sensitive to very specific wavelength intervals (e.g., centered at 1.4 and 2.2 μm).

Soil Organic Matter

Plants and animals decompose and become organic humus in the upper portions of the soil horizon. The amount of organic matter in the soil has a significant impact on the spectral reflectance characteristics of exposed soils. Generally, the greater the amount of organic content in the upper portions of the soil, the greater the absorption of incident energy and the lower the spectral reflectance. Figure 13-8 summarizes the curvilinear relationship well for soils with 0 to 100 percent organic matter.

Iron Oxide

In the southeastern United States, southeast Asia, and several other parts of the world, iron oxides are present in the soil. The existence of iron oxides generally causes an increase in reflectance in the red portion of the spectrum (600 – 700 nm), and hence its reddish color (Figure 13-9). There is also a noticeable decrease in the blue and green reflectance in the iron-oxide soil. The iron-oxide soil also exhibits an absorption band in the 850 – 900 nm region when compared with a sandy loam soil with no iron oxide.

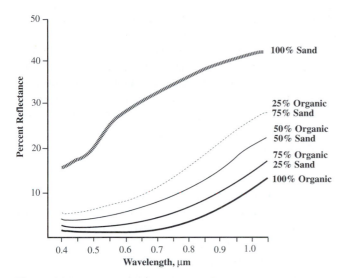

Figure 13-8   The greater the amount of organic content in a soil, the greater the absorption of incident energy and the lower the spectral reflectance.

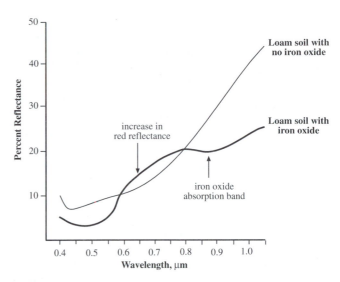

Figure 13-9   Iron oxide in a sandy loam soil causes an increase in reflectance in the red portion of the spectrum (0.6 – 0.7 μm) and a decrease in reflectance in the near-infrared region (0.85 – 0.90 μm).

Surface Roughness

As we learned in Chapter 9 (Active Microwave Remote Sensing), the smaller the local surface roughness relative to the size of the incident radiation, the greater the specular spectral reflectance from the terrain. We may utilize this information to make some general statements about the amount of spectral reflectance that *should* be exiting certain soil textures, assuming they contain no moisture, organic content, or iron oxides. Dry, fine-texture clayey soil should produce a higher spectral response throughout the visible and near-infrared portions of the spectrum due to the near-specular reflection that should take place from its surface versus silt or sand surfaces. Unfortunately, this is often not the case because as moisture, organic content, or iron oxide is added to the clayey soil, it begins dramatically to absorb incident radiant flux and quite possibly to appear like silt or perhaps even sand on remote sensor data. This can cause interpretation problems.

Conversely, dry sand with its well-drained large grains should diffusely scatter the incident wavelengths of visible and near-infrared energy more than clayey and silt soils. Therefore, it is common for more coarse-grained sand beaches to be among the brightest terrain in the landscape while clayey soils are among the darkest.

 ## Remote Sensing of Rocks and Minerals

Rocks are assemblages of *minerals* that have interlocking grains or are bound together by various types of cement (usually silica or calcium carbonate). When there is minimal vegetation and soil present and the rock material is visible directly by the remote sensing system, it may be possible to differentiate between several rock types and obtain information about their characteristics. Most rock surfaces consist of several types of minerals. Clark (1999) suggests that it is possible to model the reflectance from an exposed rock consisting of several minerals or a single mineral based on Hapke's (1993) equation:

$$r_\lambda = \left[\frac{w'}{4\pi} \times \frac{\mu}{\mu + \mu_o} \right] \times [(1 + B_g)P_g + H_\mu H_{\mu_o}^{-1}] \qquad (13\text{-}3)$$

where r_λ is the reflectance at wavelength λ, μ_o is the cosine of the angle of incident light onto the rock or mineral of interest, μ is the cosine of the angle of emitted light, g is the phase angle, w' is the average single scattering albedo from the rock or mineral of interest, B_g is a backscatter function, P_g is the average single particle phase function, and H is a function for isotropic scatterers.

Armed with this advanced reflectance theory and the known optical constants of the minerals involved, it is possible to compute the theoretical reflectance spectra for 1) pure minerals that have a single grain size, 2) a pure mineral with a variety of grain sizes, and 3) mineral mixtures with varying grain sizes (Clark, 1999). This is important because it means that we may be able to predict what the reflectance curves of specific types of minerals or rocks should look like at various wavelengths. This can be important when we attempt to interpret the information content of imaging spectroscopy remote sensor data of mineralized terrain.

Imaging Spectroscopy of Rocks and Minerals

Imaging spectrometry instruments are useful for obtaining quantitative information about rock type and mineral composition. Chapter 7 reviewed the fundamental characteristics of imaging spectroscopy sensor systems such as the Airborne Visible Infrared Imaging Spectrometer (AVIRIS). This section reviews the fundamental characteristics that impact our ability to determine rock type and mineral composition remotely using imaging spectrometry.

Imaging Spectrometers

Imaging spectrometers may be used in the lab, field, or in an airborne remote sensing mission. Reflectance and emittance spectroscopy of natural surfaces are sensitive to specific chemical bonds in materials. In the past, one of the problems associated with spectroscopy was that it was too sensitive to small changes in the chemistry and/or the structure of the material. This resulted in the creation of very complex spectral reflectance curves that were often unintelligible. Fortunately, significant strides have been made in 1) the quality of the imaging spectrometer sensors, and 2) our ability to understand and disentangle the information content of the spectroradiometer information. Thus, what was once a drawback (i.e., very complex spectral reflectance curves) is now an advantage as it allows scientists to extract more valuable information about the chemistry of the natural environment.

Energy-Matter Interactions — Reflection and Absorption Processes

As previously discussed, photons of light incident on a mineral or rock are 1) reflected from grain surfaces onto other grain surfaces, 2) passed through the grain onto other grains, and/or 3) absorbed within a grain. The photons that are reflected from the grain surfaces or refracted through a particle are said to be *scattered*. The scattered photons of light may encounter another grain or be scattered away from the

surface, perhaps toward a remote sensing system where the amount and properties can be detected and measured. We are primarily concerned here with photons of light that originated at the Sun and then interacted with the mineral of interest. It is important to remember, however, that photons may also be *emitted* from a mineral because (as discussed in Chapter 8) all objects above absolute zero emit radiation.

If every mineral absorbed and scattered the incident photons of light in an identical manner, then there would be no basis for mineralogical remote sensing. The amount of energy leaving each type of mineral at each specific wavelength would be identical. Fortunately, certain types of minerals absorb and/or scatter the incident energy differently.

There are a number of processes that determine how a mineral will absorb or scatter the incident energy. Also, the processes absorb and scatter light differently depending on the wavelength (λ) of light being investigated. The variety of absorption processes and their wavelength dependence allow us to derive information about the *chemistry* of a mineral from its reflected or emitted energy. The ideal sensor for us to use is the imaging spectrometer because it can record much of the absorption information, much like using an *in situ* spectroradiometer. For example, consider the spectral reflectance curves for alunite shown in Figure 13-10. The most detailed spectral reflectance information is obtained using an *in situ* spectroradiometer. Spectral reflectance data obtained using the 63-channel Geophysical and Environmental Research Imaging Spectrometer (GERIS) also retains much of the spectral information. Unfortunately, much of the spectral absorption information is lost when the spectral data are obtained from six bands of simulated Landsat Thematic Mapper data (Kruse et al., 1990).

All materials have a complex *index of refraction*. For example, the vacuum of outer space, the atmosphere, quartz, and water all have different indexes of refraction. If we illuminate a plane surface with photons of light from directly overhead, the light, *R*, will be reflected from the surface according to the Fresnel equation:

$$R = \frac{(n-1)^2 + K^2}{(n+1)^2 + K^2} \tag{13-4}$$

where n is the index of refraction, and K is the extinction coefficient. Both the index of refraction and the extinction coefficient of quartz are shown in Figure 13-11a (Spitzer and Klienman, 1960; Clark, 1999). From this illustration it is clear that the optical constants of n and K for quartz vary strongly with wavelength. Note that the index of refraction reaches a minimum just before 8.5 µm and 12.6 µm. The rel-

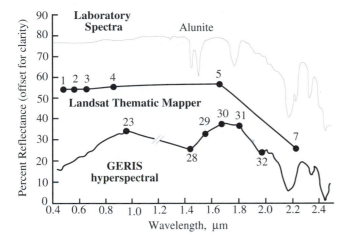

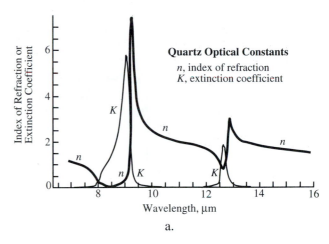

a.

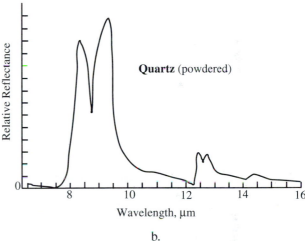

b.

Figure 13-10 Comparison of a laboratory spectra of alunite (an aluminum sulfate), a simulated Landsat Thematic Mapper spectra (resampled from the lab spectrum), and spectra obtained using an airborne 63-channel Geophysical and Environmental Research Imaging Spectrometer (GERIS) at Cuprite, NV. Symbols and channel numbers on the TM and GERIS spectra identify band centers. Most characteristic absorption band information is lost with the TM spectrum (with the exception of low reflectance in TM band 7 at 2.2 μm) while much of the spectral information is preserved in the GERIS spectrum. The spectra are offset vertically for clarity (after Kruse et al., 1990).

Figure 13-11 a) The index of refraction and extinction coefficient of quartz for the wavelength interval 6 – 16 μm. b) The spectral reflectance characteristics of powdered quartz obtained using a spectroradiometer (after Clark, 1999).

ative reflectance of powdered quartz measured by a spectrometer for the wavelength interval from 6 – 16 μm is shown in Figure 13-11b (Clark, 1999). The reflectance spectra of quartz throughout the visible and near-infrared region is effectively zero and is therefore not shown. However, from 8 – 9.5 μm and at 12.6 μm there is a dramatic increase in reflectance. If quartz is to be detected at all using imaging spectrometry, it may be necessary to sense in the region from 8 – 10 μm as shown. But what causes the reflectance spectra of quartz to appear as it does? Why are certain parts of the spectrum absorbed more completely than others? The answer lies at the heart of using imaging spectrometry for mineral analysis. It is because of the specific types of absorption that take place within the minerals.

Absorption Processes

As demonstrated in the previous illustration, a typical spectral reflectance curve obtained by an imaging spectrometer exhibits various maxima and minima. The minima are caused by strong absorption bands. For example, laboratory and AVIRIS remote sensing derived spectra for three minerals, kaolinite clay, aluminum sulfate (alunite), and budding-

tonite (an ammonium feldspar) are shown in Figure 13-12 (Van der Meer, 1994). Scientists have documented that specific minerals exhibit relatively unique absorption spectra. For example, key absorption features associated with kaolinite are typically found at 2.17, 2.21, 2.32, and 2.38 μm. If a spectra exhibits minima at these locations, it may well be kaolinite. It is important to point out here that only a hyperspectral sensor with a spectral bandwidth resolution of approximately 10 nm could capture such information. Spectroradiometers with 20 nm bandwidths might miss the important minima or maxima entirely. This diagram also suggests that the differences in maxima, minima, and the slope between maxima and minima might allow these three minerals to be differentiated one from another using hyper-

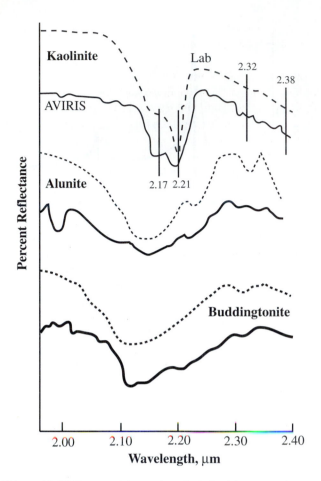

Figure 13-12 Spectra of three minerals derived from the Airborne Visible Infrared Imaging Spectrometer (AVIRIS) and a laboratory spectroradiometer (after Van der Meer, 1994). The vertical lines identify key absorption features useful for identifying kaolinite. The discrepancy between the laboratory and AVIRIS curves for each mineral is due to 1) the lab spectra are produced from pure samples whereas the AVIRIS data are extracted from relatively large 20 x 20 m areas that contain a mixture of materials within the IFOV of the sensor system, 2) the AVIRIS data is recorded through an atmosphere, and 3) the lab samples are dry whereas the real-world terrain may exhibit slight differences in soil moisture. The spectra are offset vertically for clarity.

spectral remote sensor data. The absorption bands in these minerals are caused by electronic and vibrational processes.

Electronic Processes: The most common electronic process revealed in the spectral reflectance curves of minerals is due to unfilled electron shells of transition elements such as Ni, Cr, Co, Fe, etc. (Clark, 1999). This is called a crystal field effect. Absorption bands can also be caused by

charge transfer absorptions, conduction bands, and *color centers.* Please refer to Clark (1999) or the *Manual of Remote Sensing – Geosciences* (Rencz, 1999) for detailed information about these electronic absorption processes.

Vibrational Processes: The bonds in a molecule or crystal lattice are like springs with weights attached that cause the system to vibrate. The frequency of vibration depends on the bonds in the molecule and their masses. For a molecule with N atoms, there are 3N–6 normal modes of *vibrations* called *fundamentals.* Each vibration can also occur roughly at multiples of the original fundamental frequency. The additional vibrations are called *overtones* or *combinations.* Hunt (1977) devised a spectral signature diagram that summarizes the impact of both *electronic* and *vibrational* absorption processes for selected minerals (Clark, 1999). Water, hydroxyl, carbonates, phosphates, borates, arsenates, and vanadates have diagnostic vibrational absorption bands.

Spectral Reflectance Libraries

The U.S. Geological Survey has compiled a Digital Spectral Library (Clark et al., 1993). The California Institute of Technology Jet Propulsion Laboratory compiled a spectral library that includes the Salisbury et al. (1991) library. These are accurate spectral libraries. One should only use spectral libraries that have been created using rigorous imaging spectroscopy calibration standards.

Creating Mineral Maps Using Hyperspectral Data

If we obtain high spectral resolution remote sensing spectra for an unknown surficial rock material, remove the atmospheric effects and convert the brightness values to percent reflectance (or exitance), then it may be possible to search a spectral library and identify the type of mineral that has an identical or very similar spectra. In this manner, imaging spectroscopy can be used to derive significant mineralogical information with little or no *in situ* field work. In certain instances, mineralogical maps can be made.

For example, consider the two mineral maps shown in Color Plate 13-1 derived from low- and high-altitude AVIRIS data obtained over Cuprite, NV (Swayze et al., 1999). USGS scientists first removed atmospheric effects in the data using a computer program called ATREM. They then used a computer program (Tetracorder) that compared the calibrated spectra obtained for each pixel in the AVIRIS data with the spectra contained in a mineral spectral reflectance library. Twenty-four mineral categories were found in the 2 – 2.5 μm region using Tetracorder. Note the detail present in the 2.3 x 7 m low-altitude AVIRIS data when compared with the results from the coarse 18 x 18 m high-altitude AVIRIS data (Swayze et al., 1999).

EOS Terra ASTER Reflectance and Absorption Characteristics of Value for Rock and Mineral Discrimination

Many of the spectral diagnostic characteristics of rocks and minerals are found throughout the thermal infrared portion of the spectrum. The most important satellite sensor that provides such information is ASTER — the Advanced Spaceborne Thermal Emission and Reflection Radiometer launched onboard EOS *Terra*. Six ASTER bands (4 – 9) cover the short-wavelength infrared (SWIR) range. Band 6 is centered on a clay-absorption feature often associated with hydrothermal alteration and mineral potential. Band 8 is centered on a carbonate-absorption feature, allowing global discrimination of limestones and dolomites from other rocks. Five ASTER bands (10 – 14) cover the thermal infrared (TIR) range. Bands 10, 11, and 12 are designed to detect sulfates and silica spectral features. Evaluating reflectance in SWIR band 6 with the thermal infrared band 10 allows discrimination between common minerals such as alunite (a sulfate mineral important to precious metal deposits) and anhydrite (an evaporative sulfate common in arid regions). Band 14 is centered on a carbonate-absorption spectral feature complementing the SWIR band 8 (Ellis, 1999).

 Geology

The Earth is not rock solid. It is constantly changing, moving, and being rearranged. *Geology* is the science of rocks. It reveals the immense history of the Earth and explains its geological formations. From a geological viewpoint, humanity is a relatively recent arrival, and it is one species among many millions that share an Earthly heritage (Busby et al., 1996).

Paleontology is the study of fossils, the remains of ancient organisms that have been turned to stone. William Smith's (1769 – 1839) work marked the beginning of geology. He studied English coal mines and observed that the same layers of sedimentary rocks were revealed in cuttings over large geographic areas. He identified and correlated the strata by their fossil content. He discovered that the order in which rock units were deposited did not vary across their geographic extent. He concluded that rock units at the bottom were older than those above.

Scottish geologist James Hutton (1726 – 1797) built upon Smith's findings and concluded that the history of rocks occurs in *cycles* (Trefil and Hazen, 1995). Rocks are decomposed into sediment by *weathering*. The sediment is often moved by forces of *erosion* and *mass transport* to accumulate in new locations. The sediment is often consolidated into a new type of rock that could be buried under more rock, until, heated to its melting point, it may flow back under pressure to the surface as extrusive lava or to some interior location as intrusive magma. The lava or magma cools into rock. Weathering begins immediately on extruded lava, whereas it may take millennia to expose the intrusive magma at the surface, where it may be weathered. In either case, the cycle begins again. Hutton raised the possibility that the Earth was much older than previously thought, and was continually changing and recycling itself (Selby, 1989).

Charles Lyell (1797 – 1875), the founder of modern geology, introduced *uniformitarianism* — the concept that past events occurred at the same rate as they do today. For example, sedimentary rock was formed by the same processes of sediment deposition and cementation, and at the same rate as those that can be observed today. He was the first to suggest that the Earth was millions of years old.

Finally, Alfred Wegener (1880 – 1930) proposed the theory of *continental drift,* which was hotly debated for decades. Then, during the cold war of the 1950s, U.S. and Russian scientists set up sensitive vibration detectors to monitor each other's atomic tests. Such instruments also allowed them to identify the epicenter of thousands of earthquakes throughout the world. These epicenters usually occurred along distinctive lines. It became clear to geologists that these lines were edges of enormous plates that covered the Earth's surface and that earthquakes were the result of friction between the plates. Continents on the plates were slowly moving on the surface of the Earth. This is the concept of *plate tectonics* which is related but not identical to continental drift. Today, most geological phenomena can be explained by these two concepts.

Most of our geologic information comes from detailed *in situ* investigation by geologists and paleontologists. They excavate, bore holes in the Earth to extract geologic core samples, and interpret fossil remains. In addition, some of these scientists use airborne and satellite remote sensing technology to supplement their *in situ* investigations. Remotely sensed images are routinely interpreted to identify lithology, structure, drainage-pattern characteristics, and

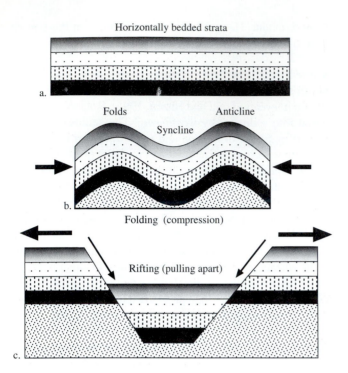

Figure 13-13 a) Horizontally bedded strata. b) Horizontally bedded strata that has been compressed, creating folds in the rock, including anticlines and synclines. c) An example of faulting where a block has been downthrown relative to two stable blocks of rock. This is referred to as rifting.

landforms. In addition, remote sensing is often used in mineral and petroleum exploration.

Lithology

Geologists often use remote sensing in conjunction with *in situ* observation to identify the *lithology* of a rock type, i.e., its origin. The different rock types are formed by one of three processes:

- *igneous* rocks are formed from molten material;

- *sedimentary* rocks are formed from the particles of pre-existing rocks and plant and animal remains; or

- *metamorphic* rocks are formed by applying heat and pressure to previously existing rock.

Rocks are weathered from the parent material and transported by erosion and mass transport. The unconsolidated sedimentary materials are called *surficial deposits*. The surficial deposits may be transported and eventually deposited by water, in which case we call them *alluvial deposits*. Examples include alluvial fans, sandbars, spits, and river terraces. If the unconsolidated material is transported by ice, we have glacial till, including eskers and drumlins. When the unconsolidated material is transported by the wind, we encounter eolian landscapes, including sand dunes and glacial loess deposits. Finally, the mass wasting of rocks by gravity can produce talus or scree deposits. Under certain favorable circumstances, all these materials may be visible on remote sensor data.

Structure

The major mountain ranges of the world are of volcanic or folded origin. The mountains in the ocean (islands) are usually volcanic. The vast majority of mountains found on the continents were created by folding. Basically, most of the *orogenesis* (mountain building) takes place at the margins of the continents where continental plates push against one another and where rock is compressed and forced to move upwards, perhaps thousands of meters.

The type of rock determines how much differential stress (or compression) it can withstand. When a rock (Figure 13-13a) is subjected to compression, it may experience 1) *elastic* deformation in which case it may return to its original shape and size after the stress is removed, 2) *plastic* deformation of rock called *folding,* which is irreversible (i.e., the compressional stress is beyond the elastic limit) (Figure 13-13b), or 3) *fracturing* where the plastic limit is exceeded and the rock breaks into pieces (the pieces can be extremely large!) (Figure 13-13c). Basalt has an average compressive rupture or fracture strength (kg/cm^2) of approximately 2,750, quartzite (2,020), granite (1,480), slate (1,480), marble (1,020), limestone (960), and sandstone (740). Thus, if more than 2,750 kg/cm^2 of compression is applied to a piece of basalt, it may fracture. Sandstone fractures under far less compressional stress.

Folding

Folding takes place when horizontally bedded materials are compressed. The compression results in wavelike undulations imposed on the strata. There are four basic types of folds. A *monocline* is a single fold on horizontally bedded material. It is like a rounded ramp (Figure 13-14b). Monoclines are usually asymmetrical. Archlike upfolds are called *anticlines* (Figure 13-14c). Anticlines typically have a convex upward fold with the oldest rocks in the core or center. The beds of sedimentary strata dip in opposite directions away from the central axis of the anticline. The downward

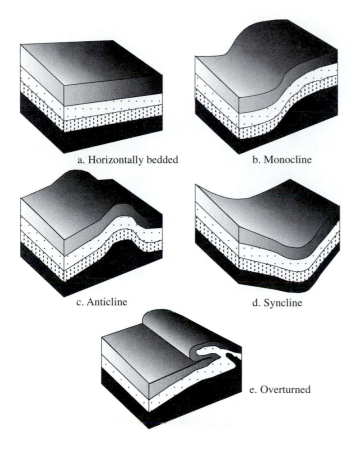

Figure 13-14 a) Horizontally bedded terrain. b) A gently sloping monocline. c) A gently sloping anticline. d) A syncline. e) An overturned fold.

a. Horizontally bedded

b. Monocline

c. Anticline

d. Syncline

e. Overturned

counterpart of an anticline is a *syncline* (a troughlike downfold) (Figure 13-14d). Synclines have a concave downward fold that typically has the youngest rocks in the center. A syncline is a fold in which the sedimentary beds dip inward from both sides toward the axis, forming a synclinal valley.

Anticlines and synclines can be symmetrical or asymmetrical. They can also plunge dramatically below the surface. In this case we have a plunging anticline or syncline. Sometimes the compressive forces are so great that the sedimentary beds are *overturned* on top of one another (Figure 13-14e). Anticlines and synclines are continuously weathered, giving rise to a variety of landforms such as hogbacks, etc. Examples of folded strata in remotely sensed images are found in the section on "Landforms Developed on Folded Strata."

Faults

Rock fractures may be divided into two categories: joints and faults. A *joint* is a crack in rock along which there is no appreciable displacement (i.e., sideways or vertical movement). A *fault* is a crack along which displacement has occurred. Faulting typically involves the movement of massive blocks of rock and usually generates earthquakes. It is only the fractures along which rock bodies actually move relative to one another that we call *faults*. This displacement can be local in nature or occur for thousands of kilometers. In certain instances, the amount of movement (displacement) can be recorded and interpreted on remotely sensed imagery.

There are three major types of displacement in faults, including: dip-slip, strike-slip, and oblique-slip, as shown in Figure 13-15. Displacement occurring up or down the fault plane walls produces a *dip-slip* fault (Figure 13-15a). If the displacement occurs parallel with the fault line, a *strike-slip* fault is created (Figure 13-15b). Displacement up and down and along the fault line creates a hybrid *oblique-slip* fault (Figure 13-15c). The diagram also highlights several parameters that may be identified using a combination of field work and remote sensor data. The *fault scarp* and the *fault line* may be identifiable. They may trend in a certain direction on the terrain, called a *strike direction*. The angle that the surface is tilted from a horizontal plane is the *dip angle* (θ). It is measured downward from a horizontal plane and ranges between 0° and 90°. The *dip direction* (0° to 359°) is measured orthogonal to the strike direction. There are several types of dip-slip faults. These are discussed in the section on "Fault-Controlled Landforms."

Drainage Density and Pattern

Earth landscapes exhibit varying stream densities and patterns that can be identified using remote sensor data.

Drainage Density and Texture

Drainage density (D) is the total length (L) of n stream channels in a drainage basin divided by the surface area (A) of the basin:

$$D = \frac{\sum_{1}^{n} L}{A}. \tag{13-5}$$

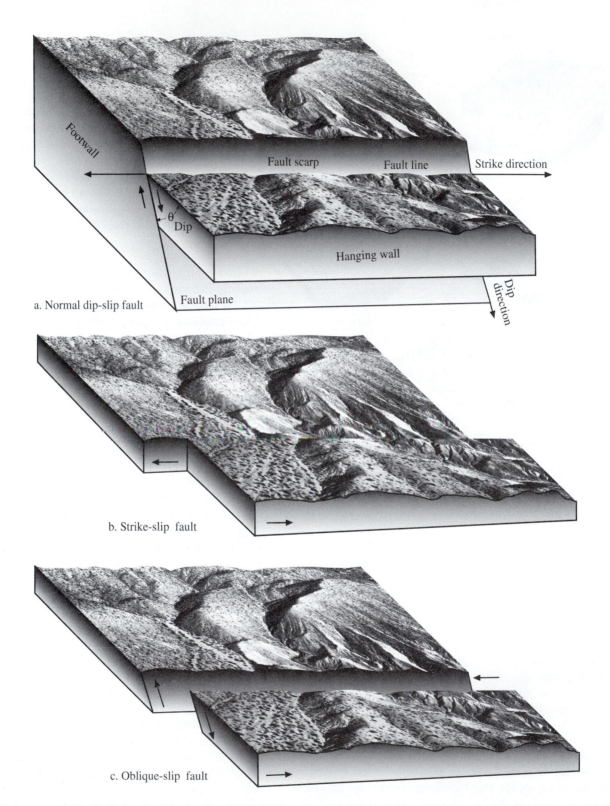

Figure 13-15 a) A normal dip-slip fault, where the hanging wall is displaced from the footwall. Note that the angle of the fault plane is called the dip and that the fault line trends across the terrain in a certain azimuth direction called the strike. The displacement causes a fault scarp. b) A strike-slip fault. c) An oblique-slip fault.

The permeability of a soil or rock (i.e., how easily water passes through the material) has a great deal to do with drainage density. Permeability is inversely proportional to runoff. Where permeability is low and runoff is high, many gullies typically form. Conversely, when permeability is high and runoff is low, much of the water infiltrates into the ground and a larger surface area is required to provide sufficient runoff for the creation and maintenance of a channel (tributary). Weak, relatively impermeable clays and shales produce the highest drainage density. Regions with extremely high drainage density are called badlands, e.g., the Badlands National Monument in South Dakota.

The drainage patterns discussed below are often adjectivally described as having fine, medium, or coarse textures. These textures generally refer to how dense the drainage network is per unit area (e.g., per km^2 or per watershed). A *fine-textured* drainage pattern exhibits relatively short distances between the first-order tributaries, i.e., it has a high density of closely spaced channels per unit area. This texture indicates high runoff from easily eroded formations such as those composed of shale and clay-rich soils of low permeability. A *medium-textured* drainage pattern typically has moderate tributary spacing and density. It is produced from moderate runoff from relatively permeable soils or bedrock. Thin beds of sandstone can produce such a texture. A *coarse-textured* drainage pattern has a low drainage density with widely spaced channels. It suggests the presence of hard, resistant rock types (e.g., granite, gneiss) and/or very permeable soils. Such materials typically absorb a great deal of water, and there is little runoff. Also, the bedrock may be fractured.

Drainage Pattern

The drainage pattern developed through time on a landscape provides clues about the bedrock lithology (e.g., igneous, sedimentary, metamorphic), topography (slope, aspect), the texture of the soil and/or bedrock materials, the permeability of the soil (how well water percolates through it), and the type of landform present (e.g., alluvial, eolian, glacial). While *in situ* observations are essential, physical scientists often use the synoptic bird's-eye-view provided by remote sensing to appreciate a regional drainage pattern. Therefore, it is important to be able to recognize the major drainage patterns present in remote sensor data.

Sometimes there may be a variety of drainage patterns within a watershed or region, in which case it is called a *mixed* pattern. Whenever possible, it is better to stratify the mixed pattern into multiple areas with unique drainage patterns. Also, it is important to remember that the drainage pattern may be composed of wet and dry channels, especially in arid environments.

Dendritic: The most common type of drainage pattern is dendritic (Figure 13-16a). It is characterized by a treelike or fernlike pattern (Busby et al., 1996) with branches that intersect primarily at acute angles (i.e., < 90°). There are few abrupt bends in the stream channels. Dendritic drainage patterns are developed by random headward erosion of insequent streams on rocks of uniform resistance with little or no structural control caused by folding or faulting. It is often found on landforms composed of relatively homogeneous, horizontally bedded sedimentary rock (e.g., shale), glacial till (e.g., loess), volcanic tuff, on sandy coastal plains, on tidal marshes (Figure 13-16a), or on glaciated outwash plains.

Pinnate: A pinnate drainage pattern is a variation of the dendritic pattern where the streams or gullies have a featherlike branching pattern that typically intersects at acute angles. This pattern indicates that the landform has a high silt content – usually loess, silty alluvium, or very soft erodible sedimentary materials.

Trellis: The trellis drainage pattern is a modified dendritic pattern that resembles a vine on a garden trellis. It often exhibits straight, parallel primary tributaries and shorter secondary tributaries that join the larger branches at right angles (Avery and Berlin, 1992). This drainage pattern often indicates that the bedrock structure is tilted, folded, or faulted. It is often found on interbedded sedimentary rocks. Trellis drainage is especially common in areas of folded sedimentary beds of differing resistance such as the Ridge and Valley topography of the Appalachian Mountains from New York to Alabama or associated with the hogbacks of the Uinta Mountains of Utah (Figure 13-16b). The primary parallel channels follow the less resistant beds. The shorter tributaries flow down the sides of the more resistant upturned beds.

Rectangular: A rectangular drainage pattern also exhibits a treelike pattern; however, the main channels have more abrupt bends. Streams often join at approximately right angles. It develops in areas where joints or faults have developed in the bedrock. Rectangular drainage is present where the bedrock is fractured, jointed, or foliated and is likely to occur on metamorphic slate, schist, or gneiss. It can also form in hard, resistant sandstone in arid climates (Figure 13-16c). The stronger and more rectangular the pattern, generally the thinner the soil covering the bedrock (Rasher and Weaver, 1990). Although both trellis and rectangular drainage patterns have right-angle tributary junctions, individual

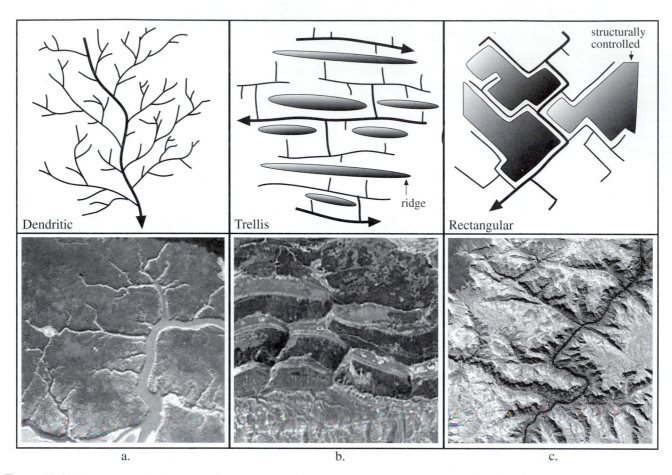

Figure 13-16 a) Top: Hypothetical dendritic drainage pattern. Bottom: Vertical aerial photograph of a dendritic drainage pattern found on Tivoli North Bay of the Hudson River National Estuarine Research Reserve in New York. b) Trellis drainage pattern found on the hogback ridges on the edge of the Uinta Mountains in Utah (Landsat MSS data). c) Structurally controlled rectangular drainage pattern of the Colorado River in the Grand Canyon (Thematic Mapper band 4 data; courtesy Space Imaging, Inc.).

streams in rectangular patterns generally have right-angle bends in their channels.

Parallel: A parallel drainage pattern consists of tributaries that flow nearly parallel to one another. All the tributaries join the main channel at approximately the same angle. Parallel drainage suggests that the area has a gentle, uniform slope. The tributaries follow beds in the less resistant rock. Parallel and radial drainage on a volcano on Maui, HI, are shown in Figure 13-17a.

Radial (Centrifugal) and Centripetal: A radial drainage pattern forms when water flows downward or outward from a hill or dome. The radial drainage pattern of channels produced can be likened to a wheel consisting of a circular network of parallel channels (i.e., the spokes) flowing away from a central high point or dome (i.e., the hub). In

this case, however, the hub of the wheel is elevated above the spokes. Radial drainage is often found on the slopes of volcanic cones such as Diamond Head Crater on Oahu, HI, or on steep mountainous terrain (Figure 13-17b). It is also referred to as *centrifugal* radial drainage.

A radial drainage pattern can also develop on circular areas that drain into a common, enclosed central basin or depression. This is referred to as *centripetal* radial drainage. In this case, the hub of the wheel is lower than the spokes. A cinder cone volcano may have centrifugal radial drainage on the exterior sides of the volcano and centripetal radial drainage on the interior walls of the crater (Figure 13-17b).

Annular: An annular drainage pattern is similar to the radial pattern except that ringlike tributaries intercept radial streams at right angles (Rasher and Weaver, 1990). These

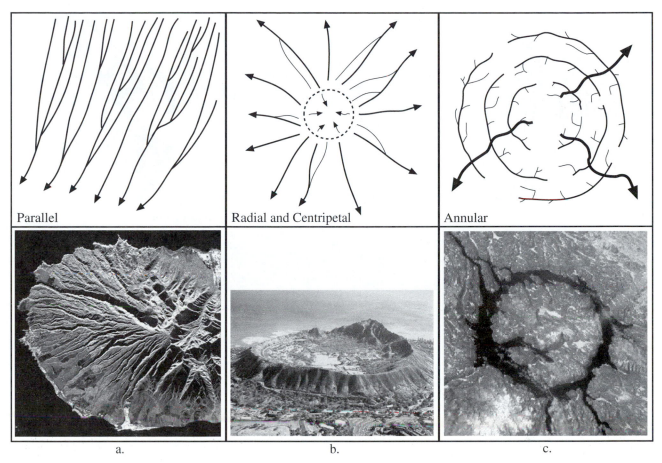

Figure 13-17 a) SIR-C radar image of parallel and radial drainage patterns found on the volcanic cone in western Maui, HI (courtesy NASA Jet Propulsion Laboratory). b) Low-oblique aerial photograph of Diamond Head Crater on Oahu, HI, revealing both radial and centripetal interior drainage patterns. c) Space Shuttle photograph of the annular drainage pattern on the Manicouagan Crater on the Canadian Shield in Quebec Province, Canada (courtesy NASA Johnson Space Center; STS009-48-3139).

develop when stream courses adjust to flow around resistant domes. Granite or sedimentary domes often develop this pattern, as do some meteorite craters (Figure 13-17c).

Dichotomic: A dichotomic drainage pattern may be found on alluvial fans or on alluvial deltas at the mouth of streams or rivers (Figure 13-18a). The water and suspended sediment usually enter the alluvial fan or deltaic area through a single channel. The flow is then redistributed throughout the fan or delta via a number of *distributaries*. On an alluvial fan, the coarse materials (cobbles, gravel, etc.) are deposited at the apex of the fan while the smaller and lighter minerals are deposited toward the terminus of the fan. In certain instances, the dichotomic drainage pattern may completely disappear near the terminus of the fan, i.e., the material is so porous that all of the water percolates underground. Similar dichotomic distributaries may form on river deltas (e.g., the Mississippi and Nile). The dichotomic drainage patterns can

only form on deltas when the suspended sediment is very fine-grained and can be transported great distances.

Braided: A braided stream pattern may develop on floodplains where stream velocity is not sufficient to move the suspended sediment load downstream. The result is the deposition of suspended material in intertwined channels that appear much like braided hair. For example, consider the braided stream pattern found at the mouth of the Betsiboka River in Madagascar (Figure 13-18b). The sands and gravels that are systematically sorted and deposited in a braided stream drainage network may eventually have significant economic importance.

Deranged: A deranged drainage pattern exhibits streams that wander in disorder in swamps and among water bodies (not shown). It is mainly found on very young landscapes that have almost level topography and a high water table. It

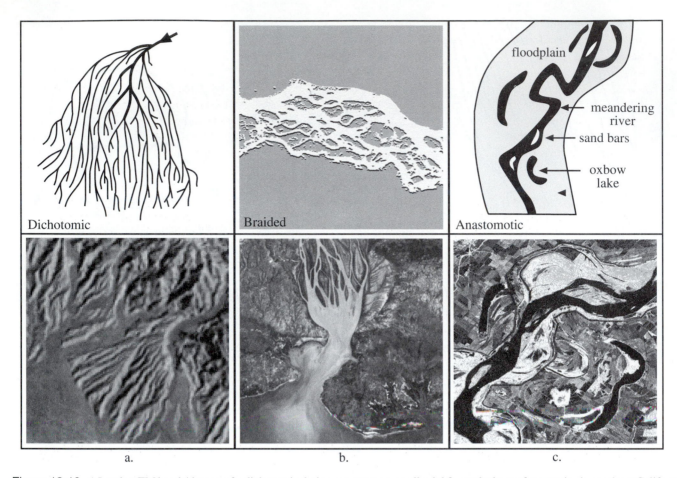

Dichotomic   Braided   Anastomotic

floodplain
meandering
river
sand bars
oxbow
lake

a.   b.   c.

Figure 13-18 a) Landsat TM band 4 image of a dichotomic drainage pattern on an alluvial fan at the base of mountains in southern California (courtesy Space Imaging, Inc.). b) Space Shuttle photograph of the braided stream pattern on Betsiboka River mouth in Madagascar (courtesy Kamlesh Lulla; NASA Johnson Space Center). c) SIR-C/X-SAR L-band radar image of Mississippi River anastomotic drainage pattern (courtesy NASA Jet Propulsion Laboratory).

can occur in young glacial till plains, moraines, low coastal plains, and floodplains. Usually swamps, marshes, bogs, and lakes or ponds are present.

Anastomotic: An anastomotic drainage pattern may be found on mature floodplains where there is ample homogeneous sediment and reduced stream flow or velocity. Basically, the hydrologic system does not have sufficient energy to remove the sediment from the area. The major stream channel adjusts to the decrease in energy by depositing its sediment load and increasing the length of the stream channel. Typical diagnostic features include meandering streams and remnant meander scars and/or oxbow lakes such as those found on the Mississippi River (Figure 13-18c).

Sinkhole (doline): A sinkhole drainage pattern consists of isolated lakes or ponds that do not appear to be connected by any systematic surface drainage. If short stream segments

are present, they often end abruptly or disappear into karst topography sinkholes or depressions. This drainage pattern develops on sedimentary limestone ($CaCO_3$) where the sinkholes and depressions have formed by chemical dissolution (Trefil and Hazen, 1995). Sometimes the subterranean drainage network that is not visible in the imagery emerges to form springs and lakes. An example of sinkhole topography in Florida is shown in the section on karst topography.

 Geomorphology

A *landform* is a three-dimensional feature on the Earth's surface formed by natural processes. Typical landforms include volcanoes, plateaus, folded mountain ranges, stream channels, etc. *Geomorphology* is the science that studies the nature and history of landforms and the processes of weathering, erosion, and deposition that created them (Selby, 1989). At one time it was known as the science of *physiography*. The study of geomorphology involves an appreciation for five major *processes* at work that constantly erode or deposit materials on the Earth's crust, including: running water (fluvial), glacial ice (glacial), wind (eolian), ground water, and wind-driven water waves. Geologists, physical geographers, and other scientists routinely use the synoptic view associated with remotely sensed data to identify and interpret geomorphic features on the Earth's surface. In fact, identifying, understanding, and appreciating the nature of landforms present on remotely sensed imagery is one of the great benefits of remote sensing science. One should take time to appreciate the tremendous beauty and variety of landforms on the Earth and how ecosystems associated with the various landforms interact with one another.

A word of caution is in order. In the late nineteenth and early twentieth century principles of landform evolution were put forth such that any current landform could be interpreted as being in a stage of evolution (e.g., that stream valleys progress from youth, maturity, to old age). Landforms do evolve, but the complexity of climatic change and its influences, the effects of tectonism, and man's impact as a geomorphological agent may be so overriding that few geomorphologists now believe that there is a simple and direct sequence of landforms that can be recognized as developing according to a particular *in*variable pattern (Selby, 1989; James, 1999).

Fortunately, many landforms do exhibit observable, repeatable attributes such as size, shape, height, three-dimensional topography, composition, slope, aspect, etc. These attributes may be used to classify landforms into the following classes (Rasher and Weaver, 1990):

- igneous landforms,

- landforms developed on horizontal strata,

- landforms developed on folded strata,

- fault-controlled landforms,

- fluvial landforms,

- karst landforms,

- shoreline landforms,

- glacial landforms, and

- eolian landforms.

The reader is referred to the Army Map Service Guide for a description of the major landforms of the world as well as quality diagrammatic examples (Army Map Service, 1954). The Soil Conservation Service produced an excellent manual that summarized the major landforms and how they appear on aerial photography (Rasher and Weaver, 1990). Several of their examples are used in this chapter. Short and Blair (1986; 1999) provided a thorough review of how to perform geomorphological analysis from space especially for obtaining information about large, regional landforms. They provide examples from around the world.

One might ask, What is the ideal type of remote sensor data and/or image presentation format to study, appreciate, and classify the Earth's landforms? First, no single sensor is ideal for the study of all landforms. Aerial photography may be ideal for studying landforms in the relatively cloud-free arid western United States, while radar data may be ideal for studying landforms in cloud-shrouded tropical environments. Many analysts can extract landform information by viewing a single aerial photograph or image. However, many scientists prefer to study landforms while viewing the terrain in a three-dimensional presentation using 1) stereoscopic photography or imagery (Caylor and Lachowski, 1988), 2) draping a monoscopic single image over a digital elevation model and then manipulating the observer's viewing position to enhance topography, and/or 3) obtaining oblique imagery that generally enhances the third-dimension. Also, imagery obtained at a relatively low Sun angle often enhances the analyst's ability to identify landforms. Finally, one has to be very careful when interpreting images in the Northern Hemisphere because shadows in images fall toward the north, away from the viewer. Therefore, experienced image analysts often rotate the Northern Hemisphere images so that south is toward the top to prevent pseudoscopic illusion from taking place (i.e., confusing mountain ridges with valleys). Some of the illustrations in this chapter are oriented in this manner to facilitate interpretation.

It is beyond the scope of this single chapter to provide examples of all the possible geomorphic landscapes present on the surface of the Earth. Therefore, only representative examples of some of the most important landforms are presented. Please refer to books on geomorphic analysis from aerial platforms such as Shelton (1966), Short and Blair (1986; 1999), Rasher and Weaver (1990), Strain and Engle (1993), and Way and Everett (1997).

Figure 13-19 Panchromatic stereopair of the southern Menan Butte cinder cone volcano in Idaho obtained on June 24, 1960. It is one of the world's largest tuff cone volcanoes, with a volume of 0.07 mi³ (0.30 cubic km³). The tuff is made of volcanic glass that has been altered by the addition of water. The cone is late Pleistocene in age. The Menan Butte volcano also extruded lava on the surface to the west (bottom of the photographs) for many kilometers. There is sunglint on Henry's Fork in the left photograph as it flows into the Snake River. Note the radial and parallel drainage on the flanks of the crater and the centripetal drainage in the interior of the crater.

Igneous Landforms

The Earth's upper mantle is partly molten, with many minerals in it that were forced deep by tectonic forces and eventually melted. Igneous rocks are produced from this molten, fluid magma. If the magma is extruded onto the Earth's surface, *extrusive* igneous (volcanic) rock is created. If the magma is extruded into some subsurface portion of the Earth's crust, *intrusive* igneous rock is formed. If the molten lava flows or is exploded out of a central vent and produces a mound or cone-shaped feature, we have a *volcano*.

Cinder cone volcanoes are built entirely of pyroclastic materials that are ejected into the air and fall to Earth nearby. Their accumulation eventually builds the cone. For example, consider one of the Menan Butte volcanic craters (Figure 13-19). It was created when a dike of magma intruded into a shallow aquifer. The water in the aquifer turned to steam and explosively fragmented the basaltic magma into volcanic glasslike particles called tuff. The volcano was built up layer by layer, by aerial deposits of tuff and larger pyroclastic (bomb) material. The two Menan Butte volcanoes also extruded lava onto the surface covering a large portion of eastern Idaho. Diamond Head Crater on Oahu, HI is also a tuff cone volcano.

Lava dome (shield) volcanoes are created when the lava pours out onto the landscape. Generally this results in more gently sloping volcanoes depending upon the viscosity of the lava (i.e., the more highly fluid the lava, the more gentle the slope; the more viscous the lava, the steeper the slope produced). Common extrusive rocks are basalt, andesite, dacite, and rhyolite. Basalt is the most widely distributed volcanic rock. It is also the most common rock on the Earth's surface. It flows out from volcanic vents over other rock in great lobes or thin sheets. It can travel hundreds of miles. It is fine-grained, black, rich in silica and the ferromagnesian minerals. Basalt normally appears dark on panchromatic film because it contains feldspar, hornblende, and micas. Dacite and rhyolite appear in light tones on panchromatic images.

The Shuttle SIR-C/X-SAR radar images in Color Plate 13-2 reveal several components of the Kilauea shield volcano located on Hawaii. The massive volcano continues to be created by the extrusion of two types of lava: *a'a* is extremely rough and spiny and will quickly tear shoes when walking on it, and *pa'hoehoe* is a ropey lava that looks like burned pancake batter. The volcano is enlarged by the extrusion of lava onto the surface. Numerous individual flows are easily seen in the three-dimensional radar-perspective view in Color Plate 13-2b. Note especially the dark materials from

Figure 13-20 A three-dimensional perspective view of Isla Isabela, one of the Galapagos Islands. This is a Space Shuttle SIR-C/X-SAR image draped over a digital elevation model. Rough *a'a* lava flows produce a bright radar-return on the side of the dome volcano. Ash deposits and relatively smooth *pa'hoehoe* lava appear as dark areas on the side of the volcano (courtesy NASA Jet Propulsion Laboratory).

the Kupaianaha crater. Kilauea has been erupting almost constantly since 1983.

Isla Isabela is one of the Galapagos Islands located off the west coast of Ecuador, South America. It is also an active lava dome (shield) volcano. Figure 13-20 depicts a Space Shuttle SIR-C/X-SAR image of Isla Isabela draped over a digital elevation model. The bright areas are very rough lava flows.

Composite cone or strato volcanoes are created from both pyroclastic materials and extruded lava. The world's most impressive volcanoes are composite cones. For example, Mount St. Helens in Washington (8,360 ft; 2,548 m) is a composite cone volcano (Figure 13-21 and Color Plate 13-3). It erupted on May 18, 1980, at 8:32 a.m. Pacific time. A series of moderate-to-severe earthquakes preceded the eruption, sending the north side of the mountain cascading downward toward Spirit Lake. This avalanche, the largest ever observed in the Western Hemisphere, weakened the magma chambers within the volcano, causing a northward lateral and vertical explosion that destroyed over 270 mi^2 (7,000 km^2) of forest in five seconds and sent a billowing cloud of ash and smoke 70,000 ft (21,000 m) into the atmosphere. A pumice plain was produced, consisting of volcanic mud, ash,

and debris that buried the original Toutle River Valley to a depth of 1,000 ft (300 m). The stereoscopic photography obtained on August 6, 1981, after the eruption revealed another lava dome developing in the center of the crater.

Intrusive igneous rock is formed when the molten magma cools and crystallizes within the Earth's crust. The material lying above this plutonic rock may eventually be eroded. Large igneous intrusive rock bodies, or *batholiths,* often form the foundation for entire mountain systems such as the Sierra Nevada in the western United States or the Andes in South America. Smaller dome-shaped intrusive rock bodies are called *lacoliths* or *stocks*.

The most common intrusive (plutonic) rocks include granite, diorite, diabase, and gabbro. Granite is a light-colored igneous rock, rich in quartz and feldspar but poor in the ferromagnesian (iron and magnesium) minerals. Granite usually appears in light tones on panchromatic imagery, whereas diabase and gabbro are dark. Granite cools for an exceptionally long time underground; therefore, it has large crystals that are visible to the naked eye. When granite is exposed at the surface, exfoliation may take place where concentric shells of rock material break off due to chemical or thermal weathering and from the release of internal stress when over-

a.

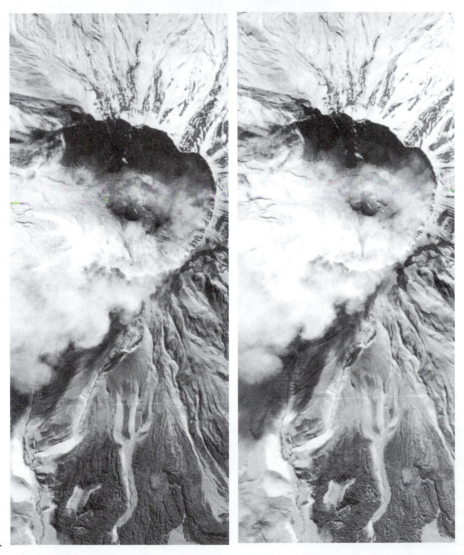

b.

Figure 13-21 a) High-oblique photograph of Mount St. Helens erupting on May 18, 1980 (courtesy of U.S. Geological Survey). b) USGS High Altitude Photography (HAP) stereopair of Mount St. Helens in Washington on August 6, 1981 (U.S. Geological Survey photos 109-84, 85). The active lava dome in the center of the cone is visible. Steam is rising from within the crater. A sediment choked radial drainage pattern has developed. North is to the left. Please refer to Color Plate 13-3.

 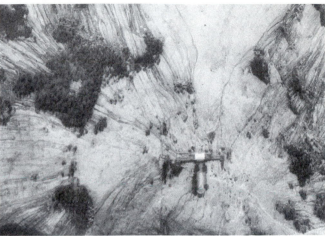

a.   b.

Figure 13-22 a) Panchromatic aerial photograph of Stone Mountain, GA. It is a granite exfoliation dome created by intrusive magma solidifying at great depth. It is believed that more than 10,000 ft of less resistant material was eroded to expose the intrusive rock. b) Large-scale photograph of the top of Stone Mountain. Note the gondola facility, linear striations in the rock, and the parallel and annular drainage pattern present.

burden is removed. For example, Figure 13-22 depicts Stone Mountain in Georgia, which is a relatively small intrusive stock. It is approximately 360 m high and 11 km in circumference. An image of the top of the dome reveals linear striations that radiate away from the apex (Figure 13-22b).

Sometimes the less resistant material surrounding the intrusive volcanic neck (or plug) of a volcano erodes, exposing the more resistant volcanic neck materials. For example, Devil's Tower in Wyoming is an intrusive volcanic neck (Figure 13-23). Occasionally the magma is extruded away from the main body via cracks and fissures in the Earth's crust to form dikes and sills. Later the material above these dikes and sills may be eroded, exposing wall-like resistant dikes.

If the granite at the surface is largely unfractured, a dendritic drainage pattern will result on the domelike hills. Conversely, if the granite has been extensively fractured, this can result in a rectangular or trellis drainage pattern.

Landforms Developed on Horizontal Strata

Many landforms have developed on flat, horizontally layered strata. A layer is considered to be horizontal when the dip (inclination from a horizontal plane) of the strata has not significantly affected the development of the topography. The type of rock, its erodability, amount of precipitation available to the region, and local stream gradient control landform development.

Horizontally bedded sedimentary rocks are formed by cementing the loose particles produced by weathering. The most common cementing agents are quartz, calcite, and iron oxide. Most sediments settle in water in relatively horizontal layers called *strata*. Sediments can collect in slow-moving streams, in swamps, lakes, and shallow seas. Sediments also collect in oceans, where they are deposited on continental shelves or flood down submarine canyons and are deposited. The flows first deposit coarse sand, then finer clays and silts, depending upon grain size. The most common rock outcrop formations on the continents are composed of mudstone, sandstone, shale, and limestone. Greater water force is required to move larger pebble-sized particles that are deposited as *conglomerate*.

Mudstone has no visible grains, being composed of extremely fine sediments deposited on quiet floodplains, in lakes, or deep oceans. Thick deposits of mudstone are present in most deltas, where rivers enter still water. Layering occurs in thick mud deposits because the clay particles, being flat, align themselves horizontally. They are usually gray to shades of red in color. The Mississippi delta contains mudstone.

Sandstone has visible grains like coarse sugar, up to almost 1/12 in. (2 mm) in diameter. Most sandstone is composed of rounded particles of quartz, but it can contain feldspar and fragments of unweathered rock. To move particles of this size, the water or wind must have force. In arid environments, sandstone is a weather-resistant, hard rock. Outcrops often appear in light tones on black-and-white photography

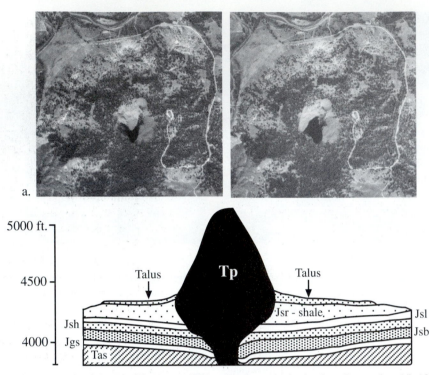

Figure 13-23 a) Panchromatic stereopair of the Devil's Tower, WY, volcanic neck obtained on September 15, 1953, with south at the top. Vertical joints divide the rock mass into polygonal columns that extend from just above the base to the top of the Tower. Devil's Tower is 865 ft tall, with an elevation of 5,117 ft. The Tower is more resistant to erosion than the surrounding sedimentary rock. It was formed by the intrusion of magma into sedimentary rock. Weathered rock (talus) lies at the base. b) Geologic cross-section: Tp = volcanic neck consisting of crystalline intrusive black igneous rock – phonolite porphyry with crystals of white feldspar; Tas = red siltstone and sandstone with interbedded red shale; Jgs = gypsum with interbedded red mudstone; Jsb = gray-green shale with sandstone; Jsh = yellow sandstone; Jsr = gray-green shale with sandstone and limestone.

unless covered by desert varnish, in which case they may have dark tones. Sandstone in arid and semiarid environments often produces *plateaus*, *mesas*, and *buttes*. Cliffs may occur where sandstone caps less resistant sedimentary rock such as shale. Sandstone is not as plastic (pliable) as shale and other sedimentary rocks. Therefore, it is broken easily when stressed. The fractures become visible and may produce angular, dendritic, trellis, or rectangular drainage patterns.

Limestone is formed from calcium carbonate. Lime precipitates onto shallow sea floors in warm tropical waters where almost no other sediments cloud the water. Limestone rocks formed in the broad centers of flooded ancient continents and along the edge of continental shelves where reefs were built. As a result, many limestones are composed of the skeletons of reef-forming animals. Limestone (and related dolomites) are susceptible to solution, meaning that they can be dissolved. In humid environments, where there is plenty of water, we may find karst topography, named after the Karst region of the former Yugoslavia, created by limestone dissolution. The undulating to hummocky terrain often contains no visible surface drainage pattern but instead may exhibit oval sinkholes called dolines. Limestone in arid environments behaves quite differently. Because there is little water for dissolution, limestone may be very resistant to weathering. In fact, like sandstone, they may form caprocks with vertical faces when underlain by less resistant sedimentary rock such as shale.

Shale sedimentary rock is produced by the deposition and compaction of silt and clay particles. It is relatively impervious to water moving through it; therefore, more water is available for surface erosion. In arid environments the soft impervious shale is easily eroded. It generally does not produce as steep slopes as sandstone or limestone. When no caprock is present, it can yield a rugged badlands topography consisting of a fine-textured pinnate dendritic drainage pattern with sharp ridgelines and steeper slopes. Generally, shale is more pliable than sandstone and therefore, is not as affected by geologic stress. This results in a greater number of fine- to medium-textured dendritic drainage patterns. Shale generally appears in light tones on panchromatic imagery.

The Grand Canyon in Arizona is a good example of various landforms developed on horizontally bedded sedimentary rock. A diagram of the bedding structure is found in Figure 13-24a. This region contains a nearly continuous sedimentary record of the Earth's history, with approximately 5,000 ft (1,500 m) of bedded horizontal layers, all stacked in sequence with the oldest at the bottom. One of the top layers, the Kaibab limestone, formed from the remains of corals, sponges, and other marine animals and is one of the youngest in the sequence – about 240 million years old. At the bottom of the canyon lies the Vishnu Schist which is about 1.7 billion years old, some of the planet's oldest rock. The in-between layers of rock reveal a turbulent history of mountain upheaval, lava deposits, erosion by wind and water, and past environmental changes as deserts were replaced by rivers, then lakes and shallow inland seas. Finally, the land was uplifted, forming the Colorado Plateau. Five to six million years ago, the young Colorado River began to cut into the layers of rock, and the Grand Canyon was created.

A panchromatic stereopair of the area near the South Rim of the Grand Canyon and Bright Angel shale butte is found in Figure 13-24b. The South Rim is composed of sandstone caprock of the Coconino Plateau and is resistant to weathering, forming a flat dissected plateau (mesa). Both sandstone and limestone form steep cliffs in arid environments. The less resistant shale is easily eroded, creating more gently sloping eroded surfaces (Figure 13-24b). Also present in the stereopair is the entrenched Colorado River with its deep, V-shaped inner gorge and rapids.

Another example of horizontally bedded strata is found a few hundred miles downstream. Figure 13-25a depicts the Colorado as it dissects the Shivwits Plateau as recorded by the Landsat Thematic Mapper. This band 4 near-infrared image provides significant detail but not nearly the information content of the color composites found in Color Plate 13-4ab, where the more resistant sandstone caprock and the more gently sloping shale sedimentary rocks are evident from their vibrant colors. Image analysts often find it very useful to have information about the elevation and slope of an area when interpreting landforms. Figure 13-25b is a representation of USGS digital elevation data. Note the higher elevation of the Shivwits Plateau and the lower elevation of the Colorado River. An analytical shaded relief image of the digital elevation data is shown in Figure 13-25c. It provides additional information of value to the image analyst about the three-dimensional nature of the terrain. Finally, the slope map (Figure 13-25d) makes it clear that the terrain with the greatest slope occurs adjacent to the Colorado and its major tributaries. Plateaus and mesas are flat, having little slope; therefore, they appear dark in the slope image.

Another excellent example of a plateau or mesa is found in the Phang Hoei Range of north central Thailand about 40 km northeast of the city of Lom Sak (Figure 13-26). The resistant plateau rises majestically above the surrounding countryside in this Shuttle Imaging Radar C-band HV polarized image.

Sometimes we encounter horizontally bedded metamorphic rocks that were created by exposure to heat and pressure. This *metamorphism* causes sandstone to become quartzite, which is very resistant to weathering and can form sharp-crested ridges in all climates. Similarly, metamorphosed shale becomes very resistant slate. Orthogneiss is metamorphosed granite and has a similar appearance. A variety of paragneiss rocks may be developed from metamorphosed sandstone and shale and other sediments. Schists are medium-grained crystalline rocks derived from sedimentary rocks. They are easily weathered in humid climates but resistant in arid environments.

Landforms Developed on Folded Strata

Diastrophism (mountain building) can eventually tilt or fold the horizontal sedimentary or metamorphic rock layers. The folded terrain may produce monocline domes (often ramp shaped), anticlines, and synclines, as previously illustrated in Figure 13-14. The folded strata are eroded differentially, depending upon the type of sedimentary rock present. For interbedded sandstone and shale in arid environments, the more resistant sandstone forms *anticlinal ridges,* whereas the shales are eroded to *synclinal valleys.* This can form nearly parallel systems of resistant ridges separated by eroded valleys. The resistance of carbonate rocks such as limestone and dolomite depends on the climatic conditions. In arid regions with little water, limestone is resistant to weathering and tends to create ridges. Conversely, in humid regions limestone may be dissolved and therefore is more easily eroded, forming valleys or lowlands.

Depending upon the steepness of the dipping terrain, the ridges may be asymmetrical or symmetrical. A *hogback* is a sharp-crested ridge formed by differential erosion of a resistant bed of steeply dipping rock. For example, a folded landscape near Maverick Spring, WY, is shown in Figure 13-27a. It is a dissected asymmetrical dome with a series of resistant hogback ridges surrounding the structure. Figure 13-27b depicts an Appalachian Mountain range synclinal valley bounded by the Brush Mountain ridgeline to the west and the Canoe Mountain ridgeline to the east near Tyrone, PA. In this environment, the sandstone is more resistant to erosion than the soluble limestone. Figure 13-28 depicts the San

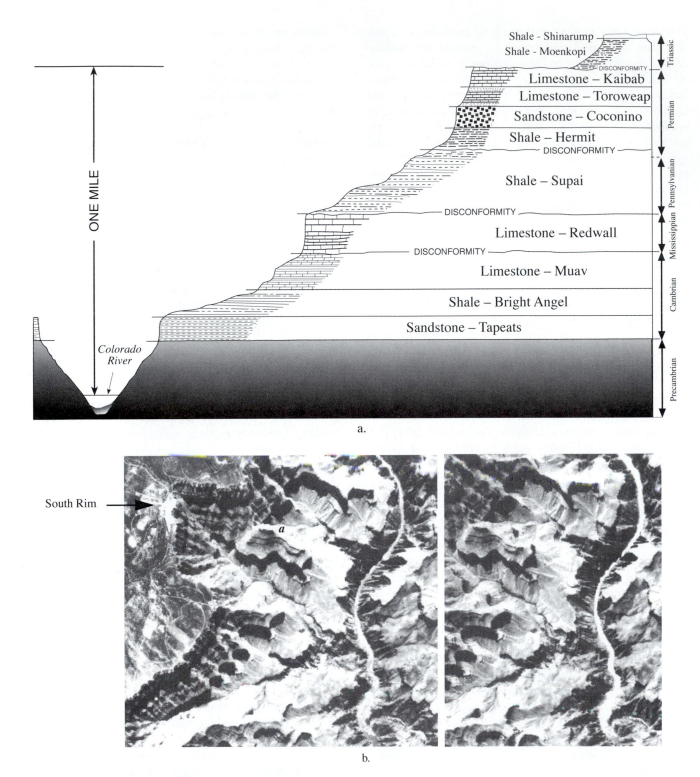

Figure 13-24 a) Geologic cross-section of the Grand Canyon in Arizona. b) Panchromatic stereopair of the Grand Canyon. Viewing and camping facilities are located on the sandstone caprock of the Coconino Plateau on the South Rim. The images were rotated so that south is to the left to avoid pseudoscopic illusion. The resistant sandstone and limestone form steep cliffs. The less resistant shale is easily eroded, causing more gradual slopes to form. The Bright Angel butte is located at "*a*".

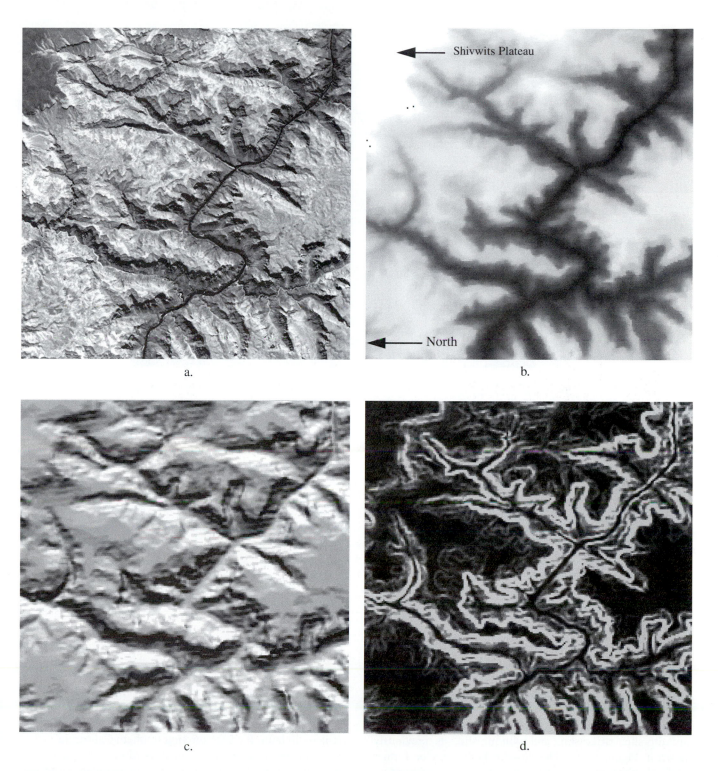

Figure 13-25 a) Another section of the Grand Canyon on the Colorado River in Arizona, as recorded by the Landsat Thematic Mapper (band 4; 30 x 30 m). North is to the left to improve interpretability. Color Plate 13-4 provides additional information. b) Digital elevation model derived from 30 x 30 m USGS data (displayed as 3 arc second, 93 x 93 m cells). Note the higher elevation Shivwits Plateau. c) Shaded relief map. d) Slope map where the brighter the pixel, the steeper the slope (after Mojave Desert Ecosystem Program, 1998; Landsat imagery courtesy of Space Imaging, Inc.).

Figure 13-26 Shuttle Imaging Radar C-band image of the Phang Hoei Range of north central Thailand on October 3, 1994. The resistant plateau is part of the Phu Kradung National Park (courtesy NASA Jet Propulsion Laboratory).

Rafael swell in southern Utah. This is a flat-topped upwarped monocline that slopes to the east/southeast. The more resistant sandstone ridges form hogbacks on the southern and eastern flanks of the monocline.

The angle of dip of a resistant bed such as a hogback may be calculated using photogrammetric measurement techniques and the formula:

$$\tan\theta = \frac{h}{d} \qquad (13\text{-}6)$$

where θ is the dip angle, h is the vertical distance, and d is the horizontal distance, as shown in Figure 13-29.

Fault-Controlled Landforms

There are three major types of displacement in faults, including: dip-slip, strike-slip, and oblique-slip, as previously shown in Figure 13-15. The dip-slip faults may be further subdivided into normal, reverse, thrust, graben, and horst, as shown in Figure 13-30a-e). In a *normal fault*, the hanging wall is moved (displaced) downward from the footwall, exposing a *fault scarp* on the footwall (Figure 13-30a). The Wasatch fault along the Wasatch Mountains in Utah is a good example (Figure 13-31). The Wasatch Mountain Range

of the Rocky Mountains is the upthrown block, and the downthrown block is the valley floor. Fault scarps at the base of the Wasatch Range generally face west.

A *reverse* fault is created when the displacement takes place in the opposite direction (Figure 13-30b). If the reverse faulting causes the slab of strata to move horizontally on top of the landscape as shown in Figure 13-30c, we have a *thrust* or *over-thrust* fault. A fault can also involve two fault planes. A *graben* is produced when a block of material gets displaced downward between two normal faults (Figure 13-30d). These often produce rift valleys, such as the great East African rift system. A *horst* is created when a block of material is thrust upward between parallel fault planes (Figure 13-30e).

When the displaced rock material occurs parallel with the fault line, a *strike-slip* fault is created (Figure 13-15b). For example, the San Andreas fault in California is a classic strike-slip fault. It passes through southern California east of Los Angeles, through the southern portion of the great Central Valley, runs through San Francisco, and then out to sea. A portion of the fault line in southern California is shown in Figure 13-32. The west side of the San Andreas fault is moving northwest, while the east side of the fault is moving southeast; hence the oft quoted observation that Los Angeles and parts of San Francisco, because they are on the western

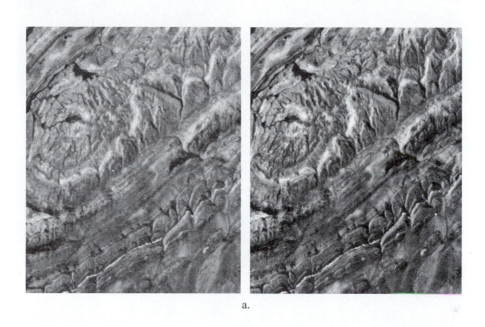

a.

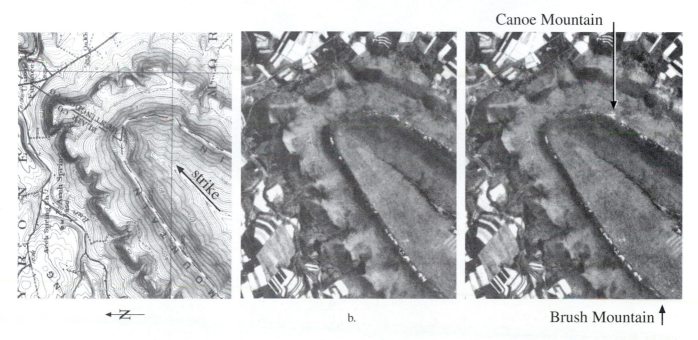

Canoe Mountain

Brush Mountain

b.

Figure 13-27 a) USGS High Altitude Photography (HAP) of an eroded folded landscape near Maverick Spring, WY. It is a dissected asymmetric dome, which is an erosional remnant of a plunging anticline. Note the fine-textured topography, the strike of the ridges and valleys, and the radial and trellis drainage controlled by the anticlinal structure (north is to the right). Also note the prominent hogback ridges. b) HAP of a synclinal valley in the Appalachian Mountains near Tyrone, PA, obtained on May 4, 1981. The ridge lines of Brush Mountain on the west and Canoe Mountain on the east are composed of more resistant sandstone, while the less resistant, soluble limestone sedimentary rock has been eroded. The interbedding of the sandstone and limestone results in hogbacks at the periphery of the syncline with a trellis drainage patter (courtesy U.S. Geological Survey).

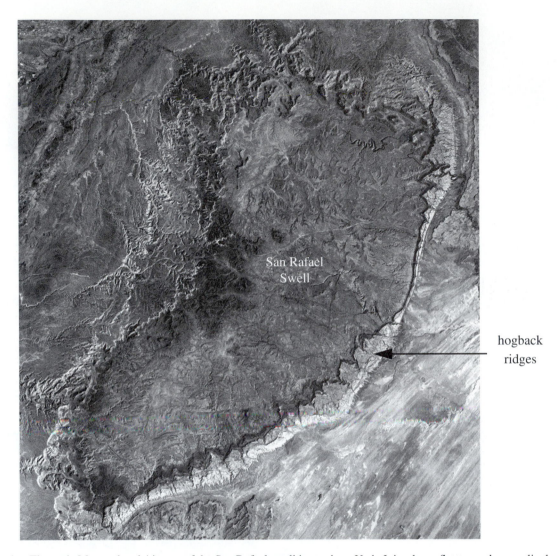

Figure 13-28 Landsat Thematic Mapper band 4 image of the San Rafael swell in southern Utah. It is a large flat-topped monoclinal upwarp bounded by hogback ridges on the southern and eastern flanks. The more resistant hogback ridges are composed of sandstone, while the less resistant shale beds have been eroded (courtesy Doug Ramsey, Utah State University and Space Imaging, Inc.).

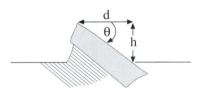

Figure 13-29 The angle of dip, θ, of a monoclinal ridge such as a hogback may be calculated from stereoscopic imagery using Equation 13-6.

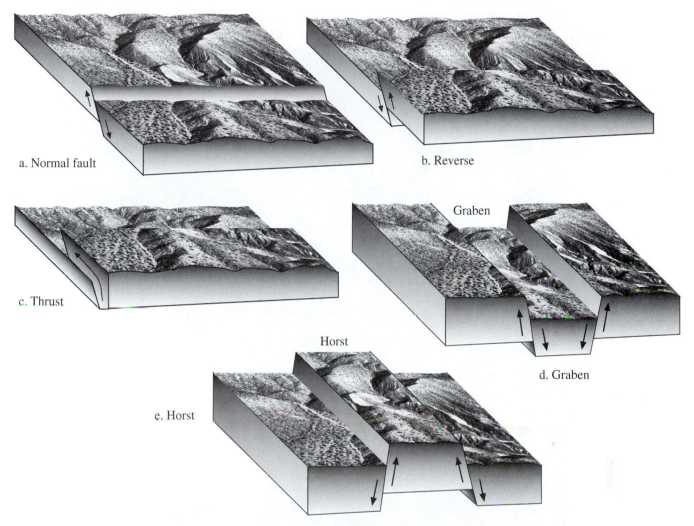

a. Normal fault

b. Reverse

c. Thrust

Graben

d. Graben

Horst

e. Horst

Figure 13-30 The major types of dip-slip faults: a) Normal fault. b) Reverse fault. c) Thrust fault. d) Graben associated with a downthrown block between two normal faults. e) Horst associated with an upthrown fault block.

side of the fault, are slowly moving out to sea albeit ever so slowly. Figure 13-32 also identifies the Garlock normal fault. The intersection of the Garlock and San Andreas faults is a very interesting place to live with minor tremors occurring almost every day. Notice that the San Andreas fault has created a small rift valley that is relatively flat when compared with the terrain on either side of the fault line. This is most evident when viewing the shaded relief version of the digital elevation model (Figure 13-32d). Natural lakes often occur in these rift valleys where the drainage pattern has been disrupted.

Fluvial Landforms

Fluvial landforms are created by the weathering, erosion, transportation, and deposition of materials by flowing water.

Fluvial processes create erosional and depositional landforms in virtually every region of the globe. Consequently, there are a tremendous variety of fluvial landforms.

It is important when interpreting fluvial landforms in an image to understand the characteristics of the rock types within the drainage basin, the drainage pattern, its density, and the gradient (slope) of the watershed. Streams or rivers flowing over terrain with steep gradients have greater velocity (energy) and hence greater bedload and suspended sediment carrying capacity. They can even entrain pebbles, cobbles, and boulders in the flow. Conversely, when the stream gradient decreases, streams or rivers may no longer be able to hold the sediment in suspension and then deposit the material in the channel bottom or as bars, spits, deltas, or alluvial fans. In effect, the velocity of the stream determines what kind of materials (sand, silt, clay, gravel, cobbles) can

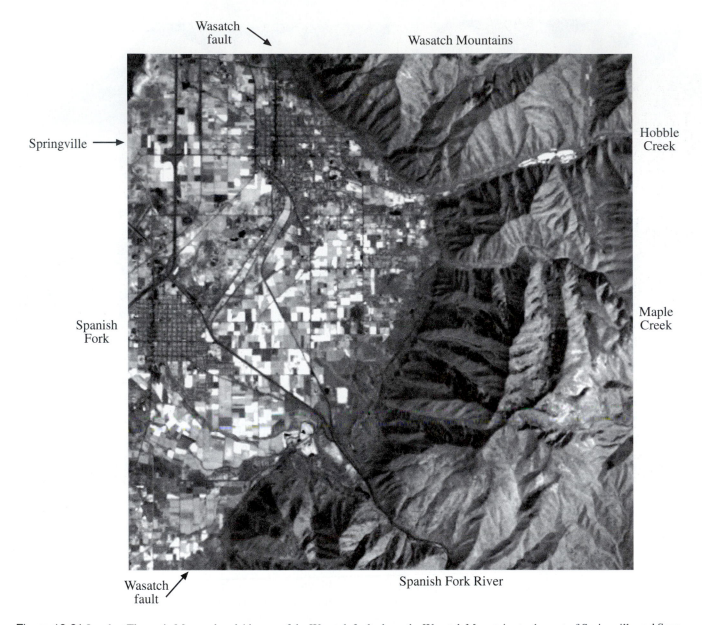

Figure 13-31 Landsat Thematic Mapper band 4 image of the Wasatch fault along the Wasatch Mountains to the east of Springville and Spanish Fork, UT. This is a classic normal fault, with the downthrown block being the valley floor and the upthrown block being the Wasatch Mountain Range. There are numerous fault escarpments all along the fault line (courtesy Space Imaging, Inc.).

be moved along by the stream. Even in deserts where storms are infrequent, sheet-wash with its intense velocity can induce massive erosion and transport the eroded materials great distances. It is important to also point out that the greater the velocity, the greater the amount of in-stream or river abrasion, corrosion, and quarrying that will take place. Thus, streams and rivers are active agents of geomorphic change that produce both erosional and depositional landforms. Some of the more important landforms include

stream-cut valleys, floodplains, terraces, deltas, alluvial fans, pediments, and playas.

Stream Valleys, Floodplains, and Relic Terraces

Uplifted terrain may be eroded by sheet erosion, creating gullies. Over time and given continued precipitation, these gullies may turn into small tributary channels or streams that have greater erosional power, especially if the terrain has

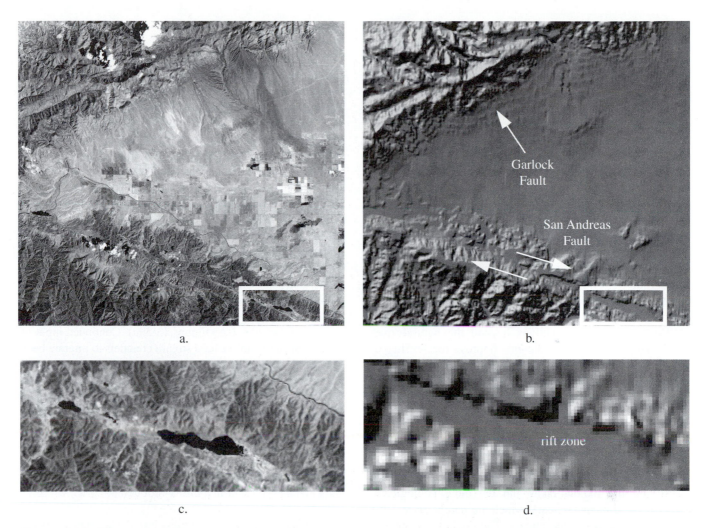

a.

b.

c.

d.

Figure 13-32 a) Intersection of the San Andreas and Garlock faults in southern California recorded on a mosaic of Landsat Thematic Mapper band 4 images. The San Andreas fault is a northwest-southeast trending strike-slip fault. b) Digital elevation model derived from USGS data (3 arc second, 93.218 m cell). c) Enlargement of the Landsat data highlighting a portion of the San Andreas rift zone in the lower right portion of (a). d) The relatively narrow rift zone as portrayed by the shaded relief model (after the Mojave Desert Ecosystem Program, 1998; courtesy Space Imaging, Inc.).

substantial gradient (slope). If the source of water continues unabated, the stream or river may incise and erode both the base and walls of the rock material creating deep, steep-sided V-shaped valleys with minimal floodplains and fairly straight water courses.

If the gradient is sufficient, the stream or river may continue to deepen and widen the river channel. However, at some point the river may erode the river bottom to the point that the river gradient is substantially reduced. If this occurs, the river loses velocity and begins to deposit some of its suspended sediment load. This can be the beginning of a river valley with floodplains, wide meanders, and occasional oxbow lakes. There are a number of factors that can cause

the river or stream to have a reduced gradient including land subsidence.

Sometimes a stream or river loses a substantial amount of its source water or its gradient is diminished greatly. When this occurs, even more sediment may be deposited. This can result in extremely broad and shallow valleys that have meanders, meander scars, cut-off meanders, point bars, numerous oxbow lakes, and natural levees. It is important to point out, however, that this process does not always happen. Unusual climatic events and man-induced impacts can intervene and dramatically alter what was once thought of as the youth-maturity-old-age stream-valley erosion cycle (Selby, 1989).

Way and Everett (1997) suggest that there are three types of river floodplains: meander, covered, and composite. *Meander floodplains* are caused by a low stream or river gradient and the deposition of some of the suspended sediment load. Thus, it is a low energy hydrologic system. Moreover, most of the sediment load in a meander floodplain stays within the confines of the river banks, even during flood stage. The greatest erosion in the floodplain takes place at the outer edges of the meanders. The greatest deposition takes place along inside edges of the meanders. This process can lead to the creation of sand and gravel point bars in the river. *Covered floodplains* are created when additional energy is provided during flooding and the floodwaters routinely overflow the riverbanks. This may create natural levees consisting of coarse sediments (usually sand) which are formed adjacent to the original stream channel. Over time, the natural levees are built up by continued accretion. Any water flowing over the natural levees has less velocity and can usually only entrain fine particle silts and clays that are then transported to distant slackwater locations beyond the natural river levees. Hopefully, these sediments improve the fertility of the soil. A *composite floodplain* contains features common to both meander and covered floodplains. Composite floodplains are the most common type of floodplain.

A small section of the Mississippi River shown in Figure 13-33 provides examples of several composite floodplain features. Both natural and man-made levees are present. The Landsat TM near- and middle-infrared bands are ideal for identifying the oxbow lakes and the more geologically recent flooded meander scars, which are barely discernible in the green and red band images. The sand and point bars are visible on all images. The middle-infrared band 5 provides more detailed soil moisture information than any of the other bands. As expected, vegetated areas are dark in the green and red bands and much brighter in both the near- and middle-infrared bands. A color-composite of TM bands 4, 3, and 2 (RGB) is found in Color Plate 13-5.

Floodplains sometimes have relic terraces associated with them. *A terrace* is a gently inclined, elevated benchlike remnant floodplain that lies at a higher elevation than the present day floodplain. Terraces may parallel the existing floodplain on both sides, although this is not always the case. A terrace typically has an escarpment on the side facing the present-day floodplain. Terraces may be produced by glacial activity or by dramatic changes in climate where more water and energy is made available to cut through the existing floodplain. Several layers of terraces present in a region indicate that dramatic erosional-depositional changes have occurred through time.

Deltas

When a stream or river with a substantial sediment load enters a standing body of water such as a lake, reservoir, sea, or ocean, its velocity is slowed dramatically. This causes the stream or river to deposit its base and suspended sediment load, with the more coarse materials deposited at the apex and the silt and clay particles moving farther out into the water body. The continued deposition of these materials over time may create a delta. These tend to have triangular shapes in planimetric view somewhat similar to the Greek letter Delta (Δ), but many other shapes also occur. There are five common delta forms.

Elongated or *digitate* deltas are created when the system is dominated by the continuous input of a large volume of water and sediment that is *not* impacted severely by waves. This can produce what is called a *bird's-foot* delta, such as the Mississippi River delta as it enters the Gulf of Mexico near New Orleans, LA (Figure 13-34a). Continued input of sediment causes the deltaic stream channels to migrate back and forth across the delta. Sometimes, however, the sediment deposited on the banks of the input channel create natural levees similar to that on the covered floodplain previously discussed. Intense flooding may cause the input stream to break free from its confinement within the natural levee and migrate to a nearby location. This process can lead to the creation of an entirely new lobe of sediment deposition in the delta. There may be several main input distributary channels present in a bird's-foot delta.

On a *lobate* delta the river builds into the sea, but relatively intense wave action redistributes much of the sediment along coastal barriers, causing it to have a convex edge or fan shape facing the water body. It is a wave-dominated system. The Niger River emptying into the Gulf of Guinea in Africa has created a classic lobate delta (Figure 13-34b). The Nile River emptying into the Mediterranean Sea is a lobate delta (Figure 12-34c).

A *crenulate* delta forms where tidal currents help create numerous sandy islands separated by tidal channels along the delta front. It is a tide-dominated system. The Irrawaddy River delta in Burmah is a good example (Figure 13-34d). Note the extensive mangroves visible as bright areas on the Irrawaddy delta.

If the waves or along-shore currents are extremely strong, they may move the sediment away from the mouth of the river, and only cuspate sand ridges will be formed that parallel the beach. This creates a *cuspate* delta.

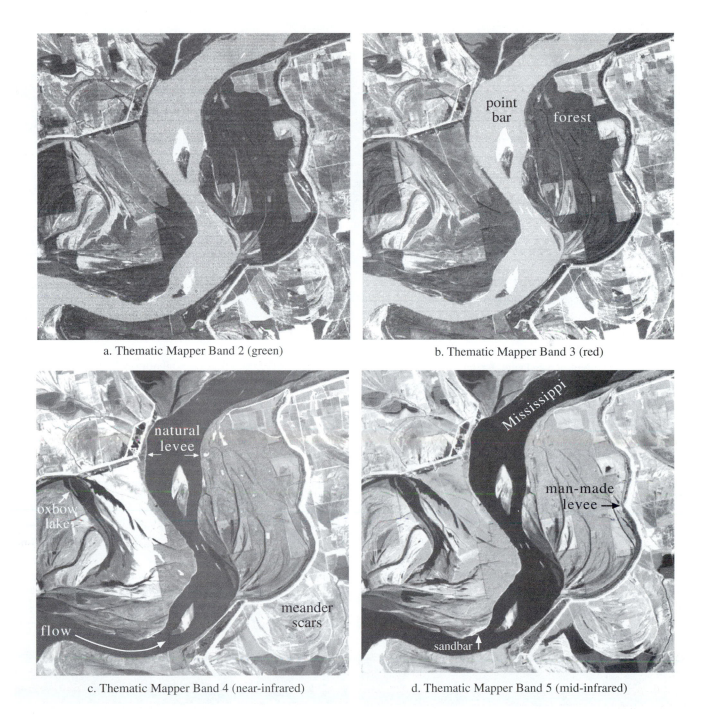

a. Thematic Mapper Band 2 (green)

b. Thematic Mapper Band 3 (red)

c. Thematic Mapper Band 4 (near-infrared)

d. Thematic Mapper Band 5 (mid-infrared)

Figure 13-33 Floodplain landforms on the Mississippi River recorded by the Landsat 4 Thematic Mapper on January 13, 1983. A color composite is found in Color Plate 13-5 (courtesy NASA Observatorium and Space Imaging, Inc.).

Mississippi River, United States

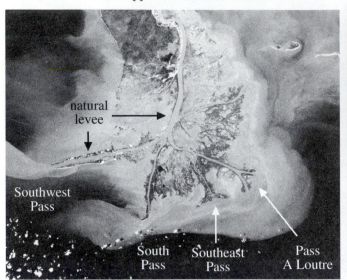

a.

Niger River Delta, Africa

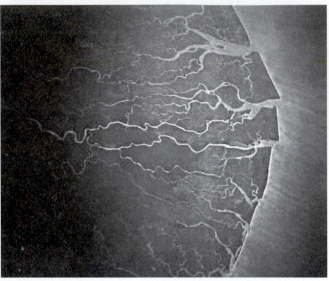

b.

Nile River Delta, Egypt

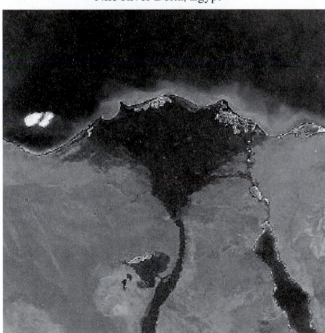

c.

Irrawaddy River Delta, Burmah

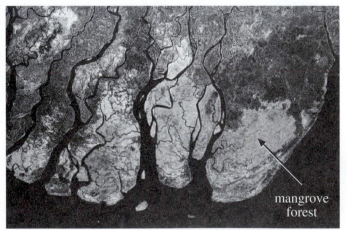

d.

Figure 13-34 a) Space Shuttle photography of the Mississippi River *elongated* bird's-foot delta, consisting of several outlets and their associated natural levees (courtesy K. Lulla, NASA Johnson Space Center). b) Space Shuttle photograph of the Niger River *lobate* delta emptying into the Gulf of Guinea in Africa (NASA Johnson Space Center; STS61C-42-0072). c) SeaWiFS image of the Nile River lobate delta in Egypt (courtesy NASA Goddard Space Flight Center and ORBIMAGE, Inc.). d) Irrawaddy River *crenulate* delta in Burmah, recorded on a Thematic Mapper band 4 near-infrared image (courtesy Space Imaging, Inc.).

Streams entering coastal estuaries also experience a reduced gradient, causing them to deposit much of their suspended sediment load. An *estuarine* delta will form if these deposits are not removed by currents in the estuary or by diurnal tidal flushing.

Alluvial Fans and Bajadas

Streams emerging from a mountain valley into a lowland area often encounter reduced slopes and are forced to deposit material in order to maintain an adequate gradient. An *alluvial fan* is a low, conical-shaped deposit with the apex of the cone at the mouth of the valley from which the fan-building stream issues. The more coarse materials such as gravel, cobbles, and even boulders are deposited near the apex while the sand, silt, and clay particles are transported to more distant parts of the fan. These materials are collectively called alluvium, hence the term alluvial fan.

The conical mound of material is deposited as distributary streams swing back and forth across the fan. This can create dichotomic or braided drainage patterns. Multiple alluvial fans that coalesce at the base of a mountain range create a *bajada*.

A Landsat Thematic Mapper band 4 image of an alluvial fan adjacent to the Salton Sea in California is shown in Figure 13-35a. Note the input stream channel in the upper right and the conical-shaped alluvial fan that has resulted due to millions of years of suspended sediment deposition from very sporadic rainfall events. It is clear that the bright-tone materials deposited on this fan had their origin in a different location when compared with the darker materials found at the base of the nearby mountain range. Also note the dichotomic drainage pattern. Another large alluvial fan in the White Mountains is shown in Figure 13-35b.

Pediments

A *pediment* is a gently inclined erosional surface carved in bedrock, thinly veneered with gravel, and developed at the base of mountains. Pediments are most prevalent in very arid environments. A pediment may look like an alluvial fan on an image, but it is dramatically different. A pediment is an erosional surface, while an alluvial fan is a depositional surface. In fact, many of the conical features in the Great Basin of North America, in southern Africa, and in Australia are actually pediments and not alluvial fans. Examples of pediments at the base of the White Mountains in California are shown in Figure 13-35b. The headward erosion of the bedrock strews sorted material downslope. The mountain may eventually drown in its own eroded debris.

Playas

Playas are shallow lake basins formed in arid desert regions that are intermittently filled with water that evaporates relatively quickly, leaving a residue of fine-textured surface materials. Many playas are saline and exhibit bright tones in imagery. Most playas are barren with little vegetation. A playa is present at the base of the alluvial fan in the White Mountains of California (Figure 13-35b).

Karst Landforms

Landforms created in limestone are generally referred to as *karst topography*. To be a true limestone at least half of the rock consists of carbonate minerals of which calcite ($CaCO_3$) is the most common (Selby, 1989). Dolomite ($CaMg$)CO_3, another carbonate rock, is also susceptible to dissolution but is not as readily affected by slightly acidic precipitation as limestone. This lowers solubility and increases resistance to weathering and erosion. Therefore, dolomite rock does not produce the same karst topography as limestone rock in humid climates.

Surface water derived from precipitation is actually somewhat acidic. When acidic surface water percolates downward through limestone, the chemical process of carbonation takes place along joints and bedding planes, gradually enlarging the openings by dissolution, until subsurface caverns, caves, and subterranean channels are created. Collapse of the roof of a subsurface cavern or downward dissolution of limestone from the surface produces depressions at the surface commonly referred to as *dolines* (the English term is *sinkhole*). *Dolines* are formally defined as closed hollows or depressions that are cone or bowl-shaped, with rocky or vegetated sides, circular or elliptical in shape, 2 – 100 m in depth, and 10 – 1000 m in diameter. The major karst topography is found in the former Yugoslavia along the Adriatic Coast, southern France, Greece, Cuba, Jamaica, the Yucatan Peninsula in Mexico, Tennessee, Kentucky, southern Indiana, northern Florida, and Puerto Rico.

Limestone with > 40 percent calcium carbonate is very susceptible to weathering in humid climates and may result in the development of an extensive network of doline features. For example, consider the black-and-white infrared photograph of karst topography found near Orlando, FL, shown in Figure 13-36. These dolines (sinkholes) were produced by dissolution occurring at favorable sites such as the intersection of major joints in the limestone. The limestone is dissolved and the solution and residue move downward,

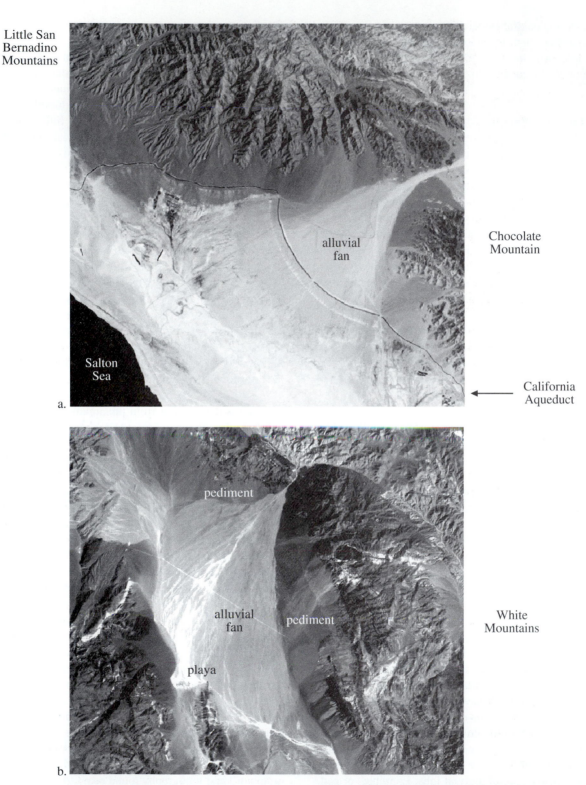

Figure 13-35 a) Alluvial fan issuing from between the Little San Bernadino Mountains and Chocolate Mountain northeast of the Salton Sea, CA. The California Aqueduct is also visible in this Landsat Thematic Mapper band 4 image. Its bowed shape gives some indication of how the fan has increased in volume over time. b) A large alluvial fan, several pediments, and a playa associated with an area in the White Mountain Range northwest of Death Valley, CA. There are several locations in these images where alluvial fans are coalescing creating a bajada (courtesy Mojave Desert Ecosystem Program and Space Imaging, Inc.).

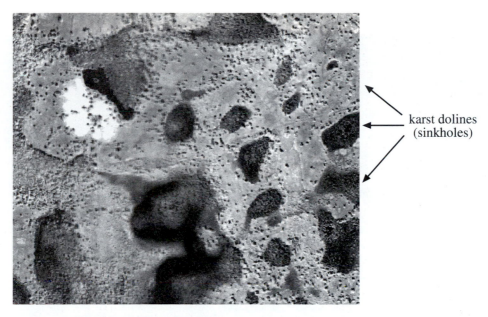

Figure 13-36 Calibrated Airborne Multispectral Scanner (CAMS) near-infrared 3 x 3 m data of karst doline (sinkhole) topography north of Orlando, FL. The depressions are created by dissolution of limestone rock. There is no surface drainage pattern in this area because all drainage is subterranean. The darkest areas are deep water. The bright area is sand. Many of these dolines are heavily vegetated.

constantly widening the fissures until a surface depression is created. This permits the collection of more of the surface water runoff and therefore the progressive enlargement of the doline. Sometimes, the dolines collapse, leaving large surface depressions. In this example, there are few surface drainage channels, only subterranean ones.

In areas of high local relief and abundant tropical precipitation such as in Puerto Rico, the weathering is greatly accelerated, causing the sinkholes to coalesce, creating haystack karst topography. Conversely, dolomitic limestone such as that found in parts of Kentucky is less susceptible to erosion. Consequently, it develops a hill and valley topography along with a surface drainage pattern.

Limestone is especially difficult to identify in extremely arid deserts or cold Arctic environments because the small amount of precipitation does not produce the weathering features discussed above. Instead, limestone is likely the resistant caprock on escarpments or plateaus.

Shoreline Landforms

More than two-thirds of the world's population lives within 100 miles of the coast. Humankind exerts great impacts on coastal resources. Consequently, scientists are very inter-ested in developing an understanding of the coastal processes associated with the dynamically changing coastal plain, beaches, spits, bars, coastal terraces, estuaries, etc. Three major agents are responsible for the creation of specific types of landforms in the coastal zone, including 1) energy from the Sun, which produces waves and ocean currents, 2) the gravitational pull of the moon and the Sun combine to create tides, and 3) man is a very active geomorphic agent.

Radiant energy from the Sun differentially heats the Earth's surface, creating variations in air pressure, which produce weather disturbances. Wind-generated surface waves are the main source of energy along a coast and are responsible for many of the erosional landforms along the coast. Conversely, waves also generate coastal currents that are responsible for the longshore movement of sediment along beaches (Selby, 1989). A wave of height (h) is produced by a sustained wind blowing over a waterbody for a certain *fetch* distance (F) where $h = 0.36\sqrt{F}$. Wave height may also be estimated if only the sustained wind velocity (V) is known using the relationship $h = 0.0024 \cdot V^2$. Sustained prevailing winds blowing over great expanses of water may create ocean currents that are also influenced by variations in water temperature and salinity. These oceanic currents such as the Gulf Stream can impact shoreline landform development because they influence the temperature and amount of sus-

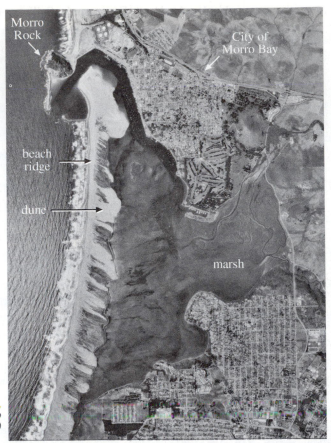

a. Portion of Morro Bay, CA, digital orthophoto quarter quadrangle (DOQQ) May 28, 1994

b. Landsat Thematic Mapper Band 4

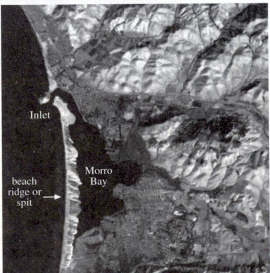

c. Landsat Thematic Mapper Band 7

Figure 13-38 a) Panchromatic USGS DOQQ of Morro Bay, CA. The bay or lagoon is separated from the ocean by an extensive beach ridge (courtesy U.S. Geological Survey). b) The beach bar also has numerous dunes, some of which are vegetated. Wave activity in the shore zone is visible in Landsat TM band 4. c) Band 7 data can be used to clearly identify the beach ridge shoreline. Wave-cut terraces are also present in this area (courtesy NASA Observatorium and Space Imaging, Inc.). Two color-infrared color composites of this area are shown in Color Plate 13-6ab.

Figure 13-39 Panchromatic version of the USGS digital orthophoto quarter quadrangle of Sullivan's Island, SC. Numerous coastal land forms are present, including a barrier island created by accretion of multiple beach ridges, sandbars, tidal inlets, and tidal marsh (*Spartina alterniflora*). A color-infrared composite of the area is found in Color Plate 13-7.

uplifts occurred. The dramatic marine terraces along the southern California coast near Palos Verdes were produced in this manner.

Coral Reefs

Coral reefs are created by living coral polyps and calcareous algae. Coral reefs develop best in seawater between 77° and 86°F with normal levels of salinity. Thus, coral is generally restricted to the tropical regions of the world between 30° N and 25° S latitude. Coral requires water that is <100 m deep and generally free of sediment so that photosynthesis can take place. Coral reefs often grow best on the windward side of islands where wave energy provides food and oxygen for the coral polyps. However, significant reef development can also take place on the leeward side of islands if a strong

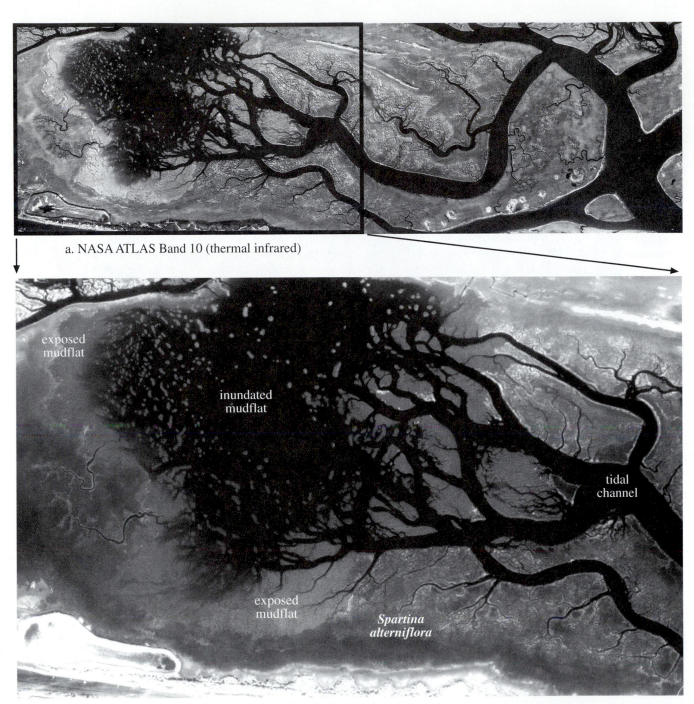

a. NASA ATLAS Band 10 (thermal infrared)

b. NASA ATLAS Band 6 (near-infrared)

Figure 13-40 a) NASA ATLAS band 10 thermal infrared 3 x 3 m image of the marsh behind Isle of Palms, SC, obtained in August, 1998.
b) Enlarged ATLAS Band 6 near-infrared image of a portion of the area. The images were obtained at mid-tide. Some of the tidal mudflats (which have no vegetation) are exposed, while others are inundated by the incoming tide flowing through the tidal channels. The more dense patches of smooth cordgrass (*Spartina alterniflora*) show up as bright areas along the natural levees in the near-infrared image. The dendritic drainage pattern associated with homogeneous soil materials is evident. It is much easier to discriminate between the tidal flats, tidal water, and smooth cordgrass when viewing a color-infrared color composite of the same area in Color Plate 13-7.

ocean current is present. For example, the Palancar reef system on the leeward side of the Island of Cozumel, Mexico, is bathed by a strong, clear upwelling ocean current. Fresh water and significant amounts of sediment can prevent coral reef growth. Coral cannot grow out of sea water; therefore, mean low-tide determines the upper limit of coral growth. Sea levels may rise due to global warming. Consequently, we may expect a commensurate growth in coral reef building during the coming century if pollution does not destroy the coral first.

The coral polyps literally create a built-up limestone surface anchored to hardbottom materials. There are three primary types of reefs. A *barrier reef* lies offshore and is separated from the land by a lagoon. The Great Barrier Reef is an excellent example (Figure 13-41a). It extends for more than 1,200 miles along the north coast of Australia. If a reef is attached to the land and extends out into the sea, it is called a *fringing reef*. A portion of the Palancar fringing reef on Cozumel is shown in Figure 13-41b. There may be gaps in fringing reefs due to the input of fresh water or suspended sediment. Finally, we may encounter an *atoll*, which is a ring-like reef enclosing a lagoon with a central island (which may or may not be present either due to subsidence or sea-level rise). The Mururoa atoll in the Pacific Ocean does not have a central island (Figure 13-41c).

Coastal Erosional Landforms

Coastal erosional landforms are created when the combined transporting capacity of wave and longshore current energy exceeds the supply of sediment. Most coastal erosion takes place during storm events due to the destructive force of wave energy. Depending upon how a coastline is oriented, wave refraction may cause either a spreading out or a convergence of wave energy. Headlands or points typically experience a convergence of wave energy and are therefore more susceptible to erosion. Conversely, adjacent bays generally experience a divergence of wave energy and less erosion.

The erosive power of water may produce *beach cliffs* with varying gradients. Rock fragments in the surf and the tremendous air pressure produced when surf breaks may combine to erode beach cliffs, creating *caves*, *arches*, and even *blowholes*.

Glacial Landforms

A *glacier* is defined as a body of ice, firn (compacted granular snow), and snow, originating on land and showing evi-

dence of past or present flow. Glacial landforms are created by the moving ice, nivation (a combination of freezing and thawing processes), and glacial meltwater. There are two general categories of glaciation: *continental* and *alpine*. During the Pleistocene Epoch (sometimes referred to as the Great Ice Age), layer upon layer of snow accumulated from year to year, creating vast glaciers that covered much of the Northern Hemisphere. There were approximately four major advances and retreats of the continental glaciers, with the most recent continental glacier retreating (disappearing) approximately 10,000 years ago from the northern United States. Only portions of Antarctica and Greenland currently have continental glaciation.

Alpine glaciation is still present in many of the great mountain ranges of the world, such as the Alps, Himalayas, Rocky Mountains, Sierra Nevada, Andes, etc. Both alpine and continental glaciation produce unique landforms based on erosion and deposition.

We focus our attention first on the major landforms that can be identified through analysis of remotely sensed data. It should be stressed that the optimum method of studying glacial landforms is often to evaluate them in three-dimensions, i.e., stereoscopically or in pseudo three-dimensions as will be demonstrated. However, some glacial features are so distinctive that they can be identified using single images.

Erosional Glacial Landforms: Cirques, Tarns, Aretes, Horns, U-Shaped Valleys, Hanging Valleys

A *cirque* is a semicircular steep-walled amphitheater-shaped valley in the upper reaches of an alpine glacier. An alpine glacier erodes headward, downward, and along the sides of the original V-shaped stream valley using the immense pressure of the overlying ice, nivation, and glacial quarrying of rock. Interestingly, after the alpine glacier retreats, there may be relatively little material in the bottom of the cirque. In fact, the base of the cirque may have been eroded more severely in certain areas creating a depression. After glacial retreat, the depression may fill with runoff, creating a glacial lake or *tarn*. Numerous cirques in the Uinta Mountains of Utah are shown in Figure 13-42ab. A color image of a Landsat MSS image of the area draped over a digital elevation model is found in Color Plate 13-8a.

An *arete* is formed when cirques on opposite sides of a divide erode headward creating a narrow, serrated mountain ridge. If the two cirques continue to erode headward, they may eventually produce a *col*, which is a sag in the serrated ridge. Numerous aretes and cols are visible in the oblique

Great Barrier Reef, Queensland, Australia

Palancar Reef, Cozumel, Mexico

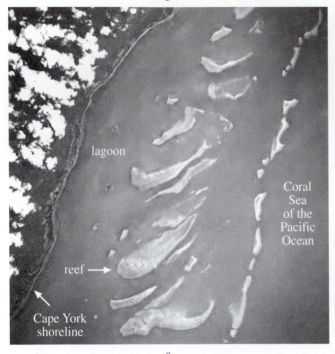

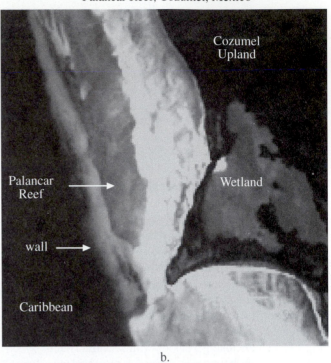

a.

b.

French Mururoa Atoll in the South Pacific Ocean

c.

Figure 13-41 a) The Great Barrier Reef of Australia extends for approximately 1,200 miles (2,000 km) along the Queensland coast. The reefs captured on this Shuttle photograph (STS 046-077-031) are located along the eastern coast of Cape York Peninsula (courtesy Kamlesh Lulla, NASA Johnson Space Center). b) SPOT HRV XS band 1 (green) image of a portion of the Palancar fringing reef on the Island of Cozumel, Mexico (© SPOT Image, Inc.). c) Landsat TM band 2 (green) image of the Mururoa coral atoll in the Pacific Ocean (courtesy Space Imaging, Inc.).

Uinta Mountains
in Utah

a. Landsat Thematic Mapper band 3 (red) image

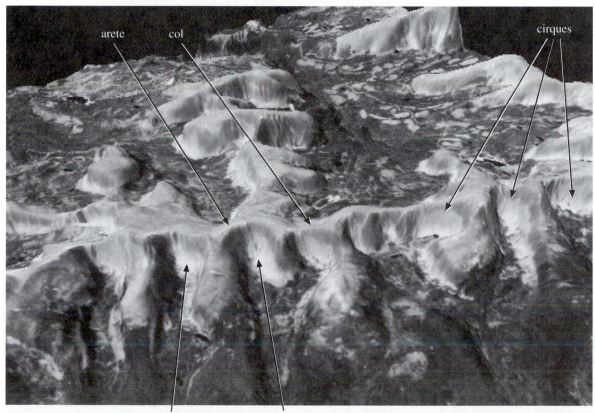

b. Landsat TM band 3 image draped over a USGS 30 x 30 m digital elevation model

Figure 13-42 a) Landsat Thematic Mapper band 3 (red) image of a portion of the Uinta Mountain Range of the Rocky Mountains in Utah. b) Landsat TM band 3 image draped over a 30 x 30 m USGS digital elevation model (vertical exaggeration is 4x) and viewed obliquely. A color version is found in Color Plate 13-8a (courtesy Space Imaging, Inc.).

Figure 13-43 High-oblique aerial photograph of the Matterhorn in Switzerland formed by the headward erosion of more than two cirques. The cirque on this side of the Matterhorn still contains a glacier and to the left is a hanging glacier clinging to the side of the peak. Glacial crevasses are readily apparent (courtesy American Geographical Society Collection archived at the National Snow and Ice Data Center, University of Colorado, Boulder, CO).

image of the Uinta Mountains shown in Figure 13-42b. A *horn* is formed when more than two cirques erode headward, leaving only a spire-shaped pinnacle. The most famous is the Matterhorn in Switzerland (Figure 13-43). Horns often appear triangular-shaped when viewed from above in vertical remote sensor data.

During glaciation in mountainous terrain, the glaciers do not normally produce new valleys. Rather, the accumulation and movement of glacial ice in existing valleys erodes headward, erodes the sides of the valley, and scours and deepens the valley floor. This process creates a *U-shaped glacial trough* or valley that is very impressive once the glacier retreats. Wallsburg Canyon in the Wasatch Mountains of Utah is a good example of a glaciated U-shaped valley (Figure 13-44; Color Plate 13-8b). Some glaciers scour extremely deep valleys that become lakes after the glacier retreats. For example, Lake Tahoe shown in Figure 13-45 and Color Plate 13-8c was created in this manner. Lake Tahoe is 23 miles long (37 km) and 12 miles wide (19 km). It lies 6,228 feet (1,898 m) above sea level and is 1,640 feet (500 m) deep; it is one of the deepest lakes in the continental United States.

If the major glacial U-shaped valley erodes downward more extensively than the tributary glacier flows, it may create *hanging valleys*. Hanging valley streams often enter the main valley as waterfalls.

Depositional Glacial Landforms: Moraines, Morainal Lakes, Till Plains, Eskers, Kettles, and Drumlins

Impressive glacial landforms are also created by the deposition of glacially eroded material. Rock material that accumulates on the side or edge of a glacier is called a *lateral moraine*. If two glaciers meet, a *medial* (or middle) *moraine* may be created at their confluence. This often appears as a dark ribbon on aerial photography or other imagery. For example, Figure 13-46a depicts the Barnard Glacier in Alaska with several tributary glaciers. Numerous lateral moraines are present. Several medial moraines are created by the confluence of the lateral moraines.

If a glacier reaches equilibrium between accumulation and ablation (melting), it produces a *terminal moraine* at the end of the glacier (Figure 13-46b). If the equilibrium lasts for a considerable time, the glacier continues to move rock and debris downslope, causing the terminal moraine to increase in volume. If ablation exceeds accumulation, the glacier retreats, leaving the terminal moraine as a remnant landform. If the glacier reaches equilibrium once again, a *recessional moraine* may be produced. It is possible for a number of recessional moraines to be present. Sometimes water is impounded behind the terminal moraine, forming a *morainal lake*.

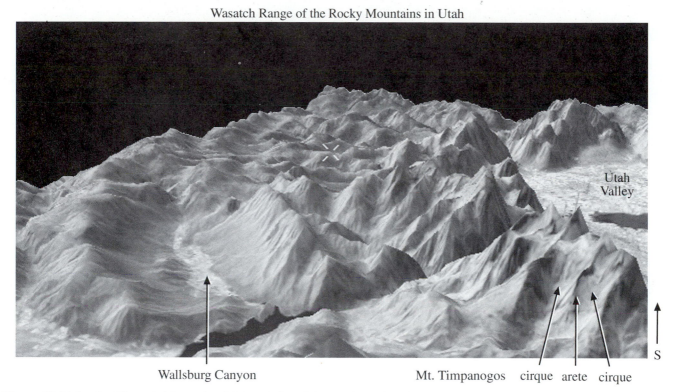

Wasatch Range of the Rocky Mountains in Utah

Utah Valley

Wallsburg Canyon   Mt. Timpanogos   cirque arete cirque

S

Figure 13-44 Landsat Thematic Mapper band 4 (near-infrared) image of a portion of the Wasatch Mountain Range of the Rocky Mountains in Utah draped over a 30 x 30 m USGS digital elevation model (vertical exaggeration is 5x) highlighting the Mt. Timpanogos glacial cirques and aretes and U-shaped Wallsburg Canyon. Several cols are present on the serrated Mt. Timpanogos ridgeline. A color version is found in Color Plate 13-8b (courtesy Space Imaging, Inc.).

Sometimes continental or even alpine glaciation deposits a mixture of clay, sand, gravel, and even large boulders over the entire countryside. After retreat, the material is referred to collectively as *glacial till* and the landform as a *glacial till plain*. There are typically no major landforms with any significant elevation in the till plain, only gently rolling terrain. Much of the midwestern United States, the bread basket of the world, is covered with glacial till deposits of varying thickness. Such till often has a mottled texture, with the lighter areas being composed of well-drained sandy soil and the darker area being composed of poorly drained clayey soil.

Several glacial landforms are produced directly within, upon, and even under the glacial ice. For example, a subglacial stream running within stagnant (non-moving) glacial ice may deposit well-sorted sand and gravel within the stream tunnels. After glacial retreat, the deposited material appears as a sinuous, linear ridge or *esker* on the terrain, often reaching heights from 5 – 30 m high and 50 – 60 m wide. Eskers may run for many kilometers. For example, Figure 13-47a is a low-oblique aerial photograph of a sinuous, gravel esker

trending across the terrain in Northern Manitoba, Canada. This esker also deposited sandy alluvium into the glacial lake as the glacier retreated.

Sometimes a large block of glacial ice that contains an assortment of glacial debris becomes detached from the main glacier. The block of ice and materials may be subsequently covered by other debris. After glacial retreat, the ice in the block melts, leaving a depression in the landscape. These depressions or *kettles* fill with water, creating lakes and swamps. A relatively large kettle is shown in Figure 13-47a.

A *drumlin* is a smooth, oval-shaped (elliptical) hill of glacial till material with its long axis parallel to the direction of flow of the former glacier. It is generally more blunt and steep on the glacier-facing end and more elongated down-glacier with a thinning tail. Drumlins often occur in clusters, referred to as drumlin swarms. Drumlins consist of unsorted gravel and sand mixed with some clay. There is much controversy as to the processes that produce drumlins. It is generally believed that a drumlin develops beneath active ice and is formed by

Sierra Nevada Mountain Range in California

Lake Tahoe

N

Figure 13-45 Landsat MSS band 3 (near-infrared) image of Lake Tahoe in the Sierra Nevada of California draped over Defense Mapping Agency Level-1 Digital Terrain Elevation Data (vertical exaggeration is 5x). In addition to Lake Tahoe, there are also several stair-stepped glaciated valleys in the southern portion of the image that contain lakes (courtesy USGS NALC database). A color version is found in Color Plate 13-8c.

high water pressure that molds the saturated till into a streamlined form. Drumlins may be 10 – 50 m high, 100 – 300 m wide and 2 – 3 km long. A small drumlin field in and around Grafton Lake in the Kejimkujik National Park of Canada is shown in Figure 13-47b.

Eolian Landforms

When wind erodes, transports, and/or deposits material it is called *eolian* (aeolian) activity. Wind typically creates both erosional and depositional landforms in coastal areas and in deserts. Eolian landforms may develop when there is an ample supply of sand and enough wind velocity to move individual soil particles. Good sources of sand include coastal and lacustrine beaches, alluvial river deposits, and material from glacial till plains. Given a strong wind and ample supply of sand, the individual grains of sand bounce along the terrain through the process of *saltation* (Selby, 1989). Most sand grains moved during saltation only rise 1 – 2 cm above the ground while a few bounce 1 – 3 m. The repeated collision of particles causes them to become increasingly rounded and smooth (and slightly smaller), which makes it even easier for the particles to become entrained in the wind field. Also, some of the energy of sal-

tating grains of sand is transferred directly to particles on the ground upon impact. This causes those grains to move forward slightly or creep. Extremely fine-grained silt smaller than 200 μm may be captured in the first 1 – 2 km of a turbulent atmosphere and moved great distances as dust. For example, approximately 10 cm of silt has been deposited in southern France from sources in Africa in the last 300 years. Thus, particles of sand are transported in the wind by suspension and saltation and on the ground by creep.

Sometimes the blowing, saltating sand encounters vegetation or other phenomena that impede its forward progress, or wind velocity decreases, causing the sand particles to be deposited. When sufficient sand is accumulated, we have a *sand dune* which is defined as a mound, ridge, or hill of windblown sand-sized particles. This mound of sand affects the flow of the air over the mound. Generally, the wind speed is greatest on the windward (upwind) side of the mound and decreases on the leeward (slipface), wind-shadow side (Figure 13-48a). The zone of reduced air velocity causes the sand to be deposited just beyond the top of the mound onto the leeward back-slope (slipface). Every size of sand grain has a different angle-of-repose, i.e., the maximum slope at which loose soil material remains stable. When this angle is reached for a particular sand (e.g., 30 – 40˚), the sand parti-

lateral
moraine →

medial moraines

lateral moraine

a. Lateral and medial moraines on Barnard Glacier, AK

terminal
moraine

lateral
moraine

b. Lateral and terminal glacial moraines on Bylot Island, Baffin Island, NW Territories, Canada.

Figure 13-46 a) High-oblique aerial photograph of the Barnard Glacier, AK. Each of the tributary glaciers has lateral moraines. The confluence of the tributary glaciers creates medial moraines. The tributary glacier on the immediate right receives material from at least one other tributary glacier, as evidenced by the two medial moraines within the center of its flow (courtesy American Geographical Society Collection archived at the National Snow and Ice Data Center, University of Colorado, Boulder, CO). b) Low-oblique photograph of lateral and terminal glacial moraines on Bylot Island, Baffin Island in the Canadian Northwest Territories.

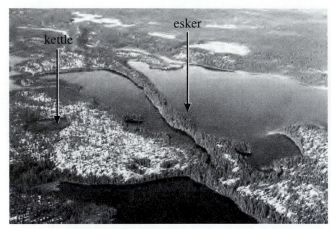

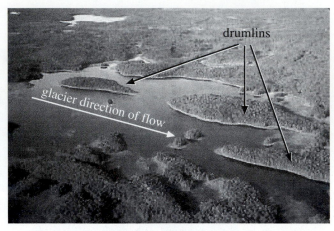

a. An esker and kettle in Northern Manitoba, Canada.   b. Drumlins in Kejimkujik National Park, Canada.

Figure 13-47 a) Low-oblique photograph of a sinuous, gravel esker and oval-shaped kettle in the boreal forest of Northern Manitoba. The light-toned area to the left and right of the esker in the foreground are sand deltas deposited by the esker into the glacial lake that was formed as the glacier retreated. b) Low-oblique aerial photograph of several drumlins in the Kejimkujik National Park of Canada (courtesy Terrain Sciences Division, Geological Survey of Canada).

cles slide down the leeward slipface causing the mound to gradually move forward in the direction of the prevailing wind.

Sand deposits may be shaped into a variety of bedforms including *ripples* (between 5 cm and 3 m apart and 0.1 – 5 cm high), *dunes* (3 – 600 m apart and 0.1 – 20 m high), and *megadunes* (300 m – 3 km apart, 200 – 400 m high, and perhaps many kilometers long). Of course, small ripples may exist on individual sand dunes and megadunes. This discussion focuses on dunes and megadunes.

Depositional Eolian Landforms

Most dunes are composed of medium-grain-sized quartz sand; however, they may also be composed of volcanic ash, calcite, and even gypsum particles. There are a variety of depositional sand dune types that have characteristic developmental cycles and diagnostic shapes that can be recognized in aerial photography or other imagery. Scientists have compiled inventories of the major types of dunes found throughout the world and how the dunes appear in remotely sensed data. For example, see the excellent work by Rinker et al. (1991), Dokka et al. (1997), and Walker (1986; 1998).

The earliest significant work on the physics, origin, and evolution of eolian sand dunes was performed by Ralph Bagnold (1941), who worked in Egypt prior to World War II. He recognized two basic dune types: the crescentic dune, which he called "barchan," and the linear dune which he called longitudinal, or "seif" (Arabic for "sword"). Unfortu-

nately, his terminology was then applied to many other types of dunes, resulting in a bewildering terminology. This section utilizes the following dune classification system adapted from several sources (Walker, 1986; 1998; Rinker et al., 1991): crescentric, linear, dome, star, and parabolic.

All dune types may occur in three forms: simple, compound, and complex. *Simple* dunes are basic forms with a minimum number of slipfaces that define the geometric type. *Compound* dunes are large dunes on which smaller dunes of similar type and slipface orientation are superimposed. *Complex* dunes are combinations of two or more dune types (Walker, 1998).

Crescentric: The crescentric dune is the most common dune form on Earth (and Mars). They are generally wider than they are long. Based on the work of Bagnold (1941), these dunes are often referred to as barchan, barchanoid, or transverse dunes. Barchan crescentric dunes such as those shown in Figure 13-48b,c may be produced when 1) the prevailing wind comes from a uniform, consistent direction, 2) there is little vegetation present to anchor any portion of the dune, and 3) there is a relatively sparse supply of sand. The two horns of the crescent point downwind. The gentle windward slope is usually < 15°. The leeward, slipface side of the barchan dune is steepest with an angle-of-repose of approximately 34 – 40° for dry sand. Individual barchans or extremely large megabarchans commonly occur in elongate chains or trains that may merge with coalesced dunes in fields or ergs. Small barchans move several meters per year, but megabarchans move more slowly (Walker, 1998).

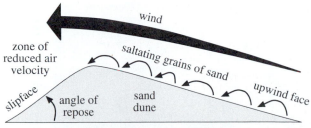

a. Sand movement on a dune.

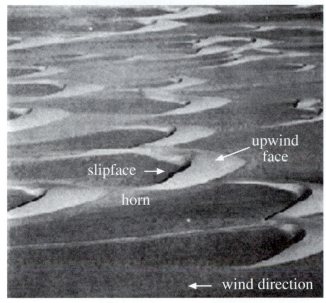

b. Oblique view of crescentric dunes
(also called barchan or transverse dunes).

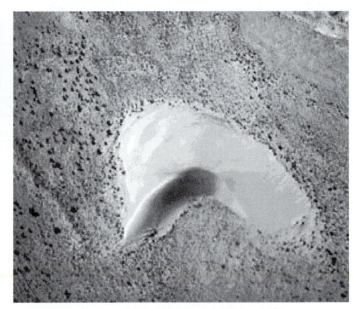

c. View of a single crescentric dune.

Figure 13-48 a) General characteristics of how sand is moved through saltation and creep on a sand dune. This illustration is keyed to the wind direction and features in (b). b) Several crescentric (barchan) dunes on the coast of Peru migrating from right to left, parallel with the wind direction (courtesy U.S. Geological Survey; Walker, 1998; and John McCauley). c) View of a single crescentric, barchanoid dune (courtesy Desert Processes Working Group; Rinker et al., 1991; Dokka et al., 1997).

Linear: Linear (or longitudinal) dunes may be straight or irregularly sinuous. A low-oblique photograph of a linear dune in the south end of Soda Lake is shown in Figure 13-49a. These dunes may be many kilometers long and hundreds of meters high. They are generally composed of loose, well-sorted very fine to medium grain sand. Extremely straight, linear dunes are often referred to as *sand ridges,* while the linear, sinuous varieties are often called *seifs.* The length of a linear dune is much greater than its width. While the crescentric dune is the most commonly occurring dune, linear dunes cover more desert area, especially in central Australia, southern Africa, the Arabian Peninsula, and parts of the Sahara.

The origin of linear dunes is controversial. Some suggest that the linear dunes are produced by unidirectional winds and that the alignment of the linear dunes is parallel with the wind direction. Others suggest that multidirectional winds also play a role in their development. It may be that the length of the dune is enlarged when the wind blows in the predominant direction. The width of the dune is enlarged when the wind blows strongly at right angles to the prevailing wind direction for relatively brief periods of time. An extensive sea of linear dunes is shown in the Space Shuttle photograph of the Marzuq Desert of Libya in Figure 13-49b.

Dome Dunes: Oval or circular mounds of sand that generally lack a slipface are called *dome dunes.* They are relatively rare. Figure 13-49c depicts a stereopair of dome dunes in the Western Shield of Saudi Arabia. The domes are partly separated by interdune hollows that are enclosed by subsidiary dunes of hard-packed sand.

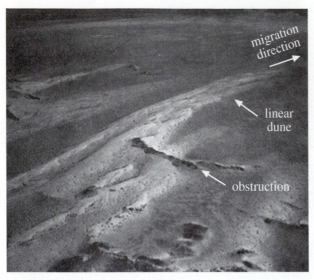

a. Oblique view of a linear (longitudinal) dune on Soda Lake.

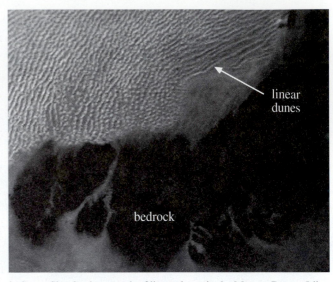

b. Space Shuttle photograph of linear dunes in the Marzuq Desert, Libya.

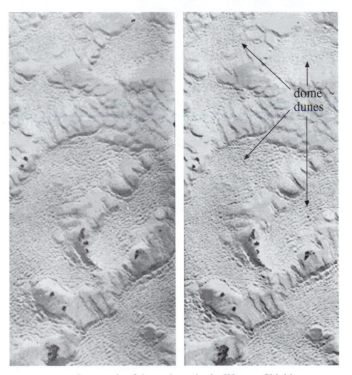

c. Stereopair of dome dunes in the Western Shield
of Saudi Arabia (Army Map Service Photos).

Figure 13-49 a) Example of a linear (longitudinal) dune. Note the physical obstruction causing the majority of sand to deviate from its course while some spills over the top (courtesy of Desert Processes Working Group; Dokka et al., 1997). b) Black-and-white version of Space Shuttle photograph STS054-152-189 obtained in January 1993, depicting an extensive sea of linear dunes. The dark areas are part of the Hammadat Marzuq (rocky desert) consisting of bedrock outcrops (courtesy NASA Johnson Space Center). c) Panchromatic stereopair of dome dunes in the Western Shield of Saudi Arabia (aerial photography obtained on October 11, 1956, by the Army Map Service; courtesy of Desert Processes Working Group; Rinker et al., 1991; Dokka et al., 1997).

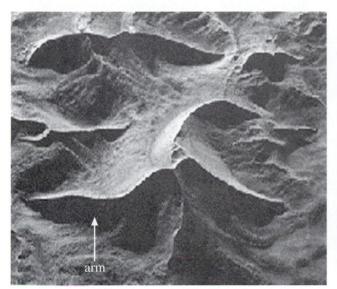

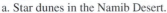

a. Star dunes in the Namib Desert.

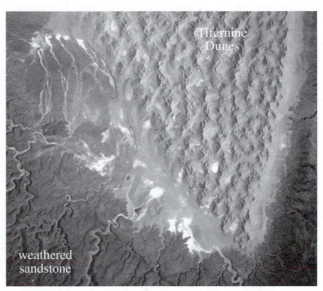

b. Space Shuttle photograph of the complex linear Tifernine Dunes in the Tassili N'Ajjer of southeastern Algeria superimposed with star dunes.

Figure 13-50 a) Large-scale vertical aerial photograph of a large star dune in the Namib desert. Each star dune consists of multiple arms each with their own slipfaces radiating from a central area. Star dunes are produced by winds coming from multiple directions (courtesy U.S. Geological Survey; Walker, 1998). b) Black-and-white version of Space Shuttle photograph STS070-705-094 of the Tifernine Dunes of Algeria in July, 1995. Extremely long, linear dunes are present. The linear dunes have star dunes superimposed on portions of them. Resistant, weathered sandstone lies to the south and east and appears in dark tones. Several white playas are present in the image (courtesy NASA Johnson Space Center).

Star Dunes: Star dunes may develop where there is an ample amount of sand and the wind blows in many directions. They are also called pyramid or radial dunes. Three or more arms of sand typically radiate from a high central mound. The arms can vary in length, width, number, and shape, but each has a slipface. Star dunes accumulate in areas with multidirectional wind regimes. Star dunes tend to increase in height rather than migrating horizontally across the landscape, causing them to be one of the tallest dunes found. They may reach 200 – 300 m in the great deserts of the world. Star dunes also occur as secondary elements on or in combination with other dunes. A single star dune in the Namib Desert is shown in Figure 13-50a, and a Space Shuttle photograph of star dunes superimposed on linear dunes in Algeria is shown in Figure 13-50b.

Parabolic Dunes: A parabolic dune has a crescent, U-shape like a barchan dune *except* that its horns point in the opposite direction into the wind. They are also called U-shaped, blowout, or hairpin dunes. Parabolic dunes are created when wind is sufficiently intense to literally blow out

the center of a dune but a relatively small amount of vegetation on either side causes the edges of the dune to be stabilized. Sometimes the horns of the parabolic dune remain attached to the vegetation for some time as the main body of the dune migrates down-wind. When this occurs, the parabolic dune resembles a hairpin lying on its side instead of a neatly shaped barchan crescent. Parabolic dunes are always associated with vegetation — grasses, shrubs, and occasional trees that anchor the trailing arms. They usually form in areas with strong unidirectional winds.

Loess: As a continental glacier retreats, it may deposit glacial till in an outwash plain. Strong winds blowing for millennia over the outwash plain may extract the small, silt-size particles. These windborne, silt-size particles may be deposited downwind from the till plain as a layer of *loess* soil. For example, the thickest known deposit of loess is 335 m on the Loess Plateau of China. Loess accumulations in Europe and in the Americas resulting from the Wisconsin glaciation are generally 20 – 30 m deep. Loess is a very highly erodable soil that is usually just a few feet thick. Drainage channels in loess soil often have very steep banks. Dunes are not generally formed on loess deposits, because of increased cohesion between the loess soil particles.

Erosional Eolian Landforms

Blowing sand can also produce impressive erosional features. The kinetic energy of wind acts upon the ground surface. Some of the energy is transferred to the ground as heat while some of it detaches soil particles from the ground and moves them along through saltation. When material is removed from a surface via wind, we have *deflation*. The continuous removal of both sand and silt from areas with an alluvial deposit leaves only a combination of coarse sand and pebbles and some clay material. This can result in the creation of a *desert pavement* landscape which is a sheet-like surface of rock fragments that remain after wind and water have removed the fine particles. Almost half of the Earth's desert surfaces are stony deflation zones. Desert pavement regions are called *regs* in Algeria, *serir* in Libya, and *gibbers* in Australia. These deflated regions may be very extensive in large deserts. A deflated basin is called a *blowout*. Sometimes a dark, shiny desert varnish is found on the desert pavement surfaces produced from manganese, iron oxides, hydroxides, and clay minerals.

The surfaces of rock bodies may also be eroded by abrasion and sandblasting by windborne particles. In extreme cases the wind may be so strong that it cuts channels in the less resistant rock, leaving residual streamlined spines called *yardangs,* which may be tens of meters high and kilometers long.

 References

Avery, T. E. and G. L. Berlin, 1992, *Fundamentals of Remote Sensing and Airphoto Interpretation*, New York: Macmillan, 377–404.

Bagnold, R. A., 1941, *The Physics of Blown Sand and Desert Dunes*, London: Methuen, 265 pp.

Busby, R. B., R. R. Coenraads, P. Willis and D. Roots, 1996, *Rocks & Fossils*, New York: Time Life Books, 288 pp.

Butler, D. R. and S. J. Walsh, 1997, "The Application of Remote Sensing and Geographic Information Systems in the Study of Geomorphology," *Geomorphology*, 21 (3-4):179–182.

Campbell, J. B., 1997, *Introduction to Remote Sensing*, New York: The Guilford Press.

Caylor, J. A. and H. M. Lachowski, 1988, *How to Use Aerial Photographs for Natural Resource Applications*, Washington: USDA Forest Service.

Clark, R. N., 1999, "Spectroscopy of Rocks and Minerals, and Principles of Spectroscopy," in the public domain at http://speclab.cr.usgs.gov/PAPERS.refl-mrs/refl4.html and published in *Manual of Remote Sensing – Geosciences,* A. Rencz (Ed.), 1999, New York: John Wiley & Sons.

Clark, R. N., G. A. Swayze, A. Gallagher, T. V. King and W. M. Calvin, 1993, "The USGS Digital Spectral Library: Version 1: 0.2 - 3.0 µm," *U.S. Geological Survey Open File Report*, 93-592, 1362 pp.

Davis, B. A. and J. R. Jensen, 1998, "Remote Sensing of Mangrove Biophysical Characteristics," *Geocarto International*, 13(4):55-64.

Dokka, R. K., L. Muyer, J. Rinker, J. Watts, G. Wakefield, J. F. McCauley and C. Breed, 1997, *MDEP Surface Materials Mapping Group Database Design Document Version 2 (MDEI-2000).*

Ellis, J., 1999, "NASA Satellite Systems for the Nonrenewable Resource Sector," *Earth Observation Magazine*, 8(3):46–48.

Goetz, A. F. H., B. N. Rock and L. C. Rowan, 1983, "Remote Sensing for Exploration: An Overview," *Economic Geology*, 78(4):573–588.

Hapke, B., 1993, *Introduction to the Theory of Reflectance and Emittance Spectroscopy*, New York, NY: Cambridge University Press.

Hoffer, R., 1978, "Biological and Physical Considerations," *Remote Sensing: The Quantitative Approach*, 241–251.

Hunt, G. R., 1977, "Spectral Signatures of Particulate Minerals in the Visible and Near-Infrared," *Geophysics*, 42: 501–513.

James, A., 1999, correspondence, geomorphologist in the Dept. of Geography, University of South Carolina, Columbia, SC.

Jensen, J. R., 1996, *Introductory Digital Image Processing: A Remote Sensing Perspective,* Upper Saddle River: Prentice-Hall, 318 pp.

Julien, R. Y., 1995, *Erosion and Sedimentation*, New York: Cambridge University Press, 279 pp.

Kruse, F. A., K. S. Kierein-Young and J. W. Boardman, 1990, "Mineral Mapping at Cuprite, Nevada with A 63-channel Imaging Spectrometer," *Photogrammetric Engineering and Remote Sensing*, 56(1):83–92.

Kunickis, S. H., 1999, *Using a Soil Survey*, Raleigh: North Carolina State Univ., http://www2.bae.ncsu.edu/courses/bae572/Special-Reports/kunickis/Soil_survey.html.

Loynachan, T. E., K. W. Brown, T. H. Cooper and M. H. Milford, 1999, *Sustaining Our Soils and Society*, Alexandria: American Geological Institute, Soil Science Society of America, and USDA, 66 pp.

Marsh, W. M. and J. Dozier, 1981, *Landscape: An Introduction to Physical Geography*, Reading: Addison Wesley, 637 pp.

Mojave Desert Ecosystem Program, 1998, *Mojave Desert Ecosystem Program Electronic Database*, Washington: U.S. Department of Defense and U.S. Department of Interior, 8 CDs and maps.

Morrissey, L. A., K. J. Weinstock, D. A. Mouat and D. H. Card, 1984, "Statistical Analysis of Thematic Mapper Simulator Data for the Geobotanical Discrimination of Rock Types in Southwest Oregon," *IEEE Transactions on Geoscience and Remote Sensing*, GE-22(6):525–529.

Petersen, G., 1999, correspondence, President of the Soil Science Society of America, Pennsylvania State University.

Pickup, G. and V. H. Chewings, 1988, "Forecasting Patterns of Soil Erosion in Arid Lands from Landsat MSS Data," *International Journal of Remote Sensing*, 9:69–84.

Rasher and Weaver, 1990, *Basic Photo Interpretation*, Washington: Soil Conservation Service with photographs and slides.

Rencz, A., (Ed.), 1999, *Manual of Remote Sensing – Geosciences*, New York: John Wiley & Sons.

Rinker, J. N., C. S. Breed, J. F. McCauley and F. A. Cord, 1991, *Remote Sensing Field Guide - Desert*, Ft. Belvoir: U.S. Army Engineer Topographic Lab, 568 pp., http://www.tec.army.mil/terrain/desert/coverpg.htm.

Salisbury, J. W., L. S. Walter, N. Vergo and D. M. D'Aria, 1991, *Infrared (2.1 - 25 μm) Spectra of Minerals*, Baltimore: John Hopkins University Press, 267 pp.

Schwaller, M. R., 1987, "A Geobotanical Investigation Based on Linear Discriminate and Profile Analyses of Airborne Thematic Mapper Simulator Data," *Remote Sensing of Environment*, 23:23–34.

Selby, M. J., 1989, *Earth's Changing Surface: An Introduction to Geomorphology*, Oxford: Clarendon Press, 607 p.

Shelton, J. S., 1966, *Geology Illustrated*, San Francisco: W. H. Freeman, 434 pp.

Short, N. M. and R. W. Blair, 1986 and 1999, *Geomorphology from Space, A Global Overview of Regional Landforms*, NASA: Washington, DC, 716 pp. An Internet version of the book became available in 1999 at http://daac.gsfc.nasa.gov//DAAC_DOCS/daac_ed.html.

Siegel, B. S. and A. R. Gillespie, 1980, *Remote Sensing in Geology*, New York: John Wiley & Sons, 702 pp.

Spitzer, W. G. and D. A. Klienman, 1960, "Infrared Lattice Bands of Quartz," *Physical Review*, 121:1324–1335.

Strain, P. and F. Engle, 1993, *Looking at Earth*, Atlanta: Turner Publishing, 304 pp.

Swayze, G. A., R. N. Clark, A. F. H. Goetz, K. E. Livo and S. S. Sutley, 1999, "AVIRIS 1998 Low Altitude Versus High Altitude Comparison Over Cuprite, Nevada," Denver: U. S. Geological Survey, Speclab Home Page, http://speclab.cr.usgs.gov/cuprite98-low/cuplow+high.html.

Trefil, J. and R. M. Hazen, 1995, *The Sciences: An Integrated Approach*, New York: John Wiley & Sons, 634 pp.

USDA, 1998, *Keys to Soil Taxonomy*, Washington: U.S. Department of Agriculture, Natural Resources Conservation Service, 8th ed., 371 pp.

Van der Meer, F., 1994, "Extraction of Mineral Absorption Features from High-Spectral Resolution Data Using Non-Parametric Geostatistical Techniques," *International Journal of Remote Sensing*, 15:2193–2214.

Vincent, R. B., 1997, *Fundamental of Geological and Environmental Remote Sensing*, Upper Saddle River: Prentice-Hall, Inc., 366 pp.

Walker, A. S., 1986, "Eolian Landforms," in Short, N. M. and R. W. Blair, *Geomorphology from Space, A Global Overview of Regional Landforms*, NASA: Washington, DC, 447–520.

Walker, A. S., 1998, *Deserts: Geology and Resources*, Washington, DC: USGS, http://pubs.usgs.gov/gip/deserts/contents.

Walsh, S. J., D. R. Butler and G. P. Malanson, 1997, "An Overview of Scale, Pattern, Process Relationships in Geomorphology: A Remote Sensing and GIS Perspective," *Geomorphology*, 21 (3-4):183–206.

Way, D. S. and J. R. Everett, 1997, "Chapter 3: Landforms and Geology," *Manual of Photographic Interpretation*, 2nd Ed., Bethesda: American Society for Photogrammetry & Remote Sensing, 117–165.

Index

Absolute zero, 31, 243, 246
Absorption
    bands,
        atmospheric, 42–43, 247, 334–335, 340
            extinction coefficient, 42
        rocks and minerals, 44–46, 471
            carbonate, 433
            clay, 433
            hydroxyl, 478
        vegetation
            carotene, 337–338, 388
            chlorophyll *a* and *b*, 335–339, 388–391
            phycoerythrin, 337–338
            phycocyanin, 337–338
        water, 381–384, 393
Across-track scanners, see Scanning
Active microwave, see Radar
ADAR 5500 sensor system, 12, 14, 233–235
Advanced Spaceborne Thermal Emission and Reflection Radiometer (ASTER), 221–222, 246, 254, 483
Advanced Very High Resolution Radiometer (AVHRR), 205–208, 244
Aerial perspective, 120
Aerial photography cameras
    focal plane, 88
    focal length, 88
    f/stop, 88–89
    image motion compensation, 93
    intervalometer, 91
    lens angle of view, 91–93
    shutter, 87–88
    shutter speed, 89–90
    types of
        digital, 95, 97
        multiple-lens, multi-band, 93, 95–96
        panoramic, 95
        single-lens mapping, 91–93
Aerial photography films, 102–116
    black-and-white
        infrared, 104, 106
        panchromatic, 103–104, 106
        orthochromatic, 103
    color
        color, 105, 114–115
        color-infrared, 114–116
    history of, 53–57
    negative/positive process, 104–107
    silver halide crystals, 102–104
    speed, 109
Aerial photography filtration, 97–102
    band-pass filtering, 99

haze filter, 115
  Kodak Wratten, 97–99, 115
  minus-blue, 99
polarizing filters, 99–102
Aerial photography platforms
  aircraft, 64, 66
    Wright brothers, 64, 66
    WWI and WWII, 64–69
  balloons, 58, 70
    Genetrix, 70
  civil war, 58
  cold war, 69–74
    U-2 aircraft, 71–73
    SR-71 aircraft, 72–74
    Corona satellite, 74–78
  gliders, 64, 66
  kites, 61–63
  ornithopters, 57
  pigeons, 63–65
  rockets, 63, 69
  unmanned aerial vehicles, 79–80
Aerosols, remote sensing of, 396–397
Agriculture NonPoint Source (AGNPS) pollution model, 402
Agriculture, remote sensing of, 350–352, 356–361
Air base, 157
Airborne Terrestrial Applications Sensor (ATLAS), 244, 254, 269, 270, 278–280
Airborne Visible/Infrared Imaging Spectrometer (AVIRIS), 228–230
Aircraft, see Aerial photography platforms
Alluvial deposits, 484
  fans, 509–510
  bajadas, 509–510
Alpine glaciation, see Glaciation
Altitude above-ground level (AGL)
  scale related to, 146
American Planning Association, 413
Analog-to-digital conversion, 1, 110
Ancillary information, 133
Angle of view
  oblique
    low, 86
    high, 87
  vertical, 85–86, 143
Anastomotic drainage pattern, 490
Annular drainage pattern, 488–489
Antenna, see Radar system components
Anticlinal ridges, 497
Area arrays, 112–113, 265
Area measurements, 174–178
  dot grid, 176–177
  on-screen digitization, 177–178
  polar planimeter, 175
Area weighted average resolution (AWAR), 411
Arete, 517, 519
Atmosphere
  absorption and transmission in, 42-43
  windows, 42–44, 246–247
  radar penetration of, 316
  scattering in, see Scattering

Backscatter, see Radar, terrain

Badland topography, see Drainage patterns, pinnate
Balloon photography, 58, 70
Bands of the electromagnetic spectrum, 12, 34, 37
  infrared
    reflective, 34, 114–116, 246–247, 382
    middle, 186, 192, 194–200, 342-344
    thermal, 34–36, 246–247
  microwave, see Radar system, wavelength (or frequency)
  ultraviolet, 34–36, 98, 382
  visible, 12, 34–36, 98, 103, 114, 382
Bar
  parallax, 159, 161–162
  shoreline, 513
Barrier island, 513, 515
Base, film, 103–104
Batholiths, 493
Beach, 513, 517
  arches, 517
  cliffs, 517
Bedrock, 472–473
Biodiversity, 368–369
Biophysical variables, 10–12
Black-and-white photography, 103–106
Blackbody, 30, 247
Block of aerial photography, 138–139
Braided drainage pattern, 489–490
Brightness temperature, see Passive microwave
Building information, 410, 422–423

Cadastral information, 410, 422–423
Cameras, see Aerial photography cameras
Camouflage detection film, see Aerial photography films
Canopy,
  closure percent, 1–2, 344–345
Cardinal effect, see Radar, terrain
Centrifugal drainage pattern, 488–489
Centripetal drainage pattern, 488–489
Change detection, 121, 370–373, 324–325
Characteristic curve, 108–109
Chlorophyll, see Vegetation and Water
Chloroplasts, 334
Cinder cone volcanoes, 492
Cirque, 517, 519
Classification
  land use and land cover, 409–413
  schemes, 413–414
Close range photogrammetry, 164
Clouds and Earth's Radiant Energy System (CERES), 20, 401
Clouds, remote sensing of,
  height, 399–400
  in thermal infrared imagery, 399–401
  in visible imagery, 397–399
  type, 401
Coastal Zone Color Scanner (CZCS), 244
Collateral information, 133
Col, a glacial landform, 517, 519
Color
  additive theory, 97–98
  balance shift in color-infrared film, 116
  complimentary, 97, 114
  false, see Aerial photography films, color-infrared
  history of, 53, 97

perception of, 128
photography, see Aerial photography films
primary colors, 97
subtractive theory, 98–99
Color photography, see Aerial photography films
Commercial land use, interpretation of, 426–438
automotive/boat, 427
central business district (CBD), 426–427
finance and construction, 427–428
food and drug, 429–430
funeral and cemetery, 430–431
house and garden, 431–433
housing (temporary), 431
other commercial, 434–439
recreation, 433–434
warehousing/shipping, 434
Communications and utilities, interpretation of, 410, 460–463
Compact Airborne Spectrographic Imager-2 (CASI-2), 228–231
Conduction, 29, 246
Cone, volcano
cinder, 492
composite, 493–494
Conglomerate, 495
Conjugate principal point, 140–142
Continental drift, 483
Convection, 29, 246
Convergence of evidence, 134
Coral reef, 515, 517–518
atoll, 517–518
barrier, 517–518
fringing, 517–518
Corona satellite, 74–78
Critical environmental areas, interpretation of, 410, 465
Crop calendars, see Vegetation phenological cycles

Data Analysis
analog (visual image processing), 21–22
digital image processing, 21–23
Data Collection
in situ, 1-2
remote sensing, 2-8
Defense Meteorological Satellite Program (DMSP), 425
Deformation of rocks, 484–485
Deltas, 506, 508
crenulate, 506, 508
cuspate, 506
elongated or digitate, 506, 508
lobate, 506, 508
Dendritic drainage pattern, 487–488
Densitometer, 110
Density, 107–108
Depression angle, see Radar system components
Depth perception, 120
Detectors, 184, 244, 256–257
Diastrophism, 497
Dichotomic drainage pattern, 489–490
Dielectric constant, complex, 311–312
Differential parallax, see Parallax, stereoscopic
Digital elevation model (DEM), 86, 138, 167, 463–464
Digital frame camera, 231
Positive Systems, Inc., 233–235
Litton Emerge Spatial, Inc., 235–236

Digitizing, 110–111
linear or area array, 112–113
microdensitometer, 110-112
video, 112
Digital image processing, 8
digital photogrammetry, 8, 22
expert systems and neural networks, 8, 23, 418–422
hyperspectral data analysis, 8, 22–23
modeling, 8, 23–24
land use/land cover classification, 408–426
pattern recognition, 8, 22
rectification, 122
terminology, 182–183
Dip, 485–486, 500, 502
Disaster emergency response, 410, 465–466
Distributaries, 489
Doline (sinkhole), 509, 511
Dome
lava, 492–493
sand dune, 525–526
Doppler frequency in radar, 303–304
Dot grids, 176–177
Drainage patterns
anastomotic, 490
annular, 488–489
braided, 489–490
dendritic, 487–488
deranged, 489
dichotomic, 489–490
parallel, 488–489
pinnate, 487
radial (centrifugal) and centripetal, 488–489
rectangular, 487–488
sinkhole (doline), 490
trellis, 487–488
Drumlin, 521, 524
Dunes, see Sand dunes

Earth Observation Satellite (EOSAT) Corporation, 20, 186
Earth Observer (EO-1) NASA
Advanced Land Imager (ALI), 200–201
Hyperion hyperspectral sensor, 200–201
LEISA atmospheric corrector, 200–201
Earth Observing System (EOS) - NASA
Landsat 7 Enhanced Thematic Mapper Plus, 197–201, 246
science plan, 18-20
Terra sensor systems
ASTER, 13, 20, 221–222, 254
CERES, 20, 401
MISR, 13, 20, 222–224
MODIS, 13, 20, 231–232
MOPPETT, 20
Earth Resource Technology Satellite (ERTS), see Landsat
EarthWatch, Inc., Earlybird and Quickbird, 224–224
Eastman, George, 57
Elastic deformation, 484
Electromagnetic radiation, 29–30, 246
absorption
by atmospheric gases, see Atmosphere
by surface features, see Energy-matter interactions
budget equation, 44
creation of, 31

flux density, 46
   exitance, 47
   irradiance, 47
frequency, 30-32
interactions, see Energy-matter
models
   particle, 35–39
   wave, 30–35
radiance, 47–48
radiant flux, 44, 246
refraction, 39
speed of light, 30–31
units of measurement, 33–35
wavelength, 30-32
Elements of image interpretation, 8, 121–133
   association, 121, 132, 418
   height/depth, 120, 131–132, 138, 152
   pattern, 121, 130–131
   shadow, 121, 125–126
   shape, 121, 124–125
   site, 121, 132, 418
   situation, 121, 132, 418
   size, 121–123
   slope/aspect, 121, 170
   texture, 121, 128–130
   tone/color, 120–121, 126–128, 140
   x,y location, 121–122
Elevation data, see Digital elevation model
El Nino, sea-surface temperature, 393–394
Emissivity, 248–253, 277, 249
Emulsion, see Aerial photography films
Endlap, stereoscopic, 138
Energy, 29
Energy-matter interactions
   hemispherical absorptance, 45–46
   hemispherical reflectance, 44, 46
   hemispherical transmittance, 44, 46
Eolian features, 522–528
Esker, 521, 524
Estuaries, 513
Exfoliation, dome, 493–495
Exposure
   film, 88, 107–110
   stations, 138-139
Eye base, 152

False color, see Aerial photography films, color-infrared
Fan, alluvial, 509–510
Faults, geologic, 485–486, 500, 503
Fiducial marks, 140
Field of view, see Instantaneous field of view
Flightline, 138, 142–143
Film, see Aerial photography films
Filters, see Aerial photography filtration
FLIR (forward looking infrared systems), 243–244, 266
Floodplains, 504–506
   composite, 506
   covered, 506
   meander, 506
Fluvial landforms, 503–509
Focal length, see Aerial photography cameras
Focal plane, see Aerial photography cameras

Folding, 484–485
Foreshortening in radar images, 300–301
Forestry, see Vegetation
Fracturing of rock, 484
f/stop, see Aerial photography cameras

Gap analysis, 368–369
Geographic Information Systems (GIS) , 7, 10, 23
Geology, 483–490
   drainage
      density, 485
      pattern, 487–490, see Drainage patterns
      texture, 487
   history, 483–484
   lithology, 484
   structure, 484
      faulting, 485–486, 500, 503
         normal, 486, 500, 503
         oblique-slip, 486
         reverse, 500, 503
         scarp, 500
         strike-slip, 486, 500
         thrust, 503
      folding, 484–485
         anticline, 484
         monocline, 484
         overturned, 485
         syncline, 485
Geometry of a vertical aerial photograph, 140–144
Geomorphology, 490–528
Geostationary Operational Environmental Satellite (GOES), 201
   GOES Imager, 201–204, 397–398, 464–465
   GOES Sounder 203–205, 397
Glacial till, 521
Glaciation
   alpine, 517
   continental, 517
Global Positioning Systems (GPS), 122, 269
Graben, 500, 503
Graybody, 249
Ground-range, see Radar, geometry
Ground reference information, 2

Hanging valley, 520
Haze, see Atmosphere
Heat Capacity Mapping Mission (HCMM), 254
Heat loss, thermal infrared images of
   buildings, 274–278
   buried heating lines, 261
Height measurement using
   air photo relief displacement, 148–150
   shadow length, 149–150
   stereoscopic parallax, 151, 157–163
Highways, see Transportation infrastructure
Hogback, 497
Hook, 512–513
Horn, 520
Horst, 500, 503
Hotspots in aerial photographs, 346–347
Housing, see Residential
Hyperspectral remote sensing, 181, 226–232,
Hyperstereoscopy, 152

535

Igneous rocks, 484, 492
 extrusive, 492
 intrusive, 492
IKONOS, 224–226
Image
 annotation, 93
 enhancement, 274–275
 interpretation
  analog or visual, 8, 119–136
  digital, see Digital image processing
Imaging spectroscopy, 228–230, 480–483
 imaging spectrometers, 480–483
  see AVIRIS, CASI-2, MODIS
 reflection and absorption processes, 480–483
  electronic processes, 482
  vibrational processes, 482
 spectral reflectance libraries, 482
Incident angle, see Radar system components
Index of refraction, 39-40
Indian Remote Sensing (IRS) Satellite program, 220–221
Industrial land use, interpretation of, 443–456
 classification logic, 443
  manuals, 443
 extraction industries, 443–444
 fabrication industries, 453–456
  heavy, 455-456
  light, 456
 processing industries, 444–453
  chemical, 450–451
  heat, 451–453
  mechanical, 444–449
Infrared
 reflective, 34, 114–116, 246–247, 382
 middle, 186, 192, 194–200
 thermal, 34–36, 246–247
Infrared images,
 photography, see Aerial photography films, infrared
 thermal, see Thermal infrared remote sensing
In situ data collection, 1–2, 263–264, 271, 273, 379, 480
Instantaneous field of view, 111, 255–259
Intervalometer
Inverse-square law, 257–258

Karst topography, 490, 509–511
Kettle, 521, 524
Kinetic temperature, 246, 251–252, 263
Kirchoff's radiation law, 250–251
Kodak aerial films, see Aerial photography films
Kodak Wratten filters, see Aerial photography filtration

Lacoliths, 493
Lagoons, 513,518
Lambertian surface, 44
Land cover, see Land use/land cover
Landforms, interpretation of, 490–528
 developed on horizontal strata, 495–497
 developed on folded strata, 497–500
 eolian, 522–528
 fault-controlled, 500–503
 fluvial, 503–509
 glacial, 517–522

 igneous, 492–495
 karst, 490, 509–511
 shoreline, 511–517
Landscape ecology, 366–369
 indicators, 367
 patch metrics, 367–369
Land use/land cover, 409-411, 413
 classification schemes, 413–414
  APA Land-Based Classification Standard, 413
  U.S. Geological Survey, 413–418, 426
 definition of, 413
 developmental cycles of, 408, 412
 form and function of, 418
Landsat, 184–201
 Enhanced Thematic Mapper Plus (ETM+), 197–201, 246
 history of, 184–186, 197–201
 Multispectral Scanner (MSS), 188–192
 Return Beam Vidicon (RBV) camera, 188
 Thematic Mapper, 192–197, 244
 Worldwide Reference System (WRS), 188
La Nina, sea-surface temperature, 393–394
Latent image, 55
Leaf, see vegetation
Leaf area index (LAI), 1–2, 345
LIDAR (Light Detection and Ranging), 285, 326-329
 accuracy of measurements, 328–329
 canopy penetration, 329
 sensor, 327–328
Limestone, 496
Line of flight, see Flightline
Linear Imaging Self-Scanning Sensors (LISS), 220–221
Linear arrays, 112–113, 183, 212, 265
Litton Emerge Spatial, Inc., 235–236
Loess, 527–528
Look angle, see Radar
Look direction, see Radar

Maps
 distances on, compared to photo distances, 142–144
 of land use/land cover, see Land use/land cover
 orthophoto and imagemaps, 22, 24, 164–174
 transfer of photointerpreted information to, 155–157
Marine terraces, 513–515
Mesophyll, see Vegetation
Metamorphic rocks, 484
Metamorphism, 497
Microdensitometer, 110
Mission Planning, 116–117
Moderate Resolution Imaging Spectrometer, 231–232, 392
Moisture content of
 soil, 249
 plant (turgidity), 342
Monoclines, see Geology, folding
Moraines, 520
 lateral, 520, 523
 medial, 520, 523
 recessional, 520
 terminal, 520, 523
Motion, compensation of in images, 93
Mudstone, 495
Multiband imaging, see Multispectral remote sensing systems
Multi-angle Imaging Spectroradiometer (MISR), 222–224, 349

Multi-concept in remote sensing, 134–135
    multidisciplinary, 134–135
    multiscale, 134
    multispectral, 134
    multitemporal, 134
Multispectral remote sensing systems, 181–241
Multispectral scanners, aircraft
    Daedalus DS-1260, DS-1268, and AMS, 209–212
    see Airborne Terrestrial Applications Sensor (ATLAS)

National Aerial Photography Program (NAPP), 113, 411
National High-Altitude Photography (NHAP), 113
Newton, Sir Isaac, 53
Niepce, M. Joseph Nicephore, 54
NIIRS (National Image Interpretation Rating Scale), 411, 414
Nonselective scattering, see Scattering
Normalized Difference Vegetation Index (NDVI), 362
North American Industrial Classification Standard, 413

Oblique view, see Angle of view
Opacity, 107–108
Orbits
    geosynchronous, e.g., GOES, 201–204
    Sun-synchronous, e.g., Landsat, 188–192
ORBIMAGE, Inc., Orbview-3 and Orbview-4, 226
Orthophotos/orthoimages, 86, 122
    differential correction, 166–167
    history, 164–165
    improvement in digital elevation model, 167–168
    soft-copy photogrammetry production of, 167–169
    traditional, 172–173,
    true, 173–174
Overlapping photographs
    stereoscopic viewing of, 153–157
    uncontrolled mosaics of, 139–140

Paleontology, 483
Panchromatic images, see Aerial photography films
Parallax, stereoscopic, 151
    absolute, 149
    differential, 151, 158–159
    measurement of
        direct, 160–161
        parallax bar (stereometer), 159, 161–162
    parallactic angle, 152
    principle of the floating mark, 161–164
Parallel drainage pattern, 488–489
Passive microwave, 325–326
    brightness temperature, 325
    radiometers, 325
        Advanced Microwave Scanning Radiometer, 325
        Special Sensor Microwave/Imager (SSM/I), 325–326
        Tropical Rainfall Measuring Mission (TRMM), 326-327
Path radiance 48-51, 380
Pattern recognition, 22
Pediments, 509
Photography
    history, 53–57
        additive color combining, 57
        calotype, 56
        camera obscura, 54
        collodion wet-plate, 56

        color film, 57
        daguerreotype, 54–55
        heliograph, 54
        latent image, 55
        positive-negative process, 56-57
Photographic interpretation
    definition of, 2, 119
    elements of, 121–133
Photogrammetry, 137–180
    analog, 137
    definition of, 2, 137
    digital, 137
Photo mensuration, 137
Photomaps (imagemaps), 24
Photons, 37
Photosynthesis, 333-334
Pinnate drainage pattern, 487
Pixels, 182–183
Planimetric map, 165
Plankton, 388–390
    bacterioplankton, 388
    phytoplankton, 388, 390
    zooplankton, 388
Plastic deformation, 484
Plate tectonics, 483
Plateaus, 496, 499–500
Polarization, 99–102, 290–293
Pollution
    nonpoint source, 271–274, 402–404
Population estimation, 422–424
Positive Systems, Inc., 233–235
Precipitation, remote sensing of
    active and passive microwave techniques, 395–396
    visible-infrared techniques, 395
Primary colors, 97
Principal point of aerial photographs, 140–142
Pseudoscopic illusion, 126
Pushbroom linear array detectors, see Linear arrays

Quality of
    life indicators, 424–425
    orthophotos, 170–174
Quantization, 182

Radar, 285–331
    advantages of, 289
    atmospheric penetration, 316
    equation, 305–308
    geometry
        foreshortening, 300–301
        ground-range display, 292–294
        layover, 300–302
        relief displacement, 299–300
        shadows, 300–302
        slant-range display, 292-294
    history, 285–288
    interferometry, 323–325
        topographic mapping, 323–324
        velocity mapping, 324–325
    radargrammetry, 286
    side-looking airborne radar (SLAR), 286, 288
    system components, 288–289

antenna, 286
azimuth flight direction, 290–291
azimuth resolution, 295–299
depression angle, 290–291
Doppler principle, 303–304
frequency, 287–290
hologram, 305
incident angle, 290–292
polarization (HH, VV, VH, HV), 290–293
pulse length, 289, 295
range or look direction, 290–291
range resolution, 295–297
real aperture, 286, 288
speckle, 302–303
synthetic-aperture, 286, 288, 302–308
    correlation (optical and digital), 305
wavelength, 287–290
terrain
    backscatter coefficient, 308, 313–314
    complex dielectric constant, 311–312
    cross-section, 307
    surface roughness response to radar energy, 308–310
    urban structure response to microwave energy, 317–319
        cardinal effect, 317
    vegetation response to microwave energy, 312–317
        penetration, 312–315
        surface scattering, 312–315
        volume scattering, 312–315
    water response to microwave energy, 317
        soil moisture, 317
Radar sensors
ALMAZ-1, 287, 323
ERS-1,2, 287, 322
JERS-1, 287, 322
RADARSAT, 287, 319–321
SEASAT, 287, 319
SIR-A, -B, and SIR-C, 287, 319
STAR 3i, 288, 324
Radial drainage pattern, 488–489
Radiance,
    path, 48–51
    target, 46–51
Radiant energy, 39,
Radiant flux, 44, 246, 334
Radiant temperature, 246
Radiation, see Electromagnetic radiation
Radiometers
    handheld, 1–2, 263
    see Advanced Very High Resolution Radiometer (AVHRR)
    see Passive microwave
Radiometric variables, 48–49
Radiosonde, 263
Railroads, see Transportation infrastructure
Rain, see Precipitation
Rayleigh criterion, see Radar terrain surface roughness
Rectangular drainage pattern, 487–488
Reflectance
    diffuse, 44
    hemispherical, 44–46
    near-perfect, 44
    specular, 44
Refraction, 39, 480–481

Relief displacement on
    aerial photographs, 148–149
    scanning systems, 258–259
Remote sensing
    advantages/limitations, 7-8
    art or science, 4-5
    biophysical variables to be sensed, 10-11
    data collection, 1-8
    definitions of, 2-4
        ASPRS, 2-3
        maximal and minimal, 3,4
    distance, how far is remote, 7
    history of, 1-3
    instruments, 5, 13
    logic, 4-5
    milestones in, 5-7 (Table 1-1)
    process, 8-25
    resolution
        radiometric, 16, 210
        spatial, 15-16, 192, 194–195, 200, 203, 210, 409-413
        spectral, 12-15, 194–195, 200, 203, 206–209, 408-411
        temporal, 16, 194–195, 200, 408–410
    statement of the problem, 9-10
Representative fraction, 142-144
Residential land use, interpretation of, 410, 418–426
    energy demand and conservation, 410, 425–426
    energy surveys, 274–278
    single- and multi-family, 410, 418–422
Return Beam Vidicon (RBV) camera, see Landsat
Rocks, lithology of
    igneous, 484
    metamorphic, 484
    sedimentary, 484
Rocks and minerals, remote sensing of, 479
    imaging spectroscopy, 228–230, 480–483
        imaging spectrometers, 228–230, 480–483
            see AVIRIS, CASI-2, MODIS
        reflection and absorption processes, 480–483
            electronic processes, 482
            extinction coefficient, 481
            index of refraction, 480–481
            vibrational processes, 482
    spectral reflectance libraries, 482
Russian SPIN-2 TK-350 and KVR-1000 Cameras, 235–238

Saltation, 522
Sand dunes, 522–528
    types
        dunes, 522–528
        crescentric, 524–525
        dome, 525–526
        linear, 525, 527
        megadunes, 524
        parabolic, 527
        ripples, 524
        star, 527
    radar penetration of, 311–312
Sandstone, 495–496
Scale of a vertical aerial photograph, 142–145
    terrain
        level, 144–145
        variable, 147–148

representative fraction, 142–144
verbal scale, 142–144
Scanning
across-track, 256
Scattering
atmospheric
mie, 42
non-selective, 43
rayleigh, 41-42
surface
in radar imagery, 308–310
in water column, 380–381
SEASAT, 287, 319
Sea-viewing Wide Field of View Sensor (SeaWiFS), 208–209, 391
Sedimentary rocks, 484
Selectively radiating bodies, 248–249
Services land use, interpretation of
public buildings and facilities, 439–442
Shadow, 121, 125–126
Shale, 496–497
Shape, 121, 124–125
Shuttle Imaging Radar (SIR), 287, 319
Sidelap, stereoscopic, 138-139
Side-looking airborne radar, see Radar
Sigmoid or s-shape distortion, 262
Silver halide, see Aerial photography films
Sinkholes, see Landforms, karst
Size, 121–123
Slant-range, see Radar
Space Imaging, Inc., IKONOS, 224–226
Space Shuttle photography, 239
Special Sensor Microwave/Imager (SSM/I), 325–326
SPOT Image Corporation, 212–220, 370–373
history, 212–214
SPOT 1,2,3 Resolution Visible (HRV) sensors, 213–217
SPOT 4 and 5, 217–220
Snow, remote sensing of
microwave region, 402
middle-infrared region, 401–402
visible spectrum, 401
Socioeconomic information, 422–426
Soil
definition of solum, 472
grain size (sand, silt and clay), 473–474
horizons, 472–473
bedrock, 472–473
zone of eluviation, 472–473
zone of illuviation, 472–473
reflectance dominant factors,
air-soil interface scattering/reflectance, 475, 477
iron oxide, 478–479
moisture content, 476–478
organic matter, 478–479
subsurface volume scattering/reflectance, 475
surface roughness, 479
texture, 473–476
hydroxyl absorption bands, 478
interstitial air spaces, 477
taxonomy, 474
Soils and rocks
active microwave response to surface roughness, 308–310
thermal response of, 266–269

SONAR, 285
Spectral bands, see Bands of the electromagnetic spectrum
Spectral reflectance, 126–127
Spectral signature, 126–127
Spectroradiometers
airborne, see AVIRIS, CASI-2, MODIS
handheld, 1–2
Spectrum
electromagnetic, 35–36, 39
Spit, 512
SR-71 aircraft, 72–74
Standard Land Use Coding Manual, 413-414
Star dunes, 527
Stefan-Boltzmann law, 31, 247–248
Stereoscopy
methods of viewing, 152–157
principles, 151-152
Stereometer, see Parallax bar
Stereoscopic model, 138, 153, 156
Stereoscopic viewing
alignment of photographs in, 155–157
anaglyphic or polarizing glasses, 154
crossed-eyes, 153–154
parallel-eyes, 153–154
using a stereoscope, 153–154
Stereopair, 153
Stereoplotters, 163
Stereoscopes
lens, 154–157 163
mirror, 154–156
Stereoscopic aerial photography, 137-138
Strike, 486
Sunglint, 380
Sun-synchronous orbit, 186
Surface roughness, 249
Surficial deposits, 484
Synclinal valleys, 497
Synthetic-aperture radar, see Radar

Tangential scale distortion, 260–263
Temperature
cross-over periods in, 268–270
diurnal, 266–268
images of, see thermal infrared remote sensing
kinetic (true), 246, 251–252, 263
radiant, 246, 251–252
Temporal resolution, 16–18, 352–361, 408–410
Terra, see Earth Observing System (EOS)
Texture
drainage, see Drainage patterns
photographic, see Elements of image interpretation
in radar images, see Radar, terrain, surface roughness
Thematic Mapper (TM), see Landsat
Thermal infrared multispectral scanner (TIMS), 254
Thermal infrared remote sensing, 243–283
angular field of view, 256
atmospheric transmission models
LOWTRAN and MODTRAN, 264–265
calibration
radiometric, 263–264
sources, 256
split-window, 265

detectors, 256–257
emissivity, 248–250, 252, 277,
forward looking infrared (FLIR), 243–244, 266–267
ground
    resolution cell size, 258
    swath width, 258
history, 244–246
inverse-square law, 257–258
instantaneous field of view, 255–256
radiometric resolution, 257
relief displacement, 258–259
scanning
    across-track, 256
tangential scale distortion, 260–263
thermal response number (TRN), 280–281
Thermal properties
    capacity, 253, 269
    conductivity, 253
    inertia and apparent inertia, 253–254
Thermography, 243
Tidal flats, 513, 515–516
Tidal marshes, 513, 515
Tornado damage, 466
Tournachon, Gaspard Felix (NADAR), 58–59
Tracking data and relay satellites (TDRS), 181
Transmittance, 107
Transportation infrastructure, interpretation of, 410, 456–462
    boats and ships, 459–460
    roads and highways, 410, 456–459
    railroads, 459
Trellis drainage pattern, 487–488
Tropical Rainfall Measuring Mission (TRMM), 395–396

U-2 aircraft, 71–73
Ultraviolet radiation, see Bands of the electromagnetic spectrum
Uncontrolled photomosaic, 139–140
Uniformitarianism, 483
United States Geologic Survey (USGS)
    land use land cover classification system, 413–418, 426
Unmanned aerial vehicles, 79–80
Universal soil loss equation, 404, 471
Urban
    developmental cycle, 408
    heat island, 278
    land use/land cover, see Land use/land cover
    meteorological data, 464–465
Urbanization, 407

Verbal scale expression, 142–144
Vegetation
    agriculture, remote sensing of 350-352, 356–361
    biodiversity and gap analysis of, 368–369
    biomass/leaf area index (LAI) of, 1-2, 345, 353–356
    canopy modeling of, 23–24, 350
    change detection of, 370–373
    classification of, 351, 356–361, 370–373
    forestry thermal infrared sensing of, 278–279
    imaging spectrometry of, 350–352
    indices, 361–366
    landscape ecology metrics of, 366–368
    phenological cycles, 352–361

managed, 354–361
    natural, 353–354
photosynthetically active radiation (PAR), 338
reflectance dominant factors
    chlorophyll a and b pigments in, 337
        lower epidermis, 337
        palisade mesophyll cells, 337-339
        spongy mesophyll cells, 339–342
        upper epidermis, 337
    other pigments
        anthocyanin, 338
        carotenes, 337–338
        xanthophyll, 337
        phycoerythrin, 337–338
        phycocyanin, 337–338
    leaf additive reflectance, 340–341
    leaf water content, 342–344
    Bidirectional reflectance distribution function, 344–350
        field goniometer system, 346–348
    canopy closure, 345
chlorotic, 339
    photosynthesis, 333–334
microwave response of, 312–317
thermal response of, 266–269
wetland, 353–356, 370–373
Volcanoes, 484, 492
Volume scattering, see Radar

Water
    optical properties, 381–383
    penetration, 382–384, 393
    quality modeling, 402–404
    reflectance dominant factors
        bottom radiance, 381
        chlorophyll, 388–392
            in coastal and inland water, 392
            in ocean water, 391-392,
        dissolved organic material, 392
            Gelbstoffe, 392
        free-surface layer radiance, 380
        subsurface volumetric radiance, 380–381
        suspended minerals, 385–368
        wavelength, 381–383
    surface extent mapping, 382–384
    temperature, 393–395
    vapor, 398, 401
Wavelength
    dominant, 248
    units of measurement, 33–35, 39
Wien's Displacement law, 31, 248
Weathering, 483
Wetlands, remote sensing of, 353–356, 370–373
Whiskbroom scanning, 184
Windows, atmospheric, 42–44, 246–247

x-parallax, see Parallax, stereoscopic

Zoom stereoscopes, 156
Zoom-transfer-scope, 157

Appendix A - Sources of Remote Sensing Information

Textbooks

Avery, T. E. and G. L. Berlin, 1992, *Fundamentals of Remote Sensing and Airphoto Interpretation*, NY: Macmillan.

Bukata, R. P., J. H. Jerome, K. Y. Kondratyev and D. V. Pozdnyakov, 1995, *Optical Properties and Remote Sensing of Inland and Coastal Waters*, NY: CRC Press.

Campbell, J. B.,1996, *Introduction to Remote Sensing*, NY: The Guilford Press.

Colwell, R. N., (Ed.), 1983, *Manual of Remote Sensing*, 2nd Ed., Falls Church, VA: American Society for Photogrammetry and Remote Sensing.

Cracknell, A. P. and L. W. B. Hayes, *Introduction to Remote Sensing*, NY: Taylor & Francis.

Egan, W. G., 1985, *Photometry and Polarization in Remote Sensing*, NY: Elsevier.

Engman, E. T. and R. J. Gurney, 1991, *Remote Sensing in Hydrology*, NY: Chapman and Hall.

Greve, C., (Ed.), 1996, *Digital Photogrammetry: An Addendum to the Manual of Photogrammetry*, Bethesda, MD: American Society for Photogrammetry & Remote Sensing.

Henderson, F. M. and A. J. Lewis, (Eds.), 1998, *Principles & Applications of Imaging RADAR: Manual of Remote Sensing,* Vol. 2, NY: John Wiley & Sons.

Ikeda, M. and F. W. Dobson, 1995, *Oceanographic Applications of Remote Sensing*, NY: CRC Press.

Jensen, J. R., 1996, *Introductory Digital Image Processing: A Remote Sensing Perspective*, 2nd Ed., Saddle River, NJ: Prentice-Hall, Inc.

Kidder, S. Q. and T. H. VanderHaar, 1995, *Satellite Meteorology: An Introduction*, NY: Academic Press.

Lillesand, T. M. and R. W. Kiefer, 2000, *Remote Sensing and Image Interpretation*, NY: John Wiley & Sons, Inc., 4th Ed., in press.

Logicon, 1995, *Multispectral Users Guide*, Fairfax, VA: Logicon Geodynamics.

Logicon, 1997, *Multispectral Imagery Reference Guide*, Fairfax, VA: Logicon Geodynamics.

Lunetta, R. S. and C. D. Elvidge, 1998, *Remote Sensing Change Detection: Environmental Monitoring Methods and Applications*, Ann Arbor, MI: Ann Arbor Press.

Philipson, W. R., (Ed.), 1997, *Manual of Photographic Interpretation*, 2nd Ed., Bethesda, MD: American Society for Photogrammetry and Remote Sensing.

Quattrochi, D. A. and M. F. Goodchild, 1997, *Scale in Remote Sensing and GIS*, Boca Raton: CRC Press.

Rasher, M. E. and W. Weaver, 1990, *Basic Photo Interpretation*, Washington: USDA Soil Conservation Service, with hardcopy aerial photographs and slides.

Rencz, A., (Ed.), 1999, *Manual of Remote Sensing: Geosciences*, NY: John Wiley & Sons.

Sabins, F. F., 1997, *Remote Sensing: Principles and Interpretation*, 3rd Ed., NY: W. H. Freeman & Company.

Schott, J. R., 1997, *Remote Sensing — The Image Chain Approach*, NY: Oxford University Press.

Strain, P. and F. Engle, 1993, *Looking at Earth*, Atlanta: Turner Publishing.

Vincent, R. K., 1997, *Fundamentals of Geological and Environmental Remote Sensing*, Upper Saddle River, NJ: Prentice-Hall.

Warner, W. S., R. W. Graham and R. E. Read, 1996, *Small Format Aerial Photography*, Scotland: Whittles Publishing.

White, L., 1995, *Infrared Photography Handbook*, Buffalo, NY: Amherst Media.

On-line Tutorials

Fundamentals of Remote Sensing Tutorial by Canada Centre for Remote Sensing, http://www.ccrs.nrcan.gc.ca/ccrs/eduref/tutorial/tutore.html

Remote Sensing Core Curriculum sponsored by ASPRS, NASA, NCGIA, and ISPRS, http://research.umbc.edu/~tbenja1/index.html

The Remote Sensing Tutorial - An On-line Handbook by Nicholas M. Short, NASA Goddard Space Flight Center, http://rst.gsfc.nasa.gov/

NASA Observatorium Education-Reference Module, http://observe.ivv.nasa.gov/nasa/education/reference/main.html

Remote Sensing Tutorial by the Virtually Hawaii RSD project, http://hawaii.ivv.nasa.gov/space/hawaii/vfts/oahu/rem_sens _ex / rsex.spectral.1.html

WWW Virtual Library: Remote Sensing, http://www.vtt.fi/aut/ava/rs/virtual/

NASA Global Change Master Directory contains information about worldwide Earth science and global change data, http://gcmd.gsfc.nasa.gov/

Societies

Alliance for Marine Remote Sensing Association — http://www.amrs.org/

American Geophysical Union — http://www.agu.org/

American Society for Photogrammetry & Remote Sensing — http://www.asprs.org/

Canadian Remote Sensing Society — http://www.casi.ca/remote.htm

European Association Remote Sensing Laboratories — http://www-earsel.cma. fr/

Federation of American Scientists, image intelligence section — http://www.fas.org/

IEEE Geoscience and Remote Sensing Society — http://www.igarss.org/

International Commission on Remote Sensing — http://hydrolab.arsusda. gov/~jritchie/

International Society for Photogrammetry and Remote Sensing — http://www.isprs.org/isprs.html

National Stereoscopic Association — http://www.nas-3d.org/

The Remote Sensing Society (UK) — http://www.the-rss.org/

Remote Sensing Specialty Group of the Association of American Geographers — http://www.earthsensing.com/rssg/index.html

SPIE - Intl. Soc. for Optical Engineering — http://www.spie.org/

National Space Agencies

Canada Centre for Remote Sensing — http://www.ccrs.nrcan.gc.ca/common/comndexe.html

Centre National d'Etudes Spatiales — http://www.cnes.fr/ (in French)

CSIRO – Scientific and Industrial Research for Australia — http://www.csiro.au/

European Space Agency — http://www.esa.int/

National Aeronautics and Space Administration — http://www.nasa.gov/

National Remote Sensing Agency India — http://www.nrsa.gov. in/

National Space Development Agency of Japan — http://www.nasda.go.jp/index_e.html

Journals

Advances in Space Research — http://www.elsevier.com:80/inca/publications/store/6/4/4/

Atmospheric Research — http://www.elsevier.com:80/inca/publications/store/5/0/3/3/2/3/

Backscatter – marine remote sensing community — http://www.amrs.org/magazine.html

Computers & Geosciences — http://www.elsevier.nl:80/inca/publications/store/3/9/8/

Earth Observation Magazine — http://www.eomonline.com/index 3.htm

Geocarto International: A Multi-disciplinary Journal of Remote Sensing & GIS — http://www.geocarto.com/geocarto.html

GeoInformatics — http://www.geoinformatics.com/

Geo Info Systems – http://www.geoinfosystems.com/welcome.htm

IEEE Transactions on Geoscience and Remote Sensing — http://www.ieee.org/organizations/pubs/pub_preview/grs_toc.html

International Journal of Remote Sensing — http://www.tandf.co.uk/journals/tf/01431161.html

ISPRS Journal of Photogrammetry and Remote Sensing — http://www.elsevier.nl:80/inca/publications/store/5/0/3/3/4/0/

Photogrammetric Engineering & Remote Sensing — http://www.asprs.org/asprs/publications/pe&rs/index.html

Remote Sensing of Environment: An Interdisciplinary Journal — http://www.elsevier.nl/inca/publications/store/5/0/5/7/3/3/

Remote Sensing Reviews — http://www.gbhap.com/journals/315/315-top.htm

Stereo World — National Stereoscopic Association — http://www.nsa-3d.org

Aerial Photography Sources

Kite aerial photography — http://www-archfp.ced.berkeley.edu/kap/kaptoc.html

NHAP and NAPP— U.S. Geological Survey, National High Altitude Photography (NHAP) & National Aerial Photography Program (NAPP), http://edcftp.cr.usgs.gov/glis/hyper/guide/napp

U.S. Department of Agriculture Photography — http://www.fsa.usda.gov/dam/APFO/airfto.htm. Order from USDA, FSA Aerial Photography Field Office, Sales Branch, 2222 W. 2300 S., Salt Lake City, UT, 84119. Phone: (801) 975-3500, Email: sales@apfo.usda.gov.

Multispectral Remote Sensing Data Sources

ADEOS-II — NASDA Advanced Earth Observing Satellite-II, http://titan.eorc.nasda.go.jp/test/GLI/adeos2.html

AMSR — Advanced Microwave Scanning Radiometer (onboard ADEOS-II), http://titan.eorc.nasda.go.jp/test/AMSR/amsr.html

ASTER — NASA *Terra* Advanced Spaceborne Thermal Emission and Reflection Radiometer, http://asterweb.jpl.nasa.gov/aster-home/

ATLAS — NASA Airborne Terrestrial Applications Sensor, http://www.crsp.ssc.nasa.gov/aits/ATLAS.GIF

Autometric Inc., http://www.autometric.com

AVHRR — NOAA Advanced Very High Resolution Radiometer, http://edcwww.cr.usgs.gov/landdaac/1KM/avhrr.sensor.html. Land Pathfinder — http://daac.gsfc.nasa.CAMPAIGN_DOCS/LAND_BIO/GLBDST_main.html, AVHRR Oceans Pathfinder — http://podaac-www.jpl.nasa.gov/sst/

AVIRIS — NASA Jet Propulsion Lab Airborne Visible/Infrared Imaging Spectrometer, http://makalu.jpl.nasa.gov/html/aboutav.html

CASI-2 — Compact Airborne Spectrographic Imager-2, ITRES Research Limited of Canada, http://www.itres.com.

CERES — NASA *Terra* Clouds and the Earth's Radiant Energy System, http://asd-www.larc.nasa.gov/ceres/ASDceres.html

CORONA, ARGON, LANYARD declassified satellite photography, http://edcsgw9.cr.usgs.gov/glis/hyper/guide/disp

CZCS — NOAA Nimbus-7 Coastal Zone Color Scanner, http://daac.gsfc.nasa.gov/SENSOR_DOCS/CZCS_Sensor.html

DAEDALUS — Airborne Multispectral Scanner (AMS), Daedalus Enterprises, Inc., P.O. Box 1869, Ann Arbor, MI 48106, Phone: (313) 769-5649.

DMSP — Defense Meteorological Satellite Program, http://www.ngdc.noaa.gov/dmsp/dmsp.html

EarthWatch, Inc., http://www.digitalglobe.com

EO-1 — NASA New Millennium Program, Earth Observing-1, Advanced Land Imager (ALI), http://eo1.gsfc.nasa.gov

EOS PM-1 — http://mtpe.gsfc.nasa.gov/eos-pm1/PM1PROJ1.htm

GOES — NOAA/NASA Geostationary Operational Environmental Satellite, http://rsd.gsfc.nasa.gov/goes/

i^3 (Information, Integration, & Imaging, LLC), http://www.i3.com

IKONOS — Space Imaging, Inc., http://www.spaceimaging.com/

IRS-1A,B,C,D — Indian Remote Sensing System, Linear Imaging Self-scanning Sensor (LISS), http://www.spaceimaging.com/aboutus/satellites/IRS/IRS.html

LANDSAT — http://geo.arc.nasa.gov/sge/landsat/landsat.html

LANDSAT MSS — Multispectral Scanner, http://edcwww.cr.usgs.gov/glis/hyper/guide/landsat

LANDSAT TM — Thematic Mapper, http://edcwww.cr.usgs.gov/glis/hyper/guide/landsat_tm. Order from Space Imaging, Inc. (if acquired after October 1992), or USGS EROS Data Center (prior to October 1992)

LANDSAT 7 (ETM+) — Landsat 7 Enhanced Thematic Mapper Plus, http://landsat7.usgs.gov

Litton Emerge Spatial, Inc., Digital Frame Camera — http://www.espatial.wsicorp.com

Mercury, Gemini, Apollo, Skylab, http://images.jsc.nasa.gov/

METEOSAT — http://www.eumetsat.de/en/

MISR — NASA *Terra* Multi-angle Imaging Spectroradiometer, http://www-misr.jpl.nasa.gov/

MODIS — NASA *Terra* Moderate Resolution Imaging Spectrometer, http://modarch.gsfc.nasa.gov/MODIS

MOPITT — NASA *Terra* Measurements of Pollution in the Troposphere, http://eos.acd.ucar.edu/mopitt/home.html

MOS-1/MOS-1b — Marine Observation Satellite, http://yyy.tksc.nasda.go.jp/Home/Earth_Obs/e/mos_e.htm

OCTS — Ocean Color and Temperature Scanner, http://www.eoc.nasda.go.jp/guide/satellite/sendata/octs_e.html

ORBVIEW-3 and ORBVIEW-4 (Warfighter) — ORBIMAGE, Inc., http://www.orbimage.com/satellite/satellite.html

POES — NOAA/NASA Polar-orbiting Operational Environmental Satellites, http://poes2.gsfc.nasa.gov/

Positive Systems, Inc., ADAR 5500 — Airborne Data Acquisition and Registration system, http://www.possys.com

QUICKBIRD — EarthWatch, Inc., http://www.digitalglobe.com/company/spacecraft/quickbird.html

RESURS-01 — Russian Satellite, http://www.eurimage.com/Products/RESURS_O1.html

SeaWiFS (ORBVIEW-2) — NASA and ORBIMAGE, Inc., Sea-viewing Wide Field-of-View Sensor, http://seawifs.gsfc.nasa.gov/SEAWIFS.html or http://www.orbimage.com/satellite/orbview2/orbview2.html

SOHO Mission — Solar and Heliospheric Observatory, http://sohowww.nascom.nasa.gov/instrument.htm

Space Shuttle Photography — http://www.nasm.edu/ceps/RPIF/SSPR.html

SPIN-2 TK-350 and KVR-1000 — http://www.spin-2.com. SOVINFORMSPUTNIK (http://www.sovinformsputnik.com) has an agreement with Aerial Images, Inc., Raleigh, NC, and Central Trading Systems, Inc., Huntington Bay, NY, to market the data. Most U.S. coverage is served on Microsoft's Terra Server (http://terraserver.microsoft.com).

SPOT-1,2,3,4 — SPOT 1, 2, and 3 High Resolution Visible (HRV) Sensors and SPOT 4 High Resolution Visible Infrared (HRVIR) and Vegetation Sensor, http://www.spot.com/spot/home/system/introsat/welcome.htm

SSM/I — NOAA/NASA Special Sensor Microwave/Imager, http://daac.gsfc.nasa.gov/SENSOR_DOCS ssmi_sensor.html

TERRA — formerly NASA EOS AM-1, http://terra.nasa.gov

TOMS — Total Ozone Mapping Spectrometer, http://jwocky.gsfc.nasa.gov/

TOPEX/POSEIDON — http://topex-www.jpl.nasa.gov/

TRIANA — http://triana.gsfc.nasa.gov/home/

TRMM — NASA/NASDA Tropical Rainfall Measuring Mission, http://trmm.gsfc.nasa.gov/

UARS — NASA Upper Atmosphere Research Satellite, http://umpgal.gsfc.nasa.gov/uars-science.html

Radar Sensor Systems

AIRSAR — NASA/JPL aircraft SAR, http://airsar.jpl.nasa.gov/

ALMAZ — Russian SAR satellite, http://www.neosoft.com/Almaz/default.html

ENVISAT-1 — 10 instruments, including SAR, http://envisat.estec.esa.nl/

ERS-1,2 — European Remote Sensing Satellite, European Space Agency, http://earth1.esrin.esa.it/ERS

JERS-1 — Japanese Earth Resource Satellite, http://yyy.tksc.nasda.go.jp/Home/Earth_Obs/e/jers_e.html

LightSAR — NASA Spaceborne Radar (proposed), http://lightsar.jpl.nasa.gov/lightsar/

RADARSAT — http://radarsat.space.gc.ca/info/description/specifications.html

SEASAT — http://southport.jpl.nasa.gov/scienceapps/seasat.html

SIR-A, B, C/X-SAR — NASA Space Shuttle Imaging Radar

SIR-A — http://southport.jpl.nasa.gov/scienceapps/sira.html

SIR-B — http://southport.jpl.nasa.gov/scienceapps/sirb.html

SIR-C/X-SAR — http://southport.jpl.nasa.gov/sir-c/

SLAR — NASA Side-Looking Airborne Radar, http://edcwww.cr.usgs.gov/glis/hyper/guide/slar

SRTM — Shuttle Radar Topography Mission, http://www.jpl.nasa.gov/srtm/

STAR-3i — Sea Ice and Terrain Assessment Radar (STAR-1, 2, 3i), Intermap Technologies Airborne Radar, http://www.intermaptechnologies.com/HTML/mapp_star3i.htm

On-line Remote Sensing Laboratory at the University of South Carolina

Internet Site Updates

The reader is encouraged to visit the Remote Sensing Laboratory in the Department of Geography at the University of South Carolina at the following Internet website:

http://www.cla.sc.edu/geog/rsbook/links/

The Internet addresses for the previously mentioned sites will be updated often at this location because they change so rapidly. New remote sensing sites will be posted on a periodic basis. If you have more accurate information about a listed Internet site or new information about a book, society, journal, or source of remote sensing data, please contact the author at jrjensen@sc.edu.

Remote Sensing Exercises

The following Internet website provides introductory remote sensing exercises that may be used in conjunction with this book:

http://www.cla.sc.edu/geog/rsbook/exercises/

You are welcome to use the exercises and adapt them to your requirements. Several of the exercises require access to the ERDAS Imagine digital image processing system (or equivalent) to perform fundamental operations such as viewing a color-composite image. In most instances, the imagery used in the exercises is in the public domain unless otherwise noted. If you use the remote sensing exercise materials, please send the author information about innovative uses, methodological refinements that could be made, and improved datasets that might be used.